# Biochemistry

# of

# Chloroplasts

## Organizing Committee

B. H. Davies

T. W. Goodwin (Director)

J. T. O. Kirk

E. I. Mercer

L. J. Rogers

D. R. Threlfall

Marie R. Wasdell

# Biochemistry of Chloroplasts

Proceedings of a NATO Advanced Study Institute held at Aberystwyth, August 1965

*Edited by*

**T. W. GOODWIN**

Department of Biochemistry and Agricultural Biochemistry
University College of Wales, Aberystwyth, Wales

VOLUME II

1967

ACADEMIC PRESS · LONDON AND NEW YORK

ACADEMIC PRESS INC. (LONDON) LTD

Berkeley Square House

Berkeley Square

London, W.1

*U.S. Edition published by*

ACADEMIC PRESS INC.

111 Fifth Avenue

New York, New York 10003



*Library of Congress Catalog Card Number:* 66–25085

PRINTED IN GREAT BRITAIN BY

SPOTTISWOODE, BALLANTYNE & CO. LTD

LONDON AND COLCHESTER

# Introduction

Outstanding developments in our knowledge of the biochemistry of photosynthesis have been made in the last ten years following Calvin's classical work on the fixation of $^{14}CO_2$ by algae and the discovery of photosynthetic phosphorylation. Important observations are still being made in these areas but with the advent of new and powerful techniques and biochemists' growing appreciation of structure—function relationships, the emphasis has shifted towards studies designed to provide information on the chloroplast itself.

Studies with the electron microscope are revealing the intimate details of chloroplast structure and, in particular, the developmental morphology of this organelle is gradually being elucidated. Parallel with these investigations the chemical components of the chloroplast are now being accurately defined and new and interesting compounds are being discovered. The biosynthesis of many of these compounds is also being actively studied in numerous laboratories. Furthermore, a start has also been made on the exciting problem of correlating morphological development with biosynthetic activities.

The aim of the NATO Study Institute was to bring together leaders in these various cognate fields to present not only their most recent investigations but the significant background from which they evolved; and then together to hammer out as far as possible an integrated view of "The Biosynthesis of Chloroplasts".

The present volume records the formal activities of this Study Institute and presents what appears to us an authoritative assessment of the "Biochemistry of the Chloroplast" in the second half of 1965. It is hoped that it will serve as a basic text for some years for recruits into the field of chloroplast biochemistry.

The idea of this Study Institute arose from very informal talks at Aberystwyth with Dr. Paul Stumpf, and later at Davis he helped in formulating the first drafts of the programme.

Thanks are due to many people for making the Study Institute a reality: to The Scientific Division of NATO, in particular Dr. B. Coleby; to all my colleagues on the Organizing Committee; to the staff of the Department of Biochemistry and Agricultural Biochemistry at Aberystwyth; to the college authorities for excellent hostel facilities; and, not least, to the eminent scientists who considered it worth their while to make the long trip to Aberystwyth to take part in the Study Institute.

*May* 1966 T. W. GOODWIN

## List of Contributors to Volume II

S. AARONSON, *Haskins Laboratories, New York, New York, U.S.A.*

J. AMESZ, *Biophysical Laboratory, State University, Leiden, Netherlands*

C. M. A. ANDERSON, *Botany Department, University College of Wales, Aberystwyth, Wales*

D. I. ARNON, *Department of Cell Physiology, University of California, Berkeley, California, U.S.A.*

D. J. BAISTED, *The University of Mississippi, Oxford, Mississippi, U.S.A.*

H. BALTSCHEFFSKY, *Kungl, Universitetet I Stockholm, Wenner-Grens Institut, Stockholm, Sweden*

E. S. BAMBERGER, *Department of Biology, Brandeis University, Waltham, Massachusetts, U.S.A.*

D. S. BENDALL, *Department of Biochemistry, University of Cambridge, England*

C. C. BLACK, *Charles F. Kettering Research Laboratory, Yellow Springs, Ohio, U.S.A.*

K. BLOCH, *Department of Chemistry, Harvard University, Cambridge, Massachusetts, U.S.A.*

P. BÖGER, *Charles F. Kettering Research Laboratory, Yellow Springs, Ohio, U.S.A.*

L. BOGORAD, *Department of Botany, University of Chicago, Chicago, Illinois*

C. BOVÉ, *Service de Biochimie, IFAC, Station de Physiologie Végétale, Centre National de Réchèrches Agronomiques, Versailles, France*

J. M. BOVÉ, *Service de Biochimie, IFAC, Station de Physiologie Végétale, Centre National de Réchèrches Agronomiques, Versailles, France*

J. W. BRADBEER, *Botany Department, University College of Wales, Aberystwyth, Wales*

J. BROOKS, *Department of Biochemistry and Biophysics, University of California, Davis, California, U.S.A.*

E. CAPSTACK JR., *The University of Mississippi, Oxford, Mississippi, U.S.A.*

A. H. CASWELL, *Department of Biochemistry, and Agricultural Biochemistry, University College of Wales, Aberystwyth, Wales*

C. O. CHICHESTER, *Department of Food Science and Technology, University of California, Davis, California, U.S.A.*

H. CLAES, *Max-Planck-Institut for Biologie, Tübingen, Germany*

G. CONSTANTOPOULOS, *Department of Chemistry, Harvard University, Cambridge, Massachusetts, U.S.A.*

J. COOMBS, *Botany Department, Imperial College, University of London, England*

H. V. DONOHUE, *Department of Food Science and Technology, University of California, Davis, California, U.S.A.*

H. D. DORRER, *Organische-chemisches Institut der Technischen Hochschule München und Pflanzenphysiologisches Institut der Universität Göttingen, Abt. Biochemie der Pflanzen, Göttingen, Germany*

R. C. DOUGHERTY, *Argonne National Laboratory, Argonne, Illinois, U.S.A.*

K. EGLE, *Department of Botany, University of Frankfurt, Germany*

J. M. EISENSTADT, *Yale University School of Medicine, New Haven, Connecticut, U.S.A.*

H. ELBERTZHAGEN, *Institut für Angewandte Botanik der Techn. Hochschule München, Germany*

P. W. ELLYARD, *Department of Biology, Brandeis University, Waltham, Massachusetts, U.S.A.*

E. ELSTNER, *Pflanzenphysiologisches Institut der Universität Göttingen, Abt. Biochemie der Pflanzen, Göttingen, Germany*

R. G. EVERSON, *Department of Biology, Brandeis University, Waltham, Massachusetts, U.S.A.*

H. FOCK, *Department of Botany, University of Frankfurt, Germany*

G. FORTI, *Laboratorio di Fisiologia Vegetale, Istituto di Scienze Botaniche, Milano, Italy*

J. FRIEND, *Department of Botany, University of Hull, Yorkshire, England*

T. GALLIARD, *Department of Biochemistry and Biophysics, University of California, Davis, California, U.S.A.*

H. GHOSH, *Department of Biochemistry and Biophysics, University of California, Davis, California, U.S.A.*

M. GIBBS, *Department of Biology, Brandeis University, Waltham, Massachusets, U.S.A.*

A. GIBOR, *The Rockefeller University, New York, New York, U.S.A.*

T. W. GOODWIN, *Department of Biochemistry and Agricultural Biochemistry, University College of Wales, Aberystwyth, Wales*

A. GORCHEIN, *Medical Research Council Research Group, Department of Chemical Pathology, St. Mary's Hospital Medical School, London, England*

S. GRANICK, *The Rockefeller University, New York, New York, U.S.A.*

B. R. GRANT, *Department of Cell Physiology, University of California, Berkeley, California, U.S.A.*

W. T. GRIFFITHS, *Department of Biochemistry and Agricultural Biochemistry, University College of Wales, Aberystwyth, Wales*

B. E. S. GUNNING, *Department of Botany, Queen's University of Belfast, Northern Ireland*

I. HABERER-LIESENKÖTTER, *Institut für Angewandte Botanik der Techn. Hochschule, München, Germany*

P. HARRIS, *Unilever Research Laboratories, Colworth House, Sharnbrook, Bedford, England*

R. V. HARRIS, *Unilever Research Laboratories, Colworth House, Sharnbrook, Bedford, England*

J. C. HAWKE, *Department of Biochemistry and Biophysics, University of California, Davis, California, U.S.A.*

U. W. HEBER, *Institut für Landwirtschaftliche Botanik der Universität Bonn, Germany*

K. W. HENNINGSEN, *Institute of Genetics, University of Copenhagen, Denmark*

R. HILL, *Department of Biochemistry, University of Cambridge, England*

S. H. HUNTER, *Haskins Laboratories, New York, New York, U.S.A.*

M. P. JAGOE, *Department of Botany, Queen's University of Belfast, Northern Ireland*

A. T. JAMES, *Unilever Research Laboratories, Colworth House, Sharnbrook, Bedford, England*

B. E. JUNIPER, *Department of Biochemistry and Agricultural Biochemistry, University College of Wales, Aberystwyth, Wales*

O. KANDLER, *Institut für Angewandte Botanik der Techn. Hochschule München, Germany*

C. KENYON, *Department of Chemistry, Harvard University, Cambridge, Massachusetts, U.S.A.*

J. T. O. KIRK, *Department of Biochemistry and Agricultural Biochemistry, University College of Wales, Aberystwyth, Wales*

J. J. KATZ, *Argonne National Laboratory, Argonne, Illinois, U.S.A.*

R. J. MANS, *Botany Department, University of Maryland, College Park, Maryland, U.S.A.*

A. F. H. MARKER, *Botany Department, Imperial College, University of London, England*

G. MOREL, *Service de Biochimie, IFAC, Station de Physiologie Végétale, Centre National de Réchèrches Agronomiques, Versailles, France*

A. MOYSE, *Laboratoire de Photosynthèse du C.N.R.S., Gif-sur-Yvette, France*

J. NAGAI, *Department of Chemistry, Harvard University, Cambridge, Massachusetts, U.S.A.*

T. O. M. NAKAYAMA, *Department of Food Science and Technology, University of California, Davis, California, U.S.A.*

W. R. NES, *The University of Mississippi, Oxford, Mississippi, U.S.A.*

A. NEUBERGER, *Medical Research Council Research Group, Department of Chemical Pathology, St. Mary's Hospital Medical School, London, England*

W. W. NEWSCHWANDER, *The University of Mississippi, Oxford, Mississippi, U.S.A.*

B. W. NICHOLS, *Biosynthesis Unit, Colworth House, Sharnbrook, Bedford, England*

R. OLSSON, *Department of Botany, University of Hull, Yorkshire, England*

J. PREISS, *Department of Biochemistry and Biophysics, University of California, Davis, California, U.S.A.*

E. R. REDFEARN, *Department of Biochemistry, University of Leicester, England*

L. J. ROGERS, *Department of Biochemistry and Agricultural Biochemistry, University College of Wales, Aberystwyth, Wales*

M. RONDOT, *Service de Biochimie, IFAC, Station de Physiologie Végétale, Centre National de Réchèrches Agronomiques, Versailles, France*

P. T. RUSSELL, *The University of Mississippi, Oxford, Mississippi, U.S.A.*

A. SAN PIETRO, *Charles F. Kettering Research Laboratory, Yellow Springs, Ohio, U.S.A.*

S. P. J. SHAH, *Department of Biochemistry and Agricultural Biochemistry, University College of Wales, Aberystwyth, Wales*

H. SIMON, *Organische-chemisches Institut der Technischen Hochschule München und Pflanzenphysiologisches Institut der Universität Göttingen, Abt. Biochemie der Pflanzen, Göttingen, Germany*

R. SIMONI, *Department of Biochemistry and Biophysics, University of California, Davis, California, U.S.A.*

R. M. SMILLIE, *Plant Physiology Unit, School of Biological Sciences, University of Sydney, Sydney, Australia*

H. H. STRAIN, *Argonne National Laboratory, Argonne, Illinois, U.S.A.*

J. M. STUBBS, *Biosynthesis Unit, Colworth House, Sharnbrook, Bedford, England*

P. K. STUMPF, *Department of Biochemistry and Biophysics, University of California, Davis, California, U.S.A.*

G. H. TAIT, *Medical Research Council Research Group, Department of Chemical Pathology, St. Mary's Hospital Medical School, London, England*

D. R. THRELFALL, *Department of Biochemistry and Agricultural Biochemistry, University College of Wales, Aberystwyth, Wales*

A. TREBST, *Pfanzenphysiologisches Institut der Universität Göttingen, Abt. Biochemie der Pflanzen, Göttingen, Germany*

W. J. VREDENBERG, *Biophysical Laboratory, State University, Leiden, Netherlands*

D. A. WALKER, *Botany Department, Imperial College, University of London, England*

B. WALLES, *Institutionen för skogsgenetik, Skogshogskolan, Stockholm, Sweden*

F. R. WHATLEY, *Department of Cell Physiology, University of California, Berkeley, California, U.S.A.*

C. P. WHITTINGHAM, *Botany Department, Imperial College, University of London, England*

S. WIĘCKOWSKI, *Department of Plant Physiology, Jagellonean University, Cracow, Poland*

S. G. WILDMAN, *Department of Botany and Plant Biochemistry, University of California, Los Angeles, California, U.S.A.*

J. WITTKOP, *Department of Biochemistry and Biophysics, University of California, Davis, California, U.S.A.*

A. C. ZAHALSKY, *Haskins Laboratories, New York, New York, U.S.A.*

G. ZANETTI, *Laboratorio di Fisiologia Vegetale, Istituto di Scienze Botaniche, Milano, Italy*

J. ZURZYCKI, *Laboratory of Plant Physiology, University of Cracow, Cracow, Poland*

# Contents of Volume II

## PART IV. BIOSYNTHESIS IN CHLOROPLASTS—PROTEINS AND NUCLEIC ACIDS



## PART V. BIOSYNTHESIS IN CHLOROPLASTS—PIGMENTS

## PART VI—BIOCHEMISTRY OF PHOTOSYNTHETIC PHOSPHORYLATION



## PART VII. BIOSYNTHETIC MECHANISM IN RELATION TO MORPHOGENESIS

# Contents of Volume I

## PART I. STRUCTURE AND MORPHOGENESIS OF CHLOROPLASTS



## PART II. CHLOROPLAST COMPONENTS-LIPIDS

## PART III. CHLOROPLAST COMPONENTS—PROTEINS, LIPOPROTEINS



## PART IV. CHLOROPLAST COMPONENTS—NUCLEIC ACIDS



## PART V. CHLOROPLASTS COMPONENTS—PIGMENTS

# Part I. Biosynthesis in Chloroplasts—$CO_2$ Fixation

# Assimilation of Carbon Dioxide by Chloroplast Preparations[1]

Martin Gibbs, Elchanan S. Bamberger, Peter W. Ellyard
and R. Garth Everson

*Department of Biology, Brandeis University,
Waltham, Massachusetts, U.S.A.*

## Introduction

About 100 years ago, Sachs and his pupils established the close connection between photosynthesis, starch and the chloroplast. They grasped the fact that starch was an end product of photosynthesis and that this polysaccharide was a starting point for the production of essentially all other organic compounds in the plant. They also clearly recognized that the site of the photochemical act in the cell was the chloroplast.

The first attempt to demonstrate $CO_2$ fixation by cell-free systems containing chloroplasts was carried out by Fager (1952 a, b). His spinach preparations could fix $^{14}CO_2$ into glyceric acid 3-phosphate and phosphoenolpyruvate. Additionally, the uptake of isotope was stimulated by light. However, no radioactivity was detected in the free sugars or phosphorylated sugars. A decisive development was provided by Allen *et al.* (1955) when they proved that under suitable conditions isolated chloroplasts were capable of reducing $CO_2$ to the level of carbohydrates. The sole source of energy supply was apparently visible light. Respiratory mechanisms which could also give rise to the reductant were minimized by conducting the incubations under nitrogen. Elimination of aerobic mechanisms could not be completely eliminated since the preparations were evolving $O_2$. A physical separation of the light and dark phases of photosynthesis was claimed by fractionating isolated chloroplasts (Trebst, Tsujimoto and Arnon, 1958).

Subsequent to these fundamental findings by Fager and by Arnon and his colleagues, $CO_2$ assimilation by isolated chloroplasts has been studied in numerous laboratories. The purpose of this article is to summarize recent findings on photosynthetic $CO_2$ metabolism as observed in isolated chloroplasts.

## Isolation of Chloroplasts

### Aqueous Methods

#### *Fager (1952)*

In this procedure, a chloroplast suspension was obtained from leaves of spinach (presumably *Spinacia oleracea*) by grinding a frozen powder in sorbitol-

[1] This investigation was aided by grants from the National Science Foundation and the United States Air Force through the Air Force office of Scientific Research of the Air Research and Development Command, under Contract AF 49(638)789.

borate buffer and filtering through heavy canvas. The pellet obtained after centrifuging at 13,000 × ***g*** for 15 min was denoted as "chloroplasts" and the supernatant fraction as "enzymes". Neither particles nor the supernatant fraction exhibited $CO_2$-fixing power. Combination of the two fractions resulted in light-stimulated $CO_2$ uptake.

*Allen, Arnon, Capindale, Whatley and Durham (1955)*

Most of our knowledge concerning $CO_2$ assimilation has been obtained using spinach chloroplasts isolated by grinding the de-ribbed leaf blades in 0·35 M NaCl. The resultant homogenate is squeezed through cheesecloth and the green juice centrifuged for 1 min at 0° to remove cell debris and abrasive. The chloroplast fraction is the pellet obtained after recentrifuging the juice for 7 min at 1000 × ***g***. The pellet resuspended in 0·35 M NaCl is referred to as "whole" chloroplasts.

Suspension of the whole chloroplasts in water or dilute buffer results in a rupturing of the organelles. The water-treated chloroplasts are referred to as "broken" chloroplasts. The "broken" chloroplasts are centrifuged at 20,000 × ***g*** for 10 min. The supernatant fraction is designated "chloroplast extract" and is probably equivalent to "enzymes" of the Fager procedure.

*McClendon (1952)*

Spinach leaves are blended in a 0·5 M sucrose solution buffered in 0·1 M potassium phosphate and 0·01 M ethylenediamine tetra-acetic acid, pH 7·2. The pellet obtained between 200 × ***g*** and 600 × ***g*** is resuspended in 0·5 M sucrose containing 0·002 M phosphate and 0·0002 M ethylenediamine tetra-acetic acid, pH 7·2. With respect to $CO_2$ assimilation, the chloroplasts isolated in sucrose appear to be similar to those prepared in NaCl (Holm-Hansen *et al.*, 1959).

*Walker (1964)*

In Walker's experiments, a chloroplast suspension was prepared from pea leaves (*Pisum sativum*) that had been kept in the dark 12 hr before use. The chloroplasts were isolated by blending the leaves in 0·33 M sorbitol solution containing $Na_2HPO_4$–$KH_2PO_4$ (0·1 M), NaCl (0·1 %), $MgSO_4$ (0·1 %) and sodium iso-ascorbate (0·1 %). The pH of the blending fluid was 6·8. The pellet obtained by centrifuging the slurry was resuspended in 0·33 M sorbitol containing NaCl (0·1 %) and $MgSO_4$ (0·1 %).

*Gee* et al. *(1965)*

In contrast to other aqueous procedures, homogenates containing chloroplasts were obtained by grinding leaves without the addition of water, buffer or sand. The resulting mixture was centrifuged at 200 × ***g*** to remove cell debris and whole cells. The supernatant fluid (homogenate) contained chloroplasts and mitochondria among other particles and cytoplasm. Gee *et al.* (1965) estimate that isolation of chloroplasts in 0·35 M NaCl results in the destruction of 80 % of the grana while 90 % of the grana remain intact in their homogenate preparations.

## NON-AQUEOUS METHODS

Chloroplasts prepared by one of the methods described in a previous section are exposed to leaching of enzymes, and ferredoxin among other water soluble components. The non-aqueous methods were developed in order to prevent such losses. After freeze-drying the leaves, chloroplasts are isolated in the dry state with a non-polar solvent. The components of the lyophilized leaves are separated on the basis of the density gradient. While the use of a solvent composed of carbon tetrachloride and hexane minimizes the leaching of polar substances (Heber, Pon and Heber, 1963), the procedure suffers from the fact that the particles prepared in non-aqueous solvents lost their ability to perform the photochemical act of photosynthesis (Havir and Gibbs, 1963). It was possible to prepare a reconstituted system which could consume $^{14}CO_2$ by adding a water extract from chloroplasts prepared in non-polar solvents to "broken" chloroplasts prepared as detailed in a previous section.

### *Stocking (1959)*

The lyophilized leaves are blended in a solution of carbon tetrachloride and hexane of density 1·305. After centrifugation of the suspension, the precipitate is suspended in a solution of 1·405 density. The chloroplast fraction usually constitutes the material precipitating between density 1·305 and 1·405. After removal of organic solvents the powder is extracted with buffered solutions.

### *Thalacker and Behrens (1959)*

Chloroplasts are prepared in mixtures of petroleum ether and carbon tetrachloride from lyophilized leaves. This method is essentially that of Stocking (1959).

## RATE OF $CO_2$ UPTAKE

The important fact emerges from the data recorded in Table I that the rate of $CO_2$ fixed by chloroplasts is of the order of about 1 to 2% of intact material. The low fixation rate appears to be a fundamental property of the isolated organelle. The rates observed are not vigorously affected by the isolating medium. Even the combination of non-aqueous with broken particle does not markedly influence the basic rate. Most of these values are not taken from time course curves and therefore include the low rates observed during the initial lag phase (see Fig. 5). Kinetic curves for $CO_2$ fixation have only been reported in a few papers (Arnon, Allen and Whatley, 1954; Bamberger and Gibbs, 1965; Walker, 1964).

The question has been raised as to what is limiting the ability of the isolated chloroplast to assimilate $CO_2$. In their chloroplast preparations, Smillie and Fuller (1959) have demonstrated that ribulose 1,5-diphosphate carboxylase is the pacemaker. Heber and Tyszkiewicz (1962) indicate that the reduction of glycerate 3-phosphate is the rate limiting step. Gibbs, Black and Kok (1961) attributed the low rate to the loss of ferredoxin during isolation of the chloroplast. Gibbs, Black and Kok (1961) argued that it was misleading to compare the observed rates of $CO_2$ uptake with NADPH and ATP formation by chloroplasts since each is determined under a different set of conditions (compare Table I with Table II). While phosphate esterification and NADPH fixation are assayed in the presence of various additives, phenazine methosulfate, riboflavin 5′-phosphate or spinach ferredoxin, $CO_2$ uptake values for

Table I. Carbon dioxide fixation by various chlorophyll-containing preparations

| Material[1] | μmoles $CO_2$/mg chl/hr | Investigator |
|---|---|---|
| Whole chloroplasts, 0·35 M NaCl | 0·25 | Allen *et al.* (1955) |
| Whole chloroplasts, 0·5 M sucrose | 1·7 | Holm-Hansen *et al.* (1959) |
| Whole chloroplasts, 0·35 M NaCl | 2·5 | Gibbs and Calo (1960) |
| Pea chloroplasts, 0·33 M sorbitol | 1·1 | Walker (1964) |
| Homogenate, 0·35 M *tris* | 2·0 | Gee *et al.* (1965) |
| Broken chloroplasts | 2·0 | Whatley *et al.* (1956) |
| Broken chloroplasts, sugar beet | 3·5 | Whatley *et al.* (1960) |
| Non-aqueous chloroplasts[2] | 24 | Heber and Tyszkiewicz (1962) |
| *Chlorella pyrenoidosa* | 180 | Hill and Whittingham (1955) |
| *Helianthus annuus* | 300 | Willstätter and Stoll (1918) |

[1] Unless stated, chloroplasts isolated from *Spinacia oleracea*.
[2] Supplemented with particles isolated in 0·5 M sucrose.

intact chloroplasts (Table I) are measured without these cofactors. Cofactors including spinach ferredoxin are omitted since they either inhibit or have no effect on assimilation of $CO_2$.

The rate of $NADP^+$ reduction catalysed by chloroplast particles without additives is of the order of 10 μmoles/mg chl/hr (Table II). Assuming as postulated by the Calvin group that 2 moles of NADPH or its equivalent are

Table II. Rate of $NADP^+$ reduction

| | $NADP^+$ reduced (μmoles/mg chl/hr) | |
|---|---|---|
| | − Ferredoxin | + Ferredoxin[1] |
| Control[2] | 7·8, 11·6 | 40·5 |
| ADP, $P_i$, $Mg^{2+}$ | 15·6, 18·4 | 93·0 |

Taken from Gibbs, Black and Kok (1961).
[1] 4·7 units of spinach ferredoxin.
[2] Spinach chloroplasts isolated in 0·4 M sucrose and assays carried out in 0·4 M sucrose.

consumed per mole of $CO_2$ reduced to the carbohydrate level, then maximum fixation rates of the order of 5 μmoles $CO_2$/mg chl/hr can be supported with preparations of this type. Values approaching these figures are recorded in Table I.

It should therefore be stressed that data reported with chloroplasts and artificial acceptors resulting in high rates of phosphate esterification and NADPH formation indicate the unusual capabilities of the chloroplast, but have little value as a basis of comparison with $CO_2$ fixation rates. The question as to what is limiting the ability of the isolated unfortified (see p. 10 for a discussion of the effect of additives to the endogenous $CO_2$ rates) chloroplast to assimilate $CO_2$ at a rate approaching that of the intact cell can best be answered by stating that this may be due to the removal of material involved in the photochemical act and the carbon assimilation cycle.

## Factors Affecting $CO_2$ Fixation

### ISOLATION TECHNIQUES

Since light-induced $CO_2$ fixation by chloroplasts isolated in 0·35 M NaCl has been considered low (0·2–3 μmoles $CO_2$/mg chl/hr), there have been

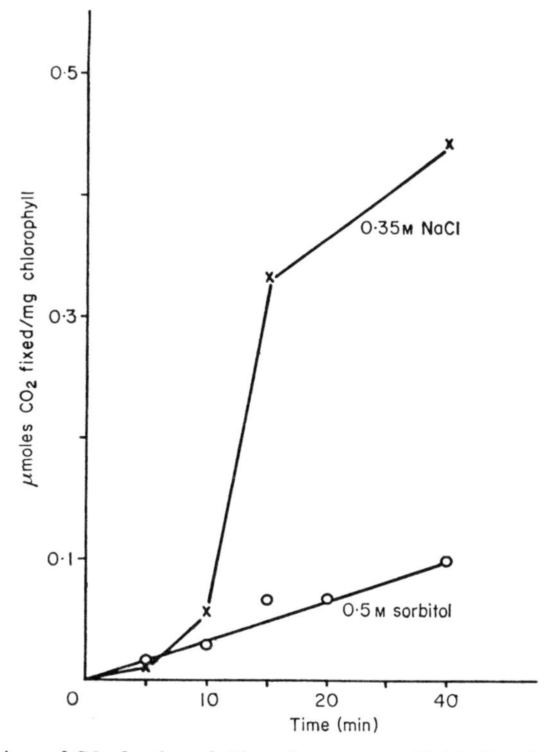


FIG. 1. Comparison of $CO_2$ fixation of chloroplasts prepared in NaCl and sorbitol from the same spinach leaves. Twenty g of leaf blades and 40 ml either *tris* (0·02 M)-NaCl (0·35 M) or *tris* (0·02 M)-sorbitol (0·5 M) were ground in a chilled mortar with sand. The mixture was passed through four layers of cheesecloth. The slurry was centrifuged for one min at 200 × *g*. The supernatant fraction was spun for 10 min at 2000 × *g*. The pellet was resuspended in 5 ml of grinding medium. The reaction mixture contained per ml: *tris*, 66 μmoles; NaCl, 250 μmoles or sorbitol, 360 μmoles; $MnCl_2$, 1 μmole; $KH_2PO_4$, 0·2 μmoles; $NaHCO_3$, 1·25 μmoles and approximately 75 μg of chlorophyll. The light intensity was 1000 foot candles and the temperature was 16°. The atmosphere was nitrogen.

repeated attempts to improve upon the rate by altering the composition of the isolating fluid.

The time course of $CO_2$ fixation of spinach chloroplasts prepared in 0·35 M NaCl and 0·5 M sorbitol are compared in Fig. 1. While the uptake in sorbitol is linear with time, the rate of $CO_2$ uptake in NaCl is considerably higher during the linear portion of the curve. In our laboratory, chloroplasts prepared in 0·35 M NaCl fix $CO_2$ at a rate roughly four times that of chloroplasts prepared by the homogenate technique of Gee *et al.* (1965). This finding contrasts

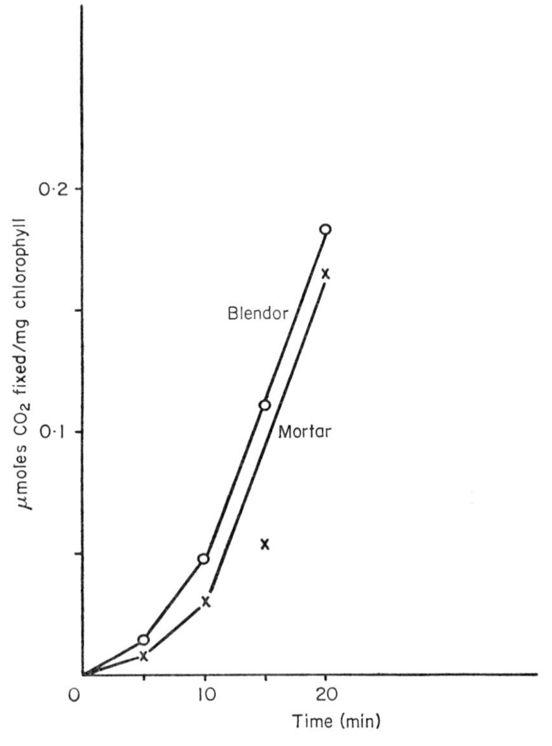


FIG. 2. Isolation of spinach chloroplasts in *tris*-NaCl: two methods compared. Spinach laminae (100 g) were added slowly to 200 ml of *tris*-NaCl in a chilled Waring Blendor cup while blades were rotated sufficiently fast to draw leaves under the solution. The speed was increased to maximum for 5 sec. Contents of cup was poured through eight layers of cheesecloth. See Fig. 1 for reaction mixture and standard mortar method.

with the data recorded by them. In addition, the dark fixation rates of chloroplasts prepared in NaCl are consistently less than 3% of the rate induced by light. This contrasts sharply with the mixture obtained by grinding spinach leaves without an isotonic salt solution.

Some procedures call for grinding the leaves in a mortar and pestle while others use the blender. A comparison (Fig. 2) between chloroplasts isolated in 0·35 M NaCl indicates little difference between the two methods.

It should be noted that little attention has been given thus far to varying the concentrations or ionic strengths of the grinding media or of the incubation mixtures. The composition of these fluids, age or source of the plant (Table III), presence of heavy metals within the leaf tissue (Gibbs and Calo, 1960) have not been documented thoroughly. These areas of investigation may lead to an improvement in the fixation rate.

Table III. Rate of $CO_2$ fixation by spinach chloroplasts isolated from various sources

| Source | $\mu$moles $CO_2$/mg chl/hr |
|---|---|
| Local market, A and P | 0·6 |
| Potted plants, soil | 0·9 |
| Field grown, Long Island, New York | 0·8 |
| Hydroponic, Hoagland solution | 2·2 |
| Hydroponic { unwashed particles | 2·5 |
| Hydroponic { washed particles | 1·3 |

STABILITY OF CHLOROPLAST

Table IV shows the effect of aging the spinach chloroplast on $CO_2$ fixation and $NADP^+$ reduction. In comparison to endogenous $NADP^+$ reduction, chloroplasts prepared and stored in 0·35 M NaCl at 0° rapidly lose their ability of assimilate $CO_2$.

Table IV. Effect of aging on stability of $CO_2$ fixation and $NADP^+$ reduction by spinach chloroplasts

| Expt. | Time after grinding (min) | $NADP^+$ reduced $\mu$moles/mg chl/hr | $^{14}CO_2$ fixed cpm/aliquot |
|---|---|---|---|
| 1 | 30 | 22·4 | 4800 |
| | 265 | 22·9 | 215 |
| 2 | 110 | 19·1 | 2200 |
| | 320 | 18·1 | 100 |
| | 420 | 16·4 | 70 |

Taken from Gibbs, Black and Kok (1961).

pH DEPENDENCE

The incorporation rate of $CO_2$ fixation by chloroplasts isolated in 0·35 M NaCl proceeds at maximal velocity in the vicinity of pH 7·4 (Elbertzhagen and Kandler, 1962; Gibbs and Calo, 1959). The fixation rate falls off slowly at neutrality but declines sharply on the alkaline side. In contrast, the pH optimum for the broken preparation is pH 8·1 (Elbertzhagen and Kandler, 1962). These authors concluded that the reaction mechanism is light induced fixation of $CO_2$ and reduction in whole chloroplasts is different from that in broken chloroplasts (see p. 30).

## KINETICS OF $CO_2$ FIXATION

The effect of white light intensity on $CO_2$ fixation by whole spinach chloroplasts prepared in 0·35 M NaCl was first observed by Turner *et al.* (1962). In-

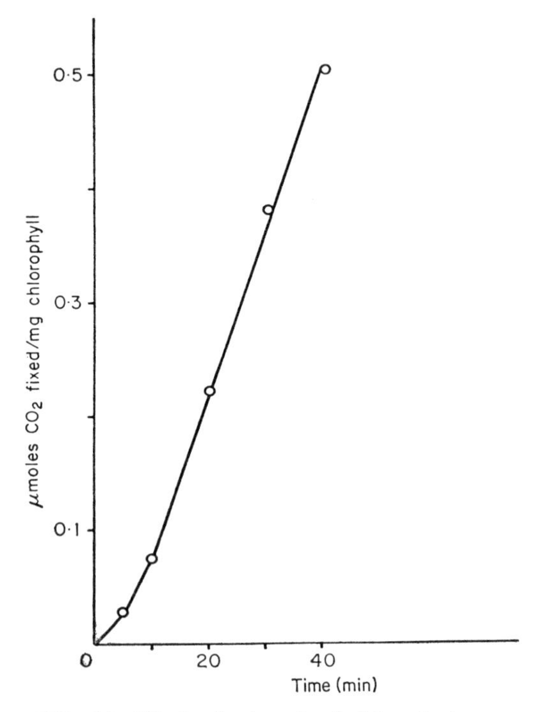


FIG. 3. Presence of "lag" in $CO_2$ fixation by spinach chloroplasts prepared in *tris*-sorbitol. The reaction was run at 6°. See Fig. 1 for comparison with 16°.

creasing light intensity brought about a higher rate of $CO_2$ fixation but at low light intensities there was a lag in the rate. The lag in $CO_2$ fixation was thought to be linked to low ATP production since the light intensity curves for the two reactions were similar. In contrast, NADPH production was linear at low light intensities.

Spinach chloroplasts prepared in 0·35 M NaCl also display a lag in $CO_2$ fixation with respect to time (Figs. 1, 2 and 5). On the other hand, chloroplasts isolated in 0·5 M sorbitol show a linear response to time when the light induced uptake is carried out at 16° (Fig. 2 and also Walker, 1964) but a lag is observed at 6° (Fig. 3). Apparently the presence of NaCl is required for the lag since chloroplasts isolated in 0·5 M sorbitol and kept in sorbitol display linear kinetics while the lag reappears in chloroplasts transferred from 0·5 M sorbitol to 0·35 M NaCl (Fig. 4).

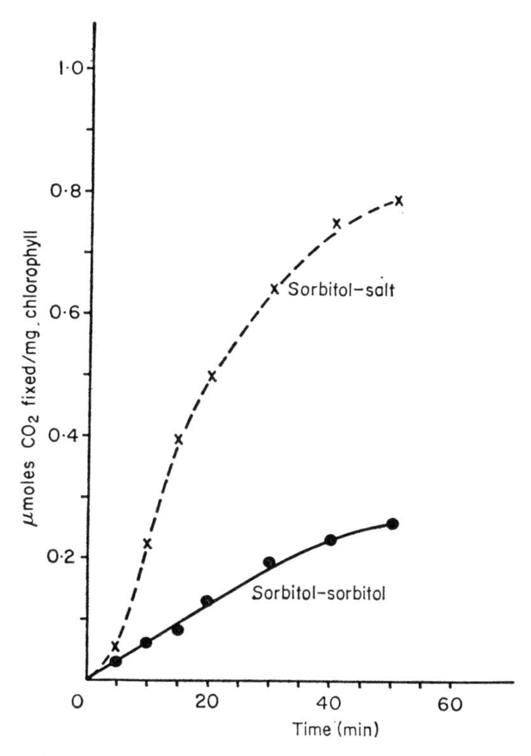


FIG. 4. Influence of resuspension in *tris*-NaCl on $CO_2$ fixation by chloroplasts prepared in *tris*-sorbitol. Leaf blades (20 g) and 40 ml 0·2 M *tris*-0·5 M sorbitol were ground in a mortar with sand. Half of the chloroplast fraction was suspended in *tris*-0·35 M NaCl and half was resuspended in *tris*-sorbitol.

## PHOSPHORYLATED COMPOUNDS

A two-fold increase (Table V) in the rate of $CO_2$ fixation following the addition of fructose 1,6-diphosphate, ribose 5-phosphate, glucose 1-phosphate, glucose 6-phosphate or glyceric acid 3-phosphate to a fragmented spinach chloroplast system (chloroplast extract and broken chloroplasts) was demonstrated by Whatley *et al.* (1956). Recently these studies have been extended to whole spinach chloroplasts (Table V) by Bamberger and Gibbs (1963, 1965) and to whole pea chloroplasts by Walker (1964).

Chloroplasts prepared in sorbitol respond more vigorously than those isolated in salt to the added phosphorylated compounds. The reason for the greater response is unknown.

There is some difficulty in explaining the contrasting effects of glycerate 3-phosphate, fructose 1,6-diphosphate and ribose 5-phosphate with that of the hexose phosphates and erythrose 4-phosphate (Table V, last column) with respect to enhancing $CO_2$ uptake by whole spinach chloroplasts. The important intermediate of the photosynthetic carbon cycle and the one readily formed from each of the phosphorylated compounds causing enhancement seems to be glyceraldehyde 3-phosphate. The availability of the aldotriose is a pacemaker in the oxidative pentose phosphate cycle and may well play a similar role in the reductive pentose phosphate pathway. Although there is some variation, whole spinach chloroplasts preserved in 0·35 M NaCl generally do not appear capable of converting hexose phosphates into triose phosphates. Therefore, the hexose monophosphates could enter the mainstream of the carbon cycle only through a transketolase cleavage of fructose 6-phosphate provided glyceraldehyde 3-phosphate was present to act as an acceptor molecule. Thus, we have noted that the addition of 0·1 mM glyceraldehyde 3-phosphate together with 1 mM glucose 6-phosphate to a chloroplast preparation resulted in a higher enhancement of $CO_2$ uptake than that achieved by the triose alone. In this respect, erythrose 4-phosphate should act similarly to the hexose monophosphates because it can enter the cycle only when triose phosphate is available.

Table V. Effect of phosphorylated compounds on photosynthetic carbon dioxide fixation

| Additions | Whatley[1] | Walker[2] | Bamberger and Gibbs[3] |
|---|---|---|---|
| None | 2·0[4] | 1·1[4] | 1·3[4] |
| Glyceric acid 3-phosphate | 3·7 | — | 2·6 |
| Glucose 1-phosphate | 4·4 | 4·7 | — |
| Glucose 6-phosphate | 3·8 | — | ±[5] |
| Fructose 6-phosphate | 3·1 | — | ±[5] |
| Glycerate 2-phosphate | — | 6·2 | — |
| Fructose 1,6-diphosphate | 5·2 | 15·2 | 2·6 |
| Ribose 5-phosphate | 4·0 | 24·3 | 3·0 |
| Sedoheptulose 7-phosphate | — | — | 0·3 |
| Sedoheptulose 1,7-diphosphate | — | — | 0·3 |
| Erythrose 4-phosphate | — | — | 1·3 |

[1] Whatley *et al.* (1956), fragmented spinach chloroplasts.
[2] Walker (1964), whole pea chloroplasts in 0·33 M sorbitol.
[3] Bamberger and Gibbs (1965), whole spinach chloroplasts in 0·35 M NaCl.
[4] $\mu$moles $CO_2$ fixed/mg chl/hr.
[5] Results variable.

An unexpected observation was the inhibition of $CO_2$ fixation by sedoheptulose 7-phosphate and sedoheptulose 1,7-diphosphate. More recent work suggests glyceraldehyde 3-phosphate dehydrogenase as the site of inhibition (Gibbs, 1963). This finding is the first indication that an intermediate of the photosynthetic carbon cycle may function in a regulatory manner in $CO_2$

reduction. In the broken chloroplast preparations, these seven carbon sugars could prevent the operation of the photosynthetic carbon cycle.

Some of the phosphorylated compounds overcome the lag in $CO_2$ fixation with respect to time. Fructose 1,6-diphosphate (Fig. 5), glycerate 3-phosphate, ribose 5-phosphate, dihydroxyacetone phosphate and glyceraldehyde 3-phosphate were found to respond similarly. However, as seen in Fig. 5, the maximal rate observed after addition of the phosphorylated compound is essentially equal to that obtained in the control reaction mixture. This fact

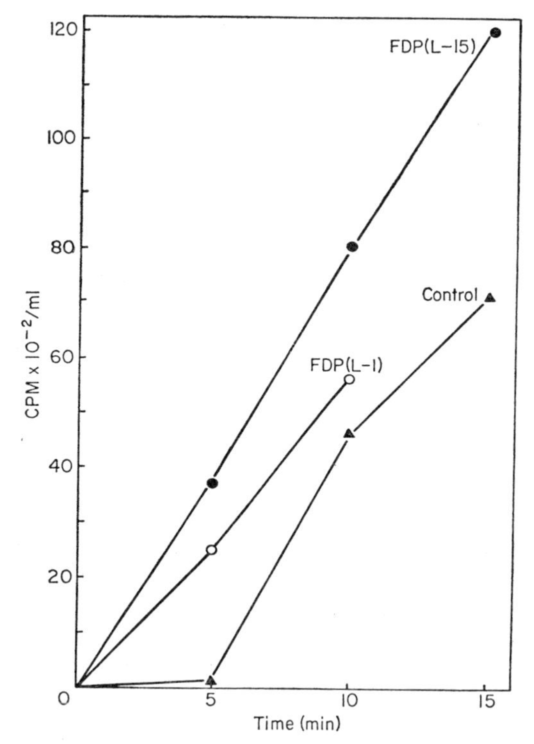


FIG. 5. Effect of 0·1 mM fructose 1,6-diphosphate (FDP) on $CO_2$ fixation by whole spinach chloroplasts. The fructose 1,6-diphosphate was added 15 min (L−15) or 1 min (L−1) prior to illumination.

suggests a rate limiting step common to both systems after the initial lag phase.

In a recent paper, a most striking effect of ribose 5-phosphate was recorded by Walker (1964). Pea chloroplasts in 0·33 M sorbitol showed a seventeen-fold increase in their ability to fix $CO_2$ on addition of the aldopentose 5-phosphate.

The lag in $CO_2$ fixation observed in 0·35 M NaCl chloroplasts may be due to a lack of ATP or intermediates of the carbon cycle required during the induction phase. The endogenous stores of these substances are apparently leached away during isolation of the chloroplast in 0·35 M NaCl. Sorbitol-prepared chloroplasts, however, appear to be less permeable with respect to these materials. The fact that glycerate 3-phosphate which in contrast to glyceraldehyde 3-phosphate, dihydroxyacetone phosphate, or fructose 1,6-diphosphate is not capable of yielding ATP through substrate phosphorylation in these preparations is as effective in enhancing $CO_2$ fixation justified the conclusion that ATP is not limiting in the lag phase. Since these experiments are carried out under nitrogen and in the presence of glucose oxidase to remove traces of $O_2$, oxidative phosphorylation would seem to be eliminated. To account for these findings, we propose that the phosphorylated compounds affect $CO_2$ fixation by filling up depleted pools of the photosynthetic carbon cycle. On the other hand, the possibility that the phosphorylated compounds may function as a ready source of ATP through a transphosphorylation reaction has not been ruled out. Many workers have proposed a direct transfer of phosphate attached to carbon 1 of fructose 1,6-diphosphate or sedoheptulose 1,7-diphosphate to ribulose 5-phosphate rather than a dephosphorylation of these compounds as called for in the Calvin cycle. Thus far, attempts to demonstrate such transphosphorylation reactions in chloroplast extracts have been negative.

Losada, Trebst and Arnon (1960) have claimed that rates obtained with broken preparations fortified with fructose 1,6-diphosphate (see p. 30) are approximately 25% those of the parent leaves. Their spinach leaves fixed $CO_2$ at the extremely low rate of 20 $\mu$moles/mg chl/hr. When compared to values of roughly 200 (Table I) obtained by others, it would appear that the photosynthetic rate of their intact leaves was limited by $CO_2$ supply among other factors.

Comparisons of this kind can be misleading. Photosynthesis is a coupled system. It involves the uptake of $CO_2$ and the release of $O_2$. There are no reports in the literature that the addition of phosphorylated compounds to broken or whole chloroplasts stimulates *both* $CO_2$ assimilation *and* $O_2$ evolution. Until an amount of $O_2$ evolved equivalent to that of $CO_2$ assimilated is demonstrated, calculations presented by Losada, Trebst and Arnon (1960) must be considered tentative and shaky.

## INHIBITORS OF $CO_2$ FIXATION

According to current chloroplast dogma, ATP and NADPH formed in the light phase of photosynthesis are consumed to reduce $CO_2$ to the carbohydrate level. It should be possible to classify the substances which block $CO_2$ assimilation into three groups, namely, those which preferentially inhibit the photochemical act or the carbon cycle and those which inhibit both processes to an equal extent.

### COMPOUNDS WHICH AFFECT THE PHOTOCHEMICAL ACT

Substances which inhibit either NADP formation (electron transport) or phosphorylation should prevent $CO_2$ uptake.

#### *Electron transport*

2-Heptyl-4-hydroxyquinoline-*N*-oxide, 2-nonyl-4-hydroxyquinoline-*N*-oxide and 3-(*p*-chlorophenyl)-1,1-dimethylurea block $NADP^+$ reduction

(Avron, 1961, Jagendorf and Margulies, 1960). Their effect could be overcome by ascorbate (Fig. 6) and by reduced lipoic acid (Table VI) but not by oxidized

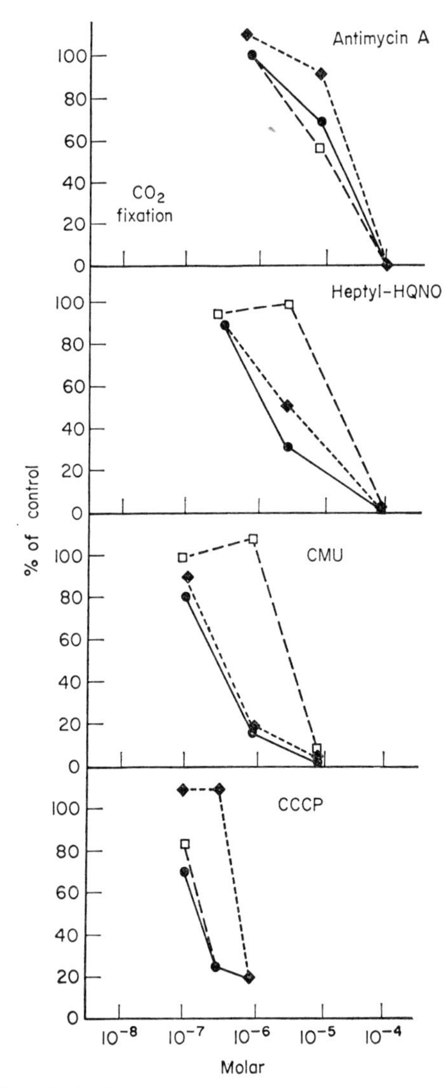


FIG. 6. Effect of CMU (3-(*p*-chlorophenyl) 1,1-dimethylurea), heptyl-HQNO (2-heptyl-4-hydroxy-quinoline-*N*-oxide), CCCP (*m*-chlorocarbonyl cyanide phenylhydrazine) and antimycin A on $CO_2$ fixation by whole chloroplasts. $CO_2$ fixation in the presence of 10 mM ascorbate (□), fructose 1,6-diphosphate (◆) and control (●).

lipoic acid. On the other hand fructose 1,6-diphosphate (Fig. 6), GSH, cysteine, or BAL could not alleviate the inhibition. Reduced lipoic acid, like ascorbic acid (Vernon and Zaugg, 1960), can apparently bypass one of the two

Table VI. Effect of sulfhydryl compounds on DCMU inhibition of $CO_2$ fixation by spinach chloroplasts[1]

| DCMU 0·5 μM | Reduced lipoate 1 mM | GSH[2] 1 mM | Cysteine 1 mM | BAL[3] 0·1 mM | cpm[4] |
|---|---|---|---|---|---|
| –[5] | – | – | – | – | 2150 |
| – | + | – | – | – | 2200 |
| – | – | + | – | – | 2200 |
| – | – | – | + | – | 1900 |
| – | – | – | – | + | 1975 |
| + | – | – | – | – | 100 |
| + | + | – | – | – | 825 |
| + | – | + | – | – | 210 |
| + | – | – | + | – | 170 |
| + | – | – | – | + | 125 |

[1] Whole chloroplasts in 0·35 M NaCl, 15 min photosynthesis.
[2] Reduced glutathione.
[3] 2,3-Dimercaptopropanol.
[4] Counts per minute of an aliquot.
[5] Compound omitted, –; compound added, +.

photochemical reactions and the oxygen-evolving mechanism and donate electrons to acceptors which yield the reductant of $CO_2$ (Table VII). Reduced lipoic acid overcomes DCMU with respect to $CO_2$ inhibition but appears to act as an uncoupler of ATP formation when broken chloroplasts are used (Table VII). Petrack (1965) has established that reduced lipoic acid is required for photohydrolysis but inhibits photophosphorylation.

Table VII. Effect of reduced lipoate and ascorbate on inhibition of $NADP^+$ reduction and ATP formation in the presence of DCMU[1]

| DCMU 0·5 μM | Ascorbate 5 mM | Reduced lipoate 5 mM | NADPH | ATP | ATP/NADPH |
|---|---|---|---|---|---|
| – | – | – | 146[2] | 138[2] | 0·95 |
| + | – | – | 22 | 12 | 0·54 |
| – | + | – | 146 | 150 | 1·03 |
| – | – | + | 109 | 66 | 0·61 |
| + | + | – | 41 | 21 | 0·51 |
| + | – | + | 34 | 34 | 0·09 |

[1] Broken spinach chloroplasts fortified with saturating amounts of spinach ferredoxin.
[2] μmoles/mg chl/hr.

### *Uncouplers of Phosphorylation*

*m*-Chlorocarbonyl cyanide phenylhydrazone (CCCP) is known to be an uncoupler of photophosphorylation without affecting $NADP^+$ reduction (Heytler and Prichard, 1962; Avron, 1964). $CO_2$ fixation is sensitive to this reagent but this effect is overcome by fructose 1,6-diphosphate. Ascorbate does not alleviate the inhibition (Fig. 6).

Low concentrations of arsenate (0·35 μM) were shown by Gibbs and Calo (1959) to inhibit $CO_2$ uptake. Avron and Jagendorf (1959) concluded that arsenate is a competitive inhibitor of light-induced phosphorylation. In agreement with their suggestion, phosphorylated intermediates of the carbon cycle could bypass arsenate inhibition (Table VIII) apparently by providing the acceptor for $CO_2$.

Table VIII. Effect of phosphorylated compounds on inhibition of $CO_2$ fixation by arsenate[1]

| Arsenate μM | Fructose 1,6-diphosphate 1 mM | Glycerate 3-phosphate 1 mM | Phosphoenol-pyruvate 1 mM | cpm[2] |
|---|---|---|---|---|
| –[3] | – | – | – | 3200 |
| 35 | – | – | – | 1600 |
| 100 | – | – | – | 910 |
| – | + | – | – | 3600 |
| – | – | + | – | 3700 |
| – | – | – | + | 1000 |
| 100 | + | – | – | 2500 |
| 100 | – | + | – | 1700 |
| 100 | – | – | + | 400 |

[1] Whole chloroplasts in 0·35 M NaCl, 15 min photosynthesis.
[2] Counts per minute of an aliquot.
[3] Compound omitted, –; compound added, +.

Arsenate is a much more effective inhibitor of $CO_2$ assimilation in whole chloroplasts (Gibbs and Calo, 1959) than in broken preparations (Losada *et al.*, 1960). $CO_2$ fixation in the broken preparation is inhibited only 28% by 10 mM arsenate. This fact again points to a difference in the two chloroplast systems. In the broken preparations, arsenate appears not to uncouple photophosphorylation but rather to inhibit ribulose 1,5-diphosphate carboxylase.

An interesting inhibitor of $CO_2$ fixation of whole chloroplasts is peroxyacetyl nitrate, a component of photochemical smog (Dugger *et al.*, 1963). Apparently it is an uncoupler of photophosphorylation since addition of ascorbate to the chloroplast preparations restored $CO_2$ fixation.

Desaspidin, a phlorobutyrophenone derivative has been shown by Baltscheffsky and de Kiewiet (1964) to uncouple light-induced phosphorylation. Low concentrations (0·1 μM) of desaspidin-inhibited phosphorylation linked to the ascorbate-dichlorophenol indophenol donor system while the same concentration had little effect on phosphorylation that is coupled with an electron flow from water to $NADP^+$. The data in Table IX show a similar trend. Where electrons flow from water to $NADP^+$, 0·25 $\mu$M desaspidin has little effect on $CO_2$ uptake. In contrast, when ascorbate is added to the control system, sensitivity to desaspidin increases and the same concentration inhibits $CO_2$ uptake completely. In the presence of DCMU, the inhibitory effect of desaspidin is bypassed to some extent. These data have been considered by others (Baltscheffsky and de Kiewiet, 1964) as support for two sites of phosphorylation. The $CO_2$ data can be discussed in these terms. If there is a site of phosphate esterification linked to ascorbate as donor then this ATP as well as

Table IX. Effect of ascorbate on inhibition of $CO_2$ fixation by desaspidin[1]

| DCMU[2] 0·5 $\mu$M | Ascorbate 1 mM | Desaspidin $\mu$M | cpm[3] |
|---|---|---|---|
| –[4] | – | — | 1540 |
| – | + | — | 2350 |
| – | – | 0·25 | 1375 |
| – | – | 0·35 | 75 |
| – | + | 0·25 | 35 |
| – | + | 0·35 | 0 |
| + | – | — | 160 |
| + | + | — | 1025 |
| + | + | 0·25 | 700 |
| + | + | 0·35 | 155 |

[1] Whole chloroplasts in 0·35 M NaCl, 15 min photosynthesis.
[2] 3-(3,4-dichlorophenyl)-1,1-dimethylurea.
[3] Counts per minute of an aliquot.
[4] Compound omitted, –; compound added, +.

that formed at the other site is required for $CO_2$ reduction. It follows that ascorbate must effectively block electrons derived from water since the recent studies of Avron (1964) and Wessels (1964) strongly support the hypothesis that the site of ATP formation during the light-induced reduction of $NADP^+$ lies between water as the electron donor and the point of entry of electrons from the ascorbate plus dichlorophenol indophenol couple. Measurement of oxygen evolution is required to determine whether the addition of ascorbate blocks the oxygen evolving system in the whole spinach chloroplast.

However, it should be kept in mind that the electron carriers comprising the photochemical chain may be in different levels of reduction depending on whether the source of the electrons is water or ascorbate. There may be one site of photophosphorylation but the sensitivity of the site to desaspidin may depend on the strength or potential of the electron donor.

## CARBON METABOLISM

Iodoacetamide inhibits $CO_2$ fixation (Whatley *et al.*, 1956; Calo and Gibbs, 1960) but has little effect on the photochemical act. The site of inhibition is glyceraldehyde 3-phosphate dehydrogenase and ribulose 5-phosphate kinase (Calo and Gibbs, 1960). The inhibition is relieved by reduced glutathione.

Cyanide is a potent inhibitor of $CO_2$ fixation and is an effective inhibitor of ribulose 1,5-diphosphate carboxylase. It may exert its effect by forming a cyanohydrin addition product with ribulose 1,5-diphosphate. The hydrolysis of the addition compound is hamamelonic acid (Kandler, 1957).

D-Threose 2,4-diphosphate has been reported to inhibit muscle and yeast glyceraldehyde 3-phosphate dehydrogenase specifically (Racker, Klybas and Schramm, 1959; Fluharty and Ballou, 1959). Park *et al.* (1960) observed that threose 2,4-diphosphate inhibited $CO_2$ fixation by sonically ruptured spinach chloroplasts and assigned the site of inhibition as the NADP associated glyceraldehyde 3-phosphate dehydrogenase. Bamberger and Gibbs (1963) have repeated their experiments using the whole spinach chloroplast. Threose 2,4-diphosphate did not inhibit $CO_2$ uptake. The diphosphate may not penetrate the chloroplast but other diphosphates, sedoheptulose and fructose, can affect $CO_2$ fixation.

Arsenite at low concentrations is a potent inhibitor of $CO_2$ fixation (Gibbs and Calo, 1960) however, the site of inhibition has not been assigned. Arsenite does not affect ATP formation or $NADP^+$ reduction (Whatley *et al.*, 1956) nor does it appear to affect any enzyme of the carbon cycle (Gibbs and Calo, 1960). The inability of arsenite to affect $CO_2$ fixation demonstrates that lipoic acid is not directly involved in the primary conversion of light into chemical energy (Calvin, 1954).

Antimycin A unlike arsenite affects a photochemical process but does not affect $CO_2$ fixation significantly (Fig. 6). Phosphorylation linked to $NADP^+$ reduction was inhibited slightly at 0·1 mM and $CO_2$ was affected some 32% at 10 $\mu$M (Arnon, Allen and Whatley, 1956; Baltscheffsky, 1960; Bamberger *et al.*, 1963). A possible site of antimycin A inhibition was demonstrated by Fewson *et al.* (1962). They demonstrated that spinach ferredoxin catalysed the aerobic production of 105 $\mu$moles ATP/mg chl/hr. Antimycin A was 50% inhibitory at about 5 $\mu$M, in contrast to photophosphorylations catalysed by other electron acceptors for which the corresponding value is roughly 50 $\mu$M. The requirement for $O_2$ suggests that some endogenous substance, presumably spinach ferredoxin, was acting in a cyclic manner. These observations have also been recorded by others (Horio and Yamashita, 1962; Tagawa, Tsujimoto and Arnon, 1963).

## EFFECT OF LIGHT AND PHOSPHORYLATED COMPOUNDS ON INHIBITORS

Arsenite and iodoacetamide inhibit $CO_2$ uptake by intact chloroplasts but have little effect on the photochemical processes. It became of interest to determine whether the phosphorylated sugars could bypass this inhibition. Additionally, the time sequence of addition of the inhibitor was investigated and found to be critical.

With respect to arsenite, when 0·1 mM arsenite was added to a 0·35 M NaCl prepared chloroplast mixture either 1 min (L − 1) or 15 min (L − 15) before the lights were turned on, $CO_2$ fixation was almost abolished (Fig. 7). In sharp contrast when arsenite was tipped into the incubation mixture in the light (L + 1) the inhibition was partially overcome and no effect was observed at (L + 5). Similar observations were made when arsenite was replaced by 0·1 mM iodoacetamide. However, inhibition of $CO_2$ uptake by 0·1 mM arsenate was not relieved by light or by any of the phosphorylated compounds.

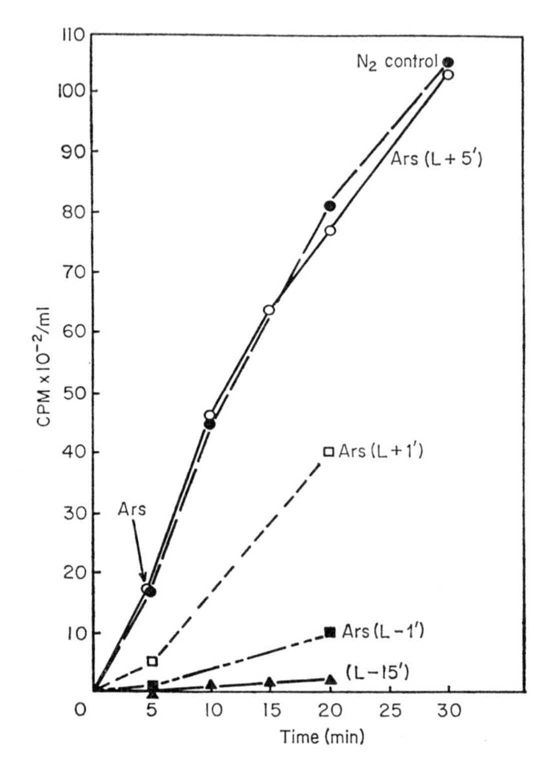


FIG. 7. Effect of incubation time on the degree of inhibition of $CO_2$ fixation by 0·1 mM arsenite.

The effect of 0·1 mM arsenite on $CO_2$ uptake was investigated in the presence of the phosphorylated compounds which enhanced or had no effect on the fixation rate. Whole chloroplasts in 0·35 M NaCl were incubated with 1mM of either fructose 1,6-diphosphate, glycerate 3-phosphate, ribose 5-phosphate or glucose 6-phosphate respectively for 15 min in the dark. At (L − 1), 0·1 mM arsenite was added. In the control, 0·1 mM arsenite brought about 91 % inhibition (Fig. 8). In the presence of fructose 1,6-diphosphate arsenite caused only a 46 % inhibition with respect to the fructose 1,6-diphosphate control (a rate of 5·7 μmoles $CO_2$/mg chl/hr). The same was found in the experiments with ribose 5-phosphate and glycerate 3-phosphate. However, arsenite inhibition was not relieved by glucose 6-phosphate (Fig. 9).

The rather broad effect exerted by the phosphorylated compounds capable of enhancement against inactivation by arsenite and iodoacetamide support the earlier work of Gibbs and Calo (1960) that in addition to the well-known sulfhydryl-containing enzymes like ribulose 5-phosphate kinase, chloroplasts possess a structural unit in which vicinal sulfhydryl groups play a central part. The structural unit can be visualized as similar to the one proposed by Lynen

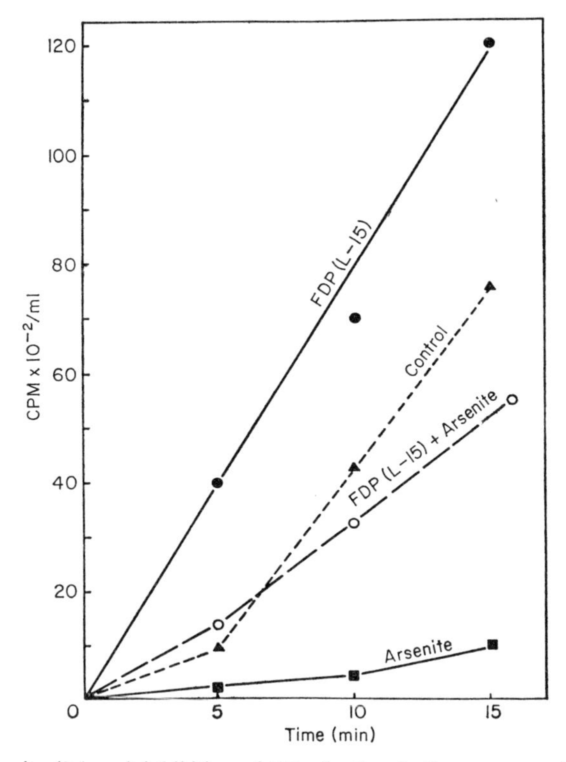


FIG. 8. Arsenite (0·1 mM) inhibition of $CO_2$ fixation in the presence of 1 mM fructose 1,6-diphosphate. Fructose 1,6-diphosphate (FDP) was added in the dark 15 min (L−15) before the lights were turned on. Arsenite was added 1 min prior to illumination (L−1).

(1961) for the multi-enzyme fatty acid synthetase system or to the highly organized and multi-functional enzyme system envisaged by Bassham (1963) for the photosynthetic carbon cycle. The formulation of such a structure could explain the apparent lack of an enzymic site for arsenite inhibition once the enzyme complex was disintegrated. The affinity of the structural group for the phosphorylated compounds capable of enhancing $CO_2$ fixation or of an intermediate derived from them would explain their protection against the inhibitors. It may also explain the inability of arsenite to inhibit $CO_2$ fixation when added to the chloroplasts during a period of photosynthesis. Thus, while the light-dependent $CO_2$ fixation was in progress, a substrate was firmly bound to the sulfhydryl group affording protection against the inhibitor. Apparently this kind of "endogenous" protection occurred during a subsequent dark period of 5 min.

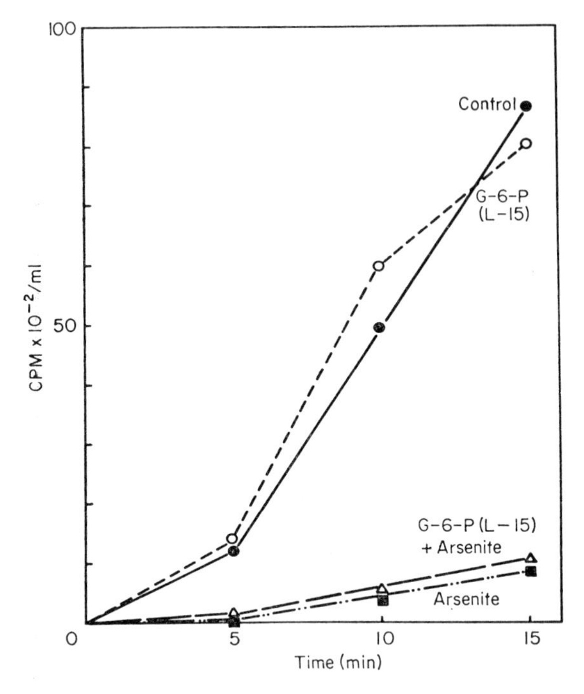


FIG. 9. Arsenite (0·1 mM) inhibition of $CO_2$ fixation by whole chloroplasts in the presence of glucose 6-phosphate. Glucose 6-phosphate (G 6-P) was added in the dark 15 min (L−15) prior to illumination. Arsenite was added 1 min before illumination.

## WARBURG EFFECT

### INTACT CELLS

The inhibition of photosynthesis by oxygen was discovered by Warburg (1920). It has therefore been referred to as the Warburg effect. The Warburg effect has been demonstrated to apply to $O_2$ evolution as well as $CO_2$ fixation. The effect has been studied periodically in many laboratories but a rigorous explanation for the effect is still wanting. The subject has been elegantly reviewed by Turner and Brittain (1962).

The effect is dependent on both light intensity and $CO_2$ concentration. At high light intensities, Tamiya and Huzisige (1949) demonstrated that the evolution of $O_2$ by *Chlorella* was inhibited by 100%, $O_2$ about 5% at high $CO_2$ concentration and about 85% at low $CO_2$ tensions. McAlister and Myers (1940) found that in the presence of 0·03% $CO_2$ and at light saturation an inhibition of 25% was recorded for $CO_2$ uptake when the $O_2$ content was increased from 0·5 to 20%. While evidence in this field is conflicting, all investigators have noted that at very low $CO_2$ concentration, the inhibition of photosynthesis by $O_2$ is strongly affected by light, being maximal at light saturation.

Since the Warburg effect was increased at low carbon dioxide concentration, Tamiya and Huzisige (1949) proposed that there was competition between oxygen and carbon dioxide for the carboxylating enzyme (ribulose 1,5-diphosphate carboxylase). More recently Tamiya and his associates have modified their earlier stand and propose that the reducing substance, R, produced as a result of pre-illumination, is itself reoxidized by molecular oxygen (Miyachi, Izawa and Tamiya, 1955). R was shown not to be a pyridine nucleotide. Turner (1962) favors the hypothesis that oxygen acts on photosynthesis primarily by reversibly inactivating one or more of the sulfhydryl enzymes of the carbon cycle. Endogenous reductants would reactivate the enzyme.

## CHLOROPLAST

Arnon, Allen and Whatley (1954) noted briefly a limited Warburg effect for $^{14}CO_2$ fixed by whole spinach chloroplasts in air as opposed to nitrogen. We have now fully documented the Warburg effect in the whole spinach chloroplast.

In contrast to intact tissue, fixation of radiocarbon by the whole chloroplast is sensitive to oxygen even at saturating light intensities and high carbon dioxide concentrations. Figure 10 illustrates that a 50 % inhibition is brought about by

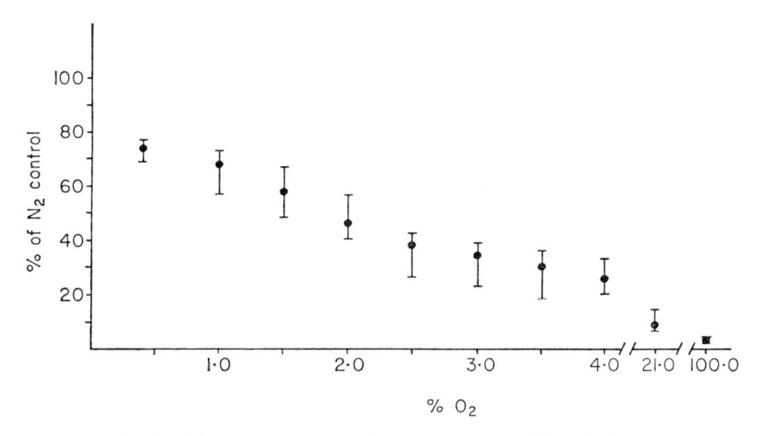


FIG. 10. Effect of concentration of oxygen on $CO_2$ fixation.

an oxygen concentration of 2 %. In air, inhibition is of the order of 80 to 90 %. Similar to intact plant tissue (Tamiya and Huzisige, 1949), the effect is rapidly reversible in the chloroplast. Therefore, aerobic conditions do not cause permanent damage to the isolated organelle. Figure 11 shows the progress curves for the reversibility experiments. After $CO_2$ fixation was run in nitrogen and a mixture of 1·5 % $O_2$ : 98·5 % $N_2$ for 10 min, the lights were turned off and the atmospheres of the Warburg vessels were reversed during a 5 min dark period. Where nitrogen was exchanged with 1·5% oxygen the rate of fixation dropped dramatically and when nitrogen replaced the 1·5% oxygen the rate of fixation was vigorously enhanced. In either condition the lag in $CO_2$ fixation was reintroduced after the dark phase.

The rapid reversibility of the process makes it unlikely that the oxidation of —SH component of a protein is involved. Rather it would appear that $O_2$ interferes with components of the electron carrier chain.

The time course of carbon dioxide fixation as a function of $O_2$ concentration is presented in Fig. 12. At each concentration of $O_2$, there is a lag in fixation for approximately 5 min prior to a linear uptake with time. A lag period independent of the $O_2$ concentration indicates that the Warburg effect is not related to

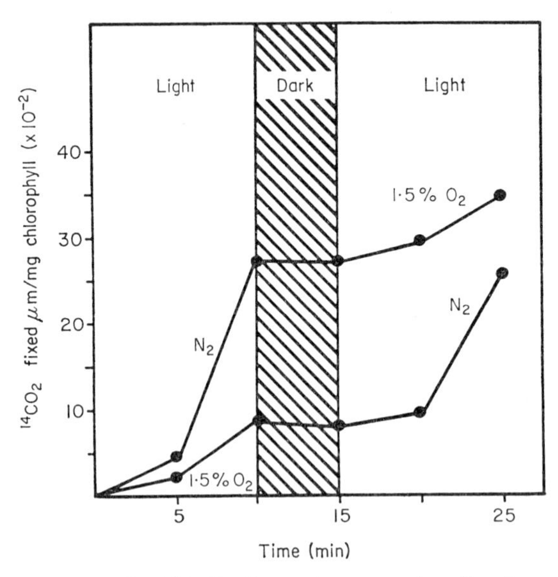


FIG. 11. Progress curves for $CO_2$ fixation under successive alteration of the gas phase between nitrogen and nitrogen containing 1·5% oxygen.

the lag period. The lag phase can be eliminated by certain intermediates of the carbon cycle which appear to fill up pools depleted during isolation of the chloroplast. Since both sugar phosphates and glycerate 3-phosphate are active in this respect, $O_2$ does not appear to affect the pertinent dehydrogenase or transfer reactions. It is not probable that an enzyme of the carbon cycle is inhibited by $O_2$.

The enzyme glyceraldehyde 3-phosphate dehydrogenase has been considered as a possible site for the inhibition of photosynthesis by $O_2$ (Turner *et al.*, 1958). This group demonstrated that at high concentrations of $O_2$, the purified sulfhydryl-containing enzyme was slowly and irreversibly inactivated. If the rate limiting step in photosynthesis is triose phosphate dehydrogenase, which it might be at saturating concentrations of light and carbon dioxide, then its inhibition by oxygen would reduce the reoxidation of reduced pyridine nucleotides and carbon dioxide assimilation. However, the Warburg effect is at a maximum when the carbon dioxide is low and under this set of conditions an enzyme of the carbon cycle would not be expected to set the pace.

A direct test of the enzyme-inactivating hypothesis was conducted by permitting chloroplasts to photosynthesize 30 min in nitrogen gas or 100% $O_2$ (Table X). Samples of chloroplasts were removed periodically, extracted with *tris* buffer and the resulting extracts assayed for the glyceraldehyde 3-phosphate dehydrogenases linked to NADP and NAD. The enzymes were determined in the back-reaction using glyceric acid 3-phosphate kinase in excess. The so-called non-arsenate dependent dehydrogenase could not be detected

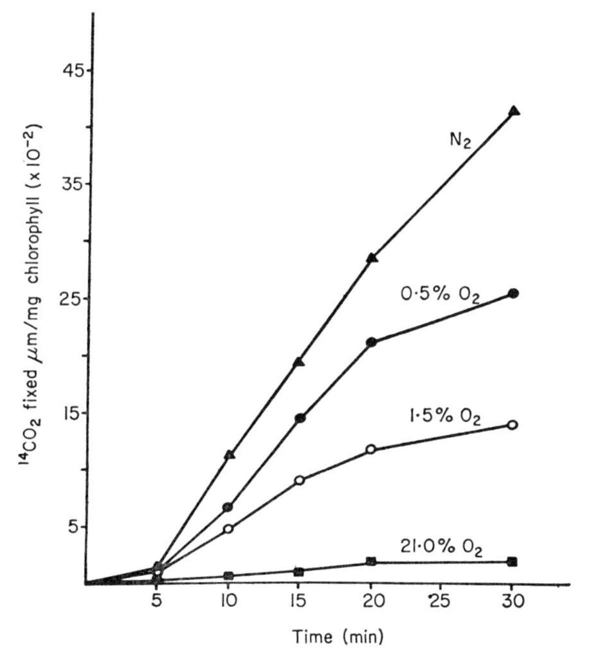


FIG. 12. Time course of $^{14}CO_2$ fixation by whole chloroplasts as a function of oxygen concentration.

in these chloroplasts. Aldolase, another sulfhydryl-containing protein, was also assayed. The rate of $CO_2$ fixation under nitrogen was roughly 2 $\mu$moles/mg chl/hr. Therefore at time zero, the endogenous (assayed in the absence of cysteine) rate of the NADP-dependent enzyme is some forty-fold in excess of the $CO_2$ fixed. The corresponding value for fructose 1,6-diphosphate aldolase is 11. While dehydrogenase activity fell off approximately 50 to 75% in 100% $O_2$ in the 30-min period of photosynthesis there was limited inactivation under the atmosphere of nitrogen. The aldolase activity remained constant. Under nitrogen, $CO_2$ fixation was linear during the 30-min period while in 100% $O_2$ the amount of radiocarbon assimilated was less than 5% of the nitrogen control (Fig. 12). Therefore during the full 30-min period at saturating light intensity

Table X. Effect of 100% $O_2$ on glyceraldehyde 3-phosphate dehydrogenase and aldolase isolated after photosynthesis by spinach chloroplasts

| Time | Atmosphere | Glyceraldehyde 3-phosphate dehydrogenase NADP+ | NAD+ | NADP+ + cysteine | Aldolase |
|---|---|---|---|---|---|
| 0 | $N_2$ | 79[1] | 33[1] | 186[1] | 22[1] |
| 0 | $O_2$ | 93 | 22 | 186 | 24 |
| 5 | $N_2$ | 74 | 19 | 219 | 19 |
| 5 | $O_2$ | 56 | 19 | 204 | 22 |
| 15 | $N_2$ | 84 | 31 | 298 | 20 |
| 15 | $O_2$ | 56 | 17 | 219 | 19 |
| 30 | $N_2$ | 56 | 19 | 242 | 21 |
| 30 | $O_2$ | 28 | 14 | 139 | 23 |

[1] μmoles of substrate utilized/mg chl/hr.

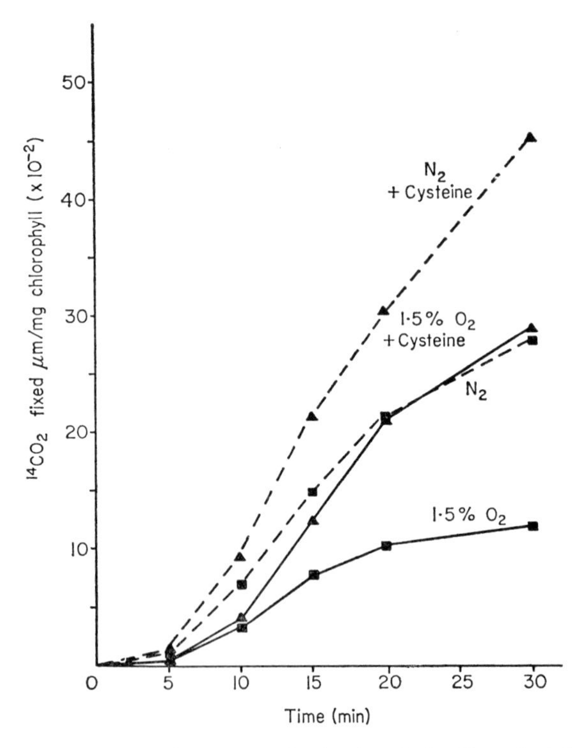


FIG. 13. Effect of 1·5 mM cysteine on $CO_2$ fixation by whole chloroplasts in nitrogen or in nitrogen containing 1·5% oxygen.

and high carbon dioxide concentration triose phosphate dehydrogenase did not appear to be the rate-limiting step. Whether both or only one of the glyceraldehyde 3-phosphate dehydrogenases are involved in the reductive pentose phosphate cycle is at present an unresolved question. Most evidence favors the NADP-linked enzyme, on the basis of the specificity of the flavoprotein transferring electrons from reduced spinach ferredoxin to $NADP^+$ rather than to $NAD^+$.

We had proposed on p. 14 that the lag in $CO_2$ uptake was not related to the Warburg effect (see also Fig. 12) but was due to a filling up of pools depleted of intermediates of the carbon cycle during isolation of the chloroplast. If this is

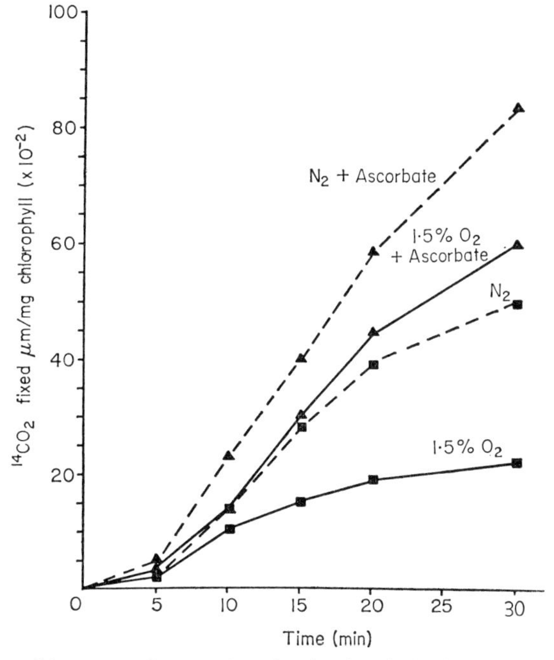


FIG. 14. Effect of 1 mM ascorbate on $CO_2$ fixation by whole chloroplasts in nitrogen or in nitrogen containing 1·5% oxygen.

so, then intermediates of the cycle like fructose 1,6-diphosphate but not electron donors like ascorbate should influence the lag. The lag was overcome by fructose 1,6-diphosphate (Fig. 5) but as illustrated in Figs. 13 and 14, ascorbate and cysteine do not bypass the lag, regardless of the composition of the atmosphere. The two contrasting experiments confirm our notion that the lag is not caused by a lack of reductant which is the result of a block in the photochemical act. The fact that ascorbate and cysteine in contrast to fructose 1,6-diphosphate (Table XI) do increase the rate of $CO_2$ uptake during the linear portion of the time course plot when the $O_2$ content is increased from 0% to 1·5% strongly suggests that there is an interference by molecular oxygen with some component of the photochemical act. The time course plots given in the two figures are essentially identical. The addition of the electron donor enhances the rate in 1·5% $O_2$ to that of the nitrogen control. On the other hand, ascorbate and cysteine also enhance the basic nitrogen rate. It would seem that the rate-limiting step in the basic nitrogen control and in the 1·5% $O_2$ incubations fortified with either ascorbate or cysteine is common to both. The question arises

Table XI. Effect of concentration of $O_2$ on $CO_2$ fixation and response to cysteine, fructose 1,6-diphosphate and glycerate 3-phosphate[1]

| | Cysteine 1·5 mM | Fructose 1,6-diphosphate 1 mM | Glycerate 3-phosphate 1 mM |
|---|---|---|---|
| $N_2$ | | | |
| 0·5% $O_2$ | 1·4[2] | 3·7[2] | 4·5[2] |
| 1·5% $O_2$ | 2·7 | 3·4 | 3·4 |
| 21% $O_2$ | 5·4 | — | 4·0 |
| 100% $O_2$ | 5·5 | — | — |

[1] Whole chloroplasts in 0·35 M NaCl, Photosynthesis of 15 min.

[2] Ratio of $\frac{\text{substrate}}{\text{control}}$

as to why the electron donors enhance the rates obtained under nitrogen. The atmosphere of a whole chloroplast carrying out photosynthesis always contains traces of oxygen. The component of the electron chain which is responsible for the Warburg effect is apparently easily re-oxidized. The presence of cysteine or ascorbate increases the flow of electrons and even under an atmosphere of

Table XII. Effect of reduced and oxidized lipoic acid and NADH on $^{14}CO_2$ fixation[1]

| Addition | μmoles/$CO_2$/mg chl/hr $N_2$ | 1·5% $O_2$·98·5% $N_2$ |
|---|---|---|
| Control | 2·0 | 0·9 |
| Reduced lipoate, 1 mM | 2·7 | 1·6 |
| Oxidized lipoate, 1 mM | 0·3 | 0·1 |
| NADH, 0·1 mM | 1·8 | 1·0 |
| Oxidized lipoate, 1 mM and NADH 0·1 mM | 0·3 | 0·1 |

[1] Whole spinach chloroplasts in 0·35 M NaCl. Photosynthesis was 15 min.

apparently "100% $N_2$" keeps the carrier in the reduced state. Additionally the transfer of electrons from water to the carrier which is reduced by ascorbate may be rate-limiting.

While 5 mM reduced lipoic acid can substitute for ascorbate or cysteine (Table XII), NADH or oxidized lipoic acid with NADH has no effect on the inhibition by molecular oxygen. It is of interest to note that oxidized lipoic acid is a potent inhibitor of $CO_2$ uptake. Petrack (1965) has demonstrated that oxidized lipoic acid is an inhibitor of photophosphorylation.

The observations suggest that the inhibition of photosynthesis by oxygen is in the formation of the reductant. The reduction of $CO_2$ is thought to be catalysed by the combined action of ATP and a reduced pyridine nucleotide. Present knowledge of the chain transferring electrons (or hydrogens) from water to $NADP^+$ points to spinach ferredoxin as the carrier most readily influenced by oxygen. In the absence of $NADP^+$, it is rapidly oxidized by molecular $O_2$. On the other hand, NADPH is quite stable in the presence of chloroplasts. Little is known about the stability of other reduced components of the photochemical act in the presence of oxygen.

Table XIII illustrates the fact that in the presence of saturating amounts of ferredoxin and large concentrations of $NADP^+$, air does not inhibit the formation of ATP or of NADPH. It has been established that reduced ferredoxin is auto-oxidizable but that in the presence of $NADP^+$ and molecular $O_2$, electrons are transferred to the former rather than to the latter. Therefore, in the presence of sufficient acceptor, the formation of $NADP^+$ is not affected by oxygen. Experiments are now in progress to determine the effect of oxygen on the formation of $NADP^+$ with chloroplasts depleted of ferredoxin and where the pyridine nucleotide is limiting.

Table XIII. Effect of aerobiosis, anaerobiosis and concentration of NaCl on photophosphorylation in the presence of ferredoxin and NADP[1]

| | μmoles/mg chl/hr | | | |
|---|---|---|---|---|
| | NADPH | ATP | $\frac{ATP}{NADPH}$ | Ferredoxin[2] −NADPH |
| Air, no NaCl added | 70 | 53 | 0·76 | 57 |
| $N_2$, no NaCl added | 70 | 56 | 0·80 | 21 |
| Air, 0·5 M NaCl | 8 | 19 | 2·4 | 47 |
| $N_2$, 0·5 M NaCl | 10 | 10 | 10 | 28 |

[1] Three units of spinach ferredoxin and 2 μmoles NADP/2 ml of reaction mixture. Spinach chloroplasts prepared in 0·35 M NaCl, then transferred to water or to 0·5 M NaCl.
[2] Thirty units of spinach ferredoxin.

Pertinent to this discussion is the fact that the amount of stimulation of $CO_2$ uptake by fructose 1,6-diphosphate in contrast to cysteine or ascorbate is independent of the $O_2$ concentration (Table XI). In order for this to occur, the phosphorylated sugars require a source of ATP. While oxygen may inhibit NADPH formation, nevertheless, ATP may be formed. Chloroplast fragments have been shown by Fewson, Black and Gibbs (1963) to catalyse phosphorylation in the presence of large amounts of spinach ferredoxin and no other added electron acceptor. Rates of photophosphorylation up to 108 μmoles

ATP/mg chl/hr were obtained. The phosphorylation is inhibited by anaerobic conditions.

In conclusion, there is no agreement as to the cause of the Warburg effect. Our data suggest that oxygen does not act on photosynthesis primarily by reversibly inactivating an enzyme of the carbon cycle. We favor the hypothesis that oxygen interferes with the formation of NADPH by reoxidizing reduced ferredoxin.

## Compounds Formed During Assimilation of $^{14}CO_2$

### whole chloroplasts

After $^{14}CO_2$ fixation by spinach chloroplasts kept in 0·35 M NaCl, Allen *et al.* (1955) found among the water-soluble products the following radioactive compounds: phosphate esters of fructose, glucose, ribulose, dihydroxyacetone, and glyceric acid; glycollic, malic and aspartic acids; alanine, glycine, free dihydroxyacetone and glucose. About 20% of the $^{14}C$ assimilated was found in an insoluble polysaccharide, apparently starch.

Table XIV. Rate and distribution of $^{14}C$ following 1 hr photosynthesis of $^{14}CO_2$ by various spinach leaf preparations

| Compounds | Whole chloroplasts plus cell sap[1] 1 min | Whole chloroplasts[1] 30 min | Homogenate 0·35 M NaCl[2] 30 min | Intact leaf[2] 30 min |
|---|---|---|---|---|
| Diphosphates | 13 | 3·4 | 13[3] | 6·0[3] |
| UDPGlucose | — | 4·9 | — | — |
| Monophosphates | 11 | 46 | 26 | 2·5 |
| Phosphoglycerate | 61 | 78 | 5·2 | 0·5 |
| Aspartic acid | 7·9 | 4·5 | 26·5 | 6·6 |
| Glutamic acid | — | — | 2·0 | 4·5 |
| Alanine | — | 1·8 | 2·5 | 2·5 |
| Glycine | — | 7·8 | 2·0 | 1·0 |
| Glyceric acid | — | — | 4·5 | 1·5 |
| Glycollic acid | — | — | 2·5 | 7·0 |
| Malic acid | 4·2 | 4·0 | 0·0 | 0·6 |
| Sucrose | — | — | 1·1 | 44·0 |
| Glucose | — | — | 6·0 | 3·0 |

[1] Holm-Hansen *et al.* (1962).
[2] Gee *et al.* (1965).
[3] Values are given as per cent total activity found on the paper chromatogram.

Using the isolation technique of McClendon (1952), whole spinach chloroplasts were found by Holm-Hansen *et al.* (1959) to distribute isotope into similar compounds after a 30-min period of photosynthesis in $^{14}CO_2$ (Table XIV). The addition of "cell sap" (extracts of whole leaves) stimulated $CO_2$ uptake about four-fold. This increase was probably due to phosphorylated sugars present in the cell sap (Dugger *et al.*, 1963). These authors (Holm-Hansen *et al.*, 1959) also presented a time-course study with the intact chloroplasts combined with sap. The uptake of $CO_2$ was linear with time and glyceric acid 3-phosphate was the dominant radioactive material, especially for the shorter exposure periods to radioactive carbon dioxide. A plot of $^{14}C$ with time in each compound as a percentage of the total soluble radioactivity showed that glyceric acid 3-phosphate and the diphosphate areas had negative slopes but of these the glyceric acid 3-phosphate contained the most activity, having 61% of the soluble activity after 1 min of fixation time.

Gee *et al.* (1965) have also measured $^{14}CO_2$ uptake by chloroplasts prepared in 0·35 M NaCl (Table XIV). They compared the distribution of compounds found in intact leaves with that found in various spinach preparations. They found that the distribution of radioactive products in the NaCl-free preparation more closely resembled that of the intact tissue than does the homogenate prepared in 0·35 M.

Experiments similar to Allen *et al.* (1955) have been carried out recently in this laboratory. The incubations were conducted in 0·35 M NaCl for a period of 20 min. Compounds were separated on a Dowex 1-chloride column (Table XV). Roughly 80% of the total fixed was found in the water soluble fraction. The remainder was located as glucose obtained after a 1 hr hydrolysis of the insoluble material. The rate ($\mu$moles $CO_2$/mg chl/hr) of the control was 1·4 during the 5 min lag phase and 5·8 during the 15 min period after the lag phase. In the presence of 0·1 mM fructose 1,6-diphosphate the rate of fixation was 9·5. In Walker's experiments (1964), the major products were triose phosphates together with some pentose phosphates, hexose phosphates, malate, aspartate, and glycollate.

Table XV. Distribution of $^{14}C$ among water-soluble products after $CO_2$ fixation in the presence and absence of fructose 1,6-diphosphate

| Compound | Control[1] cpm × $10^{-3}$ | FDP[2] 0·1 mM cpm × $10^{-3}$ |
|---|---|---|
| Amino acids | 435 | 782 |
| Glucose 6-phosphate and fructose 6-phosphate | 749 | 670 |
| Triose phosphate and organic acids | 708 | 662 |
| Glyceric acid 3-phosphate | 1244 | 2366 |
| Fructose 1,6-diphosphate | 45 | 361 |

[1] Total fixation was 4,347,800 cpm.
[2] Total fixation was 5,882,300 cpm.

One important conclusion can be drawn from all these experiments, namely, that the products formed by whole chloroplasts after photosynthesizing for short times in $^{14}CO_2$ are similar to those found with intact leaves.

## BROKEN PREPARATIONS

This system, the reconstituted chloroplast preparation, has been documented thoroughly by Losada, Trebst and Arnon (1960). Here, the reductive phase (NADPH and ATP formation) is catalysed by broken chloroplasts and the

Table XVI. Effect of time on pattern of $CO_2$ fixation in light by isolated chloroplasts[1]

| Time min | Total $^{14}CO_2$ fixed cpm | Total $^{14}CO_2$ fixed as PGA % | Total $^{14}CO_2$ fixed as Sugar phosphates % |
|---|---|---|---|
| 1 | 108,000 | 100 | |
| 2 | 196,000 | 75 | 24 |
| 5 | 450,000 | 54 | 38 |
| 10 | 600,000 | 40 | 57 |
| 20 | 750,000 | 13 | 86 |

Taken from Losada, Trebst and Arnon (1960).
[1] Primer is ribulose 1,5-diphosphate. Other additives are ADP, FMN, GSH, ascorbate and phosphate.

enzymes of the carbon cycle is a water-soluble extract of chloroplast. In contrast to the whole chloroplast, the reconstituted chloroplast preparation *must* be fortified with a primer. In the experiments of Losada, Trebst and Arnon

Table XVII. Effect of phosphorylated compounds on photosynthetic $CO_2$ fixation by broken chloroplasts[1]

| Additions | $CO_2$ fixed $\mu$moles |
|---|---|
| None | 2·0 |
| Glyceric acid 3-phosphate | 3·7 |
| Ribose 5-phosphate | 4·0 |
| Glucose 1-phosphate | 4·4 |
| Glucose 6-phosphate | 3·8 |
| Fructose 6-phosphate | 3·1 |
| Fructose 1,6-diphosphate | 5·2 |

Taken from Whatley *et al.* (1956).
[1] Three $\mu$moles of each phosphorylated compound and 10 $\mu$moles of $CO_2$ added.

(1960), the primers used have been ribulose 1,5-diphosphate, glucose 1-phosphate, ribose 5-phosphate among others. In the absence of a primer, there is no fixation by the reconstituted system. The fixation products differ somewhat from those observed with the whole chloroplasts. Tracer does not appear in an insoluble polysaccharide or in amino and organic acids. All isotope appears to be located in glyceric acid 3-phosphate and sugar phosphates (Table XVI).

A cycle to be a true cycle like the citric acid cycle of Krebs and the photosynthetic carbon cycle of Calvin and Benson must regenerate an acceptor. In the citric acid cycle the regenerated acceptor is oxaloacetate while ribulose 1,5-diphosphate plays a similar role in photosynthesis. Mitochondria can consume large quantities of acetate when fortified with catalytic amounts of oxaloacetate. On a *molar* basis the amount of $CO_2$ fixed by the broken preparations has *never* been demonstrated to exceed the primer added (Table XVII). It if can be demonstrated that the reconstituted chloroplast system like the whole chloroplast can assimilate more than one mole of $CO_2$ per mole of primer added then one critical criterion of a complete carbon cycle will be satisfied.

## INTRAMOLECULAR SPREAD OF ISOTOPE DURING $^{14}CO_2$ ASSIMILATION

Losada, Trebst and Arnon (1960) have concluded that fragmented chloroplasts as well as whole chloroplasts have a complete and functioning photosynthetic carbon cycle. This conclusion was based essentially on the appearance of $^{14}C$ in some compounds of the cycle after light-induced $^{14}CO_2$ uptake. Since the tracer can spread to a number of compounds by a series of reactions which do not necessarily constitute a cycle, its appearance in a few compounds of the reductive pentose phosphate cycle is only suggestive evidence that a functioning reductive $CO_2$ cycle is present in the isolated systems. For this reason, it is important to determine whether during assimilation of $^{14}CO_2$, isotope spreads throughout the molecules in a manner similar to that found in the intact algal or higher plant cell.

In a preliminary study, Gibbs and Cynkin (1958) reported a similarity in photosynthetic $^{14}CO_2$ assimilation between the whole chloroplast and the whole plant. They observed an asymmetrical and subsequent uniform distribution of $^{14}C$ in the glucose residue of spinach chloroplast polysaccharide formed from $^{14}CO_2$. A similar observation was made earlier with intact algal and high plant tissues by Gibbs and Kandler (1957). More recently Havir and Gibbs (1963) degraded glucose 6-phosphate, the polysaccharide glucose, and glycerate 3-phosphate isolated from whole spinach chloroplasts (Table XVIII). In contrast, other workers (Trebst and Fiedler, 1961, 1962; Havir and Gibbs, 1962, 1963) could observe only a limited spread of tracer in glucose 6-phosphate, fructose 1,6-diphosphate, glycerate 3-phosphate, and dihydroxyacetone phosphate formed with a broken spinach chloroplast preparation fortified with each of the following substrates: ribose 5-phosphate and $^{14}CO_2$, ribulose 1,5-diphosphate and $^{14}CO_2$ or [1-$^{14}C$]glycerate 3-phosphate (Table XIX).

The striking aspect of the data (compare Table XVIII with XIX) is the contrast between the rapidity with which isotope spreads to all the carbon atoms of polysaccharide glucose and glycerate 3-phosphate in the intact chloroplast and the slowness with which the tracer spreads to these compounds in the reconstituted systems. It is of significance that this difference in extent of distribution of tracer could not have been predicted on the basis of the chromatography

Table XVIII. Distribution of $^{14}C$ in polysaccharide glucose and glycerate 3-phosphate during photosynthetic $^{14}CO_2$ fixation by whole spinach chloroplasts

| Compound degraded | $^{14}C$ content after reaction time of 4 min | 10 min | 20 min |
|---|---|---|---|
| Carbons of polysaccharide glucose | | | |
| C-1 | 72 | 78 | |
| C-2 | 62 | 61 | |
| C-3 | 86 | 94 | |
| C-4 | 100 | 100 | |
| C-5 | 40 | 60 | |
| C-6 | 45 | 67 | |
| Carbons of glycerate 3-phosphate | | | |
| COOH | 100 | 100 | 100 |
| CHOH | 55 | 63 | 73 |
| $CH_2OP$ | 62 | 63 | 79 |

Taken from Havir and Gibbs (1963).

Table XIX. Distribution of $^{14}C$ in hexose phosphates isolated from reconstituted chloroplast systems

| Compound degraded | $^{14}C$ content with substrate: Ribose 5-phosphate 5 min | 20 min | 40 min | Ribulose 1,5-diphosphate 5 min | 20 min | 40 min |
|---|---|---|---|---|---|---|
| Carbons of glucose 6-phosphate | | | | | | |
| C-1 | 6[1] | 22 | 9 | 5 | 18 | 17 |
| C-2 | 7 | 29 | 6 | 5 | 30 | 12 |
| C-3 | 87 | 98 | 56 | 88 | 83 | 70 |
| C-4 | 100 | 100 | 100 | 100 | 100 | 100 |
| C-5 | 3 | 8 | 1 | 0·4 | 8 | 5 |
| C-6 | 8 | 7 | 2 | 0·6 | 5 | 4 |
| Carbons of glycerate 3-phosphate | | | | | | |
| COOH | | | | | | 100 |
| CHOH | | | | | | 4 |
| $CH_2OP$ | | | | | | 3 |

The 5 and 20 min incubations are those of Trebst and Fiedler (1961); the 40 min incubations—Havir and Gibbs (1963).

[1] The carbon content is based on C-4 or COOH=100.

(paper or column) since label appeared in essentially the same compounds in the experiments with both whole chloroplasts and reconstituted systems.

The rapid approach toward a uniformly labeled polysaccharide hexose in the whole chloroplast preparations was comparable to that observed in whole cells (Gibbs and Kandler, 1957). The inability of the reconstituted systems to bring about the tracer distribution observed with the intact preparations remains to be explained. In the experiments of Havir and Gibbs (1963), the preparations were fortified with spinach ferredoxin and with extracts of lyophilized chloroplasts (non-aqueous procedure) to insure sufficient reductant. It would appear that the regeneration of the $CO_2$-acceptor from hexose phosphate is the rate-limiting step in the broken preparations.

It is concluded that the whole chloroplasts possess a complete reductive pentose phosphate cycle for converting $CO_2$ into carbohydrate. In sharp contrast, this cycle is operating at a rather limited rate, if at all, in the broken preparations.

## Effect of Fructose 1,6-Diphosphate and Glycerate 3-phosphate on Spreading of Isotope in Whole Chloroplasts

From the distribution of $^{14}C$ among the products of $^{14}CO_2$ by intact chloroplasts (Table XVI), the major increases in $^{14}C$, caused by the addition of 1 mM fructose 1,6-diphosphate, is in the glycerate 3-phosphate fraction. The appearance of excess glycerate 3-phosphate indicates that the regenerative cycle, i.e. the formation of ribulose 1,5-diphosphate from fructose 1,6-diphosphate and its subsequent carboxylation proceeds at a faster rate than the reductive phase of the carbon cycle. An appreciable amount of labeled fructose 1,6-diphosphate was obtained apparently by a "trapping" mechanism involving the carrier fructose 1,6-diphosphate.

The question whether the phosphorylated compounds which enhance $CO_2$ fixation do or do not contribute their carbon skeletons to the regenerative phase of the carbon cycle, could not be satisfactorily answered until the extent of intramolecular distribution of $^{14}C$ would have been examined in at least one product of the cycle. This was a procedure used to test the lack or presence of the complete cycle in the reconstituted chloroplast system. A criterion for the operation of the complete cycle is the rate of spread of radiocarbon from the carboxyl group to the $\alpha$ and $\beta$ carbons of glycerate 3-phosphate. The degradation data in Table XX indicate that isotope was initially highest in the carboxyl carbon and spread rapidly into the $\alpha$ and $\beta$ carbons (see Table XIX and contrast with spread in reconstituted system). The spread of isotope was slowed by addition of 1 mM fructose 1,6-diphosphate or 1 mM glycerate 3-phosphate. This effect was less pronounced when the concentration of the additive was reduced to 0·1 mM.

Two mechanisms can be suggested to account for the apparent decrease in the spread of isotope with the glycerate 3-phosphate: (1) dilution of the [$^{14}C$]-ribulose 1,5-diphosphate by unlabeled ribulose 1,5-diphosphate formed from the added phosphorylated compound whereby on subsequent carboxylation with $^{14}CO_2$ the $\alpha$ and $\beta$ carbons could have their specific activities kept lower with respect to the carboxyl carbon and (2) inhibition of the carbon cycle by the added substrate. Since the rate of $CO_2$ fixation is not inhibited by fructose 1,6-diphosphate or glycerate 3-phosphate and since ribulose 1,5-diphosphate carboxylase is not affected by any of the phosphorylated compounds tested (unpublished observations), the first mechanism is likely to occur even though the major accumulation of tracer is in the glycerate 3-phosphate fraction. We interpret these findings as additional evidence that a major effect of the added phosphorylated compound on $CO_2$ fixation by whole chloroplasts is that of supplying intermediates of the photosynthetic carbon cycle.

Table XX. Distribution of $^{14}C$ in glycerate 3-phosphate during photosynthetic $^{14}CO_2$ fixation by whole spinach chloroplasts. Effect of fructose 1,6-diphosphate and glycerate 3-phosphate

| | $^{14}C$ content after reaction time of | | | | | | | | |
|---|---|---|---|---|---|---|---|---|---|
| | 1 min | | | 3 min | | | 2 min | | |
| Treatment | COOH | CHOH | $CH_2OH$ | COOH | CHOH | $CH_2OH$ | COOH | CHOH | $CH_2OH$ |
| $CO_2$ | 100 | 25 | 25 | 100 | 73 | 73 | | | |
| $CO_2$+1 mM PGA | 100 | 16 | 16 | 100 | 48 | 43 | | | |
| $CO_2$+1 mM FDP | 100 | 7 | 7 | 100 | 8 | 7 | | | |
| $CO_2$+0·1 mM PGA | | | | | | | 100 | 27 | 23 |
| $CO_2$+0·1 mM FDP | | | | | | | 100 | 40 | 44 |

## Factors That Do not Affect $CO_2$ Fixation in Whole Chloroplasts

### FERREDOXIN

While there is substantial evidence that spinach ferredoxin is involved in the photochemical act of the broken preparations unequivocal evidence of its participation in the whole chloroplast is missing. Whole chloroplasts show no response when fortified with spinach ferredoxin.

### ATP AND REDUCED PYRIDINE NUCLEOTIDES

ATP in low amounts (0·2 $\mu$M) inhibits photosynthetic $CO_2$ fixation by whole chloroplasts in 0·35 M NaCl. Oxidized and reduced pyridine nucleotides, likewise, have not been reported to stimulate $CO_2$ fixation by whole chloroplasts.

## References

Allen, M. B., Arnon, D. I., Capindale, J. B., Whatley, F. R. and Durham, L. J. (1955). *J. Am. chem. Soc.* 77, 4149.
Arnon, D. I., Allen, M. B. and Whatley, F. R. (1954). *Nature, Lond.* **174**, 394.

Arnon, D. I., Allen, M. B. and Whatley, F. R. (1956). *Biochim. biophys. Acta* **20**, 449.
Avron, M. (1961). *Biochem. J.* **78**, 735.
Avron, M. (1964). *Biochem. Biophys. Res. Commun.* **17**, 430.
Avron, M. and Jagendorf, A. T. (1959). *J. biol. Chem.* **234**, 967.
Baltscheffsky, H. (1960). *Acta chem. scand.* **14**, 264.
Baltscheffsky, H. and De Kiewiet, D. Y. (1964). *Acta chem. scand.* **18**, 2406.
Bamberger, E. S., Black, C. C., Fewson, C. A. and Gibbs, M. (1963). *Pl. Physiol.* **38**, 438.
Bamberger, E. S. and Gibbs, M. (1963). *Pl. Physiol.* **38** X.
Bamberger, E. S. and Gibbs, M. (1965). *Pl. Physiol.* **40**, 919.
Bassham, J. A. (1963). *In* "Photosynthetic Mechanisms of Green Plants" (A. T. Jagendorf and B. Kok, eds.) p. 635. National Academy of Sciences, Washington, D.C.
Calo, N. and Gibbs, M. (1960). *Z. Naturf.* **15**b, 287.
Calvin, M. (1954). *Fed. Proc.* **13**, 697.
Dugger, Jr., W. M., Koukol, J., Reed, W. O. and Palmer, R. L. (1963). *Pl. Physiol.* **38**, 468.
Elbertzhagen, H. and Kandler, O. (1962). *Nature, Lond.* **194**, 312.
Fager, E. W. (1952a). *Archs Biochem. Biophys* **37**, 5.
Fager, E. W. (1952b). *Archs Biochem. Biophys.* **41**, 383.
Fewson, C. A., Black, C. C. and Gibbs, M. (1963). *Pl. Physiol.* **38**, 680.
Fluharty, A. and Ballou, C. E. (1959). *J. biol. Chem.* **234**, 2517.
Gee, R., Joshi, G., Bils, R. F. and Saltman, P. (1965). *Pl. Physiol.* **40**, 89.
Gibbs, M. (1963). *In* "Photosynthetic Mechanisms of Green Plants" (A. T. Jagendorf and B. Kok, eds.) p. 666. National Academy of Sciences, Washington, D.C.
Gibbs, M., Black, C. C. and Kok, B. (1961). *Biochim. biophys. Acta* **52**, 474.
Gibbs, M. and Calo, N. (1960). *Biochim. biophys. Acta* **44**, 341.
Gibbs, M. and Calo, N. (1959). *Pl. Physiol.* **34**, 318.
Gibbs, M. and Cynkin, M. A. (1958). *Nature, Lond.* **183**, 1241.
Gibbs, M. and Kandler, O. (1957). *Proc. natn. Acad. Sci. U.S.A.* **43**, 446.
Havir, E. A. and Gibbs, M. (1962). *Fed. Proc.* **21**, 92.
Havir, E. A. and Gibbs, M. (1963). *J. biol. Chem.* **238**, 3183.
Heber, U., Pon, N. G. and Heber, M. (1963). *Pl. Physiol.* **38**, 355.
Heber, U. and Tyszkiewicz, E. (1962). *J. exp. Bot.* **13**, 185.
Heytler, P. G. and Prichard W. N. (1962), *Biochem. biophys. Res. Commun.* **7**, 272.
Hill, R. and Whittingham, C. P. (1955). "Photosynthesis." Methuen and Co. Ltd., London.
Holm-Hansen, O., Pon, N. G., Nishida, K., Moses, V. and Calvin, M. (1959). *Physiol. Plantarum* **12**, 475.
Horio, T. and Yamashita, T. (1962). *Biochem. Biophys. Res. Commun.* **9**, 142.
Jagendorf, A. T. and Margulies, M. M. (1960). *Archs Biochem. Biophys.* **90**, 184.
Kandler, O. (1957). *Naturwissenschaften* **21**, 562.
Losada, M., Trebst, A. V. and Arnon, D. I. (1960). *J. biol. Chem.* **235**, 832.
Lynen, F. (1961). *Fed. Proc.* **4**, 941.
McAlister, E. D. and Myers, J. (1940). *Smithsonian Inst. Pub. Misc. Collections* **99**, 6.
McClendon, J. H. (1952). *Am. J. Bot.* **39**, 275.
Miyachi, S., Izawa, S. and Tamiya, H. (1955). *J. Biochem.* (*Tokyo*) **42**, 221.
Park, R. B., Pon, N. G., Louwrier, K. P. and Calvin, M. (1960). *Biochim. biophys. Acta* **42**, 27.
Petrack, B. (1965). *J. biol. Chem.* **240**, 906.
Racker, E., Klybas, V. and Schramm, M. (1959). *J. biol. Chem.* **234**, 2510.
Smillie, R. M. and Fuller, R. C. (1959). *Pl. Physiol.* **34**, 651.
Stocking, C. R. (1959). *Pl. Physiol.* **34**, 56.
Tagawa, K., Tsujimoto, H. Y. and Arnon, D. I. (1963). *Proc. natn. Acad. Sci. U.S.A.* **49**, 567.
Tamiya, H. and Huzisige, H. (1949). *Stud. Tokugawa Inst.* **6**, 83.
Thalacker, R. and Behrens, M. (1959). *Z. Naturf.* **14**b, 443.
Trebst, A. and Fiedler, F. (1961). *Z. Naturf.* **16**b, 284.
Trebst, A. and Fiedler, F. (1962). *Z. Naturf.* **17**b, 553.
Trebst, A. V., Tsujimoto, H. Y. and Arnon, D. I. (1958). *Nature, Lond.* **182**, 351.
Turner, J. F. and Brittain, C. G. (1962). *Biol. Rev.* **37**, 130.
Turner J. F., Black, C. C. and Gibbs, M. (1962). *J. biol. Chem.* **237**, 577.

Turner, J. S., Turner, J. F., Shortman, K. D. and King, J. E. (1958). *Aust. J. Biol. Sci.* **11**, 336.
Vernon, L. P. and Zaugg, W. S. (1960). *J. biol. Chem.* **235**, 2728.
Walker, D. A. (1964). *Biochem. J.* **92**, 22c.
Warburg, O. (1920). *Biochem. Z.* **103**, 188.
Wessels, J. S. C. (1964). *Biochim. biophys. Acta* **79**, 640.
Whatley, F. R., Allen, M. B., Rosenberg, L. L., Capindale, J. B. and Arnon, D. I. (1956). *Biochim. biophys. Acta* **20**, 462.
Whatley, F. R., Allen, M. B., Trebst, A. V. and Arnon, D. I. (1960). *Pl. Physiol.* **35**, 188.
Willstätter, R. and Stoll, A. (1918). "Untersuchungen über Die Assimilation der Kohlensäure." J. Springer, Berlin.

## Discussion

*W. W. Hildreth:* Will you please comment on the present status of the quantum efficiency of $CO_2$ fixation in chloroplast preparations.

*M. Gibbs:* As far as I am aware, there are no reports in the literature referring to the quantum efficiency of $CO_2$ fixation in chloroplast preparations. We attempted such measurements a few years back. The values were ridiculously bad.

---

*D. Arnon:* Much has been said about the use of salt in the isolation of chloroplasts for biochemical work. New students of the biochemistry of chloroplasts may wonder why salt is used at all in the isolation of chloroplasts. We introduced the method of isolating chloroplasts with isotonic sodium chloride solutions in 1954 when procedures for isolating chloroplasts in isotonic sucrose or glucose solutions had already been used by other investigators. In 1954, our aim was to investigate ATP formation and carbohydrate synthesis by chloroplasts under such experimental conditions that light would be the sole source of added energy. We wanted, therefore, a system as free as possible from added carbohydrates and other energy-rich substrates that might conceivably contribute chemical energy for synthetic purposes. We selected NaCl as the least harmful salt to provide a proper osmotic environment for the isolated chloroplasts. By observing isolated spinach chloroplasts under a light microscope we found that 0·35 M NaCl was the concentration which was reasonably iso-osmotic for them. The techniques of electron microscopy for chloroplasts were not available to us in 1954. We were happy, therefore, to see now from the electron micrographs of Prof. von Wettstein that spinach chloroplasts at 0·35 M NaCl preserve reasonably well their structural integrity.

# Rate-limiting Factors in the Photosynthesis of Isolated Chloroplasts

O. Kandler, H. Elbertzhagen and I. Haberer-Liesenkötter

*Institut für Angewandte Botanik der Techn. Hochschule München, Germany*

## Introduction

The rate of $CO_2$ fixation in broken chloroplasts reported so far in the literature amounts to maximally 3 to 5 μmoles/mg chl/hr (Gibbs and Calo, 1959; Whatley *et al.*, 1960; Heber and Tyszkiewicz, 1962). This is about 5% of the optimal rate of photosynthesis in intact cells at a corresponding temperature. As a result of this low rate the sugar phosphates are labelled only after several minutes of photosynthesis in $^{14}CO_2$ (Holm-Hansen *et al.*, 1959), whereas in intact cells a few seconds are sufficient.

The label spreads also very slowly within the hexose molecule, as indicated by the results of Trebst and Fiedler (1962). They find only 1–2% of the radioactivity in each terminal position of fructose 1,6-diphosphate (FDP), compared to more than 90% still left in C-3+4, after 11 min photosynthesis of broken chloroplasts in $^{14}CO_2$. Even after 20 min photosynthesis in $^{14}CO_2$ or in [1-$^{14}$C]-3-phosphoglyceric acid (PGA) only 1–3% of the label in glucose phosphate were contained in C-6 or C-5 while C-1 or C-2 showed 8–11% (Trebst and Fiedler, 1961). This finding suggests that practically no newly formed sugar passed through the regeneration cycle in spite of the long reaction times. Obviously, the label was spread only in the upper half, probably caused by a shuttle of the top $C_2$ unit between pentose and hexose monophosphates. Such a mechanism was postulated by Gibbs and Kandler (1957) under the designation "exchange and dilution pathway" as one of the possible explanations of the asymmetric labelling of the hexose, and its existence was demonstrated by Trebst and Fiedler (1962) in broken chloroplasts. The effect of this mechanism is especially pronounced if the turnover in the regeneration cycle is inhibited.

It seems that in isolated chloroplasts two steps are limiting: (1) The regeneration of the $CO_2$ acceptor, which was pointed out especially by Holm-Hansen *et al.* (1959). (2) The reduction of the carboxylation product, which is emphasized by Heber and Tyszkiewicz (1962). According to Holm-Hansen *et al.* (1959) the limitation at these two sites results from a too low photosynthetic phosphorylation caused by a lack of cofactors or unfavourable pH, whereas Heber and Tyszkiewicz (1962) regard dilution of the enzymes as the decisive factor. In fact, they were able to obtain an increase in $CO_2$ fixation by the addition of protein-containing extracts from chloroplasts prepared in non-aqueous media. However, only an increase of PGA formation and not of complete photosynthesis was achieved.

It is clear that the dilution caused during the isolation of chloroplasts not only concerns the enzymes but also the cofactors and intermediate compounds. In the intact cell photosynthesis, including starch formation, takes place in compartments which are protected by osmotic barriers, namely the chloroplasts, as suggested already by the observations of Sachs. At least some of the intermediate compounds and cofactors involved in photosynthesis show a rather slow exchange with the corresponding cytoplasmic pools (Heberer and Willenbrinck, 1964).

In contrast, in broken chloroplasts the osmotic barriers, which are necessary for the maintenance of the high concentrations of intermediates, are destroyed and the structurally bound enzymes are partly or entirely dissolved or inactivated.

To elucidate some of the questions still outstanding on the limiting factors in the photosynthesis of isolated chloroplasts, the influence of PGA, NADP and ADP concentrations on the rate of $CO_2$ fixation and reduction of broken chloroplasts was studied. Later the changes in the distribution of $^{14}CO_2$ caused by the addition of large amounts of unlabelled PGA during photosynthesis in $^{14}CO_2$ were determined. This was intended to show if the PGA, newly formed by $CO_2$ fixation, is preferentially reduced to sugar phosphates as compared to the unlabelled PGA in the solution. This would indicate the participation of structurally bound enzymes and intermediates in the remaining lamellar structures of the chloroplasts.

## Materials and Methods

Spinach was bought at the local market or harvested freshly from the field. After picking over and removing the midribs and petioles, the leaves were used to prepare isolated chloroplasts according to the method of Arnon *et al.* (1956) and Gibbs and Calo (1959). To grind the chilled leaves a 0·35 M sodium chloride solution with 0·015 M *tris* buffer, pH 7·5, and 0·01 M sodium ascorbate was used.

In order to obtain broken chloroplasts and chloroplast extract the whole chloroplasts were washed once, suspended in a 0·01 M sodium ascorbate solution and centrifuged after 15 min.

The reaction was carried out in specially designed flat glass vessels with a plane basal surface of 4 $cm^2$, and two openings which were closed during the reaction by rubber caps. The vessels were fixed to a shaker, kept in a water bath of 20°, and illuminated from below by three 250 W tungsten lamps. All experiments were performed under light saturation (25,000 lux).

After flushing with purified $N_2$ for 5 min in the light, the reaction was started by injecting the substrate with a syringe. The reaction was terminated by adding a double volume of 96% alcohol + 1% acetic acid, or trichloroacetic acid to a final concentration of 5%. Trichloroacetic acid was preferred when determinations of intermediate compounds were carried out.

For measurement of the $CO_2$ fixation, Na $H^{14}CO_3$ of known specific activity was used. The total tracer fixed was determined by plating duplicate aliquots on aluminum planchets and counting at dryness with a gas-flow counter.

For paper chromatography the chloroplasts were extracted with 80 and 20%

alcohol. Two–five% of the total amount was spotted on filter paper Whatman No. 1. The separation was carried out one-dimensionally in butanol–propionic acid–water, or two-dimensionally using the same solvent and water-saturated phenol (Benson *et al.*, 1950). After radioautography on X-ray film the radioactivity of the spots was determined by a gas-flow thin window counter. For identification the spots were eluted, treated with phosphatase and re-chromatographed together with authentic substances.

The various intermediate compounds were determined enzymatically as described previously (Kandler *et al.*, 1961). The chlorophyll content was determined by the method of Arnon (1949).

## RESULTS

### EFFECT OF PGA CONCENTRATION ON $CO_2$ FIXATION IN BROKEN CHLOROPLASTS

Different quantities of PGA together with $H^{14}CO_3^-$ were added to several samples of broken chloroplasts after 5 min flushing with $N_2$ in the light and then incubated for 15 min under continued illumination. As presented in Table I $CO_2$ fixation was slightly inhibited by increasing PGA concentrations in three experiments (a–c), whereas it was increased in other experiments (d, e).

Table I. Effect of PGA concentration on $CO_2$ fixation of broken chloroplasts. Numbers express $\mu$mole $CO_2$ mg chl/hr

| | $\mu$mole PGA added | | |
|---|---|---|---|
| Experiment | 0 | 0·34 | 3·4 |
| a | 3·0 | 2·6 | 2·4 |
| b | 2·8 | 2·6 | 2·4 |
| c | 2·6 | 2·2 | 2·0 |
| d | 2·2 | 2·4 | 3·7 |
| e | 2·52 | 2·63 | 2·9 |

The reaction mixture (total volume 2·0 ml) contained: 200 $\mu$g (a–c) or 400 $\mu$g (d, e) chlorophyll and chloroplast extract equivalent to 800 $\mu$g chlorophyll; in $\mu$mole: sodium chloride 700; *tris* pH 8·2; manganese chloride 0·8; magnesium chloride 2·0; $Na_2HPO_4$ 2; sodium ascorbate 20; $NaH^{14}CO_3$ 6; specific activity 24·5 $\mu$c/$\mu$mole; ADP 0·2; $NADP^+$ 0·12; FMN $2\times10^{-3}$; R 5-P 0·4; GSH 4·0. pH=8·0–8·2; $t$=20°; gas phase nitrogen; light intensity 25,000 lux; time of exposure 15 min.

As discussed more fully below, the increase is not caused by photosynthetic $CO_2$ fixation but by an increasing Wood-Werkman reaction[1] leading to the formation of aspartic acid. The ability of the preparations to perform this additional $CO_2$ fixation is rather different with the different batches. Probably

[1] The term Wood-Werkman reaction is used here since it was not determined which of the many 3+1 addition reactions known today is actually involved here (presumably PEP-carboxylase).

the variations depend on the purity of the preparations. In literature too, the statements on the occurrence of aspartic acid after photosynthesis of broken chloroplasts are varying. Allen *et al.* (1955), Holm-Hansen *et al.* (1959) and Heber and Tyszkiewicz (1962) found significant amounts of aspartic acid regularly, while it is not mentioned in the work of Trebst *et al.* (1959, 1960), Whatley *et al.* (1960) or Losada *et al.* (1960).

## REDUCTION OF PGA AND FORMATION OF DIHYDROXYACETONE PHOSPHATE (DAP), FDP AND GLUCOSE 6-PHOSPHATE (G 6-P)

### *Kinetics*

Practically no PGA is present in broken chloroplasts. The determination of PGA in the extract of a large amount of broken chloroplasts yielded only about $3 \times 10^{-4}$ $\mu$moles PGA/mg chlorophyll. This means that merely 6 to $12 \times 10^{-5}$ $\mu$mole per sample are present. Thus, at the beginning of $CO_2$ fixation, PGA reduction will most probably be limited by the low concentration of PGA. At a rate of fixation of 3 $\mu$moles/mg chl/hr, as determined in the more efficient of our preparations, 0·2 $\mu$mole PGA per sample is formed within the first 10 min, thus leading to a gradual increase in the rate of reduction.

The actual reduction capacity may be determined by adding PGA and measuring the amounts left after different intervals. In order to be sure that the consumed PGA was really reduced and not transformed otherwise (aspartic acid!) determinations of some sugar phosphates were carried out simultaneously. As shown in the two experiments of Table II, PGA reduction amounted to about 15 $\mu$moles/mg chl/hr at both PGA concentrations. This would correspond (10 min value) to a photosynthetic $CO_2$ fixation of 7·5 $\mu$moles. After short reaction times the quantity of the three sugar phosphates determined is equivalent to about 50% of the PGA used up.

Table II. PGA reduction and sugar phosphate production by broken chloroplasts. Numbers express $\mu$mole/sample

| Exp. | $\mu$mole PGA added | Time min | PGA | DAP | G 6-P | FDP |
|---|---|---|---|---|---|---|
| A | 0·8 | 3 | 0·23 | 0·05 | 0·03 | 0·007 |
| | | 10 | 0·45 | 0·17 | 0·09 | 0·017 |
| | | 10 dark | 0·09 | 0·01 | 0·043 | 0·0 |
| | 0·0 | 10 | — | 0·03 | 0·07 | 0·003 |
| B | 3·76 | 5 | 0·30 | 0·18 | 0·02 | 0·005 |
| | | 10 | 0·47 | 0·23 | 0·04 | 0·025 |
| | | 20 | 1·20 | 0·26 | 0·18 | 0·021 |
| | | 20/without $NADP^+$ | 0·37 | 0·028 | 0·05 | 0·0 |
| | 0·0 | 20 | — | 0·06 | 0·09 | 0·003 |

Conditions as in Table I, but 200 $\mu$g chlorophyll; 10 $\mu$moles $NaHCO_3$; 0·36 $\mu$mole $NADP^+$.

In the dark the PGA consumption is only 1/5 of the amount in the light and also the omission of $NADP^+$ from the mixture of cofactors leads to a considerable decrease in the PGA consumption (Table II). However, there is still a significant consumption of PGA in the light without $NADP^+$. Presumably, minimal amounts of $NADP^+$ are sufficient to allow a slight reductive activity. Without the addition of PGA, but under otherwise normal conditions, the concentration of the sugar phosphates is clearly lower.

During the first 10 min the rate of PGA consumption is linear as shown by

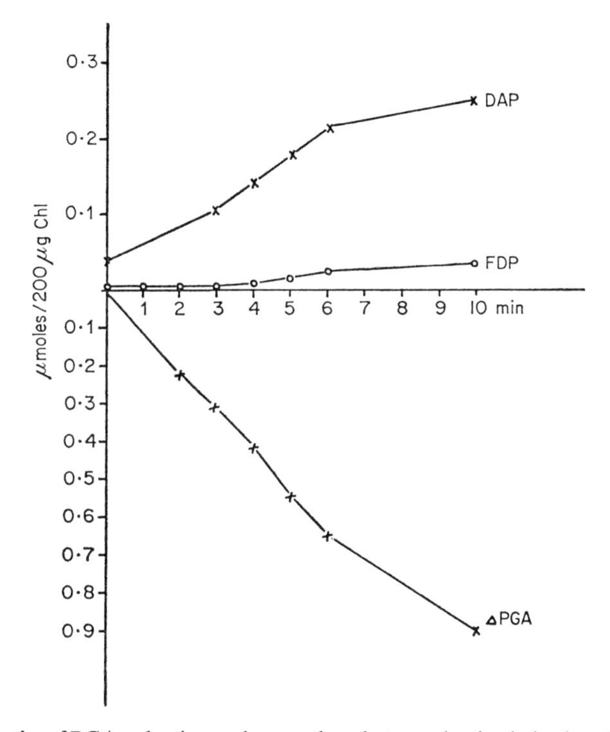


FIG. 1. Kinetics of PGA reduction and sugar phosphate production in broken chloroplasts. 2.4 μmoles PGA added; other conditions as in Table III.

Fig. 1. DAP shows a linear increase during the first few minutes, but then gradually levels off. The same is true for FDP but on a much lower level.

When the reaction time was more than 10 min the rate of PGA reduction frequently fell off very strongly as indicated by Tables III and IV. Accordingly, the concentrations of DAP and FDP decreased after 10 or 20 min. G 6-P behaves differently as it even continues to increase. There is also a significant formation of G 6-P in the dark as shown in Table II. Probably G 6-P originates not only from reduction of PGA but also from residual polysaccharides in the chloroplasts.

Table IV presents the results of an experiment in which PGA reduction as well as $CO_2$ fixation is measured. It shows that the former runs at a rate which

Table III. Kinetics of PGA reduction and sugar phosphate production in broken chloroplasts. 4 μmoles PGA added

| Time (min) | ΔPGA | DAP | G 6-P | FDP |
|---|---|---|---|---|
| 0 | — | 0·014 | 0·014 | 0·0 |
| 5 | 0·56 | 0·17 | 0·015 | 0·01 |
| 10 | 0·92 | 0·33 | 0·03 | 0·04 |
| 30 | 1·00 | 0·25 | 0·06 | 0·02 |
| 40 | 1·26 | 0·22 | 0·13 | 0·015 |

Conditions as in Table II, but 0·71 μmole $NADP^+$.

is about five times that of the latter. It is also remarkable that $CO_2$ fixation does not fall off simultaneously with PGA reduction after 20 min but continues at about the same rate. In other experiments the same or also the opposite,

Table IV. Kinetics of PGA reduction, $CO_2$ fixation and sugar phosphate production in broken chloroplasts. 6 μmoles PGA added

| Time (min) | $CO_2$ fixed | ΔPGA | DAP | FDP |
|---|---|---|---|---|
| 0 | — | — | 0·06 | 0·01 |
| 5 | 0·027 | 0·32 | 0·44 | 0·085 |
| 10 | 0·047 | 0·73 | 0·55 | 0·12 |
| 20 | 0·09 | 1·05 | 0·38 | 0·07 |
| 30 | 0·14 | 1·14 | 0·30 | 0·06 |

Conditions as in Table III, but 10 μmoles $NaH^{14}CO_3$ specific activity 0·28 μc/μmole.

namely an earlier drop of $CO_2$ fixation, was observed. Obviously the conditions are rather different under which the enzymes necessary for $CO_2$ fixation and PGA reduction are inactivated.

## *Effect of pH*

In earlier experiments (Elbertzhagen and Kandler, 1962) it was shown that $CO_2$ fixation is optimal at pH 8·2 under the conditions used in our assays. To check if the same is true for PGA reduction the pH of the reaction mixture was varied between pH 7·0 and pH 8·4. The other conditions were the same as given in Table III. The highest rates of PGA reduction were found between pH 7·7 and 8·2, with a flat maximum around pH 8·0. At pH 8·4 the PGA reduction rate was already 40% lower than at pH 8·2.

## *Effect of ADP concentration*

As illustrated in Table V, the usually applied amount of 0·2 μmole ADP/sample is already optimal for the rate of PGA reduction. Higher concentrations show an inhibitory effect. $CO_2$ fixation behaves similarly as observed in other experiments, but the maximum is much less pronounced.

Table V. Effect of ADP concentration on PGA reduction and ATP, DAP and FDP production in broken chloroplasts. 8 μmoles PGA added. 10 min light

| μmole ADP added | ΔPGA | ATP | DAP | FDP |
|---|---|---|---|---|
| 0 | 0·70 | 0·028 | 0·18 | 0·025 |
| 0·1 | 0·91 | — | 0·28 | 0·057 |
| 0·2 | 1·23 | 0·15 | 0·37 | 0·06 |
| 0·4 | 0·73 | 0·3 | 0·44 | 0·055 |
| 0·6 | 0·55 | 0·42 | 0·29 | 0·047 |
| 0·8 | 0·38 | 0·54 | 0·33 | 0·066 |
| 1·0 | 0·1 | 0·66 | 0·36 | 0·058 |

Conditions as in Table III.

### *Effect of $NADP^+$ concentration*

The amount of $NADP^+$ originally used (0·12 μmole/sample) is far from being optimal for PGA reduction, as demonstrated by the results of the experiments recorded in Table VI. Saturation is only reached at 0·72 μmole. At this concentration also the two sugar phosphates showed the highest values.

Table VI. Effect of $NADP^+$ concentration on PGA reduction and sugar phosphate production in broken chloroplasts. 6 μmoles PGA added. 10 min light

| μmole $NADP^+$ added | ΔPGA | DAP | FDP |
|---|---|---|---|
| 0 | 0·0 | 0·012 | 0·0 |
| 0·12 | 0·5 | 0·13 | 0·020 |
| 0·24 | 0·78 | 0·24 | 0·025 |
| 0·36 | 0·95 | 0·27 | 0·028 |
| 0·48 | 1·08 | 0·33 | 0·040 |
| 0·60 | 1·30 | 0·33 | 0·042 |
| 0·72 | 1·38 | 0·37 | 0·047 |
| 0·84 | 1·34 | 0·37 | 0·048 |

Conditions as in Table II, but $NADP^+$ as indicated above.

At an optimal $NADP^+$ concentration a rate of reduction of 42 μmoles PGA/mg chl/hr is calculated from the PGA decrease. This value has to be increased by the amount of PGA formed by $CO_2$ fixation. If a value of

3 μmoles $CO_2$/mg chl/hr is taken as a basis, 6 μmoles PGA have to be added and a maximal rate of reduction of 48 μmoles PGA/mg chl/hr is obtained, i.e. a value which is eight times higher than in $CO_2$ fixation without the addition of PGA.

### THE ADP/ATP RATIO AT DIFFERENT ADDITIONS OF ADP

In experiments with varying ADP concentrations the amount of ATP was also determined in order to check if photosynthetic phosphorylation is functioning well in our preparations. Even without ADP addition a relatively large amount of ATP, obviously originating from the broken chloroplasts, could be detected. Correspondingly, a reasonable reduction of PGA took place even without ADP. With smaller ADP additions about three-quarters of the amount of ADP appeared as ATP compared to two-thirds with the higher ADP additions. This is in agreement with the observations of Heber and Tyszkiewicz (1962). Hence the ATP supply is certainly not a limiting factor for the rate of PGA reduction or the regeneration of the $CO_2$ acceptor in these preparations.

### DISTRIBUTION OF $^{14}C$ AFTER PHOTOSYNTHESIS WITH $^{14}CO_2$ OR [1-$^{14}C$]-PGA

In the experiments described so far only a selection of easily determinable products resulting from PGA reduction were studied. To get a complete picture of the fate of the added PGA and the $CO_2$ fixed, to one part of the samples $^{14}CO_2$+PGA, to the other part $CO_2$+[1-$^{14}C$]-PGA was added. Within 10 min 0·06 μmole $CO_2$/sample had been fixed and 1·8 μmoles out of the added 3·76 μmoles PGA had been consumed. So in this example the rate of PGA reduction exceeded by about fifteen times that of PGA formation by $CO_2$ fixation (based on the assumption that 2 PGA are formed from each $CO_2$ fixed).

Figure 2 illustrates the distribution of radioactivity. Evidently the pattern is very different depending on whether the label was administered as [1-$^{14}C$]-PGA (A) or as $^{14}CO_2$ (B). In case (A) about half the radioactivity is found in PGA, and the other half in the sugar phosphates, whilst only a very small percentage appeared in aspartic or other acids.

With $^{14}CO_2$ about equal amounts of the label were contained in PGA and the sugar phosphates, but in contrast with case (A), one-third of the radioactivity is found in aspartic acid and 4% in other acids.

This observation may be readily explained by the fact that carboxylation is necessary for the formation of aspartic acid from PGA. Hence the specific activity of the newly formed aspartic acid is practically identical with that of the added $^{14}CO_2$, whereas the specific activity of $^{14}CO_2$ entering PGA and the resulting sugar phosphates is strongly diluted by the large quantity of inactive PGA added. If the portion of PGA transformed to sugar phosphates on the one hand, and aspartic acid on the other hand, is judged only by the chromatograms of experiments with $^{14}CO_2$, a wrong impression of the actual processes can easily be formed.

Only in some experiments where large amounts of PGA were added the portion of aspartic acid was observed to be as great as in Fig. 2. The dependence of the formation of aspartic acid on the PGA concentration is demonstrated

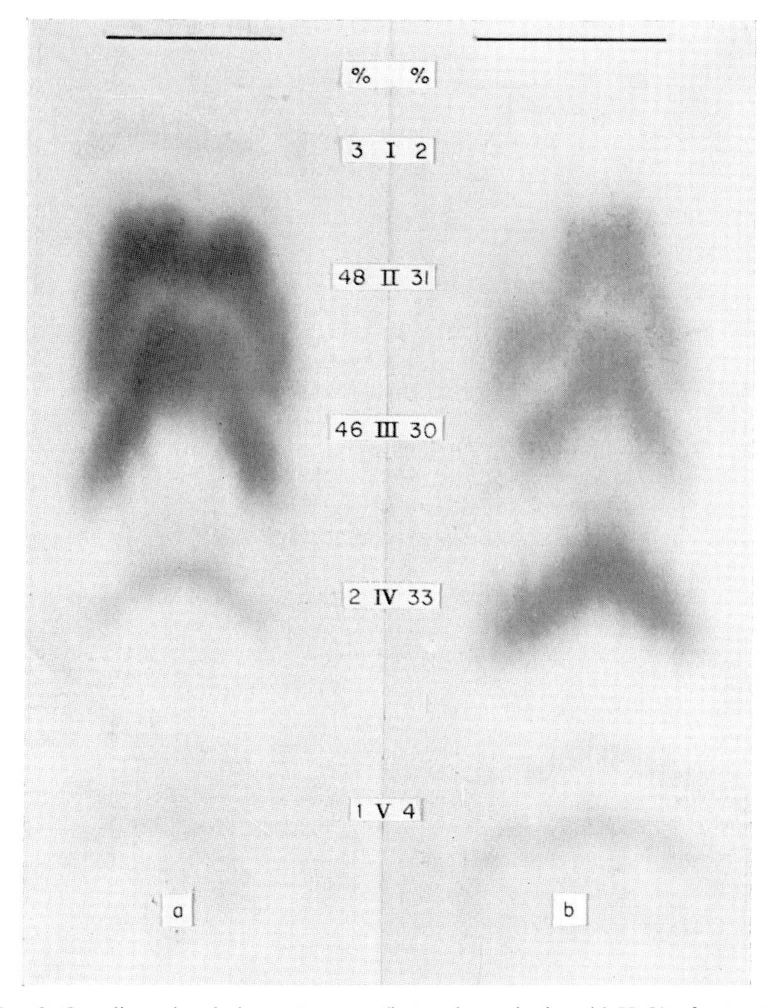


FIG. 2. One-dimensional chromatograms (butanol–propionic acid–$H_2O$) of extracts of broken chloroplasts. 10 min photosynthesis in the presence of 3·76 $\mu$moles [1-$^{14}$C]-PGA + 6 $\mu$moles $NaHCO_3$ (a) or 6 $\mu$moles $NaH^{14}CO_3$ (24·5 $\mu$c/$\mu$mole) + 3·76 $\mu$moles PGA (b); 400 $\mu$g chl/sample; other conditions as in Table III. I = sugar diphosphates, II = sugar monophosphates, III = PGA, IV = aspartic acid, V = malic acid + alanine + glutamic acid + citric acid.

clearly by the results of three experiments compiled in Table VII. For example, without the addition of PGA, 7·6% of the fixed $^{14}CO_2$ was found in aspartic acid after 30 min in experiment a, whereas 25% was obtained when 3·4% $\mu$mole

PGA had been added. In this experiment the total fixation did not increase with increasing PGA concentration, but even dropped, suggesting that in this case high PGA concentrations inhibited photosynthetic $CO_2$ fixation but promoted fixation by a Wood-Werkman reaction.

In the two other experiments (b and c) the formation of aspartic acid was considerably slower. Even at high concentrations of PGA it amounted only to a few % of the total fixation.

The distribution of $^{14}C$ in experiments with chloroplast extract without the addition of broken chloroplasts showed a very similar pattern. Here, the "assimilatory power" was added in the form of ATP and NADPH and the extract was incubated in the dark.

## DISCUSSION

### WHICH STEP IS RATE-LIMITING?

The experiments described above show that neither low enzyme activity nor a slow production of ATP and NADPH are limiting factors for PGA reduction. in assays with broken chloroplasts, but rather the low concentration of PGA at the beginning of the experiment. Only if the PGA concentration is increased to about 1 $\mu$mole/sample (i.e. roughly $10^{-3}$ M, a concentration found in intact *Chlorella* cells, Kandler and Haberer-Liesenkötter, 1963a), optimal reduction rates amounting to many times that of $CO_2$ fixation, are obtained. However, they are still lower than the rates of photosynthesis *in vivo.*

In spite of the acceleration of a part of the photosynthetic cycle, $CO_2$ fixation does not increase. Hence, either the carboxylation reaction or the regeneration of the $CO_2$-acceptor have to be considered as the rate-limiting factor for the complete photosynthesis. When carboxylation is inhibited in intact cells RuDP piles up, as repeatedly demonstrated (Bassham and Calvin, 1957; Kandler and Liesenkötter. 1963b). Correspondingly, one would expect an accumulation of RuDP in broken chloroplasts too, if carboxylation is the limiting factor.

However, the diphosphate fraction was always very weakly labelled no matter if $^{14}CO_2$ or [$^{14}C$]-PGA was fed, and most of the radioactivity was accumulated in the monophosphate area. This indicates that the regeneration of RuDP rather than carboxylation is rate-limiting. This view is also supported by the findings of Trebst and Fiedler (1961, 1962) and Gibbs (1963) that the label spreads very poorly in the hexose molecule.

### LOCALIZATION OF THE PGA REDUCTION

Experiments with intact cells like those on pool saturation with $^{14}C$ (Bassham and Kirk, 1963) and those on the effect of cyanide on the distribution of $^{14}C$ (Kandler and Liesenkötter, 1963b) indicate that the PGA formed by $CO_2$ fixation is not in equilibrium with the total pool of PGA in the cell, but is preferentially reduced to sugar phosphate.

This separation of pools might be brought about not only by osmotic barriers of the chloroplasts, but also by a specific arrangement of the enzymes in the lamellar structure, thus forming a multi-enzyme to which the carboxylation product might be firmly attached. In fact, Rabin and Trown (1964) were able to show that the $\beta$-ketoacid resulting from the carboxylation of RuDP is bound to carboxydismutase by a thioether-bond which hydrolyses very slowly.

Some of the experiments reported here allow conclusions on the question if, also in broken chloroplasts, the PGA formed by carboxylation of RuDP is more rapidly reduced than the one added from the outside.

(a) Under the two conditions of the experiment shown in Fig. 2, the quotient [$^{14}$C] sugar phosphate: [$^{14}$C]-PGA should never be higher but always lower or, at best, equal in sample B as compared to sample A, if the newly formed PGA is freely mixed with the PGA added. Vice versa, a preferential reduction of the newly formed PGA to sugar phosphates would lead to a higher quotient in case B. The observed values are very similar, namely 1·11 for case A and 1·10 for case B.

Table VII. Distribution of $^{14}$C after $^{14}CO_2$ fixation of broken chloroplasts (light) and chloroplast extract (dark+ATP and NADPH) at different PGA concentrations. Numbers express % of total activity

| | Time (min) | μmole PGA added | Exp. | μmole $^{14}CO_2$ fixed | PGA | Sugar phosphates | Aspartic acid |
|---|---|---|---|---|---|---|---|
| Broken | | | a | 0·075 | 53·5 | 42·0 | 5·0 |
| chloroplasts | | 0·0 | b | 0·072 | 60·0 | 32·0 | 1·3 |
| | | | c | 0·063 | 68·0 | 30·0 | 1·0 |
| | | | a | 0·065 | 70·7 | 21·4 | 8·0 |
| | 15 | 0·34 | b | 0·063 | 75·2 | 16·4 | 1·6 |
| | | | c | 0·055 | 75·0 | 23·0 | 1·5 |
| | | | a | 0·06 | 80·9 | 4·0 | 15·0 |
| | | 3·4 | b | 0·06 | 90·7 | 4·5 | 2·0 |
| | | | c | 0·05 | 89·0 | 7·0 | 3·0 |
| | | | a | 0·16 | 43·0 | 45·4 | 7·6 |
| | | 0·0 | b | 0·10 | 53·8 | 37·0 | 1·8 |
| | | | c | 0·11 | 56·0 | 40·0 | 2·0 |
| | | | a | 0·17 | 57·3 | 24·0 | 12·3 |
| | 30 | 0·34 | b | 0·105 | 61·8 | 28·5 | 3·2 |
| | | | c | 0·09 | 60·0 | 35·0 | 2·5 |
| | | | a | 0·14 | 64·5 | 4·2 | 25·0 |
| | | 3·4 | b | 0·095 | 80·0 | 12·0 | 4·4 |
| | | | c | 0·07 | 78·5 | 16·0 | 4·0 |
| Chloroplast | | | a | 0·19 | 67·0 | 23·7 | 5·5 |
| extract | | 0·0 | b | 0·06 | 76·2 | 17·5 | 4·0 |
| | 30 | | c | 0·035 | 68·0 | 26·0 | 2·0 |
| | | | a | 0·13 | 72·9 | 6·8 | 15·8 |
| | | 3·4 | b | 0·05 | 85·3 | 5·0 | 6·0 |
| | | | c | 0·025 | 75·0 | 18·0 | 3·0 |

Conditions: Samples contain broken chloroplasts as in Table I a–c. Samples contain chloroplast extract only, as Table I, but 3 μmoles ATP (instead of ADP) 1 μmole NADPH; 3 μmoles G 6–P, 10 μg glucose-6-phosphate dehydrogenase.

(b) The addition of increasing amounts of unlabelled PGA during photosynthesis with $^{14}CO_2$ results in an increasing dilution of the newly formed PGA if the two "species" of PGA exchange readily. Hence after a given reaction time an increasingly smaller portion of the fixed $^{14}C$ will reach the sugar phosphates and the distribution is shifted in favour of PGA. The accumulation of $^{14}C$ in PGA depending on the addition of PGA is clearly shown by the data in Table VII.

(c) Finally, the distribution of $^{14}C$ during photosynthetic $CO_2$ fixation should be compared to that during chemosynthetic $^{14}CO_2$ fixation by chloroplast extract to which ATP and NADPH were added. In this case the participation of enzymes and intermediate compounds bound to the lamellar structure of the chloroplasts is excluded. The effect of the addition of PGA on the distribution of $^{14}C$ found (Table VII) is very similar to the one obtained in broken chloroplasts.

The three observations mentioned, support the view that the reduction of PGA in preparations of broken chloroplasts occurs in the solution, independent of the "species" of PGA. There is no indication for a structurally bound multi-enzyme which hands the carboxylation product directly on to the site of reduction. This statement does of course not necessarily apply to intact cells.

## Summary

After addition of substrate amounts of PGA to broken chloroplasts, the rate of photosynthetic PGA reduction and formation of DAP, FDP and G 6-P as well as the fixation of $CO_2$ were measured.

The distribution of $^{14}C$ after fixation or $^{14}CO_2$ or reduction of [1-$^{14}C$]-PGA was determined by paper chromatography.

At optimal conditions rates of PGA reduction up to 48 $\mu$moles/mg chl/hr were measured, while only up to 4 $\mu$moles $CO_2$/mg chl/hr were fixed.

The PGA consumed was transformed mainly to sugar phosphates, while only a few per cent were detectable in aspartic or other organic acids.

After the fixation of $^{14}CO_2$ the label was found chiefly in PGA, sugar phosphates and, to a smaller and more varying extent, also in aspartic and other acids. The portion incorporated into the latter increased when higher concentrations of PGA were used. This indicates that a Wood-Werkman reaction is enhanced, whereas the $CO_2$ fixation by carboxydismutase is slightly inhibited.

When increasing amounts of PGA were added, a rising proportion of radioactivity was found in PGA and less in the sugar phosphates. This shows that the newly formed PGA mixes very well with the PGA added and is not preferentially reduced to sugar phosphates by structurally bound enzymes, as is likely to happen in intact cells.

The view is held that the enzyme activity and the supply with light-generated "reducing power" are high enough to drive the PGA reduction many times faster than the $CO_2$ fixation. The lag in the reduction of PGA formed by $CO_2$ fixation in the absence of exogenous PGA results from the low concentration of residual PGA in the broken chloroplasts (about $6 \times 10^{-5}$ mole/sample) at the beginning of the experiment.

The regeneration of the $CO_2$ acceptor is considered to be the true bottleneck of photosynthetic $CO_2$ fixation in broken chloroplasts.

ACKNOWLEDGEMENT

This research has been sponsored by the Office of Scientific Research through the European Office, Aerospace Research United States Air Force, Grant No. AF 61(052)-244 and AF EOAR 62-42, as well as by the Deutsche Forschungsgemeinschaft. We are indebted to Miss L. Weidermann and E. Seidel for perfect technical assistance.

## References

Allen, M. B., Arnon, D. I., Capindale, J. B., Whatley, F. R. and Durham, L. J. (1955). *J. Am. Chem. Soc.* **77**, 4149.
Arnon, D. I. (1949). *Pl. Physiol.* **24**, 1.
Arnon, D. I., Allen, M. B. and Whatley, F. R. (1956). *Biochim. biophys. Acta* **20**, 449.
Bassham, J. A. and Calvin, M. (1957). "The Path of Carbon in Photosynthesis." Prentice-Hall, Inc., Engelwood Cliffs, New Jersey.
Bassham, J. A. and Kirk, M. (1963). "Studies on Microalgae and Photosynthetic Bacteria" (Tokyo) 493.
Benson, A. A., Bassham, J. A., Calvin, M., Goodale, T. C., Haas, V. A. and Stepka, W. (1950). *J. Am. chem. Soc.* **72**, 1710.
Elbertzhagen, H. and Kandler, O. (1962). *Nature, Lond.* **194**, 312
Gibbs, M. (1963). "Photosynthesis Mechanisms in Green Plants", 663.
Gibbs, M. and Kandler, O. (1957). *Proc. natn. Acad. Sci., U.S.A.* **43**, 446.
Gibbs, M. and Calo, N. (1959). *Pl. Physiol.* **34**, 318.
Heber, U. and Tyszkiewicz, E. (1962). *J. exp. Bot.* **13**, 185.
Heberer, U. and Willenbrinck, J. (1964) *Biochim. biophys. Acta* **82**, 313.
Holm-Hansen, O., Pon, N. G., Nishida, K., Moses, V. and Calvin, M. (1959). *Physiol. Plantarum* **12**, 475.
Kandler, O., Liesenkötter, I. and Oaks, B. A. (1961). *Z. Naturf.* **16**b, 50.
Kandler, O. and Haberer-Liesenkötter, I. (1963a). *Z. Naturf.* **18**b, 718.
Kandler, O. and Liesenkötter, I. (1963b). "Microalgae and Photosynthetic Bacteria" (Tokyo) 513.
Losada, M., Trebst, A. V. and Arnon, D. I. (1960) *J. biol. Chem.* **235**, 832.
Rabin, B. R. and Trown, P. W. (1964). *Proc. natn. Acad. Sci.*, U.S.A. **52**, 88.
Trebst, A. V., Losada, M. and Arnon, D. I. (1959). *J. biol. Chem.* **234**, 3055.
Trebst, A. V., Losada, M. and Arnon, D. I. (1960). *J. biol. Chem.* **235**, 840.
Trebst, A. and Fiedler, F. (1961). *Z. Naturf.* **16**b, 284.
Trebst, A. and Fiedler, F. (1962). *Z. Naturf.* **17**b, 553.
Whatley, F. R., Allen, M. B., Trebst, A. V. and Arnon, D. I. (1960). *Pl. Physiol.* **35**, 188

# Photosynthetic Activity of Isolated Pea Chloroplasts

D. A. Walker
*Botany Department, Imperial College, London, England*

## Isolation of Chloroplasts

### CHOICE OF PLANT

Considered in terms of subsequent impact on the study of photosynthesis, *Claytonis perfoliata* Donn (one of the first species used by Hill in 1937), would clearly warrant a place in any list of plants which have been important in the isolation of chloroplasts. It is equally clear that pride of place would go to spinach (*Spinacea oleracea* L.). For this purpose spinach has many of the qualities of the ideal plant. Its leaf is soft, its cell sap is near to neutrality, it does not contain large quantities of tannins which may be precipitated on plastids during isolation and it is readily available in local markets throughout North America.

In Britain, unfortunately, spinach is less easily obtained and the various culinary substitutes which are principally varieties of spinach beet (*Beta vulgaris* L. ssp. *vulgaris*) are at their best for a only a few months in the summer. Though largely or entirely free from starch the leaves also contain large quantities of calcium oxalate crystals which can interfere with the separation of chloroplasts.

The literature is full of conflicting reports about what effects age and pre-treatment of the leaf may have on the isolation and subsequent activity of chloroplasts (for a summary see Rabinowitch, 1956) and the plant biochemist would welcome a standard leaf as warmly as he has welcomed uniform cultures of *Chlorella*. The various types of spinach do not readily lend themselves to this purpose. Germination is often slow and uneven, growth is seldom rapid during the early stages, and the plants are often susceptible to virus infection. Peas (*Pisum sativum* L.) have none of these disadvantages. Suitable varieties germinate evenly and grow rapidly in coarse vermiculite ("micafil", heat-expanded mica). In 2 weeks at 15°, 200 g (dry wt) of seed will produce about 200 g (fresh wt) of suitable leaf and shoot material. In full daylight, starch accumulates in pea chloroplasts. The grains are more dense than the chloroplast so that centrifugation may then lead to considerable disruption (Smillie and Fuller, 1959). Grown under low light (150–200 ft candles) however, starch is almost entirely absent. The pea chloroplast would also appear to retain its integrity more readily than those of some other species (see Section IIA).

### METHOD OF SEPARATION

#### *Composition of Grinding Medium*

Hill's chloroplasts were first isolated in sucrose (see e.g. Hill, 1965) and a variety of sugars and other compounds have been used since to maintain the osmotic pressure of the medium at a level which it was hoped would not cause undue contraction or expansion of the chloroplast. Following the work of Arnon and Whatley and their colleagues in the field of photosynthetic phosphorylation, isolation in *tris*-NaCl (usually 0·35 M NaCl M/15 *tris*hydroxymethylaminomethane, pH 7·5) became fashionable and much of our present knowledge of photosynthesis by isolated chloroplasts derives from work in which media of this nature (see e.g. Whatley and Arnon, 1963) have been employed. For many purposes it is not necessary that chloroplasts should retain a high degree of structural integrity and for some purposes structural change may prove to be an advantage (see Section IIIA). It now seems reasonably well agreed however that the changes which occur in *tris*-NaCl can be extensive and can include considerable loss of the outer membrane and even separation of the thylakoids (Gee *et al.* 1965). It is possible that not all this damage need be attributed to the use of NaCl. Both "*tris*" and the alkaline pH may also contribute to the structural changes and it may be noted that Kahn and von Wettstein (1961) found that 65% of spinach chloroplasts retained their outer membranes when prepared in 0·35 M NaCl, 0·02 M $KH_2PO_4$-$Na_2HPO_4$ pH 6·7. Varietal differences and differences in growing conditions and pretreatment of the leaf will undoubtedly add to the variety of changes seen in different laboratories but it does seem clear that "whole" chloroplasts in *tris*-NaCl may prove to be much less "whole" than some workers may have suspected. For some uses therefore there are advantages in returning to sucrose or other sugars in which structural modification following isolation may be less pronounced (see e.g. Figs. 2 and 3). There is evidence that for the Hill reaction chloroplasts may be more stable at acid than alkaline pH's (Holt *et al.*, 1951) so that it also may be an advantage to prepare chloroplasts at slightly acid pH even if they are subsequently to be used at pH 7·5 or above. It has been shown (Good, 1962) that in some respects "tricine" (*tris*hydroxymethylglycine) is markedly superior to "*tris*". Irreversible uncoupling can occur in cation-deficient media (Jagendorf and Smith, 1962) and this may be prevented by the inclusion of small quantities of magnesium chloride. Ascorbate is known (Arnon, 1958) to slow deterioration of chloroplasts on storage and might also be assumed to minimize some oxidation during separation if added to the grinding medium.

### *Disruption of Tissue*

The pestle and mortar, with or without an added abrasive, is still favoured for the initial disruption of the cell. The Waring Blendor is mostly thought to be too destructive, though it is evident that short term blending can yield chloroplasts in which the structural integrity (Figs. 2 and 3) and activity (Walker, 1964, 1965) are in no way inferior to those obtained by other methods. The motor-driven blender has the advantage of offering a more reproducible technique and in general it permits more adequate control of temperature during maceration. Its main disadvantage is probably the strength of the disruptive forces induced by cavitation effects and the frothing caused by air drawn into the macerated tissue by the rapidly rotating blades. These defects may be minimized by reducing the duration of blending. Five seconds may be adequate if the leaf is soft (see Section IV).

### *Centrifugation*

Centrifugation is usually preceded by straining or filtering through muslin or cheese-cloth to remove coarse cell-debris. Finer cell-debris is frequently removed by a short initial centrifugation and the chloroplasts themselves are then usually separated from the supernatant by low speed centrifugation. This process may be shortened by more extensive filtration to replace the initial centrifugation and by a faster, briefer spin to bring down the chloroplasts. Chloroplasts prepared by any simple variant of these procedures will inevitably be contaminated by small amounts of other sub-cellular particles such as nuclei and mitochondria together with small pieces of cell-wall debris and even traces of occluded cytoplasm (see e.g. Figs. 2 and 3). These may be sufficiently insignificant to be detected only with difficulty under the light microscope but it would be unwise to suppose that they were ever completely absent except possibly where density gradient techniques have been employed. These elegant methods (e.g. James and Das, 1956) are of great value in some areas of cytological analysis though at present, purity and activity are often inversely related, if only because of the time which must elapse during preparation by relatively complex techniques. It should also perhaps be emphasized that while it is possible to prepare intact chloroplasts to a high degree of purity in sucrose gradients (Leech, 1964), glycerol gradients yield chloroplasts which have lost their external bounding membranes and most if not all of their stroma (James and Leech, 1964) so that any conclusions regarding their enzyme complement must be interpreted accordingly.

### *Summarized Procedure*

The pea chloroplasts used in the work described in this article were prepared in 0·33 M sugar solution (mostly sorbitol or sucrose) containing 0·1 M sodium and potassium phosphate at pH 6·8 (Walker, 1964, 1965). They were resuspended in 0·45 M sugar solution from which the phosphate was omitted. The grinding and resuspending media also contained small quantities of $MgCl_2$ and NaCl (1 mg/ml) and a small amount of sodium isoascorbate was usually employed as an anti-oxidant. Fifty g of leaf and shoot were blended for 5 sec in 200 ml of ice-cold grinding medium and the juice squeezed from the homogenate through two layers of muslin and filtered through a further eight layers. The filtrate was poured into four 50 ml tubes, placed in a high-speed angle-head centrifuge and the chloroplasts separated by allowing the centrifuge to accelerate to 6000 rev/min and then bringing it to rest as rapidly as possible by manual braking (the total spinning time was approximately 90 sec). The pellet was washed and then resuspended using a glass rod and cotton-wool.

## MICROSCOPY

### PHASE CONTRAST AND ELECTRON MICROSCOPY

When examined with the light microscope, chloroplast suspensions are found to contain a proportion of plastids which are distinctly granulated and a proportion which are more opaque. These differences may be emphasized by a combination of phase contrast and dark ground illumination. Under these

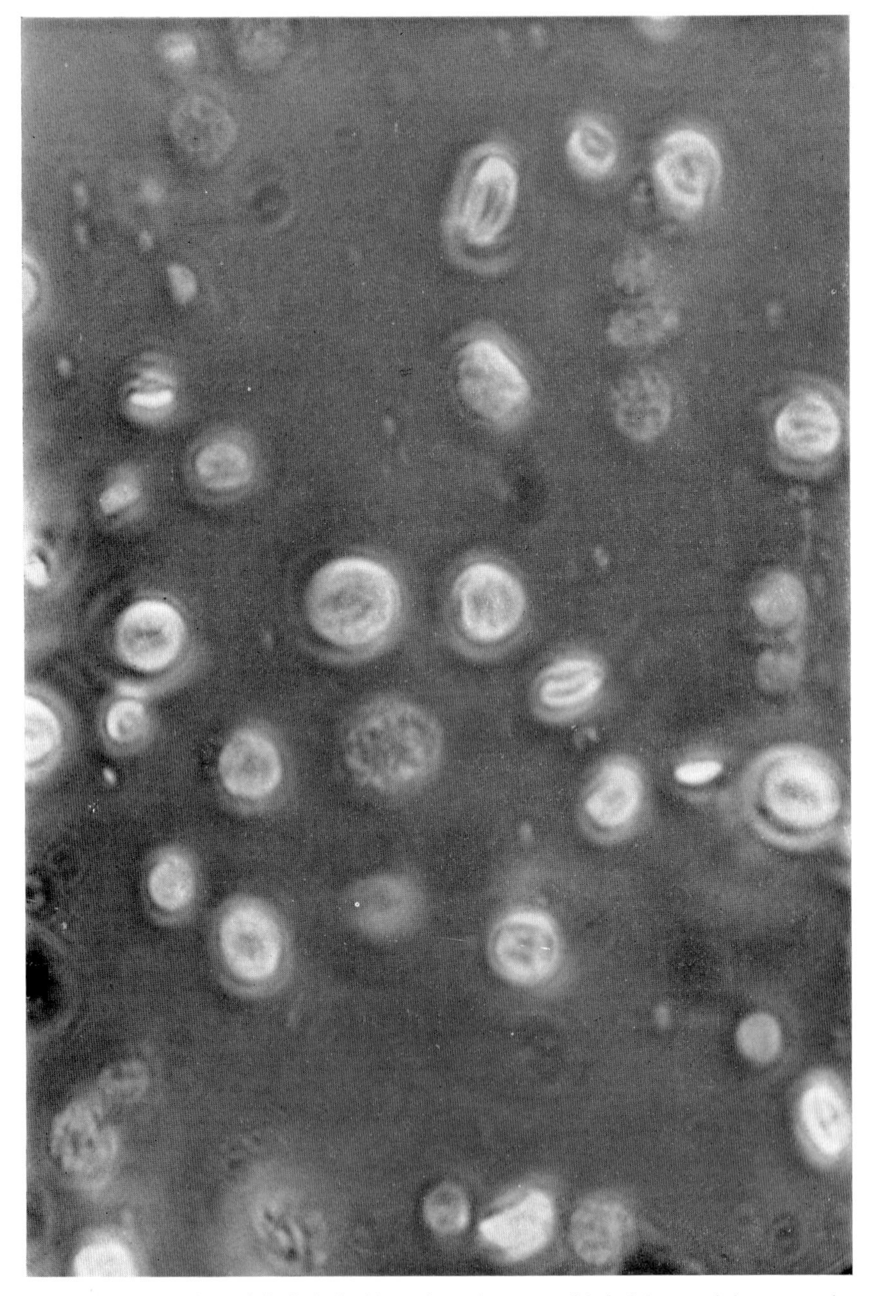

FIG. 1. Suspension of "whole" chloroplasts (opaque with halo) containing a number of membrane-free chloroplasts (dark, granulated) seen under phase contrast.

conditions (cf. Spencer and Unt, 1965) the opaque chloroplasts have a halo, they shine and appear to stand out from the surface (Fig. 1). The granulated chloroplasts are darker, flatter and (on average) a little larger, as though they are no longer rounded but have flattened against the surface of the slide (Fig. 1). In resuspending medium, from which the sucrose has been omitted so that it is now strongly hypotonic, all the chloroplasts are dark and granulated. There is now ample evidence (Leech, 1964; James and Leech, 1964) which supports the view (Kahn and von Wettstein, 1961) that the granulated chloroplasts have

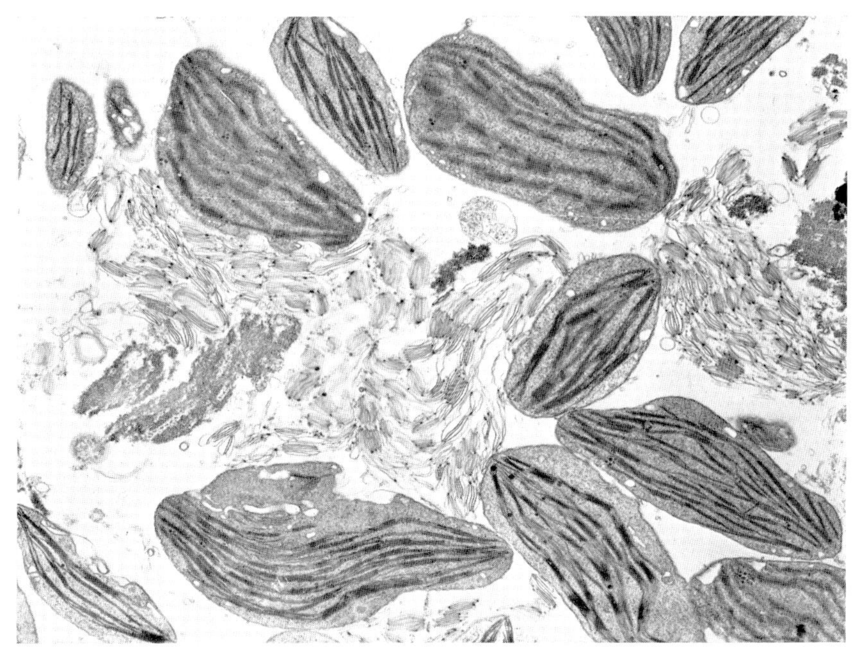

FIG. 2. Sample field ($\times$ 17,000) from a thin section of a pellet recovered from resuspending medium to show the two structural states of the chloroplasts. A majority have retained all the normal components of the stroma and possess unbroken envelope membranes. A minority have lost their envelopes, and their stroma has been almost completely dispersed, liberating the internal free-lamella systems. The free-lamella systems become crushed and distorted in the pellet. Micrograph and legend by A. D. Greenwood.

lost their external membrane and part or all of their stroma whereas the opaque chloroplasts are still, in this sense, intact. By this criterion, pea chloroplasts isolated by the method described in Section B and resuspended in isotonic or hypertonic media may contain as many as 90% plastids with entire membranes and only rarely as few as 50%. The correctness of this interpretation is supported by electron microscopy (Figs. 2 and 3). Chloroplasts isolated from spinach by the same method contained rather fewer intact chloroplasts and these appeared to lose their membranes more readily than pea chloroplasts while standing at room temperature. Chloroplasts isolated from *Stellaria adime* (L.) Vill. were entirely membrane-free, thus implying that there might be differences in the relative "toughness" of chloroplasts from different species. It should, perhaps, be emphasized that while it is possible to say with some confidence that a preparation contains virtually all membrane-free chloroplasts other estimates of the proportion of whole to membrane-free chloroplasts are necessarily very approximate because of the size of the samples which can be counted in relation to the number of plastids in a given preparation.

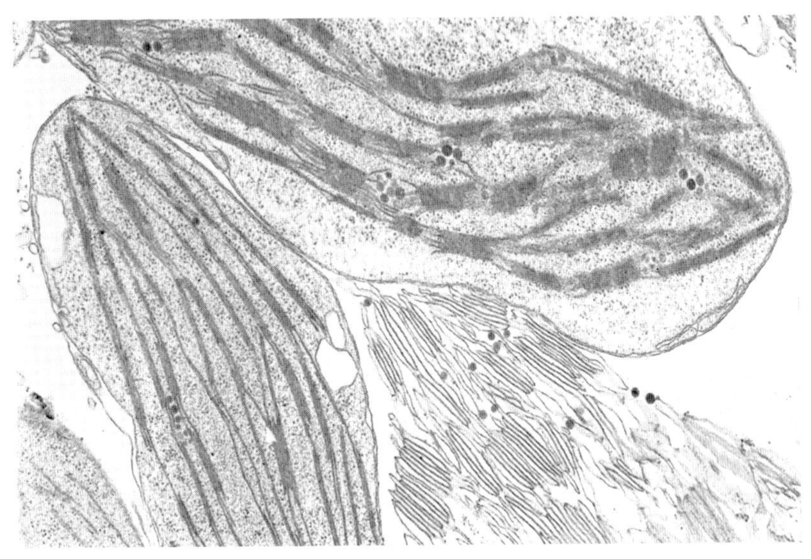

FIG. 3. At high magnification (× 38,000) the double nature of the envelope is clearly discernible and in some places the innermost membrane is invaginated to form flattened or expanded vesicles in the adjacent stroma. The external surfaces of the envelopes are notably free from adherent debris. The stroma contains densely stained lipid globules and smaller ribosome-like particles sometimes arranged in rows adjacent to granum or intergranum lamellae and other particles and finely fibrillar material, mainly of lesser density. In the free-lamella systems continuity of the granum and intergranum membranes is maintained and the intrathylakoid spaces remain closed although somewhat distended. All but traces of the granular and fibrillar fine stroma has dispersed but lipid droplets remain trapped between the lamellae. Fixed in osmium tetroxide, ethanol dehydrated, embedded in Epon, and stained with uranyl acetate followed by lead citrate. Micrograph and legend by A. D. Greenwood.

## LOSS OF OUTER MEMBRANE IN DILUTE MEDIA

Pea chloroplasts appear to retain their membranes when suspended in media containing 0·2 M sucrose and to lose them in media containing less than 0·1 M. Examined immediately after resuspension in media containing 0·09 M sucrose many plastids were membrane-free but there were also a large number of plastids in which an external membrane was distended to form a "bubble" (greater than the volume of the original chloroplast), prior to bursting.

When added to reaction mixtures in which the final osmotic pressure would be approximately equivalent to that exerted by 0·2 M sucrose and illuminated strongly (6000 ft candles) at 20° for 10 min there was no apparent decrease in the proportion of chloroplasts with entire outer membranes.

## Cyclic Photophosphorylation

### CORRELATION BETWEEN MEMBRANE INTEGRITY AND RATE

Chloroplasts which had been exposed to hypotonic media were used from an early stage in the study of cyclic photosynthetic phosphorylation when it became clear that they gave rather better rates than those maintained in hypertonic media (Whatley *et al.*, 1956; Whatley and Arnon, 1963). The superiority of "broken" or "disrupted" chloroplasts was often not large nor did it excite much comment though it was commonly accepted that the "whole" chloroplast presented some barrier to the entry of exogenous reactants or cofactors. Recent work with pea chloroplasts has shown that this effect may be substantial and that there is a correlation between the proportion of membrane-free chloroplasts and the rate of cyclic photosynthetic phosphorylation. The smaller stimulations observed in the past could conceivably reflect the relatively large proportion of membrane-free chloroplasts present in "whole" chloroplast preparations isolated from spinach leaves in *tris*-NaCl. This interpretation (see Section IB) would be in accord with recent electron microscope studies (Gee *et al.*, 1965).

Table I summarizes some of the results obtained with pea chloroplasts (Walker, 1965). "Whole" chloroplasts were prepared according to the method outlined in section II and contained a proportion of membrane-free chloroplasts which in some preparations was as much as 50% and in others as little as 10%. These, when resuspended in hypotonic media containing only small quantities of $MgCl_2$, NaCl and isoascorbate, became entirely membrane-free. Membrane loss was very rapid and had apparently gone to completion within 1 min of resuspension in this medium.

Table I. μMoles phosphate esterified/mg chlorophyll/hr

| "Whole" chloroplast preparations containing between 10% and 50% membrane-free plastids | Membrane-free chloroplast preparations containing no whole plastids | Percentage rate (whole/membrane-free) |
|---|---|---|
| 195 | 420 | 46·4 |
| 155 | 485 | 31·9 |
| 162 | 578 | 27·8 |
| 102 | 864 | 11.8 |

Full experimental details of most of these experiments can be found elsewhere (Walker, 1965). Photophosphorylation was measured by the disappearance of inorganic phosphate. The majority of the values are maximum rates taken from time course experiments. Pyocyanine was used as a cofactor. In most of the experiments sucrose was added to reaction mixtures containing chloroplasts suspended in hypotonic medium so that the final concentration was the same as that in reaction mixtures containing "whole" chloroplasts (though it was also shown that omission of this sucrose did not apparently affect the rate).

Results such as those in Table I suggest that pea chloroplasts with intact membranes support pyocyanine catalysed (Hill and Walker, 1959) cyclic photophosphorylation of *exogenous* ADP at low rates, if at all, and indeed it is clearly possible that the disappearance of inorganic phosphate could be entirely attributed to membrane-free chloroplasts. There are of course many possible explanations. The intact membranes might prevent entry of a reactant such as ADP or a cofactor such as pyocyanine. Alternatively the "whole" chloroplasts could utilize ATP as rapidly as it is produced. Retention of an active ATP-ase by whole chloroplasts is a possibility that cannot be discounted. Utilization of ATP in carbon dioxide fixation in sufficient quantities to account for the observed differences seems unlikely because the initial rates of esterification are considerably in excess of the rates of $CO_2$ fixation by these chloroplasts in the absence of sugar phosphates and in the presence of pyocyanine.

The results in Table I do not contradict those of Jagendorf and Smith (1962) or of Spencer and Unt (1965) in which water-treatment of cation-deficient chloroplasts caused uncoupling. Jagendorf and Smith recorded an increase in activity following water treatment of chloroplasts not exposed to EDTA which is in accord with the present observations (cf. also Whatley *et al.*, 1956).

### EFFECT OF MEMBRANE LOSS ON RATE OF $CO_2$ FIXATION

It has been shown on a number of occasions (see e.g. Whatley *et al.*, 1956) that chloroplasts exposed to hypotonic media lose much of their ability to fix $CO_2$. This is also true of the present work where membrane-free chloroplasts with enhanced rates of cyclic photophosphorylation showed diminished rates of $CO_2$ fixation. In such an experiment a maximum rate of fixation of 36·9 $\mu$moles $CO_2$/mg chl/hr was achieved by "whole" chloroplasts whereas membrane-free chloroplasts gave a rate of only 2·7 (Walker, 1965). Electron microscopy (Figs. 2 and 3) showed that the "whole" chloroplasts contained a substantial proportion with intact outer membranes and stroma whereas the membrane-free chloroplasts had lost most but not all of their stroma. It is assumed therefore that the decrease in fixation rate is largely caused by dilution of the Calvin cycle enzymes.

## CARBON DIOXIDE FIXATION BY PEA CHLOROPLASTS

### MAXIMUM RATE

The rates of carbon dioxide fixation by isolated chloroplasts achieved recently (Gibbs and Bamberger, 1962; Walker, 1964, 1965) are an improvement on those obtained in earlier work (e.g. Gibbs and Cynkin, 1958b) but they still fall considerably short of those supported by the intact plant. In a number of experiments with isolated pea chloroplasts rates in excess of 30 $\mu$moles/mg chl/hr have been obtained but despite attempts to standardize materials and procedure there is a large day to day variation in rate. In exploratory experiments in which the best possible rate was not of primary importance and in which sub-optimal conditions were often used for various reasons, most of the values fell between 10 and 20. These compare unfavourably with the much

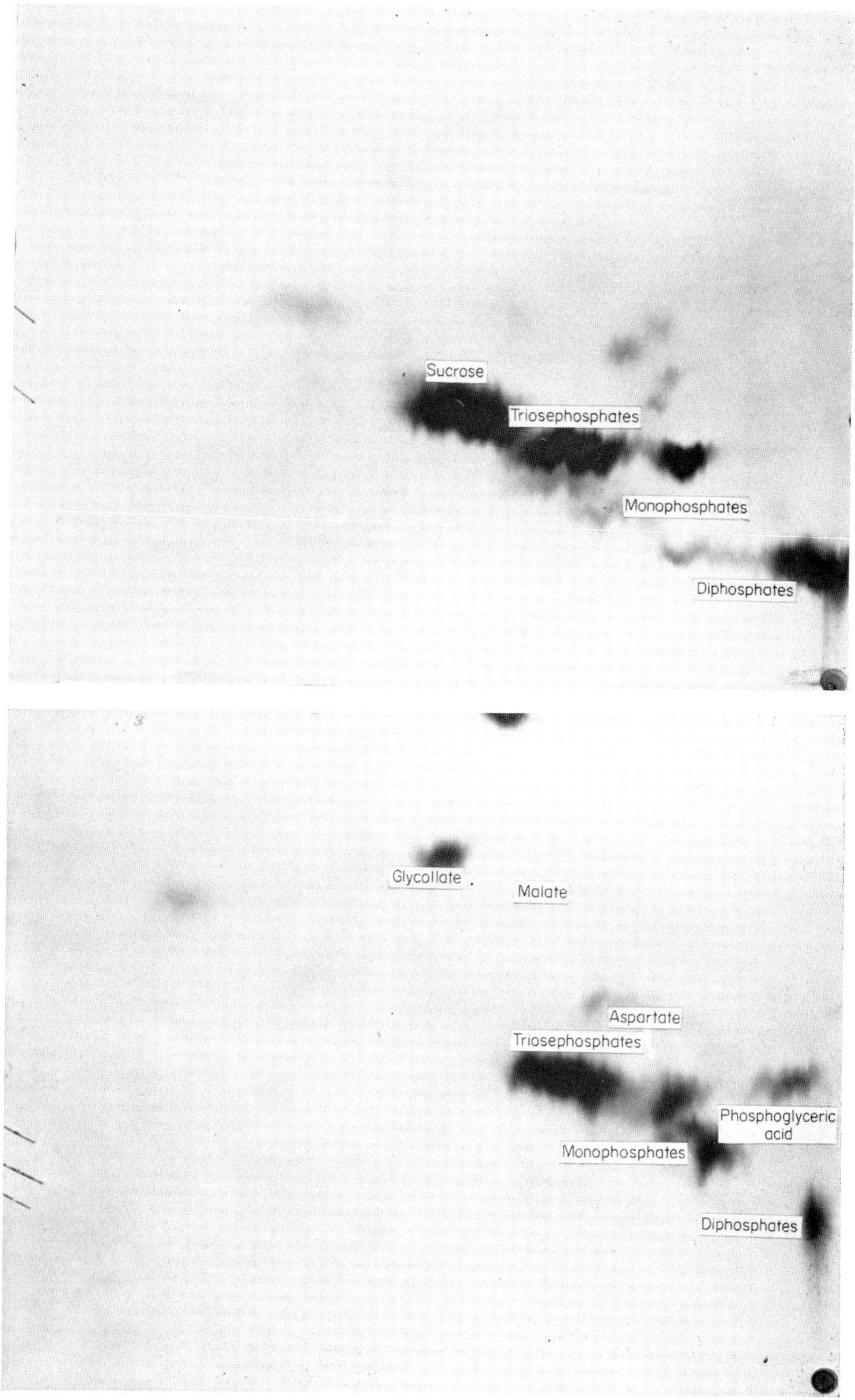

FIG. 4. Radioautographs of products of light $^{14}CO_2$ fixation by pea leaf discs (top) and isolated pea chloroplasts (bottom). Phenol-water (←), butanol-acetic (↑).

quoted maximum of 180 $\mu$moles/mg chl/hr for the intact leaf (Rabinowitch, 1956). Though it is difficult to compare fixation by chloroplasts and whole tissues under closely similar conditions, fixation by pea leaf *discs* under approximately similar conditions (Fig. 4) gave rates of 150 $\mu$moles/mg chl/hr. It should perhaps be added that, for many years, the *best* rates for the Hill reaction by higher plant chloroplasts (see Rabinowitch, 1956) were approximately 100 $\mu$moles $O_2$ evolved/mg chl/hr.

## RELATIONSHIP TO WHOLE CELL PHOTOSYNTHESIS

Gibbs and Cynkin (1958a) have shown that whole spinach chloroplasts support $CO_2$ fixation in which the intramolecular distribution of radioactivity

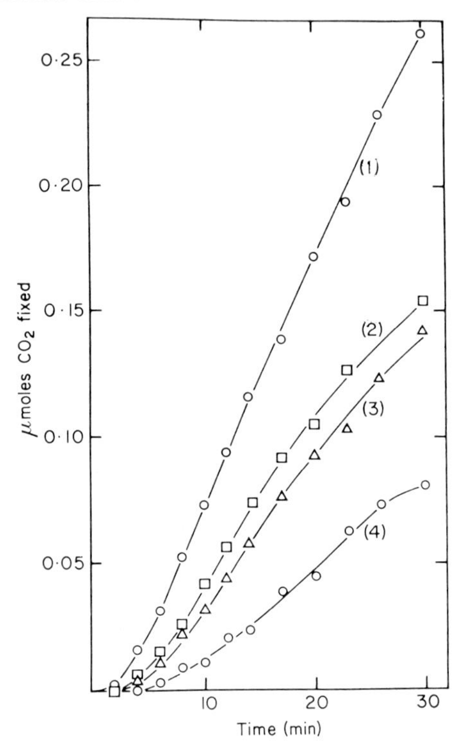


FIG. 5. Progress curves of $^{14}CO_2$ fixation by isolated pea chloroplasts in the presence and absence of ribose 5-phosphate: curves (1) 2·0 $\mu$moles ribose 5-phosphate; (2) 0·1 $\mu$mole ribose 5-phosphate; (3) as for curve 2; (4) no added ribose 5-phosphate. Curves 1 and 2 53 $\mu$g chlorophyll; curves 3 and 4 67 $\mu$g chlorophyll. In addition to chloroplasts (0·1 ml in 0·45 M sucrose), each reaction mixture contained in a final volume of 0·3 ml, $Na_2{}^{14}CO_3$, 1·4 $\mu$moles; $NaHCO_3$, 1·4 $\mu$moles; inorganic phosphate, 0·25 $\mu$mole; $MgCl_2$, 0·25 $\mu$mole; $MnCl_2$, 0·25 $\mu$mole; EDTA, 0·25 $\mu$mole; sucrose, 30 $\mu$moles; Na isoascorbate, 0·50 $\mu$mole; tricine-NaOH, pH 7·5, 7·5 $\mu$moles; glutathione, 1 $\mu$mole. The reactions were carried out at 15° in saturating light (> 6,000 ft candles). Except where indicated, the usual procedure was to add the chloroplasts in the dark to a reaction mixture at 15° followed, after 1 min temperature equilibration, by the addition of the carbonate–bicarbonate immediately prior to illumination. In (1) the chloroplasts were added at 0°, 6 min prior to illumination.

is characteristic of that in whole leaves. Isolated *pea* chloroplasts give their best rates of carbon dioxide fixation when provided with substrate levels of ribose 5-phosphate (authentic ribulose 1,5-diphosphate has not been tested). It might be inferred that these chloroplasts, while retaining an ability to carboxylate, lack a cyclic regenerating system. This inference is made untenable by the following observations.

### *Formation of Phosphoglycerate*

Phosphoglyceric acid is apparently not a major product. Most of the radioactivity from $^{14}CO_2$ appears to be located in triose phosphates and hexose or pentose monophosphates and diphosphates (Fig. 4). Radioactivity also appears in these compounds if $^{32}P$ is substituted for $^{14}C$.

### *Catalytic Effect of Added Sugar Phosphates*

When supplied with catalytic quantities of ribose 5-phosphate the maximum rate is only slightly less than with substrate amounts of ribose 5-phosphate (Section V). Moreover fixation continues at an unchanged rate when sufficient $CO_2$ has been fixed to account for the stoicheiometric conversion of the added catalytic ribose 5-phosphate to ribulose 1,5-diphosphate and its subsequent carboxylation (Fig. 5).

### *Induction Phenomena*

Pea chloroplasts will fix carbon dioxide after an initial induction period in the light (Section V) in the absence of added sugar phosphates (cf. Gibbs and Bamberger, 1962). The rate of fixation is rather lower than in the presence of added sugar phosphates but then by the time the maximum rate is achieved the chloroplasts may have been incubated in bright light at 15° for 20 min or more, by which time thermal inactivation is probably well advanced.

## ABSENCE OF SUCROSE

Though it would seem that isolated pea chloroplasts can catalyse much of the complete Calvin cycle it is of interest that sucrose has not been detected on any of a hundred or more chromatograms (representing a variety of experimental conditions) though it is a major product of photosynthesis by pea leaf discs (Fig. 2). Since the "whole" chloroplasts can apparently fix and reduce $CO_2$ at rates equivalent to at least 20% of those achieved by the whole tissue and since this rate falls to less than 2% in membrane-free chloroplasts it seems most unlikely that the complete absence of label in sucrose can be entirely attributed to enzyme loss during isolation. Failure to detect labelled sucrose among the products of fixation by chloroplasts seems to be a common experience of workers in the field and it is notable that where it has been detected (Tolbert and Zill, 1954; Gee *et al.*, 1965) the plastids have been supplemented with a cytoplasmic fraction.

## Induction Periods in Carbon Dioxide Fixation by Isolated Pea Chloroplasts

### THE EFFECT OF ADDED SUGAR PHOSPHATES ON LAG PERIODS

An initial lag period in $CO_2$ fixation by isolated chloroplasts has been reported by Gibbs and Bamberger (1962). Pea chloroplasts behave in a similar fashion. Figure 5 combines the results from two experiments in which curves 1 and 2 were obtained with one chloroplast preparation and curves 3 and 4 with another. From the first experiment, a time course obtained after dark pre-

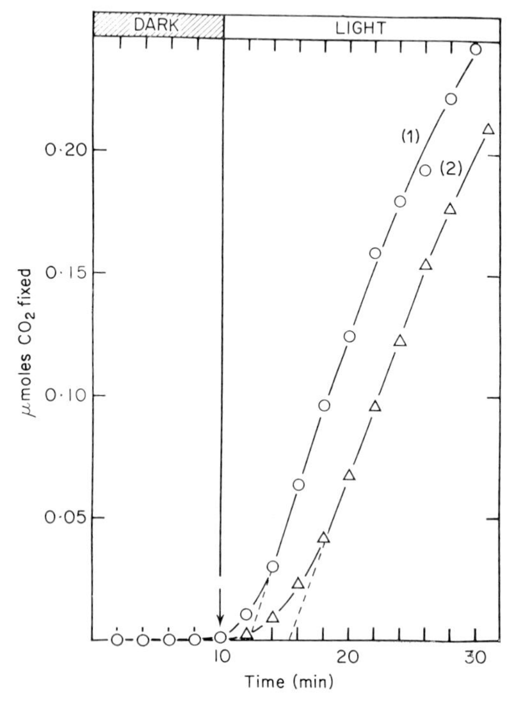


FIG. 6. Progress curves of $^{14}CO_2$ fixation by isolated pea chloroplasts with and without dark pre-incubation: curves (1) reaction started by the addition of chloroplasts at zero time; (2) reaction started by the addition of chloroplasts added immediately before illumination at 10 min. Chlorophyll 72 μg; reaction mixture: see Fig. 5.

incubation with a substrate quantity of ribose 5-phosphate (curve 1) is compared with that obtained when a catalytic quantity of ribose 5-phosphate was added immediately before illumination at zero time (curve 2). Curve 3 was obtained in the same way as curve 2 and is included to permit direct comparison with curve 4 which was obtained in the absence of any added sugar phosphate. These curves illustrate what further experiments showed to be reproducible behaviour, that chloroplasts which were not provided with exogenous sugar phosphates displayed a prolonged lag or induction period before attaining

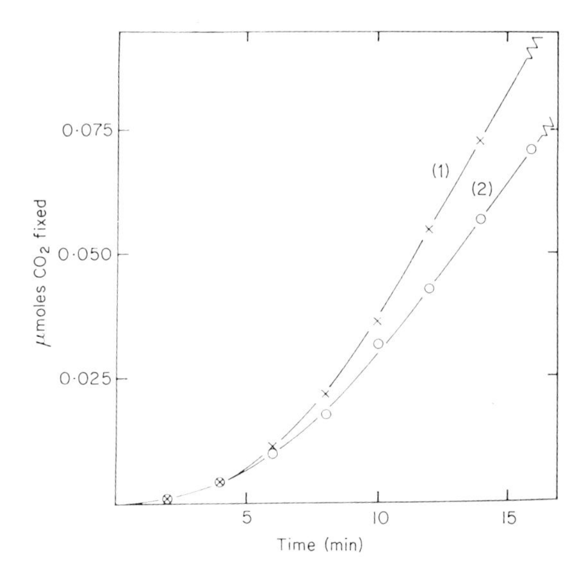


FIG. 7. Progress curves of $^{14}CO_2$ fixation by isolated pea chloroplasts in the presence of fructose 1,6-diphosphate; curves (1) 2·0 μmoles fructose diphosphate; (2) 0·1 μmole fructose disphosphate. Chlorophyll 68 μg; reaction mixture: see Fig. 5.

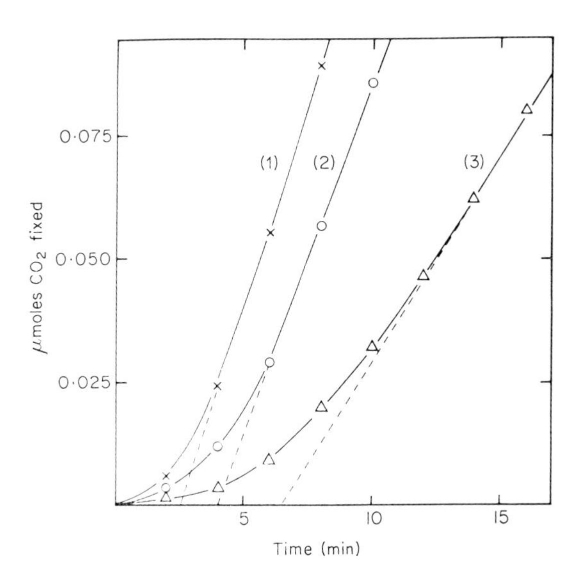


FIG. 8. Progress curves of $^{14}CO_2$ fixation by isolated pea chloroplasts in the presence and absence of sugar phosphates: curves (1) 2·0 μmoles ribose 5-phosphate; (2) 2·0 μmoles fructose 1,6-diphosphate; (3) 2·0 μmoles fructose. Chlorophyll 60 μg; reaction mixture: see Fig. 5.

their maximum rate of $CO_2$ fixation. The lag could be shortened by the addition of "catalytic" quantities of sugar phosphate and brought to minimum by pre-incubation with substrate levels of sugar phosphate. The effect of dark pre-incubation in a complete reaction mixture is best seen in Fig. 6. This shows that the lag period which was found when the ribose 5-phosphate was added immediately before illumination (curve 1) was shortened but not eliminated by 10 min dark pre-incubation with ribose 5-phosphate and $CO_2$ (curve 2) at 15°.

Figure 7 compares substrate (curve 1) and catalytic (curve 2) levels of fructose 1,6-diphosphate and it can be seen here that the time taken to reach the maximum rate is not lengthened and the difference in maximum rates is not proportional to the twenty-fold difference in sugar phosphate concentration. Figure 8 shows part of the rate curves obtained with substrate levels of ribose 5-phosphate (curve 1), fructose 1,6-diphosphate, (curve 2) and free fructose (curve 3). In all these figures it is evident that the fastest rate followed the shortest lag. It is possible that the different rates obtained with different additives could therefore be attributed at least in part. to loss of activity during the lag periods and that in the absence of such losses the slower starters might have eventually achieved the same rates as the others.

## THE EFFECT OF CHLOROPLAST CONCENTRATION

In Figure 9, curve 1 was obtained with a reaction mixture otherwise identical to that used for curve 2 except that it contained twice the quantity of chloro-

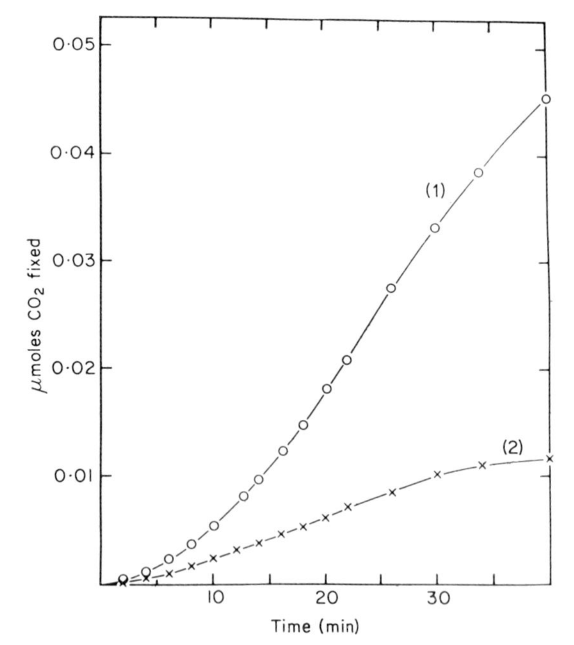


FIG. 9. Progress curves of $^{14}CO_2$ fixation by isolated pea chloroplasts in sorbitol without added sugar phosphates: curves (1) 83 μg chlorophyll; (2) 41 μg chlorophyll. Reaction mixture: see Fig. 5.

plasts. It may be noted that in this experiment sucrose was replaced by sorbitol at all stages, and that in the experiments of Gibbs in which induction periods were reported the osmotic pressure was maintained with NaCl rather than a sugar or a sugar alcohol. It will be seen that for the first 8 min the reaction mixture with twice the chlorophyll fixed $CO_2$ at almost exactly twice the rate, but subsequently the curves progressively diverged and the reaction mixture with more chloroplasts continued to fix $CO_2$ at an increasing rate. It is possible that the immediate cause of this divergence is a more rapid inactivation of the chloroplasts in more dilute suspension. Such an effect has been recorded by Holt, Smith and French (1951) in relation to the ability of chloroplasts to catalyse the Hill reaction.

## AUTOCATALYSIS

The results illustrated in Figs. 5–10 suggest that isolated pea chloroplasts initially lack enough of a component (or components) of the photosynthetic cycle needed to give maximal rates of $CO_2$ fixation but that this is built up during illumination in an autocatalytic fashion. Ribose 5-phosphate or fructose

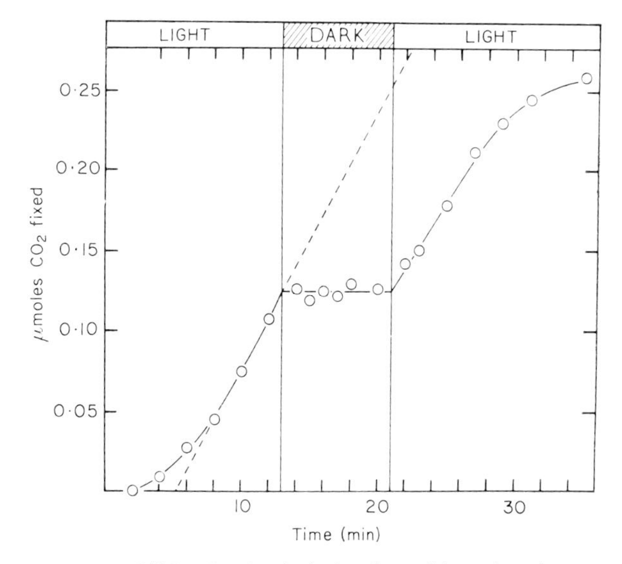


FIG. 10. Progress curve of $^{14}CO_2$ fixation by isolated pea chloroplasts interrupted by a dark interval. The reaction mixture contained 2·0 μmoles of ribose 5-phosphate and 80 μg chlorophyll. In this experiment sorbitol was substituted for sucrose both in the reaction mixtures and in the preparation of the chloroplasts. Reaction mixture: see Fig. 5.

1,6-diphosphate can apparently shorten but not entirely eliminate this induction period. In this respect the present results differ somewhat from those of Gibbs and Bamberger (1962) who found that with spinach chloroplasts the lag in the absence of sugar phosphates could be entirely eliminated by their addition.

If it is assumed that the residual lag in the presence of ribose 5-phosphate is a consequence of its conversion in the light to ribulose 1,5-diphosphate, the further shortening of the lag which follows dark pre-incubation with ribose 5-phosphate would imply that the chloroplast membrane offers some resistance to the penetration of the metabolite. Unfortunately the only ribulose 1,5-diphosphate available during this work was of doubtful authenticity and it has not yet been possible to test this further. However Fig. 10 shows that when illumination of chloroplasts, provided with ribose 5-phosphate, was interrupted by a dark period of 8 min carbon dioxide fixation was resumed at the same rate following re-illumination without any detectable lag. This would imply that in these circumstances some rate-limiting reactant could be made immediately available in sufficient quantity to allow resumption of fixation at the maximum rate.

### THE EFFECT OF DARK PRETREATMENT OF LEAVES

In the experiments described above, pea seedlings (13–14 days old) grown under low light (150 ft candles) with an 11 hr day and a 13 hr night were taken from the dark and illuminated in full daylight for approximately 1 hr before the chloroplasts were separated. Preliminary experiments indicate that if the plants were kept in the dark for 72 hr before use (cf. Arnon, 1958) the lag period in the presence and absence of sugar phosphates was extended and the maximum rate decreased. This effect could be partly reversed by illuminating the plants in full daylight for 4 hr.

## CONCLUSIONS

The results discussed in this article add to existing evidence that it is possible to isolate chloroplasts with intact membranes in aqueous media by conventional techniques and that these retain the ability to carry out photosynthesis in a manner which is not substantially different in many respects from that in the intact cell. It is true that the best rates of $CO_2$ fixation obtained with isolated chloroplasts are at present only 20% of those achieved by the intact leaf but 5 years ago they were only 2% and it seems unlikely that these figures could not be improved by relatively minor changes in techniques or reaction mixtures. The larger problem is perhaps not the isolation of intact functioning chloroplasts but rather the prevention of subsequent inactivation. The decline in the ability of isolated chloroplasts to catalyse several photochemical reactions is well documented (see e.g. Rabinowitch, 1956) but its causes are only poorly understood.

It is possible that the lag periods described in Section V are partly artifacts of an *in vitro* system but it is perhaps more probable that they relate to corresponding induction phenomena in intact tissues (Rabinowitch, 1956).

## ACKNOWLEDGEMENTS

This work was supported by grants from the S.R.C. and the University of London Central Research Fund.

The Gillet and Sibert photomicroscope used for the examination of chloroplasts was provided by a grant from the Royal Society.

I am most grateful to Mr. A. D. Greenwood for permission to reproduce his electron micrographs of pea chloroplasts; to Professor C. P. Whittingham for his advice and criticism, and to Carl Baldry without whose skilled technical assistance this work could not have been undertaken.

## References

Arnon, D. I. (1958). *In* "Symposium on the Photochemical Apparatus its Structure and Function", p. 181. Upton, New York.
Gee, R., Joshi, G., Bils, R. F. and Saltman, P. (1965). *Pl. Physiol* **40**, 89.
Gibbs, M. and Bamberger, E. S. (1962). *Pl. Physiol.* **37**, (supp.) lxiii.
Gibbs, M. and Cynkin, M. A. (1958a). *Nature, Lond.* **182**, 1241.
Gibbs, M. and Cynkin, M. A. (1958b). *Pl. Physiol.* **33**, xviii.
Good, N. E. (1962). *Archs Biochem. Biophys.* **96**, 653.
Hill, R. (1965). *In* "Essays in Biochemistry", **1**, p. 121. Academic Press, London and New York.
Hill, R. and Walker, D. A. (1959) *Pl. Physiol.* **34**, 240.
Holt, A. S., Smith, R. F. and French, C. S. (1951). *Pl. Physiol.* **26**, 164.
Jagendorf, A. T. and Smith, M. (1962). *Pl. Physiol.* **37**, 135.
James, W. O. and Das, V. S. R. (1956). *New Phytol.* **56**, 325.
James, W. O. and Leech, R. M. (1964). *Proc. R. Soc.* B **160**, 13.
Kahn, A. and von Wettstein, D. (1961). *J. Ultrastr. Res.* **5**, 557.
Leech, R. M. (1964). *Biochim. biophys. Acta* **79**, 637.
Rabinowitch, E. I. (1956). "*Photosynthesis*", Vol. II, p. 2. Interscience Publishers, Inc., New York.
Smillie, R. M. and Fuller, R. C. (1959). *Pl. Physiol.* **34**, 651.
Spencer, D. and Unt, H. (1965). *Aust. J. biol. Sci.* **18**, 197.
Tolbert, N. E. and Zill, L. P. (1954). *J. gen. Physiol.* **37**, 575.
Walker, D. A. (1964). *Biochem. J.* **92**, 22c.
Walker, D. A. (1965). *Pl. Physiol.* **40**, 1157.
Whatley, F. R., Allen, M. B., Rosenburg, L. L., Capindale, J. B. and Arnon, D. I. (1956). *Biochim. biophys. Acta* **20**, 462.
Whatley, F. R. and Arnon, D. I. (1963). *In* "Methods in Enzymology", p. 308. Academic Press, New York and London.

# Transport Metabolites in Photosynthesis

U. W. Heber
*Institut für Landwirtschaftliche Botanik der Universität Bonn, Germany*

## Introduction[1]

Isolated intact chloroplasts possess the complete enzymic machinery for the uptake and reduction of $CO_2$ and the regeneration of the $CO_2$ acceptor molecule (Losada *et al.*, 1960; Havir and Gibbs, 1963). However, information is scarce as to whether efficient photosynthesis requires the cooperation of cytoplasmic constituents. It is well known that isolated chloroplasts exhibit only low rates of photosynthesis (cf Gee *et al.*, 1965). The aim of the experiments to be reported was to establish where reactions brought about by photosynthesis occur and whether there is an intracellular transport of photosynthetic intermediates. Three different types of experiments have been performed:

1. $^{14}CO_2$ was fed, during steady state photosynthesis, to intact leaves, which were, after different times, immersed in liquid nitrogen and freeze-dried. Chloroplasts were separated from the remainder of the cellular components by a non-aqueous isolation method. From the analysis of the fractions it was possible to locate intracellular sites of incorporation of label and to follow its spread into other parts of the cell. It has been shown that the non-aqueous isolation technique effectively prevents leaching of water-soluble compounds from their original location in the cell (Heber and Willenbrinck, 1964; Urbach *et al.*, 1965).
2. Leaf cells exhibit very pronounced changes in the levels of a number of metabolites in the dark/light or the light/dark transition (Bassham *et al.*, 1956; Kandler and Haberer-Liesenkötter, 1963). Again the non-aqueous chloroplast isolation technique was employed to explore whether these changes are confined to the chloroplasts or whether they occur both in the chloroplasts and in the cytoplasm.
3. Intact and broken chloroplasts, which had been isolated in aqueous buffer containing sucrose, convert added intermediates of photosynthesis into other products at markedly different rates. This effect has been used as an indicator of whether the chloroplast membrane is permeable to a particular compound or not.

[1] Abbreviations: RuDP, ribulose 1,5-diphosphate; SuDP, sedoheptulose 1,7-diphosphate; FDP, fructose 1,6-diphosphate; FMP, fructose monophosphate; GMP, glucose monophosphate; DHAP, dihydroxyacetone phosphate; GAP, glyceraldehyde 3-phosphate; PGA, 3-phosphoglyceric acid; *tris*, *tris*(hydroxymethyl)aminomethane; NADP(H), (reduced) triphosphopyridine nucleotide.

## RESULTS

### LABELLING EXPERIMENTS

Figure 1 shows the distribution of some labelled metabolites between the chloroplasts and the remainder (cytoplasm) of leaf cells from *Elodea* after photosynthesis for various times in the presence of $H^{14}CO_3^-$. The distribution data from this steady state experiment are very similar to those obtained previously under non-steady state conditions (Heber and Willenbrinck, 1964). From Fig. 1 it is evident that labelled RuDP is not, or only to a small extent,

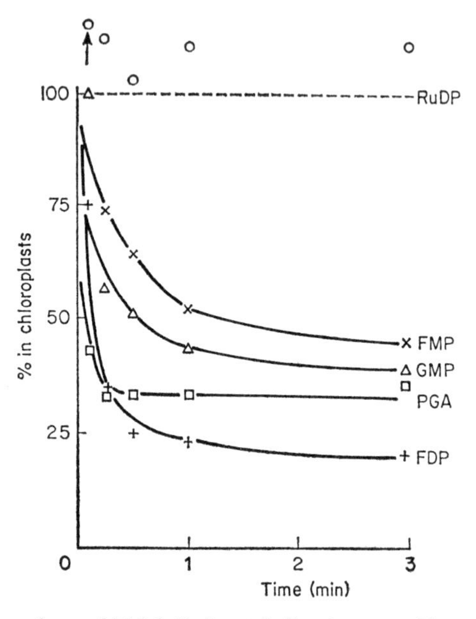


FIG. 1. Distribution of some $^{14}$C-labelled metabolites between chloroplasts and cytoplasm of leaf cells from *Elodea densa*. Ordinate: % of the total amount present in the chloroplasts; Abscissa: time in min. Conditions: Pre-illumination of *Elodea* shoots for 10 min in $2 \times 10^{-3}$ M $H^{12}CO_3^-$, subsequent transfer, under illumination, in $H^{14}CO_3^-$ for the indicated times; killing in liquid nitrogen; non-aqueous chloroplast isolation; further procedures as in Heber and Willenbrinck (1964).

present in the cytoplasm, even after some minutes of photosynthesis. This suggests that the chloroplast membrane is impermeable to RuDP. Very similar results have been obtained for labelled SuDP. In contrast, percentages approaching 100 of labelled PGA, FDP, FMP and GMP are to be found in the chloroplasts after only very brief exposure to $^{14}CO_2$. These compounds then show a rapid decrease with time until an equilibrium in the distribution between chloroplasts and cytoplasm is reached. Labelled PGA, FDP, FMP and GMP thus appear rapidly in the cytoplasm. This can only be explained by a transfer from the chloroplasts, where labelling has taken place. Whether this transfer is direct or occurs via transport metabolites different from the compounds under investigation cannot readily be decided from these experiments.

INTRACELLULAR LEVELS OF PGA, FDP AND DHAP

A possible approach to the problem as to which compounds serve as transport metabolites in photosynthesis is the study of fluctuations in the pool sizes of metabolites. A compound capable of traversing the chloroplast membrane should be expected to exhibit concomitant changes on both the chloroplasts and the cytoplasm during the dark/light and the light/dark transition. The same may not be true for a compound confined to either chloroplasts or cytoplasm. In fact, during illumination the levels of $NADP^+$ and NADPH change in the chloroplasts, but not in the cytoplasm (Heber and Santarius, 1965). Very similar results have also been obtained for chromatographically

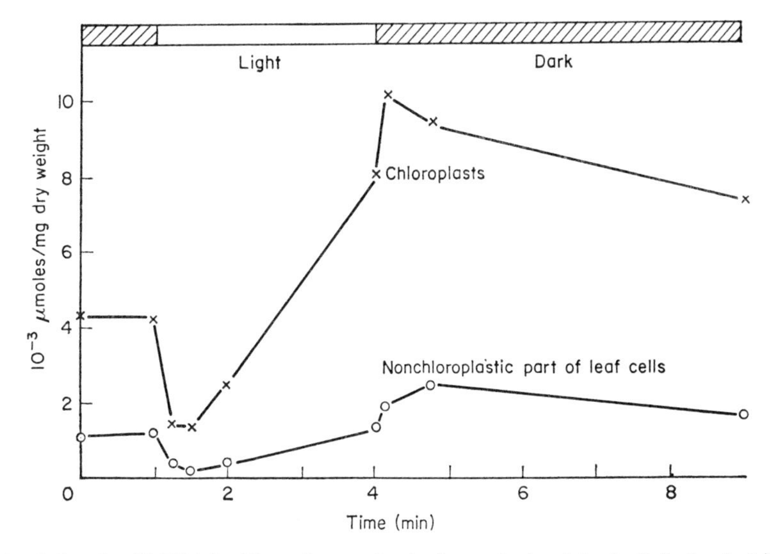


FIG. 2. Levels of 3-PGA in chloroplasts and cytoplasm of spinach leaf cells in the dark/light and the light/dark transition. Ordinate: PGA content of chloroplasts and non-chloroplastic part of leaf cells (cytoplasm including cell walls and vacuolar constituents); Abscissa: time in min. Conditions: Leaves were kept in the dark for several hours and then illuminated or darkened as indicated. Killing in liquid nitrogen; non-aqueous chloroplast isolation; enzymatic determination of PGA (Czok and Eggert, 1962); calculations according to Heber and Willenbrinck (1964).

separated labelled sugar diphosphates, which have, however, not been resolved into the individual components (unpublished experiments). It appears likely that the response to illumination in the chloroplasts exceeding that in the cytoplasm is caused by RuDP.

In contrast, changes in the levels of PGA are not restricted to the chloroplasts (Urbach *et al.*, 1965). In both the chloroplasts and the cytoplasm PGA at first decreases after illumination and then increases slowly (Fig. 2). On darkening a further rapid increase in PGA occurs in chloroplasts and in the cytoplasm. The percentage distribution of PGA between chloroplasts and cytoplasm is very similar to that of the [$^{14}$C]PGA after one or more minutes photosynthesis in the presence of $^{14}CO_2$. In *Spinacia* and *Elodea* approximately 70% and 35% respectively of the total PGA of the cells is found in the chloroplasts. In addition to the concomitant changes in chloroplasts and cytoplasm, the similarity in the distribution of total and of labelled PGA between chloroplasts and cytoplasm strongly suggests that PGA is directly transferred across the chloroplast membrane.

The levels of FDP and DHAP are, in the dark, very low in the chloroplasts as well as in the cytoplasm (Fig. 3). Illumination results in a large increase and darkening in a rapid decrease in the levels in both the chloroplasts and the cytoplasm. These concomitant changes may be interpreted in one of the following ways:

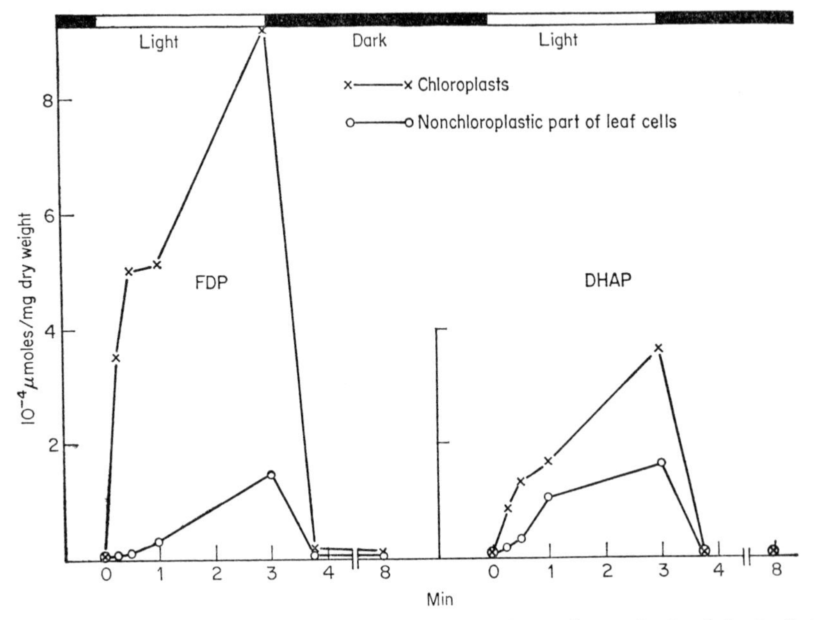


FIG. 3. Levels of FDP and DHAP in chloroplasts and cytoplasm of spinach leaf cells in the dark/light and the light/dark transition. Ordinate: FDP and DHAP content of chloroplasts and the nonchloroplastic part of leaf cells; Abscissa: time in min. Conditions: As indicated for Fig. 3; enzymatic determination of DHAP and FDP (Bücher and Hohorst, 1962).

1. The transfer from the chloroplasts into the cytoplasm may be indirect and may be mediated by a third intermediate. PGA can be ruled out as a mediator, since its reduction in the cytoplasm does not appear possible to any major extent owing to lack of reducing power.

2. Predominantly one of the two intermediates may be transferred from the chloroplasts to the cytoplasm. Aldolase and triosephosphate isomerase are available in high activities in chloroplasts and cytoplasm (Stocking, 1959; Heber, 1960; Peterkofsky and Racker, 1961) to convert the transport metabolite into the other intermediate.

3. Both intermediates may easily pass the chloroplast membrane.

On the basis of the data presented in Section C the second (or the third) possibility is favoured.

## RATES OF THE CONVERSION OF EXOGENOUS METABOLITES CATALYSED BY INTACT AND BROKEN CHLOROPLASTS

Another approach to the transport problem is offered by the observation that enzymes present in intact isolated chloroplasts are not freely accessible

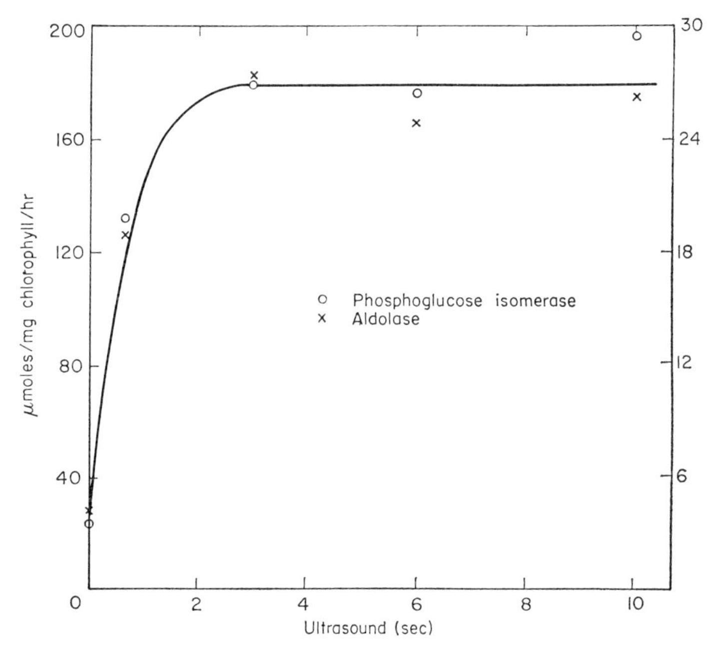


FIG. 4. The stimulation of the chloroplast aldolase and phosphoglucose isomerase reaction by ultrasound. Ordinate (left): μmoles of DHAP formed from FDP/mg chl/hr; Ordinate (right): μmoles of glucose 6-phosphate formed from fructose 6-phosphate/mg chl/hr; Abscissa: exposure of the chloroplasts to ultrasound. Conditions: Chloroplast isolation as in Table I. The reaction mixture contained *tris* (35 mM), sucrose (0·3 M), $MgCl_2$ (1 mM) and FMP (4 mM) or FDP (5 mM), pH 7·6; reaction time 5 min; determination of the reaction products by optical tests (Bücher and Hohorst, 1962; Hohorst, 1962).

to exogenous substrates. Breaking the chloroplast structure by ultrasound or by osmotic shock results in markedly increased reaction rates. Only a brief exposure to ultrasound is sufficient to afford maximal stimulation (Fig. 4). Very large stimulation rates characterize the light-dependent reduction of $NADP^+$ and the carboxydismutase reaction (Tables I and II). This agrees with the observations mentioned in Sections A and B, according to which there is no or only little transfer of RuDP and of $NADP^+$ or NADPH between chloroplasts and cytoplasm. Other reactions show considerably less or even no stimulation. This is especially true for the reduction in light of PGA to DHAP.

Table I. Rates of enzymic reactions in a system containing intact and broken spinach chloroplasts in sucrose solution

| | Rate of the reaction in | | |
|---|---|---|---|
| | "Intact" chloroplasts | Supernatant of "intact" chloroplasts | Broken chloroplasts |
| | | cts/mg chl/hr | |
| $RuDP + CO_2 \rightarrow 2\ PGA$ | $9{\cdot}8 \times 10^6$ | $9{\cdot}9 \times 10^6$ | $42 \times 10^6$ |
| | | μmoles/mg chl/hr | |
| $NADP^+ \xrightarrow{h\nu} NADPH$[1] | 0·4 | — | 13 |
| GAP → DHAP | 188 | 181 | 592 |
| FDP → DHAP | 70 | 25 | 168 |
| $PGA \xrightarrow{h\nu} DHAP$[2] | 7 | — | 4·4 |

[1] No addition of ferredoxin.
[2] No addition of pyridine nucleotide.
Conditions: Chloroplast isolation in a buffer containing *tris* (50 mM), sucrose (0·4 M), phosphate (10 mM), ascorbate (10 mM), cysteine (3 to 5 mM) and, sometimes, mercaptoethanol (20 mM); pH 7·8. The chloroplasts were washed once in a buffer containing *tris* (35 to 50 mM) and sucrose (0·3 M). Ultrasonication (3 sec) in a MSE disintegrator (20,000 c/s). Since washing of the chloroplasts did not remove all exogenous enzyme and since leaching of enzymes occurred during the procedure, exogenous enzyme was determined immediately prior to the test by centrifuging the intact chloroplasts and assaying the supernatant. Modifications of the test systems given by Arnon *et al.* (1959), Pon (1960), Bücher and Hohorst (1962) and Hohorst (1962) were used. All reaction mixtures contained 0·3 M sucrose in addition to substrates and cofactors.

Isolated chloroplasts reduce added PGA much faster than added $NADP^+$. However, while $NADP^+$ reduction is enhanced by ultrasound, sometimes more than 100-fold, PGA reduction is not stimulated (Urbach *et al.*, 1965). The low absolute rate of PGA reduction (Table I) is easily explained by the fact that isolated chloroplasts contain, owing to losses during the isolation procedure, very little pyridine nucleotide (Krogmann, 1958; Heber and Santarius, 1965).

Table II. The stimulation of reaction rates by ultrasound

| | Rates in broken chloroplasts[1] / Rates in intact chloroplasts |
|---|---|
| $RuDP + CO_2 \rightarrow 2\ PGA$ | 25/1 to ∞ |
| $NADP^+ \xrightarrow{h\nu} NADPH$ | 25/1 to ∞ |
| GAP → DHAP | 20/1 to ∞ |
| FDP → DHAP | 3/1 to 8/1 |
| $PGA \xrightarrow{h\nu} DHAP$ | <1/1 |

[1] The rate for the broken chloroplasts was divided by the rate for the intact chloroplasts, after the rate for the supernatant enzyme (cf. Table I) was subtracted from both values.

The results indicate that PGA, as well as the reaction product DHAP, function as transport metabolites. It has long been recognized that the latter compound is capable of passing the mitochondrial membrane (Bücher and Klingenberg, 1958).

FDP is converted to DHAP in chloroplasts owing to the presence of aldolase and an excess of triose phosphate isomerase. This reaction is much less stimulated by ultrasound than is the carboxylation of RuDP (Tables I and II). This indicates that FDP can also pass the chloroplast membrane. From the extent of the stimulation it appears likely that the inflow of FDP to the reaction sites rather than the outflow of DHAP is the rate-limiting step of the reaction in intact chloroplasts.

In contrast to the reduction of PGA and the splitting of FDP, the conversion of GAP to DHAP and vice versa is stimulated by ultrasound about as much as is the carboxydismutase reaction. Probably the chloroplast membrane is impermeable to GAP.

The results discussed in this section have been obtained with chloroplasts isolated in an isotonic buffer system containing sucrose. The isolation of chloroplasts in buffer containing 0·35 M NaCl does not yield suitable preparations. Stimulations in the reaction rates induced by ultrasound are then very small, even in the case of the carboxydismutase reaction, indicating that the mass of the chloroplasts did not retain their permeability properties.

## Discussion

The rapid spread of label from the chloroplasts to the cytoplasm (Fig. 1) raises the question which compounds are engaged in the transport of carbon across the chloroplast membrane. There are a great number of possibilities. However, during the first 20 or 30 sec of photosynthesis, when intracellular transport can already be demonstrated, label is essentially confined to phosphorylated intermediates and a few amino acids. Thus phosphorylated intermediates of metabolism are among the compounds which function as transport metabolites. This leads to the preservation of energy, since no re-phosphorylation is necessary for subsequent transformations, which occur in the cytoplasm. The experiments with intact leaves (Sections A and B) suggest that PGA and one or both of DHAP and FDP can leave the chloroplasts. Additional information stems from experiments with aqueously isolated chloroplasts. While RuDP is not converted into PGA by isolated intact chloroplasts —an indication that these chloroplasts are impermeable to RuDP—PGA and DHAP can enter and leave such intact chloroplasts. The penetration of FDP may be slow. There is some uncertainty as to whether the permeability properties of aqueously isolated chloroplasts remain essentially unaltered during the isolation procedure. Since the results are in line with related observations from *in vivo* experiments, it may be assumed that this is indeed the case.

## Summary

1. After feeding $^{14}CO_2$ to intact leaves under continuous illumination PGA, FDP, GMP and FMP are labelled in the chloroplasts and then appear rapidly

in the cytoplasm. Labelled RuDP and SuDP are retained in the chloroplasts.

2. The levels of PGA, FDP and DHAP show large concomitant changes in chloroplasts and cytoplasm in the dark/light and the light/dark transition.

3. Aqueously isolated chloroplasts suspended in isotonic sucrose buffer convert some exogenous substrates at a rate lower than that found when chloroplasts are previously exposed to ultrasound or osmotic shock. The stimulation of the reaction rate induced by ultrasound is very large in the case of $NADP^+$ reduction and of the reactions catalysed by carboxydismutase and triosephosphate isomerase. The reactions catalysed by aldolase and glyceraldehyde phosphate dehydrogenase/phosphoglycerate kinase are less or not stimulated.

4. It is concluded that PGA, DHAP and, perhaps to a much smaller extent, FDP function as transport metabolites in photosynthesis.

## References

Arnon, D. I., Whatley, F. R. and Allen, M. B. (1959). *Biochim. biophys. Acta* **32**, 47.

Bassham, J. A., Shibata, K., Steenberg, K., Bourdon, J. and Calvin, M. (1956). *J. Am. chem. Soc.* **78**, 4120.

Bücher, Th. and Klingenberg, M. (1958). *Angew. Chem.* **70**, 552.

Bücher, Th. and Hohorst, H.-.J (1962). *In* "Methoden der enzymatischen Analyse" (H. U. Bergmeyer, ed.) p. 247. Verlag Chemie. Weinheim.

Czok, R. and Eggert, L. (1962). *In* "Methoden der enzymatischen Analyse" (H. U. Bergmeyer, ed.) p. 224. Verlag Chemie. Weinheim.

Gee, R., Joshi, G., Bils, R. F. and Saltman, P. (1965). *Pl. Physiol.* **40**, 89.

Havir, E. A. and Gibbs, M. (1963). *J. biol. Chem.* **238**, 3183.

Heber, U. (1960). *Z. Naturf.* **15**b, 100.

Heber, U. and Willenbrinck, J. (1964). *Biochim. biophys. Acta* **82**, 313.

Heber, U. and Santarius, K. A. (1965). *Biochim. biophys. Acta* **109**, 390.

Hohorst, H.-J. (1962). *In* "Methoden der enzymatischen Analyse" (H. U. Bergmeyer, ed.) p. 134. Verlag Chemie, Weinheim.

Kandler, O. and Haberer-Liesenkötter, I. (1963). *Z. Naturf.* **18**b, 718.

Krogmann, D. (1958). *Archs Biochem. Biophys.* **76**, 75.

Losada, M., Trebst, A. V. and Arnon, D. I. (1960). *J. biol. Chem.* **235**, 832.

Peterkofsky, A. and Racker, E. (1961). *Pl. Physiol.* **36**, 409.

Pon, N. G. (1960). "Studies on the carboxydismutase system and related materials." Thesis. University of California UCRL-9373.

Stocking, C. R. (1959). *Pl. Physiol.* **34**, 56.

Urbach, W., Hudson, M. A., Ullrich, W., Santarius, K. A. and Heber, U. (1965). *Z. Naturf.* **20**b, 890.

# Light Respiration—Correlations Between $CO_2$ Fixation, $O_2$ Pressure and Glycollate Concentration

K. Egle and H. Fock
*Department of Botany, University of Frankfurt, Germany*

## Introduction

The effect of visible light on respiration of photo-autotrophic plants has so far not been explained satisfactorily. A few experiments which resulted in an improvement of respiration with light are contradicted by other findings, which show an inhibition or no effect at all of light on the respiratory metabolism (Egle, 1960; Rosenstock and Ried, 1960; Ducet and Rosenberg, 1962; Krotkov, 1963; Zelitch, 1964). The basis for these various contradictory results, apart from other factors, appears to be that the term "respiration" has without limitations frequently been identified with the rate of the $CO_2$ output and with the $O_2$ intake.

The team of Krotkov at Kingston, Ontario, Canada (Krotkov *et al.*, 1958; Lister *et al.*, 1961; Tregunna *et al.*, 1961) as well as the team in the Department of Botany, Frankfurt a.M., Germany (Egle and Döhler, 1963, 1964; Döhler and Egle, 1964; Fock *et al.*, 1964; Fock, 1965) in recent years have described transitory effects in $CO_2$ exchange during the light–dark change in green leaves as well as in suspensions and sediments of unicellular green algae, which within the first few minutes of the dark period produce a considerably increased $CO_2$ liberation. It had to be confirmed whether the observed more intense liberation of carbon dioxide is dealt with a process, which begins only at the end of the light phase or whether the subsequently increased $CO_2$ output already occurs during the light phase and fades away only during the first minutes of the dark phase.

## Materials and Methods

We used attached leaves of *Phaseolus vulgaris* L. and *Helianthus annuus* L., thalli of the liverwort *Conocephalum conicum* and suspensions of *Chlorella vulgaris* Beyerinck (strain 211-f of the Pringsheim collection, Göttingen, Germany). An equipment of three gas regulator pumps allowed us to change the carbon dioxide and oxygen concentrations in a gas mixture (together with nitrogen) over a very wide range within a few seconds. The $CO_2$ exchange has been measured by an infrared gas analyser (URAS) with a sensitivity of 100 ppm $CO_2$ over the whole scale range (25 cm) of a compensation recorder (Linecomp). Details of the method will be given elsewhere (Fock and Egle, 1965).

## Results

The results obtained with leaves and thalli of liverwort are almost similar, but as the curves for thalli are more pronounced, only these are given in the following figures. Figure 1 shows the $CO_2$ exchange of *Conocephalum conicum* during the light and dark period with the same carbon dioxide concentration of the gas mixture of 300 ppm but with different oxygen concentrations (1 %, 25 %, 75 %). The curves demonstrate that the apparent photosynthesis decreases considerably with constant partial pressure of $CO_2$ and with increas-

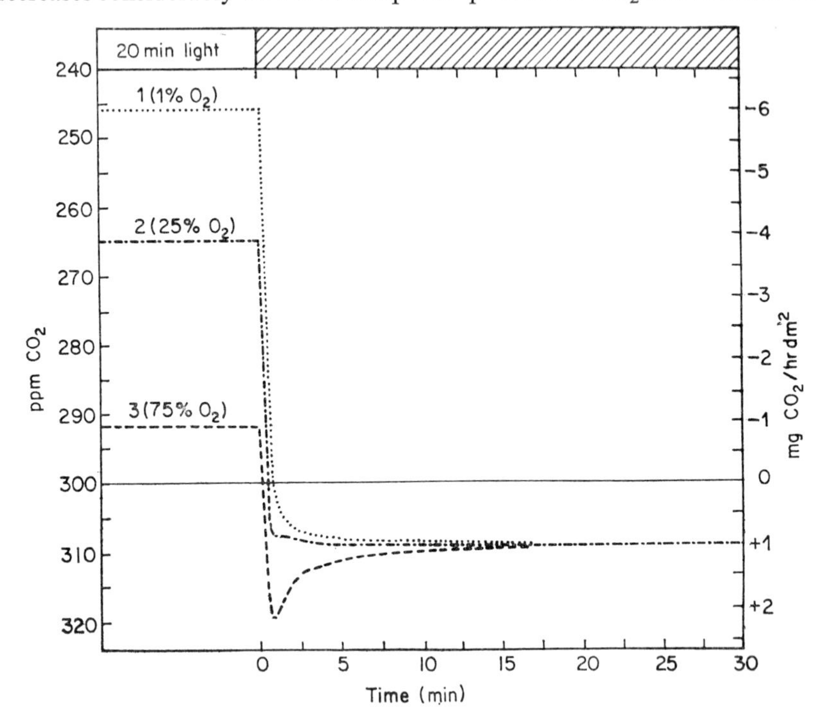


FIG. 1. $CO_2$ uptake and liberation of *Conocephalum conicum* during the light and dark phases. The gas stream (43 l/hr) introduced into the assimilation chamber consists of a constant $CO_2$ concentration of 300 ppm and of a different $O_2$ concentration of 1 %, 25 % and 75 %, together with nitrogen. Surface of the thalli = 71 $cm^2$, light intensity = 14,000 lux, temp = 20°, rel. humidity = 100 %.

ing oxygen concentrations. After darkening the $CO_2$ output of the thalli with very low $O_2$ concentration increases continuously, until after the filling up of the $CO_2$ deficiency the stationary state of dark respiration is reached. This transient effect takes about 5 min. With high $O_2$ pressure, however, within the first minute of darkening a strong $CO_2$ outburst occurs followed by a decrease of the $CO_2$ liberation, very rapidly at first and then more slowly, down to the stationary level of dark respiration, which is reached after 15 min. These effects are reversible by changing the $O_2$ pressure and can be repeated frequently with the same plant material. It should be mentioned here, that the rate of steady state dark respiration is not influenced by $O_2$ pressures between 1% and 99%.

With high carbon dioxide concentration of 1200 ppm (Fig. 2) the $CO_2$ uptake during illumination similarly depends on the $O_2$ pressure in the gas mixture, but the depression of the apparent photosynthesis with high oxygen concentrations does not reach those extreme values, even when we use a gas mixture with 99% oxygen, 1% nitrogen together with 0·12% carbon dioxide. The "outburst" of carbon dioxide in the first minute of the dark period does not occur under these conditions with relatively high $CO_2$ pressure; there are only small intervals in the time which is needed to reach the stationary dark respira-

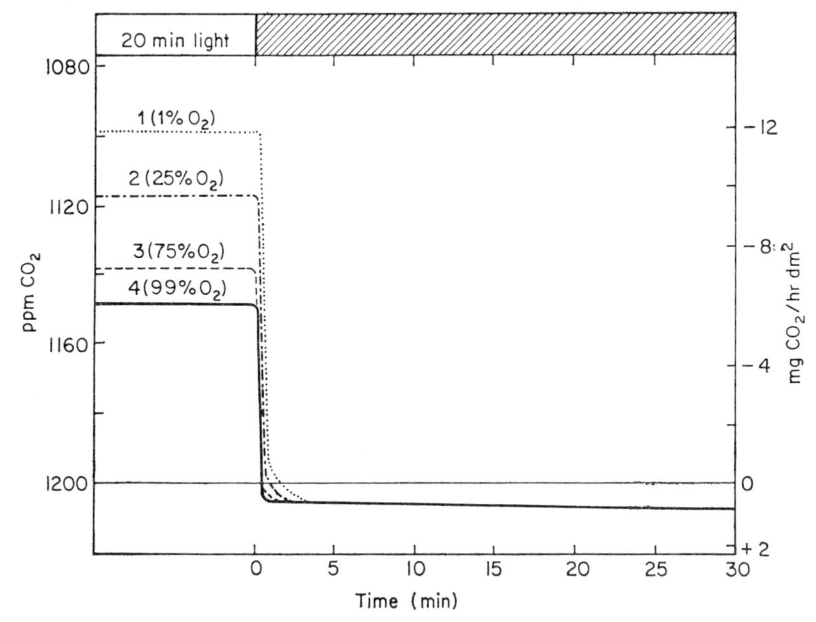


FIG. 2. $CO_2$ uptake and liberation of *Conocephalum conicum* during the light and dark phases. The gas stream (45 l/hr) introduced into the assimilation chamber consists of a constant $CO_2$ concentration of 1200 ppm and of a different $O_2$ concentration of 1%, 25%, 75% and 99% together with nitrogen. [For other data see Fig. 1.]

tion rate between very low and very high $O_2$ concentrations in the gas phase surrounding the plant material.

Apart from the dependence of this rapid and strong carbon dioxide liberation in the first minute of the dark period on the high level of $O_2$ pressure and on the low level of $CO_2$ pressure there exists a correlation between this $CO_2$ outburst and the intensity of the illumination during the preceding light period. Figure 3 shows the $CO_2$ exchange of liverwort in a gas stream consisting of 300 ppm $CO_2$ and 75% $O_2$ in nitrogen, with light intensities varying between 14,000 and 2500 lux. With increasing light intensity the $CO_2$ outburst at the beginning of the next dark period becomes stronger and more pronounced. At lower light intensities the smaller $CO_2$ liberation within the first minutes after darkening is followed by a small second wave until the steady state dark respiration is reached.

With these results alone we cannot decide exactly whether this observed $CO_2$ outburst immediately after the end of the light period is an effect that starts at the beginning of the dark period or an occurrence that lasts during the illumination period and fades away in the first minutes of darkness. In case this $CO_2$ outburst only sets in with darkness the decrease of the carbon dioxide uptake depending on $O_2$ pressure, as observed for the first time by Warburg (1920), would be dealing with a true inhibition of photosynthesis (cf. Kessler, 1960, p. 942). If this $CO_2$ outburst, however, is the result of a fading effect of a continuous occurrence of $CO_2$ liberation during the light period, then the considerably different $CO_2$ uptake depending on the oxygen partial

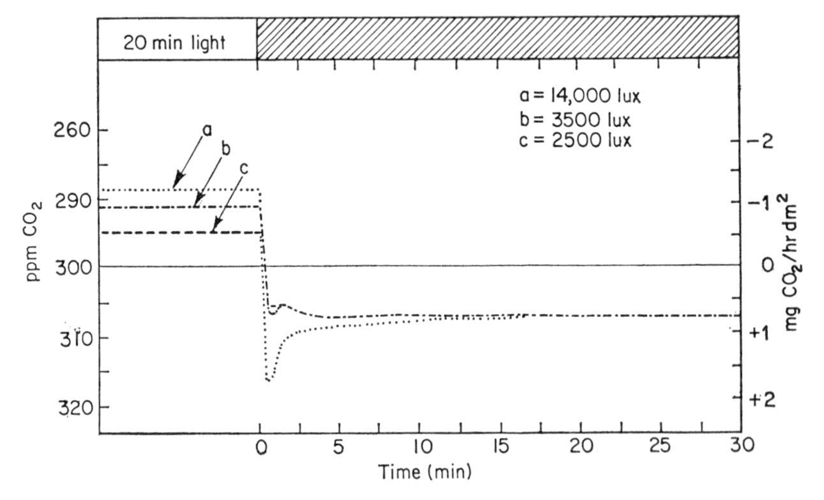


FIG. 3. $CO_2$ gas exchange of *Conocephalum conicum* with different light intensities. The gas stream (41 l/hr) consists of 300 ppm $CO_2$ and 75% $O_2$ in nitrogen. Surface of the thalli $=70$ cm². [For other data see Fig. 1.]

pressure predominantly cannot deal with an inhibition of photosynthesis. In this case the measured photosynthetic rate could be considered as a resultant between $CO_2$ liberation in light, which is strongly influenced by the oxygen concentration, and the almost constant total photosynthesis. The following results on the carbon dioxide liberation of the plant material in an atmosphere free from carbon dioxide may contribute to the solution of this problem.

Figure 4 demonstrates the $CO_2$ output of the liverwort in a $CO_2$ free gas stream, consisting of nitrogen together with an oxygen content of 1%, 25% and 75%. During the steady state dark respiration the $CO_2$ level of the gas stream passing through the assimilation chamber is increased from zero to approximately nine ppm $CO_2$. During illumination with very low oxygen pressure (1%) an amount of nearly 70% of the respiratory carbon dioxide is reassimilated; after darkening the steady rate dark respiration is continuously reached again after 5 min. With a higher oxygen pressure (25%) only a small percentage of the respiratory carbon dioxide is reassimilated. The time course of the $CO_2$ liberation reaches a minimum rapidly within the first light minute and then increases gradually to a higher level almost constantly throughout the following light period. At the end of the illumination period a small but very rapid $CO_2$ outburst occurs followed by a second wave until the steady state dark respiration is reached gradually after about 30 min. With the high oxygen pressure of 75% the curve declines to a weaker minimum within 30 sec after the illumination begins, but then follows a $CO_2$ liberation throughout the entire light period, which reaches a level almost double that of the steady state dark respiration. When the light is turned off, an additional rapid $CO_2$ outburst occurs within 30 sec. Then the $CO_2$ liberation is reduced at first rapidly and passing over a second wave the steady state dark respiration is reached after 30 min.

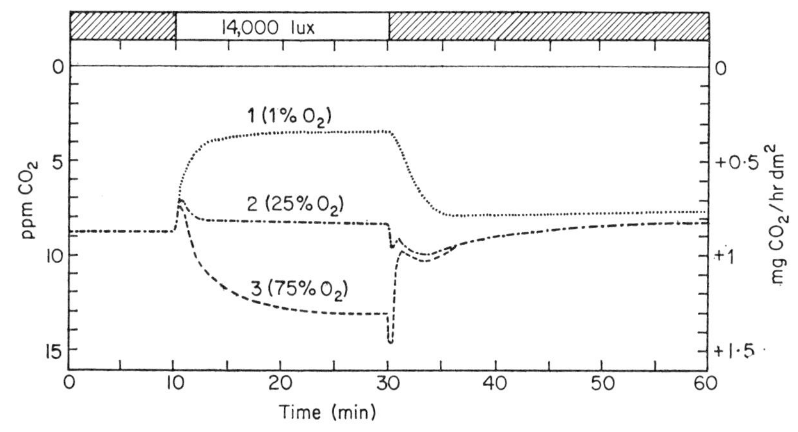


FIG. 4. $CO_2$ liberation of *Conocephalum conicum* in a gas stream (39 l/hr) free of carbon dioxide with different $O_2$ concentrations of 1%, 25% and 75% in nitrogen. Surface of the thalli = 72 $cm^2$. [For other data see Fig. 1.]

In case there occurs an inhibition of photosynthesis, the plants during the light period should not liberate under any circumstances more carbon dioxide than during the steady state dark respiration. However, as the thalli actually liberate considerably more carbon dioxide in light at a high $O_2$ pressure than in the dark, the hypothesis for an inhibition of photosynthesis does not arise. The small peak which is reached in the course of the curve with the higher $O_2$ pressures shortly after the beginning of illumination as well as the short but strong $CO_2$ outburst immediately after the end of illumination forces us to conclude that in this case we are dealing with two superimposed processes during the light period, one of these consumes carbon dioxide and the other one liberates carbon dioxide. The $CO_2$ liberation sets in later on and with a high $O_2$ pressure it can considerably exceed the photosynthetic $CO_2$ uptake. When photosynthesis ceases at the end of illumination there may be observed for another few seconds the $CO_2$ liberating process in the form of a short strong $CO_2$ outburst like a fading effect, which superimposes the dark fixation of carbon dioxide. The second waves in the time course of the curves which occur later on may be attributed to the longer lasting dark fixation of carbon dioxide and to the more rapid decomposition of intermediates from the preceding light period (Egle and Döhler 1963).

As the $CO_2$ output observed during light is closely linked with photosynthesis, therefore carbohydrates or one or more intermediates of the photosynthetic carbon reduction cycle have to be considered as substrates of $CO_2$ liberation in "light respiration". The development and further metabolism of some intermediates in the course of this decomposition, at least to some extent, should be motivated by similar external factors such as the $CO_2$ liberation observed in light. With the help of specific inhibitors for certain steps of metabolism it should be possible to pursue the accumulation of substrates or intermediates of light respiration depending on external factors.

It is known, that in submerged algae the metabolism of glycollic acid is affected in a similar manner as light respiration (Warburg and Krippahl, 1960; Tolbert, 1963; Bassham, 1964). It was thus considered important to investigate the glycollate level in higher plants and to bring this level into closer relationship with the light respiration. As a specific inhibitor accumulating glycollate in the tissues and inhibiting glycollate oxidase disodium sulphoglycollate was used (Zelitch, 1958, 1959, 1964).

Table I shows the glycollate concentration of bean leaves depending on the preliminary treatment of the testing material. The values obtained demonstrate

Table I. Glycollate concentration of bean leaves depending on pre-treatment

| Pre-treatment | | Glycollic acid (μmole/g dry weight) | |
|---|---|---|---|
| Inhibitor | Gas atmosphere | 14,000 lux | Darkness |
| Stalk in water | 75% $O_2$+ 0·03% $CO_2$ in $N_2$ | <2 | <2 |
| | 1% $O_2$+ 0·03% $CO_2$ in $N_2$ | <2 | <2 |
| Stalk in solution of 0·01 M Disodium sulphoglycollate | 75% $O_2$+ 0·03% $CO_2$ in $N_2$ | 0·5 | 0·6 |
| | | 0·6 | 1·3 |
| | | 1·3 | 1·6 |
| | 1% $O_2$+ 0·03% $CO_2$ in $N_2$ | 8·6 | 1·1 |
| | | 9·2 | 1·0 |
| | | 8·1 | 0·8 |

that glycollate can accumulate in the green tissues only in light and only after the inhibition of glycollate oxidase and only with very reduced $O_2$ pressure. Among all other conditions—particularly with light and with high oxygen concentrations—we are able to find only a small amount of glycollate. With the explanation of the relationships between glycollate metabolism and light respiration our experiments so far have shown that the time course of $CO_2$ gas exchange cannot be influenced essentially by partly inhibiting the glycollate oxidase.

According to the investigations of Warburg and Krippahl (1960), Bassham and Kirk (1962), Tolbert (1963) and Whittingham and co-workers (1963) on the biogenesis of glycollic acid, low $CO_2$ pressure, high $O_2$ pressure and high illumination intensities are necessary. Particularly with a high $O_2$ concentration in the chloroplasts the enzyme oxidizes the glycollic acid to a considerable extent through glyoxylate to carbon dioxide (Tolbert *et al.*, 1949; Zelitch, 1953). This mechanism according to Zelitch (1958, 1959, 1964) should play a very important part with the light dependent "end oxidation" (light respiration). Probably this mechanism is the main source of the $CO_2$ liberation observed by us in leaves and liverwort thalli during the light period.

It is to be particularly emphasized here, that concerning the glycollate metabolism there exists an essential difference between the submerged water plants, particularly the algae, and leaves or thalli of emerged plants (Fock, 1965). Land plants are able to oxidize glycollic acid to a much larger extent, because high $O_2$ concentrations from outside can reach directly the locations of photosynthetic reactions. Moreover, the leaves have no opportunity to excrete the glycollate into the surrounding medium. In leaves the residue of glycollate oxidase, which is still active after the addition of inhibitors, is sufficient to oxidize the glycollate considerably, which is produced continuously in light with a high $O_2$ pressure in the surrounding air. With a low $O_2$ concentration around the leaves, by the action of photo-oxygen a large quantity of glycollate is still produced. Glycollate accumulates only under these conditions, as the quantities of photo-oxygen contained in the chloroplasts are insufficient for the further oxidation of glycollate if the glycollate oxidase is partly blocked.

Suspensions and sediments of *Chlorella* do not show the $CO_2$ outburst immediately after darkening, even when we use a very high $O_2$ concentration of 99% in the air (Fig. 5). The depression of apparent photosynthesis with high $O_2$ pressures does not reach those extreme values as shown with liverworts and leaves. After darkening the $CO_2$ output passes over a broad maximum and reaches the steady state rate after 15 min as shown already by Egle and Döhler (1963). We can compare this broad maximum with the second wave in the curves with liverworts and leaves. With a gas stream free from carbon dioxide (Fig. 6) we observe the almost total reassimilation of carbon dioxide in light; after darkening the short and strong $CO_2$ outburst does not appear, but the continuous $CO_2$ output reaching the steady state dark respiration is more pronounced with high oxygen pressure (4, Fig. 6). We ascribe these different effects with *Chlorella* to the buffering effect of the system in suspensions and to the different solubility of carbon dioxide and oxygen as well as to the fact that submerged plants excrete large amounts of glycollate into the surrounding medium.

Although with the observed $CO_2$ liberation in light, carbon dioxide is similarly formed, as in the dark respiration, the so-called light respiration does not show a respiration process in the sense of an oxidative decomposition of assimilates to obtain energy. According to recent investigations,

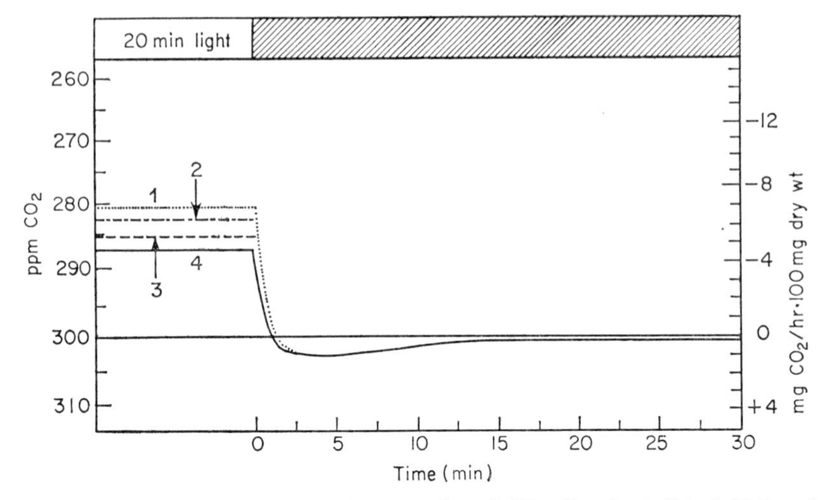


FIG. 5. $CO_2$ uptake and liberation of a suspension of *Chlorella vulgaris* (26 ml, 12·7 mg dry weight). The gas stream (25 l/hr) introduced into the assimilation chamber consists of a constant $CO_2$ concentration of 300 ppm and of a different $O_2$ concentration of 1% (1), 25% (2), 75% (3) and 99% (4), together with nitrogen. Light intensity = 14,000 lux, temp = 20°.

partial processes of dark respiration are inhibited by light (Weis and Brown, 1959; Hoch *et al.*, 1963; Kandler and Haberer-Liesenkötter, 1963). However, the $CO_2$ liberation in light is closely dependent on such external factors, which

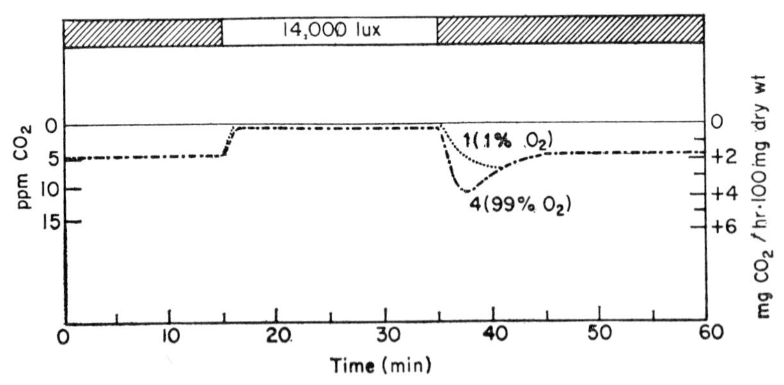


FIG. 6. $CO_2$ liberation of a suspension of *Chlorella vulgaris* in a gas stream free of carbon dioxide with different $O_2$ concentrations of 1% (1) and 99% (4). [For other data see Fig. 5.]

have no influence at all on the dark respiration. Moreover the light respiration of old leaves in contradiction to the dark respiration is more pronounced than in young leaves (Zelitch and Barber, 1960; Fock, 1965). The most convincing, however, are those results of Zelitch and Barber (1960) contradicting the $CO_2$ liberation in light from a respiration process, which show clearly that the oxidation of glycollate is not linked with phosphorylation. Thus in future in this connection we should not use the term "light respiration" at all.

Possibly the $CO_2$ liberation in the light phase is only a side reaction of a metabolism chain, in which the surplus of oxygen is reduced, which generally causes photo-oxidations (Zelitch, 1953). The significance of $CO_2$ liberation during the light phase probably can be seen in that the $CO_2$ partial pressure increases in the locations of the photosynthetic reactions, in which the carbon dioxide concentration is greatly reduced with higher light intensities. Thus the damage to the cells caused with light by $CO_2$ deficiency may probably be avoided. Considering the fact that the quantity of carbon dioxide liberated in light increases considerably with the reduced $CO_2$ content in the surrounding air, it may be assumed that green plants have developed this mechanism when in the course of the history of the earth the carbon dioxide concentration in the atmosphere has been reduced to the low level of today.

## References

Bassham, J. A. (1964). *Annu. Rev. Pl. Physiol.* **15**, 101.
Bassham, J. A. and Kirk, M. (1962). *Biochem. Biophys. Res. Commun.* **9**, 376.
Döhler, G. and Egle, K. (1964). *Z. Naturf.* **19**b, 137.
Ducet, G. and Rosenberg, A. J. (1962). *Annu. Rev. Pl. Physiol.* **13**, 171.
Egle, K. (1960). "Encyclopedia of Plant Physiology" (W. Ruhland, ed.) Vol. V/1, p. 182. Springer-Verlag, Berlin, Göttingen, Heidelberg.
Egle, K. and Döhler, G. (1963). *Beitr. Biol. Pflanzen* **39**, 295.
Egle, K. and Döhler, G. (1964). *Z. Naturf.* **19**b, 773.
Fock, H. (1965). Thesis, Science Faculty, University of Frankfurt/M.
Fock, H., Schaub, H., Ziegler, R. and Egle, K. (1964). *Beitr. Biol. Pflanzen* **40**, 293.
Fock, H. and Egle, K. (1966). *Beitr. Biol. Pflanzen* **42**, (in press).
Hoch, G., Owens, O. v. H. and Kok, B. (1963). *Archs Biochem. Biophys.* **101**, 171.
Kandler, O. and Haberer-Liesenkötter, I. (1963). *Z. Naturf.* **18**b, 718.
Kessler, E. (1960). "Encyclopedia of Plant Physiology" (W. Ruhland, ed.) Vol. V/1, p. 935. Springer-Verlag, Berlin, Göttingen, Heidelberg.
Krotkov, G. (1963). "Photosynthetic Mechanisms of Green Plants", p. 452. National Acad. Sci., Washington.
Krotkov, G., Runeckles, V. C. and Thimann, K. V. (1958). *Pl. Physiol.* **33**, 289.
Lister, G. R., Krotkov, G. and Nelson, C. D. (1961). *Can. J. Bot.* **39**, 581.
Rosenstock, G. and Ried, A. (1960). "Encyclopedia of Plant Physiology" (W. Ruhland, ed.) Vol. XII/2, p. 259. Springer-Verlag, Berlin, Göttingen, Heidelberg.
Tolbert, N. E. (1963). "Photosynthetic Mechanisms of Green Plants", p. 648. National Acad. Sci., Washington.
Tolbert, N. E., Clagett, C. O. and Burris, R. H. (1949). *J. biol. Chem.* **181**, 905.
Tregunna, E. B., Krotkov, G. and Nelson, C. D. (1961). *Can. J. Bot.* **39**, 1045.
Warburg, O. (1920). *Biochem. Z.* **103**, 188.
Warburg, O. and Krippahl, G. (1960). *Z. Naturf.* **15**b, 197.
Weis, D. and Brown, A. H. (1959). *Pl. Physiol.* **34**, 235.
Whittingham, C. P., Hiller, R. G. and Bermingham, M. (1963). "Photosynthetic Mechanisms of Green Plants", p. 675. National Acad. Sci., Washington.
Zelitch, I. (1953). *J. biol. Chem.* **201**, 719.
Zelitch, I. (1958). *J. biol. Chem.* **233**, 1299.
Zelitch, I. (1959). *J. biol. Chem.* **234**, 3077.
Zelitch, I. (1964). *Annu. Rev. Pl. Physiol.* **15**, 121.
Zelitch, I. and Barber, G. A. (1960). *Pl. Physiol.* **35**, 205.

# Part II. Biosynthesis in Chloroplasts—Carbohydrates

# Les Chloroplastes: Activité Photosynthétique et Métabolisme des Glucides

Alexis Moyse
*Laboratoire de Photosynthèse du C.N.R.S.*
*et Faculté des Sciences Orsay, France*

## Introduction

On sait, depuis les premières observations de Sachs en 1862, que l'amidon peut se déposer dans les chloroplastes de diverses feuilles (*Tropaeolum majus*, *Nicotiana tabacum*, etc.) lorsqu'elles sont éclairées, alors qu'il disparaît complètement après 48 hr d'obscurité. Sachs généralisa cette observation et remarqua que les feuilles qui ne font pas la synthèse de ce polysaccharide (*Allium cepa*, *Tulipa sp.*, par exemple) peuvent, à sa place, renfermer du saccharose (voir Sachs, 1868).

Des lévanes jouent le même rôle dans les feuilles de nombreuses Graminées. Des glucides peuvent également s'accumuler dans les algues; ainsi diverses variétés d'amidon se rencontrent dans les plastes d'algues vertes ou d'algues rouges.

Cette présence permet de poser plusieurs questions. En voici quelques unes.

— Quelle est la participation des glucides eux-mêmes à l'assimilation du $CO_2$, à la formation d'autres glucides, et par quel mécanisme?

— Jusqu'à quel degré de complexité, les glucides s'élèvent-ils dans les chloroplastes? Quelle diversité y atteignent-ils?

— Quelles sont les conséquences de l'accumulation des glucides dans les chloroplastes pour la cinétique même de la photosynthèse?

Bien que toutes les réactions concernant les glucides soient des réactions sombres, leur cinétique *in vivo* est étroitement subordonnée au déroulement des

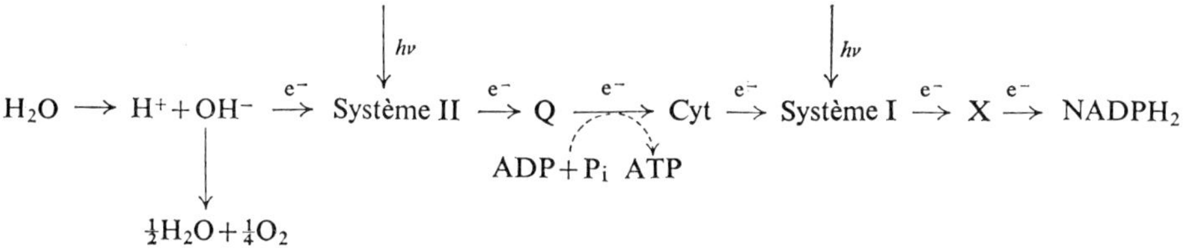

$$H_2O \longrightarrow H^+ + OH^- \xrightarrow{e^-} \text{Système II} \xrightarrow{e^-} Q \xrightarrow{e^-} \text{Cyt} \xrightarrow{e^-} \text{Système I} \xrightarrow{e^-} X \xrightarrow{e^-} NADPH_2$$

($h\nu$ → Système II; $h\nu$ → Système I; $ADP + P_i \rightarrow ATP$ couplé à Q → Cyt; $OH^- \rightarrow \frac{1}{2}H_2O + \frac{1}{4}O_2$)

FIG. 1. Schéma du transfert d'électrons dans la photosynthèse, inspiré de Duysens et Amesz (1962). Q=probablement plastoquinone; Cyt=cytochrome *f* (dans les feuilles); X=ferredoxine, d'après Tagawa, Tsujimoto et Arnon (1963).

réactions photochimiques, puisqu'au cours de celles-ci l'énergie électromagnétique de la lumière est convertie en énergie chimique utilisable dans les

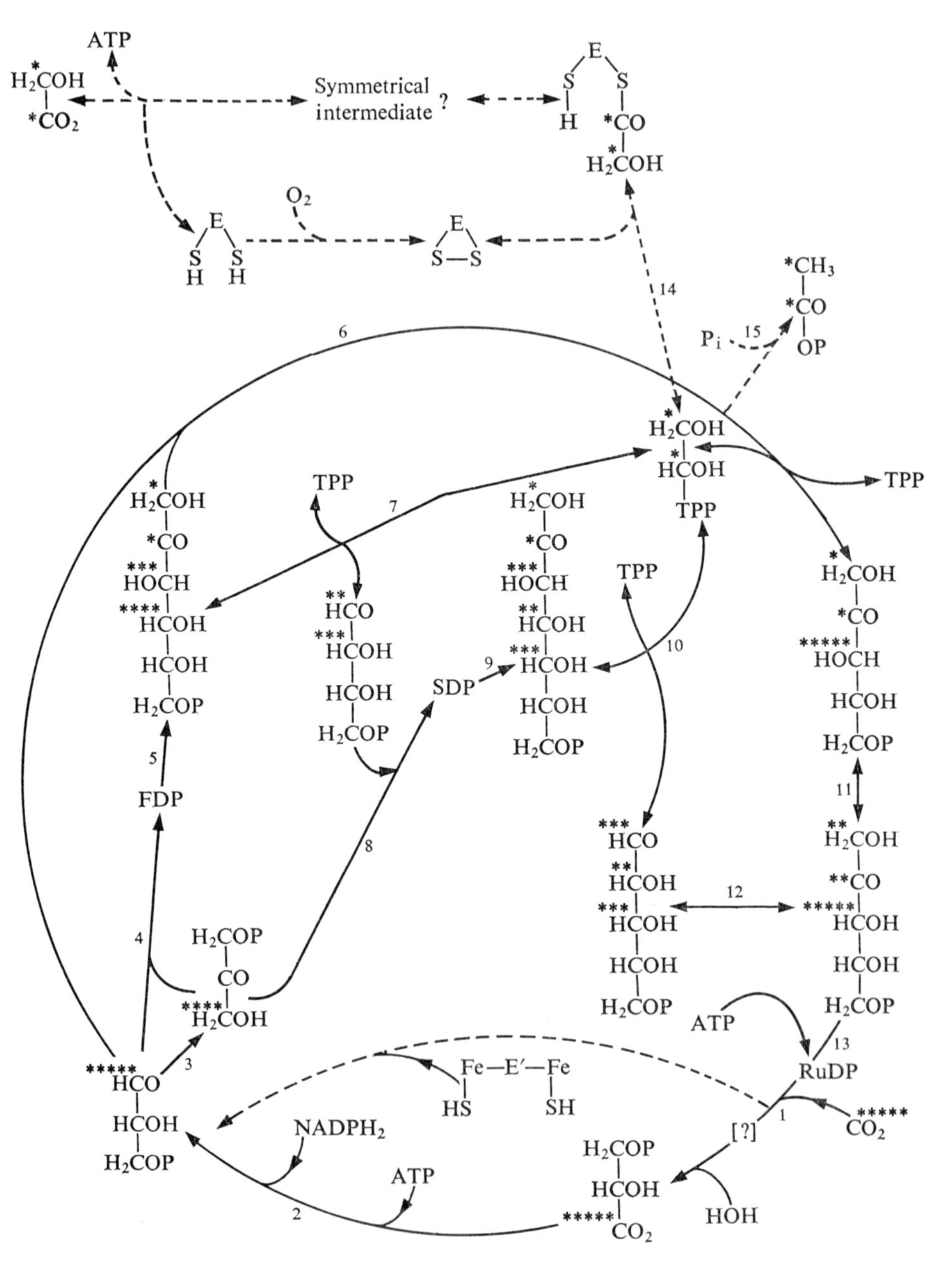


FIG. 2. Le cycle photosynthétique de la réduction de $CO_2$ ou cycle de Calvin, d'après Bassham (1964).

(1) carboxylation du ribulose 1,5-diphosphate (RuDP)

(2) réduction de PGA par $NADPH_2 + ATP$ ou réduction directe d'un intermédiaire de carboxylation par un transporteur d'électrons du type de la ferredoxine symbolisée par $E'(Fe\text{–}SH)_2$

(3) isomérisation du glycéraldéhyde 3-phosphate en dihydroxycétone phosphate

(legend continued on facing page)

réactions sombres. L'activité des deux systèmes photochimiques auxquels participent deux complexes différents de la chlorophylle *a* est résumée dans la Fig. 1, inspirée des interprétations de Duysens et Amesz (1962).

Ce qui est important pour notre propos est la genèse de NADPH, de NADH et d'ATP indispensables à la formation des glucides à partir de $CO_2$. L'inhibition de la formation de l'un ou l'autre de ces transporteurs entrave l'assimilation de $CO_2$ par les chloroplastes (Bamberger *et al.*, 1963).

## La Participation des Glucides a l'Assimilation de $CO_2$

Le premier mécanisme complet de la photosynthèse des glucides a été proposé par Calvin et ses collaborateurs (voir Bassham et Calvin, 1960). Critiqué par Stiller (1962), revu par Bassham (1964), il constitue une bonne base de départ pour tenter de répondre à notre première question.

Ce mécanisme a été établi à l'aide d'expériences utilisant le plus souvent des tensions partielles ou des concentrations de $CO_2$ plus élevées que celles qui sont présentes dans les conditions naturelles. Ces conditions ont pu masquer d'autres voies possibles.

Plusieurs mécanismes de carboxylation sont connus. La carboxylation du ribulose 1,5-diphosphate est l'un des plus rapides, bien qu'elle ne constitue pas la seule voie d'entrée du $CO_2$.

L'inhibition de la fixation photosynthétique de $CO_2$ *in vivo* par l'acide lipoïque et l'acide 8-méthyllipoïque (Bassham, 1963), composés dont on ne connaît pas la spécificité d'action à l'égard des enzymes mises en jeu dans le cycle de Calvin, montre également que le dernier mot n'est pas dit à ce propos.

Nous allons cependant essayer d'en dégager la généralité et ses limites, tel qu'il est présenté dans la Fig. 2.

---

(4) condensation des trioses phosphates en fructose 1,6-diphosphate
(5) déphosphorylation du fructose 1,6-diphosphate en fructose 6-phosphate
(6) transcétolisation, équilibre entre le fructose 6-phosphate, le glycéraldéhyde 3-phosphate, le xylulose 5-phosphate et l'érythrose 4-phosphate
(7) transcétolisation, équilibre entre le fructose 6-phosphate, le glycolyle-thiamine pyrophosphate (TPP) et l'érythrose 4-phosphate
(8) transaldolisation, formation du sédoheptulose 1,7-diphosphate par condensation du dihydroxycétone phosphate et de l'érythrose 4-phosphate
(9) déphosphorylation du sédoheptulose 1,7-diphosphate en sédoheptulose 7-phosphate
(10) transcétolisation, équilibre entre glycolyle-TPP, sédoheptulose 7-phosphate et ribose 5-phosphate
(11) épimérisation entre le xylulose 5-phosphate et le ribulose 5-phosphate
(12) isomérisation entre le ribose 5-phosphate et le ribulose 5-phosphate
(13) phosphorylation du ribulose 5-phosphate en ribulose 1,5-diphosphate
(14) équilibre entre le glycolyle-TPP et un transporteur de glycolyle du type de l'acide lipoïque
(15) oxydoréduction conduisant à l'acétyl-phosphate et aux lipides (?) Les relations entre le radical glycolyle et l'acide glycolique sont supposées avoir lieu par un intermédiaire symétrique assurant le marquage égal des atomes de C.

Les astérisques indiquent l'ordre de grandeur relatif du marquage des différents atomes de C, à partir de $^{14}CO_2$. (Reproduit avec l'autorisation de J. A. Bassham que je remercie vivement.)

## LA FIXATION DE $CO_2$ PAR CARBOXYLATION DU RIBULOSE 1,5-DIPHOSPHATE

La carboxylation du ribulose 1,5-diphosphate est le mécanisme de fixation de $CO_2$ qui paraît quantitativement le plus important parmi les divers mécanismes connus de carboxylation.

L'emploi du $^{14}CO_2$ et les études cinétiques effectuées en photosynthèse avec des feuilles, des algues (voir Bassham et Calvin, 1960; Doman *et al.*, 1964), des bactéries phototrophes (*Chromatium*: Fuller *et al.*, 1961; *Chlorobium thiosulfatophilum*: Smillie *et al.*, 1962; *Chromatium* et *Thiopedia*: Hurlbert et Lascelles, 1963) ou encore en chimiosynthèse avec des Bactéries chimiotrophes (*Thiobacillus denitrificans*: Trudinger, 1956; Aubert *et al.*, 1957; *Hydrogenomonas facilis*: Bergmann *et al.*, 1958) montrent la généralité de la formation du 3-phosphoglycérate, l'un des premiers produits de la fixation du $CO_2$.

Il est possible qu'il soit précédé dans diverses feuilles, dont celles de céléri, par un composé relativement stable, insoluble dans l'alcool à 80°, mais soluble dans l'eau chaude, livrant de l'acide phosphoglycérique rapidement (Trip *et al.*, 1964).

La genèse de l'acide 3-phosphoglycérique est liée à la carboxylation du ribulose 1,5-diphosphate, réaction thermodynamiquement spontanée, entraînant une très faible libération d'énergie libre. L'enzyme de carboxylation, la ribulose 1,5-diphosphate carboxylase ou carboxydismutase n'est guère présente que dans les cellules des organismes chlorophylliens. On la trouve cependant dans quelques bactéries hétérotrophes (*Escherichia, Pseudomonas*). Chez les plantes, elle est localisée, au moins en majeure partie, dans les chloroplastes (Heber *et al.*, 1963) et à l'extraction elle se trouve dans la fraction I des protéines foliaires (épinard, tabac). Elle n'est active qu'en présence d'ions $Mg^{2+}$ (ou $Mn^{2+}$) et son maximum d'activité est obtenu après incubation avec un sel de Mg en présence de $CO_2$ (Pon, 1962). Il se formerait dans ces conditions un complexe Enz-$CO_2$-$Mg^{2+}$ qui réagirait activement avec le ribulose 1,5-diphosphate.

La forte affinité du complexe pour $CO_2$ et le caractère irréversible de la réaction de carboxylation assureraient la fixation du gaz carbonique, même lorsqu'il n'est présent qu'à faible concentration dans les cellules.

La saturation par l'ion bicarbonate est obtenue avec une concentration de $3 \times 10^{-2}$ M, aussi bien pour les feuilles entières que pour les homogénats foliaires d'épinard. Les chlorelles et les extraits de chlorelles sont saturés pour $2 \times 10^{-2}$ M. Par contre, les extraits d'*Euglena* ne sont saturés que pour $3 \times 10^{-2}$ alors que les organismes intacts le sont pour $1 \times 10^{-3}$ M (Peterkofsky et Racker, 1961).

L'inhibition de la réaction *in vitro* par le *p*-chloromercuribenzoate et l'iodoacétamide laisse supposer l'intervention de groupes sulfhydryles dans le complexe enzymatique actif, ce que suggère tout particulièrement la levée d'inhibition due à l'iodoacétamide par l'addition de ribulose 1,5-diphosphate.

Rabin et Trown (1964) ont donné l'interprétation suivante de la série des réactions *in vitro*:

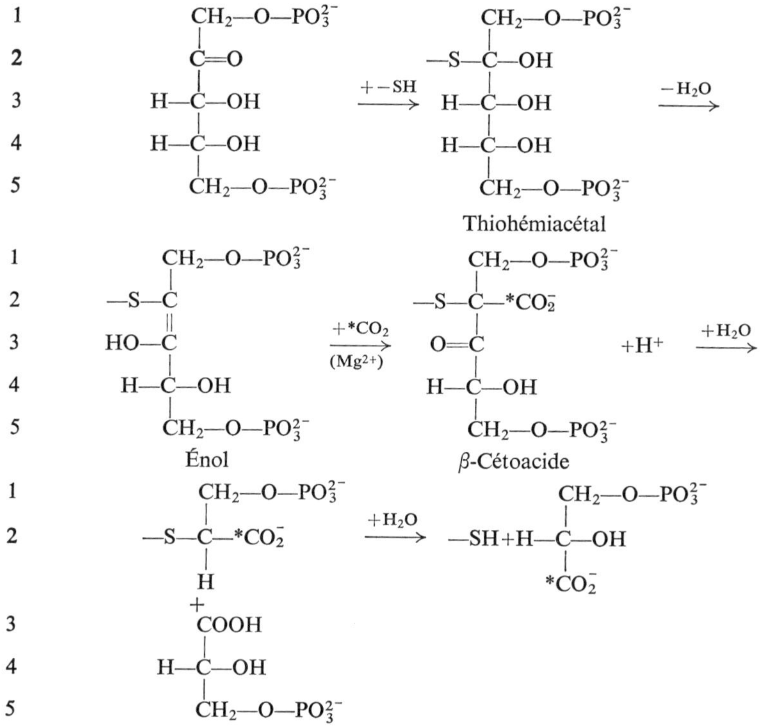

1 $CH_2—O—PO_3^{2-}$ | $CH_2—O—PO_3^{2-}$
2 $C=O$ | $—S—C—OH$
3 $H—C—OH$ $\xrightarrow{+-SH}$ $H—C—OH$ $\xrightarrow{-H_2O}$
4 $H—C—OH$ | $H—C—OH$
5 $CH_2—O—PO_3^{2-}$ | $CH_2—O—PO_3^{2-}$

Thiohémiacétal

1 $CH_2—O—PO_3^{2-}$ | $CH_2—O—PO_3^{2-}$
2 $—S—C$ | $—S—C—{}^*CO_2^-$
3 $HO—C$ (double liaison C=C entre 2 et 3) $\xrightarrow[(Mg^{2+})]{+{}^*CO_2}$ $O=C$ $+H^+$ $\xrightarrow{+H_2O}$
4 $H—C—OH$ | $H—C—OH$
5 $CH_2—O—PO_3^{2-}$ | $CH_2—O—PO_3^{2-}$

Énol | $\beta$-Cétoacide

1 $CH_2—O—PO_3^{2-}$ | $CH_2—O—PO_3^{2-}$
2 $—S—C—{}^*CO_2^-$ $\xrightarrow{+H_2O}$ $—SH+H—C—OH$
$H$ | ${}^*CO_2^-$
$+$
3 $COOH$
4 $H—C—OH$
5 $CH_2—O—PO_3^{2-}$

Antérieurement, des études de la cinétique du marquage des différents atomes de C de l'acide phosphoglycérique et du ribulose 1,5-diphosphate ont été effectuées avec des chlorelles intactes par Bassham et Kirk (1960) et Bassham et Calvin (1961).

Elles ont conduit ces auteurs à proposer *in vivo*, à la lumière, la libération d'une seule molécule d'acide phosphoglycérique et d'une molécule d'un composé réduit qui pourrait être un triose. Lorsque la fixation de $CO_2$ se déroule à l'obscurité par contre, deux molécules d'acide phosphoglycérique sont libérées. Les réactions suivantes résument les 2 modes:

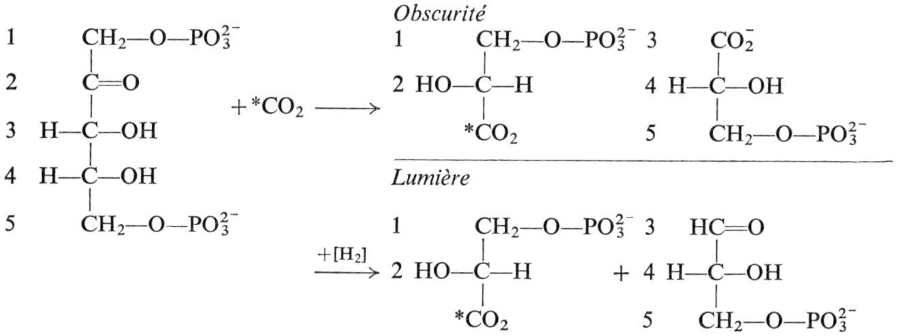

1 $CH_2—O—PO_3^{2-}$
2 $C=O$
3 $H—C—OH$ $+{}^*CO_2 \longrightarrow$
4 $H—C—OH$
5 $CH_2—O—PO_3^{2-}$

*Obscurité*

1 $CH_2—O—PO_3^{2-}$ | 3 $CO_2^-$
2 $HO—C—H$ | 4 $H—C—OH$
${}^*CO_2$ | 5 $CH_2—O—PO_3^{2-}$

*Lumière*

1 $CH_2—O—PO_3^{2-}$ | 3 $HC=O$
$\xrightarrow{+[H_2]}$ 2 $HO—C—H$ $+$ 4 $H—C—OH$
${}^*CO_2$ | 5 $CH_2—O—PO_3^{2-}$

Avec des chloroplastes isolés de feuilles et dont l'activité de fixation ne dépasse pas 5 % de l'activité réalisée *in vivo*, ce sont 2 moles d'acide phosphoglycérique qui sont libérées (Pon, 1962).

## LA RÉDUCTION DE L'ACIDE PHOSPHOGLYCÉRIQUE. LA FORMATION DES TRIOSES

Quoiqu'il en soit, tous les auteurs sont d'accord pour admettre que la réduction de l'acide phosphoglycérique constitue l'étape énergétique essentielle du métabolisme carboné photosynthétique.

La réaction:

$$\text{3-PGA} + \text{NADPH} + \text{ATP} + \text{H}^+ \rightarrow \text{3-phosphoglycéraldéhyde} + \text{ADP} + \text{P}_i + \text{NADP}^+$$

est très légèrement exergonique lorsqu'elle se déroule de gauche à droite.

Elle comprend d'abord la phosphorylation en 1 de l'acide 3-phosphoglycérique, catalysée par une phosphoglycérique kinase et dans laquelle intervient l'ATP. La présence de l'enzyme dans les chloroplastes de l'épinard a été confirmée (Kahn, 1964). Une triose phosphate déshydrogénase intervient ensuite.

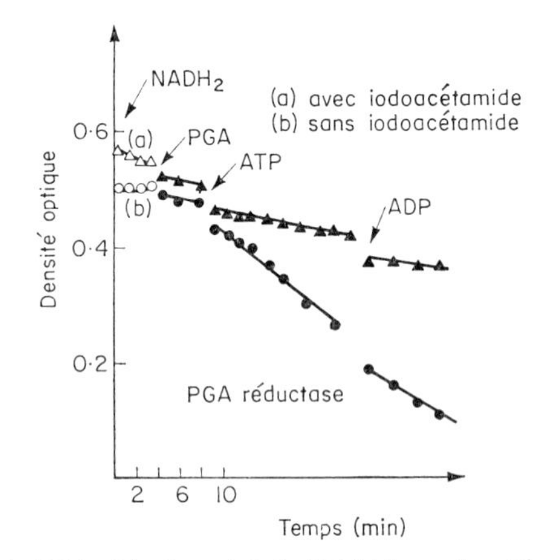


Fig. 3. Activités de la PGA réductase et de la PGA kinase des chloroplastes isolés de *Bryophyllum daigremontianum*, mesurées par la diminution de la densité optique à 340 m$\mu$. Inhibition par l'iodoacétamide (Garnier-Dardart, 1965b).

Le thréose 2,4-diphosphate inhibiteur de la triose phosphate déshydrogénase bloque la synthèse des glucides dans les préparations de chloroplastes isolés et provoque l'accumulation d'acide phosphoglycérique (Park *et al.*, 1959).

Curieusement, le sédoheptulose 7-phosphate exerce également un effet inhibiteur (Gibbs, 1963). Ce dernier fait est très intéressant, car on peut y trouver un point de départ de l'étude des mécanismes de régulation du cycle de la réduction du $CO_2$.

Les chloroplastes renferment à la fois la totalité de la triose phosphate déshydrogénase dépendant du $NADP^+$ présente dans les cellules des feuilles et environ le tiers de la triose phosphate déshydrogénase dépendant du NAD

(PGA-réductase), ainsi que l'ont montré Heber *et al.* (1963). Les Figs 3 et 4 en donnent une confirmation, effectuée par J. Garnier-Dardart (communication personnelle).

Les deux enzymes sont sensibles à l'iodoacétamide.

La synthèse de la première serait accélérée, dans les feuilles étiolées de *Phaseolus vulgaris*, par les radiations rouges, indépendamment de toute genèse de chlorophylle (Margulies, 1965). Il faut observer que les chloroplastes d'épinard renferment plus de $NAD^+$ que de $NADP^+$, d'après Ogren et Krogmann

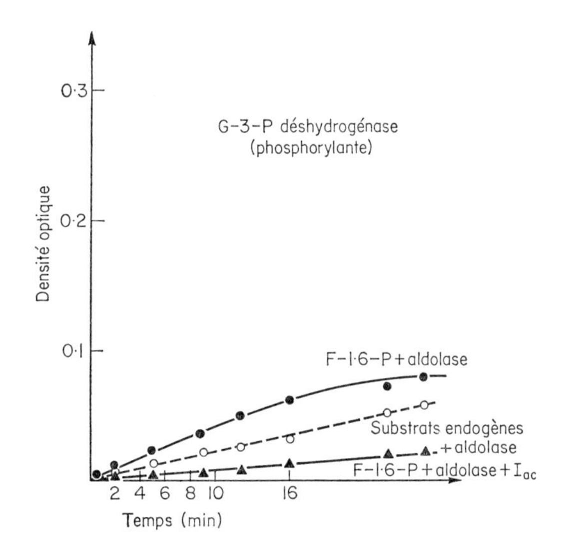


FIG. 4. Activité de la glycéraldéhyde 3-phosphate déshydrogénase des chloroplastes isolés de *Bryophyllum daigremontianum*, mesurée par l'augmentation de la densité optique à 340 m$\mu$, en présence de fructose 1,6-diphosphate et d'aldolase. Inhibition par l'iodoacétamide. G 3-phosphate: 3-phosphoglycéraldéhyde. F 1,6-diphosphate: fructose 1,6-diphosphate.

(1963). La pyridine nucléotide transhydrogénase, présente dans les chloroplastes (Keister *et al.*, 1960) catalyserait la réduction du $NAD^+$ par le NADPH. Chez les bactéries phototrophes et chimiotrophes, le NADH est généralement le transporteur d'électrons.

## LA FORMATION DU FRUCTOSE 1,6-DIPHOSPHATE

Parmi les oses phosphates formés, le fructose 1,6-diphosphate occupe une place à part dans les conceptions classiques. On peut en effet le considérer à la fois comme un intermédiaire dans la régénération du ribulose 1,5-diphosphate et comme l'origine des autres hexoses et de leurs produits de condensation (saccharose, lévanes, amidon). Son apparition rapide et son marquage symétrique dans les algues et les feuilles, après 15 s de photosynthèse en présence de $^{14}CO_2$ (Bassham *et al.*, 1954), avec prédominance égale de la radioactivité en ses atomes de C n$^{os}$ 3 et 4 ont depuis longtemps suggéré la séquence de réactions suivantes:

$$\text{Ⓟ-Glycéraldéhyde} \underset{}{\overset{\text{triose phosphate isomérase}}{\rightleftharpoons}} \text{Dihydroxycétone-Ⓟ}$$

$$\left.\begin{array}{r}\text{Ⓟ-Glycéraldéhyde}\\ \text{dihydroxycétone-Ⓟ}\end{array}\right\} \overset{\text{aldolase}}{\rightleftharpoons} \text{Fructose 1,6-diⓅ}$$

La triose phosphate isomérase et l'aldolase ont été caractérisées aussi bien dans les chloroplastes isolés de feuilles que dans les chlorelles.

Deux difficultés se sont présentées à propos de cette séquence: la non généralité du marquage symétrique des hexoses dérivés du fructose 1,6-diphosphate et la non généralité de la présence de l'aldolase.

La deuxième est en voie de résorption. Quelques organismes ont bien semblé dépourvus d'aldolase: diverses algues rouges (paraissant également dépourvues de triose phosphate déshydrogénase) (Jacobi, 1957), diverses algues bleues-vertes dont *Anacystis nidulans* (Richter, 1959; Fewson *et al.*, 1962). Il en a paru de même pour la Bactérie pourpre *Rhodopseudomonas spheroides.* Par contre, cette Bactérie, de même que la Cyanophycée *Anacystis,* renferme une transcétolase clivant les pentoses phosphates avec formation de trioses phosphates (Richter, 1961). Mais, récemment, l'activité de l'aldolase a été constatée dans plusieurs de ces organismes. L'addition de réducteurs (cystéine; $SO_4Fe$, $7H_2O$) permet de retrouver dans les extraits d'*Anacystis* et d'autres Cyanophycées une activité comparable à celle que l'on constate dans les organismes verts (Van Baalen, 1965).

Ces résultats ont été aussitôt confirmés par Willard *et al.* (1965) pour *Anacystis nidulans.* Les mêmes auteurs ont également trouvé une forte activité aldolase chez *Rhodopseudomonas spheroides.* La séparation rapide des débris cellulaires et de l'extrait enzymatique est nécessaire pour conserver une activité importante. L'enzyme est sensible au *p*-chloromercuribenzoate, à l'*o*-phénanthroline et à l'$\alpha$-$\alpha'$-dipyridyle. Elle diffère de l'aldolase des plantes supérieures et des algues vertes par son appartenance au groupe des enzymes métallo-sulfhydrylées.

On peut conclure de ces résultats que la présence d'aldolase en particulier et d'enzymes en général ne peut être exclue même lorsque des essais de caractérisation dans des extraits bruts ont échoué.

Plus complexe apparaît la première difficulté. En étudiant la répartition du $^{14}C$ dans le glucose 1-phosphate, dans l'uridine diphosphoglucose et dans le glucose libéré à partir de l'amidon, après que ces substances aient été synthétisées par des chlorelles ou des feuilles en présence de $^{14}CO_2$, Kandler et Gibbs (1956), Gibbs et Kandler (1957) ont constaté une distribution non symétrique dont il sera discuté plus loin, après l'examen des faits relatifs aux transformations possibles du fructose 1,6-diphosphate.

### LA RÉGÉNÉRATION DU RIBULOSE 1,5-DIPHOSPHATE

Les réactions d'équilibre entre les trioses, les pentoses, le sédoheptulose et le fructose, empruntées à la voie des phosphopentoses, ont servi de guide, d'une manière justifiée en raison du marquage précoce de ces composés, dans les expériences utilisant soit des algues, soit des feuilles mises en présence de $^{14}CO_2$. Elles constituent l'un des meilleurs exemples de cycle rapide) (voir Fig. 2).

### *Réactions Catalysées par des Phosphatases*

Fructoses 1,6-diⓅ → Fructose 6-Ⓟ + $P_i$

Sédoheptulose 1,7-diⓅ → Sédoheptulose 7-Ⓟ + $P_i$

Les phosphatases spécifiques de chacun des deux substrats ont été mises en évidence dans les chloroplastes d'épinard (Racker et Schroeder, 1958).

### *Réactions Catalysées par la Transcétolase*

Fructose 6-Ⓟ + glycéraldéhyde 3-Ⓟ ⇌ Xylulose 5-Ⓟ + érythrose 4-Ⓟ

Sédoheptulose 7-Ⓟ + glycéraldéhyde 3-Ⓟ ⇌ Ribulose 5-Ⓟ + xylulose 5-Ⓟ

Au cours de la première réaction, les 4 atomes de C n$^{os}$ 3, 4, 5 et 6 du fructose forment l'érythrose (Racker *et al.*, 1954), tandis que les atomes n$^{os}$ 1 et 2 sont transférés sur le glycéraldéhyde 3-phosphate pour former le xylulose 5-phosphate. La réaction inverse est possible, le xylulose ayant en 1, 2 et 3 la même configuration que le fructose (Horecker *et al.*, 1956). Une situation parallèle se retrouve dans la deuxième réaction.

La transcétolase a été trouvée dans les feuilles (Horecker et Smyrniotis, 1953a; Axelrod *et al.*, 1953). Son activité requiert la présence de thiamine pyrophosphate et de $Mg^{2+}$ (Racker *et al.*, 1953). L'érythrulose peut servir de donneur de groupe glycolaldéhyde et l'érythrose peut jouer, comme le glycéraldéhyde, le rôle d'accepteur (Horecker *et al.*, 1953).

Pour la transcétolase isolée de la levure de bière, le glucose 6-phosphate comme le glycéraldéhyde 3-phosphate et l'arabinose 5-phosphate, peut servir d'accepteur (Datta et Racker, 1961a). En présence d'hydroxypyruvate comme donneur et de glucose 6-phosphate, il se forme de l'octulose 8-phosphate, composé dont la présence a été signalée dans le fruit d'avocatier (Charlson et Richtmyer, 1959).

Le complexe glycolaldéhyde-transcétolase peut également transférer le "glycolaldéhyde actif" au glycolaldéhyde en solution dans les milieux réactionnels, avec formation d'érythrulose, et au ribose 5-phosphate avec genèse d'un autre glucide phosphorylé qui est peut-être du sédoheptulose phosphate mais n'a pas été identifié avec certitude. Il peut également l'échanger contre le groupe glycolaldéhyde du fructose 6-phosphate (Datta et Racker, 1961b).

La spécificité de la (ou des) transcétolase est donc assez lâche à l'égard des aldoses accepteurs comme à l'égard des cétoses donneurs du groupe glycolaldéhyde.

Il faut cependant observer que les donneurs, mis à part le glycolaldéhyde lui-même, doivent avoir la même configuration que le fructose, pour leurs atomes de C n$^{os}$ 1, 2 et 3. En raison de sa configuration différente pour le n° 3, le ribulose 5-phosphate ne peut servir de donneur, ce qui peut limiter d'une manière heureuse la transformation des pentoses en heptulose.

*Réactions Catalysées par la Transaldolase*

Érythrose 4-Ⓟ + dihydroxycétone-Ⓟ ⇌ Sédoheptulose 1,7-diⓅ

à laquelle il faut ajouter:

Fructose 6-Ⓟ + érythrose 4-Ⓟ ⇌ Sédoheptulose 7-Ⓟ + glycéraldéhyde 3-Ⓟ

La transaldolase présente dans les feuilles assurerait ainsi la genèse rapide du sédoheptulose, par la réaction subordonnée à la formation antérieure d'érythrose 4-phosphate (Horecker et Smyrniotis, 1953b).

L'ensemble des réactions catalysées par la transaldolase et la transcétolase peut avoir lieu à partir du fructose 6-phosphate comme seul substrat (Horecker *et al.*, 1963). La séquence est alors la suivante:

Fructose 6-Ⓟ + transcétolase ⇌ Glycolaldéhyde-transcétolase + érythrose 4-Ⓟ

Fructose 6-Ⓟ + transaldolase ⇌ Dihydroxycétone-transaldolase + glycéraldéhyde-Ⓟ

Erythrose 4-Ⓟ + dihydroxycétone-transaldolase ⇋ Sédoheptulose 7-Ⓟ + transaldolase

Glycéraldéhyde 3-Ⓟ + glycolaldéhyde-transcétolase ⇌ Xylulose 5-Ⓟ + transcétolase

Ces réactions ont bien été démontrées *in vitro*. Mais la formation de l'érythrose 4-phosphate *in vivo* n'a pas été, à ma connaissance, décelée avec une certitude complète. Parmi les produits intermédiaires synthétisés par les chlorelles et *Scenedesmus*, en présence de $^{14}CO_2$, il a bien été signalé, après déphosphorylation et séparation chromatographique, l'acide érythronique. Ce composé résulterait de l'oxydation à l'air de l'érythrose dont la formation est donc très probable (Moses et Calvin, 1958).

Il faut ajouter que le mannoheptulose phosphate accompagne souvent le sédoheptulose phosphate. La cinétique de sa formation est très régulièrement croissante pendant plusieurs minutes, malgré sa faible abondance (Bean *et al.*, 1963), aussi se trouve-t-il peut-être plutôt sur une voie latérale au cycle.

Une autre possibilité de synthèse du sédoheptulose a été mise en évidence par Pontremoli *et al.* (1960) en faisant agir un mélange de transcétolase (d'épinard) et de transaldolase (de levure) sur du fructose 6-phosphate.

Les réactions suivantes se réaliseraient, en stoechiométrie:

4 Fructose-Ⓟ ⇌ 2 Heptulose-Ⓟ + 2 pentose-Ⓟ

2 Pentose-Ⓟ ⇌ Heptulose-Ⓟ + triose-Ⓟ

La présence initiale de triose-Ⓟ n'est pas nécessaire.

Enfin, la transaldolase extraite de la levure de bière peut catalyser une réaction d'échange entre le fructose 6-phosphate et le glycéraldéhyde 3-phosphate (Ljungdahl *et al.*, 1961). Une réaction du même type n'est pas à exclure dans les chloroplastes.

*Réactions Catalysées par la Phosphocétopentose Épimérase, la Phosphopentose Isomérase et la Phosphoribulokinase*

La première:

Xylulose 5-Ⓟ ⇌ Ribulose 5-Ⓟ

est catalysée par la xylulose 5-Ⓟ épimérase.

La seconde:

$$\text{Ribose 5-Ⓟ} \rightleftharpoons \text{Ribulose 5-Ⓟ}$$

est catalysée par la ribose 5-Ⓟ isomérase (Tabachnick *et al.*, 1958).

La présence des enzymes requises a été démontrée dans les feuilles (Horecker *et al.*, 1956).

La nécessité de la phosphoribulokinase a été très tôt reconnue (Calvin, 1955), puisqu'une phosphorylation est requise pour régénérer le ribulose 1, 5-diphosphate. L'ATP, toujours présent dans les chloroplastes, à la lumière, permet de l'assurer. La présence de l'enzyme, dans les feuilles d'épinard, a été vérifiée (Racker, 1957).

Du point de vue énergétique, les réactions de transcétolisation et de transaldolisation se passant entre oses-monophosphates entraînent une perte d'énergie liée à l'intervention des phosphatases.

Enfin, le ribulose peut provenir de l'oxydation partielle du glucose 6-phosphate par la voie des pentoses phosphates, mais cette modalité entraîne une perte de $CO_2$ non compatible avec le gain photosynthétique. D'autres modalités de genèse sont également possibles, mais non réalisées en raison même du manque de précurseurs dans les conditions habituelles: il en est ainsi de la formation de ribulose à partir du L-arabitol fourni comme aliment aux feuilles de Tabac (Kocourek *et al.*, 1964).

## LE MARQUAGE ASYMÉTRIQUE DU GLUCOSE ET LA COMPLEXITÉ DU CYCLE DES PENTOSE PHOSPHATES

Les premières observations d'une asymétrie dans le marquage des atomes de C du glucose sont dues à Kandler et Gibbs (1956). Après de courtes durées d'expérience, des chlorelles illuminées en présence de $^{14}CO_2$ ont synthétisé du

Tableau I. Distribution du $^{14}C$ dans le glucose phosphate, d'après Kandler et Gibbs (1956)

| Composés | Impulsions/mg C selon la position des atomes de C, dans la molécule de glucose | | | | | |
|---|---|---|---|---|---|---|
| | 1 | 2 | 3 | 4 | 5 | 6 |
| Monophosphate | 51 | 43 | 186 | 243 | 19 | 35 |
| Uridinediphosphate de glucose | 88 | 72 | 468 | 530 | 43 | 59 |
| | 83 | 80 | 730 | 811 | 12 | 20 |
| Glucose phosphate de nature non précisée | 216 | 200 | 838 | 1074 | 117 | 141 |

glucose possédant un marquage plus élevé en 4 qu'en 3 d'une part, et plus élevé en 1 et 2 qu'en 5 et 6 d'autre part, ainsi que l'indiquent les résultats du Tableau I.

Ces observations contrastent avec celles qu'avaient faites antérieurement les chercheurs du groupe de Calvin (voir Bassham *et al.*, 1954) à propos du fructose isolé des *Scenedesmus*.

Pour le fructose, la radioactivité est à peu près également distribuée en 3 et 4 d'une part, également distribuée encore (mais faible) entre 1, 2, 5 et 6 d'autre part, après des durées brèves d'expérience.

A propos du marquage excédentaire en 4 par rapport à 3, Bassham et Calvin (1960) ont supposé que la condensation catalysée par l'aldolase pouvait concerner des molécules de phosphoglycéraldéhyde fortement marquées dans leur carbonyle avec des molécules de dihydroxycétone phosphate non marquées, provenant du pool du métabolisme de fond. En effet, l'équilibre entre les deux trioses, catalysé par la triose phosphate isomérase, peut ne pas s'établir immédiatement devant l'afflux de molécules d'aldéhyde phosphoglycérique nouvellement synthétisées.

Le métabolisme de fond antérieur aura pu laisser, par contre, une plus forte proportion de cétotriose, puisque, lorsque l'équilibre est réalisé, le dihydrocétone phosphate l'emporte en concentration sur l'aldéhyde.

Cet effet initial devra s'effacer avec le temps, les nouvelles molécules de cétose formé étant elles-mêmes marquées.

Le schéma suivant (qui n'est pas reporté dans la Fig. 2) exprime cette interprétation, en supposant que 2 molécules de phosphoglycéraldéhyde sont marquées pour une de dihydroxycétone phosphate.

| | | | | | | | | | |
|---|---|---|---|---|---|---|---|---|---|
| C | | C | | C | | C | ⎫ | C | 1 |
| C=O | | C=O | | C=O | | C=O | ⎪ | C=O | 2 |
| *C | | C | | *C | | C | ⎪ | *C | 3 |
| | + | | → | | + | | ⎬ | | |
| *C | | *C | | *C | | *C | ⎪ | **C | 4 |
| C | | C | | C | | C | ⎪ | C | 5 |
| C | | C | | C | | C | ⎭ | C | 6 |

Un marquage excédentaire en 4 par rapport à 3 pourrait être également dû à l'activité de la transaldolase. La transaldolase extraite de la levure peut en effet catalyser l'échange de la moitié inférieure de la molécule de fructose contre du glycéraldéhyde phosphate. Si ce dernier est fortement marqué en 1, il communiquera un marquage élevé à l'atome de C n° 4 des nouvelles molécules de fructose (Ljungdahl *et al.*, 1961).

Il faut admettre alors que l'excédent de marquage en 4 dans le fructose phosphate passe inaperçu dans les chloroplastes, ce glucide phosphorylé n'existant guère qu'à l'état de traces, tandis que ses conséquences vont se manifester dans les glucose phosphate et l'amidon, plus abondants. Ou bien faut-il estimer, avec Stiller (1962), que le fructose initialement formé est marqué d'une manière strictement symétrique et ne participe pas au cycle de la régénération du ribulose diphosphate? Seules les molécules de fructose synthétisées ensuite, à partir du pentose accepteur régénéré seraient alors marquées asymétriquement en 3 et 4 par le jeu des transcétolisations et transaldolisations, ainsi qu'on va le voir maintenant.

Comme l'ont observé dès l'origine Kandler et Gibbs (1956), l'asymétrie concerne également les autres atomes de C du glucose, les $C_1$ et $C_2$ étant plus marqués que les $C_5$ et $C_6$. Pour expliquer ce fait, Schroeder et Racker (1959) ont suggéré de faire appel à l'intervention des transcétolase et transaldolase, ce qu'illustre assez bien le schéma de la Fig. 2 proposé par Bassham (1964). Ce schéma tient compte également de l'intervention possible de l'acide glycolique dérivé du clivage des cétoses.

D'après cette interprétation, l'érythrose 4-phosphate issu du fructose 6-phosphate (réaction 7) est marqué essentiellement en 1 et 2. Ces atomes deviennent les $C_4$ et $C_5$ du sédoheptulose, puis les $C_2$ et $C_3$ des pentoses. Le $C_3$ du sédoheptulose est fortement marqué puisqu'il peut provenir du C le plus marqué du dihydroxycétone phosphate. Il fournit le $C_1$ des pentoses. En résumé, les pentoses vont donc se trouver nettement marqués en 1 et 2, plus fortement encore en 3 ($C_4$ du fructose→$C_2$ de l'érythrose→$C_3$ des pentoses) et par contre faiblement en 4 et 5 ($C_5$ et $C_6$ du fructose).

Le marquage excédentaire du glucose en 1 et 2 peut également être obtenu par condensation d'un radical glycolyle marqué dans ses 2 atomes de C (dérivé d'un cétose: ribulose, xylulose ou fructose déjà marqués en 1 et 2) et de phosphoglycéraldéhyde. Comme ce dernier est fortement marqué en 1, il maintiendra le marquage prépondérant des pentoses en 3.

D'autres possibilités peuvent certainement être considérées et, comme l'écrit Stiller: *The enzymatic implementation of this early approach to the problem of hexose synthesis makes an intriguing subplot to the intellectual romance that is biochemistry.*

Mais le roman dispose dès maintenant de quelques précisions supplémentaires d'origine expérimentale.

Trebst et Fiedler (1961, 1962) et Gibbs (1963) ont retrouvé, avec des chloroplastes isolés d'épinard, la distribution asymétrique dans les molécules de glucose 6-phosphate s'opposant à la symétrie du marquage des molécules de fructose 1,6-diphosphate, en donnant simultanément $^{14}CO_2$ et soit du ribose 5-phosphate, soit du ribulose 1,5-diphosphate, ou bien encore du [1-$^{14}$C]-3-phosphoglycérate en absence de $^{14}CO_2$.

Dans des conditions de fixation plus importantes, Havir et Gibbs (1963) ont effectué des recherches analogues avec des préparations de chloroplastes isolés de feuilles d'épinard ou de pois et les mêmes précurseurs.

Le marquage du fructose 1,6-diphosphate est symétrique ou bien très proche de la symétrie, quel que soit le composé organique offert comme précurseur. Par contre, le marquage du glucose est doublement asymétrique, ainsi qu'il a été indiqué plus haut et l'asymétrie est plus grande en présence de pentoses phosphates et de $^{14}CO_2$ qu'en présence de [1-$^{14}$C]-3-phosphoglycérate.

Alors que la répartition de la radioactivité entre les C n$^{os}$ 4, 5 et 6 du glucose est toujours très voisine de celle qui est présente dans les C 1, 2 et 3 du 3-phosphoglycérate ou du dihydroxycétone phosphate, la répartition est différente dans l'autre moitié de la molécule, avec un enrichissement en C 1 et C 2 dont peuvent rendre compte la transcétolisation et la transaldolisation précédemment décrites.

Les liens possibles entre le glycolate ou le groupe glycolyle intervenant dans les transcétolisations compliquent encore le problème.

## LES RELATIONS DU GLYCOLATE AVEC LE MÉTABOLISME DES GLUCIDES

Le mécanisme de la formation du glycolate et ses relations avec l'activité photosynthétique ne sont pas encore clairement élucidés, ainsi que l'a souligné Tolbert (1963).

Si aucun mécanisme n'est connu pour sa synthèse directe à partir de $CO_2$, le fait qu'il soit très rapidement marqué dans les expériences utilisant $^{14}CO_2$ et ses variations quantitatives considérables selon les conditions expérimentales, établissent l'intérêt que son étude peut présenter.

Le glycolate fourni comme aliment aux cellules ne semble cependant pas être incorporé tel quel dans les substances synthétisées. Les résultats obtenus autrefois par Schou *et al.* (1950) avec des *Scenedesmus* assimilant du [1-$^{14}$C]- ou [2-$^{14}$C]glycolate montrent que les deux atomes de C de l'acide glycolique sont randomisés d'une manière complexe dans les molécules d'acide phosphoglycérique. Le [2-$^{14}$C]glycolate donne de l'acide phosphoglycérique également marqué en 2 et 3 et beaucoup plus faiblement en 1. L'inverse se produit avec le [1-$^{14}$C]glycolate.

Jimenez *et al.* (1962) ont trouvé des résultats analogues par incorporation de glycolate dans les feuilles de blé ou de soja. Lorsqu'il est marqué en 2, les hexoses (obtenus à partir du saccharose) sont marqués en 1, 2, 5 et 6.

Lorsque le glycolate est marqué en 1, les hexoses le sont en 3 et 4.

La randomisation permet de penser que le glycolate est pour une grande part au moins dégradé avant que ses atomes de C ne soient assimilés. Il est par suite difficile de le considérer comme un accepteur plus ou moins direct de $CO_2$, comme l'a proposé Stiller (1962).

L'acide glycolique formé dans des conditions physiologiques normales peut être excrété par les chlorelles (Tolbert et Zill, 1956).

L'excrétion par ces algues a lieu lorsque la concentration en $CO_2$ est faible et la lumière forte (Pritchard *et al.*, 1962). Elle n'a pas lieu en absence de Mn dans le milieu de culture (Miller *et al.*, 1963). Nous avons également constaté son marquage rapide à la lumière, avec une activité spécifique au moins aussi élevée que celle de l'acide phosphoglycérique dans les feuilles de *Bryophyllum daigremontianum* en présence de $^{14}CO_2$ (Moyse et Jolchine, 1957); il s'accumule lorsque les plantes reçoivent un sel d'$NH_4$ comme aliment azoté (Blondon-Ragage *et al.*, 1963).

La formation de glycolate par des chloroplastes d'épinard, en présence de fructose 6-phosphate (Bradbeer et Racker, 1961), son excrétion très importante par les chloroplastes isolés ayant fixé $^{14}CO_2$, alors que dans les cellules intactes il est rapidement métabolisé avec formation d'hexoses (Kearney et Tolbert, 1961), la présence d'acide phosphoglycolique dans les algues, celle de la glycolique phosphatase dans les feuilles de tabac (Richardson et Tolbert, 1961) ont également attiré l'attention sur ce composé dicarboné et ses possibilités d'intervention dans le métabolisme photosynthétique normal.

L'acide glycolique radioactif se forme également lorsque les chlorelles sont cultivées à la lumière, en présence de glucose uniformément marqué et de $CO_2$ à faible tension partielle. Lorsque la tension partielle de $CO_2$ est élevée, l'acide glycolique est peu abondant, par contre le saccharose est fortement marqué. Tous ces faits établissent les relations directes entre la formation du glycolate et le déroulement des réactions du mécanisme photosynthétique (Whittingham *et al.*, 1963a,b).

La formation d'acide glycolique en diverses circonstances où la photosynthèse est déprimée: faible tension partielle de $CO_2$ (Wilson et Calvin, 1955), action d'inhibiteurs: alcools (Lefrançois et Ouellet, 1958), isonicotinyl hydrazide (Whittingham *et al.*, 1962), forte tension partielle d'$O_2$ (Warburg et Krippahl, 1960; Whittingham *et al.*, 1964), montre également que la genèse de ce composé est liée au mécanisme de la photosynthèse des glucides, ainsi que l'ont proposé, de manière différente d'ailleurs, Stiller (1962) et Bassham (1964).

Ces résultats ont été interprétés souvent comme les conséquences d'une déficience partielle de la carboxylation du ribulose 1,5-diphosphate ou comme celles d'une libération du radical glycolyle lors des réactions de transcétolisation.

Les observations de Bradbeer et Anderson (1964) semblent être en accord avec cette conception: en présence d'$\alpha$-hydroxy-2-pyridine méthane sulfonate, inhibiteur de la glycolate oxydase, évitant la dégradation ultérieure du glycolate, les chloroplastes isolés d'épinard accumulent de l'acide glycolique à la lumière. Cette accumulation requiert la présence d'oses phosphates (fructose 6-phosphate ou glucose 6-phosphate, fructose 1,6-diphosphate, xylulose 5-phosphate ou sédoheptulose 7-phosphate) participant habituellement aux réactions de transcétolisation.

Cependant, Stiller (1962) a attribué au glycolate un rôle direct dans la fixation initiale de $CO_2$, par un mécanisme différent de celui des carboxylations connues.

Récemment, Zelitch (1965) a étudié la formation d'acide glycolique par des disques de feuilles de tabac exposés à $^{14}CO_2$ en présence de lumière pendant 5 ou 10 min et flottant sur une solution d'acide $\alpha$-hydroxy-2-pyridine méthane sulfonique.

Lorsque la tension partielle de $CO_2$ est faible, bien que supérieure à celle qui existe dans l'air (0·11 ou 0·18 % en volume au début des expériences), l'acide glycolique s'accumule 2 à 3 fois plus qu'en présence de concentrations plus élevées de $CO_2$ (1·1 %). Ces résultats confirment ceux qu'avaient obtenus Warburg et Krippahl (1960) avec les chlorelles.

La radioactivité spécifique des atomes de C du glycolate est voisine de celle du $^{14}CO_2$ fourni et elle est plus élevée que celle de l'atome de C n° 1 de l'acide 3-phosphoglycérique.

L'auteur en conclut que l'acide glycolique est formé à partir de $CO_2$ par un mécanisme non connu, sans passer par l'acide phosphoglycérique ni par l'acide phosphoglycolique dont la radioactivité spécifique est faible.

Il pourrait ensuite être intégré dans les glucides, soit par transcétolisation, soit par l'intermédiaire de la glycine et de la sérine. On sait en effet que ce dernier corps peut être transformé en glucides sans remaniement de ses atomes de C. En effet, les hexoses formés à partir de la [3-$^{14}$C]sérine sont marqués en 1 et 6 (Jimenez *et al.*, 1962). Il faut remarquer cependant que l'atome de C n° 1 du glucose 6-phosphate a une activité spécifique proche de celle du C n° 1 du phosphoglycérate.

Lorsque la concentration en $CO_2$ est élevée, l'acide 3-phosphoglycérique l'emporte sur l'acide glycolique.

Zelitch suppose donc qu'aux concentrations de $CO_2$ voisines de celle qui est présente dans l'air, l'assimilation peut se faire d'abord par la formation de glycolate et le 3-phosphoglycérate ne se formerait alors que secondairement. Quand la concentration de $CO_2$ est élevée par contre, la carboxylation du ribulose diphosphate serait favorisée, comme c'est le cas dans de nombreuses conditions expérimentales.

Aucun mécanisme actuellement connu ne permet de rendre compte de la genèse directe du glycolate. Les mécanismes enzymatiques de la formation des autres composés en $C_2$ (acétate, glyoxylate) partent généralement de composés plus riches en atomes de C. Par ailleurs, aucun mécanisme de carboxylation du glycolate n'est connu non plus (Voir Wood et Utter, 1965).

De nouvelles recherches seraient nécessaires pour confirmer ces résultats, préciser la spécifité de l'inhibiteur employé et les réactions intermédiaires responsables, enfin élucider les processus de régulation de la formation du glycolate.

## L'Importance Quantitative de la Voie Glucidique dans la Photosynthèse

L'étude de la cinétique d'apparition des différents composés marqués, synthétisés à la lumière par les chlorelles en présence de $^{14}CO_2$, a permis à Bassham et Calvin (1961), de préciser l'importance du cycle qu'ils ont proposé.

— après 10 s, la quasi-totalité du $^{14}C$ fixé dans les composés stables, non volatils, l'a été par le ribulose 1,5-diphosphate.

— de 10 à 30 s, les différents composés isolés par chromatographie sur papier et qui ont pour la plupart permis d'établir le cycle contiennent au moins 85 % du $^{14}C$ fixé.

— après de plus longues durées, on peut estimer qu'au moins 70 % du $^{14}C$ incorporé passe par l'intermédiaire de l'acide phosphoglycérique, les trioses phosphates et le cycle précédemment décrit. Il s'agit d'une valeur minimum puisqu'elle ne tient pas compte des pertes de substances au cours des extractions et des séparations analytiques.

Parmi les autres réactions de carboxylation, la $\beta$-carboxylation de l'acide phosphoénolpyruvique est quantitativement la plus importante. Chez les chlorelles, 3 % du $^{14}C$ fixé peuvent l'être par cette voie, l'acide oxaloacétique et ses dérivés possèdent en effet 12 % de la radioactivité fixée lorsqu'ils sont uniformément marqués. Quand la vitesse de fixation est maximum, ce pourcentage peut être plus élevé.

Dans les feuilles de Crassulacées, comme le *Bryophyllum*, ce mode de fixation est plus important puisque pour un éclairement de 5000 lux, il représente, après 6 min, 20 à 25 % de la totalité du $^{14}C$ fixé; il atteint encore 20 % à 20,000 lux (Moyse et Jolchine, 1957).

Il est probable que la faiblesse de la photophosphorylation aux faibles intensités lumineuses défavorise la réduction du phosphoglycérate.

Ce mode serait également très actif dans les feuilles de canne à sucre chez qui l'acide malique et l'acide aspartique apparaîtraient relativement plus marqués que l'acide phosphoglycérique décelable pendant les temps inférieurs à 60 s (Kortschak *et al.*, 1965). La formation du phosphoénolpyruvate et sa $\beta$-carboxylation seraient, au début de l'illumination, plus rapides que la réduction de l'acide phosphoglycérique en aldéhyde. Il faut encore remarquer qu'aux faibles éclairements, la genèse des acides aminés liés aux premiers produits de fixation de $CO_2$ ($\alpha$-alanine, acide aspartique, sérine, glycocolle) est, en valeur relative, favorisée par rapport à celle des glucides qui reste cependant prépondérante (Champigny, 1960).

Très rapidement, une fraction très importante des produits photosynthétisés émigre des chloroplastes dans le cytoplasme. Dans les feuilles d'épinard, après 5 ou 10 min de photosynthèse en présence de $^{14}CO_2$, les chloroplastes ne renferment plus que la moitié des substances organiques radioactives (Heber et Willenbrinck, 1964).

## Le Métabolisme des Glucides dans les Chloroplastes Isolés

Si l'on a pu douter, il y a une vingt ans encore, de la capacité des chloroplastes de réaliser seuls, sans la participation du cytoplasme, la synthèse des glucides, il n'en est plus de même depuis une dizaine d'années (Allen *et al.*, 1955).

Actuellement, la lecture de la bibliographie consacrée à la description de l'équipement enzymatique des chloroplastes conduit plutôt à poser la question: de quelles enzymes les chloroplastes sont-ils dépourvus ? Il serait en effet trop long de répondre à la question: quel est l'équipement enzymatique des chloroplastes ?

A vrai dire, il n'est pas toujours facile de préciser si des enzymes présentes dans les feuilles sont réellement présentes dans les chloroplastes eux-mêmes, les risques de pollution des préparations chloroplastiques d'une part, les risques de pertes d'enzymes au cours de l'extraction d'autre part, étant considérables. Mais Thomas (1960) en signale plus d'une centaine comme pouvant être dans les chloroplastes eux-mêmes.

En raison de leur taille, et compte tenu de leur richesse en protéines de structure constituant la majeure partie des lamelles et des *grana*, les chloroplastes peuvent contenir plusieurs centaines de protéines-enzymes de nature différente.

Il est bien établi que les premiers échecs dans l'étude de l'assimilation de $CO_2$ par les chloroplastes isolés *in vitro* ont été dus aux pertes d'enzymes subies par les organites au cours de leur isolement (voir Arnon, 1960).

La fixation et la réduction de $CO_2$ confirment la présence dans les chloroplastes des enzymes requises, notamment de la ribulose 1,5-diphosphate carboxylase, de la phosphoglycérate kinase, de la triose phosphate déshydrogénase, de l'aldolase, de la fructose diphosphatase, de la transaldolase et de la transcétolase. L'action de divers inhibiteurs: cyanure pour la carboxylation du ribulose 1,5-diphosphate, arséniate pour la réduction du phosphoglycérate, iodoacétamide pour la formation de ribulose 1,5-diphosphate à partir de ribose 5-phosphate apporte également un argument pour la présence de ces enzymes, dans les chloroplastes (Trebst *et al.*, 1960).

La restitution partielle d'un équipement enzymatique permettant la synthèse des glucides à partir de $CO_2$ a été obtenue par l'addition de protéines solubles extraites des chloroplastes isolés en milieu aqueux (Whatley *et al.*, 1956) et de différents glucides phosphorylés (ou substances voisines): glucose 1- ou 6-phosphate, fructose 1,6-diphosphate, ribose 5-phosphate, 3-phosphoglycérate, etc. L'extrait de chloroplastes est destiné en particulier à apporter de la ribulose 1,5-diphosphate carboxylase. Les chloroplastes après leur isolement en milieu aqueux ne retiennent guère en effet que 10% de la ribulose 1,5-diphosphate carboxylase présente dans les feuilles et cette faible rétention rend compte en partie de leur faible activité photosynthétique (Smillie et Fuller, 1959). L'addition de l'extrait de chloroplastes apporte également de la pyridine nucléotide réductase photosynthétique (PPNR) de San Pietro et Lang (1958).

Les glucides sont ajoutés comme précurseurs possibles du ribulose 1,5-diphosphate. Dans les meilleurs cas, avec de telles préparations, la fixation de $CO_2$ ne dépasse guère le trentième de la fixation maximum réalisée par les feuilles intactes (Losada *et al.*, 1960) (voir Tableau II).

Tableau II. Fixation de $CO_2$ par les chloroplastes isolés
($\mu$moles fixées/hr/mg de chlorophylle)

| | Fixation maximum | Références |
|---|---|---|
| Chloroplastes *aqueux* d'épinard, intacts | 2·5 | Gibbs et Calo (1959) |
| Chloroplastes *aqueux* d'épinard, brisés + fructose 1,6-diphosphate | 6 | Losada *et al.* (1960) |
| Chloroplastes *aqueux* + extrait de chloroplastes *organiques* d'épinard | 10 | Heber et Tyszkiewicz (1962) |
| —D°, extrapolé à une concentration de protéines voisine de celle du cytoplasme | 30 | —d°— |
| Chloroplastes *in vivo* (cellules intactes) | 180 | |

Avec des homogénats de feuilles d'épinard, Gee *et al.* (1965) ont obtenu une fixation plus élevée, atteignant à la lumière, en présence de ribulose 1,5-diphosphate, de $NADP^+$ et d'ADP, 7·6 $\mu$mol de $CO_2$ par mg de chlorophylle et par heure. A l'obscurité, l'addition de ribulose 1,5-diphosphate, d'ATP, de NADPH et de NADH permettrait d'atteindre 17·2 $\mu$mol, soit autant que la fixation réalisée par des feuilles semblables à celles qui ont servi à préparer les homogénats. Il faut observer cependant que la capacité de ces feuilles était faible. Avec des chloroplastes de pois, isolés en solution de saccharose et pourvus de leur membrane, Walker (1964) a obtenu la fixation de prés de 25 $\mu$moles de $CO_2$ par mg de chlorophylle et par heur.

L'élévation du pouvoir de fixation et de réduction de $CO_2$ par les chloroplastes isolés *in vitro* a également été obtenue par l'addition de protéines solubles extraites d'autres chloroplastes isolés de feuilles préalablement lyophilisées.

Ces derniers chloroplasts sont séparés par centrifugation dans des mélanges de substances organiques ($CCl_4$+hexane) constituant des gradients de densité (Stocking, 1959). S'ils ont perdu la majeure partie de leur activité photochimique par suite de la perte de substances liposolubles du groupe des plastoquinones (Ogren et Krogmann, 1963), ils ont conservé les enzymes hydrosolubles présentes dans la matrice (*matrix*) chloroplastique et sont 2 à 3 fois plus riches en protéines que les chloroplastes primitivement extraits en milieu aqueux (Heber et Tyszkiewicz, 1962).

Aussi l'addition de ces enzymes à des suspensions de chloroplastes préparées initialement en milieu aqueux, et qui ont gardé la majeure partie de leur activité photochimique, permet-elle d'obtenir des préparations également actives dans la réduction de $CO_2$.

Par cette méthode mixte, Heber et Tyszkiewicz (1962) ont obtenu des préparations chloroplastiques de *Tetragonia expansa*, *Vicia faba*, *Nicotiana tabacum*, qui fixent 7 à 8 fois plus de $CO_2$ que les chloroplastes extraits simplement en milieu aqueux. Ces résultats, obtenus sans addition de glucides, atteignent la fixation réalisée par les chloroplastes extraits en milieu aqueux et enrichis en glucides précurseurs du ribulose 1,5-diphosphate (Tableau II). L'examen comparé du marquage de l'acide phosphoglycérique et des glucides phosphorylés, à l'aide de $^{32}P$ d'une part et à l'aide de $^{14}C$ d'autre part, montre qu'avec les préparations aqueuses seules, le rapport $^{32}P$ fixé/$^{14}C$ fixé est toujours plus grand que 1, traduisant un réel déficit de la fixation de $^{14}CO_2$ par rapport à la phosphorylation des substrats (Tableau III). L'acide phosphoglycérique formé, peu marqué dans ses atomes de C, dériverait en partie des glucides présents dans les organites.

Tableau III. Rapport molaire de $^{32}P/^{14}C$ incorporés à la lumière dans différentes substances par des chloroplastes isolés, après 20 min

| | Chloroplastes *aqueux* | Chloroplastes *aqueux+extrait de* chloroplastes *organiques* | Chloroplastes *aqueux+extrait de* chloroplastes *organiques*+NADPH |
|---|---|---|---|
| Ac.phosphoglycérique | 1/0·61 | 1/1·1 | 1/2·1 |
| Glucides monophosphates | 1/0·31 | 1/0·65 | 1/0·25 |
| Glucides diphosphates | 1/0·07 | 1/0·08 | |

d'après Heber et Tyszkiewicz (1962).

L'addition d'un extrait de chloroplastes séparés en milieu organique provoque une nette diminution du rapport qui devient voisin de 1. La formation d'acide phosphoglycérique à partir des produits photosynthétisés doit mieux fonctionner dans ces conditions et il se trouve plus abondamment marqué dans ses atomes de C n$^{os}$ 1 et 2, par suite de l'apport des enzymes du cycle des pentoses phosphates.

L'addition de NADPH, élève encore le rapport qui devient égal à 2. On remarque cependant qu'il n'atteint pas la valeur théorique de 3 qui correspondrait au marquage rapide par le $^{14}C$ des 3 atomes de C de l'acide phosphoglycérique. La déficience est encore plus prononcée pour les glucides phosphorylés. Bien que les extraits de chloroplastes isolés en milieu organique aient enrichi les préparations en triose phosphate déshydrogénase, la lenteur de la réduction de l'acide phosphoglycérique en aldéhyde, et par suite la lenteur des réactions du cycle des pentoses phosphates sont dues à celle de la réduction du $NADP^+$. Il en résulte que la majeure partie du $CO_2$ fixé reste à l'état de carboxyle, sans être réduit.

Ces résultats expliquent également l'efficacité des glucides phosphates ajoutés comme précurseurs de ribulose 1,5-diphosphate. Par rapport aux activités des cellules chlorophylliennes intactes, la faiblesse des organites isolés *in vitro* est également due à la dilution des enzymes et des substrats au sein des liquides de suspension qui comprennent à la fois des sucs vacuolaires normalement séparés des organites *in vivo* et les solutions de sels et de tampons appropriés.

L'élévation de la concentration en protéines solubles de 2 à 12 mg par ml de suspension permet de multiplier par 3 à 5 la vitesse de fixation de $CO_2$. Si l'on extrapole jusqu'à des concentrations en protéines comparables à celles qui existent dans le cytoplasme *in vivo* (100 mg de protéines environ par ml), on peut estimer que des préparations *in vitro*, de même concentration, permettraient d'atteindre une fixation de 30 $\mu$mol $CO_2$/hr/mg chlorophylle. Encore cette fixation ne s'accompagne-t-elle pas d'une réduction au niveau glucidique de tout le $CO_2$ retenu.

*Remarque.* La faiblesse encore très grande de cette valeur, comparée aux 180 $\mu$moles de $CO_2$ qui peuvent être fixées par hr et par mg de chlorophylle, lorsque des feuilles intactes se trouvent placées dans les meilleures conditions possibles de photosynthèse (voir Rabinowitch, 1956) est due certainement en partie à l'inhibition exercée par les constituants non chloroplastiques de la cellule. Ces constituants se trouvent mis en contact après le broyage avec les chloroplastes et leurs fragments.

Aux divers facteurs dépressifs bien connus: concentrations en sels ou en substances organiques et pH non optimum, effet des produits d'oxydation libérés à la suite de l'intervention intempestive de phénoloxydases, inhibition partielle des enzymes à groupe SH, etc., il faut ajouter sans doute l'effet éventuel de tannins (Maslow, 1964).

Enfin, il est bien certain que les liens topographiques entre les divers pigments, les divers systèmes photochimiques et les complexes enzymatiques solubles ne sont plus aussi bien assurés du point de vue fonctionnel dans les préparations isolées *in vitro*, alors qu'ils sont beaucoup plus efficacement coordonnés entre eux *in vivo*.

## L'Oxydation du Glucose 6-Phosphate dans les Chloroplastes

Les chloroplastes isolés en milieu aqueux, lorsqu'ils sont débarrassés des pollutions mitochondriales, n'ont que peu d'activité oxydasique. S'ils possèdent l'enzyme de condensation, ils sont dépourvus des enzymes des décarboxylations oxydatives du cycle des acides tricarboxyliques et des systèmes catalytiques des oxydations terminales, notamment de la cytochrome oxydase (voir Moyse, 1961). Cette absence d'enzymes oxydatives permet de bien comprendre l'intervention essentielle sinon exclusive de ces organites dans les réductions.

La question restait cependant posée au sujet des enzymes qui catalysent les oxydations à l'origine du cycle oxydatif des pentoses phosphates.

Au cours d'une étude des chloroplastes de *Bryophyllum* isolés par centrifugation en milieu organique selon la méthode de Stocking (1959), Garnier-Dardart (1965) a mis en évidence une activité très nette de la glucose 6-phosphate déshydrogénase et de la 6-phosphogluconate déshydrogénase (Tableau IV). La première est indifférente à l'iodoacétamide, alors que la seconde est inhibée presque complètement en sa présence à la concentration de $3 \times 10^{-5}$ M. L'activité glucose 6-phosphate déshydrogénase des chloroplastes représente 20 % de celle des cellules entières. Les deux réactions d'oxydation précédentes laissent supposer une activité intermédiaire de la gluconolactonase.

Dans le cas présent, la forte activité de ces enzymes est certainement en relation avec la genèse de l'acide malique dans les chloroplastes eux-mêmes, à l'obscurité.

Tableau IV. Activités enzymatiques des chloroplastes de *Bryophyllum* isolés en milieu organique ($\mu$moles NAD(P)$H_2$ formé ou oxydé par hr)

| Réactions | Par mg N protéique | Par mg chlorophylle |
|---|---|---|
| G-6-Ⓟ + $NADP^+$ → glucono-lactone-6-Ⓟ + $NADPH^+$ + $H^+$ | 5·1 | 6·8 |
| glucose-6-Ⓟ: NADP oxydoréductase | (2·5)[1] | |
| 6-Ⓟ-gluconate + $NADP^+$ → RuDP + $CO_2$ + NADPH + $H^+$ | 5·1 | 6·8 |
| 6-Ⓟ-gluconate: NADP oxydoréductase décarboxylase | | |
| PEP + $CO_2$ + $H_2O$ → $P_i$ + OAA | 7·7 | 10·3 |
| PEP carboxylase | | |
| OAA + NADH + $H^+$ ⇌ malate + $NAD^+$ | 1280 | 1706 |
| malate déshydrogénase | (640)[2] | |
| Malate + $NADP^+$ → pyruvate + $CO_2$ + NADPH + $H^+$ | 6·6 | 8·8 |
| enzyme malique | | |

[1] Corrigée de l'activation possible par les ions $Cl^-$.
[2] Corrigée de l'activation possible par NaCl présent dans les préparations.
D'après Garnier-Dardart (1965).

Ainsi que le montre le Tableau IV, les enzymes requises pour assurer la fixation de $CO_2$ sur le phosphoénolpyruvate et la réduction de l'oxaloacétate y sont présentes. La formation du phosphoénolpyruvate à partir de phosphoglycérate et de ce dernier par carboxylation du ribulose 1,5-diphosphate, aussi bien *à l'obscurité* qu'à la lumière, explique le double marquage de l'acide malique dans ses atomes de C n^os^ 4 et 1 (Jolchine, 1959). L'ensemble du mécanisme exige *à l'obscurité* la formation de ribulose 1,5-diphosphate qui peut ainsi avoir lieu à la suite de l'oxydation du glucose issu de l'amylolyse. Bien qu'il s'agisse d'un cas un peu spécial puisque seules les plantes grasses (Crassulacées et autres plantes crassulentes) accumulent de l'acide malique en grande quantité à l'obscurité, la généralité du mécanisme ne peut être exclue, l'acide étant très généralement synthétisé par les chloroplastes. Les activités glucose 6-phosphate déshydrogénase, 6-phosphogluconate déshydrogénase, phosphoénolpyruvate carboxylase et malate déshydrogénase, mesurées dans les chloroplastes, sont suffisantes pour expliquer l'accumulation de l'acide malique dans les feuilles à l'obscurité.

La présence de la glucose 6-phosphate déshydrogénase et de la 6-phosphogluconate déshydrogénase pose le problème de la régulation de la genèse des pentoses phosphates dans les chloroplastes par la voie oxydative à l'obscurité et par la voie inverse à la lumière.

On peut estimer que cette régulation est gouvernée par la quantité de NADPH et de NADH.

## La Diversification des Glucides

Parmi les composés glucidiques formés dans les chloroplastes ou à leur voisinage, tel qu'il est schématisé dans la Fig. 5, trois catégories peuvent être distinguées, selon leur devenir.

(a) *Les oses-phosphates qui constituent le pool du métabolisme photosynthétique* assurant, en rétroaction, la régénération de l'accepteur de $CO_2$.

(b) *Les glucides engagés dans l'édification de l'appareil photosynthétique* pendant la période de croissance des chloroplastes.

Parmi eux, on peut distinguer des pentoses: ribose et ribodésose, engagés dans la synthèse des acides nucléiques chloroplastiques.

Il faut y ajouter les chaînes glucidiques des galactolipides et sulfolipides (Benson, 1961; Wintermans, 1962). Le transfert du galactose dans les galactolipides des chloroplastes d'épinard dépend de l'intervention de l'UDP-D-galactose (Neufeld, 1963).

(c) *Les glucides condensés* dont le plus important est l'amidon. Le rôle que jouent les nucléotides dans sa synthèse a été récemment exposé (voir Cabib, 1963; Northcote, 1964; Ashwell, 1964; Nordin et Kirkwood, 1965). Il s'accumule fréquemment dans les chloroplastes des feuilles des Dicotylédones (voir Porter, 1962) et parfois même de Monocotylédones (Maïs); on le trouve également dans les plastes d'algues vertes (Giraud, 1963). Sa formation est liée à l'activité d'une transglycosylase assurant la condensation de glycosyles de l'UDPG (voir Leloir et Cardini, 1962) ou plus efficacement de l'adénosine diphosphate de glucose, selon la réaction:

$$\text{ADPG} + (\alpha\text{-D-glucose } 1\text{–}4)n \rightarrow (\alpha\text{-D-glucose } 1\text{–}4)n+1 + \text{ADP}$$

Le transfert de glycosyle peut se faire également à partir du saccharose, tout au moins dans les grains de maïs en maturation (De Fekete et Cardini, 1964) et de riz (Murata *et al.*, 1964).

L'ADPG a été trouvé non seulement dans ces organes riches en amidon dans lesquels il intervient très efficacement dans la synthèse de l'amylose et de l'amylopectine (Akazawa *et al.*, 1964), mais il existe aussi dans les chloroplastes. Ainsi, dans les plastes de *Phaseolus aureus*, il intervient en présence d'une transglycosylase liée aux grains d'amidon eux-mêmes (Murata et Akazawa, 1964). La présence d'ADPG a été également constatée dans les chlorelles (Kauss et Kandler, 1962).

On peut supposer, comme l'a fait Badenhuizen (1963), que l'amylose serait formé en premier, puis transformé ensuite en amylopectine à la suite de l'inter-

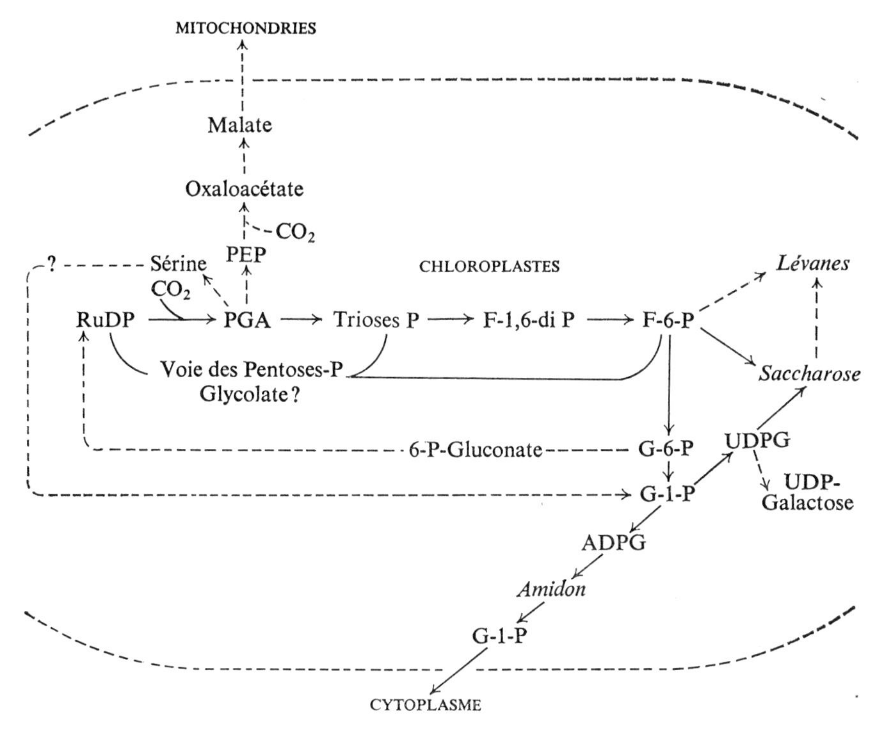


FIG. 5. Schéma des relations entre les principaux glucides et composés apparentés participant à la fixation de $CO_2$ dans les chloroplastes.

vention d'un système enzymatique de branchement comparable à la Q enzyme de la pomme de terre décrite par Peat *et al.* (1959).

Les grains d'amidon chloroplastiques ont aussi une activité phosphorylase qui, à un faible degré, peut également participer à la synthèse de l'amidon, bien que l'enzyme catalyse essentiellement sa solubilisation.

Chez les algues rouges, l'amidon floridéen formé par l'union de résidus glycosyles liés entre eux en 1–3 et non en 1–4 comme dans l'amylose de l'amidon habituel, se dépose contre le pyrénoïde des chloroplastes (Giraud, 1963). Ces algues renferment également du floridoside, de l'isofloridoside et du floridoside-α-mannoside, dont les mécanismes terminaux de synthèse sont mal connus.

Les algues brunes élaborent fréquemment des quantités élevées de mannitol (Bidwell *et al.*, 1958) et des polysaccharides formés de $\beta$-glycosyles liés en 1–3 (ex: laminarine). Il en est de même pour les Euglènes qui synthétisent le paramylon, polymère de $\beta$-glycosyles liés en 1–3. L'UDPG intervient dans la synthèse du paramylon (Goldemberg et Maréchal, 1963). Quant aux algues bleues, elles synthétisent des polysaccharides solubles analogues au glycogène.

Le saccharose est aussi un des premiers glucides formés dans les cellules chlorophylliennes (voir Calvin, 1955). Bien qu'il prenne naissance semble-t-il hors des chloroplastes, tout au moins dans les feuilles, le mécanisme de sa genèse sera rapidement décrit.

Si, jusqu'à une époque récente, différents organismes, dont les algues rouges, semblaient ne pas en former, il ne faudrait pas en conclure à leur incapacité réelle. Ainsi diverses d'entre elles sont-elles capables de le synthétiser (Champigny, 1959; Quillet, 1965). Plusieurs renferment également du tréhalose.

La synthèse du saccharose et celle de l'amidon sont indépendantes l'une de l'autre. Ainsi, des chlorelles en présence d'eau tritiée forment du saccharose dont le glycosyle est rapidement marqué en position $C_2$ alors que les glycosyles de l'amidon ne le sont que lentement à la même place (Simon *et al.*, 1964). Dans les feuilles d'épinard, le saccharose semble être formé hors des chloroplastes, dans le cytoplasme lui-même, comme le montrent les analyses effectuées par Heber et Willenbrinck (1964) sur des fractions cellulaires isolées par centrifugation en gradients de densité organiques à partir de feuilles lyophilisées, préalablement exposées à $^{14}CO_2$. La synthèse de l'UDPG aurait lieu aussi principalement dans le cytoplasme.

Il est connu depuis longtemps que le saccharose se forme dans les cellules chlorophylliennes par l'action d'un système enzymatique décrit d'abord par Leloir et Cardini (1955). Ce système catalyse la réaction:

$$\text{UDPG} + \text{fructose 6-Ⓟ} \rightarrow \text{Saccharose-Ⓟ} + \text{UDP}$$

Sous l'action d'une phosphatase, le saccharose phosphate donne naissance au saccharose, tandis que l'UDPG est régénéré par la suite des réactions suivantes:

$$\text{UDP} + \text{ATP} \rightarrow \text{UTP} + \text{ADP}$$

$$\text{UTP} + \text{glucose 1-Ⓟ} \rightarrow \text{UDPG} + \text{Pyrophosphate}$$

Cette dernière réaction est en effet catalysée par des préparations enzymatiques extraites des feuilles de betterave (Burma et Mortimer, 1956). L'UDPG pyrophosphorylase qui la catalyse et l'UDPG-fructose transglycosylase qui assure la synthèse du saccharose ont été identifiées dans la canne à sucre (Hatch *et al.*, 1963). L'UDP-D-glucose est spécifiquement requis, ainsi que l'ont montré Frydman et Hassid (1963) avec des feuilles de canne à sucre. L'accumulation du saccharose dans les feuilles est fréquente chez les graminées qui synthétisent aussi du raffinose, des oligosaccharides et des lévanes. Si la formation de ces derniers composés est très active, en particulier dans les gaines des feuilles (Rocher et Bourdu, 1964), son mécanisme est très mal connu.

Si l'ADPG et l'UDPG jouent un rôle prépondérant dans les transferts de glycosyles, en raison même de l'abondance de l'amidon et du saccharose dans les cellules chlorophylliennes, il n'est pas exclu que d'autres nucléotides interviennent. Ainsi le thymidine diphosphate et, à un moindre degré, le guanosine diphosphate et le cytidine diphosphate, quoique beaucoup moins efficaces que l'ADPG et l'UDPG, peuvent cependant jouer le rôle de substrats dans la synthèse de saccharose (Milner et Avigad, 1965).

La régulation de la synthèse et de l'hydrolyse des polysaccharides commence à être connue pour le glycogène dans les cellules hépatiques (Krebs, 1965). Il n'en est pas encore de même pour l'amidon et les autres glucides condensés des végétaux, qui se forment à la lumière et représentent une réserve aisément mobilisable, émigrant facilement après hydrolyse, à l'obscurité, hors des cellules vertes.

Il est probable qu'à cette régulation participent les phosphates, les oses non-phosphorylés (Badenhuizen, 1963), les oses phosphates (comme ils le font dans la régulation de la biosynthèse du glycogène hépatique) (Steiner, 1964), les osyle-nucléotides, les transosylases et les phosphorylases. Peut-être des effets de rétroaction comparables à ceux qui ont été mis en évidence dans le foie, à propos de la régulation de la synthèse des osyle-nucléotides (Kornfeld *et al.*, 1964), interviennent-ils également. Il y a, à ce propos, tout un champ de recherches encore bien peu exploré.

## L'Accumulation des Glucides dans les Chloroplastes et le Fonctionnement Photosynthétique. Le Cas des Feuilles Carencées en Azote

Il est bien connu que les cellules chlorophylliennes peuvent accumuler plus ou moins de glucides ou de protéines ou bien encore de lipides, selon les conditions de nutrition. Ces faits sont particulièrement nets chez les chlorelles et les *Scenedesmus* qui, en nutrition azotée normale, synthétisent surtout des protéines et des glucides, tandis qu'après carence en azote ils accumulent des lipides.

Pour les plantes supérieures, l'accumulation des glucides a souvent lieu dans les chloroplastes eux-mêmes. Elle concerne souvent l'amidon (voir Porter, 1962). Mais les sucres solubles peuvent aussi s'accumuler et, comme l'a signalé Burström (1943) à propos des feuilles de blé carencées en N, les glucides constituent alors, par rapport aux protéines foliaires, un excédent de chaînes carbonées.

L'accumulation peut également avoir lieu dans les cellules des algues bleues. Ainsi, *Agmenellum quadriplaticum*, Cyanophycée marine, accumule-t-elle un polymère du glucose dont la quantité atteint 45% de la masse de substance sèche lorsque les cultures sont faites en présence d'acide urique comme unique source d'N (Van Baalen, 1965).

### L'accumulation d'amidon en fonction de la carence en azote. La levée de carence et l'hétérogénéité des chloroplastes dans une même feuille

Des *Bryophyllum daigremontianum* Berger, cultivés depuis plusieurs mois sur un milieu synthétique dépourvu d'azote, ont des chloroplastes généralement riches en amidon, et par contre très appauvris en protéines et en chlorophylle

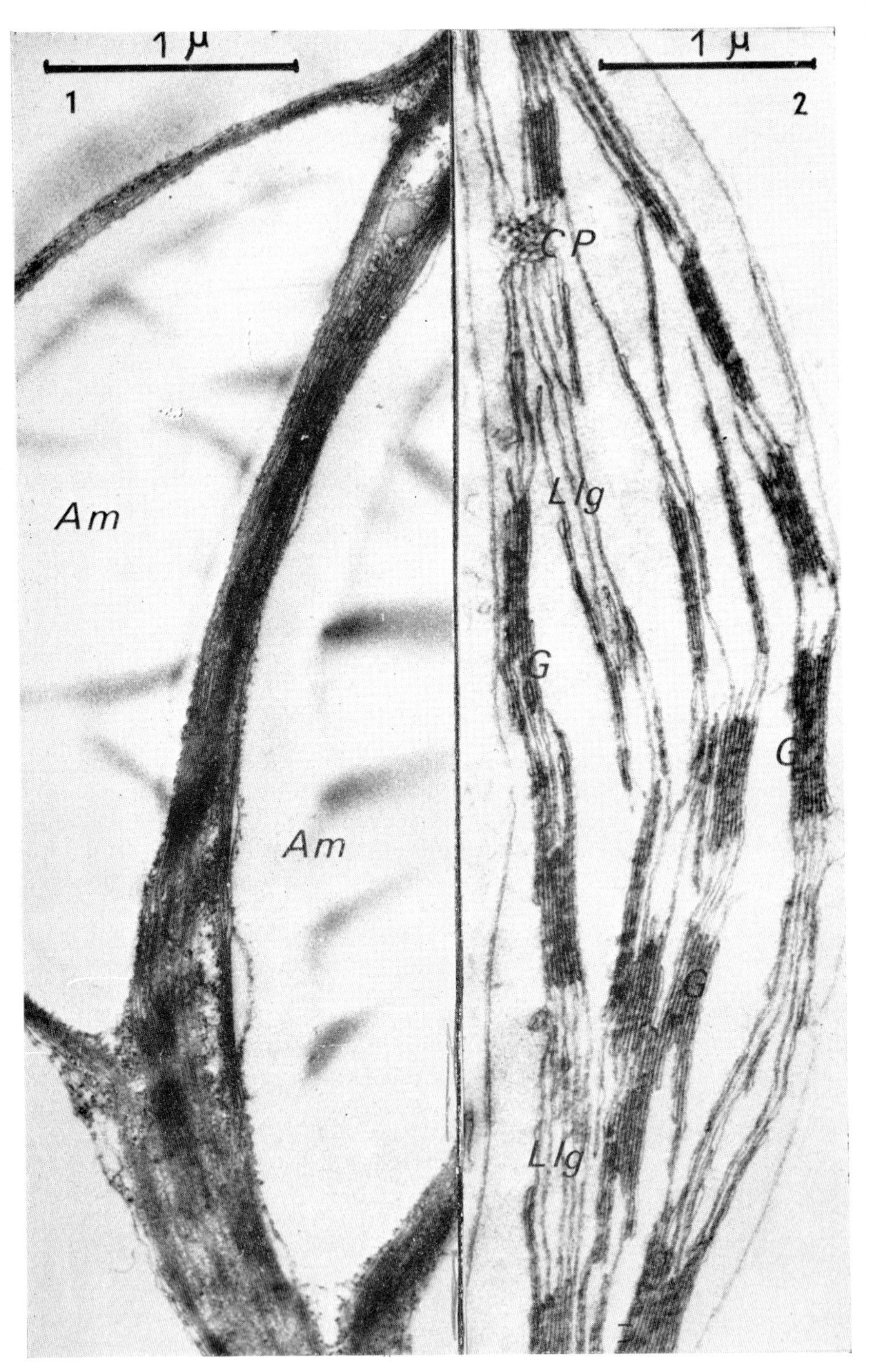


Fig. 6. (1) Feuille n° 2 de *Bryophyllum daigremontianum*. Sujet carencé en azote. Chloroplaste dont le système lamellaire réduit est écrasé par de grosses inclusions d'amidon *Am*. Fixation tétroxyde d'osmium. × 35,000. (2) Feuille n° 2. Témoin non carencé. Chloroplaste lenticulaire, sans amidon, pourvu d'une organisation lamellaire juvénile bien différenciée en grana *G* et lamelles intergranaires *Llg*, avec un centre prolamellaire *CP*. Fixation permanganate. × 30,000. (Préparation et cliché M. Lefort.)

(*Plantes sans N*). L'amidon y subsiste encore après 72 hr de mise à l'obscurité. Afin de faire une étude précise de leurs activités, nous avons suivi l'évolution des chloroplastes *toujours dans la feuille de même niveau initial*, la feuille n° 2 (le n° 1 étant attribué aux feuilles les plus jeunes situées au sommet des plantes). Cette étude a été faite principalement après la levée de carence par apport de nitrates. Comparativement, des chloroplastes des feuilles voisines, plus âgées ou plus jeunes, et des chloroplastes de plantes non carencées ont été examinés comme témoins (*Plantes* $NO_3^-$) (Bourdu *et al.*, 1965a,b; Rémy *et al.*, 1965).

Dans les feuilles n° 2 des *plantes* $NO_3^-$, les chloroplastes possèdent peu d'amidon et sont constitués normalement (Fig. $6_2$).

Les feuilles de même niveau, pour les *plantes sans N*, ont des chloroplastes fréquemment dépourvus de membrane et très riches en amidon. Les lamelles et les granums y sont plus rares (Fig. $6_1$). Les teneurs en chlorophylle et en azote protéique y sont faibles (Tableau V).

Tableau V. Teneurs en chlorophylle et en N protéique de chloroplastes totaux des feuilles "carencées" en N, avant (0 j) et après 7, 14 et 21 j de levée de carence

| | Chloroplastes des feuilles primitivement carencées | | | | Chloroplastes des feuilles témoins non carencées |
|---|---|---|---|---|---|
| | Avant levée de carence | Après levée de carence | | | |
| | 0 j | 7 j | 14 j | 21 j | |
| Chlorophylle | 2 | 6 | 27.9 | 47·4 | 46·3 |
| N protéique | 2·8 | 19·2 | 60 | 57·8 | 66 |

Les résultats sont exprimés en $\mu$g par mg de substance sèche, les chloroplastes ayant été extraits de feuilles n° 2 de *Bryophyllum daigremontianum* Berger, après lyophilisation et centrifugation en milieu organique ($CCl_4$ + hexane).

La *levée de carence* des plantes carencées, réalisée par apport de nitrates, est suivie de lents remaniements qui aboutissent cependant à une accélération très sensible de la croissance foliaire, à la formation de nouveaux chloroplastes juvéniles et à un appauvrissement en amidon d'une fraction importante des chloroplastes carencés décrits précédemment.

La Fig. 7 montre un chloroplaste dans lequel l'amidon se lyse et un autre, plus jeune, dans lequel de nouvelles lamelles se développent après 7 jours de levée de carence. Après 14 jours, de nouveaux chloroplastes apparaissent et nombre d'entre eux sont riches en lamelles et grana (Fig. 8).

Il en résulte une très grande hétérogénéité des chloroplastes dans la même feuille. S'il s'agit dans le cas présent d'un cas particulièrement net, il ne faut cependant pas oublier qu'une hétérogénéité chloroplastique très sensible existe toujours entre les cellules d'une même feuille, l'âge des cellules étant différent selon les régions foliaires. Les mesures d'activité photosynthétique,

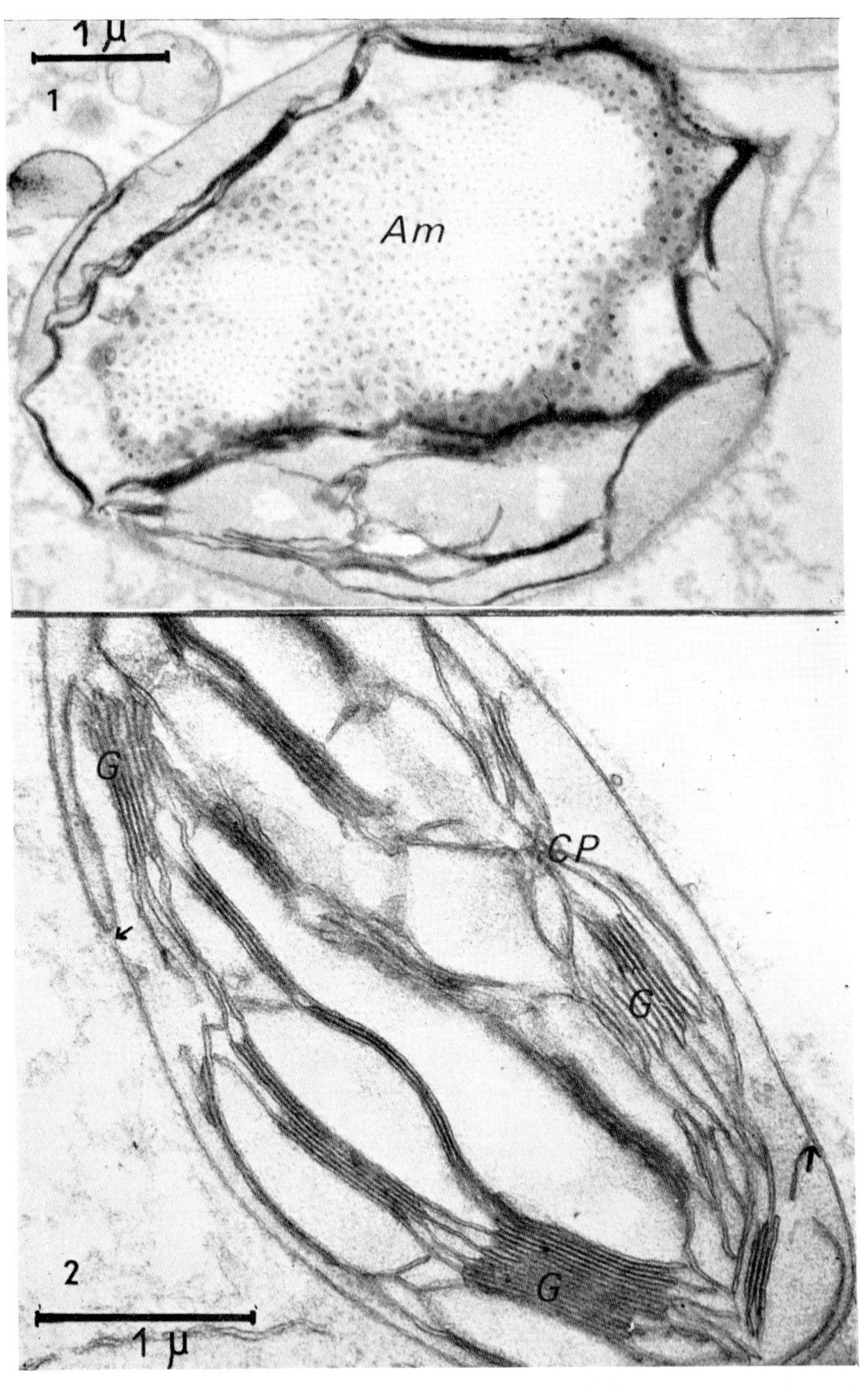


FIG. 7. (1) Feuille n° 2 de *Bryophyllum daigremontianum* initialement carencé, puis ayant reçu des nitrates pendant 7 jours. Chloroplaste "festonné" avec globule d'amidon *Am* en voie de digestion. Fixation permanganate. ×18,000. (2) Feuille n° 2 initialement carencée, puis ayant reçu des nitrates pendant 7 jours. Chloroplaste jeune où l'organisation lamellaire est en cours d'édification. On observe de nombreuses expansions du feuillet interne de la membrane plastidiale (cf. Flèches) *CP*, centre prolamellaire de vésicules lipoprotéiques. *G* grana. Fixation permanganate. ×30,000. (Préparation et cliché M. Lefort.)

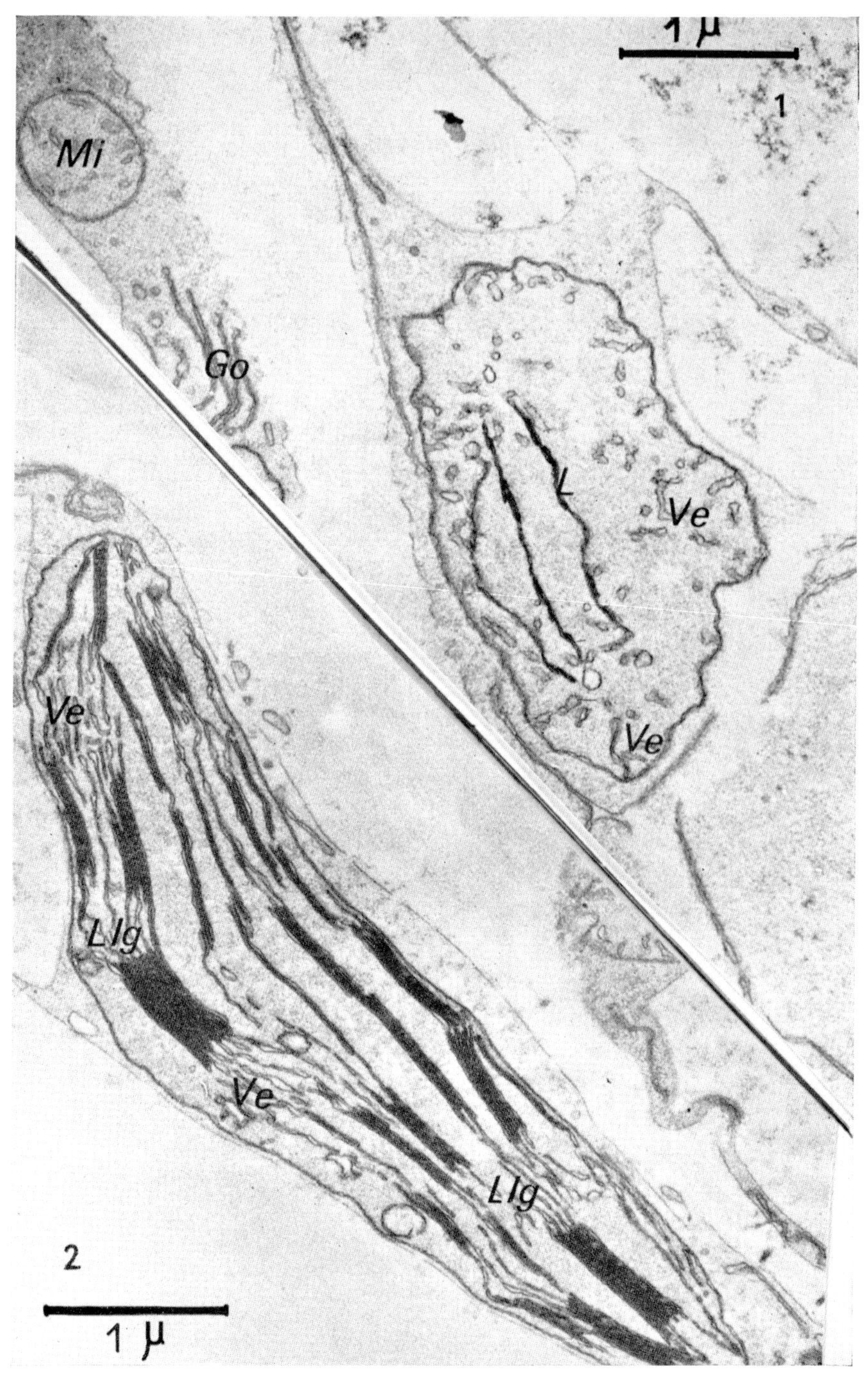


FIG. 8. (1) Feuille n° 2 de *Bryophyllum daigremontianum* initialement carencé, puis ayant reçu des nitrates pendant 14 jours. Très jeune plaste encore amiboïde pourvu de nombreuses vésicules *Ve* issues du feuillet interne de la membrane et des premières lamelles doubles *L*. *Mi*: mitochondrie, *Go*: appareil de Golgi. Fixation permanganate. ×24,000. (2) Feuille n° 2 initialement carencée, puis ayant reçu des nitrates pendant 14 jours. Dans la même feuille et dans des cellules voisines de celles de la photographie précédente, chloroplaste déjà différencié. On observe des vésicules *Ve* qui contribuent à l'élaboration des lamelles intergranaires *Llg*. Fixation permanganate. ×25,000 (Préparation et cliché M. Lefort.)

faites avec des feuilles entières ou bien avec des chloroplastes isolés, ne traduisent qu'une moyenne réalisée par des organites à des stades différents de développement.

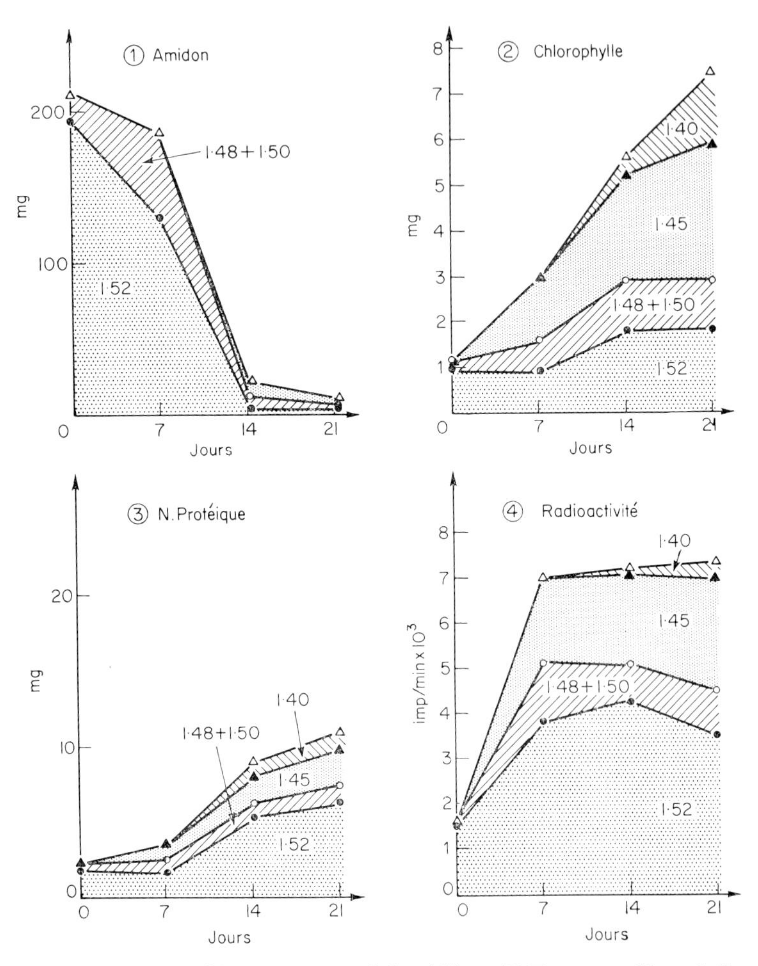


FIG. 9. (1) Teneur en amidon en mg par g de lyophilisat; (2) Teneur en chlorophylle en mg par g de lyophilisat; (3) Teneur en N protéique en mg par g de lyophilisat; (4) Radioactivité de $^{14}CO_2$ fixée par les différentes fractions contenues dans 1 g de lyophilisat. Chloroplastes des feuilles n° 2 de *Bryophyllum daigremontianum* préalablement carencé en N et après la levée de carence (7,14 et 21 jours). Les différentes fractions chloroplastiques isolées par centrifugation en gradient de densité organique sont indiquées par leur densité.

En même temps que la structure des chloroplastes subit d'importantes modifications, leur composition varie.

Le premier indice de variation est présenté par la densité des différentes couches chloroplastiques qui peuvent être séparées par centrifugation des broyats de feuilles lyophilisées, dans des gradients de densité constitués par des mélanges de tétrachlorure de C et d'hexane, selon Stocking (1959).

Dans les feuilles des plantes carencées, on peut séparer aisément 2 fractions chloroplastiques. La plus abondante a comme densité 1·52, l'autre est un peu plus légère (1·48–1·50). Après 7 jours d'alimentation nitrique la couche 1·52 a diminué d'importance, tandis qu'apparaît une nouvelle couche de densité 1·45.

Après 14 jours, une couche plus légère encore apparaît (densité 1·40).

D'une manière générale, il apparaît de nouvelles couches moins denses, tandis que les couches lourdes diminuent d'importance.

A l'origine, les couches lourdes sont très riches en amidon et elles s'appauvrissent très rapidement de telle sorte qu'après 14 jours la majeure partie de l'amidon en a disparu (Fig. $9_1$).

En même temps, les quantités de chlorophylle augmentent dans chaque couche, surtout dans celles qui sont récemment apparues (densités 1·45 et 1·40) (Fig. $9_2$).

La quantité de protéines présentes dans les différentes couches augmente également (Fig. $9_3$).

Les expériences d'assimilation de $^{14}CO_2$ montrent que les protéines néoformées sont peu marquées. Il est donc probable qu'une part importante de leurs chaînes carbonées provient de l'amidon qui disparaît.

L'examen comparé de la structure, des densités et de la composition des différentes fractions chloroplastiques ainsi séparées conduit à penser que la couche la plus lourde est faite en majeure partie de chloroplastes âgés, primitivement très riches en amidon formé pendant la période de carence. Ces chloroplastes sont pauvres en grana, en lamelles et en matrice (*stroma*).

Il en est de même, quoiqu' à un moindre degré, pour la couche de densité 1·48–1·50.

Dans les couches légères qui apparaissent ensuite, les chloroplastes sont ou bien en voie de régénération complète ou bien nouvellement formés. Tous sont riches en pigments et en protéines.

## L'ACTIVITÉ PHOTOSYNTHÉTIQUE EN FONCTION DE LA TENEUR DES CHLOROPLASTES EN AMIDON

Les plantes préalablement carencées reçoivent ensuite des nitrates et les mesures d'activité photosynthétique sont faites après 7, 14 et 21 jours succédant à la levée de carence.

Après fixation de $^{14}CO_2$ par les feuilles pendant 1 hr, en présence de lumière (9000 lux), les chloroplastes sont extraits des feuilles lyophilisées et séparés en milieu organique.

L'analyse montre que 70 à 84 % du $^{14}C$ assimilé dans les différentes fractions se trouvent dans l'amidon néoformé, le reste étant présent surtout dans les glucides phosphorylés solubles, les acides aminés libres, les acides organiques. Une faible quantité se trouve dans les protéines (0·1 à 8 %).

La Fig. $9_4$ donne les résultats relatifs à la fixation totale de $^{14}CO_2$.

Pour les chloroplastes âgés (densités 1·52 et 1·48–1·50), l'intensité de la photosynthèse croît d'abord durant la période de disparition de l'amidon antérieurement accumulé. Elle décroît ensuite après 7 ou 14 jours, tandis que les chloroplastes des fractions légères, plus jeunes ou régénérés, montrent une vitesse d'assimilation croissante.

Pour toutes les fractions, l'intensité rapportée à l'unité de chlorophylle croît d'abord après la levée de carence, pour diminuer lorsque la chlorophyllogenèse a permis aux feuilles d'obtenir la même teneur en pigments que les témoins non carencés (Tableau VI).

Tableau VI.

Intensité de la photosynthèse.

| Fractions chloroplastiques (densités) | Carencés | Après levée de carence 7 j | 14 j | 21 j |
|---|---|---|---|---|
| | 10⁶ impulsions par min et mg chlorophylle | | | |
| 1·45 | | 12·5 | 5 | 7·8 |
| 1·48–1·50 | 2·3 | 18·4 | 7·7 | 9·3 |
| 1·52 | 15·5 | 43 | 23·6 | 18·6 |
| | $10^6$ impulsions par min et mg N protéique | | | |
| 1·45 | | 15·8 | 7·6 | 9·7 |
| 1·48–1·50 | 0·9 | 13 | 8·5 | 11·8 |
| 1·52 | 7·7 | 21·5 | 8 | 5·3 |
| Rapport de l'intensité de la photosynthèse à la teneur des chloroplastes en amidon | | | | |
| | (a) $\frac{10^6 \text{ impulsions par min et mg chlorophylle}}{10^3 \text{ mg amidon par g de chaque fraction}}$ | | | |
| 1·45 | | | 20 | 60 |
| 1·48–1·50 | 3 | 27 | 50 | 70 |
| 1·52 | 80 | 300 | 3900 | 6200 |
| | (b) $\frac{10^6 \text{ impulsions par min et mg d'N protéique}}{10^3 \text{ mg amidon par g de chaque fraction}}$ | | | |
| 1·45 | | | 33 | 81 |
| 1·48–1·50 | 1 | 19 | 57 | 86 |
| 1·52 | 40 | 140 | 1300 | 1800 |

Pendant la période de croissance de l'intensité de la photosynthèse par unité de chlorophylle, il est intéressant d'observer que l'intensité est d'autant plus élevée que les chloroplastes sont moins riches en amidon.

La même remarque peut être faite à propos de l'intensité exprimée par unité d'N protéique.

L'efficacité des pigments et celle des enzymes chloroplastiques paraissent alors d'autant plus grandes que les chloroplastes s'appauvrissent en amidon.

Il est très probable que l'action inhibitrice de l'amidon est liée à la gêne topographique qu'il exerce au milieu des lamelles, des grana et de la matrice.

Les contacts nécessaires entre les systèmes pigmentaires responsables des réactions photochimiques, les enzymes du cycle du carbone, les transporteurs d'énergie et les substances intermédiaires de l'assimilation de $CO_2$ peuvent être troublés par les grains d'amidon, de telle sorte que pour une même quantité de chlorophylle ou d'enzyme, l'assimilation est plus lente.

Ce trouble s'ajoute à la déficience générale en pigments et en protéines pour affaiblir la capacité photosynthétique des feuilles carencées.

Les études suivantes le précisent à propos de la réaction de Hill et de la photophosphorylation cyclique.

## LA RÉACTION DE HILL

L'intensité de la réaction de Hill, mesurée avec des chloroplastes isolés en *milieu aqueux*, en présence de 2,6-dichlorophénolindophénol, est environ 4 fois plus faible pour les organites carencés que pour les chloroplastes de feuilles normalement alimentées en azote.

Après l'apport de nitrates, l'intensité de la réaction augmente plus rapidement que ne le fait la vitesse de synthèse de la chlorophylle (Fig. 10).

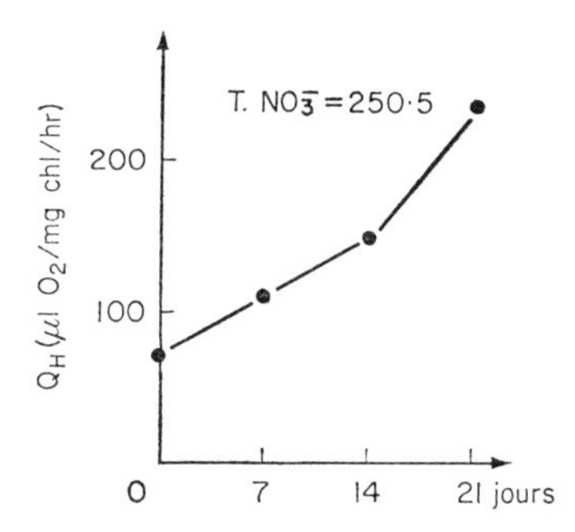


FIG. 10. Activité Hill de chloroplastes isolés en milieu aqueux à partir de feuilles de *Bryophyllum daigremontianum* carencé en N et après levée de carence (7,14 et 21 jours). Réactif: 2,6-dichlorophénolindophénol. La mesure est faite par mesure de la densité optique à 600 m$\mu$ et les résultats sont convertis en $O_2$ émis par mg de chlorophylle et par hr. T. $NO_3^-$ = Chloroplastes témoins isolés de feuilles non carencées.

On peut penser, en raison de l'enrichissement moyen des chloroplastes en N protéique, qu'ils acquièrent rapidement de nouveaux complexes protéines-pigments très actifs, accroissant l'efficacité de la chlorophylle.

## LA PHOTOPHOSPHORYLATION

Les chloroplastes des plantes carencées ont une photophosphorylation cyclique, en présence de phénazine méthosulfate ou de vitamine $K_3$, ordinairement 4 à 20 fois plus faible que celle des chloroplastes des plantes normalement alimentées en azote.

Comme pour la réaction de Hill, l'apport de nitrates aux plantes carencées est suivi d'une élévation de la vitesse de photophosphorylation, par rapport à l'unité de chlorophylle. Dans les feuilles n° 2, de même âge initial, la vitesse atteint après 10 ou 14 jours une valeur voisine de celle des chloroplastes de plantes qui ont toujours reçu des nitrates (Fig. 11). Elle diminue ensuite lors de l'apparition d'une nouvelle feuille. La même observation est faite si l'on rapporte la photophosphorylation à l'unité d'N protéique.

L'efficacité des systèmes phosphorylants dépend étroitement de la genèse des cofacteurs physiologiques ou des enzymes nouvellement formés.

L'étude de la photophosphorylation, faite comparativement avec des chloroplastes extraits de feuilles préalablement mises à l'obscurité, apporte un renseignement complémentaire en relation avec leur teneur en glucides.

Ainsi que le montrent, pour différentes feuilles, les résultats indiqués dans le Tableau VII, l'appauvrissement en amidon qui intervient à l'obscurité permet ensuite une augmentation de la vitesse de photophosphorylation.

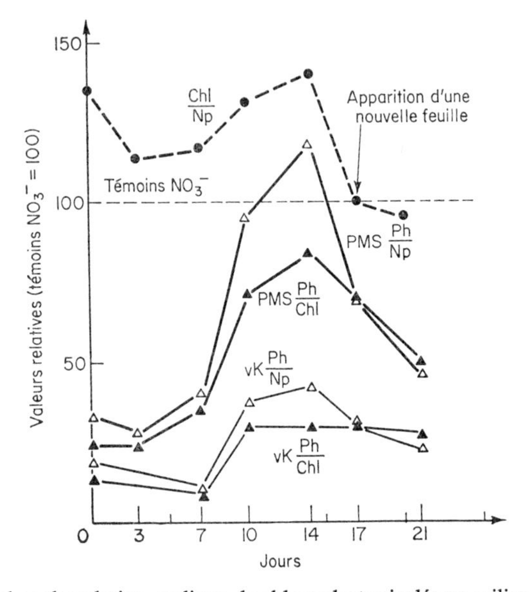


Fig. 11. Photophosphorylation cyclique de chloroplastes isolés en milieu aqueux à partir de feuilles de *Bryophyllum daigremontianum*. Les mesures sont faites par incorporation de $^{32}P$ en présence de phénazine méthosulfate (PMS) ou de vitamine $K_3$. Les résultats sont exprimés soit par rapport à la chlorophylle (Ph/Chl), soit par rapport à l'N protéique (Ph/Np) et par comparaison dans chaque cas avec des témoins extraits de feuilles non carencées (Témoins $NO_3^-$ pris arbitrairement égaux à 100). Chl/Np = rapports chlorophylle/N protéique (en masses) par comparaison avec les témoins.

Inversement, une durée excédentaire d'illumination entraîne une diminution de l'activité. Ces résultats s'accordent avec l'effet dépressif du saccharose à l'égard de la fixation du $CO_2$ dans les feuilles de canne à sucre (Hartt, 1963).

La faiblesse des activités des chloroplastes carencés en N est liée à plusieurs causes parmi lesquelles on peut distinguer:

— la faible teneur en chlorophylle,

— la faible teneur en protéines, qu'il s'agisse des protéines membranaires des lamelles et des grains ou des enzymes de la matrice,

— la désorganisation partielle des lamelles ou des grana,

— l'accumulation de l'amidon. Les résultats décrits ci-dessus précisent la suggestion faite il y a plus de 15 ans par Loomis (1949) au sujet du caractère limitant de l'accumulation des glucides vis-à-vis de la photosynthèse.

Tableau VII. Influence de l'illumination ou de l'obscurité antérieures sur la vitesse de la photophosphorylation (incorporation de $^{32}P$, par unité de chlorophylle, exprimée par rapport aux témoins correspondants)

| | | N° de la feuille | | |
|---|---|---|---|---|
| | | 1 | 2 | 3 |
| Témoins (matin) | | 100 | 100 | 100 |
| Pl. non carencées | Après 72 hr d'obscurité | 168 | 168 | 245 |
| Pl. carencées | Après 72 hr d'obscurité | | 150 | |
| | Après 72 hr d'illumination | | 25 | |

Après la levée de carence, la restauration des capacités photosynthétiques se fait par la genèse de nouveaux organites et vraisemblablement aussi par le rajeunissement de chloroplastes antérieurement formés qui perdent leur amidon et s'enrichissent en lamelles et en grana. L'efficacité de leurs appareils photosynthétiques croît à mesure qu'ils s'appauvrissent en amidon.

## Conclusion

Le ribulose 1,5-diphosphate, par sa carboxylation, constitue l'un des participants majeurs à l'incorporation du $CO_2$. La transformation des hexoses et des trioses en pentoses (*reductive pentose phosphate cycle*) assure aux chloroplastes la possibilité d'entretenir le mécanisme de fixation.

L'acide 3-phosphoglycérique, formé par la carboxylation, est pour la plus grande partie réduit en 3-phosphoglycéraldéhyde. Cette étape, la plus importante dans l'énergétique des transformations des substances intermédiaires est assurée à la fois par la 3-phosphoglycérate kinase et la 3-phosphoglycérate réductase fonctionnant avec le NADH comme donneur d'electrons et par la 3-phosphoglycéraldéhyde déshydrogénase fonctionnant avec le NADPH. Les chloroplastes isolés la réalisent avec difficulté en raison de la dispersion des enzymes et des substrats, consécutive aux lésions qu'ils subissent lors du broyage des cellules.

Le cycle des pentoses draine dans la synthèse glucidique des radicaux glycolyles appartenant déjà aux cétoses ou bien d'origine non connue, mais toujours proche du $CO_2$ assimilé.

Il en résulte une complication dans la distribution du carbone fixé, dont ne peut rendre compte une simple représentation schématique.

De plus, la $\beta$-carboxylation de l'acide phosphoénolpyruvique provenant d'une fraction non réduite de l'acide 3-phosphoglycérique, engage rapidement du $CO_2$ dans la synthèse des acides malique et aspartique. Ces réactions, et beaucoup d'autres d'ailleurs appartenant au métabolisme intermédiaire, assurent une diversification rapide des premiers produits de la fixation et de la réduction de $CO_2$.

Quelques éléments de la régulation du mécanisme de réduction sont connus: la concentration en ATP, en pyridine nucléotides réduits, l'action freinatrice du sédoheptulose 7-phosphate sur la triose phosphate déshydrogénase par exemple.

Il est fréquent, néanmoins, que la majeure partie du carbone assimilé s'accumule d'abord dans les chloroplastes sous forme de polysaccharides, produits stables dépositaires de l'énergie d'origine photochimique stockée.

Les principaux agents connus du transfert des groupes glucosyles dans leur synthèse sont l'UDPG pour le saccharose et l'ADPG pour l'amidon.

L'accumulation de l'amidon dans les chloroplastes déprime leur activité photosynthétique. La dépression est également sensible pour la réaction de Hill avec le 2-6-dichlorophénolindophénol comme oxydant et pour la photophosphorylation réalisée en présence de vitamine $K_3$ ou de phénazine méthosulfate. L'effet dépresseur est particulièrement net dans les chloroplastes de plantes carencées en azote qui accumulent de l'amidon.

Parmi les facteurs internes de la régulation de la vitesse de la photosynthèse, interviennent à la fois les complexes pigmentaires des lamelles ou des grana, les enzymes de la matrice et l'accumulation des glucides dans les chloroplastes.

## Références


Akazawa, T., Minamikawa, T. et Murata, T. (1964). *Pl. Physiol.* **39**, 371.

Allen, M. B., Arnon, D. I., Capindale, F. B., Whatley, F. R. et Durham, L. J. (1955). *J. Am. chem. Soc.* **77**, 4149.

Arnon, D. I. (1960). *In* "Encyclopedia of Plant Physiology" (W. Ruhland, éd.) Vol. 5, Part 1 p. 773. Springer Verlag, Berlin, Göttingen, Heidelberg.

Ashwell, G. (1964). *Annu. Rev. Biochem.* **33**, 101.

Aubert, J. P., Milhaud, G. et Millet, J. (1957). *Ann. Inst. Pasteur* **92**, 515.

Axelrod, B., Bandurski, R. S., Greiner, C. M. et Jang, R. (1953). *J. biol. Chem.* **202**, 619.

Badenhuizen, N. P. (1963). *Nature, Lond.* **197**, 464.

Bamberger, E. S., Black, C. C., Fewson, C. A. et Gibbs, M. (1963). *Pl. Physiol.* **38**, 483.

Bassham, J. A. (1963). *In* "Photosynthetic Mechanisms of Green Plants", p. 635. Publ. 1145, Natl. Acad. Sci., Natl. Res. Council, Washington, D.C.

Bassham, J. A. (1964). *Annu. Rev. Pl. Physiol.* **15**, 101.

Bassham J. A., Benson, A. A., Kay, L. D., Harris, A. Z., Wilson, A. T. et Calvin, M. (1954). *J. Am. chem. Soc.* **76**, 1760.

Bassham, J. A. et Calvin, M. (1960). *In* "Encyclopedia of Plant Physiology" (W. Ruhland, éd.) Vol. 5, Part 1, p. 884. Springer Verlag, Berlin, Göttingen, Heidelberg.

Bassham, J. A. et Calvin, M. (1961). *Proc. Vth Int. Congr. Biochem. Moscou*, Vol. VI, p. 285. Pergamon Press, Oxford, Londres, New York et Paris; PWN-Polish Sc. Publ., Varsovie (1963).

Bassham, J. A. et Kirk, M. (1960). *Biochim. biophys. Acta* **43**, 447.

Bean, R. C., Porter, G. G. et Barr, B. K. (1963). *Pl. Physiol.* **38**, 280.

Benson, A. A. (1961). *Proc. Vth. Int. Congr. Biochem. Moscou*, Vol. VI, p. 340. Pergamon Press, Oxford, Londres, New York et Paris; PWN-Polish Sc. Publ., Varsovie (1963).

Bergmann, F. H., Towne, J. C. et Burris, R. H. (1958). *J. biol. Chem.* **230**, 13.

Bidwell, R. G. S., Craigie, J. S. et Krotkov, G. (1958). *Science N.Y.* **128**, 776.

Blondon-Radage, F., Lebas, M., Menez, L. et Moyse, A. (1963). *Bull. Soc. Fr. Physiol. vég.* **9**, 194.

Bourdu, R., Champigny, M.-L., Lefort, M., Rémy, R. et Moyse, A. (1965a). *Physiol. vég.* **3**, 305.

Bourdu, R., Champigny, M.-L., Lefort, M., Maslow, M. et Moyse, A. (1965b). *Physiol. vég.* **3**, 355.
Bradbeer, J. W. et Anderson, C. M. A. (1964). *Xth Int. Bot. Congr. Edinburgh Abstracts,* 155.
Bradbeer, J. W. et Racker, E. (1961). *Fed. Proc.* **20**, 88.
Burma, D. P. et Mortimer, D. C. (1956). *Archs Biochem. Biophys.* **62**, 16.
Burström, H. (1943). *Annu. roy. agric. Coll., Sweden* **11**, 1.
Cabib, E. (1963). *Annu. Rev. Biochem.* **32**, 321.
Calvin, M. (1955). *Proc. 3rd Int. Congr. Biochem., Brussels*, p. 211 (C. Liébecq, éd.). Academic Press, New York et Londres.
Champigny, M.-L. (1959). *Rev. Cytol. Biol. vég.* **21**, 1.
Champigny, M.-L. (1960). *Rev. gén. Bot.* **67**, 65.
Charlson, A. J. et Richtmyer, N. K. (1959). *J. Am. chem. Soc.* **81**, 1512.
Datta, A. G. et Racker, E. (1961a). *J. biol. Chem.* **236**, 617.
Datta, A. G. et Racker, E. (1961b). *J. biol. Chem.* **236**, 624.
De Fekete, M. A. R. et Cardini, C. E. (1964). *Archs Biochem. Biophys.* **104**, 173.
Doman, N. G., Shkolnik, R. J. et Terenteva, Z. A. (1964). *Doklady Akad. Nauk. SSSR* **156**, 698.
Duysens, L. N. M. et Amesz, J. (1962). *Biochim. biophys. Acta* **64**, 243.
Fewson, C. A., Al-Hafidh, M. et Gibbs, M. (1962). *Pl. Physiol.* **37**, 402.
Frydman, R. B. et Hassid, W. Z. (1963). *Nature, Lond.* **199**, 382.
Fuller, R. C., Smillie, R. M., Sisler, E. C. et Kornberg, H. L. (1961). *J. biol. Chem.* **236**, 2140.
Garnier-Dardart, J. (1965). *Physiol. vég.* **3**, 215.
Gee, R., Joshi, G., Bils, R. F. et Saltman, P. (1965). *Pl. Physiol.* **40**, 89.
Gibbs, M. (1963). *In* "Photosynthetic Mechanisms of Green Plants", p. 663. Publ. 1145, Natl. Acad. Sci., Natl. Res. Council, Washington, D.C.
Gibbs, M. et Calo, N. (1959). *Pl. Physiol.* **34**, 318.
Gibbs, M. et Kandler, O. (1957). *Proc. natl. Acad. Sci. (U.S.A.)* **43**, 446.
Giraud, G. (1963). *Physiol. vég.* **1**, 203.
Goldemberg, S. H. et Marechal, L. R. (1963). *Biochim. biophys. Acta* **71**, 743.
Hartt, C. E. (1963). *Naturwissenschaften* **50**, 666.
Hatch, M. D., Sacher, J. A. et Glasziou, K. T. (1963). *Pl. Physiol.* **38**, 338.
Havir, E. A. et Gibbs, M. (1963). *J. biol. Chem.* **238**, 3183.
Heber, U., Pon, N. G. et Heber, M. (1963). *Pl. Physiol.* **38**, 355.
Heber, U. et Tyszkiewicz, E. (1962). *J. exp. Bot.* **13**, 185.
Heber, U. et Willenbrinck, J. (1964). *Biochim. biophys. Acta* **82**, 313.
Horecker, B. L., Cheng, T. et Pontremoli, S. (1963). *J. biol. Chem.* **238**, 3428.
Horecker, B. L. et Smyrniotis, P. Z. (1953a). *J. Am. chem. Soc.* **75**, 1009.
Horecker, B. L. et Smyrniotis, P. Z. (1953b). *J. Am. chem. Soc.* **75**, 2021.
Horecker, B. L., Smyrniotis, P. Z. et Hurwitz, J. (1956). *J. biol. Chem.* **223**, 1009.
Horecker, B. L., Smyrniotis, P. Z. et Klenow, H. (1953). *J. biol. Chem.* **205**, 661.
Hurlbert, R. E. et Lascelles, J. (1963). *J. Gen. Microbiol.* **33**, 445.
Jacobi, G. (1957). *Kieler Meeresforsch.* **13**, 212.
Jimenez, E., Baldwin, R. L., Tolbert, N. E. et Wood, W. A. (1962). *Archs Biochem. Biophys.* **98**, 172.
Jolchine, G. (1959). *Bull. Soc. Chim. biol.* **41**, 227.
Kahn, J. S. (1964). *Biochim. biophys. Acta* **79**, 421.
Kandler, O. et Gibbs, M. (1956). *Pl. Physiol.* **31**, 411.
Kauss, H. et Kandler, O. (1962). *Z. Naturf.* **17**b, 858.
Kearney, P. C. et Tolbert, N. E. (1961). *Pl. Physiol.* **36**, Suppl. XXVI.
Keister, D. L., San Pietro, A. et Stolzenbach, F. E. (1960). *J. biol. Chem.* **235**, 2989.
Kocourek, J., Ticha, M. et Kostir, J. (1964). *Archs Biochem. Biophys.* **108**, 349.
Kornfeld, S., Kornfeld, R., Neufeld, E. F. et O'Brien, P. J. (1964). *Proc. natn. Acad. Sci., U.S.A.* **52**, 371.
Kortschak, H. P., Hartt, C. E. et Burr, G. O. (1965). *Pl. Physiol.* **40**, 209.
Krebs, H. A. (1965). *Abstr. Comm. Second Meeting Fed. Europ. Biochem. Soc.* p. 351. Vienne, F 1.

Lefrançois, M. et Ouellet, C. (1958). *Can. J. Bot.* **36**, 457.
Leloir, L. F. et Cardini, C. E. (1955). *J. biol. Chem.* **214**, 157.
Leloir, L. F. et Cardini, C. E. (1962). *In* "The Enzymes" (Boyer, P. D., Lardy, H. et Myrback, K., éds.) Vol. 6, p. 317. Academic Press, New York et Londres.
Ljungdahl, L., Wood, H. G., Racker, E. et Couri, D. (1961). *J. biol. Chem.* **236**, 1622.
Loomis, W. E. (1949). *In* "Photosynthesis in Plants" (Franck, J. et Loomis, W. E., éds.) p. 1. Iowa State College Press, Ames, Iowa.
Losada, M., Trebst, A. V. et Arnon, D. I. (1960). *J. biol. Chem.* **235**, 832.
Margulies, M. M. (1965). *Pl. Physiol.* **40**, 57.
Maslow, M. (1964). *Physiol. vég.* **2**, 209.
Miller, R. M., Meyer, C. M. et Tanner, H. A. (1963). *Pl. Physiol.* **38**, 184.
Milner, Y. et Avigad, G. (1965). *Nature, Lond.* **206**, 825.
Moses, V. et Calvin, M. (1958). *Archs Biochem. Biophys.* **78**, 598.
Moyse, A. (1961). *Prov. Vth Int. Congr. Biochem. Moscou*, Vol. VI, p. 310. Pergamon Press, Oxford, Londres, New York et Paris; PWN-Polish Sc. Publ. Varsovie (1963).
Moyse, A. et Jolchine, G. (1957). *Bull. Soc. Chim. biol.* **39**, 725.
Murata, T. et Akazawa, T. (1964). *Biochem. Biophys. Res. Commun.* **16**, 6.
Murata, T., Sugiyama, T. et Akazawa, T. (1964). *Archs Biochem. Biophys.* **107**, 92.
Neufeld, E. F. (1963). *Fed. Proc.* **22**, 464.
Nordin, J. H. et Kirkwood, S. (1965). *Annu. Rev. Pl. Physiol.* **16**, 393.
Northcote, D. H. (1964). *Annu. Rev. Biochem.* **33**, 51.
Ogren, W. L. et Krogmann, D. W. (1963). *In* "Photosynthetic Mechanisms of Green Plants", p. 684. Publ. 1145, Natl. Acad. Sci., Natl. Res. Council, Washington, D.C.
Park, R. B., Pon, N., Bassham, J. A. et Calvin, M. (1959). *Proc. IXth Int. Congr. Bot., Montréal*, Vol. II, Abstracts, p. 293.
Peat, S., Turvey, J. R. et Jones, G. (1959). *J. chem. Soc.* Part II, p. 1540.
Peterkofsky, A. et Racker, E. (1961). *Pl. Physiol.* **36**, 409.
Pon, N. G. (1962). "La Photosynthèse", Colloque Intern. C.N.R.S. n°. 119, p. 597. C.N.R.S. éd., Paris (1963).
Pontremoli, S., Bonsignori, A., Grazi, E. et Horecker, B. L. (1960). *J. biol. Chem.* **235**, 1881.
Porter, H. K. (1962). *Annu. Rev. Pl. Physiol.* **13**, 303.
Pritchard, G. G., Griffin, W. J. et Whittingham, C. P. (1962). *J. exp. Bot.* **13**, 176.
Quillet, M. (1965). *C.r. hebd. Séanc. Acad. Sci., Paris* **260**, 6192.
Rabin, B. R. et Trown, P. W. (1964). *Nature, Lond.* **202**, 1290.
Rabinowitch, E. I. (1956). *In* "Photosynthesis and Related Processes" Vol. II, Part 2, p. 1262. Interscience Publ. Inc., New York.
Racker, E. (1957). *Archs Biochem. Biophys.* **69**, 300.
Racker, E., De La Haba, G. et Leder, I. G. (1953). *J. Am. chem. Soc.* **75**, 1010.
Racker, E., De La Haba, G. et Leder, I. G. (1954). *Archs Biochem. Biophys.* **48**, 238.
Racker, E. et Schroeder, E. A. R. (1958). *Archs Biochem. Biophys.* **74**, 326.
Rémy, R., et Champigny, M.-L., (1966) *Physiol. vég.* **4**, (sous presse).
Richardson, K. E. et Tolbert, N. E. (1961). *J. biol. Chem.* **236**, 1285.
Richter, G. (1959). *Naturwissenschaften* **46**, 604.
Richter, G. (1961). *Biochem. biophys. Acta* **48**, 606.
Rocher, J. P. et Bourdu, R. (1964). *Bull. Soc. Fr. Phys. vég.* **10**, 154.
Sachs, J. (1868). *In* "Physiologie végétale", trad. fr. (Masson et fils, éd.) p. 29. Paris, Genève.
San Pietro, A. et Lang, H. M. (1958). *J. biol. Chem.* **231**, 211.
Schou, L., Benson, A. A., Bassham, J. A. et Calvin, M. (1950). *Physiol. Plantarum* **3**, 487.
Schroeder, E. A. R. et Racker, E. (1959). *Féd. Proc.* **18**, 318.
Simon, von H., Dorrer, H.-D. et Trebst, A. (1964). *Z. Naturf.* **19**b, 734.
Smillie, R. M. et Fuller, R. C. (1959). *Pl. Physiol.* **34**, 651.
Smillie, R. M., Rigopoulos, N. et Kelly, H. (1962). *Biochim. biophys. Acta* **56**, 612.
Steiner, D. F. (1964). *Nature, Lond.*, **204**, 1171.
Stiller, M. (1962). *Annu. Rev. Pl. Physiol.* **13**, 151.
Stocking, C. R. (1959). *Pl. Physiol.* **34**, 56.

Tabachnick, M., Srere, P. A., Cooper, J. et Racker, E. (1958). *Archs Biochem. Biophys.* **74**, 315.
Tagawa, K., Tsujimoto, H. Y. et Arnon, D. I. (1963). *Nature, Lond.* **199**, 1247.
Thomas, J. B. (1960). *In* "Encyclopedia of Plant Physiology" (W. Ruhland, éd.) Vol. V, Part 1, p. 511. Springer Verlag, Berlin, Göttingen, Heidelberg.
Tolbert, N. E. (1963). *In* "Photosynthetic Mechanisms of Green Plants", p. 648. Publ. 1145, Natl. Acad. Sci., Natl. Res. Council, Washington, D.C.
Tolbert, N. E. et Zill, L. P. (1956). *J. biol. Chem.* **222**, 895.
Trebst, A. et Fiedler, F. (1961). *Z. Naturf.* **16**b, 284.
Trebst, A. et Fiedler, F. (1962). *Z. Naturf.* **17**b, 553.
Trebst, A., Losada, M. et Arnon, D. I. (1960). *J. biol. Chem.* **235**, 840.
Trip, P., Nelson, C. D. et Krotkov, G. (1964). *Archs Biochem. Biophys.* **108**, 359.
Trudinger, P. A. (1956). *Biochem. J.* **64**, 274.
Van Baalen, C. (1965). *Nature, Lond.* **206**, 193.
Walker, D. A. (1964), *Biochem. J.* **92**, 22c.
Warburg, O. et Krippahl, G. (1960). *Z. Naturf.* **15**b, 197.
Whatley, F. R., Allen, M. B., Rosenberg, L. L., Capindale, J. B. et Arnon, D. I. (1956). *Biochim. biophys. Acta* **20**, 462.
Whittingham, C. P., Bermingham, M. et Hiller, R. G. (1963a). *Z. Naturf.* **18**b, 701.
Whittingham, C. P., Hiller, R. G. et Bermingham, M. (1963b). *In* "Photosynthetic Mechanisms of Green Plants", p. 675. Publ. 1145, Natl. Acad. Sci., Natl. Res. Council, Washington, D.C.
Whittingham, C. P., Bermingham, M., Hiller, R. G. et Pritchard, G. G. (1962). *In* "La Photosynthèse", Colloque Intern. C.N.R.S. n° 119, p. 571. C.N.R.S. éd., Paris (1963).
Whittingham, C. P., Coombs, J. et Miflin, B. (1964). *Xth Int. Bot. Congr. Edinburgh, Abstracts*, p. 154.
Willard, J. M., Schulman, M. et Gibbs, M. (1965). *Nature, Lond.* **206**, 195.
Wilson, A. T. et Calvin, M. (1955). *J. Am. chem. Soc.* **77**, 5948.
Wintermans, J. F. G. M. (1962). *In* "La Photosynthèse", Colloque Intern. C.N.R.S. n° 119, p. 381. C.N.R.S. éd., Paris (1963).
Wood, H. G. et Utter, M. F. (1965). *In* Essays in Biochemistry (P. N. Campbell et G. D. Greville, éd.) Vol. I, 1, Academic Press, New York et Londres.
Zelitch, I. (1965). *J. biol. Chem.* **240**, 1869.

# Regulation of the Biosynthesis of Starch in Spinach Leaf Chloroplasts

Jack Preiss[1], Hara Prasad Ghosh and Judith Wittkop
*Department of Biochemistry and Biophysics, University of California, Davis, California, U.S.A.*

## Properties of Plant Starch Synthetases

The biosynthesis of starch was first shown by Leloir's group (DeFekete *et al.*, 1960; Leloir *et al.*, 1961) to occur by the transfer of the glucosyl portion of uridine diphospho-D-glucose (UDP-glucose) to an existing starch granule primer (reaction 1).

$$\text{UDP-glucose} + \text{starch primer} \rightarrow \text{UDP} + \text{glucosyl-primer} \quad (1)$$

The starch granules were obtained from dwarf beans; however starch granules from young potatoes and sweet corn also contained the transferase activity. Subsequently, Recondo and Leloir (1961) reported that adenosine diphospho-D-glucose (ADP-glucose) was a better glucosyl donor than UDP-glucose (reaction 2).

$$\text{ADP-glucose} + \text{starch primer} \rightarrow \text{ADP} + \text{glucosyl-primer} \quad (2)$$

Since then starch synthetase systems consisting of starch granules from many sources have been found. Frydman (1963) reported the presence of starch synthetase in starch granules of potato tubers, potato sprouts, from seeds of wrinkled peas, red and white corn, and waxy maize. In each case the rate of transfer of glucose from ADP-glucose was four to ten times higher than from UDP-glucose. The potato enzyme was found to be inactive toward CDP-D-glucose, ADP-D-galactose and UDP-D-galactose. Similar results were obtained by Akazawa's group with starch granules from rice grains (Akazawa *et al.*, 1964; Murata *et al.*, 1964). ADP-glucose was five times more active than UDP-glucose in the rate of transfer of its glucosyl residue to the rice starch granules. In all these systems the starch synthetases were intimately associated with the starch granule. Therefore it was not possible to show a primer requirement. However, if oligosaccharides of the maltodextrin series were added to the reaction mixtures containing ADP-[$^{14}$C]glucose (or UDP-[$^{14}$C]glucose) and the starch synthetase, glucose transfer to these oligosaccharides could be observed.

In order to facilitate the study of the starch synthetase it was desirable to obtain it in soluble form and separated from the starch granules. A soluble $\alpha$-1,4-glucan synthetase from plant material was first reported by Frydman and Cardini (1964a). This enzyme, which was isolated from sweet corn endosperm, required primer for activity and was present in the $100{,}000 \times g$ supernatant

[1] Supported in part by a USPHS Grant No. AI 05520.

fluid of the kernel extract. Amylopectin, phytoglycogen, animal glycogen and malto-oligosaccharides were glucosyl acceptors for this enzyme (Frydman and Cardini, 1965) while amylose, starch and starch granules were completely ineffective as primers. Recently four laboratories have reported the presence of soluble ADP-glucose α-1,4-glucan transferases in a variety of plant tissues. Frydman and Cardini (1964b) showed that the α-1,4-glucan synthetase of tobacco leaves chloroplasts could utilize heated starch granules, amylopectin and glycogen as acceptors of glucose from ADP-glucose. The enzyme could not

Table I. Properties of various plant α-1,4-glucan synthetases

| Source | Glucose donor | Primer specificity | Reference |
|---|---|---|---|
| I. Starch granules | | | |
| Dwarf bean | | | 6 |
| Wrinkled peas | ADP-glucose | | 3 |
| Red corn | | | 3 |
| White corn | or | | 3 |
| Waxy maize | UDP-glucose | | 3 |
| Potato tubers | | | 3 |
| Rice grains | | | 1,8 |
| II. Soluble glucan synthetases | | | |
| Waxy maize | ADP-glucose | Amylopectin, glycogen | 5 |
| Glutinous rice grains | ADP-glucose, UDP-glucose | Starch, amylose, glycogen, amylopectin | 9 |
| Potato tubers | ADP-glucose, UDP-glucose | Only starch granules | 4 |
| III. Leaf glucan synthetases | | | |
| Tobacco leaf chloroplasts | ADP-glucose | Heated starch granules, amylopectin, glycogen | 4 |
| Spinach leaf chloroplasts | ADP-glucose | | 2,10 |
| Butter lettuce chloroplasts | ADP-glucose | | 11 |
| Soy bean leaf granules | ADP-glucose | | 7 |

[1] Akazawa *et al.* (1964).
[2] Doi *et al.* (1964).
[3] Frydman (1963).
[4] Frydman and Cardini (1964b).
[5] Frydman and Cardini (1965).
[6] Leloir *et al.* (1961).
[7] Murata and Akazawa (1964).
[8] Murata *et al.* (1964b).
[9] Murata *et al.* (1965).
[10] Ghosh and Preiss (1965b).
[11] Present paper.

utilize UDP-glucose as a glucosyl-donor. The glucan synthetase from potato tubers however could only use whole starch granules as an acceptor of glucose from ADP-glucose. Murata *et al.* (1965) showed that the starch synthetase of extracts of glutinous rice grains was present in the $100{,}000 \times g$ supernatant fluid. Doi *et al.* (1964) and Ghosh and Preiss (1965a,b) both reported the presence of a soluble starch synthetase in extracts of spinach chloroplasts.

The properties of the various α-1,4-glucan synthetases are summarized in Table I. It is seen that there is some variation in the properties of the α-1,4-glucan synthetases. Whereas ADP-glucose and UDP-glucose are active with the glucan synthetases of seeds and tubers, the leaf starch synthetases are totally specific for ADP-glucose. The various glucan synthetases also differ with respect to their primer requirements. The leaf enzymes appear capable of using a number of α-1,4-glucans; heated starch granules, amylose, amylopectin and glycogen. The soluble potato tuber enzyme can only use whole starch granules as a primer, Table II shows some of the properties of the butter lettuce chloroplast enzyme. ADP-glucose but not UDP-glucose or glucose 1-phosphate was active as a glucosyl donor. Transfer of glucose was dependent on enzyme and amylose. Hydrolysis of the radioactive alcohol-insoluble product by β-amylase caused the formation of a radioactive compound that behaved similarly to maltose in three paper chromatographic solvent systems and in paper electrophoresis in borate buffer (Ghosh and Preiss, 1965b). This indicated the formation of $\alpha(1\rightarrow4)$ glucosidic linkages during the enzymic transfer. It is pertinent to note that the plant α-1,4-glucan synthetases are not stimulated by glucose 6-phosphate or any other glycolytic intermediate (Frydman, 1963; Murata *et al.*, 1965; Ghosh and Preiss, 1965b). This is in contrast to what has been reported for the mammalian UDP-glucose; glycogen α-4-glucosyl transferase (Rosell-Perez and Larner, 1964).

Table II. Properties of butter lettuce chloroplast starch synthetase

| Sugar nucleotide | Additions or omissions | [14C]glucose incorporated (mμmoles) |
|---|---|---|
| ADP-[14C]glucose | None | 4·2 |
| UDP-[14C]glucose | None | <0·02 |
| [14C]glucose 1-phosphate | None | <0·02 |
| ADP-[14C]glucose | − Enzyme | <0·02 |
| ADP-[14C]glucose | − Amylose | <0·02 |
| ADP-[14C]glucose | + α-Amylase, 20 μg | 0·04 |

Reaction mixture contained 30 mμmoles of [14C]glucose compound (listed above), 10 μmoles of glycine-NaOH buffer, pH 8·5, 5 μmoles of KCl, 2 μmoles of GSH, 1 μmole EDTA, 250 μg of soluble corn amylose and enzyme in a volume of 0·2 ml. Glucose incorporation into starch was measured as previously described (Ghosh and Preiss, 1965b). The butter lettuce enzyme was prepared exactly as the spinach leaf chloroplast enzyme (Ghosh and Preiss, 1965b).

## Synthesis and Occurrence of Adenosine Diphosphate Glucose and Uridine Diphosphate Glucose in Plants

Two other reactions are therefore pertinent to starch synthesis. They are catalysed by ADP glucose pyrophosphorylase and UDP glucose pyrophosphorylase (reactions 3 and 4).

$$\text{ATP} + \alpha\text{-D-glucose 1-P} \rightleftharpoons \text{ADP-D-glucose} + \text{PP}_i \qquad (3)$$

$$\text{UTP} + \alpha\text{-D-glucose 1-P} \rightleftharpoons \text{UDP-D-glucose} + \text{PP}_i \qquad (4)$$

Espada (1962) reported the presence of ADP-glucose pyrophosphorylase in wheat flour extracts while Ghosh and Preiss (1965a) have reported the presence of ADP-glucose pyrophosphorylase in spinach leaf chloroplasts. UDP-glucose pyrophosphorylase which was first found in yeast extracts by Munch-Peterson *et al.* (1953) has been shown to be present in a number of plant tissue extracts (Neufeld *et al.*, 1957; Ginsburg, 1958).

Both UDP-glucose and ADP-glucose have been found to occur naturally in plants. ADP-glucose has been identified in *Chlorella pyrenoidosa* (Kauss and Kandler, 1962) and isolated from sweet corn (Recondo *et al.*, 1963) and rice grains (Murata *et al.*, 1964a). UDP-glucose is usually found to be three to ten times higher in concentration than ADP-glucose. For this reason Leloir (1964) has postulated that both UDP-glucose and ADP-glucose play an equal role in starch synthesis even though the reaction rate of the starch synthetases with ADP-glucose as the glucosyl donor is three to ten times faster than with UDP-glucose. While this statement may be true for the starch synthetases present in the starch granules of many seedlings and fruits, it does not apply to the starch synthetases present in the leaf chloroplasts. In these systems the only natural glucose donor appears to be ADP-glucose. Therefore UDP-glucose does not appear to play a role in leaf starch synthesis.

## Regulation of Starch Synthesis

Starch resulting from photosynthetic carbon assimilation accumulates in the chloroplasts of leaf tissues of higher plants. This starch appears to have a higher turnover compared to starch stored in structures such as tubers, roots, grains and leaf sheaths. The starch found in the chloroplasts is known as assimilation starch, being readily consumed as metabolic fuel as compared to reservoir starch found in other parts of plants. We may then ask ourselves how is starch synthesis in the chloroplast regulated? As we have stated before there appears to be no stimulation of starch synthetase by glucose 6-phosphate or by any other glycolytic intermediate tested. Thus it appears that regulation of starch synthesis is not accomplished by controlling the activity of starch synthetase. This would then be in contrast to what has been so elegantly shown in the case of mammalian glycogen synthesis (Craig and Larner, 1964; Rosell-Perez and Larner, 1964). We then proceeded to study leaf chloroplast ADP-glucose pyrophosphorylase to see if it was affected by any glycolytic intermediates. It had already been shown in this laboratory that bacterial ADP-glucose pyrophosphorylase activities were activated by glycolytic intermediates (Shen and Preiss, 1964; Preiss *et al.*, 1965). In bacteria, ADP-glucose is the glucosyl donor for bacterial glycogen synthesis (Greenberg and Preiss, 1964) and therefore analogous to the role that ADP-glucose plays in starch synthesis in plants. Leaf chloroplast ADP-glucose pyrophosphorylase activity was indeed stimulated by a number of glycolytic intermediates (Ghosh and Preiss, 1965a). The largest stimulation was observed with 3-phosphoglyceric acid. It was thus thought that this might be of some significance in the control of starch formation during photosynthesis. In chloroplast extracts the level of ADP-glucose pyrophosphorylase is lower than the ADP-glucose: starch transglucosylase activity. Thus formation of ADP-glucose probably is the limiting reaction in the synthesis of starch in chloroplasts. During $CO_2$ fixation accumulation of 3-phosphoglycerate would cause an increase of synthesis of ADP-glucose by stimulation of ADP-glucose pyrophosphorylase. Increasing ADP-glucose concentrations in the chloroplast would then increase the rate of synthesis of starch.

In order to study this phenomenon further we proceeded to purify the ADP-glucose pyrophosphorylase. Two assays were used to measure the enzyme (Ghosh and Preiss, 1965a). The first method measured the formation of [$^{32}$P]-ATP from ADP-glucose and $^{32}PP_i$. The labelled ATP was isolated by adsorption onto Norit. The second method measured the synthesis of ADP-

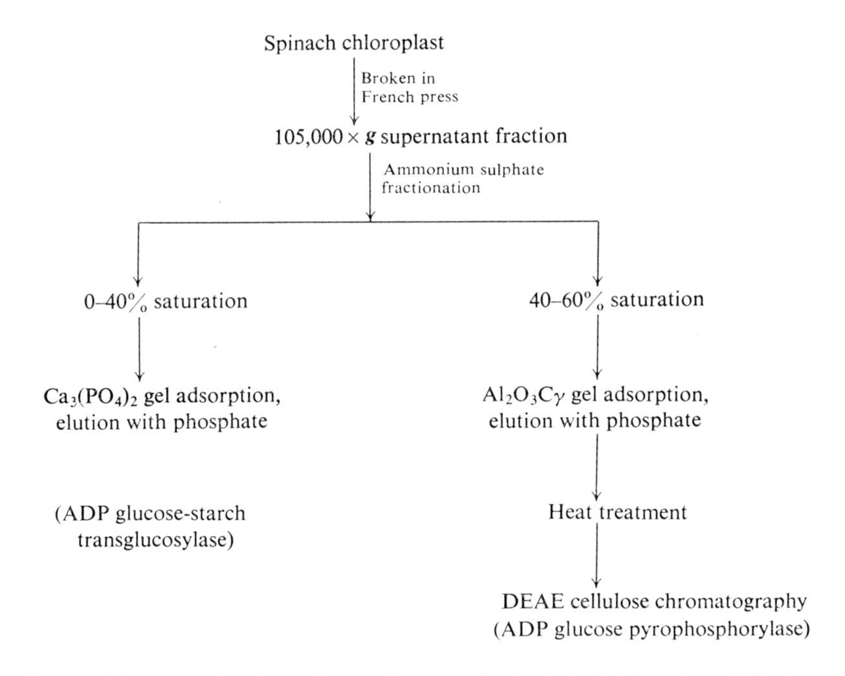


FIG. 1. Diagrammatic scheme for purification of spinach leaf chloroplast ADP-glucose pyrophosphorylase and ADP-glucose: starch transglucosylase.

[$^{14}$C]glucose from ATP plus [$^{14}$C]glucose 1-phosphate. The reaction was terminated by heat denaturation. The remaining [$^{14}$C]glucose 1-phosphate was hydrolyzed to free [$^{14}$C]glucose and inorganic phosphate by *E. coli* alkaline phosphatase. A 100 $\mu$l portion of the phosphatase-treated reaction mixture was then placed on a disc (diameter, 2·5 cm) of DEAE-cellulose paper. The discs were washed three times with 150 ml volumes of de-ionized water to remove the [$^{14}$C]glucose. The discs were then dried and counted by liquid scintillation technique as previously described (Shen and Preiss, 1964).

Figure 1 is a diagrammatic scheme showing the purification of both ADP-glucose pyrophosphorylase and ADP-glucose: starch transglucosylase. The pyrophosphorylase was purified about sixty-fold. Table III shows the specific activity of ADP-glucose pyrophosphorylase both in synthesis and pyrophosphorolysis. Various contaminating enzymes were also assayed for and their activities are also listed in Table III. A small amount of inorganic pyrophosphatase activity was observed and this activity was inhibited 100% by 0·01 M NaF.

Table III. Various enzymatic activities in the DEAE-cellulose fraction

| Enzyme | Activity ($\mu$moles product formed $min^{-1}$ $mg^{-1}$) |
|---|---|
| ADP-glucose pyrophosphorylase | 3·6 |
| ADP-glucose synthetase | 3·0 |
| ADP-glucose phosphorylase | < 0·003 |
| UDP-glucose pyrophosphorylase | < 0·003 |
| TDP-glucose pyrophosphorylase | < 0·003 |
| ADP-glucose pyrophosphatase | < 0·003 |
| Phosphoglucomutase | < 0·004 |
| Adenosine triphosphatase | < 0·007 |
| Inorganic pyrophosphatase | 0·10 |
| Phosphohexose isomerase | < 0·004 |
| Enolase | < 0·018 |
| Phosphoglyceromutase | < 0·018 |
| Adenylate kinase | < 0·012 |
| Aldolase | < 0·004 |
| Hexokinase | < 0·004 |

## Activation of Spinach Leaf ADP-Glucose Pyrophosphorylase

### Activation by 3-Phosphoglycerate

Figures 2 and 3 show the effect of enzyme concentration on the rates of ADP-glucose pyrophosphorolysis and synthesis. They also show the stimulation of these reactions by 3-phosphoglycerate. 3-Phosphoglycerate stimulates pyrophosphorolysis three to four-fold and synthesis eight to ten-fold. Figure 3 also shows the effect of yeast inorganic pyrophosphatase on the synthesis of ADP-glucose. The addition of inorganic pyrophosphatase to reaction mixtures containing no 3-phosphoglycerate resulted in a decrease of the reaction rate. Pyrophosphatase action on the reaction product, inorganic pyrophosphate, would result in the formation of inorganic phosphate. Phosphate is a potent inhibitor of the pyrophosphorylase ($K_i = 2 \times 10^{-5}$ M; see Fig. 11, p. 147). Linearity of reaction rate to protein concentration therefore would not be expected since increasing extents of reaction would increase the amount of inhibitor in the reaction. In reaction mixtures containing phosphoglycerate inhibition by phosphate is not a problem since phosphoglycerate overcomes the phosphate inhibition (see Figs. 11 and 12, p. 147–8). The addition of inorganic

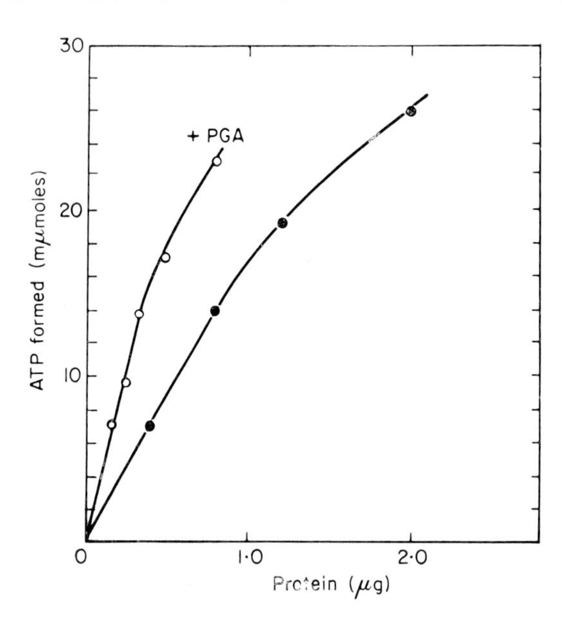


FIG. 2. Effect of protein concentration on the pyrophosphorolysis of ADP-glucose. Reaction mixtures which had no 3-phosphoglycerate (3PGA) contained 20 μmoles of glycylglycine buffer, pH 7·5, 0·5 μmole of ADP-glucose, 0·3 μmole of inorganic [$^{32}$P]pyrophosphate (specific activity, 2·0 to $2·5 \times 10^5$ cpm/μmole), 0·5 μmole of $MgCl_2$, 100 μg of bovine plasma albumin, 2·5 μmoles of KF and enzyme in a total volume of 0·25 ml. Reaction mixtures with 1·0 mM 3PGA contained 0·1 μmole of ADP-glucose, 0·5 μmole of inorganic [$^{32}$P]pyrophosphate, 1·5 μmoles of $MgCl_2$, 20 μmoles of glycylglycine buffer, pH 7·5, 2·5 μmoles of KF, 100 μg of bovine plasma albumin, and enzyme in a total volume of 0·25 ml. [$^{32}$P]ATP was assayed as described previously (Ghosh and Preiss, 1965a). Incubation time was 10 min at 37°.

Table IV. The concentration of 3-phosphoglycerate in reaction mixtures synthesizing ADP-glucose

| Time (min) | ADP-glucose formed (μmoles) | 3-Phosphoglyceric acid present (μmoles) |
|---|---|---|
| 0 | 0·0 | 0·159 |
| 10 | 0·028 | 0·165 |
| 30 | 0·048 | 0·167 |

The reaction mixtures and conditions of the experiments are described in Fig. 3. Inorganic pyrophosphatase was included in the reaction mixtures. 3-Phosphoglyceric acid was analysed by coupling it to NAD oxidation with phosphoglyceromutase, enolase, pyruvate kinase, and lactate dehydrogenase.

pyrophosphatase to reaction mixtures containing 3-phosphoglycerate causes an increase in the range of linearity of reaction rate to enzyme concentration. In the absence of the pyrophosphatase the rate of reaction decreases with higher enzyme concentration since the reaction is approaching equilibrium. During the formation of ADP-glucose the concentration of 3-phosphoglycerate remains constant (Table IV). Thus 3-phosphoglycerate is not consumed during the reaction and appears to be stimulating the ADP-glucose pyrophosphorylase activity. In another experiment [$^{14}$C]-3-phosphoglycerate was

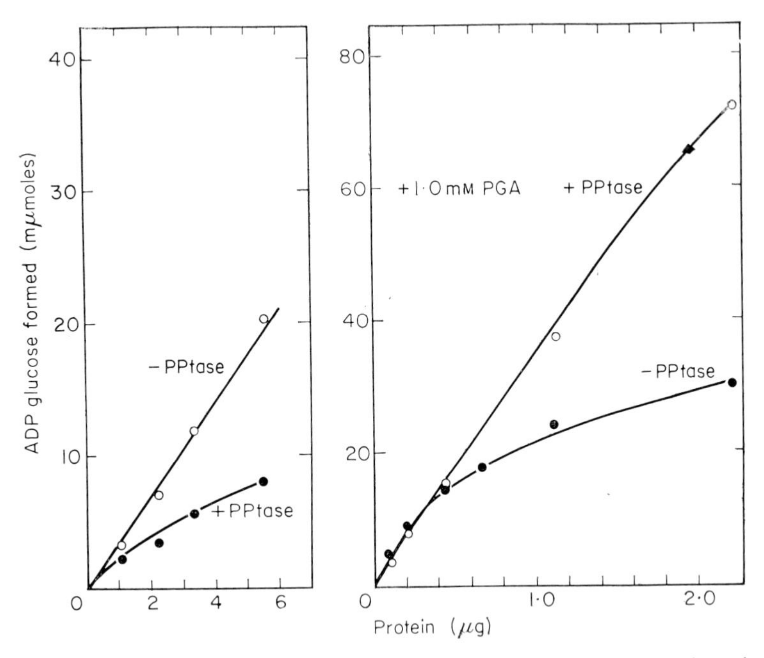


FIG. 3. Effect of protein concentration on the synthesis of ADP-glucose. Reaction mixtures contained 20 μmoles of glycylglycine buffer, pH 7·5, 0·2 μmole of ATP, 0·1 μmole of [$^{14}$C]-α-glucose 1-phosphate, 1·0 μmole of $MgCl_2$, 100 μg of bovine plasma albumin and enzyme in a total volume of 0·2 ml. Incubation time was 10 min at 37°. 3-Phosphoglycerate (3PGA) in the amount of 0·2 μmole and yeast inorganic pyrophosphatase in the amount of 0·8 μg were also added where indicated. ADP-glucose formation was measured according to Shen and Preiss (1964).

added to the reaction mixture. No incorporation of [$^{14}$C]-3-phosphoglycerate into ADP-glucose was observed in the presence or absence of α-glucose 1-phosphate. Figure 4 shows the effect of 3-phosphoglycerate concentration on the rate of ADP-glucose formation. Using the reciprocal plots method according to Lineweaver and Burk (1934) one obtains an apparent $K_m$ of activation of $2{\cdot}0 \times 10^{-5}$ M for 3-phosphoglycerate.

ADP-glucose pyrophosphorylase may then be classified as one of the group of proteins considered to be regulatory enzymes. The catalytic activities of these enzymes are affected by molecules other than their substrates. These effector molecules can either increase or decrease the activity of these enzymes. These molecules are usually related to the enzymes that they affect. For instance, CTP, an end product of pyrimidine metabolism, regulates the production of carbamyl-aspartate by inhibiting aspartate transcarbamylase (Gerhart and Pardee, 1964; Yates and Pardee, 1956). This phenomenon is known as feedback inhibition and appears to be an important physiological mechanism for regulating the synthesis of compounds by a biosynthetic pathway (Umbarger, 1961). Similarly activation of enzymes by metabolites has also been observed. Activation of the mammalian glycogen synthetase by glucose 6-phosphate

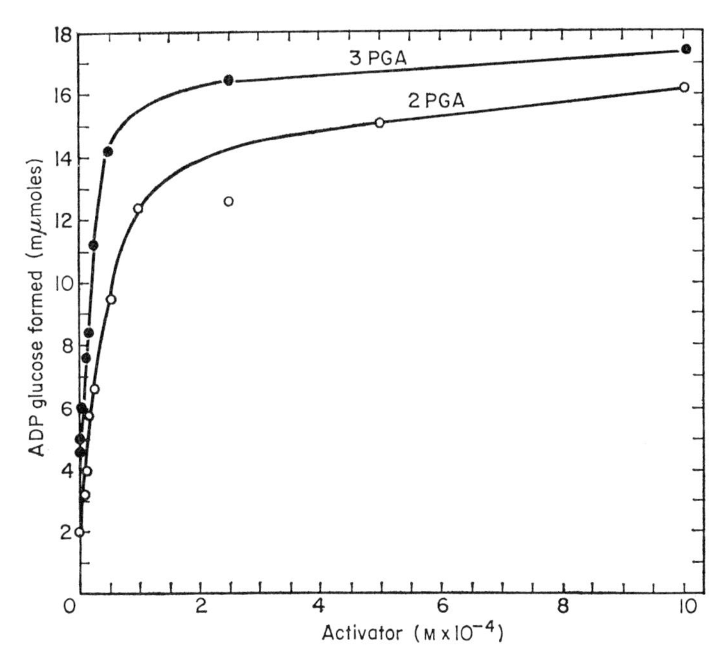


FIG. 4. Stimulation of ADP-glucose synthesis by various metabolites. The reaction mixtures and assay are described in Fig. 3. Yeast inorganic pyrophosphatase was included in the reaction mixtures.

(Leloir *et al.*, 1959; Kornfield and Brown, 1962; Robbins *et al.*, 1959; Rosell-Perez and Larner, 1964) and activation of the NAD-specific isocitrate dehydrogenase by citrate (Atkinson *et al.*, 1965; Sanwal and Stachow, 1965) are examples of this phenomenon. Activation is presumably also important in the physiological regulation of metabolic pathways. The 3-phosphoglycerate activation may be then considered as a stimulation of enzymic activity since: (1) it does not appear to participate in the reaction catalysed by the enzyme and (2) analogous activations by various metabolites of other enzymes have been observed. The formation of the primary carbon reduction product of photosynthesis would then cause stimulation of starch synthesis by increasing the formation of ADP-glucose.

## SPECIFICITY OF ACTIVATION

Figure 3 shows that at higher concentrations 2-phosphoglycerate is as active as 3-phosphoglycerate in stimulating ADP-glucose formation. Its $K_m$ of activation is $5{\cdot}0 \times 10^{-5}$ M. Table V shows the specificity of the activation with

Table V. Activation of spinach leaf ADP-glucose pyrophosphorylase by various metabolites

| Activator | ADPG formed ($\mu$moles mg$^{-1}$ hr$^{-1}$) | Activation fold |
|---|---|---|
| None | 22·6 | |
| 3-Phosphoglycerate | 210 | 9·3 |
| 2-Phosphoglycerate | 187 | 8·3 |
| Acetyl-CoA | 102 | 4·5 |
| 2,3-Diphosphoglycerate | 93 | 4·1 |
| Fructose 6-phosphate | 81 | 3·6 |
| Phosphoenolpyruvate | 79 | 3·5 |
| Deoxyribose 5-phosphate | 72 | 3·2 |
| Ribose 5-phosphate | 63 | 2·8 |
| α-Glycerol phosphate | 54 | 2·4 |
| Phosphohydroxypyruvate | 52 | 2·3 |
| Fructose diphosphate | 50 | 2·2 |

Conditions of the experiment were the same as described in Fig. 3. The effector molecules were added at a concentration of 1·5 mM.

Table VI. The effect of adding two effector molecules to the same reaction mixture

| Activators | Concentration (mM) | ADP-glucose formed (m$\mu$moles) |
|---|---|---|
| None | | 2·0 |
| 3-Phosphoglycerate | 1·0 | 20·2 |
| 3-Phosphoglycerate | 0·0075 | 5·33 |
| 2-Phosphoglycerate | 0·05 | 8·7 |
| 3-Phosphoglycerate + 2-phosphoglycerate | 0·0075 0·05 | 11·2 |
| Fructose 6-phosphate | 0·5 | 3·7 |
| Fructose 6-phosphate + 3-phosphoglycerate | 0·5 0·0075 | 7·7 |
| Fructose diphosphate | 0·02 | 3·7 |
| Fructose diphosphate + 3-phosphoglycerate | 0·02 0·0075 | 7·4 |

The conditions of the experiment were the same as those described in Fig. 3.

respect to other compounds. Compounds that gave stimulations of between 1·1 to 2-fold at a concentration of 1·5 mM were sedoheptulose 7-phosphate,[1] galactose 6-phosphate, glucose 6-phosphate, glycerate, glycollate, phosphoglycollate,[2] and glycolaldehyde phosphate. No stimulation was observed with the following compounds up to a concentration of 1·5 mM; β-glycerophosphate, lactate, pyruvate, dihydroxyacetone phosphate, 3-phosphoglyceraldehyde, bicarbonate, malate and succinate. No stimulation was observed with ribulose 1,5-diphosphate. However, the preparation of ribulose 1,5-diphosphate was contaminated with inorganic phosphate, which, as stated before, is a potent inhibitor of ADP-glucose pyrophosphorylase.

It is not known whether all these activators occupy the same site (sites) on the enzyme. However, Table VI shows that these other effector molecules can enhance the activation caused by unsaturating concentrations of 3-phosphoglycerate.

## Properties of the Pyrophosphorylase

### pH optimum

Figures 5 and 6 show the pH optimum of the enzyme in the presence and absence of 3-phosphoglycerate in various buffers. The activating effect of

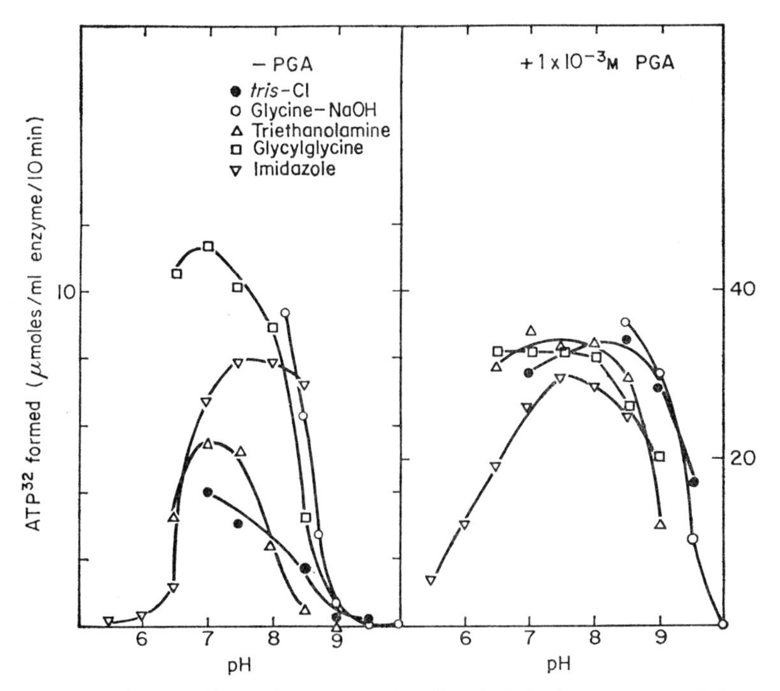


FIG. 5. pH optimum of ADP-glucose pyrophosphorolysis in the presence and absence of 3-phosphoglycerate. The conditions of the experiment are described in Fig. 2.

[1] A sample of sedoheptulose 7-phosphate was kindly provided by Dr. B. L. Horecker.

[2] A generous gift of phosphoglycollate was provided by Dr. I. Zelitch.

3-phosphoglycerate was observed in all the different buffers used. The enzyme activity was approximately the same when different buffers were used in the reaction mixtures containing 3-phosphoglycerate. However, in the reaction mixtures containing no activator, the enzyme exhibited 1·4 to 2·5 times more activity with glycylglycine than with the other buffers. It is interesting to note that in glycylglycine buffer the synthesis of ADP-glucose is activated eighty-fold at pH 8·5, twenty-three-fold at pH 8·0, eleven-fold at pH 7·5, and nine-fold at pH 7·0.

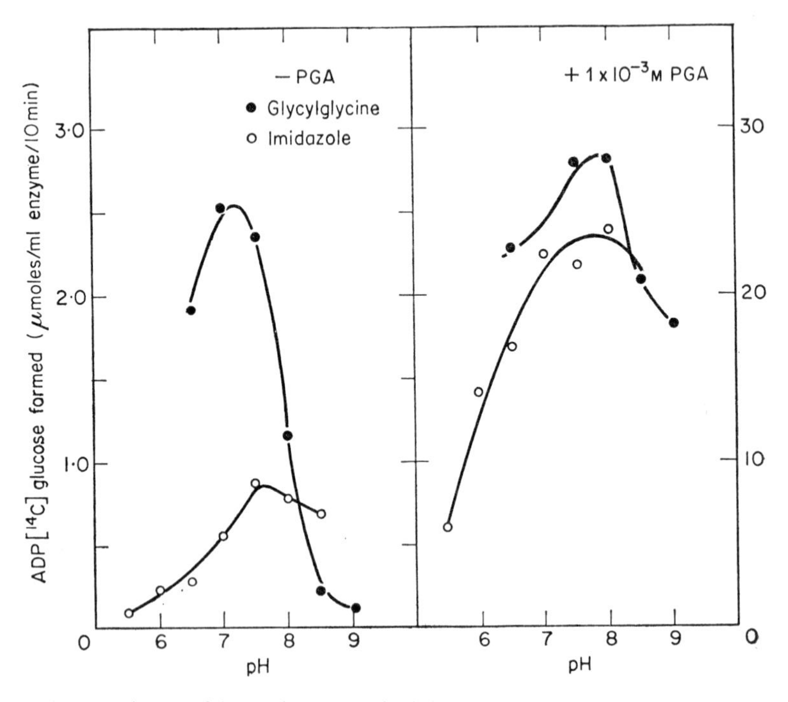


FIG. 6. pH optimum of ADP-glucose synthesis in the presence and absence of 3-phosphoglycerate. Conditions are described in Fig. 3.

## REQUIREMENTS

Table VII shows that in the presence or absence of activator the requirements for pyrophosphorolysis of ADP-glucose are $Mg^{2+}$ and enzyme. $^{32}P_i$ cannot replace pyrophosphate; UDP-glucose, and TDP-glucose cannot replace ADP-glucose. The crude extract however does contain both a UDP-glucose and a TDP-glucose pyrophosphorylase, which are not activated by 3-phosphoglycerate. Similarly synthesis of ADP-glucose from ATP and glucose 1-phosphate required $Mg^{2+}$ and enzyme (Table VIII). The enzyme cannot utilize uridine triphosphate (UTP) or thymidine triphosphate(TTP) for sugar nucleotide synthesis.

Table VII. Requirements of pyrophosphorolysis of ADP-glucose

| Reaction conditions | ATP formed (mμmoles) A | B |
|---|---|---|
| Complete | 13·4 | 19·8 |
| –Enzyme | 0·08 | 0·08 |
| –ADP-glucose | 0·18 | 0·18 |
| –NaF | 16·0 | 20·4 |
| –Bovine plasma albumin | 10·8 | 13·2 |
| –$MgCl_2$ | 0·26 | 0·18 |
| –3-phosphoglycerate | — | 4·8 |
| –ADP-glucose + UDP-glucose or TDP-glucose | 0·15 | 0·15 |

In A the complete system contained 20 μmoles of glycylglycine buffer, pH 7·5, 0·5 μmole of $MgCl_2$, 50 μg bovine plasma albumin, 2·5 μmoles NaF, 0·5 μmole of $^{32}PP$, 0·5 μmole of ADP-glucose and the purified enzyme in a total volume of 0·25 ml. In the case of B, the complete system contained 1·5 μmole of $MgCl_2$, 0·1 μmole of ADP-glucose, 0·25 μmole of 3-phosphoglycerate and the other components as present in assay A.

Table VIII. Requirements of synthesis of ADP-glucose

| Reaction conditions | ADP-[$^{14}C$]glucose formed (mμmoles) A | B |
|---|---|---|
| Complete | 9·5 | 29·3 |
| –ATP | <0·05 | 0·05 |
| –ATP + UTP or TTP | <0·05 | 0·05 |
| –$MgCl_2$ | <0·05 | 0·8 |
| –Bovine plasma albumin | 7·0 | 22·0 |
| –3-phosphoglycerate | — | 1·7 |

The complete system contained 20 μmoles of glycylglycine buffer, pH 7·5, 1 μmole of $MgCl_2$, 50 μg of bovine plasma albumin, 0·2 μmole of ATP, 0·1 μmole of [$^{14}C$]glucose 1-phosphate (specific activity $7{\cdot}5 \times 10^5$ counts/min/μmole), 0·8 μg of inorganic pyrophosphatase and the enzyme in a final volume of 0·2 ml. In the case of B, it also contained 0·2 μmole of 3-phosphoglycerate.

Table IX. Kinetic parameters of spinach leaf ADP-glucose pyrophosphorylase in glycylglycine buffer pH 7·5

| Substrate | $K_m$ (mM) | | $V_{max}$ (μmoles $mg^{-1}$ $min^{-1}$) | |
|---|---|---|---|---|
| | –PGA | +PGA | –PGA | +PGA |
| ADP-glucose | 0·93 | 0·15 | 1·9 | 5·0 |
| Pyrophosphate | 0·50 | 0·04 | 1·9 | 5·0 |
| ATP | 0·45 | 0·04 | 0·47 | 3·8 |
| Glucose 1-phosphate | 0·07 | 0·04 | 0·41 | 3·8 |
| $Mg^{2+}$ | 1·4 | 1·6 | — | — |

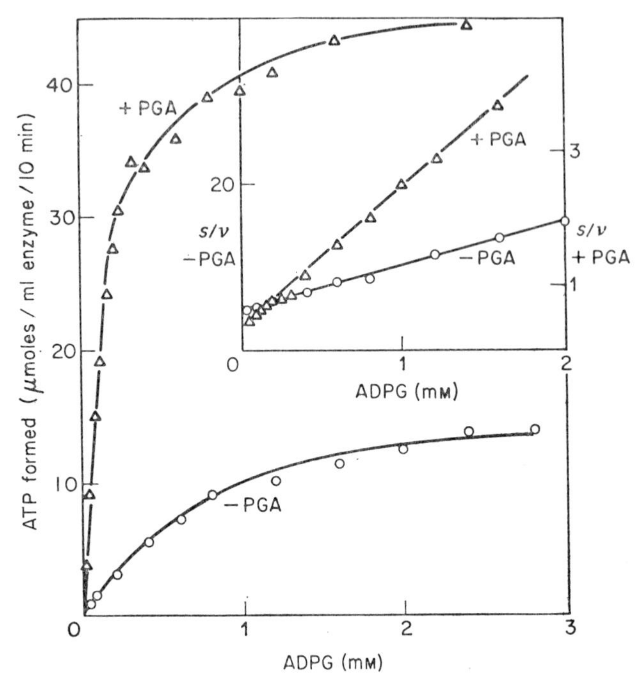


FIG. 7. Dependence of velocity of ADP-glucose pyrophosphorolysis on ADP-glucose concentration. The conditions of the experiment were the same as described in Fig. 2 except that the ADP-glucose concentration was varied.

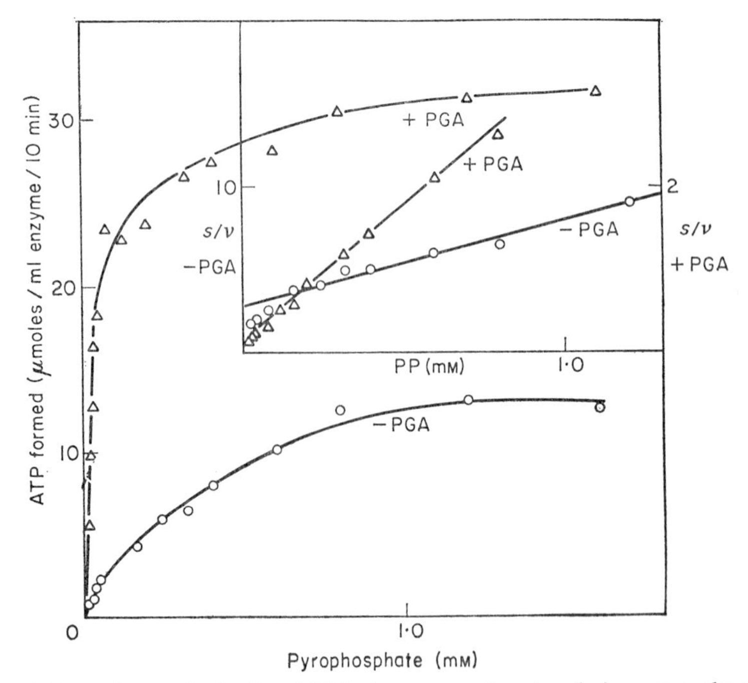


FIG. 8. Dependence of velocity of ADP-glucose pyrophosphorolysis on pyrophosphate concentration. The conditions of the experiment were the same as described in Fig. 2 except that the pyrophosphate concentration was varied.

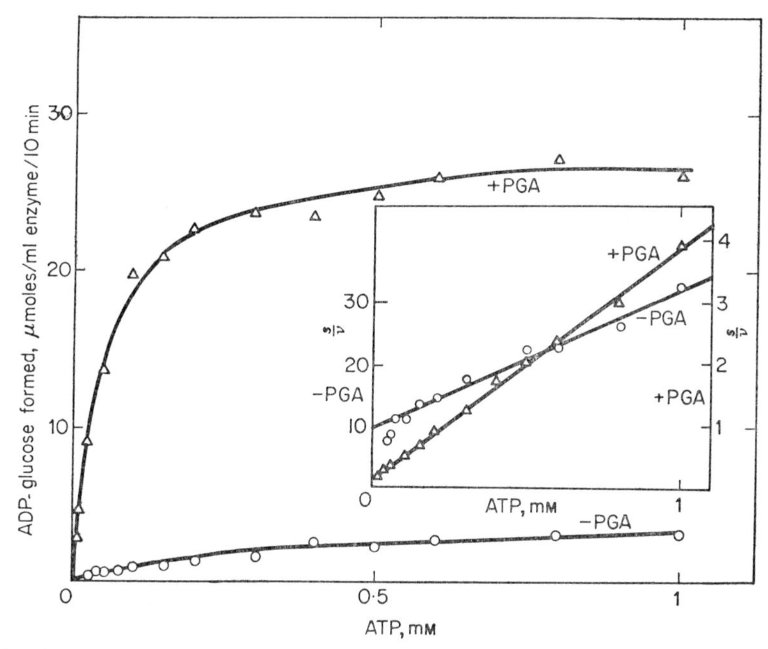


FIG. 9. Dependence of velocity of ADP-glucose synthesis on ATP concentration. The conditions of the experiment were the same as described in Fig. 3 except that the ATP concentration was varied.

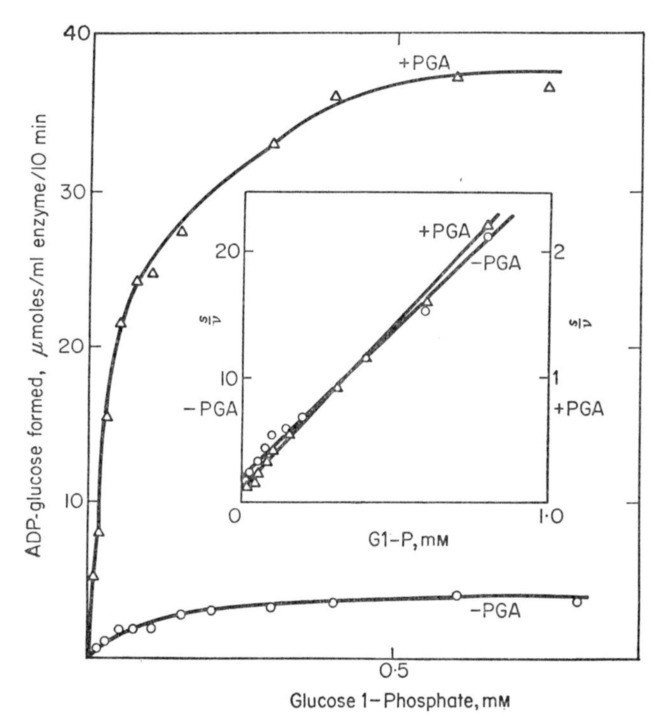


FIG. 10. Dependence of velocity of ADP-glucose synthesis on [14C]glucose 1-phosphate concentration. The conditions of the experiment were the same as described in Fig. 3 except that the glucose 1-phosphate concentration was varied.

### C. KINETICS OF THE ADP-GLUCOSE PYROPHOSPHORYLASE REACTION

Figures 7, 8, 9, and 10 show the effect of substrate concentrations on the kinetics of the reaction. In the presence of 3-phosphoglycerate, the $K_m$ values for ATP and pyrophosphate were decreased ten-fold (Table IX). The $K_m$ for ADP-glucose was decreased six-fold by the presence of 3-phosphoglycerate. The $V_{max}$ values of pyrophosphorolysis and synthesis of ADP-glucose were increased three-fold and nine-fold, respectively, by 3-phosphoglycerate. All substrates obeyed Michaelis-Menten kinetics in the presence or absence of 3-phosphoglycerate.

## Activation of ADP-Glucose Pyrophosphorylase in Other Plant Extracts

Table X shows the stimulation of the ADP-glucose pyrophosphorylases of various plant extracts by a number of glycolytic metabolites. Of all the metabolites tested either 3-phosphoglycerate or 2-phosphoglycerate was the best

Table X. Stimulation of various ADP-glucose pyrophosphorylases from plant tissues

| | Activator (mμmoles of ADP-glucose formed/min/mg protein) | | | | | |
|---|---|---|---|---|---|---|
| Extract | None | 3PGA | PEP | 2PGA | FDP | F6P |
| Wheat germ | 0·25 | 0·44 | 0·25 | 0·75 | 0·35 | — |
| Avocado mesocarp | 0·55 | 3·5 | 0·45 | 6·4 | 0·4 | 0·55 |
| Etiolated mung bean seedlings | 0·14 | 0·23 | 0·2 | 0·19 | 0·12 | 0·09 |
| Etiolated Alaska pea seedlings | 0·13 | 0·44 | 0·11 | 0·16 | 0·11 | 0·19 |
| Potato tubers | 0·34 | 1·76 | 1·1 | 0·5 | 0·41 | 0·57 |
| Carrot roots | 0·67 | 6·7 | 4·0 | 5·0 | 4·2 | 2·2 |
| Barley leaves | 3·1 | 50·0 | 3·4 | 46·0 | 10·8 | 5·7 |
| Butter lettuce | 0·15 | 13·1 | 0·83 | 2·5 | 1·1 | — |

The conditions of the experiment were described in Fig. 3 except that the buffer was glycine-NaOH, pH 8·5. The activators were added to the reaction mixture at a concentration of 1·5 mM. Mung bean and Alaska pea seedlings were germinated for 3 days in the dark. All extracts with the exception of the carrot roots and barley leaves were prepared by homogenizing the tissues in a Waring blendor with 0·05 M *tris*, pH 7·5, buffer containing 0·01 M glutathione. The carrot root and barley leaf extracts were prepared by grinding with acid-washed sand and extracting with the *tris*-glutathione buffer. Solid ammonium sulfate was then added to all the extracts up to 0·7 saturation. The precipitates, which were isolated by centrifugation, were dissolved in the above *tris*-glutathione buffer and dialysed against the same buffer. These dialysed fractions were used as the source for ADP-glucose pyrophosphorylase. The abbreviations used are as follows: 3PGA, 3-phosphoglycerate; 2PGA, 2-phosphoglycerate; PEP, phosphoenolpyruvate; FDP, fructose 1,6-diphosphate; F6P, fructose 6-phosphate.

activator. The activation always seemed to be greater with the enzymes isolated from the leaf extracts than from the other extracts under the conditions used. In each case the product of synthesis was shown to be ADP-glucose by its chromatographic behavior in three solvent systems (Ghosh and Preiss, 1965a). It thus appears that the ADP-glucose pyrophosphorylases of plant origin have the general property of being stimulated by phosphoglyceric acid.

## Inhibition of ADP-Glucose Pyrophosphorylase by Inorganic Phosphate

During the course of study it was found that phosphate was a potent inhibitor of the pyrophosphorylase. Figure 11 shows the effect of phosphate and sodium chloride concentration on the rate of ADP-glucose synthesis. The phosphate

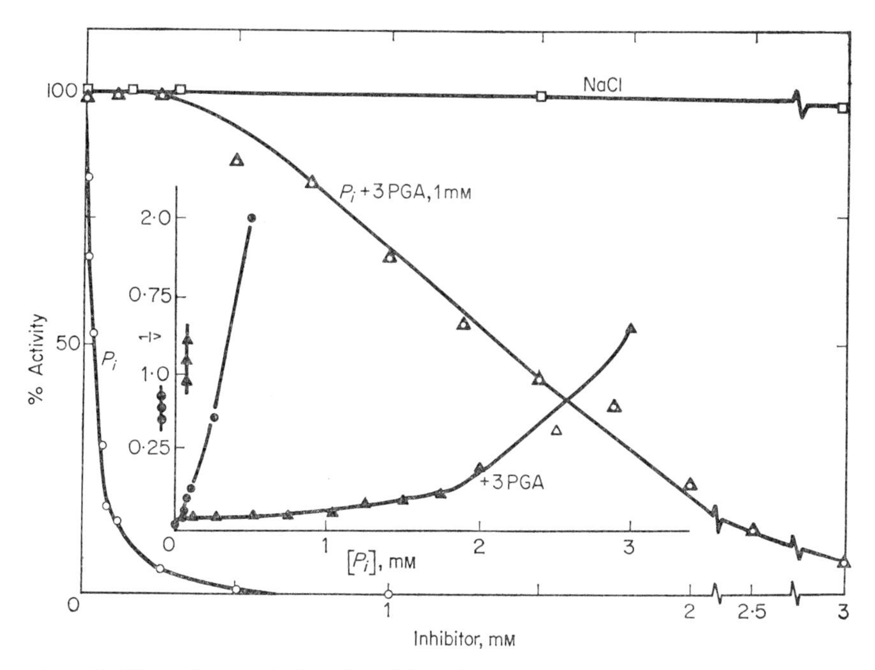


Fig. 11. Effect of inorganic phosphate ($P_i$) on the synthesis of ADP-glucose. The conditions of the experiment are described in Fig. 3. 3-Phosphoglycerate (3PGA) was added where indicated at a concentration of 1 mM. Velocity, $v$, denotes $\mu$moles of ADP-glucose formed in 10 min/ml of enzyme.

inhibition curve is sigmoid-shaped in nature while there is little inhibition with sodium chloride. When 3-phosphoglycerate is present in the reaction mixture (Fig. 11) higher phosphate concentrations are required to inhibit the enzymic activity. Fifty % inhibition occurs at $2{\cdot}2 \times 10^{-5}$ M phosphate; in the presence of

1 mM 3-phosphoglycerate, $1{\cdot}3 \times 10^{-3}$ M phosphate is required for 50% inhibition. The phosphate inhibition curves do not follow Michaelis-Menten kinetics (see insert, Fig. 11) and are sigmoid-shaped. This suggests that phosphate binds to more than one site on the enzyme and that these sites are interacting. The expression, $\log[v/(V_{max}-v)] = n \log$ (inhibitor) $- \log K$ has been used for the treatment of kinetic data whose reaction rate *vs* concentration curves were sigmoidal. $V_{max}$ is the reaction velocity of the enzyme in the absence of inhi-

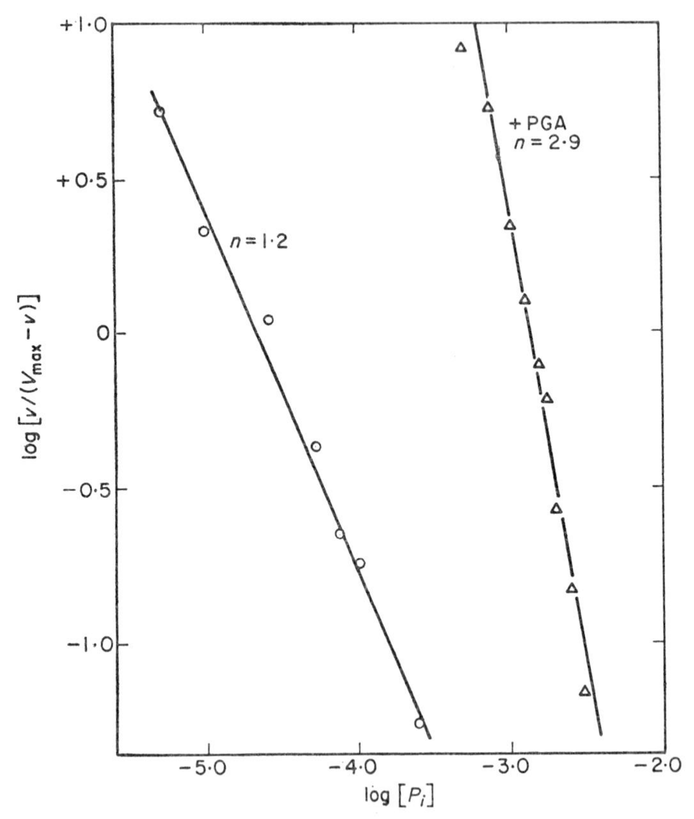


FIG. 12. Plot of log $[v/(V_{max}-v)]$ against log of inorganic phosphate ($P_i$) concentration. The conditions are the same as in Fig. 11.

bitor; $v$ is the velocity of reaction in presence of inhibitor; $K$ is the product of the $n$ dissociation constants of the $n$ binding sites. $n$ is known as an interaction coefficient and is dependent on the total number of binding sites as well as their strength of interaction. If the interactions are weak, then $n$ will be less than the number of binding sites and if there is no interaction $n$ will equal one. Changeux (1963) introduced this expression as the Hill equation. Atkinson *et al.* (1965) and Taketa and Pogell (1965) have derived similar expressions using Michaelis-Menten assumptions. Figure 12 shows that a plot of the data in Fig. 11 as

$\log[v/(V_{max}-v)]$ *vs* log of phosphate concentration gives interaction coefficients of 1·2 for phosphate in the absence of 3-phosphoglycerate and 2·9 in the presence of 3-phosphoglycerate. This can be interpreted as either 3-phosphoglycerate is causing the formation of new binding sites for phosphate on the enzyme or, more logically, 3-phosphoglycerate is causing greater interaction between the already existing phosphate sites. The effect of phosphate concentration on the velocity *vs* 3-phosphoglycerate concentration curve was next examined. Figure 13 shows that in the presence of phosphate the 3-phosphoglycerate curve is now sigmoidal. In the presence of phosphate the sites for

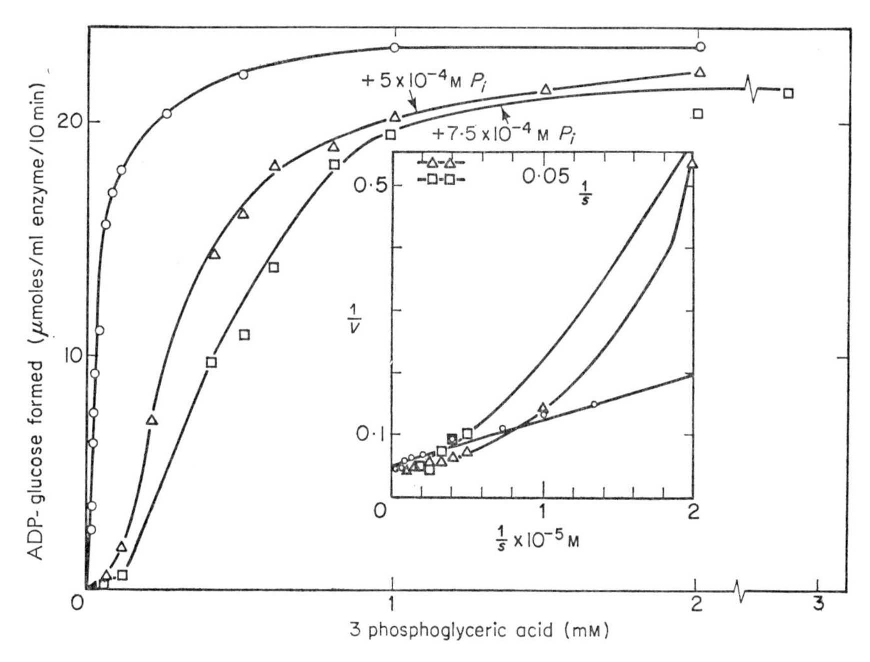


FIG. 13. Effect of 3-phosphoglycerate concentration on the phosphate ($P_i$) inhibition of ADP-glucose synthesis. The conditions are the same as in Fig. 3 except phosphate was added where indicated.

3-phosphoglycerate appear to interact while in the absence of phosphate, the 3-phosphoglycerate curve follows Michaelis-Menten kinetics. Plotting the data in Fig. 13 as $\log[v/(V_{max}-v)]$ *vs* log of 3-phosphoglycerate concentration (Fig. 14) one obtains an interaction coefficient of 1·0 for 3-phosphoglycerate in the absence of phosphate, 1·9 in the presence of $5{\cdot}0\times10^{-4}$ M phosphate, and 2·5 in the presence of $7{\cdot}5\times10^{-4}$ M phosphate. Thus phosphate causes an interaction of the binding sites of 3-phosphoglycerate, and conversely 3-phosphoglycerate causes a greater interaction of the phosphate sites. Whether the phosphate or 3-phosphoglycerate sites are equivalent is presently unknown. Recent results (H. P. Ghosh and J. Preiss, unpublished) indicate that inorganic phosphate inhibits in a non-competitive manner against the substrates of the enzyme.

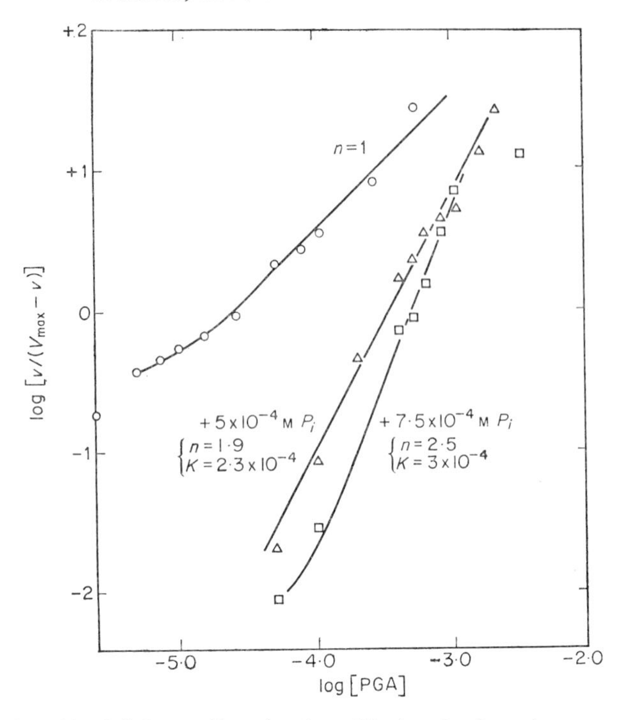


FIG. 14. Plot of $\log[v/(V_{max}-v)]$ against log of 3-phosphoglycerate concentration. The data were obtained from Fig. 13.

## DISCUSSION

It seems therefore that synthesis of ADP-glucose in spinach chloroplasts is controlled in part by two metabolites, 3-phosphoglycerate and inorganic phosphate. In the light, the level of inorganic phosphate is decreased owing to photophosphorylation and there is fixation of $CO_2$ to form 3-phosphoglycerate. Simultaneously there is also an increase of reducing power in the cell. The inorganic phosphate is converted into high energy phosphate, presumably ATP (which would be needed for starch synthesis). The newly formed 3-phosphoglycerate would be reduced and then converted into the hexose phosphates. Therefore a decrease of inorganic phosphate concentration and an increase of 3-phosphoglycerate concentration would lead to an increase in the rate of ADP-glucose synthesis and thereby direct the flow of carbon to the formation of starch.

In the dark the inorganic phosphate concentration would increase while 3-phosphoglycerate concentration would decrease. This would cause the inhibition of ADP-glucose synthesis and therefore starch synthesis. Conceivably, increased phosphate concentrations would also cause starch breakdown by the enzyme, starch phosphorylase. To support this regulatory mechanism information on the concentrations of the various metabolites in the chloroplast would be required. As yet no information of this kind is available. Fekete and Cardini

(1964) have postulated a control of ADP-glucose levels in the cell by the enzyme ADP-glucose phosphorylase (Dankert *et al.*, 1964). High levels of phosphate would cause a degradation of ADP-glucose by reaction 5.

$$\text{ADP-glucose} + P_i \rightarrow \text{ADP} + \text{glucose 1-phosphate} \quad (5)$$

This reaction has been found only in wheat germ extracts. Whether this enzyme is ubiquitous in plants is not known.

The spinach leaf enzyme appears to be similar to the bacterial ADP-glucose pyrophosphorylases (Shen and Preiss, 1964; Preiss *et al.*, 1965). All these enzymes are inhibited by phosphate and are activated by glycolytic intermediates. However, some differences are noted. The *E. coli* and *Arthrobacter* ADP-glucose pyrophosphorylases are inhibited by 5′-adenylate and adenosine diphosphate. The spinach enzyme is not inhibited by these nucleotides.

It should be mentioned that the statements on the nature of the concentration *vs* rate curves pertain only to the conditions used in the experiments. The sigmoid-shaped curves of aspartate concentration *vs* velocity of aspartate transcarbamylase activity are seen only at alkaline pH and disappear at neutral pH (Gerhart and Pardee, 1964). In the case of spinach leaf ADP-glucose pyrophosphorylase recent results have indicated that at pH 8·5 the 3-phosphoglycerate *vs* velocity curve is sigmoid-shaped and does not obey Michaelis-Menten kinetics.

The spinach leaf ADP-glucose pyrophosphorylase may then be included in the group of enzymes known as regulatory enzymes. Recent reports have postulated mechanisms consisting of subunit interaction to explain the properties of these proteins. These proteins were defined as allosteric proteins (Monod *et al.*, 1963; Changeux, 1964; Monod *et al.*, 1965) and the site of the effector molecule (inhibitor or activator) was defined as the allosteric site. At present possible subunit interactions in the spinach leaf ADP-glucose pyrophosphorylase have not been looked for. Further experiments are required to see if the activation or inhibition of ADP-glucose pyrophosphorylase is due to subunit interaction or to conformational changes as proposed by Koshland (1958, 1963, 1964). However, the simplest explanation for 3-phosphoglycerate activation and phosphate inhibition would be to assume that they are acting at sites other than the substrate sites. It would also be interesting to know if other enzymes of the photosynthetic carbon reduction cycle are regulated by 3-phosphoglycerate.

## References

Akazawa, T., Minamikawa, T. and Murata, T. (1964). *Pl. Physiol.* **39**, 371.
Atkinson, D. E., Hathaway, J. A. and Smith, E. C. (1965). *J. biol. Chem.* **240**, 2682.
Changeux, J. P. (1963). *Symposia on Quantitative Biology* **XXVIII**, 497.
Changeux, J. P. (1964). *In* "Brookhaven Symposia in Biology" **17**, 232.
Craig, J. W. and Larner, J. (1964). *Nature, Lond.* **202**, 971.
Dankert, M., Goncalves, R. J. and Recondo, E. (1964). *Biochim. biophys. Acta* **81**, 78.
Doi, A., Doi, K. and Nikuni, Z. (1964). *Biochim. biophys. Acta* **92**, 628.
Espada, J. (1962). *J. biol. Chem.* **237**, 3577.
Fekete, M. A. R. de, Leloir, L. F. and Cardini, C. E. (1960). *Nature, Lond.* **187**, 918.
Fekete, M. A. R. de and Cardini, C. E. (1964). *Archs Biochem. Biophys.* **104**, 173.
Frydman, R. B. (1963). *Archs Biochem. Biophys.* **102**, 242.
Frydman, R. B. and Cardini, C. E. (1964a). *Biochem. Biophys. Res. Commun.* **14**, 353.
Frydman, R. B. and Cardini, C. E. (1964b). *Biochem. Biophys. Res. Commun.* **17**, 407.
Frydman, R. B. and Cardini, C. E. (1965). *Biochim. biophys. Acta* **96**, 294.

Gerhart, J. C. and Pardee, A. B. (1964). *Fed. Proc.* **23**, 727.
Ghosh, H. P. and Preiss, J. (1965a). *J. biol. Chem.* **240**, PC 960.
Ghosh, H. P. and Preiss, J. (1965b). *Biochemistry* **4**, 1354.
Ginsburg, V. (1958). *J. biol. Chem.* **232**, 55.
Greenberg, E. and Preiss, J. (1964). *J. biol. Chem.* **239**, 4314.
Kauss, H. and Kandler, O. (1962). *Z. Naturf.* **17b**, 858.
Kornfeld, R. and Brown, D. H. (1962). *J. biol. Chem.* **237**, 1772.
Koshland, D. E., Jr. (1958). *Proc. natn. Acad. Sci. U.S.A.* **44**, 98.
Koshland, D. E., Jr. (1963). *Cold Spr. Harb. Symp. quant. Biol.* **28**, 473. Long Island Biol. Assoc., Cold Spring Harbor, New York.
Koshland, D. E., Jr. (1964). *Fed. Proc.* **23**, 719.
Leloir, L. F., Olavarria, J. M., Goldemberg, S. H. and Carminatti, H. (1959). *Archs Biochem. Biophys.* **81**, 508.
Leloir, L. F., Fekete, M. A. R. de and Cardini, C. E. (1961). *J. biol. Chem.* **236**, 636.
Leloir, L. F. (1964). *Biochem. J.* **91**, 1.
Lineweaver, H. and Burk, D. (1934). *J. Am. chem. Soc.* **56**, 658.
Monod, J., Changeux, J. P. and Jacob, F. (1963). *J. mol. Biol.* **6**, 306.
Monod, J., Wyman, J. and Changeux, J. P. (1965). *J. mol. Biol.* **12**, 88.
Munch-Petersen, A., Kalckar, H. M., Cutolo, E. and Smith, E. E. B. (1953). *Nature, Lond.* **172**, 1036.
Murata, T. and Akazawa, T. (1964). *Biochem. Biophys. Res. Commun.* **16**, 6.
Murata, T., Minamikawa, T., Akazawa, T. and Sugiyama, T. (1964a). *Archs Biochem. Biophys.* **106**, 371.
Murata, T., Sugiyama, T. and Akazawa, T. (1964b). *Archs. Biochem. Biophys.* **107**, 92.
Murata, T., Sugiyama, T. and Akazawa, T. (1965). *Biochem. Biophys. Res. Commun.* **18**, 371.
Neufeld, E. F., Ginsburg, V., Putman, E. W., Fanshier, D. and Hassid, W. Z. (1957). *Archs Biochem. Biophys.* **69**, 602.
Preiss, J., Shen, L. and Partridge, M. (1965). *Biochem. Biophys. Res. Commun.* **18**, 180.
Recondo, E. and Leloir, F. L. (1961). *Biochem. Biophys. Res. Commun.* **6**, 85.
Recondo, E., Dankert, M. and Leloir, L. F. (1963). *Biochem. Biophys. Res. Commun.* **12**, 204.
Robbins, P. W., Traut, R. R. and Lipmann, F. (1959). *Proc. natn. Acad. Sci.* **45**, 6.
Rosell-Perez, M. and Larner, J. (1964). *Biochemistry* **3**, 773.
Sanwal, B. D. and Stachow, C. S. (1965). *Biochim. biophys. Acta* **96**, 28.
Shen, L. and Preiss, J. (1964). *Biochem. Biophys. Res. Commun.* **17**, 424.
Taketa, K. and Pogell, B. M. (1965). *J. biol. Chem.* **240**, 651.
Umbarger, H. E. (1961). *Cold Spr. Symp. quant. Biol.* **26**, 301.
Yates, R. A. and Pardee, A. B. (1956). *J. biol. Chem.* **221**, 757.

## Addendum

Recent experimental results show that purer preparations of 2-phosphoglyceric acid activated the ADP-glucose pyrophosphorylase very slightly. This is in contradiction to the results reported in the body of this paper. The reason for this anomaly was due to the contamination of the 2-phosphoglyceric acid (Ba salt, Sigma Chemical Co. Batch No. 114B-0060) by 3-phosphoglyceric acid. Subsequent use of purer 2-phosphoglyceric acid preparation (Na salt, Calbiochem, A grade, Batch No. 52868) gave practically no activation.

The presence of about 5% 3-phosphoglyceric acid in the 2-phosphoglyceric acid preparation would, in fact, give such stimulation. Moreover, the 2-phosphoglyceric acid obtained from Calbiochem did not have any inhibitory effect on the stimulation by 3-phosphoglycerate.

## Discussion

*O. Kandler* (*Germany*): I should like to make two comments. (1) At the end of his talk Dr. Preiss said that it would be important to know how the concentrations of

PGA and inorganic phosphate change during the light–dark transition, since the regulation of starch synthesis depends on these two substances. In fact, 15 years ago, we showed [*Z. Naturf.* **5**b, 423 (1950)] that the inorganic phosphate drops by about 35% within 30 sec of illumination and increases when illumination is terminated. The steady-state level in the light is about 10–20% lower than in the dark. At least in our opinion, these data provided the first experimental evidence for the existence of photophosphorylation, for in those days photophosphorylation in chloroplasts was not yet known. During the last few years we have also studied the change of the concentrations of some of the intermediates of the photosynthetic cycle, among them also PGA. It turned out that in *Chlorella* PGA drops by about 30% upon illumination, but after a few minutes the concentration returns to about the same level as in the dark. The concentration of PGA during steady-state photosynthesis is certainly not higher than in the dark. So only the concentration of inorganic phosphate shows the behaviour as predicted by Dr. Preiss for the regulation of starch synthesis.

(2) Is there any indication that maltose could be an intermediate in polysaccharide synthesis? Of course, I am familiar with the general opinion that maltose results from the break-down of starch, but in many experiments in which we studied the kinetics of the distribution of $^{14}C$ after photosynthesis of leaves of various species in $^{14}CO_2$, we found that maltose is very early labelled. In fact, when the percentage of label fixed in the various compounds is plotted against time, maltose shows an optimum already after 1–2 min thus immediately following the optimum of the sugar phosphates. When $^{12}CO_2$ was supplied after 30 min photosynthesis in $^{14}CO_2$ the label in maltose was diluted out very quickly. Thus maltose behaves obviously differently from sucrose and starch and has to be considered as an intermediate in the sugar metabolism. I could imagine that glucose from UDPG or ADPG is transferrred at first to free glucose to form the 1–4 bond and a *trans*-glucosidase transfers this glucose further on to polysaccharide or other glucosides. The rapid labelling of maltose could of course also be due to an active exchange between maltose and nucleotide-activated glucose. So the formation of maltose would be a side-reaction and not an important process in the net-synthesis of polysaccharides.

*J. Preiss:* (1) Your first comment, of course, is applicable only to *Chlorella.* Our preliminary results indicate that the Chlorella ADP-glucose pyrophosphorylase is activated by fructose 1,6-diphosphate, possibly by fructose 6-phosphate and not by 3-phosphoglycerate. According to your results [*Z. Naturf.* **18**b, 718 (1963)] fructose diphosphate levels in *Chlorella* increase about ten-fold after illumination.

(2) We have no evidence that maltose is an intermediate in the ADP-glucose:starch transferase. Maltose is a poor acceptor of the glucosyl moiety from ADP-glucose while glucose is inactive as an acceptor.

# The Role of Glycollate in Photosynthetic Carbon Fixation

C. P. Whittingham, J. Coombs and A. F. H. Marker
*Botany Department, Imperial College, London, England*

## Introduction

Glycollic acid has been observed as a product of photosynthesis in *Chlorella* and other green plants by many workers (Tolbert and Zill, 1956; Zelitch, 1958; Warburg and Krippahl, 1960; Whittingham *et al.*, 1963). Maximal production is observed at lower partial pressures of carbon dioxide and higher light intensities. Glycollic acid has been shown to accumulate in green leaves in the presence of α-hydroxysulphonates (Zelitch, 1958) and in *Chlorella* in the presence of isoniazide (INH) (Pritchard *et al.*, 1962). It has been generally considered that glycollate is further metabolized through glyoxylate and glycine to serine. The hydroxysulphonate is thought to inhibit glycollic oxidase and isoniazide to inhibit the formation of serine from glycine. Glycollic oxidase, glyoxylate reductase and a phosphatase active on phosphoglycollate have all been observed in green tissue. The enzymes concerned in the synthesis of serine from glycine in green plants have not been isolated. In animal tissues, Richert *et al.* (1962) have shown that INH inhibits the formation of serine from glycine in avian livers. From an initial suggestion of Bassham *et al.* (1954) and later Griffith and Byerrum (1959), the origin of glycollic acid has been generally believed to be from the primary photosynthetic cycle. Bradbeer and Racker (1961) found that crystalline transketolase in the presence of ferricyanide catalysed the formation of glycollate from fructose 6-phosphate. Others have proposed that ribulose diphosphate is cleaved to give rise to glycollate and triose phosphate. There seems little evidence to support the view that glycollate arises from a carboxylation other than that in the primary Calvin cycle. The observations of Zelitch (1965) which show that in tobacco leaf discs the specific activity of glycollate is greater than that of phosphoglycerate is not, in our opinion, conclusive.

Rabson *et al.* (1962) have shown that [2-$^{14}$C]glycollate supplied to wheat leaves in air is metabolized to serine and glycine. Jiminez *et al.* (1962) have shown that such feeding also gave rise to sucrose in which the glucan component was labelled almost equally in the 1, 2, 5 and 6 carbon atoms. [3-$^{14}$C]Serine gave sucrose in which the glucose was labelled in C-1 and C-6. Wang and Waygood (1962) obtained similar data with wheat seedlings when [1-$^{14}$C]glycine, [2-$^{14}$C]glycine and [1,2-$^{14}$C]glycollate were fed. They showed further that when glycollate and glyoxylate were supplied together with radioactive glycine or serine, there was no effect on the radioactivity of the sugar formed, but addition of glycine or serine, together with [$^{14}$C]glyoxylate did diminish the radioactivity. B. Miflin in our laboratory has undertaken similar experiments with pea leaves and obtained essentially similar results. Furthermore Miflin (1965) showed that the metabolism of [1-$^{14}$C]glycollate and [2-$^{14}$C]glycine to sucrose was inhibited by either addition of INH or prior infiltration of the leaves with non-radioactive serine. There is therefore considerable evidence that in leaves, sucrose can be formed from exogenously supplied glycollate and that the route is via glycine and serine.

Tolbert (1963) has suggested that one role of glycollic acid may be as part of a permease system at the chloroplast membrane. He has proposed that phosphoglycollate can permeate the membrane and pass from the chloroplast to the cytoplasm. Here it may be oxidized to glyoxylate which can either be further metabolized or enter the chloroplast where it can be reduced and phosphorylated. If a NADP-linked glyoxylate reductase were present in the chloroplast and a NAD-linked reductase in the cytoplasm as suggested by Zelitch (1955), this mechanism would operate as a transfer mechanism allowing the oxidation of NADPH to $NADP^+$ in the chloroplast, and the concomitant reduction of NADH to $NAD^+$ in the cytoplasm. Butt and Peel (1963) have shown that hydroxysulphonates inhibit the light-activated uptake of glucose by *Chlorella*. In the absence of carbon dioxide or an alternative hydrogen acceptor for reduced co-enzymes, these authors suggest that a glycollate/glyoxylate cycle could be utilized to re-oxidize NADPH in the chloroplast permitting a continuous cyclic electron transport in the light with phosphorylation as the sole end product.

Partial pressures of oxygen greater than that in air have been shown to inhibit photosynthesis and this inhibition is most marked at lower carbon dioxide concentrations. Turner *et al.* (1958) suggested that this inhibition resulted from an inhibition by oxygen of the enzyme glyceraldehyde phosphate dehydrogenase. In the work to be described we have shown, in confirmation of the work of others, that the primary effect of higher partial pressures of oxygen is to increase the production of glycollate and decrease the pool size of the various intermediates of the photosynthetic cycle. After establishing this we have gone on to consider how far this can be utilized to demonstrate the existence of a metabolic path in which glycollate is converted into sucrose.

## Experimental

### The influence of partial pressures of oxygen

The effect of oxygen partial pressure on the distribution of radioactivity amongst different photosynthetic products at air levels of carbon dioxide was determined with *Chlorella*. The time was varied so that the same fixation took place at all oxygen partial pressures. The longest time, which was for the oxygen sample, was 3 min. For a given amount of carbon fixed the higher the oxygen partial pressure the larger the percentage activity in glycollate and the smaller that in sucrose and to a less extent in an ethanol insoluble fraction (Fig. 1). Radioactivity of the latter was shown, in the main, to be present in a glucose component of a polysaccharide.

The time course of fixation in air and with 99·97% oxygen is shown in Fig. 2. This again shows a decrease in sucrose with a corresponding rise in glycollate in oxygen which is accentuated in the presence of INH.

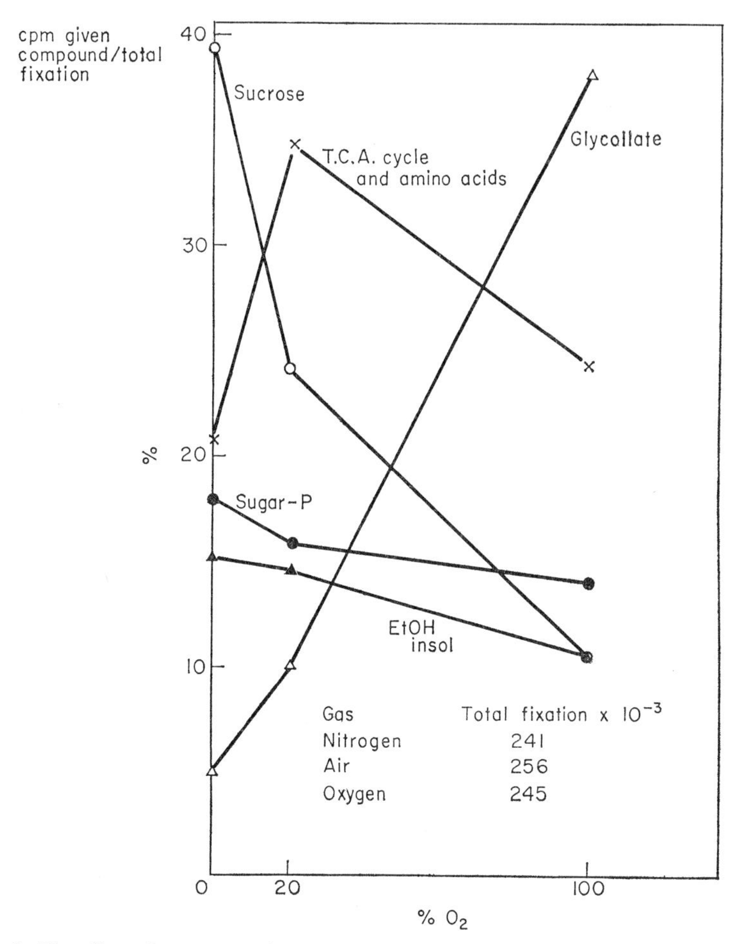


FIG. 1. The effect of oxygen partial pressures on the distribution of activity in photosynthetic products at air levels of carbon dioxide. Samples (10 ml) were run into a test tube and sealed with a "Suba Seal". $Na_2\,^{14}CO_3$ (3 μc) was injected into the tube. The oxygen sample was allowed to photosynthesize for 3 min. The other samples were allowed to fix the same amount of radioactivity. The cells were bubbled with the gas mixture prior to but not during the experimental period.

Cells were pre-incubated in air in the presence of INH for 30 min and then the gas phase was changed to either air or oxygen and samples taken up to 30 min after transfer. The products resulting from 1 min photosynthesis were determined from aliquots taken from a photosynthesizing cell suspension after transfer from air to oxygen. In the first 10 min after transfer from air to oxygen the monophosphates showed a rapid fall reaching a minimum. On return from oxygen to air a rise of smaller extent was observed (Fig. 5). Glycine and glycollate showed an increase during the first 10 min after the transition

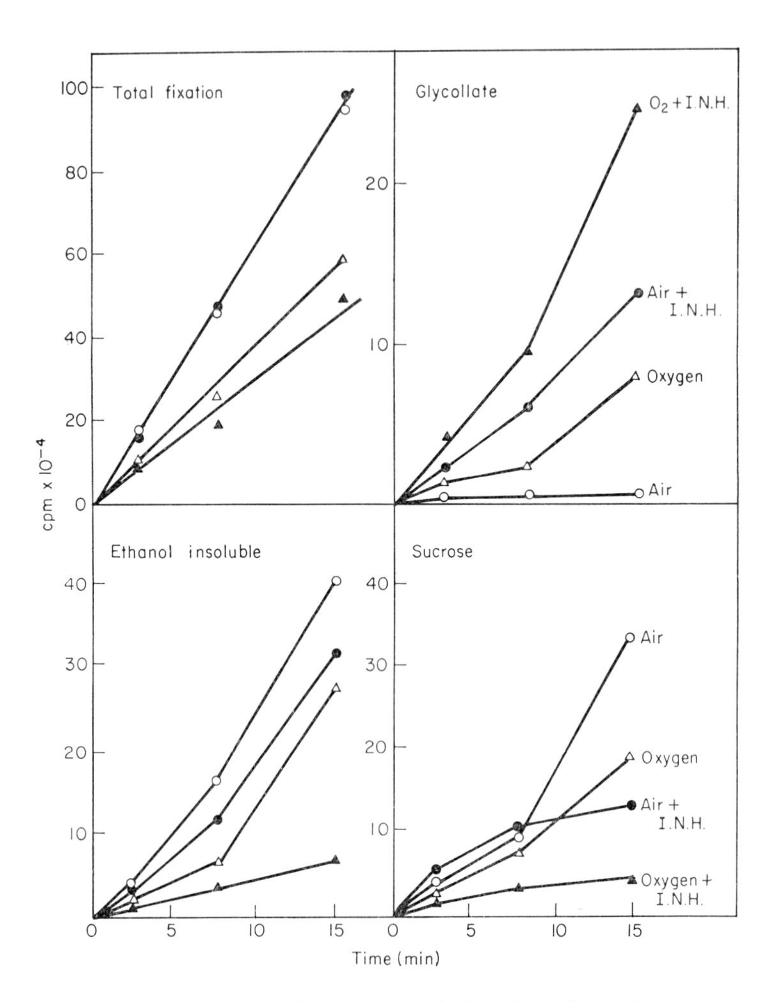


FIG. 2. Long term photosynthesis in phosphate buffer. The radioactivity was continuously injected into the base of a large perspex lollipop by means of a motor driven syringe, which delivered the solution at the rate of 1 ml every 7 min. The cell suspension was bubbled with gas containing the required level of $CO_2$. The amount of $^{14}CO_3^-$ injected was not sufficient to alter significantly the $CO_2$ concentration. $10^{-2}$ M INH was used.

with little further increase after 10 min and a marked decrease on transfer from oxygen back to air (Fig. 5). These were the major changes observed. Sucrose showed a behaviour which varied according to the light intensity.

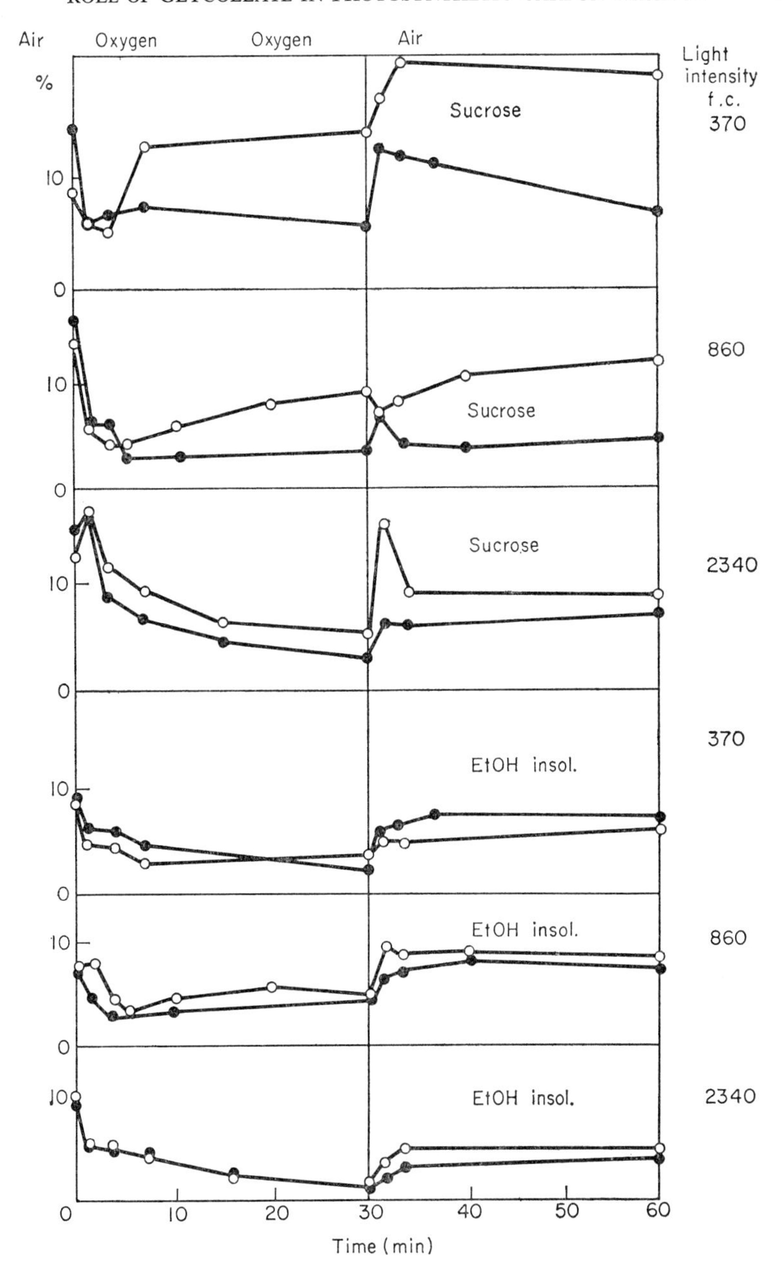


FIGS. 3, 4 and 5. The effect of light intensity on the total fixation of $^{14}CO_2$, and the distribution of radioactivity into products of photosynthesis, during transitions of gas phase, from air to oxygen and from oxygen to air. Fifty ml of *Chlorella* were suspended at a density of 5 $\mu$l/ml phosphate buffer and preincubated for 30 min in the gas mixture, at the stated light intensity. A sample (10 ml) was run into a test tube, sealed with a "Suba Seal" and injected with $Na_2{}^{14}CO_3$ (3 $\mu$c). The cells were killed after 1 min photosynthesis in the closed tube. The gas flowing through the lollipop was changed and similar samples taken at various time intervals up to 30 min after the gas stream had been changed. ○ = Control; ● = +INH.

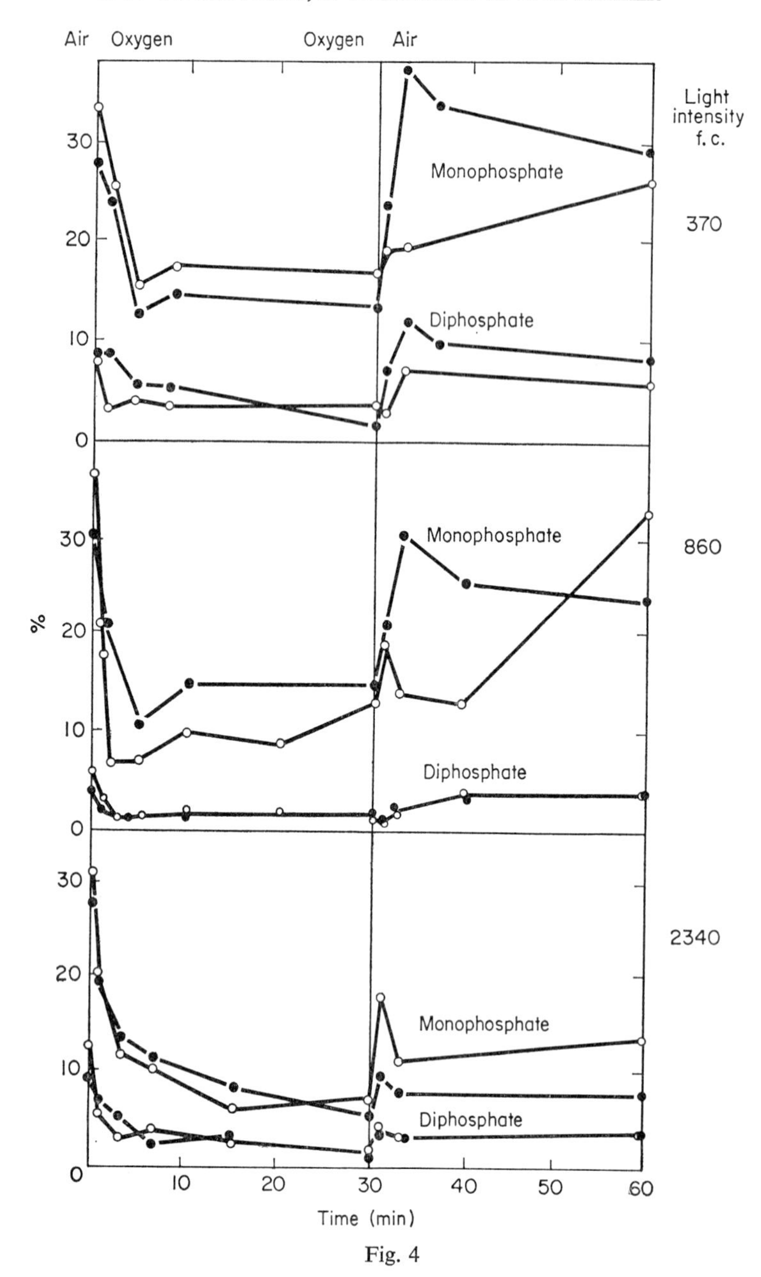


Fig. 4

After transition to oxygen at low light intensities there was an initial fall followed by an increase whereas at high light intensities there was an initial rise followed by a fall. On return to air in all cases a rise in activity was observed (Fig. 3). These results are consistent with the view that high partial pressures of oxygen cause an oxidation of some intermediate of the photosynthetic cycle resulting in a decrease in the size of the sugar phosphate pool and an increase in the production of glycollate.

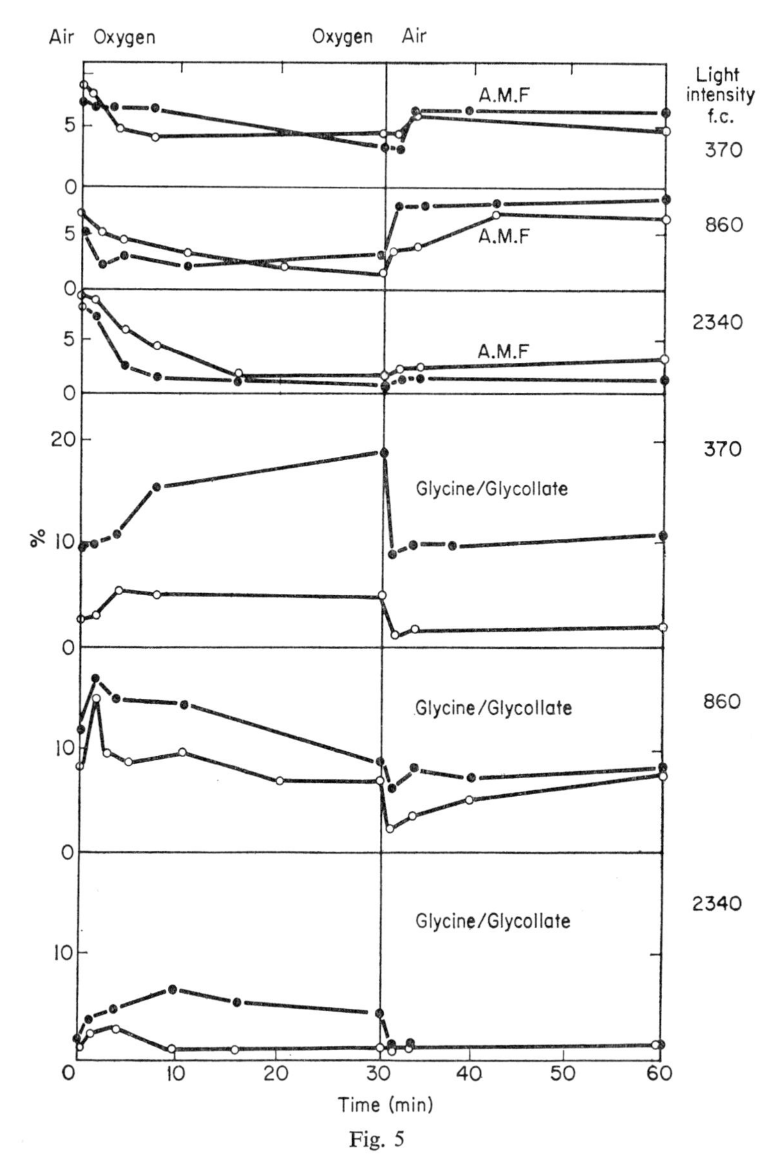


Fig. 5

[AMF=total counts in aspartate, malate and fumarate]

An experiment was undertaken to determine whether the increased production of glycollate in high partial pressures of oxygen was related to a change in the percentage of glycollate excreted from the cell. It has been widely observed that under conditions of maximum glycollate production a large fraction of the glycollate produced is excreted from *Chlorella.* The glycollate excreted from the cell was determined in the supernatant using Calkin's method and the internal and external radioactivity in glycollate also determined. There is close agreement between the radioactivity and colorimetric determination of external glycollate, indicating that glycollate is the only radioactive product

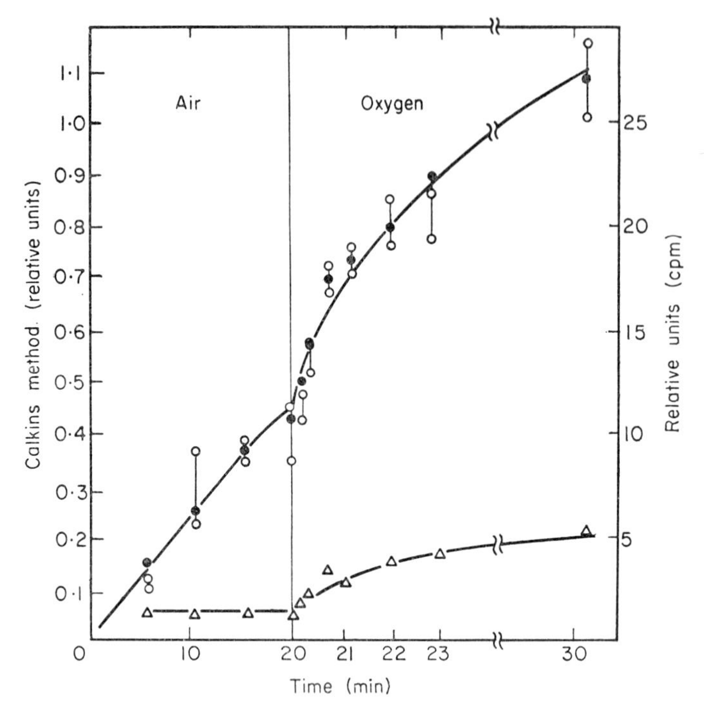


FIG. 6. Changes in the metabolism of carbon, during transitions from air to oxygen. Fifty ml of *Chlorella* suspended at a density of 5 $\mu$l/ml phosphate buffer were injected with $Na_2{}^{14}CO_3$ (10 $\mu$c) by means of the automatic syringe. Samples (1 ml) were taken at intervals of about 10 sec. External: Calkins method ●—●; Glycollate-$^{14}$C = ○—○; Internal: Glycollate-$^{14}$C = △—△.

excreted. Furthermore the data show that the proportion of glycollate excreted is not very different in air or oxygen (Fig. 6).

In a further experiment the influence of oxygen partial pressure on the rate of flow of carbon from the photosynthetic cycle intermediates to end products of photosynthesis was determined. After an initial short period of fixation of $^{14}$C after which 35% of the activity still remained in sugar monophosphates, the gas phase was rapidly changed to either air or oxygen, both free from carbon dioxide. The rate of loss of $^{14}$C in a number of intermediates was determined. During the first 3 min a marked fall in monophosphates took place which was slightly faster in air than in oxygen. There was an initial rise in diphosphates

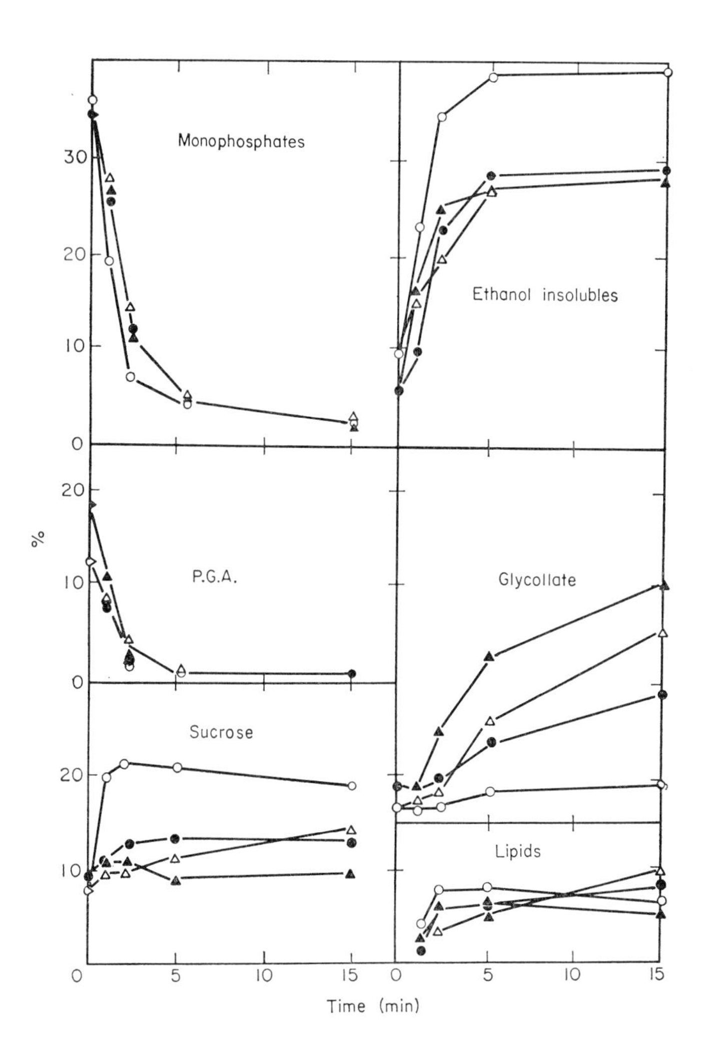


FIG. 7. The effect of INH, oxygen partial pressures and $CO_2$ concentration on the further metabolism of sugar phosphates labelled with $^{14}CO_2$ in the air. Fifty ml of *Chlorella* suspended at a density of 5 μl/ml phosphate buffer were preincubated in air for 30 min. The gas stream was stopped and $Na_2{}^{14}CO_3$ (15 μc) was injected into the lollipop. The cells were allowed to fix the $^{14}CO_2$ for 1 min, and the excess $^{14}CO_2$ was blown off by a rapid stream of gas, either air or oxygen. A sample was taken as the gas stream was changed. In this sample about 50% of the carbon fixed could be recovered from the sugar phosphate fraction. The further metabolism of these labelled sugar phosphates was studied by taking samples (10 ml) at intervals after the transition. INH was added at the beginning of illumination. ○=air; △=$O_2$; ●=air+INH; ▲=$O_2$+INH.

followed by a fall after between 3–5 min; subsequent to this glycollate showed a considerable rise (Fig. 7).

In all the experiments described the glycollate produced was degraded using the perchloratocerate method. After periods of 1 min or less fixation of carbon dioxide the ratio of C-2:C-1 varied from 1·01 in air to 0·87 in the presence of INH or in oxygen. Similarly with longer periods of photosynthesis, up to 15 min, the ratio in air was 0·93, that in INH 0·87 and that in oxygen 0·84. Conditions which favour the production of free glycollate thus tend to decrease the incorporation of activity into the C-2 of glycollate, suggesting that the C-2 precursor to glycollate may be to some extent asymmetrically labelled.

Summarizing, the effect of high oxygen partial pressure on photosynthesis in *Chlorella* can be demonstrated in two ways. Firstly during the transition from air to oxygen there is an increased formation of glycollate associated with a decrease in the size of the sugar phosphate pools of the photosynthetic cycle. Secondly, with respect to steady state photosynthesis, the percentage incorporation of carbon dioxide into glycollate is greater at high partial pressures of oxygen, which is again associated with the existence of a smaller pool size of sugar phosphate intermediates.

## DEGRADATION STUDIES

It can be seen from the experiments just described that in general, conditions which favour an increase in glycollic acid formation, result in a decrease in sucrose. Together with other evidence this has led to the view that part of the sucrose formed in photosynthesis is derived through a pathway in which glycollic acid or some related compound is an intermediate. If there are two synthetic paths for sucrose formation, one of which passes through glycollate which is approximately equally labelled in its two carbon atoms, and one which is derived from triose phosphate which is initially predominantly labelled in C-1, the distribution of labelling within the atoms of sucrose should be markedly different. To establish whether this was the case the sucrose formed under a variety of photosynthetic conditions has been degraded. Details of the methods used will be given in detail in further papers.

### *Glucose Feeding Experiments*

Experiments were undertaken in which radioactive glucose, labelled in specific carbon atoms, was fed to *Chlorella* under a variety of conditions. The radioactivity was found to be incorporated in the light into sucrose, an insoluble fraction (shown to be a polyglucan) and glycollic acid as main products. The glucose components of the sucrose and the polyglucan were then degraded to determine the distribution of isotope within the molecule. It was found that while most of the activity had stayed in the primary position after incorporation, 10–20% of the activity equilibrated with the opposite end of the molecule (i.e. C-1 equilibrated with C-6, C-2 with C-5 etc.), as previously observed by many others (for example see Maclachlan and Porter, 1959). Essentially the same behaviour was found whatever the composition of the gas mixture used to aerate the *Chlorella* suspension (0·034% $CO_2$ with either 20% or 99·97%

$O_2$). It was desirable to see if this equilibration occurred immediately on entry to the cell and was not directly linked with the photosynthetic apparatus. Specifically labelled glucose was fed to *Chlorella* aerated with a stream of air containing 0·034% $CO_2$ (nonradioactive) in the light and dark. The rate of glucose metabolism was found in the dark to be only a third of that in the light and little sucrose was formed. But polyglucan was found both in the light and the dark and the distribution of isotope within the glucose released on hydrolysis is shown in Table I. Light did not markedly affect the distribution. In a

Table I. Percentage distribution of isotope within glucose component of polyglucan after feeding [6-$^{14}$C]glucose

| Carbon atom | Light | Dark |
|---|---|---|
| 1 | 15·6 | 15·6 |
| 2 | 5·9 | 5·7 |
| 3 | 4·9 | 1·6 |
| 4 | 8·1 | 4·9 |
| 5 | 3·5 | 4·3 |
| 6 | 62·0 | 67·8 |

Length of experiment 1 hr. Atmosphere, 0·034% $CO_2$ in air.

second experiment specifically labelled glucose was fed to a suspension of *Chlorella* aerated with $CO_2$-free air in light. The removal of $CO_2$ from the air reduces the rate of glucose incorporation, little sucrose is formed ($\simeq$1%) and most of the isotope ends up as polyglucan and glycollate. The distribution of isotope within the glucose fraction of the polyglucan was determined and essentially the same pattern was found both in the presence and absence of INH (Table II). The glucose appears to equilibrate with the triose pool always to the same extent, and to an extent independent of the rate of photosynthesis. Photosynthesis may however have an indirect effect in as far as the rate at which the glucose is taken up is increased.

Table II. Percentage distribution of isotope within glucose component of polyglucan after feeding [1-$^{14}$C]glucose

| Carbon atom | C | INH |
|---|---|---|
| 1 | 65·2 | 68·6 |
| 2 | 3·8 | 0·8 |
| 3 | 3·2 | 1·7 |
| 4 | 4·4 | 6·4 |
| 5 | 4·4 | 0·8 |
| 6 | 18·9 | 21·5 |

Length of experiment, 10 min. Atmosphere $CO_2$-free air.

In all these data the effect first reported by Gibbs and Kandler (1957) of an inequality between the C-3 and C-4 is generally observed, that is to say in most cases C-4 is more active than C-3. In Table III the glucose present in the polysaccharide of the alcohol insoluble fraction has been degraded both when [6-$^{14}$C]glucose and [2-$^{14}$C]glucose were fed. The soluble fraction was hydrolysed in N $H_2SO_4$ for 6 hr and the glucose present separated. The suspension was illuminated in the presence of air, and with the period of illumination used a considerable fraction of radioactivity was present in the insoluble fraction. In this fraction more equilibration between the two halves of the glucose residue appears to have taken place and the inequality between C-3 and C-4 is more apparent. Again the effect of addition of INH was investigated and in this case surprisingly little effect was observed. There was no increase in the degree of symmetry between the two halves of the molecule as was suggested in the data for sucrose.

Table III. Percentage distribution of isotope in glucose fraction of hydrolysed "insolubles"

| | [6-$^{14}$C]Glucose | | [2-$^{14}$C]Glucose | | | |
|---|---|---|---|---|---|---|
| | 15 min | | 15 min | | 30 min | |
| | Control | INH | Control | INH | Control | INH |
| 1 | 17·0 | 15·6 | 3·8 | 2·6 | 3·8 | 3·7 |
| 2 | 4·3 | 2·2 | 54·2 | 54·6 | 41·9 | 47·2 |
| 3 | 1·7 | 2·2 | 5·6 | 7·6 | 7·1 | 3·9 |
| 4 | 8·0 | 5·6 | 9·1 | 10·0 | 15·0 | 15·1 |
| 5 | 4·7 | 1·1 | 22·4 | 21·0 | 26·7 | 26·8 |
| 6 | 64·8 | 73·5 | 4·9 | 4·1 | 5·4 | 3·1 |
| Cpm "solubles" | 13,012 | 15,644 | 9900 | 14,570 | 4360 | 10,760 |
| Cpm "insolubles" | 8860 | 7638 | 7500 | 3250 | 5088 | 2478 |

Illumination in presence of air.

*Chlorella* was fed specifically labelled glucose in the presence of 99·97% oxygen and 0·03 $CO_2$. Under these conditions the incorporation of radioactivity into the insoluble fraction is very much reduced and the decrease in radioactivity appearing in sucrose consequent upon the addition of INH is related to an increase in radioactivity in glycollate and glycine. Table IV shows data for feeding glucose when [2-$^{14}$C]glucose and [6-$^{14}$C]glucose is fed with 100% oxygen in the gas phase. In all cases the addition of INH tends to increase the exchange between an atom and its corresponding partner in the other half of the molecule, but it does not consistently affect the radioactivity in the remaining atoms.

The data reported confirm the opinion that the glucose is not metabolized to carbon dioxide and reformed in photosynthesis (Whittingham *et al.* 1963). In these experiments a relatively large addition of glucose had been made but this did not obscure the effect of a variation in conditions upon the relative

Table IV. Percentage intramolecular distribution of isotope in some of the metabolic intermediates

| | [1-$^{14}$C]Glucose 20 min | | [6-$^{14}$C]Glucose 20 min | | [2-$^{14}$C] Glucose 20 min |
|---|---|---|---|---|---|
| | Control | INH | Control | INH | Control |
| Glucose/Glycine | 4·6 | 34·9 | 5·4 | 32·9 | 4·2 |
| Sucrose | 57·7 | 28·7 | 38·1 | 16·8 | 51·5 |
| Glycollate | 20·7 | 23·9 | 26·4 | 34·6 | 23·1 |
| Cpm "solubles" | 10823 | 9557 | 9274 | 8946 | 6598 |
| Cpm "insolubles" | 1811 | 1298 | 1972 | 1580 | 1542 |

Percentage atomic distribution of isotope within the glucopyranose fraction of sucrose

| | [1-$^{14}$C]Glucose 20 min | | [6-$^{14}$C]Glucose 20 min | | [2-$^{14}$C] Glucose 20 min |
|---|---|---|---|---|---|
| Carbon | Control | INH | Control | INH | Control |
| 1 | 79·0 | 71·4 | 11·2 | 18·1 | 1·9 |
| 2 | 3·4 | 2·7 | 6·2 | 9·3 | 76·0 |
| 3 | 1·8 | 2·6 | 1·1 | 1·8 | 3·7 |
| 4 | 2·7 | 4·8 | 5·0 | 2·0 | 4·2 |
| 5 | 2·2 | 2·9 | 3·4 | 1·7 | 10·5 |
| 6 | 10·9 | 15·7 | 73·2 | 67·2 | 4·0 |

Gas stream—100% oxygen.

predominance of glycollate and glycine. Furthermore at high partial pressures of $O_2$ most of the glucose is converted into glycollate and sucrose alone and addition of INH markedly favours the former. Nevertheless, the sucrose formed has not shown changes in labelling pattern which would be consistent with the formation of sucrose from glycollate.

## *$H^{14}CO_3^-$ Feeding Experiments*

In these experiments a suspension of *Chlorella* in $10^{-4}$ M phosphate buffer was preincubated in the light and aerated with a stream of non-radioactive air containing 0·034% $CO_2$ with and without INH for 30 min. $H^{14}CO_3^-$ [5 $\mu$c] was added initially to 50 ml of the suspension and after a predetermined time samples were taken. The addition of INH led to a rise in the level of glycollic acid and glycine and a lowering in the level of serine. In the first 3 to 5 min sucrose formation was increased in the presence of INH but at longer times it was decreased. The sucrose and polyglucan was extracted and the glucose components of both molecules subjected to the same degradation procedures. After incubation with $H^{14}CO_3^-$ for only 1 min, the glucose residue of the sucrose has most of the isotope in C-3 and C-4. There is an indication that INH increases the degree of labelling in C-1, C-2, C-5 and C-6 relative to C-3 and C-4 (Table V)

Table V. General percentage distribution of isotope

| | C | INH |
|---|---|---|
| Diphosphates | 34·7 | 32·2 |
| Monophosphates | 16·7 | 27·7 |
| Phosphoglyceric acid | 9·4 | 8·3 |
| Glycine + Serine | 15·8 | 10·1 |
| Sucrose | 3·8 | 4·1 |
| Glycollate | 6·3 | 11·4 |
| Cpm insoluble/soluble | 111,000/1,107,000 | 109,000/1,021,000 |

Distribution of isotope within glucose moiety of sucrose after feeding $H^{14}CO_3^-$

| | Sucrose | |
|---|---|---|
| Carbon atom | Control | INH |
| 1 | 18 | 26 |
| 2 | 21 | 29 |
| 3 | 93 | 89 |
| 4 | 100 | 100 |
| 5 | 23 | 27 |
| 6 | 21 | 32 |

Length of experiment, 1 min. Atmosphere 0·034% $CO_2$ in gas stream.

In another experiment the algal suspension was allowed to photosynthesize in the presence of $H^{14}CO_3^-$ for 3 min. It can be seen from Table VI that the isotope has spread more evenly throughout the molecule in the glucose component of sucrose. Activity in atoms other than C-3 and C-4 was nearly equal, and the more so in the presence of INH. The polyglucan found after 1 and 3 min was also degraded. Similar to the sucrose it showed increased randomization with time and slightly greater randomization in the presence of INH than in its absence.

Table VI. General percentage distribution of isotope

| | C | INH |
|---|---|---|
| Diphosphates | 18·0 | 10·9 |
| Monophosphates | 19·7 | 14·4 |
| Phosphoglyceric acid | 1·7 | 0·9 |
| Glycine + Serine | 11·4 | 31·4 |
| Sucrose | 5·3 | 4·6 |
| Glycollate | 3·5 | 18·5 |
| Cpm insoluble/soluble | 361,000/937,000 | 405,000/965,000 |

Distribution of isotope within glucose derived from sucrose and polyglucan after feeding $H^{14}CO_3^-$

| | Sucrose | | Polyglucan | | | |
|---|---|---|---|---|---|---|
| | 3 min | | 1 min | | 3 min | |
| Carbon atom | Control | INH | Control | INH | Control | INH |
| 1 | 72 | 68 | 20 | 30 | 61 | 64 |
| 2 | 93 | 83 | 34 | 32 | 66 | 68 |
| 3 | 93 | 97 | 87 | 88 | 89 | 89 |
| 4 | 100 | 100 | 100 | 100 | 100 | 100 |
| 5 | 93 | 67 | 28 | 29 | 67 | 72 |
| 6 | 88 | 69 | 33 | 35 | 64 | 70 |

Length of experiment, 3 min. Atmosphere 0·034% $CO_2$ in gas stream.

Table VII. Distribution of isotope within glucose derived from sucrose and polyglucan after feeding $H^{14}CO_3^-$

| | Sucrose | | Polyglucan | |
|---|---|---|---|---|
| Carbon atom | Air | $O_2$ | Air | $O_2$ |
| 1 | 25 | 36 | 36 | 59 |
| 2 | 26 | 38 | 44 | 45 |
| 3 | 71 | 60 | 83 | 87 |
| 4 | 100 | 100 | 100 | 100 |
| 5 | 37 | 37 | 39 | 53 |
| 6 | 43 | 37 | 36 | 65 |

Length of experiment 1 min. Atmosphere 0·034% $CO_2$ in 99·966% of $O_2$ or air.

Previous work had shown that changes in the partial pressure of oxygen had a marked effect upon the formation of glycollate. A similar experiment was carried out with a gas stream of 0·03% $CO_2$ in 99·97% $O_2$ as compared with 0·03% $CO_2$ in air. The sucrose and polyglucan were extracted and again the relative distribution of isotope within the molecule determined as before. In the glucose residue of sucrose the activity of C-1, C-2 and C-5, and C-6 was more nearly equal and greater in oxygen than in the air samples. Again in the polyglucan increase in oxygen partial pressure increased the incorporation of isotope in C-1, C-2, C-5 and C-6 relative to that in the C-3 and C-4 positions (Table VII).

## Discussion

The labelling pattern of the glucose residue of the polyglucan formed from glucose fed to *Chlorella* is essentially the same in the light or dark. Furthermore the present work shows that whether the carbon dioxide concentration is that in air or the gas stream is maintained free of carbon dioxide, no change in labelling pattern resulted. Addition of INH also had no effect. It is thus clear that the labelling pattern of the polyglucan formed from exogenous glucose is independent of the presence or absence of photosynthesis. Similarly both in air and in air with higher partial pressures of $O_2$, the sucrose formed showed the same labelling pattern and this is also true in the presence of INH. This would suggest that most of the polyglucan and sucrose is formed from glucose in the cytoplasm. Presumably prior to or immediately after uptake the glucose is phosphorylated and the glucose 6-phosphate partly transformed to fructose 6-phosphate to form the two potential halves of the sucrose molecule.

By contrast, it should be noted that the proportion of glucose converted into glycollate is markedly dependent on the conditions of photosynthesis. For example with pure oxygen in the gas phase and in the presence of INH as much as 50% of the glucose incorporated may be converted to glycollic acid. The formation of glycollic acid is strictly light-dependent and only takes place at lower partial pressures of carbon dioxide. As we have suggested previously glycollate is presumably derived from an intermediate of the photosynthetic cycle and in that case glucose must obtain access to the cycle prior to conversion into glycollate. The glycollate formed is unequally labelled in the two carbon atoms when [1-$^{14}$C]glucose, [2-$^{14}$C]glucose or [6-$^{14}$C]glucose is fed. Thus, the glycollate cannot have been formed by liberation of carbon dioxide and its re-fixation in photosynthesis for it would then have been equally labelled. It is probable that sugar phosphates formed from the glucose in the cytoplasm have access to the photosynthetic cycle in the chloroplast, possibly penetrating in the form of triose phosphate. When [6-$^{14}$C]glucose is fed it is found that the glucose of sucrose is labelled predominantly in the C-6 but approximately 15–20% also in C-1; correspondingly when [1-$^{14}$C]glucose is fed, label appears in C-6 of the sucrose also to about 20%. Clearly randomization between the two ends of the molecule has taken place and presumably through conversion into a triose. The resulting triose will in each case be labelled C-3. If this triose moves from the cytoplasm to the chloroplast and into the photosynthetic cycle the ribulose diphosphate formed will be predominantly labelled in C-1 and C-5. If this molecule is cleaved into a C-2 fragment it will give rise to [2-$^{14}$C]glycollic acid, whilst the residual triose will be labelled in C-3. If this triose forms sucrose, it should be labelled equally in C-1 and C-6. The results show that such labelling is present to a limited extent indicating that sucrose cannot be formed wholly after entering the photosynthetic cycle in the chloroplast. It must be formed for a large part in the cytoplasm.

These results may be compared with those when radioactivity is supplied in the form of carbon dioxide. Randomization within the sucrose molecule is already well advanced after 5 min and degradation of material from samples subsequent to this show almost uniform labelling. It has been suggested by a number of workers that sucrose may arise from glycollate by a direct route. Since the glycollate formed even within 1 min in *Chlorella* is almost equally labelled in its two carbon atoms, it will presumably give rise to serine which is also equally labelled (cf. Rabson *et al.*, 1962) in all three atoms and hence ultimately to sucrose uniformly labelled.

If this process were taking place it would mean that the sucrose would be more uniformly labelled than would be otherwise the case. We have postulated that INH inhibits the conversion of glycine into serine in *Chlorella* and hence also the formation of sucrose from glycollate; addition of INH should therefore alter the labelling pattern of sucrose. However a complication arises in this apparently simple experiment. In the presence of INH and at air levels of carbon dioxide, there is after the addition of tracer carbon, an increased incorporation of radioactivity in sucrose during the first 5 min. Furthermore the distribution of radioactivity within the sucrose molecule is not changed by treatment with INH at these short times of 1 min exposure. This suggests that during this period the formation of sucrose by a route other than through glycollate has been accelerated presumably because the pools which have to be filled by the newly entering carbon before appearance in sucrose are decreased in size in the presence of INH. With longer periods of photosynthesis the formation of sucrose is inhibited in the presence of INH but at this time labelling within the sucrose is already random and it is not practicable to determine whether it is more random in the absence than in the presence of INH. With sucrose formed after 3 min exposure to radiocarbon there is some indication that addition of INH decreases radioactivity in C-1, C-2, C-5 and C-6 relative to that in C-3 and C-4.

The most informative set of conditions have been obtained in the presence of higher partial pressures of oxygen. In this case the appearance of radioactivity in sucrose is inhibited by INH even in the shortest times after injection of radiocarbon. This would indicate that in higher partial pressures of oxygen, sucrose is derived from C-2 compounds even in the short times of exposure to radioactive carbon dioxide. Consistent with this it has been shown that the sucrose formed from carbon dioxide in the presence of oxygen is more uniformly labelled than in an air control. It will be realized this was not the case for sucrose formed from glucose either in air or oxygen.

Our present hypothesis is that phosphoglycollate can give rise to phosphoserine which can then be converted into phosphoglyceric acid. This may equilibrate with the phosphoglyceric acid of the photosynthetic cycle proper whence it can be reduced to triose and ultimately to sucrose. It cannot be established at the present time whether the formation of sucrose from C-2 intermediates is totally independent of the formation of sucrose from triose phosphate of the photosynthetic cycle or not. The pools of sugar phosphate from which sucrose is largely formed during photosynthesis must be different from those from which sucrose is predominantly formed from exogenous glucose.

The C-2 compounds act as a by-pass to the carboxylation mechanisms of the photosynthetic cycle and this allows the cycle to operate even in the absence of carbon dioxide (see Fig. 8). Free glycollate is probably not itself an intermediate in the formation of sucrose but may be derived from an intermediate. If it were shown that phosphoglycollate phosphatase were a cytoplasmic enzyme or bound to the chloroplast membrane, then it would appear probable that exit from the C-2 by-pass to the cytoplasm would be determined by this enzyme. This would control the loss of carbon from chloroplast to cytoplasm and vice versa. In *Chlorella,* much of the glycollate formed is excreted from the cell; this represents an irreversible loss of carbon from the photosynthetic cycle intermediates. High partial pressures of oxygen activate the C-2 by-pass and the resulting increased loss of glycollate depletes the cycle intermediates lowering the rate of carbon fixation.

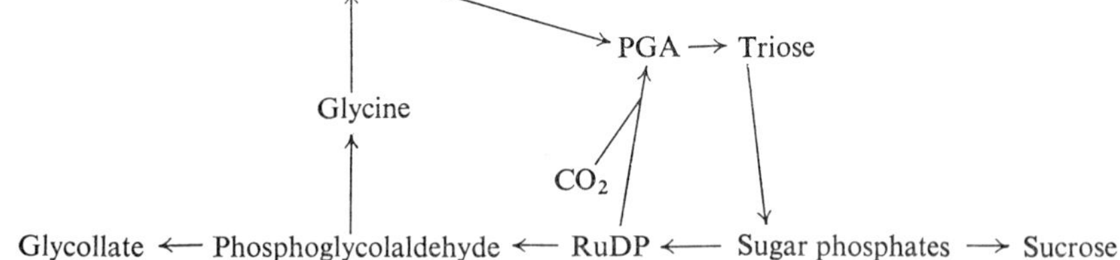


FIG. 8. Diagram of position of glycollate in relation to photosynthetic cycle.

## ACKNOWLEDGEMENTS

This work was supported in part by the United States Air Force under Grant No. AF EOAR 65-2 with the European Office of Aerospace Research (OAR), United States Air Force. The work was done while one of us (J.C.) held a D.S.I.R. Research Studentship and another (A.F.H.M.) a D.S.I.R. Fellowship.

## References

Bassham, J. A., Benson, A. A., Kay, L. D., Harris, A. Z., Wilson, A. T. and Calvin, M. (1954). *J. Am. chem. Soc.* **76**, 1760.
Bradbeer, J. W. and Racker, E. (1961). *Fed. Proc.* **20**, 88.
Butt, V. S. and Peel, M. (1963). *Biochem. J.* **88**, 31p.
Gibbs, M. and Kandler, O. (1957). *Plant Physiol.* **31**, 411.
Griffith, T. and Byerrum, R. U. (1959). *J. biol. Chem.* **234**, 762.
Jimenez, E., Baldwin, R. L., Tolbert, N. E. and Wood, W. A. (1962). *Archs Biochem. Biophys.* **98**, 172.
Maclachlan, G. A. and Porter, H. K. (1959). *Proc. roy. Soc.* B **150**, 460.
Miflin, B. (1965). "Carbon metabolism in *Chlorella.*" Ph.D. Thesis, University of London.

Pritchard, G., Whittingham, C. P. and Griffin, W. (1962). *J. exp. Bot.* **13**, 176.
Rabson, R., Tolbert, N. E. and Kearney, P. C. (1962). *Archs Biochem. Biophys.* **98**, 154.
Richert, D. A., Amberg, R. and Wilson, M. (1962). *J. biol. Chem.* **237**, 99.
Tolbert, N. E. (1963). *Phot. Mech. Grn. Plnts.* Nat. Acad. Sci. Publ. 1145, p. 648.
Tolbert, N. E. and Zill, L. P. (1956). *J. biol. Chem.* **222**, 895.
Turner, J. S., Turner, J. F., Shortman, K. D. and King, J. E. (1958). *Aust. J. biol. Sci.* **11**, 336.
Wang, D. and Waygood, E. R. (1962). *Pl. Physiol.* **37**, 826.
Warburg, O. and Krippahl, G. (1960). *Z. Naturf.* **15**, 197.
Whittingham, C. P., Hiller, R. G. and Bermingham, M. (1963). *Phot. Mech. Grn. Plnts.*, Nat. Acad. Sci. Publ. 1145, p. 675.
Zelitch, I. (1955). *J. biol. Chem.* **216**, 553.
Zelitch, I. (1958). *J. biol. Chem.* **233**, 1299.
Zelitch, I. (1965). *J. biol. Chem.* **240**, 1869.

# Glycollate Formation in Chloroplast Preparations

J. W. Bradbeer and C. M. A. Anderson
*Botany Department, University College of Wales, Aberystwyth, Wales*

In the literature there are to be found a number of suggestions as to the pathway of glycollate synthesis in green plants. We shall consider three of the pathways which have been suggested in recent years.

On the basis of tracer studies of *in vivo* systems Wilson and Calvin (1955) suggested that glycollate was formed from the pentose phosphates of the photosynthetic cycle by a reaction which involved transketolase.

Tolbert (see e.g. 1963) obtained data from labelling experiments which he interpreted as implicating phosphoglycollate in the pathway of glycollate synthesis. This conclusion was supported by the identification of a specific phosphoglycollate phosphatase in green plants. Tolbert suggested that phosphoglycollate might be formed from the diphosphates of xylulose, fructose or sedoheptulose, but not from that of ribulose.

There have been a number of suggestions that a *de novo* synthesis of glycollate from $CO_2$ may occur (Warburg and Krippahl, 1960; Tanner *et al.*, 1960; Stiller, 1962; Zelitch, 1965).

In 1961 Bradbeer and Racker reported that transketolase would catalyse the formation of glycollate in a model system. Glycollate accumulated when the following reaction mixture was incubated at 30° in the dark: 50 $\mu$moles glycylglycine buffer pH 7·4, either fructose 6-phosphate or xylulose 5-phosphate as substrate, 5 $\mu$moles ferricyanide, 1·6 units of transketolase, 15 $\mu$moles of magnesium chloride and 0·67 $\mu$moles of thiamine pyrophosphate, in a total volume of 2 ml. After 1 hr 10 $\mu$moles of fructose 6-phosphate had yielded 0·19 $\mu$moles of glycollate, and 1·2 $\mu$moles of xylulose 5-phosphate had yielded 0·38 $\mu$moles of glycollate. The mechanism of this reaction with fructose 6-phosphate appears to be as follows:

Fructose-6-phosphate + transketolase + TPP → Glycolaldehyde-TPP-transketolase complex + erythrose-4-phosphate

Glycolaldehyde-TPP-transketolase-complex + 2 ferricyanide → Glycollate + transketolase + TPP + 2 ferrocyanide

Low yields of glycollate have been obtained as a result of the enzymic action of extracts obtained by the sonication of *Chlorella pyrenoidosa*. This reaction may be mediated by transketolase since fructose 6-phosphate, but not fructose diphosphate, behaved as a substrate for glycollate formation. Furthermore a mixture of thiamine pyrophosphate and magnesium chloride, which are co-factors for transketolase, stimulated glycollate production by the *Chlorella* extracts. Both light and phenazine methosulphate stimulated the reaction, indeed occasionally in the absence of either of these factors no glycollate accumulation was detected. The results of some typical experiments are shown in Table I. Use of a glycollate oxidase inhibitor did not appear to be necessary for optimal glycollate accumulation, a finding which is in accordance with the recent failure of Hess *et al.* (1965) to detect glycollate oxidase in sonicated cell-free preparations of *Chlorella* and *Chlamydomonas.* The suggestion by these workers that glycollate is not rapidly converted into glycine and serine in illuminated green algae cannot however be supported in view of the *in vivo* $^{14}CO_2$-feeding experiments reported by Whittingham and Pritchard (1963).

Table I. Glycollate formation by a sonic extract of *Chlorella pyrenoidosa*

| Reaction mixture | Glycollate production in mμmoles/mgchl/hr | | |
|---|---|---|---|
| Complete | 1430 | 1180 | 270 |
| Omit fructose 6-phosphate | | 0 | |
| Replace fructose 6-phosphate with fructose diphosphate | 0 | | |
| Omit phenazine methosulphate | 0 | 600 | 160 |
| Omit 2-pyridylhydroxymethanesulphonate | | | 340 |
| Omit thiamine pyrophosphate/magnesium chloride | | | 70 |
| Complete in dark | 120 | 0 | 170 |

The complete reaction mixture contained the *Chlorella* extract, 5 μmoles of fructose 6-phosphate, 10 μmoles of 2-pyridylhydroxymethanesulphonate, 0·1 μmoles phenazine methosulphate, 0·5 μmoles of thiamine pyrophosphate and 15 μmoles of magnesium chloride in a total volume of 2 ml. Incubation at 20° under 1500 foot candles illumination.

We have made intensive investigations of glycollate formation in chloroplast preparations from spinach, spinach beet, meteor pea and Laxton's superb pea. We have closely followed the methods of Arnon's group (see e.g. Arnon *et al.*, 1956; Whatley *et al.*, 1956), of Zelitch and Barber (1960) and of Walker (1964) for the preparation of intact chloroplasts and chloroplast fragments, and of San Pietro and Lang (1956) for the preparation of grana. We have used non-aqueous chloroplasts kindly provided by Dr. K. J. Treharne of the Department of Biochemistry, University College of Wales. We have also made modifications to the recommended media such as the substitution of mannitol for sodium chloride and the addition of various components such as polyvinylpyrrolidone and bovine serum albumin. In addition to our use of the recommended methods of disrupting leaf cells we have used a stainless steel mill (Dr. J. D. Jones, personal communication) to grind leaves which had been infiltrated with the extraction medium. Despite our efforts none of the preparations gave high yields of glycollate from the tested substrates; see Table II for the results of some typical experiments. Fructose diphosphate gave the best yields of glycollate, but all the sugars and sugar phosphates tested gave small yields. Unlike the situation with *Chlorella* extracts, fructose 6-phosphate was a comparatively poor source of glycollate. Although a commercial preparation of ribulose diphosphate gave an appreciable yield of glycollate we have been unable to confirm this result in three recent experiments with a ribulose diphosphate preparation (Dr. E. Racker, personal communication) of our own. Tolbert (1963) has pointed out that ribulose diphosphate "does not have the necessary *trans* configuration of the hydroxyl groups between carbons 3 and 4" to make it a likely source of glycollate.

Table II. Glycollate synthesis from various substrates by higher plant chloroplast preparations

| Substrate | Glycollate accumulation in mμmoles/mg chl/hr | | | | | |
|---|---|---|---|---|---|---|
| | Intact spinach chloroplasts | | Meteor pea chloroplast fragments | | | |
| None | 5 | 10 | 0 | 90 | | |
| Fructose 6-phosphate | 25 | 24 | 20 | | | |
| Glucose 6-phosphate | 13 | 50 | | | | |
| Xylulose 5-phosphate | 38 | 170 | | | | |
| Erythrose 4-phosphate | | | | | | 20 |
| Sedoheptulose diphosphate | 27 | | | | | |
| Fructose diphosphate | 39 | 220 | 110 | 190 | 420 | 210 |
| Ribulose diphosphate | | | | | | 150 |
| Sucrose | 11 | 30 | | | | |
| Fructose | | 40 | | | | |
| Isocitrate | | | | 0 | | |
| Glyoxylate | | | | | 220 | |

The complete reaction mixtures consisted of: chloroplasts containing 0·5 to 1·0 mg chlorophyll, 100 μmoles *tris* pH 7·2, 10 to 20 μmoles 2-pyridylhydroxymethanesulphonate and 2·5 to 10 μmoles of substrate in a total volume of 2 ml. In addition the reaction mixture with intact spinach chloroplasts contained 830 μmoles of mannitol and the reaction mixture with the pea chloroplast fragments contained 0·1 μmoles FMN and 3 μmoles $MnCl_2$. The reactions were carried out at 20° under 1500 foot candles illumination for 30 to 60 min.

FMN or FAD have been found to stimulate the synthesis of glycollate from fructose diphosphate by the chloroplast preparations.

During a survey of possible sources of glycollate, hydroxypyruvate and phosphohydroxypyruvate gave good yields of glycollate with chloroplast preparations. Further investigation established that these high yields resulted from a non-enzymic reaction which required hydroxypyruvate, $Mn^{2+}$, FMN or FAD, oxygen and light. The reaction appears to be analogous to the non-enzymic decarboxylation of glyoxylate reported by Corbett and Davies (1965).

Zelitch (1965) has recently reported that glycollate can accumulate in tobacco leaf discs at a rate as high as 40 μmoles/g leaf/hr. We have been unable to obtain rates higher than one hundredth of this in experiments with higher plant chloroplast preparations with sugar phosphates or sugars as substrates. Griffith and Byerrum (1959) infiltrated tobacco leaves with [1-$^{14}$C]ribose (6·7 μmoles/g leaf) but found less than 0·1 μmoles of glycollate and related compounds/g of leaf after 3 hr illumination.

Since there appears to be no strong evidence in favour of a sugar phosphate being the starting point of the major pathway of glycollate synthesis in green plants, serious consideration must be given to the possibility of a *de novo*

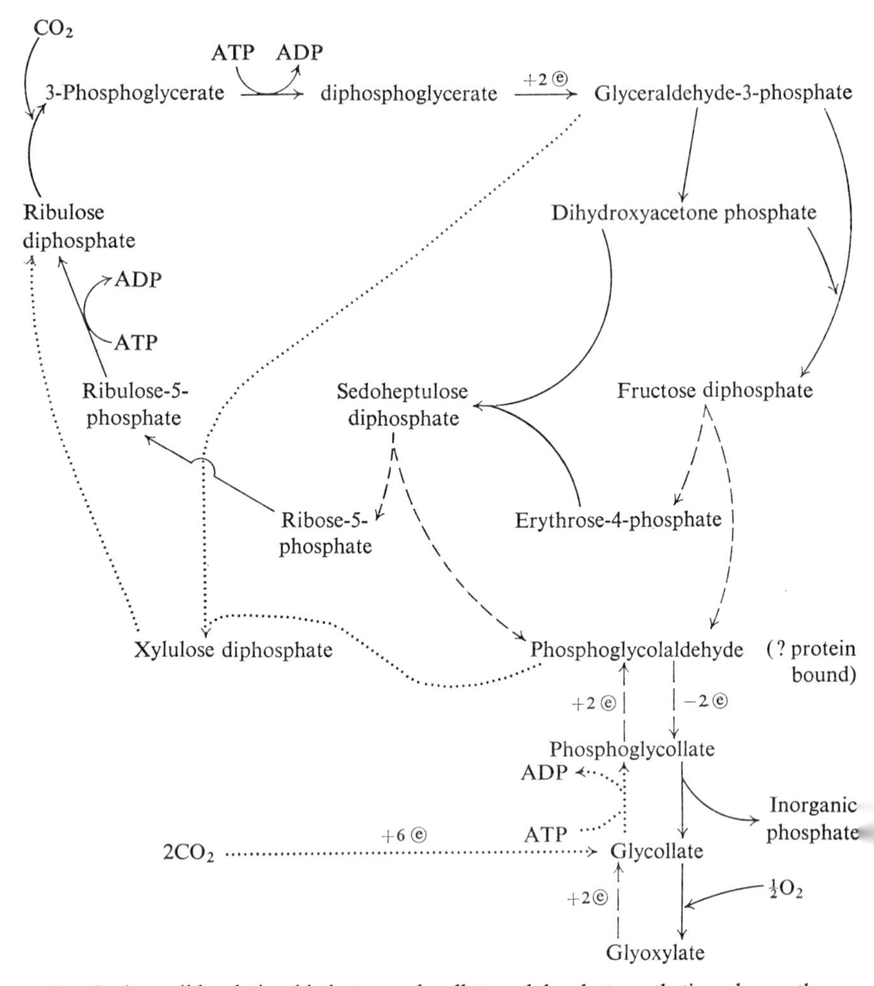


FIG. 1. A possible relationship between glycollate and the photosynthetic carbon pathway in higher plant leaves. Enzymes whose activity in cell-free leaf preparations exceeds 20 μmoles/mg chl/hr, ⟶; enzymes whose activity in leaf preparations has not been reported to exceed the above value, ---->; and enzyme reactions which have not been demonstrated in leaf preparations, ·····>.

$CO_2$ fixation being the major pathway. Zelitch (*loc. cit.*) found that the specific activity of carbon atom 1 of the accumulated glycollate in tobacco leaf discs corresponded closely to that of the supplied $^{14}CO_2$, while the specific activities of the corresponding carbon atoms of phosphoglycollate, 3-phosphoglycerate and glucose 6-phosphate were much lower. Zelitch concludes that the undefined carboxylation by which glycollate is synthesized may be as important as the ribulose diphosphate carboxylase reaction in photosynthesis in higher plants.

It does not seem possible to reach a firm conclusion about the pathways of glycollate synthesis on a basis of *in vivo* tracer experiments involving either kinetic studies or specific activity determinations. The feasibility of postulated pathways requires confirmation by *in vitro* studies of the individual enzyme reactions. In Fig. 1 are shown some pathways which may be concerned with glycollate metabolism. Since a rate of glycollate accumulation of 40 $\mu$moles/g fresh wt/hr corresponds to a rate of approximately 20 $\mu$moles/mg chl/hr, enzymic steps known to be capable of rates in excess of this value in cell-free leaf preparations are distinguished from enzymic steps of lower rates and from enzyme reactions not yet demonstrated in green leaves.

### Acknowledgement

Part of this work was supported by a grant from the Department of Scientific and Industrial Research.

## References

Arnon, D. I., Allen, M. B. and Whatley, F. R. (1956). *Biochim. biophys. Acta* **20**, 449.
Bradbeer, J. W. and Racker, E. (1961). *Fed. Proc.* **20**, 88.
Corbett, J. R. and Davies, D. D. (1965). *Pl. Physiol.* **40**, Suppl. lxviii.
Griffith, T. and Byerrum, R. U. (1959). *J. biol. Chem.* **234**, 762,
Hess, J. L., Huck, M. G., Liao, F. H. and Tolbert, N. E. (1965). *Pl. Physiol.* **40**, Suppl. xlii.
San Pietro, A. and Lang, H. M. (1956). *J. biol. Chem.* **231**, 211.
Stiller, M. (1962). *Annu. Rev. Pl. Physiol.* **13**, 151.
Tanner, H. A., Brown, T. E., Eyster, C. and Treharne, R. W. (1960). *Biochem. Biophys. Res. Commun.* **3**, 205.
Tolbert, N. E. (1963). *In* "Photosynthetic Mechanisms of Green Plants", p. 648. Nat. Acad. Sci.—Nat. Res. Council, Washington, D.C.
Walker, D. A. (1964). *Biochem. J.* **92**, 22c.
Warburg, O and Krippahl, G. (1960). *Z. Naturf.* **15**b, 197.
Whatley, F. R., Allen, M. B., Rosenberg, L. L., Capindale, J. B. and Arnon, D. I. (1956). *Biochim. biophys. Acta* **20**, 462.
Whittingham, C. P. and Pritchard, G. G. (1963). *Proc. R. Soc.* B **157**, 366.
Wilson, A. T. and Calvin, M. (1955). *J. Am. chem. Soc.* **77**, 5948.
Zelitch, I. (1965). *J. biol. Chem.* **240**, 1869.
Zelitch, I. and Barber, G. A. (1960). *Pl. Physiol.* **35**, 626.

# Triosephosphate Dehydrogenase in Plant and Microbial Photosynthesis[1]

R. C. Fuller and G. A. Hudock[2]

*Department of Microbiology, Dartmouth Medical School, Hanover, New Hampshire, U.S.A.*

## Introduction

With the possible exceptions of phosphoribulokinase and ribulose 1,5-diphosphate carboxylase, the enzymes of the pentose reductive cycle are widely distributed in nature. The photosynthetic carbon cycle shares many enzymes in common with those of both glycolysis and various carbohydrate biosynthetic pathways in animals, plants and microbial cells. The functioning of the cycle itself, however, is apparently limited to photosynthetic carbon metabolism and autotrophic $CO_2$ fixation in bacteria. One other exception that is generally found to the universality of these enzymes is that the triose-phosphate dehydrogenase (TPD) of the photosynthetic cycle is linked to NADP when photosynthesis is occurring in higher plants and algae. In both the autotrophic bacteria and the photosynthetic bacteria, the NAD-linked TPD is used as it is throughout the rest of the animal and microbial worlds.

The intracellular and phylogenetic distribution of the NAD and the NADP-dependent TPD in photosynthetic organisms has been widely examined. Research by Fuller and Gibbs (1959) and Smillie and Fuller (1960) have indicated that the NADP enzyme occurs only in the chloroplasts of higher plants and oxygen evolving photosynthetic micro-organisms. The NAD-linked enzyme also occurs in chloroplasts and in the cytoplasm of higher plant cells. In the green alga *Euglena*, studies on the kinetics of formation of the NADP enzyme by Brawerman and Konigsberg (1960) have shown that the rates of chlorophyll synthesis in re-greening cells and the increase in the NADP-linked TPD activity were approximately equal. These and other results have led to the suggestion that the NADP enzyme functions specifically in the Calvin photosynthetic cycle while the NAD-linked enzyme is operative primarily in oxidative and glycolytic systems.

Many of the enzymes of the photosynthetic cycle have been shown to be under metabolic control as a function of the environment. In both green algae

[1] This work was supported in part by Grant GB 2631 from the National Science Foundation.

[2] Post-doctoral trainee supported by Training Grant 5T1 GM 961-03 from the National Institutes of Health. Present address: Department of Zoology, Indiana University, Bloomington, Indiana.

and in bacteria, the ribulose diphosphate carboxylase reaction is repressed in the presence of organic material (Fuller *et al.*, 1961; Hudock and Levine, 1964).

Investigations were undertaken using the green alga *Chlamydomonas reinhardi* and the photosynthetic bacterium *Chromatium* strain D to elucidate further possible control mechanisms in the photosynthetic reduction cycle in both chloroplast systems and in the non-plastid bacterial systems.

## Materials and Methods

### ORGANISMS AND GROWTH CONDITIONS

Experiments on *Chlamydomonas reinhardi* mutant y-2 are reported in this paper. The mutant strain y-2 was derived from the wild type 137c. Cultures were grown in heterotrophic medium in the dark and allowed to re-green in the same medium (Hudock and Levine, 1964). *Chromatium* strain D was grown either in a photolithotrophic medium containing $CO_2$ and thiosulfate as the sole source of carbon and reductant or in an organic medium on sodium malate as a source of carbon and reductant (Fuller *et al.*, 1961).

### PREPARATION OF ENZYME EXTRACTS

Crude preparations of TPD were obtained as follows: Suspensions of cellular material were disrupted either by sonic oscillation or by liquid shear in the French press and spun at $20,000 \times g$ to remove particulate material.

TPD was extensively purified from *Chromatium* using fractionation with ammonium sulfate, and the fraction precipitating between 0·5 and 0·8 saturation was collected. This fraction contained 70–80% of the total TPD activity. The precipitate was re-suspended in a small volume of 0·1 M K phosphate buffer at pH 8·4 and centrifuged for $1\frac{1}{2}$ hr at $144,000 \times g$. The supernatant fluid contained all the remaining enzyme activity and the enzyme was further purified by passage through a column of Sephadex G-200 equilibrated in 0·1 M K phosphate buffer at pH 8·4 from which it was eluted in about 1·5 times the void volume. In order to crystallize the enzyme, the Sephadex eluant was brought to turbidity with ammonium sulfate and allowed to stand for 3 to 5 days. Twice re-crystallized enzyme had no higher specific activity than the Sephadex eluant and the latter was used for most studies. Purified enzyme was dialysed against saturated ammonium sulfate and stored in the cold in this manner until it was used.

The NAD-linked enzyme from *Chromatium* proved to be extremely unstable but could be stabilized by incubating and storing in the presence of NADH.

SH groups were measured with the nitroprusside reaction as described by Grunert and Phillips (1955).

### ENZYME ASSAY

TPD activity was assayed in both the oxidative and reductive direction as previously described (Fuller *et al.*, 1961).

## RESULTS

### *Chlamydomonas*

Since the wild type *Chlamydomonas* does not lose either its pigment or its chloroplast structure when grown in the dark, the mutant strain y-2 which does not form chloroplasts when grown in the dark was used for studies of bleaching and re-greening. The results of these experiments are shown in Fig. 1. During growth in the dark, chlorophyll synthesis ceases. NADP-linked activity per cell begins to decrease about the same time but at a lower rate than the chlorophyll

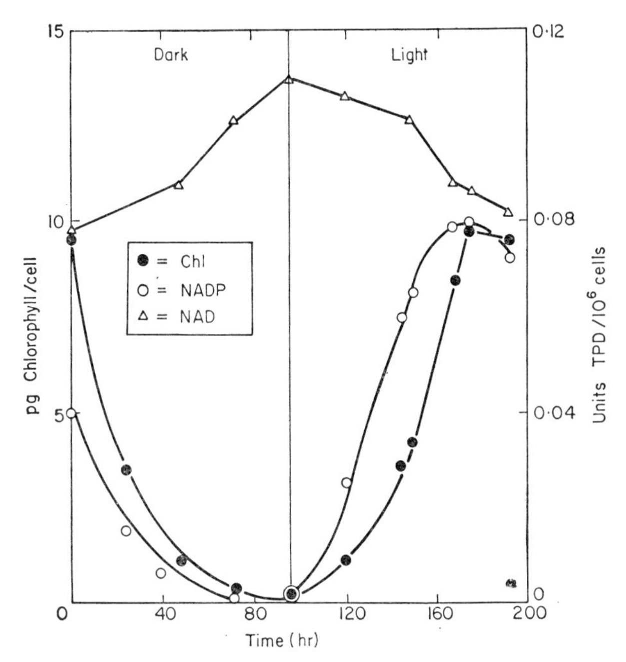


FIG. 1. Chlorophyll content, NAD-dependent TPD activity and NADP-dependent TPD activity during bleaching and re-greening of the y-2 mutant strain of *Chlamydomonas*. ● = pg chl/cell; △ = units NAD-dependent TPD per cell ($\times 10^9$); ○ = units NADP-dependent TPD per cell ($\times 10^9$).

content. NAD-linked activity increased almost two-fold during this period. On return of the cultures to the light, there is a rapid decrease in the NAD-linked activity and increases in both chlorophyll content and NADP-linked activity. It is interesting to note that the kinetics of the changes in the formation of the NADP-linked activity and the disappearance of the NAD-linked activity do not follow those of other photosynthetic enzymes in *Chlamydomonas reinhardi*. As shown in Fig. 2, the ribulose 1,5-diphosphate carboxylase shows a lag in formation relative to chlorophyll synthesis and the formation of chloroplast lamella. No such lag is seen in increase in NADP-linked activity or the decrease in the NAD-linked activity.

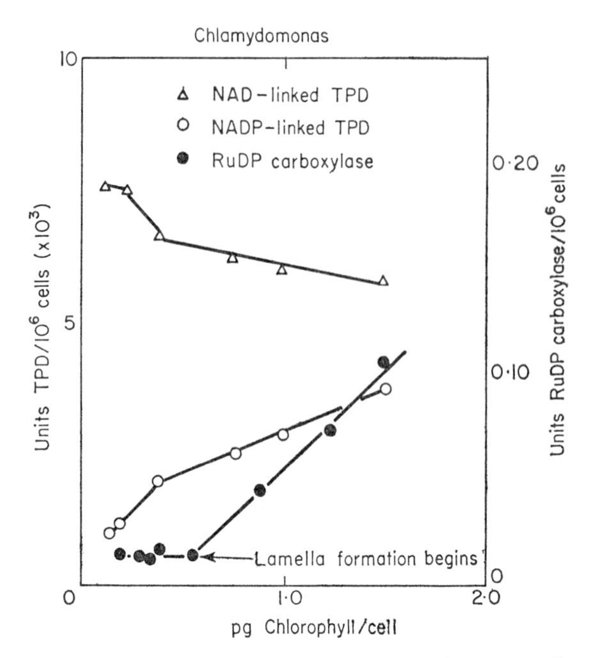


FIG. 2. Units of NAD- and NADP-dependent TPDs and RuDP carboxylase per cell plotted as a function of chlorophyll content during re-greening of y-2 *Chlamydomonas*. The arrow indicates the time at which paired lamellae are first seen in the developing chloroplast.

## EXPERIMENTS WITH THE PHOTOSYNTHETIC BACTERIUM *Chromatium*

As mentioned previously, the photosynthetic bacteria apparently lack NADP-linked TPD activity (Smillie and Fuller, 1960) whether they are grown under lithotrophic or organotrophic conditions, either in the light or in the dark. Therefore, the specificity of the NADP-linked enzyme for photosynthesis does not hold in the bacterial systems. We have studied the properties of the NAD-linked TPD of the obligately anaerobic phototrophic bacterium *Chromatium*, in order to clarify more fully the activities of this enzyme in the pentose reductive cycle in photosynthesis.

*Chromatium* strain D was grown either photolithotrophically or photo-organotrophically. Enzyme was purified from organisms grown under both sets of conditions as described previously (see Materials and Methods). The purified enzyme from both sources showed a sharp symmetrical peak of activity that was eluted from Sephadex G-200 and exhibited a monodispersed peak in the analytical ultracentrifuge, although most preparations from the ultracentrifuge show a marked tendency to dissociate into subunits very rapidly with concomitant loss in activity. However, as shown in Table I the enzyme could be stabilized and its activity maintained for a considerable period of time if stored in the presence of NADH. The purified enzyme also migrated as a single protein band during electrophoresis on acrylamide gel. The enzyme was found to have a sharp pH optimum between 8·3 and 8·5. The molecular weight was determined both by analytical ultracentrifugation and by elution from Sephadex G-100 and 200 columns and found to be 120,000 which is identical to TPD from a wide variety of other sources.

The $K_m$ values for both substrates and NAD and NADH for the purified proteins from both sources were also measured (see Table II). Measurements of both purified and crude preparations were identical, indicating that the differences in $K_m$ values were not purification artifacts. As can be seen, enzymes from cells grown under both sets of metabolic conditions had approximately the same $K_m$ values for NAD and NADH, but the $K_m$ values for DiPGA and glyceraldehyde 3-phosphate differed by a factor of three. The enzyme prepared

Table I. Effect of reduced pyridine nucleotide on stability of *Chromatium* TPD

| Time (hr) | | Enzyme (0·10 ml) $\Delta$OD/min |
|---|---|---|
| 10 | Untreated | 0·085 |
| | NADH | 0·087 |
| | NADPH | 0·080 |
| 24 | Untreated | 0·041 |
| | NADH | 0·068 |
| | NADPH | 0·040 |
| 48 | Untreated | 0·042 |
| | NADH | 0·064 |
| | NADPH | 0·045 |
| 72 | Untreated | 0·040 |
| | NADH | 0·060 |
| | NADPH | 0·039 |
| 144 | Untreated | Trace |
| | NADH | 0·020 |
| | NADPH | Trace |

from photolithotrophically grown cells had a $K_m$ for DiPGA of $3{\cdot}0 \times 10^{-3}$ M and that prepared from photo-organotrophically grown cells had a $K_m$ for DiPGA of $10^{-2}$ M. Although these differences are not large, they were consistent and did not overlap in five consecutive experiments.

The differences in these specific properties of the purified protein from cultures grown under two sets of conditions might lead one to think that two separate proteins both having a requirement for NAD might exist and be analogous to the NAD and NADP specificities in oxygen evolving organisms and chloroplasts. The major differences of the growth conditions are the source of carbon (inorganic *vs* organic) and the redox potential of the growth medium. Thiosulfate or $H_2S$ as reductant has a lower redox potential than do the organic sources of hydrogen. The effect of oxidation and reduction on the purified enzyme was measured and these results are shown in Table III. Reduction of the enzyme prepared from cells grown with sodium malate as a source of carbon and reductant was reduced by treating the enzyme with 0·1 M Na ascorbate (pH 8·5). Mild oxidation of the enzyme prepared from cells grown in the strongly reducing medium was carried out by simply dialysing against 0·1 M K phosphate buffer (pH 4·0) for 12 hr to remove cysteine which had been used to maintain the enzyme in a more stable form. Mild oxidation of the enzyme prepared from $CO_2$ grown material altered the $K_m$ of the enzyme for DiPGA from $3 \times 10^{-3}$ M to $9 \times 10^{-3}$ M, that is approximately equal to the enzyme prepared from malate grown cells. Conversely, mild reduction of the enzyme

Table II. $K_m$ values (M)[1] for purified TPD from *Chromatium*

| | Cells grown on | |
|---|---|---|
| | $H_2S$ and $CO_2$ | Malate |
| Glycerate 1,3-diphosphate | $3·0\ (\pm 1·3) \times 10^{-3}$ | $1·0\ (\pm 0·2) \times 10^{-2}$ |
| NADH | $5·0\ (\pm 1.2) \times 10^{-5}$ | $6·5\ (\pm 0·8) \times 10^{-5}$ |
| Glyceraldehyde 3-phosphate | $1·7\ (\pm 0·5) \times 10^{-4}$ | $5·0\ (\pm 1·0) \times 10^{-4}$ |
| $NAD^+$ | $3·0\ (\pm 0·9) \times 10^{-6}$ | $4·0\ (\pm 1·2) \times 10^{-6}$ |

[1] $K_m$ values, range of at least five separate determinations is given. Reaction mixtures (A) Glyceraldehyde 3-phosphate—DiPGA. The reaction mixture contained in a total of 3·0 ml: 100μmoles of *tris* (hydroxymethyl)-aminomethane, pH 8·4; 17 μmoles of Na arsenate; 12 μmoles cysteine, pH 8·4; 0·18 μmoles of NAD in standard reaction mixtures in determination of $K_m$ for G 3-P (varied from 0·002 μ mole to 0·36 μmole for determination of $K_m$ for NAD) and 0·1 ml enzyme preparation. The reaction mixture was incubated at 25° for 7 min and 1·0 μmole glyceraldehyde 3-phosphate was added (varied from 0·10 to 2·0 μmoles in determination of $K_m$ for G 3-P). After 30 sec, the optical density change at 340 mμ was followed for 3 min. (B) DiPGA—glyceraldehyde 3-phosphate. The reaction mixture contained in a total of 3·0 ml: 100 μmoles *tris*, pH 8·4; 20 μmoles of $MgSO_4, 7H_2O$; 12·0 μmoles cysteine, pH 8·4; 45 μmoles of PGA in standard assay (3·0–67·5 μmoles in determination of $K_m$ for DiPGA); two-fold excess of PGA-kinase (Sigma) in 0·10 ml; 5·0 μmoles of ATP, 0·80 μmole of NADH in standard assay (0·01–1·0 μmoles in determination of $K_m$ for NADH); 0·10 ml enzyme preparation. The optical density change at 340 mμ was followed for 3 min.

freshly prepared from malate grown cells changed the $K_m$ for DiPGA from $10^{-2}$ M to $4 \times 10^{-3}$ M, approximating the enzyme from the $CO_2$ grown material. Both of these changes were completely reversible.

The number of reactive –SH groups per mole of protein was measured in a highly purified enzyme prepared from both sources immediately after elution from Sephadex G-200. Determination of the thiol groups, therefore, was unaffected by the added cysteine. The native enzyme prepared from $CO_2$ grown material contained four available –SH groups per mole of protein. The enzyme prepared from the malate-grown material contained on average 2·4. Again, five independent determinations were carried out. Oxidation of the former reduced the number of reactive –SH groups to 2·4 and the reduction of the latter increased the number to 3·3. Changes in $K_m$ values were accompanied by change in the number of reactive –SH groups per protein molecule and again all changes were reversible. Thus, these *in vitro* changes in both the available –SH groups and the $K_m$ values parallel the *in vivo* differences, and are reversible, suggesting the existence of one protein which is altered depending upon the oxidizing and reducing conditions of the cellular environment.

Table III. Effect of oxidation and reduction on purified TPD from *Chromatium*[1]

| | Enzymes from $CO_2$ cells | Enzymes from malate cells | Enyzmes from $CO_2$ cells oxidized ($O_2$) | Enzymes from malate cells reduced (ascorbate) |
|---|---|---|---|---|
| Available –SH groups | 4·2 | 2·4 | 2·4 | 3·3 |
| $K_m$ for DiPGA | $3 \times 10^{-3}$ | $10 \times 10^{-3}$ | $9 \times 10^{-3}$ | $4{\cdot}0 \times 10^{-3}$ |

[1] Reduction of oxidized TPD was effected by adding one-tenth volume of 0·10 M Na ascorbate to the enzyme preparation. Oxidation of reduced TPD was carried out by overnight dialysis against 0·10 M K phosphate, pH 8·4; to remove cysteine and other possible reductants. In the former treatment, there was no loss of activity; in the latter treatment no more than 20% of the activity was lost. Reactions were carried out as described in Table II. Thiol groups were measured as described.

## Discussion

The procaryotic photosynthetic bacterial cell does not contain its photosynthetic system in any highly organized organelle such as the chloroplast. The bacterial cytoplasmic membrane elaborates to differing degrees as a function of species to form tubular intrusions which make up the "chromatophore fraction" of the cell. This chromatophore fraction contains the entire photochemical electron transport system. However, the enzymes of the photosynthetic reduction cycle are apparently soluble and not compartmentalized in the bacterial cell. Consequently, there is no physical separation of the function of TPD in photosynthesis from its function in glycolysis or biosynthesis. In addition, whereas green plants must support high levels of TPD activity in both glycolysis and biosynthesis, the primary activity in *Chromatium*, which does not store large amounts of reserve carbohydrates, is biosynthetic regardless of the carbon source.

During photolithotrophic growth *Chromatium* contains high ribulose 1,5-diphosphate carboxylase activity and actively reduces $CO_2$ via the Calvin cycle. TPD activity is needed for both maintenance of the Calvin cycle and for biosynthesis. Under photo-organotrophic conditions, ribulose 1,5-diphosphate carboxylase activity is reduced ten-fold and there is little Calvin cycle activity (Fuller *et al.*, 1961). The main function of TPD under such conditions would be the biosynthesis of C-6 compounds and some glycolysis depending upon the storage product formed. Therefore, although the same enzyme is used, a different function at least as to rate would be demanded of this enzyme under these two sets of metabolic conditions.

From the results obtained, it can be suggested that *Chromatium* may have developed a mechanism which combines the constitutive synthesis of TPD with control of its properties for both photosynthetic and non-photosynthetic metabolic functions. Extracts of *Chromatium* grown either photolithotrophically or photo-organotrophically contained the same amount of specific activity of TPD. Control of activity is indicated by the variation of the properties of the enzyme prepared from *Chromatium* grown under two sets of conditions reported. Studies of carbon metabolism in *Chromatium* have failed to detect any sizable pool of DiPGA or glyceraldehyde 3-phosphate while the cell is undergoing photosynthesis. With a large amount of enzyme present, minor changes in the metabolic specificities under these two metabolic conditions could strikingly alter the rates or even direction of the TPD reaction. This would also provide a very rapid control mechanism for shifts of activity in this cell, and not be dependent on protein synthesis, feedback inhibition, and other slow control mechanisms.

In *Chlamydomonas*, a true eucaryotic cell with highly organized lamella-containing chloroplast, a different situation exists. When grown in the dark, the y-2 mutant strain does not contain any chloroplast lamellae. When returned to the light, the entire re-greening process under the conditions studied requires only 8 to 10 hr. During this time, the chloroplasts containing no lamella are transformed into highly organized structures containing an elaborate internal membrane system (Hudock *et al.*, 1964).

Following an initial lag of 2 hr, and concomitant with the formation of the lamellae, various enzymes in the Calvin photosynthetic cycle are formed. However, during this lag period both an increase in the NADP-linked TPD activity and a decrease in the NAD-linked activity are observed.

The properties of the NADP-dependent TPD, found only in the chloroplasts of green plants, are extremely similar to those of the NAD-linked enzyme. They are in fact, so alike that a separation of the two enzymes in active form has not been accomplished.

The kinetics of the formation of the NADP-linked enzyme and the decrease in the amount of NAD-linked enzyme in *Chlamydomonas* suggest but clearly do not prove a possible conversion of one form of the enzyme into another. This possibility is particularly interesting in view of the findings of Ogren and Krogmann (1961) that there is a light-dependent conversion of NAD into NADP in several oxygen-evolving photosynthetic systems, and is analogous with the two forms of a single NAD-specific enzyme in *Chromatium* where such a conversion is apparently observed. Recent results with *Euglena*, which has a very different form of chloroplast, indicate that there is probably no such conversion in that cell (Brawerman and Konigsberg, 1960; Hudock and Fuller, 1965).

With the exception of the blue-green algae (Fewson *et al.*, 1962) all organisms that show NADP-linked TPD activity are true eucaryotic photosynthetic cells. That is, they contain a membrane-limited nucleus and chloroplast with carbon metabolism and all other photosynthetic reactions contained in membrane-bound form. The bacteria, however, do not show this evolved compartmentalization. The two alternate forms of TPD in the photosynthetic bacteria, might provide for a rapid direct metabolic control which might be called "biochemical compartmentalization" of the Calvin cycle in a procaryotic cell. A source of reducing power for growth, which readily alters the form of the enzyme *in vitro*, could well alter the form *in vivo*. Under the more reducing conditions, more available –SH groups would be found on the enzyme, substrate affinity would be increased and thus it would be capable of carrying out a more rapid pentose reduction cycle. Under less reducing conditions, a less reduced enzyme is formed primarily for the purpose of glycolysis and biosynthesis. The TPD reaction does not need to occur as rapidly or singularly in one direction under these conditions.

It can be suggested that as the photosynthetic apparatus of the eucaryotic cell evolved from procaryotic bacterial systems, numerous functions were included until all the enzymes in photosynthesis were included in this independent organelle. The isolation of these functions within the chloroplast would clearly constitute a strong selective agent. It is clear that the photosynthetic electron transfer system in higher plant chloroplasts is NADP-dependent and a primarily photosynthetic TPD with NADP-specificity would seem to be a logical development in such an organelle. Furthermore, the occurrence of a photosynthetic TPD which is NADP-specific and the light-dependent conversion of NAD into NADP would seem to be natural concomitants.

Figure 3 is an attempt to describe diagrammatically the situation as it might exist in both the eucaryotic and procaryotic cell. In the eucaryotic cell with the chloroplast carrying on the complete metabolic functions of photosynthesis, a Calvin cycle exists with an NADP-dependent TPD working primarily in the photosynthetic direction. Outside the chloroplast there is an NAD-dependent TPD working in both the glycolytic and biosynthetic directions. These two independently functioning systems are isolated in separate compartments of the cell although of course the chloroplast also carries on cellular functions as well. The procaryotic cell, represented from the work on *Chromatium* grown under two sets of metabolic conditions, is also diagrammed. Under metabolic conditions where carbon and hydrogen come from a weak reductant such as carboxyl of malate, the cell is carrying out primarily glycolysis and biosynthesis and the TPD exists in one form which is designated State I. However, when the cell is carrying on photosynthesis, a Calvin cycle operates. Of course, the State I functions still occur, but there is a shift of the TPD to State II in such a way that a rapidly running photosynthetic cycle could operate more efficiently.

Such a rapid control mechanism is indeed an intriguing possibility even in higher plant chloroplasts. In nature there are rapid shifts from sunlight to darkness and in the kind of metabolism that occurs in the chloroplast. The turnover of protein, and re-synthesis of new enzymes seems an awkward way for the chloroplast to handle these shifts in metabolism. The findings of Ogren and Krogmann (1961) of a light-dependent conversion of NAD into NADP, our work with *Chlamydomonas*, and the kinetics of the formation of TPD, which are at a variance with other photosynthetic systems, could suggest an alteration of a single protein to accommodate the change in pyridine nucleotide specificity. That such a shift in the protein in procaryotic bacterial cells occurs with the maintenance of pyridine nucleotide specificity already seems probable.

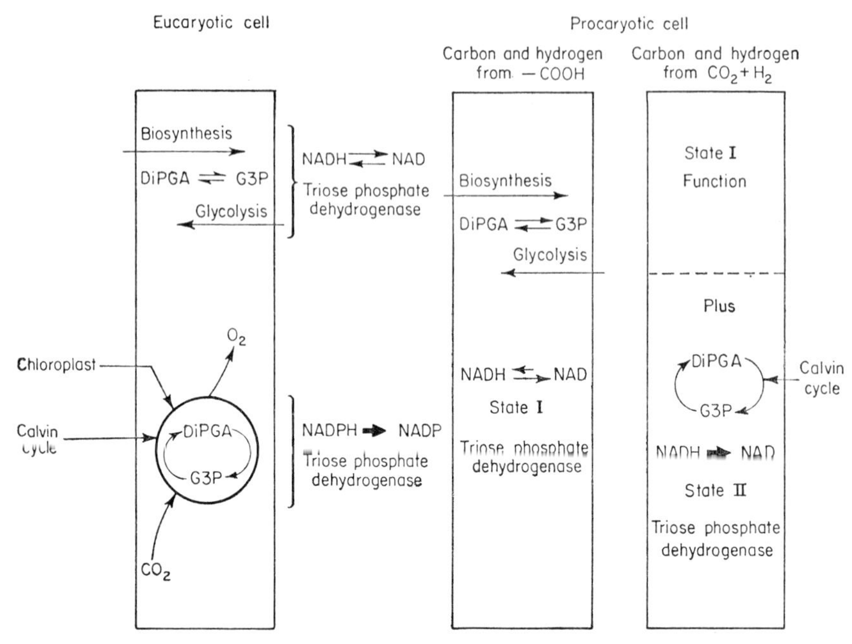


FIG. 3. Diagrammatic representation of the states of triosephosphate dehydrogenase in an hypothetical procaryotic and eucaryotic cell. As described in the test the two altered forms of the NAD enzyme in the parocaryotic cell may be the equivalent of the NAD and NADP linked enzyme in the compartmentalized eucaryotic cell.

## References

Brawerman, G. and Konigsberg, N. (1960). *Biochim. biophys. Acta* **43**, 374.
Fewson, C. A., Al-Hafidh, M. and Gibbs, M. (1962). *Pl. Physiol.* **37**, 402.
Fuller, R. C. and Gibbs, M. (1959). *Pl. Physiol.* **34**, 324.
Fuller, R. C., Smillie, R. M., Sisler, E. C. and Kornberg, H. L. (1961). *J. Biol. Chem.* **236**, 2140.
Grunert, R. R. and Phillips, P. H. (1955). *Archs Biochem. Biophys.* **30**, 217.
Hudock, G. A. and Levine, R. P. (1964). *Pl. Physiol.* **39**, 889.
Hudock, G. A. and Fuller, R. C. (1965). *Pl. Physiol.* **40**, 1205.
Hudock, G. A., McLeod, G. C., Moravkova-Kiely, J. and Levine, R. P. (1964). *Pl. Physiol.* **39**, 898.
Ogren, W. L. and Krogmann, D. W. (1961). *Fed. Proc.* **24**, 208.
Smillie, R. M. and Fuller, R. C. (1960). *Biochem. Biophys. Res. Commun.* **3**, 368.

# Distribution of $^3H$ in Photosynthetic Products after $CO_2$ Fixation in $^3H_2O$

Achim Trebst, Hans-Dieter Dorrer and Helmut Simon

*Organisch-chemisches Institut der Technischen Hochschule München und Pflanzenphysiologisches Institut der Universität Göttingen, Abt. Biochemie der Pflanzen, Göttingen, Germany*

One of the basic methods used to establish the path of carbon in photosynthesis is the determination of the intramolecular distribution of $^{14}C$ in a number of products formed after photosynthetic $CO_2$ assimilation. To investigate the path of hydrogen in carbohydrate biosynthesis the corresponding procedure is hampered from the start by the particular features encountered in work done with hydrogen isotopes such as exchange reactions, isotope effects, incorporation by isomerization reactions, etc. Moses and Calvin (1959) have shown that tritium is incorporated into the same photosynthetic products as $^{14}C$, but, for example, glycollic acid contained an unexpected high radioactivity. Vishniac (1963) studied the incorporation of tritium in a number of lipophilic compounds and Aronoff and Choi (1963) in carbohydrates in leaves.

We had determined the intramolecular distribution of tritium in carbohydrates in *Chlorella* after photosynthesis in TOH (Simon *et al.*, 1964). The results showed that isomerization reactions are mainly responsible for the incorporation of tritium into soluble sugar phosphates since the radioactivity at C-1 of fructose (6-phosphate) and at C-2 of glucose (monophosphate) was high after short incubation periods. However, a rather unexpected result was the absence of radioactivity at C-2 of glucose obtained from starch after short fixation times and the presence of very little radioactivity even after $\frac{1}{2}$ hr as compared with glucose from the glucose monophosphates. We proposed (Simon *et al.*, 1964) that this result can only be explained by a multi-enzyme complex in which: (1) a pool of hexose monophosphate is not in equilibrium, and small in comparison, with the main intracellular pool of hexose monophosphate, (2) the isomerization reaction of fructose 6-phosphate to glucose 6-phosphate is highly intramolecular and does not exchange with the protons of water, and (3) there is an isotope effect (of about 5) in the isomerization from fructose 6-phosphate to glucose 6-phosphate and on to starch and a discrimination against all fructose 6-phosphates which contain tritium at the isomerase-activated hydrogen at C-1.

Table I summarizes briefly this difference of the intramolecular distribution of tritium in glucose obtained from starch as compared with that from glucose monophosphate.

Since chloroplasts are assumed to contain the complete photosynthetic complex and since $CO_2$ fixation can be studied in isolated chloroplast systems from spinach we studied the distribution of tritium in glucose after photosynthesis in TOH of whole leaves and of isolated chloroplasts of leaves.

Since the tritium content at C-2 of glucose seems to be the most telling and

Table I. Intramolecular distribution of tritium in glucose from *Chlorella*[1]

| | Tritium content in % in glucose from G 6-Ⓟ | Starch |
|---|---|---|
| C-1 | 12 | 14·9 |
| C-2 | 30·8 | <1 |
| C-3 | 10·5 | 8·2 |
| C-4 | 22·9 | 19·1 |
| C-5 | 11·4 | 30·0 |
| C-6 | 18·3 | 24·9 |

[1] 10 min photosynthesis in TOH. Experimental details, see Simon *et al.* (1964).

unexpected result, Table II shows the tritium content at C-2 of glucose from starch obtained after photosynthesis in TOH with whole spinach leaves. Again as in *Chlorella* there is rather low tritium activity at C-2 (particularly if compared with the labelling in hexose phosphate of chloroplasts as in Table

Table II. Tritium content at C-2 of glucose from starch after photosynthesis of spinach leaves in TOH[2]

| Photosynthesis time (min) | Total T-content in starch ipm/leaf | T-content at C-2 of glucose from starch (in %) |
|---|---|---|
| 30 | $8{\cdot}6 . 10^4$ | 8·5 |
| 60 | $1 . 10^5$ | 7·6 |
| 120 | $2{\cdot}8 . 10^6$ | 4·5 |
| 240 | $1 . 10^7$ | 3·6 |

[2] Leaves in a desiccator were allowed to suck up water of 500 mc tritium/ml during illumination with 10,000 lux in an atmosphere of 5% $CO_2$/air. They were then homogenized and starch was isolated and degraded as by Simon *et al.* (1964).

III). Though the absolute tritium content in starch (and therefore at C-2 of glucose) increases with the fixation time there is a decrease in the percentage of tritium at C-2. The reason for this is not quite clear. It is different from the *Chlorella* experiments where the percentage at C-2 of glucose from starch is slowly increasing. It could be that the multi-enzyme system of spinach does not as strictly exclude cytoplasmic hexose monophosphates as that of *Chlorella*. Even after 4 hr equal distribution of tritium is not obtained in the glucose of starch since the labelling at C-2 is only 3·6% (there should be 14·3% at equal distribution).

Table III. Intramolecular distribution of tritium in glucose monophosphate and fructose monophosphate of a reconstituted chloroplast system after $CO_2$ fixation in TOH[1]

| Photosynthesis time (min) | Distribution (in %) of tritium in glucose 6-phosphate | | | fructose 6-phosphate | |
|---|---|---|---|---|---|
| | 3 | 6 | 20 | 3 | 6 |
| C-1 | 4·1 | 1·7 | 13·5 | 22·5 | 40·1 |
| C-2 | 45·0 | 28·8 | 22·5 | — | — |
| C-3 | 9·2 | 16·5 | 14·7 | 16·7 | 15·4 |
| C-4+C-5 | 33·4 | 34·3 | — | — | 34·0 |
| C-6 | 4·3 | 6·7 | 20·8 | 7·4 | 7·0 |

[1] 30 min photosynthesis; chloroplasts were prepared and incubated as by Trebst *et al.* (1959) in water containing 300 mc T; sugar phosphates were isolated and degraded as described by Simon *et al.* (1964).

Table III shows the intramolecular distribution in glucose monophosphate and fructose monophosphate after photosynthesis in a reconstituted chloroplast system. The tritium content at C-2 of glucose and at C-1 of fructose is very high and is comparable to the results obtained with soluble sugar phosphates from *Chlorella*. This shows that the assumed multi-enzyme system in intact *Chlorella* and spinach leaves is broken up during the preparation of the chloroplasts. The isomerization reaction at the level of hexose monophosphate proceeds at least partly (Rose and O'Connel, 1961) via a proton-exchange and is mainly responsible for the high tritium incorporation.

It is interesting to note the high percentage of tritium at C-6 of glucose in the chloroplast system (Table III) which becomes even higher than the tritium content at C-1 after several minutes photosynthesis. This might indicate either that the aldolase reaction has much less activity than the triose phosphate isomerase, which would lead to an exchange of the two halves of fructose diphosphate, or that there is a rapid recycling of the sugar phosphates. In any case the high tritium content at C-6 is in contradiction to the corresponding $^{14}CO_2$ experiments (Trebst and Fiedler, 1961; Havir and Gibbs, 1963). The very little $^{14}C$ activity in C-6 of glucose even after long photosynthesis in a reconstituted chloroplast system has been taken as an indication that the recycling of the sugar phosphates is the limiting or damaged part of the Calvin-cycle in isolated chloroplasts (Trebst and Fiedler, 1961; Havir and Gibbs, 1963).

## Summary

The small incorporation of tritium at C-2 of glucose from starch after photo-synthesis of *Chlorella* in TOH was interpreted as indication for a multi-enzyme system of photosynthetic $CO_2$ assimilation. This multi-enzyme system seems to be present also in the photosynthetic apparatus of spinach leaves. In the reconstituted chloroplast system, however, this multi-enzyme system is broken up and there is a high tritium content at C-2 of glucose monophosphates.

## Acknowledgement

The work described here was supported by Deutsche Forschungsgemeinschaft and Verband der Chemischen Industrie, Fonds der Chemie.

## References

Aronoff, S. and Choi, I. C. S. (1963). *Biochim. biophys. Acta* **102**, 159.
Havir, E. A. and Gibbs, M. (1963). *J. biol. Chem.* **238**, 3183.
Moses, V. and Calvin, M. (1959). *Biochim. biophys. Acta* **33**, 297.
Rose, I. A. and O'Connel, E. L. (1961). *J. biol. Chem.* **236**, 3086.
Simon, H., Dorrer, H.-D. and Trebst, A. (1964). *Z. Naturf.* **19**b, 734.
Trebst, A. and Fiedler, F. (1961). *Z. Naturf.* **16**b, 284.
Trebst, A., Losada, M. and Arnon, D. I. (1959). *J. biol. Chem.* **234**, 3055.
Vishniac, W. (1963). Symposiumsvortrag V. Int. Kongreß für Biochemie." Moskau 1961. Pergamon Press. *In* "Photosynthetic Mechanisms of Green Plants", National Academy of Sciences—National Research Council. Washington, D. C. 1963.

# Part III. Biosynthesis in Chloroplasts—Lipids

# Lipid Metabolism of Algae in the Light and in the Dark[1]

Konrad Bloch, G. Constantopoulos, C. Kenyon and J. Nagai
*Department of Chemistry, Harvard University, Cambridge, Massachusetts, U.S.A.*

The lipids of the chloroplast are remarkably complex and in several respects chemically unique (Benson, 1964). Among the lipids universally found in chloroplasts the monogalactosyl- and digalactosylglycerides are especially prominent. The unusually high content of α-linolenate and related polyunsaturated fatty acids linked to galactolipids is equally characteristic of green plant lipids. In addition, the enzymatic mechanisms for the synthesis of monounsaturated fatty acids in green plant tissues and in algae appear to differ in important detail from the mechanisms generally found in non-photosynthetic organisms. These special features of lipid chemistry and metabolism in higher and lower plants are not only of comparative interest but they also point to a specific functional or structural involvement of chloroplast lipids in the photosynthetic process. It should be mentioned at the outset that the lipid metabolism appears to be specialized only in the algae and the higher plants but not in photosynthetic bacteria.

The lipid metabolism of photosynthetic organisms first attracted our interest when we noted that the fatty acid patterns of *Euglena gracilis* differed strikingly in light-grown and dark-grown organisms (Erwin and Bloch, 1962). *Euglena gracilis* is uniquely suitable for comparing photoauxotrophic and heterotrophic patterns of metabolism and, for reasons which will become apparent, has been the organism of choice for studying lipid problems in photosynthesis in our laboratory.

## Enzymatic Syntheses of Mono-unsaturated Fatty Acids

In recent years we have described two major mechanisms for the biosynthesis of oleic acid and related mono-unsaturated fatty acids (Erwin and Bloch, 1964). One of the pathways, the oxygen-dependent desaturation of stearyl-CoA to oleyl-CoA, is widely distributed, occurring in animals, in yeast and in a variety of micro-organisms. A second, anaerobic route to unsaturated fatty acids has been found in bacteria and was shown to involve the dehydration of β-hydroxyacyl thioesters of medium length to β, γ enoates, intermediates which are then elongated to various isomeric hexadecenoic and octadecenoic acids. The latter mechanism is used by *Escherichia coli*, by Clostridia, and presumably by other anaerobes, but apparently not by higher organisms. So far neither

[1] Supported by grants-in-aid from the National Institutes of Health, the National Science Foundation and the Eugene Higgins Fund of Harvard University.

of the two mechanisms has been demonstrated in higher plants or in the Protophyta. Working with mitochondrial preparations of avocado mesocarp, Stumpf (1962) showed that oleic acid synthesis from acetyl-CoA and malonyl-CoA requires molecular oxygen as it does in the microsomal systems of yeast or liver; on the other hand, these plant preparations failed to desaturate stearate or stearyl-CoA. James (1963) has reported similar results for leaf tissue. Examining the synthesis of unsaturated fatty acids in various photosynthetic micro-organisms we were able to demonstrate the transformation of palmitate or stearate to oleate in representatives of the more primitive blue-green and red algae and in two chrysomonads (Erwin *et al.*, 1964; Levin *et al.*, 1964), but found no evidence for this conversion in euglenids and in several green algae (Table I). It therefore appeared that some of the lower plants also used a special mechanism for synthesizing oleic acid. However, since these experiments were carried out with intact cells a failure of the exogenous fatty acid to penetrate the cell membrane or to reach the chloroplast could have accounted for the negative results. That permeability is indeed one of the limiting factors is indicated by our more recent findings. For example, intact *Euglena gracilis* whether grown photoauxotrophically or heterotrophically in the dark, does not detectably convert stearate into oleate, yet, as I will show shortly, in cell-free extracts the transformations do occur provided one chooses the appropriate thioester substrate. Similar permeability barriers for the fatty acid substrate at the site of oleate formation do not seem to be limiting in *Anabaena*

Table I. Conversion of palmitate or stearate into oleate by intact cells

| | | | |
|---|---|---|---|
| *Anabaena variabilis* | +++ | *Chlorella pyrenoidosa* | − |
| *Porphyridium cruentum* | +++ | *Ankistrodesmus braunii* | − |
| *Poteriochromonas stipitata* | ++ | *Chlamydomonas reinhardi* | − |
| *Ochromonas malhamensis* | ++ | *Scenedesmus obliquus* | − |
| *Astasia longa* | +++ | *Euglena gracilis* | − |

*variabilis*, in *Porphyridium cruentum*, in the Chrysomonads or in the permanently bleached euglenid *Astasia longa*. These organisms carry out the transformation of stearate to oleate quite readily *in vivo*. From the examples cited it is clear that the failure of an exogenous substrate to be utilized *in vivo* is not necessarily valid evidence against the occurrence of a postulated transformation.

The experiments I am about to report have not solved the general problem of oleic acid synthesis in plants but they have clarified the mechanism of this reaction in at least one instance. In considering possible alternatives for the synthesis of oleic acid we felt that the evidence available for plant systems had ruled out only stearyl-CoA as precursor but not necessarily other derivatives of stearate. It seemed particularly worth exploring whether ACP, the heat-stable acyl carrier protein discovered by Vagelos and his group (Goldman *et al.*, 1963) as a component of *E. coli* fatty acid synthetase, might play a role also in the production of unsaturated fatty acids. Moreover, Overath and Stumpf (1964) have isolated from avocado mesocarp a heat-stable protein similar in properties to *E. coli* ACP, a finding which indicates a role for ACP in plant systems also. Stearyl-ACP, the presumed end-product of the elongation process, thus became a possible candidate as an oleic acid precursor. We synthesized this compound chemically and chose *Euglena gracilis* as enzyme source because a comparison of phototrophic and heterotrophic patterns of metabolism was one of the objects of this study. As shown in Fig. 1, stearyl-ACP is readily transformed to oleate (possibly oleyl-ACP) by extracts from cells grown in the light on a mineral medium (Nagai and Bloch, 1965). The enzyme from green *Euglena*, the first of this type from a plant source, has several unusual properties. When crude extracts are prepared by rupturing the cells in

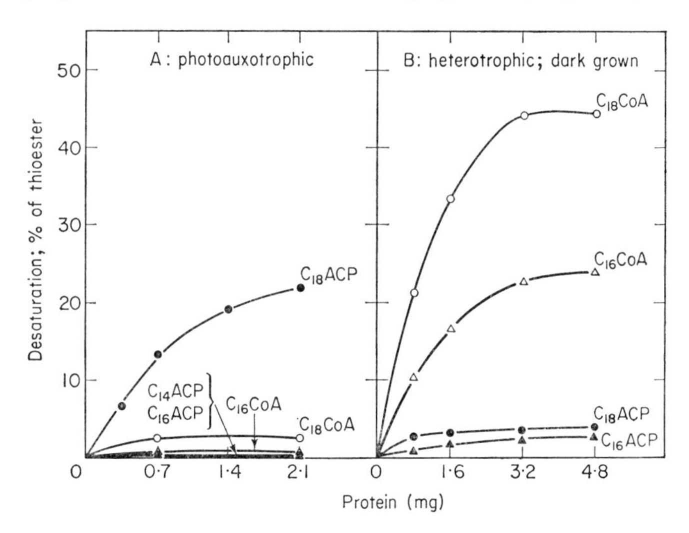


FIG. 1. For details of preparing enzymes and substrates see Nagai and Bloch, *J. biol. Chem.* **240**, PC 3702 (1965).

a French pressure cell, desaturating activity remains in the supernatant fraction on high-speed centrifugation. The solubility of the enzyme is noteworthy because the stearyl-CoA desaturases from *Mycobacterium phlei*, from yeast or from rat liver are all particle-bound enzymes. Secondly, the soluble stearyl-ACP desaturase has strict chain length and thioester specificity. Neither palmityl-ACP nor the coenzyme A esters of stearate or palmitate are converted. The soluble *Euglena* enzyme can be partly purified by ammonium sulfate fractionation and retains full activity on passage through Sephadex. At this stage cofactors other than oxygen and NADPH appear to be unnecessary. The soluble stearyl-ACP desaturase is probably a chloroplast enzyme since it is not present in etiolated cells. We have observed that the concentration of ACP is four times greater in green than in etiolated *Euglena* cells and believe, therefore, that ACP may be of special importance for the lipid metabolism of the chloroplast.

In extracts prepared from dark-grown *Euglena* the synthesis of mono-unsaturated fatty acids is of the conventional type (Nagai and Bloch, 1965). The enzyme system is particulate and desaturates either stearyl-CoA or palmityl-CoA, the substrates which are also active in a variety of non-photosynthetic systems. While *Euglena* employs one specific mechanism as a photoauxotroph, and another in the etiolated form, an intermediate situation seems to exist when the cells are grown in the light on an organic carbon source. Extracts of such cells convert both stearyl-ACP and stearyl-CoA into oleate.

Green *Euglena* cells contain an unusual variety of mono-unsaturated fatty acids (Table II). These include $\Delta^5$- and $\Delta^7$-tetradecenoic acids, $\Delta^7$- and $\Delta^9$-hexadecenoates and both oleate and *cis*-vaccenate (Hulanicka *et al.*, 1964; Korn, 1964). The origin of none of these acids, with the exception of oleate, is accounted for by the stearyl-ACP desaturase of green *Euglena*. The phenomenon of double bond isomerisms in mono-unsaturated fatty acid was originally encountered in a number of bacterial species which produce long chain monoenoic acids anaerobically by the elongation of $\Delta^3$-decenoate or $\Delta^3$-dodecenoate (Scheuerbrandt and Bloch, 1962). A biosynthetic mechanism of the bacterial type would therefore explain the structural diversity of the mono-unsaturated fatty acids found in green *Euglena*. In preliminary experiments we have indeed found certain similarities between the bacterial pathway and one of the routes that lead to mono-unsaturated fatty acids in *Euglena*. Extracts of

Table II. Mono-unsaturated fatty acids of *Euglena* lipids

| | Photoauxotrophic | Heterotrophic, etiolated |
|---|---|---|
| $C_{16}'$ | $\Delta^7$: 15–45 | 2 |
| | $\Delta^9$: 45–85 | 95 |
| | $\Delta^{11}$: 10 | 3 |
| $C_{18}'$ | $\Delta^7$: 0 | 0 |
| | $\Delta^9$: 70–90 | 55 |
| | $\Delta^{11}$: 10–30 | 45 |

Data from: Korn (1964). Hulanicka *et al.*, (1964). J. Nagai (unpublished).

photoauxotrophic *Euglena* cells elongate the ACP derivatives of the homologous series of fatty acids from $C_8$ to $C_{16}$ producing in all cases long chain saturated and unsaturated acids (Table III). The same overall transformations can be demonstrated in extracts of *Escherichia coli.* The two systems differ in that *Euglena* enzymes produce long chain unsaturated acids only aerobically (Table IV), whereas the bacterial enzyme system is oxygen-independent. A scheme rationalizing the various experimental findings and also the double bond structures of Euglena fatty acids is shown in Fig. 2. The oxidative introductions of the $\beta,\gamma$ double bond at the $C_{10}$ or $C_{12}$ stage are the key reactions of this scheme and these steps have yet to be demonstrated. In etiolated *Euglena* the desaturation-elongation pathway probably plays a lesser role for the following reason. In green *Euglena*, the $\Delta^7$- and $\Delta^9$-isomers of hexadecenoic acid occur simultaneously (Hulanicka *et al.*, 1964; Korn, 1964), but when the cells are grown in the dark the $\Delta^7$-isomer is missing (Nagai and Bloch, 1965) (Table II). No special mechanism need be postulated for the formation of

Table III. Chain elongation of acyl thioesters by extracts of photoauxotrophic *Euglena gracilis*

| Acyl thioester | Substrate d.p.m. (as hydroxamate) | Relative incorporation into | | | | | % of substrate elongated | Unsaturated / saturated |
|---|---|---|---|---|---|---|---|---|
| | | $C_{14}$ | $C_{16}$ | $C_{16}'$ | $C_{18}$ | $C_{18}'+C_{18}''$ | | |
| $C_8$-ACP | 14,000 | 3 | 3 | 8 | 100 | 174 | 14 | 1·7 |
| $C_{10}$-ACP | 22,000 | 2 | 17 | 2 | 100 | 178 | 22 | 1·5 |
| $C_{12}$-ACP | 13,000 | 12 | 11 | 4 | 100 | 241 | 17 | 2·0 |
| $C_{14}$-ACP | 3700 | – | 14 | 4 | 100 | 113 | 13 | 1·0 |
| $C_{16}$-ACP | 8300 | – | – | 0 | 100 | 118 | 5·1 | 1·2 |
| $C_{10}$-CoA | $2 \times 10^5$ | – | – | – | – | – | 2·2 | 0·6 |

palmitoleic acid, the $\Delta^9$-isomer, since the acyl-CoA desaturase of etiolated cells exhibits relatively broad chain-length specificity (Nagai and Bloch, 1965) and acts on palmityl-CoA.

Table IV. Requirement of oxygen for fatty acid synthesis in photoauxotrophic *Euglena gracilis*

| Substrate | | Relative incorporation into | | | | | % of substrate elongated | Unsaturated / saturated |
|---|---|---|---|---|---|---|---|---|
| | | $C_{14}$ | $C_{16}$ | $C_{16}'$ | $C_{18}$ | $C_{18}'+C_{18}''$ | | |
| $C_{10}$-ACP | air | 2 | 13 | 1 | 100 | 88 | 21 | 0·8 |
| 22,000 d.p.m. | $N_2$ | 2 | 12 | 1 | 100 | 9 | 19 | 0·08 |
| $C_{12}$-ACP | air | 12 | 11 | 4 | 100 | 241 | 25 | 2·0 |
| 8500 d.p.m. | $N_2$ | 5 | 4 | 1 | 100 | 9 | 30 | 0·08 |
| 8500 d.p.m. | $N_2$+ methylene blue | 5 | 3 | 3 | 100 | 13 | 24 | 0·14 |
| 8500 d.p.m. | $N_2$+PMS | 6 | 3 | 2 | 100 | 14 | 25 | 0·16 |

Chloroplast lipids have as major constituents not only α-linolenate but also a series of 16 carbon dienoic and polyunsaturated fatty acids containing double bonds in the usual divinyl pattern: $\Delta^{7,10}$-, $\Delta^{7,10,13}$-, and $\Delta^{4,7,10,13}$-$C_{16}$ (Debuch, 1962; Klenk and Knipprath, 1962). These "photosynthetic acids" are logically derived from a $\Delta^7$-monoene and it is therefore reasonable that

7-hexadecenoate is found in green but not in etiolated *Euglena*. Some photosynthetic bacteria also synthesize 7-hexadecenoate (Scheuerbrandt and Bloch, 1962) but in common with all other bacteria they lack the mechanisms for producing di- or poly-unsaturated acids.

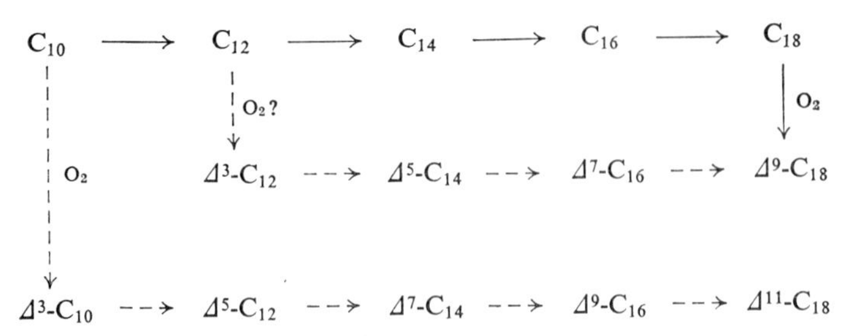


FIG. 2. Synthesis of unsaturated fatty acids in photoauxotrophic *Euglena gracilis*. The ACP derivatives are assumed to be the substrates in all the indicated transformations; from Nagai and Bloch (1965).

From our findings to date it is clear that at least in one plant system the mechanisms for the synthesis of mono-unsaturated fatty acids differ in several important details from those found in nonphotosynthetic organisms. One of these reactions is distinguished from the more widely occurring stearate desaturations by the solubility of the enzyme, the utilization of the ACP thioesters instead of the coenzyme A derivatives as substrates and the strict chain length specificity. The second pathway of phototrophic *Euglena* resembles the bacterial pathway in that it involves chain elongation of mono-unsaturated precursors. However, the introduction of the double bond which presumably occurs at the $C_{10}$ and the $C_{12}$ stage, requires oxygen in contrast to the bacterial process which is strictly anaerobic.

Exploratory experiments to demonstrate the *Euglena* pathways to mono-unsaturated fatty acids in *Chlorella* and in spinach have not been successful. It is therefore not clear whether these reactions occur generally in photosynthetic systems. At any rate, the example of *Euglena* is especially interesting because the biosynthetic patterns are fundamentally different in the green and in the etiolated form of the phytoflagellate. These differences emphasize not only the dual metabolic capacities of *Euglena* but also the apparent biosynthetic autonomy of the chloroplast.

## POLYUNSATURATED FATTY ACIDS

Several years ago our interest in comparative aspects of lipid metabolism led us to examine fatty acid patterns in photoauxotrophic and etiolated *Euglena*. The most conspicuous analytical result was the high concentration of α-linolenate in photoauxotrophic cells and the complete absence of this acid in cells that had been grown heterotrophically in the dark (Erwin and Bloch, 1962). The prominence of α-linolenate in plant lipids had long been known (Shorland, 1962) and the literature suggested that there might be a connection between α-linolenate content and photosynthetic activity. For example,

Crombie (1958) has shown that there is a marked rise of α-linolenate during the greening of developing leaves of *Citrolus vulgaris*. However, the possibility that α-linolenate might be of special significance in photosynthesis has been considered only recently (Erwin and Bloch, 1963; Rosenberg, 1963).

During the last few years the lipids of higher plants and of various representative algal groups have been analysed in great detail and the results clearly show an abundance of α-linolenate in photosynthetic tissue, the localization of this fatty acid in the chloroplast and in general a close relationship between α-linolenate levels and photosynthetic competence. Of all organisms investigated, only the photosynthetic bacteria were found to lack di- or polyunsaturated acids (Scheuerbrandt and Bloch, 1962) a deficiency which appears to be characteristic for the entire bacterial phylum. The apparently exceptional case of the photosynthetic bacteria has very much influenced our thinking and has led us to conclude that α-linolenate or lipids containing this acid might play some specific role in the photolysis of water in green plant photosynthesis rather than in photosynthesis *per se*. For various reasons our earlier view that α-linolenate might interact chemically with some components of the Hill system is no longer tenable. It now seems much more probable that lipids containing α-linolenate and related $C_{16}$ polyunsaturated acids serve as structural components of the photosynthetic apparatus facilitating, in some unknown way, optimal efficiency of electron transport in the more advanced or green plant type of photosynthesis. This, of course, is not an explanation but merely a restatement of the fact that systems carrying out the green plant type photosynthesis are associated with highly characteristic lipid structures which are not found in other energy conversion systems, such as bacterial photosynthesis or mitochondrial oxidation. In the meantime, until we have learned how membranes work and how lipid structure influences membrane properties it may serve a useful purpose to continue the structural and chemical analysis of the photosynthetic apparatus and to search carefully for relationships between structure and physiological activity.

In our earlier work we have focused attention only on α-linolenate but it has since become apparent that a second group of polyunsaturated fatty acids represented by 7,10,13-hexadecenoic acid or 4,7,10,13-hexadecenoic acid follow the α-linolenate pattern very closely, at least in the higher algae. Our results further emphasize the already well-established association of galactosylglycerides and polyunsaturated fatty acids. In all examined cases α-linolenate and the $C_{16}$ polyunsaturated fatty acids are linked preferentially to the glycolipids and these in turn are localized in the photosynthetic apparatus (Benson, 1964).

Results of special interest have been obtained by determining the fatty acid composition separately in the mono- and digalactosylglycerides (Table V). The content of polyunsaturated fatty acids in the monogalactosylglycerides is always high, α-linolenate and $C_{16}$-triene or tetraene accounting for two-thirds or more of the total fatty acids in *Chlamydomonas reinhardi*, in *Euglena gracilis* and in spinach. The digalactosylglycerides are much more saturated, at least in the algae. In *Chlamydomonas* this fraction contains essentially no α-linolenate. What is even more striking, the $C_{16}$ tri- or tetraenoic acids are virtually absent from the digalactosylglycerides of all chloroplast lipids examined so far. Since the degree of unsaturation must profoundly affect the various physicochemical properties of the lipid molecule, the two types of glycolipids are likely to make different contributions to chloroplast properties, whatever

Table V. Fatty acid composition of mono- (MGG) and digalactosylglycerides (DGG)

| | *Chlamydomonas reinhardi* | | *Euglena gracilis* (light) | | Spinach[1] | |
|---|---|---|---|---|---|---|
| | MGG | DGG | MGG | DGG | MGG | DGG |
| Palmitate | 2 | 52 | 6 | 17 | 1 | 17 |
| Stearate | 3 | 7 | 1 | 0 | 0 | – |
| Oleate | 13 | 22 | 9 | 19 | 1 | 4 |
| Linoleate | 6 | 5 | 6 | 12 | 1 | 3 |
| α-Linolenate | 32 | 2 | 41 | 26 | 25·5 | 53 |
| 4,7,10,13-$C_{16}$ | 33 | 0 | 32 | 7 | 30·0[2] | 0 |
| Total saturated | 5 | 59 | 7 | 17 | 1 | 17 |
| Total polyunsaturated | 65 | 2 | 75 | 33 | 55·5 | 53 |

[1] See also Allen *et al.* (1964).
[2] 7, 10, 13-$C_{16}$.

they are. The markedly different fatty acid profiles of the two glycolipids also pose a biosynthetic problem. A direct glycosylation of the monogalactosylglycerides to digalactosylglycerides seems highly improbable, unless one assumes that the highly unsaturated fatty acid residues are extensively reduced during this transformation. There is no evidence that reductive processes of this type occur in plant tissues.

Some of the relationships between the metabolism of galactosylglycerides and photosynthetic activity are illustrated in Figs. 3 to 10. When a culture of etiolated *Euglena* is placed in a mineral medium and then illuminated, chlorophyll synthesis and the formation of galactosylglycerides, starting from zero or near zero levels, proceed in synchronous fashion (Fig. 3). As a result the ratio of chlorophyll to glycolipid is fairly constant at all times. In *Chlorella vulgaris*, the formation of chloroplasts is not light induced. Nevertheless, the pigment-lipid relationships are qualitatively the same (Fig. 4), chlorophyll and galactolipid content increasing sharply on illumination as in *Euglena*. However, in *Chlorella* only the monogalactosylglycerides show a sharp rise on illumination, suggesting that for photosynthetic activity the level of digalactosylglycerides is less critical than the level of monogalactosylglycerides. We have already reported the striking changes in total fatty acid patterns that are seen on illumination of etiolated *Euglena* (Erwin and Bloch, 1963). The time course of these changes and the effects on mono- and digalactosylglycerides individually are shown in Figs. 5 and 6. The content of two "photosynthetic acids", α-linolenate and the $C_{16}$ tetraene, rises most steeply during the 10–25 hr illumination period, when the chlorophyll and galactolipid content of the cells also shows the greatest increase on a dry weight basis. At the same time the concentrations of palmitic, stearic or oleic acid fall sharply, changes which are consistent with precursor-product relationships between saturated and unsaturated fatty acids. Again these changes are much more pronounced in the monogalactosylglycerides than in the digalactosylglycerides.

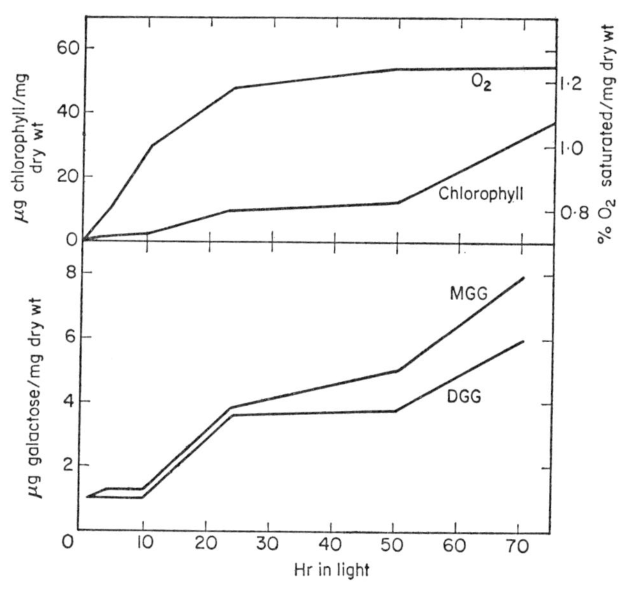


FIG. 3. *Euglena gracilis* Z were grown with shaking in organic medium in the dark at 28° and harvested in the late log phase. After centrifugation the cells were resuspended in mineral medium. Incubation was continued in the light and 5% $CO_2$-air passed through the culture. Samples were taken at the time of harvesting and 1, 2, 4, 10½, 24, 50½, and 72 hr after the transfer to mineral medium. Chlorophyll was determined by the method of Arnon (*Pl. Physiol.* **24**, 1 (1949)). Oxygen evolution was determined with an oxygen electrode. Mono- and digalactosyldiglycerides (MGG and DGG) were isolated from a $CHCl_3$–MeOH (2/1, v/v) extract of the cells by a modification of the method of Nichols (*Biochim. biophys. Acta* **70**, 1 (1963)). Galactose was determined by the anthrone method.

In *Scenedesmus* $D_3$, another alga which produces chloroplasts in the dark, light has similar but less pronounced effects on lipid patterns. Neither chlorophyll nor the content of galactolipid change significantly on illumination of dark-grown cells. In *Scenedesmus*, α-linolenate and $C_{16}$ tetraene are synthesized in the dark, at least to some extent, and this may be true for all photosynthetic organisms in which chloroplast formation does not require induction by light. Galactolipid synthesis by the chloroplast is therefore not a light-dependent process nor is photosynthetically evolved oxygen essential for the formation of polyunsaturated fatty acids even though this process requires oxygen. Still, illumination of dark-grown *Scenedesmus* markedly elevates the

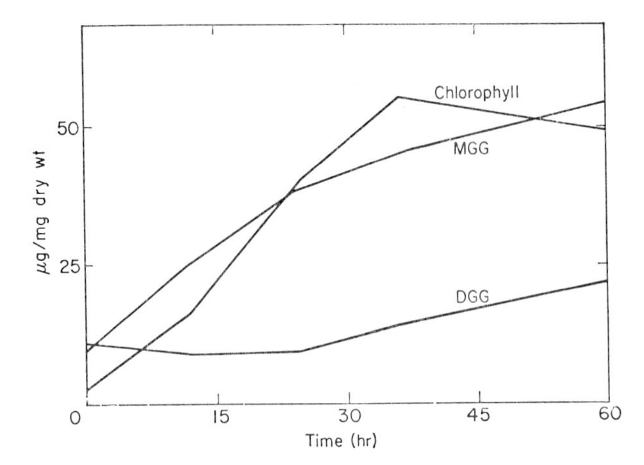


FIG. 4. *Chlorella vulgaris*, wild type, was grown in the dark in a glucose-supplemented medium. The cells were harvested in the early log phase, washed with sterile phosphate buffer pH 5·8 and resuspended in a mineral medium. At zero time the cells were exposed to light and a gas mixture of 5% $CO_2$-air passed through the culture. For the analysis of chlorophyll and of galactolipids see legend to Fig. 3. Lipid-galactose was determined by the "galactostat" method.

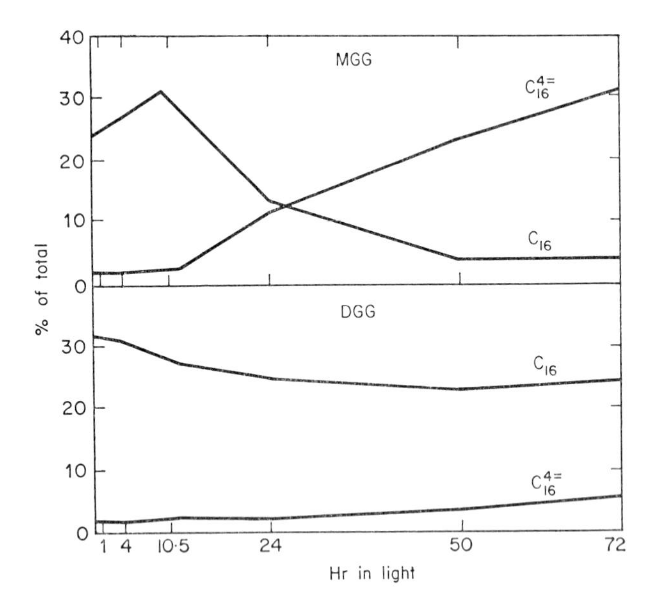


FIG. 5. The mono- and digalactosyldiglyceride fractions were isolated as described in the legend to Fig. 3. The lipids were saponified, the liberated fatty acids converted into the methyl esters and analysed by gas-liquid chromatography (Goldfine and Bloch, *J. biol. Chem.* **236**, 2596 (1961)).

proportion of α-linolenate and of the $C_{16}$ tetraene, the rise occurring again almost exclusively in the monogalactosylglyceride fraction (Figs. 7, 8). The same series of experiments extends earlier observations on the reduction of α-linolenate synthesis by CMU (Erwin and Bloch, 1963). In *Scenedesmus*, as well as in *Euglena*, the Hill reaction inhibitor prevents the light-induced stimulation of polyunsaturated fatty acid synthesis causing the galactolipids to remain as saturated as they are in dark. Again the fatty acid profile of the digalactosylglycerides is scarcely altered. In the dark CMU has no effect on the growth of *Scenedesmus*, on chlorophyll content or any of the lipid constituents of the chloroplast, in line with current views that the inhibitor impairs certain functional activities but not the integrity of the chloroplast.

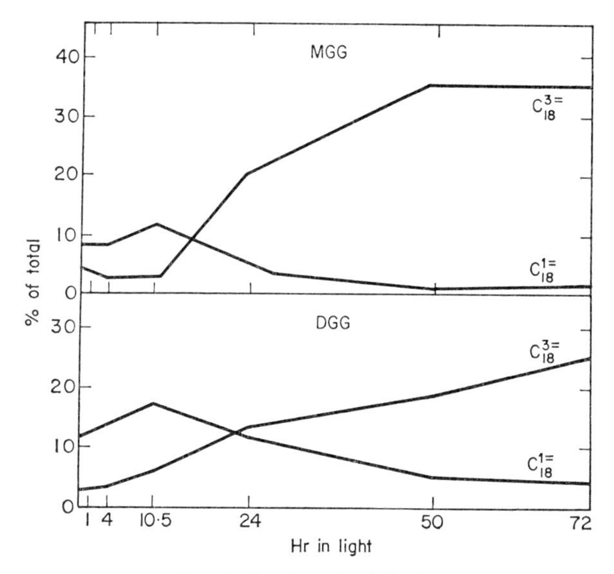


FIG. 6. See legend to Fig. 5.

Further evidence showing a relationship between the degree of unsaturation of chloroplast lipids and photosynthetic activity has been obtained by growing photoauxotrophic *Euglena* at varying light intensities (Fig. 9). As is well known from the earlier literature, the chlorophyll content of algal cells declines with increasing light intensity even though growth is markedly accelerated. At the higher light intensities the total lipid including galactosylglycerides decreases, in close parallel with the decline of the chlorophyll content. However, as the lipid concentration falls the galactolipids become markedly more unsaturated owing to the increase of α-linolenate and $C_{16}$-tetraene in the monogalactosylglyceride fraction (Fig. 10). The time course of this response and of the rise of Hill activity are remarkably similar. Light intensity has no influence on the fatty acid pattern of the digalactosylglycerides showing again the relative constancy of this fraction under a variety of conditions.

From these observations the following principal conclusions may be drawn.

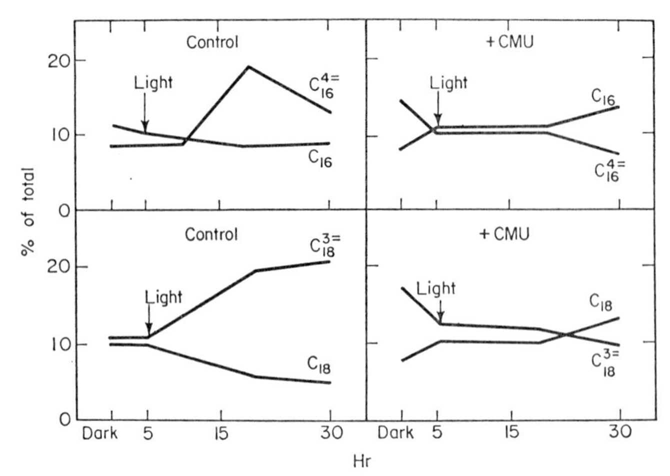


FIG. 7. Fatty acid changes in monogalactosyldiglyceride fraction. Dark-adapted *Scenedesmus* $D_3$ was grown with shaking in organic medium in the dark at 28° with or without $8 \times 10^{-6}$ M CMU. When growth had reached the late log phase, the cells were harvested by centrifugation and resuspended in mineral medium with or without $8 \times 10^{-6}$ M CMU as before. Incubation was continued in the dark for 5 hr. The cultures were then placed in the light and aerated with 5% $CO_2$-air. Samples were removed at the time of harvesting, after 5 hr in mineral medium in the dark, and after 12 and 24 hr in the light. The mono- and digalactosyldiglyceride fractions were isolated and the fatty acids analysed as described in the legends to Figs. 3 and 5.

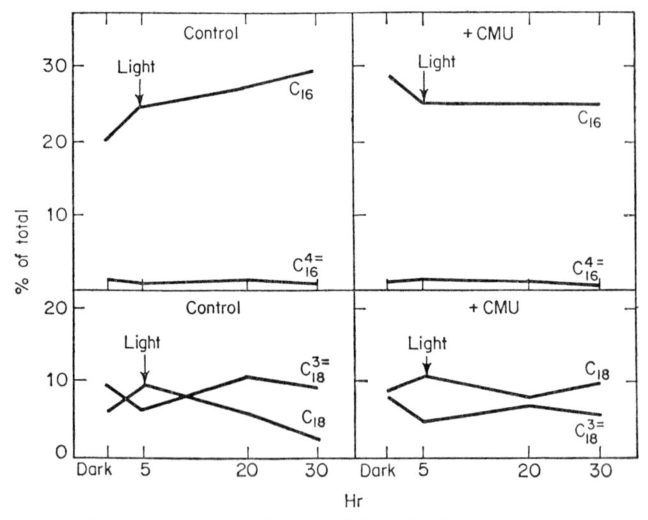


FIG. 8. Fatty acid changes in Digalactosyldiglyceride fraction. See legend to Fig. 7.

In the dark as well as in the light the morphogenesis of the chloroplast and the synthesis of galactosylglycerides occur in synchrony. The ratio of mono- to digalactosylglycerides is not constant but tends to increase in the light. The fatty acid composition of the two glycolipids differs markedly in the algae investigated, the content of polyunsaturated fatty acids tending to be much higher in the mono- than in the digalactosylglycerides. Only traces of $C_{16}$-triene or tetraene are found in the digalactosylglycerides. Algae, capable of forming chloroplasts in the dark, contain relatively saturated galactosylglycerides. The proportions of α-linolenate and of $C_{16}$ polyenoic acid rise as a function of light intensity, but only in the monogalactosylglyceride fraction.

We have already stressed the universal inability of bacteria including photosynthetic species to synthesize di- or polyunsaturated fatty acids (Erwin and Bloch, 1964). It now appears that mono- and digalactosylglycerides are also absent in photosynthetic bacteria, with the possible exception of a *Chlorobium* species (Nichols and James, 1965). In fact, the lipids of photosynthetic bacteria consist mainly of phosphatidylethanolamine and

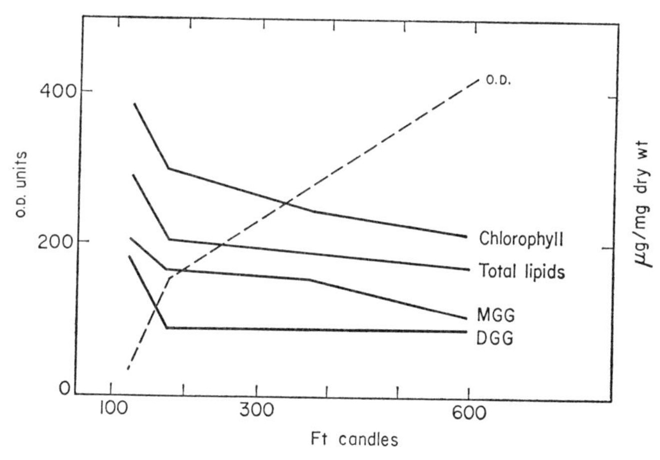


FIG. 9. Four cultures of *Euglena gracilis* Z were grown on a mineral medium at room temperature at light intensities of 120, 175, 370 and 610 ft candles respectively. A mixture of 5% $CO_2$-air was continuously passed through the culture. The rate of growth was followed by measuring O.D. at 750 mμ. Chlorophyll and lipid analyses were carried out as mentioned in the legend to Fig. 4.

phosphatidylglycerol and thus are no more complex than those of non-photosynthetic bacterial species. The more diversified lipid chemistry of algae and of higher plants may, therefore, be viewed as an expression of the more advanced type of photosynthesis. Structural complexity of the photosynthetic apparatus, i.e. chloroplast *vs* chromatophore, is probably only one of the factors, since the chromatophores of the blue-green algae contain mono- and digalactosylglycerides as well as α-linolenate (Levin *et al.*, 1964). We have in the past cited the example of the blue-green algae in support of the hypothesis that α-linolenate-containing glycolipids play some specific role in the photolysis of water, the argument being that in the blue-green algae photosynthesis is of the advanced or green plant type, even though their morphology is simple and closely related to that of the bacteria, Unfortunately the recent examination of one organism invalidates this generalization. Holton *et al.* (1964), have failed to find α-linolenate among the fatty acids of *Anacystis nidulans* and this has been confirmed in our laboratory (P. Gold unpublished). On the other hand *Anacystis* contains some linoleic acid and also mono- and digalactosyl-glycerides. Perhaps the feature distinguishing the photosynthetic systems of plants and bacteria is the presence of galactosylglycerides rather than the content of polyunsaturated fatty acids. Nevertheless, the presence of α-linolenate and $C_{16}$ polyenes in algae and higher plants is so common that it must have some physiological significance. Drs. Cohen-Bazire and Stanier have suggested to us that α-linolenate might be structure-determining, instead of serving as a component of the oxygen-evolving system. According to this view α-linolenate may be associated with lamellar but not with vesicular arrangements of the photosynthetic apparatus. We have, therefore, analysed the lipids from photosynthetic bacteria of both categories (vesicular: *Rhodospirillum*

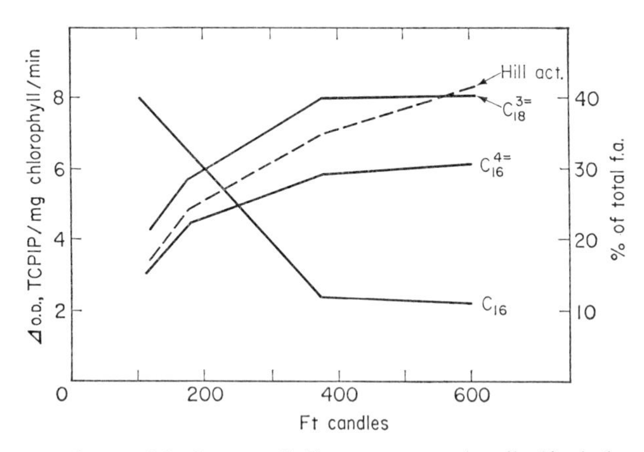


FIG. 10. Four cultures of *Euglena gracilis* Z were grown as described in the legend of Fig. 9. Chloroplasts were prepared in 0·04 M sucrose–0·01 M phosphate buffer pH 7·0. Hill activity was measured by following the reduction of TCP I P at 620 mμ. For fatty acid analyses see legend to Fig. 5.

*rubrum, Rhodopseudomonas spheroides*; lamellar: *Rhodospirillum molischianum, Rhodopseudomonas palustris*) but have found no differences in their fatty acid profiles. None of these organisms contains α-linolenate nor significant amounts of galactosylglycerides. It therefore remains an attractive working hypothesis to regard the highly unsaturated galactosylglycerides as specialized lipids which in a manner yet to be determined provide the optimal environment for the electron transport processes of higher plant photosynthesis.

One of the aims of biochemistry is to correlate chemical structure with physiological function. The structures of the polyunsaturated fatty acids suggest that they belong to two major functional types (Fig. 11). In the "plant" acids, α-linolenate, 7,10,13-$C_{16}$ and 4,7,10,13-$C_{16}$, the poly *cis* double bond system is maximally extended towards the methyl end, leaving only two saturated terminal carbons and making the distal portion of the molecule relatively rigid. Conversely, in γ-linolenate and arachidonate, acids which are essential for animals and are associated with mitochondrial electron transport systems, the terminal portion consisting of five saturated carbons is much more flexible and suitable for hydrophobic interactions. It will be a challenge

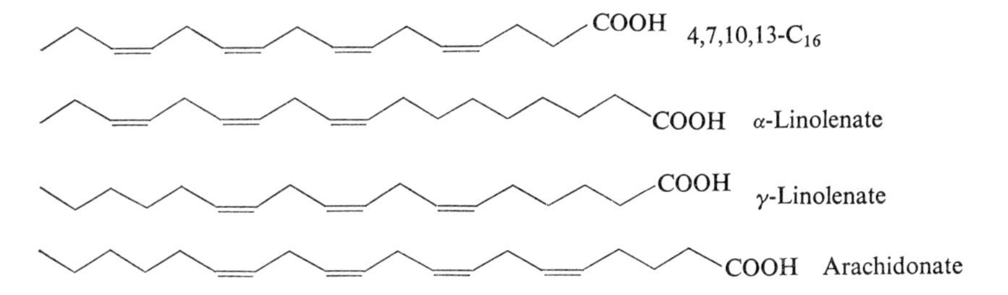


to rationalize the differences of these two structural types in terms of specific properties of the membranes of which these lipid molecules are component parts.

## References

Allen, C. F., Good, P., Davis, H. F. and Fowler, F. D. (1964). *Biochem. Biophys. Res. Commun.* **15**, 124.
Benson, A. (1964). *Annu. Rev. Pl. Physiol.* **15**, 1.
Crombie, W. M. (1958). *J. exp. Bot.* **9**, 254.
Debuch, H. (1962). *Experientia* **18**, 61.
Erwin, J. and Bloch, K. (1962). *Biochem. Biophys. Res. Commun.* **9**, 2.
Erwin, J. and Bloch, K. (1963). *Biochem. Z.* **338**, 446.
Erwin, J. and Bloch, K. (1964). *Science, N.Y.* **143**, 1006.
Erwin, J., Hulanicka, D. and Bloch, K. (1964). *Comp. Biochem. Physiol.* **12**, 191.
Goldman, P., Alberts, A. W. and Vagelos, P. R. (1963). *J. biol. Chem.* **238**, 557.
Holton, R. W., Blecker, H. H. and Ohore, M. (1964). *Phytochemistry* **3**, 595.
Hulanicka, D., Erwin, J. and Bloch, K. (1964). *J. biol. Chem.* **239**, 2778.
James, A. T. (1963). *In* "The Control of Lipid Metabolism" (J. K. Grant, ed.) p. 17. Academic Press, New York and London.
Klenk, E. and Knipprath, W. (1962). *Z. phys. Chem.* **327**, 283.
Korn, E. D. (1964). *J. Lipid Res.* **5**, 352.
Levin, E., Lennarz, W. J. and Bloch, K. (1964). *Biochim. biophys. Acta* **84**, 471.
Nagai, J. and Bloch, K. (1965). *J. biol. Chem.* **240**, 3702.
Nichols, B. W. and James, A. T. (1965). *Biochem. J.* **94**, 22P.
Overath, P. and Stumpf, P. K. (1964). *J. biol. Chem.* **239**, 4103.
Rosenberg, A. (1963). *Biochemistry* **2**, 1148.
Scheuerbrandt, G. and Bloch, K. (1962). *J. biol. Chem.* **237**, 7.
Shorland, F. B. (1962). *In* "Comparative Biochemistry" (M. Florkin and H. S. Mason, eds.) Vol. III, p. 1. Academic Press, New York and London.
Stumpf, P. K. (1962). *Nature, Lond.* **194**, 1158.

# Biosynthesis of Fatty Acids by Photosynthetic Tissues of Higher Plants

P. K. Stumpf,[1] J. Brooks,[2] T. Galliard, J. C Hawke
and R. Simoni[3]

*Department of Biochemistry and Biophysics,
University of California, Davis, California, U.S.A.*

Considerable attention has been focused in recent years on the composition of fatty acids in a wide variety of tissues. Earlier in this conference, discussion centered on the types of fatty acids in photosynthetic tissues and speculations were submitted as to their functions. In this chapter, we shall review the work conducted during the past few years on the biosynthesis of fatty acids by a variety of photosynthetic tissue preparations of higher plants.

## BIOSYNTHESIS IN THE INTACT LEAF

Surprisingly little work has been conducted with intact leaf tissue. In 1963 James reported an extensive series of experiments with castor leaf tissue. The intact leaf has a relatively simple pattern of fatty acids, namely 21 % of palmitic, 13 % of linoleic and 60 % of α-linolenic acids. Both stearic and oleic acid are minor components. Ricinoleic acid is absent in these tissue. When [$^{14}$C]acetate was added to chopped leaf preparations and $^{14}$C followed in different fatty acids as a function of time, palmitic and oleic acids were rapidly labelled; with time, the per cent incorporation declined in palmitic and oleic acids while incorporation into linoleic rose. Smaller incorporation values were noted in lauric, myristic, and stearic acids.

James then extended his observations with experiments similar to those carried out previously by Bloch and his group. Thus when octanoic, decanoic, lauric and myristic acids labelled in the carboxyl carbon with $^{14}$C were introduced via stems of intact castor leaf tissue and permitted to metabolize for 3 hr, some conversions occurred to the higher homologues. For example, [1-$^{14}$C]octanoic was elongated to small amounts of myristic, palmitic, oleic and linoleic acids; both [1-$^{14}$C]decanoic and [1-$^{14}$C]lauric gave similar elongation patterns. Interestingly both palmitic and [1-$^{14}$C]stearic acids were completely ineffective as precursors for the formation of monoenoic and dienoic acids under different conditions and varying lengths of time. Experiments were repeated with chopped leaf preparations and similar results were

[1] Supported in part by a USPHS Grant No. 10132 and a National Science Foundation Grant No. GB-2352.
[2] J.B. (NIH Predoctoral Fellowship No. 5F1-GM-15897).
[3] R.S. (NIH Predoctoral Fellowship No. 1F1-GM-24738).

obtained. In addition myristic acid was elongated to higher homologues with some conversion into oleic acid. $O_2$ was required for the formation of the unsaturated fatty acids with [$^{14}$C]acetate as substrate. These results confirmed earlier results published by Mudd and Stumpf (1962) in which they reported that both palmitic and stearic acids were ineffective as precursors of oleic acid with avocado mesocarp preparations and that oxygen was an essential component for the formation of oleic acid from [$^{14}$C]acetate.

On the basis of these experimental results, James suggested that there might be two separate pools of longer chain fatty acids in leaf tissues. In one pool myristic acid is converted into palmitic and stearic acids, which are then esterified to give galactolipids and phospholipids. In the second pool myristic acid

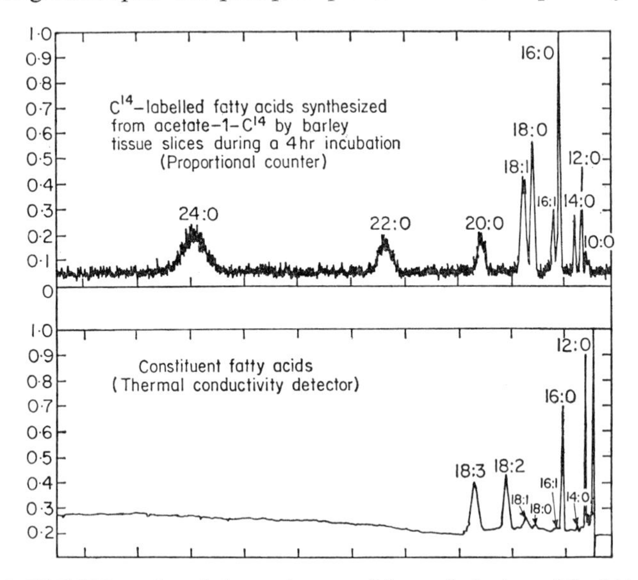


FIG. 1. A GLC-$^{14}$C monitored chromatogram of the methyl esters of the fatty acids prepared from etiolated barley tissues with [1-$^{14}$C]acetate as substrate. DEGS column at 160°. Aerograph, Model A-90P2 and Nuclear-Chicago Biospan Model 4998 as $^{14}$C detector.

is converted into palmitic and then to stearic and oleic with only oleic available for further conversions into linoleic and α-linolenic acids.

Recently Hawke and Stumpf (1965a) carried out extensive investigations with young blade tissues of five species of the *Graminae* family. Since the preliminary results with barley, wheat, oat, maize and ryegrass were quite similar, most of the experiments were carried out with both etiolated and green barley tissue. Figure 1 shows a typical GLC-$^{14}$C monitored tracing of fatty acids synthesized from [1-$^{14}$C]acetate by barley tissue slices grown aseptically in the dark. While a number of fatty acids are labelled, the principal [$^{14}$C] fatty acids include palmitic, stearic, oleic, arachidic, behenic, and lignoceric acid with smaller amounts of label in cerotic acid. Approximately 50% of the total $^{14}$C was consistently observed in the $C_{20}$–$C_{26}$ long chain saturated fatty acids. Of interest, although dynamically the leaf tissue was incorporating large amounts of $^{14}C$ into very long chain fatty acids, only 0·5–2% of the mass of fatty acid belong in this category. Thus for some reason or other, the young tissue is converting major portions of available [$^{14}C$]acetic into acids which are normally only of minor concentration in the tissues.

When green and etiolated tissues of barley were examined as to their capacity to incorporate [$^{14}C$]acetate, two points of interest were observed as illustrated in Table I. The amount of $C_{20}$–$C_{26}$ fatty acids synthesized from acetate was essentially the same although a larger proportion of lignoceric was synthesized by the etiolated than by the green tissue. Of more interest is the observation that green tissue incorporated [$^{14}C$]acetate to a rather significant extent into

Table 1. Incorporation of [1-$^{14}C$]acetate into fatty acids by green and etiolated tissue slices

| Fatty acid | Incorporation of [1-$^{14}C$]-acetate by plant tissue (mμmoles) Green | Etiolated |
|---|---|---|
| Total fatty acid | 9·2 | 5·77 |
| 10:0 | 0·08 | 0·0 |
| 12:0 | 0·12 | 0·16 |
| 14:0 | 0·12 | 0·12 |
| 16:0 | 1·14 | 0·72 |
| 16:1 | 0·08 | 0·49 |
| 18:0 | 1·70 | 0·29 |
| 18:1 | 1·43 | 0·35 |
| 18:2 | 0·73 | 0·0 |
| 18:3 | 0·28 | 0·0 |
| 20:0 | 0·57 | 0·32 |
| 22:0 | 1·0 | 0·68 |
| 24:0 | 1·49 | 2·19 |
| 26:0 | 0·40 | 0·55 |

polyunsaturated fatty acids. This observation is in line with the results of Bloch and others who have reported a high concentration of polyunsaturated fatty acid in photosynthetic organisms.

Degradation studies of the $^{14}C$ labelled fatty acids revealed two interesting points; if the $^{14}C$ label derived from [1-$^{14}C$]acetate were uniformly distributed at the odd numbered carbons along the fatty acid chain, the radioactivity located in the carboxyl carbon of 14:0, 16:0, 18:0, 20:0, 22:0, 24:0, and 26:0 fatty acids would be respectively 14·3, 12·5, 11·1, 10·0, 9·1, 8·3, and 7·7% of the total $^{14}C$. Results obtained by Hawke and Stumpf (1965a) showed that only myristic and palmitic acids synthesized from [1-$^{14}C$]acetate were consistent with such a pattern. With the saturated fatty acids from 18 to 26 carbon atoms, the decarboxylation data suggested that palmitate was the primary fatty acid onto which $C_2$ units were added to construct the higher homologues. The second observation of interest was that when the 18:1 monoenoic acid synthesized from [1-$^{14}$C]acetate was degraded by reductive ozonolysis, 50% was vaccenic acid and the remainder oleic acid. Direct evidence for palmitoleic acid conversion into vaccenic acid was not obtained, but it was noted that when label in palmitoleic acid was high the vaccenic/oleic ratio was also high. Since all the experiments of Hawke reported here were carried out under carefully controlled aseptic conditions with sterile tissue, the possibility of bacterial contamination is remote.

The types of fatty acids found in the endogeneous lipid of barley seedlings and those synthesized *de novo* from [$^{14}$C]acetate disclose sharp differences. Similar results have been observed with avocado mesocarp and with intact lettuce leaf tissue. With green barley seedlings, palmitic, linoleic and linolenic acids are the principal components of the endogeneous lipids; the newly synthesized acids from [$^{14}$C]acetate are primarily palmitic, stearic and oleic acids with smaller amounts of linoleic and linolenic acids; considerable amounts of long chain fatty acids such as arachidic, behenic, lignoceric and cerotic acids are also found. While age of the tissue and length of incubation are obvious factors to be considered, in a dynamic situation such as incorporation of [1-$^{14}$C]acetate by developing tissue for a brief period of time, there must exist additional factors which markedly affect the type of fatty acids synthesized. These intriguing problems await further investigations.

Hawke and Stumpf (1965b) further extended the type of experiments carried out earlier by Erwin and Bloch (1964) with bacteria and by James (1963) with castor leaves, in order to shed more light on the mechanism of monoene biosynthesis in higher plants. In 1963, Bloch proposed the following reactions to explain the synthesis of oleic acid in higher plants:

$$\text{Lauric} \rightarrow \text{2-dodecenoate} \quad (1)$$

$$\text{2-dodecenoate} \rightarrow \beta\text{-hydroxydodecanoate} \rightarrow \text{3-dodecenoate} \quad (2)$$

$$\text{3-dodecenoate} + 3C_2 \rightarrow \text{oleate} \quad (3)$$

Bloch postulated that lauric acid would be converted by $\beta$-oxidation into a $\beta$-OH-dodecanoic acid. $O_2$ would be required. A $\beta$-$\gamma$-elimination with subsequent elongation would lead to the final product, oleic acid. It is obvious that if 3-hydroxy decanoate were supplied as a substrate to plant tissue, a $\beta$-$\gamma$-elimination mechanism would yield an 18:1 (11) monoene; with 3-hydroxydodecanoic, 18:1 (9) and with 3-hydroxytetradecanoic, 18:1 (7). These possibilities were put to a direct test by employing both the saturated and the 3-hydroxy fatty acids appropriately labelled.

There is considerable diversity in the ability of tissue slices from 7-day-old barley seedlings to synthesize long chain fatty acids from normal lower chain saturated fatty acids of even carbon number (Table II). Only acetate is utilized for synthesis of long chain fatty acids ($C_{20}$–$C_{24}$); $C_8$–$C_{14}$ fatty acids are converted into fatty acids up to and including $C_{18}$ to the extent of 6–10% of the substrate used. A further difference to be noted in substrate-product relationships is the ratio of saturated to unsaturated fatty acids synthesized. Approximately 50% of the fatty acids synthesized from 2:0, 8:0, 10:0, and 12:0 are

Table II. Biosynthesis of fatty acids from [1-$^{14}$C] fatty acids by tissue slices prepared from green barley seedlings

| | | Products (mμmoles) | | | | | | | | | | | | | |
|---|---|---|---|---|---|---|---|---|---|---|---|---|---|---|---|
| | | | Fatty acids | | | | | | | | | | | | |
| Substrate | Atmosphere | $CO_2$ | Total | 10:0 | 12:0 | 14:0 | 16:0 | 16:1 | 18:0 | 18:1 | 18:2 | 18:3 | 20:0 | 22:0 | 24:0 |
| [1-$^{14}$C]Acetate | Air | 11·3 | 32·7 | — | 1·2 | 1·6 | 7·6 | 7·6 | 1·9 | 7·1 | 1·6 | — | 2·1 | 0·9 | 1·1 |
| | $N_2$ | 3·3 | 8·9 | — | 0·4 | 0·2 | 2·3 | 1·6 | 2·0 | 1·0 | — | — | 0·7 | 0·7 | — |
| [1-$^{14}$C]Octanoate | Air | 8·5 | 28·0 | 1·2 | 1·2 | 0·5 | 10·1 | — | 0·2 | 9·2 | 4·3 | 1·3 | — | — | — |
| [1-$^{14}$C]Decanoate | Air | 12·8 | 35 4 | — | 1 7 | 1·4 | 14·3 | — | 1·7 | 9·9 | 5·4 | 1·0 | — | — | — |
| | $N_2$ | 3·0 | 36·8 | — | 6·1 | 4·0 | 13·2 | — | 7·5 | 6·0 | — | — | | | |
| [1-$^{14}$C]Dodecanoate | Air | 7·8 | 21·3 | — | — | 1·9 | 7·4 | 0·8 | 2·0 | 5·3 | 2·6 | 1·3 | | | |
| | $N_2$ | 1·1 | 6·6 | — | — | 2·2 | 2·8 | — | 1·6 | tr | | | | | |
| [1-$^{14}$C]Tetradecanoate | Air | 12·7 | 20·8 | — | 2·3 | — | 13·9 | — | tr | 4·6 | | | | | |
| | $N_2$ | 1·9 | 6·1 | — | — | — | 35·5 | — | 0·9 | 1·7 | | | | | |
| [1-$^{14}$C]Hexadecanoate | Air | 22·1 | 0 | | | | | | | | | | | | |
| | $N_2$ | 2·6 | 0 | | | | | | | | | | | | |
| [1-$^{14}$C]Octadecanoate | Air | 15·8 | 0 | | | | | | | | | | | | |
| | $N_2$ | 1·9 | 0 | | | | | | | | | | | | |

Incubation conditions: 2·0 g (wet wt) slices prepared from seedlings grown for 7 days (final 24 hr in light), 250 mμmoles of each substrate, 0·5 μmole acetate, 400 μmoles *tris*-HCl buffer at pH 8·0, 30 μmoles $KHCO_3$ to total volume of 4·0 ml. 4 hr incubation at 30° with shaking.

unsaturated fatty acids whereas less than 25% of the total fatty acids synthesized from myristate are unsaturated. On the other hand, both palmitate and stearate are not utilized for chain elongation or synthesis of unsaturated fatty acids although the formation of $^{14}CO_2$ from these carboxyl labelled fatty acids proves that that these fatty acids enter the plant cells and are activated at least to the thioester stage. If the true substrates for aerobic mono-unsaturation of palmitic and stearic acids are the acyl thioesters of ACP, a block in the formation of these substrates may be present in higher plants by the absence of specific long chain acyl transacylases. Presumably transacylases for the shorter chain fatty acids would be present since these acids are readily converted into oleic acid. Since both palmitic and stearic acids were degraded to $^{14}CO_2$, [$^{14}$C]-acetyl-CoA must have been derived from these acids by β-oxidation. Acetyl-CoA so derived apparently is unavailable for fatty acid synthesis since no *de novo* synthesis of $C_{16}$ and $C_{18}$ fatty acids were observed with these compound.

In nitrogen, the conversion of substrates into $CO_2$ and unsaturated fatty acids decreased considerably. Difficulties in the complete removal of oxygen from tissue is the probable reason for much of the residual formation of these two products. Of interest is the sharp decrease in utilization of 2:0, 12:0 and 14:0 under anaerobic conditions.

The difference noted above between the level of utilization of the two substrates containing 12 and 14 carbon atoms directed attention to the possibility that oleate synthesis was taking place via 3-hydroxydodecanoate, followed by

Table III. Biosynthesis of fatty acids from 3-OH-fatty acids by tissue slices prepared from barley seedlings

| Fatty acids synthesized | $^3$H substrates (mμmoles) | | | | | |
|---|---|---|---|---|---|---|
| | [3-$^3$H]-3-OH-decanoate | | [3-$^3$H]-3-OH-dodecanoate | | [3-$^3$H]-3-OH-tetradecanoate | |
| | Air | $N_2$ | Air | $N_2$ | Air | $N_2$ |
| Total | 79·0 | 31·2 | 87·3 | 33·0 | 57·8 | |
| 10:0 | 3·2 | 3·1 | | | | |
| 12:0 | 3·3 | 2·6 | 30·5 | 16·7 | | |
| 14:0 | 1·6 | 1·8 | 9·9 | 6·1 | 25·4 | |
| 16:0 | 21·7 | 13·0 | 30·5 | 6·1 | 15·9 | |
| 18:0 | 3·2 | 6·6 | 4·0 | 3·7 | 0·7 | |
| 18:1 | 23·4 | 2·8 | 8·0 | 0·5 | 9·1 | |
| 18:2 | 17·7 | 0·8 | 4·5 | | 5·1 | |
| 18:3 | 4·9 | 0·5 | | | 1·1 | |

Incubation conditions were the same as in Table II.

β-γ-elimination in a manner analogous to the mechanism for the synthesis of monoenes under anaerobic conditions in bacteria. Table III gives details of the fatty acids synthesized from tritium-labelled DL,3-hydroxy derivatives of decanoate, dodecanoate and tetradecanoate. Four features of fatty acid synthesis from these substrates appear significant in a consideration of the possible role of these $\beta$-hydroxy acids as intermediates in oleate synthesis. (i) There was considerable synthesis of the corresponding saturated fatty acid from each 3-hydroxy fatty acid substrate, the synthesis being substantial from the hydroxy $C_{12}$ and $C_{14}$ derivatives. Moreover, Tables II and III show that the pattern of synthesis of fatty acids from dodecanoate and DL-3-OH-dodecanoate over a 1 to 8·5 hr period was very similar. (ii) Appreciable synthesis of $C_{18}$ monoene took place from each substrate, 3-hydroxydecanoate being the best utilized in this respect. (iii) The synthesis of unsaturated fatty acids was $O_2$-dependent. If monoenes were synthesized by $\beta$-$\gamma$-elimination of 3-hydroxy acids, no oxygen would be required. (iv) Since $\beta$-oxidation of [3-$^3$H]-3-hydroxy fatty acid would eliminate the tritium label, the initial phase of synthesis involving a $\beta$-oxidation step for acetyl-CoA production and subsequent recondensation is excluded by these data.

Acetate was the only substrate utilized for the synthesis of $C_{16}$ monoene. This monoene was essentially pure palmitoleic acid ($C_{16:1}-\Delta^{9,10}$). In addition to oleic acid ($C_{18:1}-\Delta^{9,10}$), vaccenic acid ($C_{18:1}-\Delta^{11,12}$) contributed appreciably to the total $^{14}$C-labelled 18:1 synthesized from [1-$^{14}$C]acetate. Table IV shows that the ratio of labelled oleic to vaccenic formed from acetate was

Table IV. Biosynthesis of monoenes by tissue slices from substrates of short and intermediate chain length

| Substrate | Product | % Distribution of label in degradation products | |
|---|---|---|---|
| | | Methyl fragment[1] | Carboxyl fragment[2] |
| [1-$^{14}$C]Acetate | 16:1(9) | 12·5 | 87·6 |
| | 18:1(9) $\left[\frac{(9)}{(11)}=\frac{5}{2}\right]$ | 33·4 | 66·6 |
| | 18:1(11) | 11·6 | 88·4 |
| [1-$^{14}$C]Decanoate | 18:1(9) | 0 | 100 |
| [1-$^{14}$C]Dodecanoate | 18:1(9) | 0 | 100 |
| [1-$^{14}$C]Tetradecanoate | 18:1(9) | 0 | 100 |
| DL-[3-$^3$H]-3-OH-decanoate | 18:1(9) | 100 | 0 |
| DL-[3-$^3$H]-3-OH-dodecanoate | 18:1(9) $\left[\frac{(9)}{(11)}=5\right]$ | 0 | 100 |
| | 18:1(11) | | |
| DL-[3-$^3$H]-OH-tetradecanoate | 18:1(9) $\left[\frac{(9)}{(11)}=12\right]$ | 0 | 100 |
| | 18:1(11) | — | — |

[1] *n*-Aldehyde.
[2] Methyl ester of semi-aldehyde, mono carboxylate.

approximately 5:2, although there was considerable variability in different experiments. In some experiments the radioactivity in vaccenic acid, as measured by summing up the radioactivity of the appropriate degradation fragments, almost equalled that for oleate, vaccenic/oleic ratios tending to be high when the synthesis of palmitoleic was high. Furthermore, the ratio of the radioactivity of the carboxyl to the methyl degradation fragments in both $C_{16:1}-\Delta^{9,10}$ and $C_{18:1}-\Delta^{11,12}$ were similar (6·7:1) while this ratio for oleate was in the range from 2 to 2·5. Minor amounts of radioactive fragments which would correspond to $C_{18:1}-\Delta^{6,7}$, $\Delta^{7,8}$, $\Delta^{8,9}$ fatty acids were also detected on the GLC-radiochromatograms. It should be emphasized that the tissue employed and the conditions for the reaction were aseptic and thus the possibility of bacterial contamination is excluded.

In contrast to acetate, each of the saturated fatty acids of intermediate chain length gave $^{14}$C-labelled 18:1 as the only labelled monoene which on oxidation and cleavage gave labelled azelaic semialdehyde (Table IV). This is consistent with synthesis occurring by chain elongation and by a specific desaturation between the 9–10 carbons. Minor peaks (approx. 5% of the total radioactivity) corresponding to suberic semialdehyde provides evidence for some synthesis of $C_{18:1}-\Delta^{7,8}$ from dodecanoate and tetradecanoate.

If a $\beta$-$\gamma$-elimination mechanism with subsequent elongation existed, then 3-hydroxy decanoic acid should yield a $C_{18:1}-\Delta^{11,12}$ monoene, 3-hydroxylauric acid should yield a $C_{18:1}-\Delta^{9,10}$ monoene and 3-hydroxymyristic acid should yield a $C_{18:1}-\Delta^{7,8}$ monoene. The identity of the labelled fragments obtained by reductive cleavage of the ozonides of the $C_{18}$ monoenes synthesized from these 3-hydroxy fatty acids shows that desaturation occurred to yield only $C_{18:1}-\Delta^{9,10}$ as the final product. Thus the 3-hydroxy fatty acids regardless of chain length did not influence the site of the elimination reaction. Furthermore, the non-random distribution of label between the methyl carboxyl fragments confirms a mechanism of chain elongation for the synthesis of oleate from these

Table V. Carboxyl labelling of fatty acids synthesized by tissue slices from [1-$^{14}$C]acetate

| Number of carbon atoms in fatty acid | % of Radioactivity in COOH | |
|---|---|---|
| | Saturated | Monoene |
| 12–14 | 34·0 | |
| 16 | 24·0 | 30·6 |
| 18 | 53·0 | 16·7 |
| 20 | 63·0 | |
| 22 | 42·0 | |
| 24 | 22·0 | |
| 26 | 30·0 | |

substrates. Table IV shows that small amounts of radioactive fragments, consistent with vaccenic being a minor product of biosynthesis, were also obtained. Minor amounts (approx. 5%) of $C_{18:1}-\Delta^{7,8}$ appear to be synthesized from each substrate also.

Previous work (Hawke and Stumpf 1965a) showed that stearate was synthesized by a mechanism of chain elongation from palmitate to give a very characteristic COOH-labelling for stearate (in excess of 50% of the total radioactivity) when [1-$^{14}$C]acetate was used as substrate. Since the results of the above experiments with 3-hydroxy substrates did not provide evidence for a mechanism of elimination by a $\beta$-$\gamma$-mechanism to give oleate, it was of interest to determine whether oleate had a similar distribution of label to that of stearate. Table V summarizes an experiment in which the COOH-labelling of

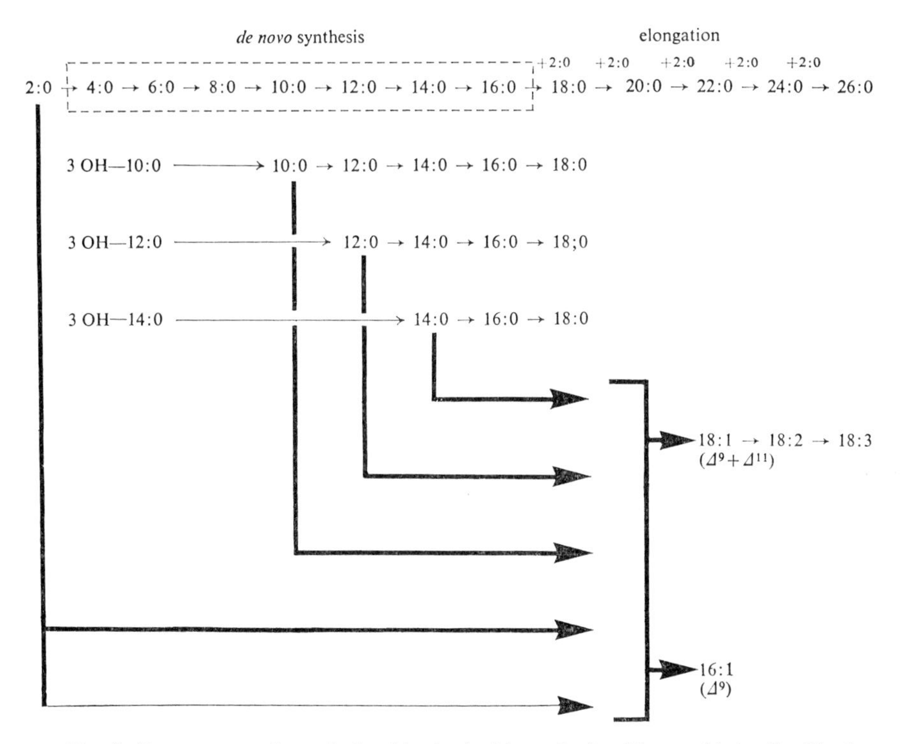


FIG. 2. Precursor-product relationships in the biosynthesis of fatty acids in sliced barley tissue. Heavy lines depict main routes and heavy bracket implies undefined precursor prior to oleic acid synthesis.

the homologous series of *n*-saturated fatty acids is compared with that of the monoenes which had been reduced to the corresponding saturated fatty acid before being subjected to Schmidt degradation. The COOH group in stearic and oleic acids contributed 53·0 and 16·7% of the total label respectively.

It is therefore obvious that the bulk of stearic acid in the plant cell is not being converted into oleic acid because of the difference in the [$^{14}$C]carboxyl labelling pattern. These data do not exclude, however, the possibility of a rapid flow of a small amount of stearyl-S-ACP (with a labelling pattern identical to the isolated oleic acid) through an aerobic unsaturation mechanism to yield oleic acid.

The data obtained with barley tissue slices are presented in summary form in Fig. 2 which depicts the relationship of acetate, shorter chain fatty acids, and 3-hydroxy fatty acid for the $C_{16}$, $C_{18}$, and longer chain fatty acids.

## BIOSYNTHESIS OF FATTY ACIDS BY ISOLATED CHLOROPLASTS

In 1960, the Russian biochemist Smirnov was the first to describe a system consisting of isolated spinach chloroplasts which readily incorporated [$^{14}$C]-acetate into long chain fatty acids. In fact he described two systems. The first consisted of spinach leaf chloroplasts which in the dark required ATP, CoA and $Mn^{2+}$ at a pH of 7·2 and incorporated acetate rather ineffectively. In the presence of light he observed a 3·5-fold stimulation of acetate incorporation which also depended on ATP, CoA and $Mn^{2+}$. When he increased the ATP concentration from 0 to 20 $\mu$moles, the light/dark (L/D) ratio of incorporation decreased from 6·4 to 2·1. Smirnov reasoned that since in the absence of ATP the L/D ratio was highest, photosynthetic incorporation of acetate was independent of ATP. Since in the absence of ATP, incorporation was small and since addition of large amounts of ATP markedly increased over-all incorporation, Smirnov concluded that a contaminating nucleotide was the actual cofactor required for photosynthetic incorporation. Analysis of ATP which he used in his experiments showed that the sample contained 69% ATP, 23% ADP, and 5% AMP. When ADP replaced ATP in the absence of inorganic phosphate, L/D ratios as high as 27·5 were achieved when 4·1 $\mu$moles of ADP were used. AMP was far less effective. Addition of inorganic phosphate depressed incorporation. The optimum concentration of ADP was 0·9 $\mu$mole/1·8 ml of reaction mixture. Smirnov therefore proposed a novel series of reactions to explain his results. The reaction:

$$\text{ADP} + \text{acetate} \xrightarrow[\text{chloroplast}]{\text{light}} \text{ADP} \sim \text{acetyl} \tag{4}$$

$$\text{ADP} \sim \text{acetyl} + \text{CoA} \rightarrow \text{acetyl CoA} + \text{ADP} \tag{5}$$

could explain the catalytic role of ADP and the lack of any requirement for inorganic phosphate and ATP. He termed reaction (4) a new photoacylation reaction which required light energy. In the dark the chloroplasts depended on a conventional acetic thiokinase type of activation with ATP, CoA and $Mn^{2+}$ as components.

In 1962 Smirnov presented further evidence in support of his earlier report. Sulfanilamide was used as a trapping agent for acyl-activated systems. Therefore, conversion of free sulfanilamide to a bound form was a direct measurement of an activated acyl group. When citrate and succinate as well as acetate were added to a chloroplast suspension, free sulfanilamide decreased in the presence of light and ADP. Since, in the presence of sulfanilamide, no CoA requirement could be shown, presumably the proposed ADP-acyl derivatives acylate sulfanilamide directly.

$$\text{ADP} \sim \text{acylate} + \text{sulfanilamide} \rightarrow \text{ADP} + \text{acyl-sulfanilamide} \tag{6}$$

In 1964 Smirnov extended his observations further. Incubation of chloroplasts from a variety of different plants will incorporate [$^{14}$C]glycine or

[$^{14}$C]alanine into lipids, RNA and proteins in the presence of light. Further characterization of the reaction products should prove of interest.

Meanwhile in 1962, Mudd and McManus initiated a series of experiments with spinach chloroplasts and Stumpf and James began work on lettuce chloroplasts. We shall first review the work from Mudd's laboratory. In essence they confirmed the over-all observations of Smirnov, namely that chloroplasts readily incorporate acetate into fatty acids and that light stimulates the incorporation. Employing broken spinach chloroplasts, they determined that ATP, CoA and NADPH were essential components. Although a light stimulation was noted, this stimulation could be replaced in the dark with increasing concentration of glucose 6-phosphate and $NADP^+$. Presumably NADPH is formed and serves as the reductant. When broken chloroplasts were separated into a pellet of grana and an amber supernatant, a considerable reduction of acetate incorporation was noted. Mixing both fractions together restored activity. Mudd also presented evidence to show that the particulate preparation he employed was devoid of fumarase, succinic dehydrogenase and NADH-cytochrome *c* reductase activity, typically found associated with mitochondrial particles. In 1964 Mudd and McManus reported the lability of their broken spinach chloroplast preparation. Thus, in 40 min at 0°, 50% of the activity was gone and after 80 min only 25% remained. Although the lipid synthetase system was markedly affected by the aging process, the incorporation of acetate into water-soluble compound was not changed. Addition of glutathione had a slight stabilizing effect on preventing deterioration of the synthetase and there seemed to be a shift in the pattern of oleic to stearic acid synthesis.

Furthermore, Mudd and McManus presented evidence suggesting that the SH group of the preparation was probably the site of the loss of activity during aging. Thus arsenite in the presence of British Anti-Lewisite (BAL) markedly inhibited lipid synthesis. $10^{-5}$ M *p*-chloromercuribenzoate (PCMB) caused 50% inhibition and this effect could be reversed by glutathione. Interpretation of the data is difficult since the investigators used 50 mμmoles of CoA–SH and free acetate in all their studies.

The same authors in 1965 in studying the relation of acetate incorporation into both lipid and water-soluble compounds by broken spinach chloroplasts noted that citric and glutamic acids were the principal water-soluble compounds. Of interest, water-soluble compounds were formed while lipid synthesis was depressed in the presence of only ATP and CoA and in the absence of NADPH. Increase of NADPH concentration restored capacity to synthesize lipids. Glutathione and mercaptoethanol stimulated incorporation of acetate into long chain fatty acids while ascorbic acid was inactive. The nature of these effects will be examined in detail in later portions of this chapter.

Earlier Stumpf and James (1963) and Stumpf, Bové and Goffeau (1963) examined extensively the properties of a lettuce chloroplast system for the incorporation of acetate into long chain fatty acids. These workers were intrigued by the earlier observations of Smirnov that ADP participated in the light-stimulated lipid synthetase system. Two important factors were examined in attempts to explain Smirnov's results.

The first factor is the preparation of the chloroplasts. Two kinds of chloroplast preparations were employed. Preparation I employed a suspending medium composed of 0·35 M NaCl, 0·01 N sodium ascorbate, 0·01 M phosphate buffer (pH 7·4), while preparation II used a suspending medium composed of 0·50 M sucrose, 0·01 M sodium ascorbate, 0·01 M NaCl, all adjusted to pH 7·4. Both media contain 0·001 M Versene. The major difference in activity between Preparations I and II is a consistent and far more pronounced requirement for adenosine nucleotides by Preparation II in contrast to Preparation I. These differences are illustrated in Table VI. Presumably a critical amount of nucleotide leaks out of chloroplasts obtained by Preparation II which necessitates a further addition to complete the nucleotide needs. Both preparations are reasonably stable when stored in an ice bath for at least 3 hr. However, if stored overnight at −10° or frozen and then promptly thawed, complete loss of acetate-incorporating activity is observed.

Table VI. Nucleotide requirement by preparations I and II

| Preparation | Nucleotide concentration ($\mu$moles) | [$^{14}$C]acetate incorporation ATP | [$^{14}$C]acetate incorporation ADP |
|---|---|---|---|
| I | — | 9800 | 9800 |
| | 1 | 11,600 | 11,000 |
| | 10 | 6000 | 9600 |
| II | — | 12,800 | 12,800 |
| | 1 | 23,300 | 26,000 |
| | 10 | 6900 | 11,000 |

Each reaction mixture contained 0·5 ml of freshly prepared chloroplast preparation (0·5 mg of chlorophyll), 100 $\mu$moles of phosphate buffer (pH 8·2), 0·5 $\mu$mole of [$^{14}$C]acetate (0·5 $\mu$c, 55,500 counts/min), 0·2 $\mu$mole of CoA, 1·0 $\mu$mole of $Mg^{2+}$, 30 $\mu$moles of sodium bicarbonate, 0·2 $\mu$mole of $NADP^+$ and varying amounts of nucleotide as indicated. Time of reaction incubation is 60 min; temperature is 20°, and 2000 ft candles of light strike the flasks. Total volume is 1·2 ml.

The second important factor is the concentration of nucleotides employed in the reaction mixture. Marked stimulation occurs at about $2 \times 10^{-3}$ M of ATP and ADP. However, with concentration of $5 \times 10^{-3}$ M of ATP inhibition takes place and increases with increasing concentrations of ATP whereas with ADP no pronounced inhibition occurs until $10^{-2}$ M ADP is attained.

Thus the preparatory conditions of the chloroplasts and a rather critical concentration of ATP must be considered in any re-evaluation of Smirnov's results. Table VII depicts the effect of increasing concentration of ATP in both the light and dark reactions. Of interest, in the dark, there is an increased response to ATP whereas in the light reaction there is an opposite or decreasing effect after a critical concentration of $2 \times 10^{-3}$ M of ATP.

Table VIII brings together the nucleotide specificity for both the light and dark reactions with respect to AMP, ADP and ATP. Table IX summarizes the cofactor requirements for the effective incorporation of acetate into long chain fatty acids.

Magnesium and manganese ions separately have a pronounced stimulatory effect with $1 \times 10^{-3}$ M being the optimum concentration. No marked effect is observed when manganese is added in increasing concentration in the presence of $10^{-3}$ M of magnesium ion.

Table VII. Effect of ATP on incorporation of acetate into lipid in whole butter lettuce chloroplasts

| ATP ($\mu$mole) | m$\mu$Atoms of $^{14}C$ incorporated into lipid | |
|---|---|---|
| | Light | Dark |
| 0 | 22·1 | 0·2 |
| 1 | 59·2 | 0·7 |
| 2 | 52·8 | 1·2 |
| 4 | 45·2 | 1·9 |
| 8 | 34·0 | 3·6 |

With intact chloroplast no effect has been observed on addition of $NAD^+$ or $NADP^+$ in the light or dark in combination or alone with or without glucose 6-phosphate.

Table VIII. The effect of AMP, ADP, and ATP on acetate incorporation into lipid in whole butter lettuce chloroplasts

| Addition, 2 $\mu$moles | m$\mu$Atoms of $^{14}C$ incorporated into lipids | |
|---|---|---|
| | Light | Dark |
| — | 8·4 | 0·18 |
| AMP | 13·5 | 0·29 |
| ADP | 15·5 | 0·35 |
| ATP | 17·0 | 1·25 |

Stumpf and James (1963) demonstrated the requirement for inorganic phosphate in the incorporation of acetate in the light, in contrast to the results reported by Smirnov. These results have recently been confirmed by Brooks in our laboratory. These results coupled with the observation that photophosphorylation inhibitors such as $NH_3$ and CMU markedly inhibit lipid biosynthesis in intact chloroplasts would strongly suggest that there is no evidence to support the photoacylation reaction proposed by Smirnov.

The early experiments by Stumpf and James were carried out with [$^{14}$C]-acetate as substrate. Later, we shall discuss the relation of substrate to fatty acid synthesis and the types of fatty acids synthesized by both intact chloroplasts and the soluble synthetase system.

Table IX. Cofactor requirements for the incorporation of [2-$^{14}$C]acetate into lipids in whole butter lettuce chloroplasts

| Sample | mμAtoms of $^{14}$C incorporated into lipid | |
|---|---|---|
| | Light | Dark |
| Complete | 54·0 | 2·2 |
| −ATP | 20·6 | 0·2 |
| −CoA | 3·2 | 1·3 |
| $-MgCl_2$ | 2·8 | 0·3 |
| $-KHCO_3$ | 2·8 | 0·8 |
| −All | 0·8 | 0·2 |

## Nature of the Light Effect

We have already shown in several tables (Table VI, VII, VIII, and IX) the pronounced photostimulation, reported first by Smirnov.

Stumpf, Bové and Goffeau (1963) studied in detail the relation of the photophosphorylation reaction to the light effect and they were able to show that non-cyclic photophosphorylation, which has as its products ATP, $O_2$ and NADPH, is an important co-reaction of lipid synthesis. Certainly the three products are essential components of lipid synthesis by chloroplasts. Cyclic photophosphorylation was considerably less effective.

These workers were about to conclude that non-cyclic photophosphorylation could explain all the results and accordingly designed an experiment to prove this point. In this experiment using intact chloroplasts first NADPH and ATP were formed in the light; after 15 min of illumination, CoA was tipped into the reaction mixtures, and acetate incorporation was allowed to proceed either in light or in dark for 60 min. Table X demonstrates that substrate amounts of ATP and NADPH which are first formed in light, do not stimulate acetate incorporation in the subsequent dark period. Sufficient amounts of NADPH and ATP are formed to supply the needs of the lipid-synthesizing system. The formation of NADPH has been measured directly on the reaction mixtures. ATP production has not been determined in these experiments; it has always been found that the reduction of 4 μmoles of $NADP^+$ is accompanied invariably by the esterification of at least 2 μmoles of ADP to ATP. These experiments prove that although chloroplasts were able to synthesize NADPH and ATP in quantities more than sufficient to fulfill all cofactors requirements, a light effect was still demonstrable; it was very transitory and could not be stabilized.

A considerable number of experiments have been conducted by the authors in an attempt to explain the photostimulation effects. What are possible explanations for this highly reproducible but transitory system? Mudd and

Table X. Acetate incorporation in the dark with NADPH and ATP formed in a preliminary light reaction

| Conditions | Gas phase | NADPH[1] ($\mu$moles) | Acetate incorporation (counts/min) |
|---|---|---|---|
| 15 min light<br>60 min dark | $N_2$ | 2·9 | 440 |
| 15 min light<br>60 min light | $N_2$ | 3·3 | 2340 |
| 15 min light<br>60 min dark | Air | 2·1 | 740 |
| 15 min light<br>60 min light | Air | 3·6 | 3260 |

CoA was tipped into the reaction mixture at the end of the 15-min preliminary light reaction. To keep certain vessels dark, they were wrapped with aluminium foil and returned into the illuminated Warburg bath.

[1] Measured at the end of the incubation period; $NADP^+$ added, 4 $\mu$moles.

McManus (1962, 1964) have offered evidence suggesting that the only role light plays is the generation of NADPH which could be easily replaced in broken chloroplasts with sufficient amounts of NADPH. The results in Table X and Table XIII are not in agreement with their conclusions.

Table XI. Incorporation of acetate, acetyl-CoA, malonyl-CoA, and malonate into long chain fatty acids in whole butter lettuce chloroplasts

| Substrate | m$\mu$Atoms of $^{14}C$ incorporated into long chain fatty acids | |
|---|---|---|
| | Light | Dark |
| [2-$^{14}$C]Acetate | 3·47 | 0·23 |
| [1-$^{14}$C]Acetyl-CoA | 0·60 | 0·18 |
| [2-$^{14}$C]Malonyl-CoA | 0·27 | 0·23 |
| [1,3-$^{14}$C]Malonate | 0·24 | 0·03 |

Another possibility is a phototransport system which permits the transport of acetate into the chloroplast particle. Permeability undoubtedly plays a limiting role in lipid synthesis. Table XI for example demonstrates that the only effective substrate is [$^{14}$C]acetate. Acetyl-CoA, malonic acid and malonyl-CoA are ineffective. The only explanation for this result is a differential permeability for acetate and none for the other three substrates. If, however, acetate enters the chloroplast particle more readily in the presence of light, then a larger amount of water-soluble compounds should be synthesized in the light than in the dark. In fact, as illustrated in Table XII, while there are

Table XII. Incorporation of acetate into lipid and water soluble material in whole butter lettuce chloroplasts

| ATP ($\mu$moles) | m$\mu$atoms of $^{14}$C in water soluble material | | m$\mu$Atoms of $^{14}$C in lipids | |
|---|---|---|---|---|
| | Dark | Light | Dark | Light |
| 0 | 1·1 | 3·6 | 0·4 | 28·7 |
| 2 | 3·4 | 12·5 | 3·0 | 50·8 |
| 4 | 6·0 | 13·9 | 3·7 | 46·8 |
| 8 | 8·7 | 14·0 | 6·7 | 44·6 |

appreciable differences in water-soluble products of acetate metabolism in the dark or light, at sufficient concentration of ATP the differences become considerably less. However the differences in lipid synthesis remain great.

Another explanation might be offered in a photoallosteric effect on a critical protein involved in lipid synthesis. However, the fact that soluble synthetase systems in the presence of grana show no light effect would argue against this possibility.

We have been intrigued for some time with the possibility that the –SS/SH ratio might be markedly influenced by a light reaction. Since a number of proteins involved in lipid synthesis have –SH groups essential for activity, in particular ADP–SH, we examined this hypothesis in some detail. A review of our recent results will now follow.

## EARLIER OBSERVATIONS

In 1954, Calvin and his coworkers suggested that there may be a transformation of electromagnetic energy from the excited states of chlorophyll to a photochemical reduction of the disulfide bond in lipoic acid to the sulfhydryl form. Although this postulate was attractive, little direct evidence became available.

In 1962, Newton working with chromatophores of *Rhodospirillum rubrum* demonstrated a photochemical disulfide reducing system. The model compound which he employed was 5,5′-dithiobis[2-nitrobenzoic acid][1] in the disulfide form. DTNB–SS is readily reduced to the intensely yellow colored DTNB–SH by thiol compounds:

[1] The oxidized disulfide form is abbreviated DTNB–SS; the reduced form DTNB–SH.

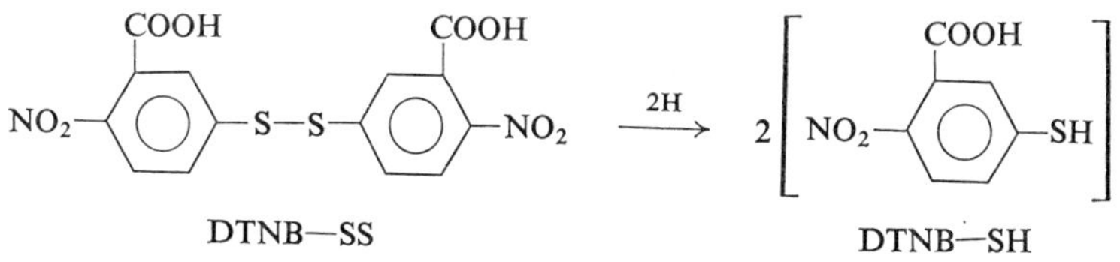

$$NO_2\text{-}C_6H_3(COOH)\text{-}S\text{-}S\text{-}C_6H_3(COOH)\text{-}NO_2 \xrightarrow{2H} 2\left[NO_2\text{-}C_6H_3(COOH)\text{-}SH\right]$$

DTNB—SS DTNB—SH

Newton observed that DTNB–SS is rapidly photoreduced to DTNB–SH under anaerobic conditions in the presence of 2,6-dichlorophenolindophenol (DCI), ascorbate, and benzyl viologen. Since the bacterial DTNB–SH system could be stored for months at 0° with little loss whereas the bacterial $NAD^+$ photoreducing system lost activity rapidly at 0° within 24 hr, a $NAD^+$ photoreducing system is probably not implicated. No correlation could be shown between the systems catalysing the sulfate reduction to sulfite and DTNB–SS reduction to DTNB–SH. The enzymatic photoreduction system consists of the following reaction:

$$\text{Ascorbate} + \text{DTNB–SS} \xrightarrow[\text{benzyl viologen}]{h\nu + \text{DCI}} \text{dehydroascorbate} + 2\text{DTNB–SH}$$

The system is highly specific since other disulfides tested such as oxidized glutathione, lipoamide, cysteine, and hydroxyethane disulfide were completely inert. Newton believes that since the standard potential of ascorbate is +0·07 and that of benzyl viologen is −0·359 and of DTNB–SS approximately −0·150, an energy input is necessary to facilitate the reduction of low potential dyes by ascorbate electrons.

Ashai in 1964 examined the capacity of spinach chloroplasts to reduce 3-phosphoadenosine-5′-phosphosulfate (PAPS) to $SO_3^-$ and 3-phosphoadenosine-5′-phosphate. He noted that although spinach chloroplasts readily photoreduce a yeast protein fraction C–SS, NADPH is not a component of the system. Chloroplast fragments rapidly photoreduced DTNB–SS without the requirement of protein C–SS and without the participation of NADPH. These fragments also photoreduced disulfide-containing proteins isolated from acetone-dried powders of spinach chloroplasts.

## REDUCTION OF DTNB–SS BY LETTUCE CHLOROPLASTS

In 1964, we undertook a reinvestigation of the capacity of lettuce chloroplasts to reduce DTNB–SS with the hope that we could correlate such a photoreduction with the light stimulated lipid synthesis system. The model system was simple and consisted of 200 μmoles of phosphate buffer at pH 8·0, 200 μmoles of sodium chloride, 0·3 μmoles of DTNB–SS and chloroplast containing 135 μg of chlorophyll in a reaction volume of 1·20 ml. This system was exposed to white light for a given period of time, centrifuged in the dark, and the clear yellow supernatant measured directly at 412 mμ. Figure 3 relates the extent of photoreduction as a function of time under aerobic and anaerobic conditions as well as in the dark.

In the dark, added NADPH or NADH did not reduce DTNB–SS; in the light, the presence of $NADP^+$ or $NAD^+$ also had no effect. When chloroplasts are exposed to light for several minutes and then DTNB–SS added in the dark, no reduction to DTNB–SH occurs, indicating that no stable SH groups were made available during light exposure. CMU markedly inhibits the reduction process. Protein C–SS[1] does not stimulate DTNB–SS reduction. In further support was the observation that after preincubation of intact chloroplasts in the light, and subsequent addition of [$^{14}$C]iodoacetamide to the reaction

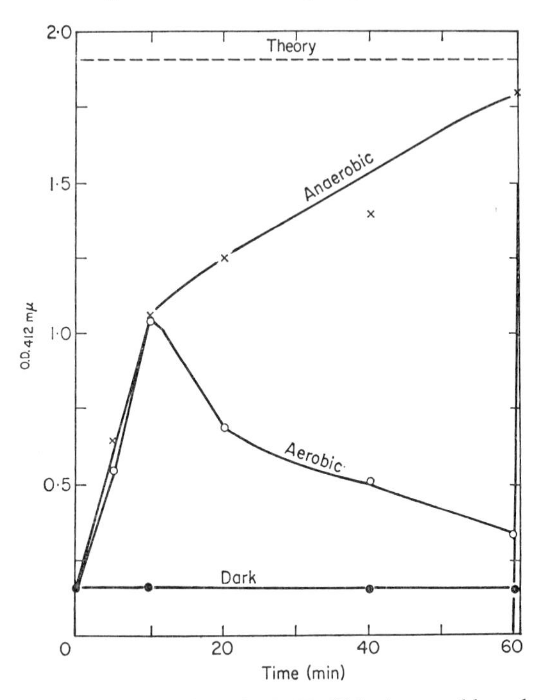


FIG. 3. The extent of photoreduction of DTNB–SS by lettuce chloroplasts as a function of time under aerobic and anaerobic conditions.

mixture, no $^{14}$C counts were fixed above that of a control experiment in which the chloroplasts were incubated throughout the experiment in the dark.

While lipid synthetase activity in lettuce chloroplasts cannot be retained overnight, chloroplast suspensions could be stored at 4° for several days with little or no loss in the capacity to reduce DTNB–SS. Broken chloroplasts readily reduce DTNB–SS but were far less stable.

The pH optimum for the photoreduction of DTNB–SS is 8·1. There was good correlation of ferricyanide reduction and DTNB–SS reduction suggesting that DTNB–SS was a Hill reagent. We confirmed Newton's earlier work that oxidized glutathione and cystine were not reduced while DTNB–SS underwent a rapid photoreduction by lettuce chloroplasts.

[1] We are indebted to Dr. Robert Bandurski for a generous sample of Protein C–SS.

Experiments were then designed to examine the capacity of chloroplasts to photoreduce known disulfide proteins. A typical protein which can be easily tested is triose phosphate dehydrogenase. The crystalline enzyme was inactivated by gently bubbling $O_2$ through a solution of the enzyme at pH 8·0. The activity could be fully restored by the addition of cysteine or glutathione. Figure 4 shows a typical experiment which would suggest that readily available disulfide groups in a protein are not photoreduced by chloroplasts. Parallel control experiments showed clearly that DTNB–SS under the same condition was rapidly reduced and that triose phosphate dehydrogenase in the presence of cysteine was fully activated.

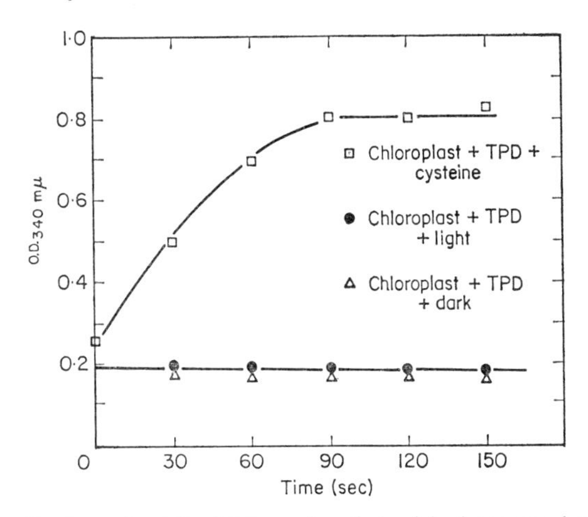


FIG. 4. The activation of oxidized triose phosphate dehydrogenase by cystine and by chloroplasts suspensions. Activity is determined by NADH formation as measured by the increase in optical density at 340 mμ. Triose phosphate serves as substrate.

The conclusion derived from these experiments were: (1) In confirmation of Newton's experiments with *Rhodospirillum rubrum* and Ashai's work with spinach chloroplasts, DTNB–SS was rapidly photoreduced to DTNB–SH. (2) The intact chloroplast system was unusually stable. The system did not require NADPH, NADH, ascorbic or the rather complicated system of Newton. (3) In agreement with Newton's results, only DTNB–SS was reduced. Cystine, oxidized glutathione and oxidized triose phosphate dehydrogenase were not photoreduced. The same chloroplast system readily photoreduced ferricyanide. These results therefore suggested that as a model compound the disulfide form of DTNB–SS was readily photoreduced but that no obvious appearance of new thiol groups occurred by exposing chloroplasts to light.

## DITHIOERYTHRITOL AND THE DARK REACTION

Recently, Beulen *et al.* (1965) demonstrated that hydrosulfite could serve as a source of electrons for a soluble nitrogen-fixing system from *Azotobacter*. It occurred to us that if the function of light in chloroplasts was simply a source of electrons for lipid synthesis, then hydrosulfite should substitute for the light requirement. Exploratory experiments to correlate lipid synthesis from acetate with hydrosulfite as a source of electrons showed that this strong reductant had no effect either in inhibiting or in stimulating lipid synthesis (Table XIII). These results therefore would tend to rule out any limiting role that either ferredoxin or NADPH might play in the light reaction.

Table XIII. Effect of hydrosulfite on incorporation of [$^{14}$C]acetate into lipids by intact lettuce chloroplasts under anaerobic conditions

| Additions | Illumination | Total cpm/90 min |
|---|---|---|
| None | Light | 9200 |
| None | Dark | 330 |
| 20 $\mu$moles $S_2O_4^=$ | Light | 10,500 |
| 20 $\mu$moles $S_2O_4^=$ | Dark | 300 |

Reaction mixture includes: CoA, ATP, $NADP^+$, Glucose 6-phosphate, $Mg^{2+}$, $CO_2$, NADH, lettuce chloroplasts, glycylglycine buffer, pH 8·2.

Dithioerythritol (DTE) was then tested in conjunction with hydrosulfite. The combination of these two reagents would provide both a source of electrons as well as the conversion of disulfide groups to thiol groups. Preliminary experiments revealed a remarkable stimulation of acetate incorporation into lipids by chloroplasts in the dark when DTE was added while hydrosulfite in the absence or presence of DTE was ineffective. The experiments to be described strongly suggest that the conversion of disulfide groups to thiol groups may be the site of the light stimulation.

Thiol groups such as those on proteins and coenzyme A are readily oxidized by air to disulfides. Although glutathione, mercaptoethanol and other thiol compounds are frequently used to reduce disulfide groups, the equilibrium constants of these reactions with these thiol reagents are nearly unity and hence high concentrations must be used to obtain a reasonable formation of the desired thiol form of the enzyme. Dithioerythritol (DTE) is a new reagent introduced by Cleland (1964). It reacts very rapidly with disulfides at pH 8 and converts these completely to the thiol form. The redox potential is −0·33 V at pH 7. The nature of the model reaction is:

$$RSSR + HS—CH_2(CHOH)_2CH_2—SH \rightleftharpoons RSH + R—SS—CH_2(CHOH)_2CH—SH \quad (6)$$

S—R
|
S
|
$CH_2$ SH
| |
CHOH $CH_2$
CHOH
→
S
$CH_2$ S
| |
CHOH $CH_2$
CHOH
+ RSH

Cyclic disulfide

DTE has a marked stimulatory effect in the dark. As indicated in Table XIV, as much as a thirty-fold stimulation has been observed with 20 μmoles of DTE. In the dark, with intact chloroplasts no requirements can be demonstrated for NADPH or NADH. This observation is rather suprising since one would assume that photoreduction is one of the principal reactions of light. Also of interest is the need for ATP in the presence of DTE. Because of the absence of photophosphorylation in the dark, a requirement for ATP becomes readily apparent.

Table XIV. Effect of DTE on incorporation of [$^{14}$C]acetate into lipids by intact lettuce chloroplasts under anaerobic conditions in the dark

| Cofactors | Total cpm/90 min |
|---|---|
| None | 379 |
| Complete | 11,180 |
| $-Mg^{2+}$, $CO_2$, CoA | 4356 |
| −NADPH, NADH | 10,180 |
| −DTE | 1928 |
| −ATP | 238 |
| +2 μmole ATP | 4820 |
| +4 μmole ATP | 7884 |
| +8 μmole ATP | 11,180 |
| +12 μmole ATP | 9936 |
| All −DTE+light | 18,470 |

The effect of a naturally occurring cyclic disulfide compound, namely lipoic acid, was examined. Lipoic acid alone at concentrations ranging from 0·5 μmole to 2·5 μmole has little effect on lipid synthesis in the dark. The addition of both DTE and lipoic acid gave a slight stimulation. The precise nature of the slight stimulation by lipoic acid in the presence of DTE is under further study.

At the present time, it is difficult to interpret the results with DTE. As noted in Table XV glutathione and mercaptoethanol had no marked stimulatory

Table XV. Effect of thiol reagents on lipid synthesis by lettuce chloroplast

| Addition | Concentration (μmole) | Dark cpm | Light cpm | L/D |
|---|---|---|---|---|
| — | — | 1560 | 15,800 | 10·0 |
| GSH | 20 | 1730 | — | — |
| Mercaptoethanol | 20 | 2460 | — | — |
| DTE | 1 | 2050 | 20,600 | 10·0 |
| DTE | 10 | 4150 | 17,000 | 4·1 |
| DTE | 20 | 7730 | 16,680 | 2·2 |
| DTE | 40 | 8200 | 17,700 | 2·1 |

effect in the dark. While the DTE activated system never attains the same capacity as does the light-activated chloroplast system, nevertheless, the light/dark ratio drops from 10 in the absence of DTE to 2·1 in the presence of DTE. On analysis of the fatty acids produced in the presence of DTE, a rather large amount of shorter chain acid is synthesized, suggesting that DTE plays two roles, one in converting a disulfide group into thiol groups in a chloroplast and secondly, a possible transacylation at an early stage of synthesis.

In reviewing the several reactions catalysed by chloroplasts we can note:

(1) a specific capacity to photoreduce DTNB–SS
(2) no activation in the dark by glutathione or mercaptoethanol
(3) a marked stimulation by DTE
(4) the lack of stimulation by hydrosulfite thereby suggesting that the light effect is not related to the flow of electrons through ferredoxin or NADPH.

Based on these observations an attractive hypothesis could be proposed. If ACP and the associated lipid-synthesizing enzymes are buried deeply between the layers of grana thylakoids of the chloroplast, and if in the dark, ACP is largely in the disulfide or inactive form, lipid synthesis would be markedly curtailed. On illumination, chloroplasts would rapidly photoreduce ACP–SS to its active thiol form. We would predict that a mechanism must be available that rapidly oxidizes the thiol form of ACP to the disulfide form in the dark. By this mechanism a chloroplast can control its lipid synthesizing capacity. Presumably a much slower system is available for lipid synthesis in the dark. Further experimentation will test the validity of this hypothesis.

## Soluble Synthetase from Chloroplasts

Stumpf and James (1963) were unsuccessful in obtaining a soluble chloroplast synthetase. Working with spinach chloroplast, Mudd and McManus

Table XVI. Cofactor requirements for fatty acid synthesis by butter lettuce chloroplast soluble enzymes

| Sample | mμAtoms of $^{14}C$ incorporated into long chain fatty acids |
|---|---|
| Complete | 13·2 |
| – ACP | 2·8 |
| – $NADP^+$ G 6-P | 1·6 |
| – GSH | 3·0 |
| – NADH | 8·4 |
| – $MgCl_2$ | 13·4 |
| – Acetyl-CoA | 14·6 |
| – All cofactors | 0·1 |

showed in 1962 that both chloroplast fragments and soluble proteins were necessary for incorporation of [$^{14}$C]acetate into long chain fatty acids when supplemented with NADPH. Recently Brooks and Stumpf (1965, 1966) defined the components of a soluble synthetase system from both lettuce and spinach chloroplasts which readily incorporated [$^{14}$C]malonyl-CoA into palmitate and stearate. Table XVI summarizes the cofactor requirements. A comparison with intact chloroplasts is of interest. The major differences are the requirements for ACP, NADPH, NADH and GSH which could not be demonstrated with intact chloroplasts. Table XVII illustrates the expected differences in cofactor requirements with acetyl-CoA and malonyl-CoA as substrates.

Table XVII. Requirement for ATP and $KHCO_3$ in fatty acid synthesis with butter lettuce chloroplast soluble enzymes

| Sample | mμatoms of $^{14}$C incorporated into long chain fatty acids | |
|---|---|---|
| | [1-$^{14}$C]acetyl-CoA | [2-$^{14}$C]malonyl-CoA |
| Complete | 25·0 | 49·9 |
| $-KHCO_3$ | 9·9 | 58·4 |
| $-$ATP | 0·7 | 51·6 |

A second striking observation is illustrated in Table XVIII which shows that unlike in intact chloroplast where only acetate is an effective substrate, in soluble systems from either spinach or lettuce leaves, malonyl-CoA is the most effective substrate. ACP is also an essential component.

Table XVIII. Incorporation of acetate, acetyl-CoA and malonyl-CoA into long chain fatty acids with broken chloroplast preparations

| Enzyme source | Substrate | mμatoms of $^{14}$C incorporated into long chain fatty acids | | |
|---|---|---|---|---|
| | | Enzyme | Enzyme +ACP | ACP |
| Lettuce | [2-$^{14}$C]Acetate | 0·3 | 0·7 | 0·0 |
| | [1-$^{14}$C]Acetyl-CoA | 1·7 | 3·4 | 0·0 |
| | [2-$^{14}$C]Malonyl-CoA | 2·2 | 8·4 | 0·05 |
| Spinach | [2-$^{14}$C]Acetate | 0·7 | 1·3 | 0·0 |
| | [1-$^{14}$C]Acetyl-CoA | 0·8 | 1·9 | 0·0 |
| | [2-$^{14}$C]Malonyl-CoA | 0·9 | 19·6 | 0·0 |

A third difference is that the soluble synthetase synthesizes primarily stearic acid whereas the intact chloroplasts form mostly oleic and palmitic

Table XIX. Products of fatty acid synthesis in butter lettuce chloroplast preparations

| Chloroplast preparation | Substrate | Gas phase | mμmoles incorporation | Relative % composition | | | | | | |
|---|---|---|---|---|---|---|---|---|---|---|
| | | | | 14:0 | 16:0 | 18:0 | 20:0 | 18:1 | 18:2 | Polar |
| Leaf tissue | acetate | air | 5·4 | 4·8 | 41·5 | 3·9 | — | 41·8 | 8 | 1·2 |
| Whole | acetate | air | 89 | 1 | 24 | — | — | 73 | — | 2 |
| Whole | acetate | nitrogen | 28 | 1 | 33 | 66 | — | — | — | — |
| Pressate | acetyl-CoA | air | 12 | — | 36 | 63 | — | — | — | 2 |
| Pressate | malonyl-CoA | air | 42 | — | 7 | 83 | 7 | — | — | 3 |
| Sol. enzymes | malonyl-CoA | air | 9 | — | — | 98 | — | — | — | 2 |

Reaction mixtures and conditions for whole chloroplast experiments were as in Table IX. Reaction mixtures for pressate and soluble enzyme experiments were as in Table XVI. In all experiments except that with soluble enzymes, incubation was in Warburg flasks with shaking and illumination. Substrates used were 0·25 μmole [2-$^{14}$C]acetate, 0·1 μmole [1-$^{14}$C]acetyl CoA and 0·1 μmole [2-$^{14}$C]malonyl-CoA.

acids. Table XIX collects together the comparative data on products formed employing different chloroplast preparations.

A fourth difference is the complete loss of the light effect. With the soluble synthetase, the extent of incorporation is not affected by light provided that all the cofactor requirements listed in Table XVI are provided. Moreover, DTE does not markedly increase activity in the soluble preparations.

The soluble synthetase remains in solution despite prolonged ultracentrifugation at 100,000 × ***g*** for 4 hr and thus this synthetase is similar to the soluble system described in both *E. coli* and avocado mesocarp and differs markedly from the mammalian and yeast enzymes which sediment at high speeds.

## The Source of Acetyl-CoA in Chloroplasts

We have shown that free acetate is the most effective substrate for incorporation into long chain fatty acids by intact chloroplasts while malonyl-CoA is the most effective substrate in the soluble system derived from the chloroplast. An interesting question now arises as to the actual source of acetic acid in the chloroplast. Acetic acid is a relatively rare acid in leaf tissue and probably does not occur in the free form.

The biochemistry of organic acids is rather poorly defined in isolated chloroplasts. Tobacco leaves contain large amounts of malic acid and smaller amounts of citric, oxalic and succinic acids. The condensing enzyme has been isolated from homogenates of bean, spinach and soybean leaves by Hiatt (1962). Aconitase activity has been detected in leaves of twelve species by Bacon *et al.* (1963). Malic dehydrogenase has been purified some 100-fold from extracts of acetone powder of spinach leaves by Hiatt and Evans (1960). The enzymes of the glyoxylate cycle, isocitritase and malate synthetase, have been looked for in a variety of leaf tissues. Extracts of tobacco, barley and bean leaves have no isocitritase activity but tomato, tobacco, and barley leaf tissues have some activity with respect to malate synthetase (Yamamoto and Beevers, 1960). Rosenberg, Capindale and Whatley (1958) have reported the presence of phosphoenolpyruvate carboxylase in isolated chloroplasts. The fixation of $CO_2$ into aspartate can take place by this reaction in the dark with isolated chloroplasts. Young and Shannon (1959) confirmed the earlier results of Giovanelli and Stumpf (1957) that malonic acid is converted into malonyl-CoA which is then decarboxylated to acetyl-CoA and $CO_2$ by leaf extracts. These results indicate a rather active organic acid metabolic machinery in leaf tissue. Considerable effort should now be directed to a precise definition of the sites for the various activities described above.

On the basis of these and other fragmentary data, one can postulate several possible reactions:

$$\text{Pyruvic acid} \xrightarrow[O_2]{\text{CoA}} \text{acetyl-CoA} + CO_2 + H_2O \tag{8}$$

$$\text{Phosphoenolpyruvate} + H_2O + CO_2 \rightarrow \text{oxaloacetic} + HPO_4^{=} \tag{8a}$$

$$\text{NADPH} + \text{oxaloacetic} \rightarrow \text{malic} + \text{NADP}^+ \tag{8b}$$

$$\text{Malic} + \text{NADP}^+ \rightarrow \text{pyruvic} + CO_2 + \text{NADPH} \tag{8c}$$

$$\text{Malate} + \text{ATP} + \text{CoA} \rightleftharpoons \text{malyl-CoA} + \text{ADP} + \text{P} \quad (9)$$

$$\text{Malyl-CoA} \rightleftharpoons \text{acetyl-CoA} + \text{glyoxylate} \quad (9a)$$

$$\text{Oxaloacetic acid} \xrightarrow{H_2O_2} \text{malonyl-CoA} + CO_2 \quad (10)$$

$$\text{Malonyl-CoA} \rightarrow \text{acetyl-CoA} + CO_2 \quad (10a)$$

$$\text{Xylulose 5-phosphate} \xrightarrow{\text{CoA}} \text{triose phosphate} + \text{acetyl-CoA} \quad (11)$$

$$\text{Glyoxylate} \rightarrow \text{acetic acid} \rightarrow \text{acetyl-CoA} \quad (12)$$

Evidence for reactions (8) to (12) in chloroplasts is very meager. Undoubtedly the most important sources of carbon compounds in chloroplasts are the photosynthetically produced sugars. Presumably phosphoenolpyruvate acid derives from the degradation of sugars. Phosphoenolpyruvate could be the pivotal point for acetyl-CoA formation. Thus, it could be further degraded to yield pyruvic acid which could enter reaction (8) or it could participate in reactions (8a), (8b), and thence to the sequence of (9) and (9a). Pyruvic acid could also be formed by the malic enzyme which catalyses (8c). Acetyl-CoA could also be formed by the series of reactions (8a), (10) and (10a). Reactions (9) and (9a) have been described by Tuboi and Kikuchi (1963) in *Rhodopseudomonas spheroides*. Reaction (9) is catalysed by a malyl-CoA synthetase and (9a) by a malyl-CoA lyase. These reactions are attractive since both malic acid and glyoxylic acid are important plant acids in leaves. Reaction (10) has been described by Shannon *et al.* (1963) in root tissue of several plants and reaction (10a) is found widespread in plant tissue (Hatch and Stumpf, 1962). Reaction (11) is catalysed by the enzyme phosphoketolase but has been limited to bacterial cells. The work of Tolbert (1963) has contributed much to to metabolism of glyoxylic acid. Perhaps this acid is further reduced to acetic acid.

It is obvious from these few remarks that further research on an enzyme level is necessary to elucidate the precise sequence of steps which must be functioning in the chloroplast to provide sufficient quantities of acetate ion. Since acetate enters the chloroplast rather easily, it is also quite possible that this acid is synthesized elsewhere in the cells of the leaf and migrates into the chloroplast. Since acetyl-CoA and malonyl-CoA appear to be rather restricted in their transport across the chloroplast membrane, these thiol esters may be formed *in situ* in chloroplasts. Free acetic acid may not be a precursor for lipid synthesis.

The earliest reference cited in this chapter which deals directly with lipid synthesis in chloroplast is that of Smirnov in 1960. Five years have elapsed since Smirnov published his first observations. Although considerable progress has already been made, much remains for further investigation. Hopefully the next five years will resolve many of the problems we have discussed here.

## References

Ashai, T. (1964). *Biochim. biophys. Acta* **82**, 58.

Bacon, J. S. D., Palmer, M. J. and De Kock, P. C. (1963). *Biochem. J.* **78**, 198.

Beulen, W. A., Burns, R. C. and Le Comte, J. R. (1965). *Proc. natn. Acad. Sci. U.S.A.* **53**, 532.

Bradley, D. F. and Calvin, M. (1955). *Proc. natn. Acad. Sci. U.S.A.* **41**, 563.
Brooks, J. L. and Stumpf, P. K. (1966). *Archs Biochem. Biophys.* **116**, 108.
Brooks, J. L. and Stumpf, J. K. (1965). *Biochim. biophys. Acta*, **98**, 213.
Calvin, M. (1954). *Fed. Proc.* **13**, 697.
Cleland, W. W. (1964). *Biochemistry* **3**, 480.
Ellman, G. L. (1959). *Archs Biochem. Biophys.* **82**, 70.
Erwin, J. and Bloch, K. (1964). *Science, N.Y.* **143**, 1006.
Giovanelli, J. and Stumpf, P. K. (1957). *Pl. Physiol.* **32**, 498.
Hatch, M. D. and Stumpf, P. K. (1962). *Pl. Physiol.* **37**, 121.
Hawke, J. C. and Stumpf, P. K. (1965). *Pl. Physiol.* **40**, 1023.
Hawke, J. C. and Stumpf, P. K. (1966). *J. biol. Chem.* **240**, 4746.
Hiatt, A. J. and Evans, H. J. (1960). *Pl. Physiol.* **35**, 662.
Hiatt, A. J. (1962). *Pl. Physiol.* **37**, 85.
James, A. T. (1963). *Biochim. biophys. Acta* **70**, 9.
Mudd, J. B. and Stumpf, P. K. (1962). *J. biol. Chem.* **236**, 2602.
Mudd, J. B. and McManus, T. T. (1962). *J. biol. Chem.* **237**, 2057.
Mudd, J. B. and McManus, T. T. (1964). *Pl. Physiol.* **39**, 115.
Mudd, J. B. and McManus, T. T. (1965). *Pl. Physiol.* **40**, 340.
Newman, D. W. (1962). *Biochem. Biophys. Res. Commun.* **9**, 179.
Newton, J. W. (1962). *J. biol. Chem.* **237**, 3282.
Rosenberg, L. L., Capindale, J. B. and Whatley, F. R. (1958), *Nature, Lond.* **181**, 632.
Rosenberg, A. and Pecker, M. (1964). *Biochemistry* **3**, 254.
Shannon, L. M., de Vellis, J. and Lew, J. X. (1963). *Pl. Physiol.* **38**, 691.
Smirnov, B. P. (1960). *Biokhimija* **25**, 419.
Smirnov, B. P. (1962). *Biokhimija* **27**, 127.
Smirnov, B. P. and Rodionov, M. A. (1964). *Biokhimija* **29**, 335.
Stumpf, P. K. and James, A. T. (1962). *Biochim. biophys. Acta* **57**, 400.
Stumpf, P. K. and James, A. T. (1963). *Biochim. biophys. Acta* **70**, 20.
Stumpf, P. K., Bové, J. M. and Goffeau, A. (1963). *Biochim. biophys. Acta* **70**, 260.
Tolbert, N. E. (1963). *In* "Photosynthesis Mechanisms in Green Plants", p. 648. National Research Council, Washington, D.C.
Tuboi, S. and Kikuchi, G. (1963). *Biochem. J.* **53**, 364.
Yamamoto, Y. and Beevers, H. (1960). *Pl. Physiol.* **35**, 102.
Young, R. H. and Shannon, L. M. (1959). *Pl. Physiol.* **34**, 149.

# Synthesis of Unsaturated Fatty Acids by Green Algae and Plant Leaves

R. V. Harris, A. T. James and P. Harris
*Unilever Research Laboratories, Colworth House, Sharnbrook, Bedford, England*

## Introduction

The work of James, and Stumpf and James established plant leaves and chloroplast preparations as an efficient fatty acid synthesizing system and provided evidence that the conventional malonyl-CoA pathway operated in this tissue. However, their work left problems concerning the origin of the polyunsaturated fatty acids which the isolated chloroplast was unable to synthesize. In addition, plant leaves were found to be unable to desaturate stearate to oleate, although oleate synthesis from acetate required molecular oxygen. This led to the proposal of an entirely new pathway for monoene synthesis in plants with an unknown mechanism, distinct from the "aerobic" and "anaerobic" pathways already elucidated by Bloch.

## General Fatty Acid Synthesis in the Photosynthetic Tissue of Higher Plants

### Whole Leaves

The efficient incorporation of [$^{14}$C]acetate into fatty acid containing lipids by plant leaves was first shown by Eberhardt and Kates (1957). The development of gas radiochromatography (James and Piper, 1961), however, allowed a detailed study of the actual fatty acids produced. Such a study (James, 1963) revealed that all component fatty acids of the leaf were synthesized from acetate. In addition, plant leaves were able to synthesize long chain saturated and unsaturated fatty acids using added $C_8$, $C_{10}$, $C_{12}$ and $C_{14}$ saturated acids as precursors. Degradation of the products showed that chain elongation of the precursor had occurred.

Palmitic and stearic acids, however, could not act as precursors for unsaturated fatty acids although they were taken up by the leaf tissue and incorporated into complex lipids. This latter finding ruled out an explanation based on impermeability of the cell wall or absence of appropriate activation enzymes. Anaerobic incubations of leaf tissue with acetate suppressed the formation of oleic acid; this oxygen requirement precluded the "anaerobic" mechanism (Bloch *et al.*, 1962). Furthermore, since the "anaerobic" pathway has only been shown to occur in micro-organisms which lack polyunsaturated fatty acids (Erwin and Bloch, 1964), the presence of these in plants also rules out this pathway. Consequently, neither of the two known mechanisms appeared to account for monoene synthesis in plant leaves and a new "plant pathway" was proposed, aerobic but not involving direct desaturation of palmitic or stearic acids (Fig. 1).

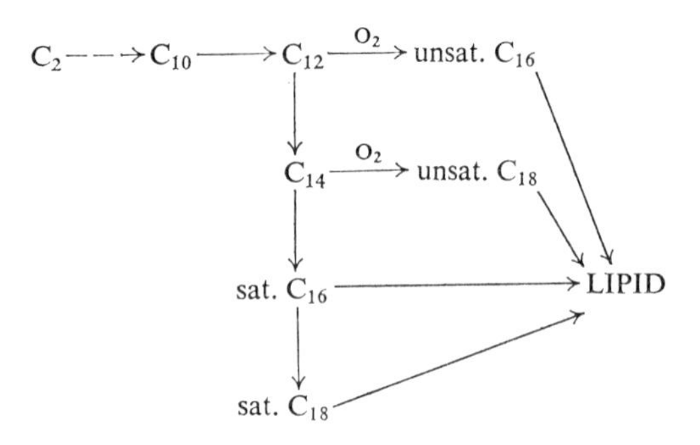


FIG. 1. Plant Pathway for unsaturated fatty acid synthesis.

## LEAF CHLOROPLAST PREPARATIONS

The major site of fatty acid synthesis in plant leaves was found by Stumpf and James (1963) to be the chloroplast and this has been confirmed by other workers (Mudd and McManus, 1962, 1964). Cofactor studies have revealed the same requirements as in many other systems and point to the conventional malonyl-CoA pathway (Stumpf and James, 1963; Mudd and McManus, 1964).

Stimulation of fatty acid synthesis by light has been found by all workers and is thought to be explicable by an increase in the supply of cofactors ATP and NADPH generated by photosynthesis. It has been claimed, in fact, that the effect is entirely due to NADPH increase (Mudd and McManus, 1962), but other workers suggest that light has an influence over and above the production of cofactors (Stumpf *et al.*, 1963). The "photoacetylation" reaction proposed by Smirnov (1960) has, however, received little support.

Brooks and Stumpf (1965) have recently reported the preparation of a soluble fatty acid synthetase system from disrupted spinach chloroplasts. Their investigations revealed a heat-stable fraction resembling the acyl carrier protein of other systems (Overath and Stumpf, 1964; Majerus and Vagelos, 1964; Pugh *et al.*, 1964), which was an absolute requirement for fatty acid synthesis. Thus there seems little fundamental difference between fatty acid synthesis in chloroplasts and in other systems.

## UNSATURATED FATTY ACID SYNTHESIS IN LEAF TISSUE

The products of fatty acid synthesis in *in vitro* chloroplast preparations differ in two important respects from those in the whole leaf. Whereas the whole leaf is able to utilize a series of acids from $C_2$ through $C_{14}$ as precursors for fatty acids, the isolated chloroplast is active only with acetate or malonate. Secondly, isolated chloroplasts are unable to synthesize linoleic or $\alpha$-linolenic acids, which are the major component unsaturated fatty acids of the organelle, linolenic acid often accounting for some 60% of the total fatty acid present. We have been most concerned with the latter problem because of the current interest in fatty acid desaturation mechanisms and the possible connection between polyunsaturated fatty acids and the primary reactions of photosynthesis (Erwin and Bloch, 1963).

More detailed examination of linoleic and α-linolenic acid synthesis in whole leaf tissue showed that oleic acid could act as a precursor. This conversion of added [$^{14}$C]oleic acid to linoleic and linolenic acids occurred without the appearance of radioactivity in shorter chain acids. A time study of this conversion revealed a typical precursor/product relationship suggesting successive desaturations of the oleate molecule (Harris and James, 1965b).

The direct desaturation of [$^{14}$C]oleate to linoleate and linolenate was confirmed by using [1-$^{14}$C]oleic acid as precursor. The radioactive linoleate and linolenate produced were isolated, hydrogenated and saponified to stearic acid

Table I. Effect of illumination on the desaturation of oleic acid

| Organism | Substrate | Condition | % conversion 18:1→18:2 | Ratio of specific activity $\frac{18:2}{18:1}$ | % of control |
|---|---|---|---|---|---|
| Chopped castor leaves | [1-$^{14}$C]oleic acid (2 μc) | Light (control) | 17·5 | 0·064 | 100 |
| | [1-$^{14}$C]oleic acid (2 μc) | Dark | 13·5 | 0·050 | 78 |
| *Chlorella vulgaris* | [1-$^{14}$C]oleic acid (1 μc) | Light (control) | 30 | 0·12 | 100 |
| | [1-$^{14}$C]oleic acid (1 μc) | Dark | 15 | 0·08 | 67 |

and degraded by a modification (James and Hitchcock, 1965) of the permanganate/acetone method of Murray (1959). After methylation, the reaction products were analysed by gas radiochromatography revealing that all radioactivity was lost on removal of the carboxyl carbon atom. Thus, the carboxyl label of the precursor was fully retained, indicating direct conversion of oleic into linoleic and linolenic acids (Harris and James, 1965a,b).

In view of the stimulatory effect of illumination on fatty acid synthesis in intact leaves and isolated leaf chloroplasts, the effect of light on linoleic acid synthesis from [$^{14}$C]oleic acid in leaves was investigated. The result (Table I) indicated that absence of light leads to a decrease in conversion but not a complete inhibition. This is in accord with the suggestion that light in photosynthetic systems stimulates fatty acid synthesis only by increasing the supply of cofactors, since these would probably persist in the dark. Any more fundamental effect might be expected to bring fatty acid synthesis to a complete halt in the dark.

The direct insertion of a double bond into long chain fatty acids has been shown (Bloomfield and Bloch, 1960; Yuan and Bloch, 1961) to require molecular oxygen. Incubation of plant leaf tissue with [$^{14}$C]oleate in the absence of oxygen completely inhibits conversion to linoleate (Table II).

Table II. Effect of anaerobiosis on the desaturation of oleic acid

| Organism | Substrate | Condition | % conversion 18:1→18:2 | Ratio specific activity $\frac{18:2}{18:1}$ | % of control |
|---|---|---|---|---|---|
| Chopped lettuce leaf | [1-$^{14}$C]oleic acid (1 μc) | Air (control) | 16·6 | — | 100 |
| | [1-$^{14}$C]oleic acid (1 μc) | Nitrogen | 0 | — | 0 |
| *Chlorella vulgaris* | [1-$^{14}$C]oleic acid (1 μc) | Air (control) | 77 | 0·58 | 100 |
| | [1-$^{14}$C]oleic acid (1 μc) | Nitrogen | 6 | 0·04 | 7 |

Water-saturated nitrogen or air (as indicated) bubbled through incubation for 1 hr prior to addition of isotope and during incubation.

The formation of linoleic and α-linolenic acids by desaturation of oleic acid is not unique to plant leaves and has been shown to occur in a variety of organisms (Erwin and Bloch, 1963, 1964). In plant tissues, the conversion has been demonstrated in safflower seeds (McMahon and Stumpf, 1964), castor seed embryo (unpublished observations), and the kinetic data of Dutton and Mounts (1964) with other oil seeds also support it. In contrast, there is no direct experimental evidence whatever for the suggestion frequently made (e.g. Hilditch and Williams, 1965) that, in plants, linolenic, linoleic, oleic and stearic acids arise, in that order, by successive *hydrogenations* of a methylene-interrupted fully unsaturated $C_{18}$ precursor.

The major piece of indirect evidence for this theory is the variation in the degree of unsaturation of the seed oils produced by plants of the same species grown at different temperatures. Here, it is found that a lower temperature invariably leads to a more unsaturated oil. Since chemical reactions occur more slowly at lower temperatures, this has lead to the belief that the saturated fatty acids are the products of the pathway, the unsaturated acids being precursors. However, it is noteworthy that Bloch and his coworkers have observed that desaturase enzyme systems derived from yeast (Meyer and Bloch, 1963) and bacteria (Fulco *et al.*, 1964) show an inverse relationship between activity and the temperature at which the cells were grown. Thus cellular control mechanisms for the desaturation reaction were able to override rate effects due to environmental temperature.

All our attempts to isolate mature leaf chloroplasts capable of synthesizing linoleic or linolenic acids have so far proved unsuccessful. Other workers in the field have likewise reported the inability of isolated chloroplasts to synthesize fatty acids more unsaturated than oleic acid (Mudd and McManus, 1962). It is possible, of course, that the enzyme involved may not be located in the chloroplast but this appears unlikely since the major proportion of the polyunsaturated fatty acid is situated there. In any case, we have found that whole leaf homogenates are also unable to synthesize linoleate, so we must conclude that wherever in the cell the enzyme or system is situated, it is extremely sensitive to the procedures generally employed for disrupting leaf tissue.

## Fatty Acid Synthesis in *Chlorella*

### WHOLE CELLS

Owing to the difficulty of preparing chloroplasts or cell-free homogenates able to carry out desaturation reactions from plant leaves, we have turned our attention to the green alga *Chlorella vulgaris*. It has been claimed that *Chlorella pyrenoidosa* (Bloch, 1964), together with the other green algae *Scenedesmus* and *Ankistrosdemus braunii* (Erwin *et al.*, 1964), follow the "plant pathway" for monoene synthesis. Studies in our laboratory (Nichols 1965) have shown

Table III. Fatty acid compositions of plant leaves and *Chlorella vulgaris*

| Acid | Castor Leaf % | *Chlorella* % |
|---|---|---|
| 14:0 | trace | 2·6 |
| 14:1 | — | trace |
| 15:0 | trace | 1·6 |
| 15:1 | — | trace |
| 16:0 | 22·0 | 19·5 |
| 16:1 | 1·5 | 6·9 |
| 16:2 | trace | 7·6 |
| 16:3 | — | trace |
| 17:0 | trace | trace |
| 17:1 | — | trace |
| 18:0 | 2·0 | 5·0 |
| 18:1 | 4·5 | 25·0 |
| 18:2 | 16·0 | 28·8 |
| 18:3 | 54·0 | 3·8 |

a close similarity between *Chlorella vulgaris* and plant leaves in terms of lipid and fatty acid composition (Table III). The fatty acid spectrum of *Chlorella* is a little more complex than that of plant leaves, including di- and tri-unsaturated $C_{16}$ acids as well as the linoleic and α-linolenic acids characteristic of higher plant leaves. The major component fatty acid in *Chlorella* is linoleic acid whilst in plant leaves it is α-linolenic acid.

*Chlorella*, a unicellular alga, offers certain advantages over plant leaves as an experimental tissue. It can be cultured under known and carefully controlled conditions and, in contrast to leaf tissue, several methods are available for disrupting algal cells for the preparation of homogenates.

Incubation of *Chlorella* in tryptone/glucose culture medium with [$^{14}$C]-acetate, $^{14}CO_2$ and the usual range of medium and long chain [$^{14}$C] fatty acids revealed good incorporation of acetate into all fatty acids but poor incorporation of $^{14}CO_2$. Unsaturated $C_{18}$ acids can be derived from the $C_8$ to $C_{14}$ saturated acids but not from palmitate or stearate (James *et al.*, 1965a,b). The poor incorporation of $CO_2$ into fatty acids suggests that under these conditions, i.e. in a medium containing abundant nutrients, *Chlorella* is growing largely heterotrophically.

Incubation of *Chlorella* with [1-$^{14}$C]oleic acid led, as in leaves, to the rapid appearance of radioactivity in linoleic and α-linolenic acids. Chemical degradation revealed full retention of labelling position indicating direct desaturation (Harris and James, 1965a,b). Incubation in the dark (Table I) or in the absence of oxygen (Table II) considerably suppressed the conversion of oleate into linoleate. Thus, biosynthesis of both saturated and unsaturated fatty acids in *Chlorella* appears to follow a pathway identical to that found in plant leaves.

## SUBCELLULAR PREPARATIONS

We have found it possible, by alteration of the culture conditions, to bring about a marked change in the fatty acid composition of *Chlorella* cells. When cells cultured in nutrient tryptone/glucose medium are transferred to a non-nutrient medium such as phosphate buffer, a rapid increase is found in the proportion of $C_{16}$ and $C_{18}$ polyunsaturated acids, accompanied by a fall in the proportion of their saturated and monounsaturated analogues (James *et al.*, 1965a]. One explanation for this, which is supported by our isotope experiments, is that the lack of nutrients in the second medium is forcing the organism to turn to a more autotrophic mode of existence, leading to rapid synthesis of polyunsaturated fatty acids for the photosynthetic apparatus.

Using cells in which polyunsaturated fatty acid synthesis is enhanced in this way, we have been able to prepare a cell-free homogenate capable of desaturating oleic acid (Harris and James, 1965a,b). The preparation is outlined in Fig. 2. After pre-incubation in phosphate buffer, the cells are disrupted using an ultrasonic disintegrator (MSE 60 W 20 kc/sec) in a medium containing sucrose (0·5 M), neutralized ascorbate (0·1 M), sodium chloride (0·1 M) and EDTA (1 mM) adjusted to pH 7·4. Centrifugation at 1000 ***g*** for 10 min serves to remove disrupted cells, and the resultant supernatant, dark green in colour, is used for incubation. The absence of viable cells in this homogenate was confirmed immediately before incubation by microscopic examination and also by plating out a small sample on a nutrient agar plate.

Using this homogenate, we have demonstrated oxygen and reduced nicotinamide nucleotide as cofactors and the CoA ester of oleic acid as preferred substrate for conversion into linoleate. These findings make more readily explicable the inhibition of conversion found in the dark since both oxygen and reduced nicotinamide nucleotides are produced by photosynthesis.

Fractional centrifugation of the homogenate yielded successive green sediments which were unable to desaturate oleic acid. Electron microscopy of these sediments showed that they contained chloroplast fragments. The supernatants from the centrifugations contained desaturase activity which, along with the green colouration, was slowly removed by increasing centrifugation

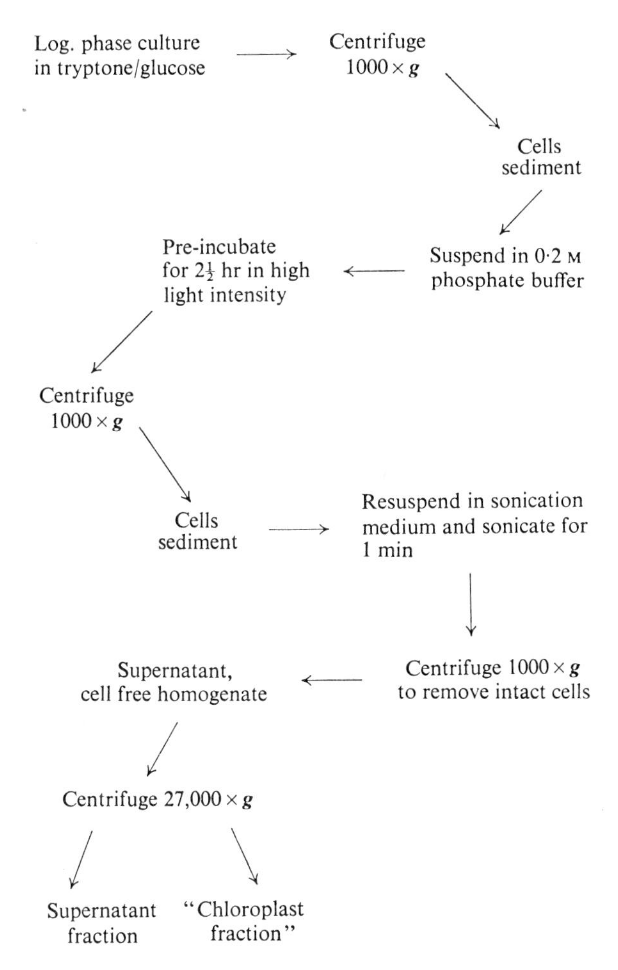


FIG. 2. Preparation of sub-cellular homogenate and fractions from *Chlorella vulgaris.*

speeds. The ability to desaturate oleic acid was thus removed by centrifugation of the homogenate without its reappearance in the chloroplast sediment. This anomaly was resolved by the use of [$^{14}$C]oleyl-CoA as substrate which was readily desaturated to both linoleic and linolenic acids by the sediment. We conclude from these results that the desaturase enzyme is located in the chloroplast but requires the CoA ester of oleic acid as substrate, there being no activating system present. An oleate-activating enzyme occurs in the supernatant fraction, which also contains chloroplast fragments (inferred from the sedimentable green colouration) sufficient to desaturate the CoA ester once formed. Increasing centrifugation speeds would remove these chloroplast fragments leaving both fractions incapable of desaturating oleic acid itself. (Harris and James, 1965a,b).

Table IV. Comparison of $C_{18}$ diene-producing sub-cellular systems

| Source | Localization | Cofactors | Isomer produced |
|---|---|---|---|
| *Chlorella vulgaris*[1] | Chloroplast | $O_2$ NADPH or NADH | $\Delta^{9,12}$ |
| Yeast (*Torulopsis utilis*)[2] | Microsomes + supernatant | $O_2$ NADPH | $\Delta^{9,12}$ |
| Safflower seeds[3] | Plastids | $O_2$ NADPH? | $\Delta^{9,12}$ |
| Rat liver[4] | Microsomes | $O_2$ NADPH or NADH | $\Delta^{6,9}$ |

[1] Harris and James (1965a,b).
[2] Meyer and Bloch (1963).
[3] McMahon and Stumpf (1964).
[4] Holloway *et al.* (1963).

A comparison with other $C_{18}$ diene-producing subcellular systems is given in Table IV. It is noteworthy that despite the dissimilarity of sources and the fact that the product in one case is a different isomer, all systems are particulate and show the same requirements for $O_2$ and NADPH or NADH as cofactors, pointing, presumably, to the same overall mechanism.

## Green Algae and Desaturation Mechanisms

As pointed out by Bloch (Fulco *et al.*, 1964) the operation of either the "aerobic "or "anaerobic" mechanism for unsaturated fatty acid synthesis appears to be mutually exclusive, no organism has yet been encountered which uses both pathways. Thus the observation that the the introduction of the second double bond into the $C_{18}$ fatty acid chain in *Chlorella* occurs by the conventional "aerobic" mechanism, prompted a re-examination of oleic acid synthesis in this organism.

With the possibility of enhanced unsaturated fatty acid synthesis produced by the culture medium changing technique referred to earlier, incorporation studies were carried out with the usual series of labelled saturated fatty acids on *Chlorella* cells, cultured in tryptone/glucose but transferred before incubation into phosphate buffer. Under these conditions, incubation with [1-$^{14}$C]-palmitate and -stearate produced considerable labelling in the $C_{16}$ and $C_{18}$ unsaturated acids. Isolation, reduction and chemical degradation of the unsaturated acids showed, in both cases, that the labelling position of the precursor was fully retained, indicating that direct desaturation of the palmitic and stearic acids had occurred (James *et al.*, 1965a,b]. This is contrary to the results reported by Bloch (1964) for *Chlorella pyrenoidosa.* Incubation of *Chlorella* with labelled palmitate or stearate in the same medium but under an oxygen-free atmosphere led to a reduction of monoene synthesis. Clearly under these conditions *Chlorella cells* are following the aerobic pathway for monoene synthesis.

Thus, *Chlorella,* when growing partly heterotrophically in nutrient medium, prefers to use carbon sources derived from the medium carbohydrate rather than added saturated acids for the synthesis of unsaturated acids. On transfer to a medium containing no nutrients, the cells are driven to produce unsaturated fatty acids for the photosynthetic apparatus. Under these conditions, the cells will directly desaturate saturated fatty acids using the "aerobic" mechanism.

Preliminary experiments with *Euglena gracilis* indicate that the same situation may exist here. Cells grown in nutrient medium and transferred to phosphate buffer for incubation produce labelled unsaturated fatty acids on incubation with [1-$^{14}$C]-stearate and -palmitate. *Euglena gracilis* had also previously been reported as following the "plant pathway" of monoene synthesis (Erwin *et al.*, 1964].

Thus for *Chlorella,* we have established the sequence of three successive desaturations: stearate$\rightarrow$oleate$\rightarrow$linoleate$\rightarrow$$\alpha$-linolenate, the insertion of each double bond requiring aerobic conditions.

## Desaturation Mechanisms in Plant Leaves

The observation that *Chlorella* can directly desaturate long chain saturated fatty acids under the conditions described, can have two explanations, either (i) *Chlorella* reverts from the "plant pathway" to direct desaturation when rapid synthesis of unsaturated fatty acids is required, or (ii) under these conditions the "plant pathway" becomes readily demonstrable as direct desaturation.

It is possible that the same situation exists in plant leaves, but that an excess of reserve carbon sources is used preferentially for unsaturated fatty acid synthesis, added stearate and palmitate not being able to enter the pathway. We have therefore attempted to prepare plant leaves low in carbon reserves. Etiolated cotyledons, very young expanding leaves and mature leaves kept in darkness for 3–10 days might all be expected to lack substantial food reserves but require rapid synthesis of the photosynthetic apparatus. None of these tissues, however, were able to utilize palmitate or stearate as precursors for the unsaturated fatty acids.

An important feature of the "plant pathway" is the depression of oleate synthesis when incubation with acetate is carried out in an oxygen-free atmosphere. Table V shows the distribution of activity when leaf discs are incubated under various conditions. When incubated aerobically (4) very little labelling appears in stearate, although the unsaturated oleic and linoleic acids become active. However, under anaerobic conditions (1) half of the total activity appears in stearate, which might suggest the accumulation of labelling in a precursor due to the omission of oxygen. This, in fact, can be regarded as a method of obtaining radioactive endogenous stearate. Now, if stearate is a precursor of oleate, exposure of a leaf containing radioactive *endogenous* stearate to aerobic conditions, should lead to a transference of this radioactivity to the oleate. If not, aerobic conditions should make no difference to the radioactive distribution.

Table V. Effect of absence of $O_2$ on oleic acid synthesis in plant leaves

| Distribution of activity in fatty acids | | (1) 6 hr An | (2) 5 hr An +1 hr A | (3) 4 hr An +2 hr A | (4) 6 hr A |
|---|---|---|---|---|---|
| Palmitic | % counts | 50 | 53 | 48 | 35 |
| | Specific activity | 0·6 | 0·7 | 0·4 | 1·3 |
| Stearic | % counts | 50 | 34 | 35 | 4·7 |
| | Specific activity | 5·2 | 3·5 | 2·3 | 1·1 |
| Oleic | % counts | 0 | 13 | 17 | 52 |
| | Specific activity | — | 0·6 | 0·9 | 6·6 |
| Linoleic | % counts | 0 | 0 | 0 | 7·9 |
| | Specific activity | — | — | — | 0·8 |

A = Aerobic conditions; An = Anaerobic conditions. For conditions of incubation, see text.

Experiments to test this were carried out as follows: leaf tissue was incubated in buffer for the indicated time with radioactive acetate under anaerobic conditions and then carefully washed, still under strictly anaerobic conditions, with dilute acetate solution, followed by buffer, and then re-incubated in air. The leaf tissue was thus exposed to radioactive acetate only during the anaerobic part of the experiment. Table V, columns (2) and (3) show that radioactivity has appeared in oleic acid. With 5 hr (2) under anaerobic conditions, there is little difference in either specific activity or % total counts of the palmitic acid compared with the totally anaerobic incubation, but both the specific activity and the % total counts of the stearic acid have fallen. With 4 hr (3) under anaerobic conditions, palmitic acid retains about the same percentage of total counts although the specific activity is less, owing to the shorter exposure to radioactive acetate; stearate again is considerably lower in both percentage total counts and specific activity. In both cases, the loss of counts from stearic acid approximately corresponds to that gained by the oleic acid.

Although not conclusive, these results certainly suggest that a proportion of the endogenous stearate can act as precursor of oleate. Previous findings, however, have shown clearly that exogenous stearate cannot exchange with this endogenous precursor stearate although the former is metabolized and incorporated into lipids.

A possible explanation to account for this situation is outlined in Fig. 3. The middle row represents the general line of saturated fatty acid synthesis leading into the lipids. These acids are in equilibrium with exogenous saturated acids. On either side are the pathways leading to the unsaturated fatty acids; these precursors are saturated acids but not in equilibrium with exogenous substrate, possibly they are bound to an acyl carrier protein analogous to that found in other systems but specific for the introduction of the first double bond into the saturated fatty acid chain. The solid arrows represent transfer systems present in all systems allowing lower chain saturated acids to pass from the saturated acid pathway to the unsaturated pathway. The dashed arrows indicate transfer systems either present or inducible in *Chlorella* when rapid

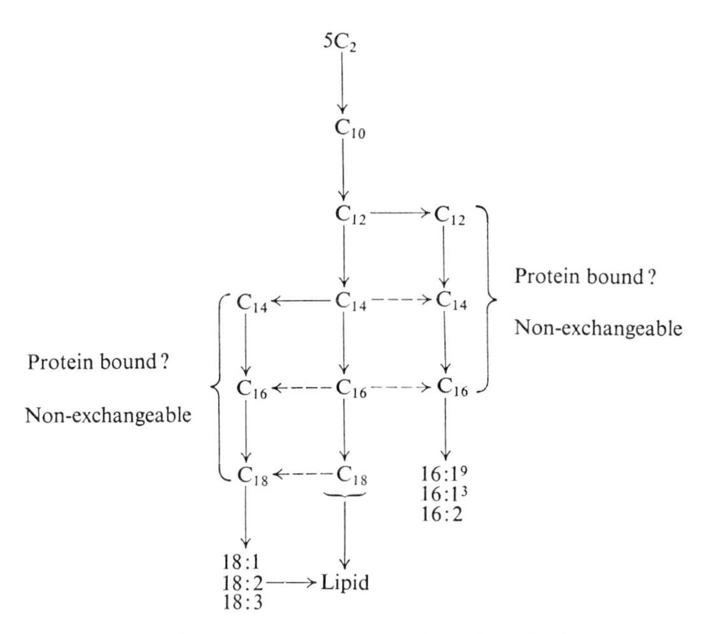


FIG. 3. Proposal for the "plant pathway" mechanism.

synthesis of unsaturated fatty acids is required. Thus the difference between the "aerobic" direct desaturation mechanism and the "plant pathway' may rest entirely upon the presence of the transfer systems represented by the dashed arrows. Ultimate confirmation of our scheme could come only with isolation from leaf tissue of the bound stearate and demonstration of its conversion into oleate.

If our hypothesis is correct then there exist only the two pathways for monoene synthesis; the "aerobic" and the "anaerobic". The "plant pathway" is fundamentally a modification of the aerobic mechanism which cannot be demonstrated as such because of the inability of exogenous stearate to enter the pathway leading to the unsaturated fatty acids.

Once oleate is formed it becomes freely interchangeable with exogenous oleate and lipid bound fatty acids, and its conversion to linoleate and linolenate becomes readily demonstrable in all intact green photosynthetic tissue and organisms examined (Erwin and Bloch, 1963, 1964; James, 1963; Harris *et al.*, 1965; James *et al.*, 1965; Nichols *et al.*, 1965). These further desaturations clearly occur by the aerobic mechanism as has been shown in subcellular preparations from *Chlorella* (Harris *et al.*, 1965; Harris and James, 1965), yeast (Meyer and Bloch, 1963) and safflower seed plastids (McMahon and Stumpf, 1964). The problem of inactivation encountered on preparation of plant leaf homogenates and isolated chloroplasts is at present hindering progress in our understanding of linoleic and linolenic acid synthesis in this tissue. However, encouraging preliminary results have been obtained using isolated proplastid preparations from etiolated castor cotyledons. These appear capable of converting oleyl-CoA into linoleic acid in the presence of NADPH (M. S. Hershfield and R. V. Harris, unpublished observations).

Another approach we have made to the problem of desaturation mechanisms in photosynthetic tissue involves a study of the fatty acid-containing lipids present in *Chlorella* and leaf tissue. *Chlorella* cells or discs of leaf tissue were allowed to metabolize [1-$^{14}$C]oleic acid and samples were withdrawn after

Table VI. Specific activity of unsaturated fatty acids in the lipids of *Chlorella vulgaris*

| Lipid | 2 hr 18:1 | 2 hr 18:2 | 4 hr 18:1 | 4 hr 18:2 |
|---|---|---|---|---|
| Neutral lipids | 2260 | 313 | 683 | 179 |
| Pigment | 0 | 0 | 0 | 0 |
| Monogalactosyl diglyceride | 160 | 41 | 296 | 138 |
| Phosphatidyl choline | 2400 | *659* | 1134 | *3856* |
| Digalactosyl diglyceride | 455 | 74 | 106 | 40 |
| Phosphatidyl ethanolamine | 1430 | 19 | 687 | 195 |
| Phosphatidyl inositol. | 802 | 0 | 179 | 51 |
| Sulpholipid | 197 | 36 | 646 | 220 |
| Phosphatidyl glycerol | 394 | 136 | 146 | 96 |
| Cardiolipin | 556 | 70 | 137 | 94 |
| Total | 992 | 243 | 1158 | 259 |

*Chlorella vulgaris* cells were incubated with [1-$^{14}$C]oleic acid in the light and samples withdrawn at 2 and 4 hr; the fatty acid containing lipids were fractionated and the specific activity of the oleic and linoleic acids esterified to each lipid determined.

varying periods of time. Fractionations of the constituent fatty acid-containing lipids were carried out (Nichols and James, 1964) and the esterified fatty acids analysed. Table VI illustratess the result of a *Chlorella vulgaris* incubation examined in this way after 2 and 4 hr. Radioactive oleic and linoleic acids are distributed generally throughout the lipids but the most significant finding is the very high specific activity of the linoleic acid in the phosphatidylcholine fraction, some fifteen times as high as in any other lipid. This is also the only lipid in which the linoleic acid has a specific activity higher than the oleic acid. Moreover, when plant leaves are allowed to metabolize [$^{14}C$] acetate in the light, phosphatidylcholine is the first lipid in which radioactive linoleic acid appears (unpublished observations).

The phosphatidylcholine molecule clearly has a very high initial specificity for biosynthesized linoleic acid. No definite conclusions can be drawn from such a general study at this stage, but this may well prove to be a fruitful line of approach and it is tempting to speculate that the phosphatidylcholine molecule may play some role in this desaturation reaction.

## References


Bloch, K. (1964). Biochemical Society Symposium No. 24, "The Control of Lipid Metabolism" (J. K. Grant, ed.), Academic Press, London and New York.

Bloch, K., Baronowsky, P., Goldfine, H., Lennarz, W. J., Light, R., Norris, A. T. and Scheuerbrandt, G. (1962) *Fed. Proc.* **20**, 921.

Bloomfield, D. K. and Bloch, K. (1960). *J. biol. Chem.* **235**, 337.

Brooks, J. L. and Stumpf, P. K. (1965). *Biochim. biophys. Acta* **98**, 213.

Dutton, H. J. and Mounts, T. L. (1964). VIth International Congress of Biochemistry, New York, VII, 43.

Eberhardt, F. M. and Kates, M. (1957). *Can. J. Bot.* **35**, 907.

Erwin, J. and Bloch, K. (1963). *Biochem. Z.* **338**, 496.

Erwin, J. and Bloch, K. (1964). *Science N.Y.* **143**, 1006.

Erwin, J., Hulanicka, D. and Bloch, K. (1964). *Comp. Biochem. Physiol.* **12**, 191.

Fulco, A. J., Levy, R. and Bloch, K. (1964). *J. biol. Chem.* **239**, 998.

Harris, R. V. and James, A. T. (1965a). *Biochem. J.* **94**, 15c.

Harris, R. V. and James, A. T. (1965b). *Biochim. biophys. Acta* **106**, 456.

Harris, R. V., Wood, B. J. B. and James, A. T. (1965). *Biochem. J.* **94**, 22P.

Hilditch, T. P. and Williams, P. N. (1965). "Chemical Constitution of Natural Fats," p. 561, 4th Ed.

Holloway, P. W., Pefullo, R. and Wakil, S. J. (1963). *Biochem. Biophys. Res. Commun.* **12**, 300.

James, A. T. (1963). *Biochim. biophys. Acta* **70**, 9.

James, A. T., Harris, R. V. and Harris, P. (1965a). *Biochim. biophys. Acta* **106**, 465.

James, A. T., Harris, R. V. and Harris, P. (1965b). *Biochem. J.* **95**, 6P.

James, A. T. and Hitchcock, C. (1965). *Kerntechnik* **7**, 5.

James, A. T. and Piper, E. A. (1961). *J. Chromatog.* **5**, 265.

Majerus, P. W. and Vagelos, P. R. (1964). *Fed. Proc.* **23**, 166.

McMahon, V. and Stumpf, P. K. (1964). *Biochim. biophys. Acta* **84**, 359.

Meyer, F. and Bloch, K. (1963). *Biochim. biophys. Acta* **77**, 671.

Mudd, J. B. and McManus, T. T. (1962). *J. biol. Chem.* **237**, 2057.

Mudd, J. B. and McManus, T. T. (1964). *Pl. Physiol.* **39**, 115.

Murray, K. E. (1959). *Aust. J. Chem.* **12**, 657.

Nichols, B. W. (1965). *Biochim. biophys. Acta* **106**, 275.

Nichols, B. W., Harris, R. V. and James, A. T. (1965). *Biochem. Biophys. Res. Commun.* **20**, 256.

Nichols, B. W., and James, A. T. (1964). *Fette Seifen AnstrMittel* **66**, 1003.

Overath, P. and Stumpf, P. K,. (1964). *Fed. Proc.* **23**, 166.

Pugh, E. L., Sauer, F. and Wakil, S. J. (1964). *Fed. Proc.* **23**, 166.

Smirnov, B. P. (1960). *Biokhimija* **25**, 545.

Stumpf, P. K., Bové, J. M. and Goffeau, A. (1963). *Biochim. biophys. Acta* **70**, 260.

Stumpf, P. K. and James, A. T. (1963). *Biochim. biophys. Acta* **70**, 20.

Yuan, C. and Bloch, K. (1961). *J. biol. Chem.* **236**, 1277.

# Biosynthesis of Terpenoid Quinones

D. R. Threlfall and W. T. Griffiths

*Department of Biochemistry and Agricultural Biochemistry,*
*University College of Wales, Aberystwyth, Wales*

## Introduction

Goodwin and Mercer (1963) whilst studying the regulation of sterol, carotenoid and phytol metabolism in germinating seedlings were led to propose that there are two sites of terpenoid biosynthesis within the photosynthetic plant cell; one is in the chloroplast for the formation of phytol and $\beta$-carotene, and the other is in the cytoplasm giving rise to the phytosterols. If this "compartmentalization" view is correct then it follows that the terpenoid side-chains of plastoquinone (PQ, I), vitamin $K_1$ (II), $\alpha$-tocopherol quinone ($\alpha$-TQ, III) and its parent chromanol $\alpha$-tocopherol ($\alpha$-T,IV) compounds which are all specifically located in the chloroplast, would be synthesized in the chloroplast. The side-chain of the mitochondria-located ubiquinone (UQ, V), on the other hand, along with the phytosterols would be formed at some extraplastidic site. Taking the premise one step further it can be argued that the entire molecules of the intraplastidic located quinones would be elaborated in the chloroplast.

The investigations outlined here were carried out firstly, to determine the relationship of quinone formation to chloroplast development and secondly to demonstrate if, in accordance with the above postulate, the chloroplastidic quinones are biosynthesized within the confines of the chloroplast. Since most of the biosynthetic work was concerned with the terpenoid portions of the quinone molecules, it is as well to emphasize that PQ and UQ both have long unsaturated polyisoprenoid side-chains whilst the other compounds studied vitamin $K_1$, $\alpha$-TQ and $\alpha$-T all have a shorter twenty carbon unit, which can formally be regarded as arising from the partly saturated polyisoprenoid alcohol, phytol (VI).[1]

## Distribution and Nature of Phytoquinones

In the course of the present work we have used a variety of photosynthetic tissues, the quinone complements of which are summarized in Table I. All the tissues, as expected, contained vitamin $K_1$, PQ, $\alpha$-TQ and $\alpha$-T; tobacco contained in addition two other quinones which we designated PQC′ and PQD′ (Threlfall, Griffiths and Goodwin, 1965); french bean also contained PQC′

[1] Some of the techniques used in these investigations have been reported briefly in previous communications (Threlfall and Goodwin, 1963, 1964; Griffiths, Threlfall and Goodwin, 1964; Threlfall, Griffiths and Goodwin, 1964, 1965) and will in due course be published *in extenso*.

and PQD′ and a further quinone having properties similar to PQB (Kegel, Henninger and Crane, 1962). Maize, barley and *Euglena gracilis* strain Z contained the UQ homologue having the same chain length as PQ, i.e. $UQ_9$

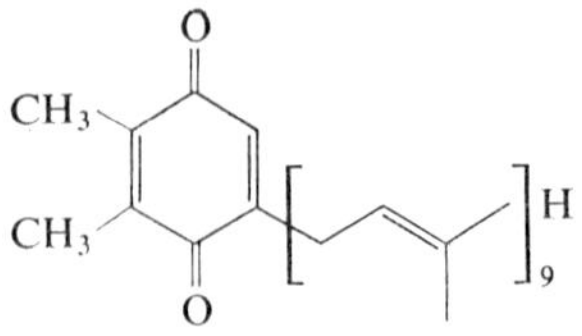

I

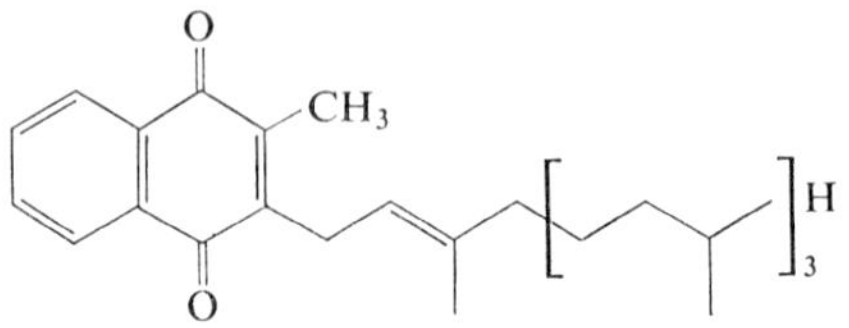

II

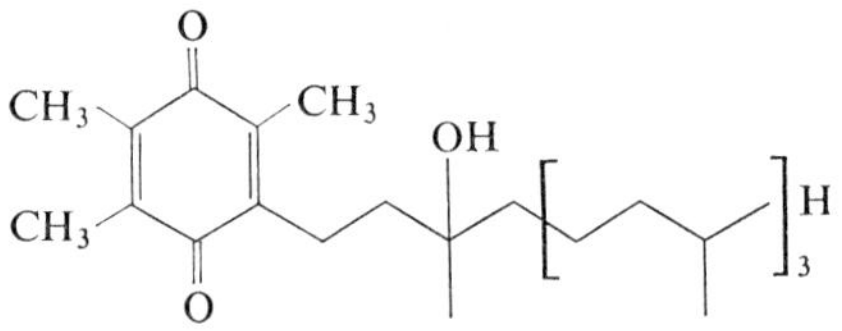

III

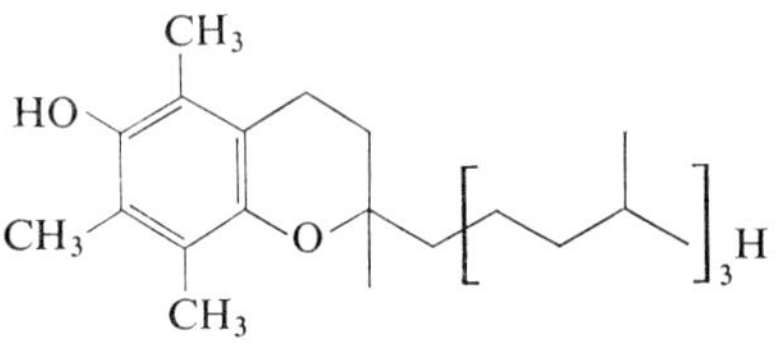

IV

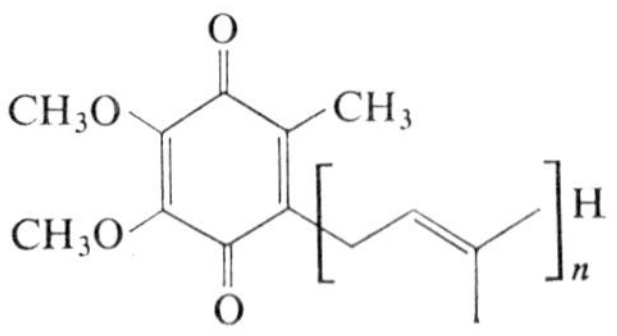

V

$n = 6 - 10$

whilst the other two tissues, tobacco and french bean, contained the homologue normally found in higher plant tissues, i.e. $UQ_{10}$. In all cases examined vitamin $K_1$, PQ, α-TQ and α-T were found localized in the chloroplast (but see later for comments on α-T). UQ on the other hand was found to be extraplastidic in distribution.

As is clear from other papers presented in this symposia, the nature and distribution of the chemically undefined plastoquinones (PQB, PQC and PQD) is confused, it may be useful here to summarize our own findings.

We have isolated from a variety of photosynthetic tissues quinones which chromatographically are identical to the PQB, C and D preparations of Crane

Table I. Quinones present in the various experimental tissues used

| Quinone | Tissue | | |
|---|---|---|---|
| Vitamin $K_1$ | Maize | Tobacco | French bean |
| α-Tocopherol quinone (and α-tocopherol) | Barley | | |
| Plastoquinone (PQ) | *Euglena* | | |
| PQC′[1] | | | |
| PQD′[1] | | | |
| PQB[1] | | | |
| Ubiquinone | $UQ_9$ in Maize, Barley and *Euglena*, $UQ_{10}$ in Tobacco and French bean | | |

(Brackets in the original: Maize, Barley, *Euglena* span Vitamin $K_1$ to Plastoquinone (PQ); Tobacco spans Vitamin $K_1$ to PQD′; French bean spans Vitamin $K_1$ to PQB.)

[1] As defined by Threlfall, Griffiths and Goodwin (1965).

N.B. This distribution takes no account of quinones present in trace amounts, e.g. Desmethyl vitamin $K_1$; McKenna, Henninger and Crane (1964).

and coworkers (Kegel *et al.*, 1962; Henninger and Crane, 1963, 1964); this has been confirmed in both Dr. Crane's laboratory and our own. However, the ultraviolet absorption characteristics of the quinones we called C and D differed from those recorded by Henninger and Crane (1964) for apparently the same quinones, this led us to designate our compounds PQC′ and PQD′; the implication being that, despite the differences in ultraviolet spectra, PQC′ and PQD′ correspond to PQC and PQD.

In a limited survey we have found that PQB, PQC′ and PQD′ do not show the same widespread distribution pattern associated with (say) vitamin $K_1$ or

H[ ]$_3$ $CH_2OH$

VI

PQ. On the basis of our findings, to date, we have suggested that in the chloroplast three types of quinone patterns can exist: (1) (basic) consists of vitamin $K_1$, PQ and α-TQ, e.g. maize, barley and *Euglena*; (2) basic + PQC′ and PQD′, e.g. tobacco; and (3) basic + PQB, PQC′ and PQD′, e.g spinach, lucerne, nettle, *Ulva* and *Polysiphonia*. The amounts of PQB, PQC′ and PQD′, relative to PQ and chlorophyll occurring in spinach, broad bean and tobacco are given in Table II.

PQB, when present appears to occur in only small amounts. From its rather anomalous chromatographic properties we have eliminated the possibility that it is a simple or partially hydrogenated homologue of PQ. PQC′ and PQD′, although showing the same restricted distribution pattern as PQB can occur in significantly larger amounts. Table II shows that when these quinones are present in a tissue, PQC′ is the second most abundant member of the quinone complement. On the basis of ultraviolet and infrared spectroscopic evidence we proposed that PQC′ and PQD′ were trialkyl substituted *p*-benzoquinones

Table II. Quinone levels in green photosynthetic tissues

| Component | μmole/100 μmole chlorophyll | | |
|---|---|---|---|
| | Spinach[1] | Broad bean[1] | Tobacco[2] |
| Chlorophyll | 100 | 100 | 100 |
| β-Carotene | 8·3 | 4·8 | 8·9 |
| Vitamin $K_1$ | 1·2 | 0·8 | 1·4 |
| PQ | 10·0 | 4·8 | 12·2 |
| PQB | 0·9 | 1·4 | ND |
| PQC′ | 4·1 | 1·5 | 4·4 |
| PQD′ | 1·8 | 0·3 | 0·5 |
| α-TQ | 0·7 | 1·6 | 1·5 |

[1] Values for chloroplast preparations.
[2] Values for deveined leaf tissue.
ND = not detected.

of the same molecular parameters as PQ, with one of the alkyl groups taking the form of a hydroxyl containing polyisoprenoid side-chain (Threlfall, Griffiths and Goodwin, 1965), i.e. monohydroxy plastoquinones.

Mass spectra determinations, carried out through the courtesy of Prof. J. W. Cornforth (Shell & Co. Ltd.), have confirmed that PQC′ and PQD′ (isolated from senescent tobacco leaves) are monohydroxy plastoquinones. In both cases the mass peak was found to be *m/e* 764, i.e. M. W. of PQ (748) + one oxygen. On the basis of this evidence and from a consideration of the ultra-

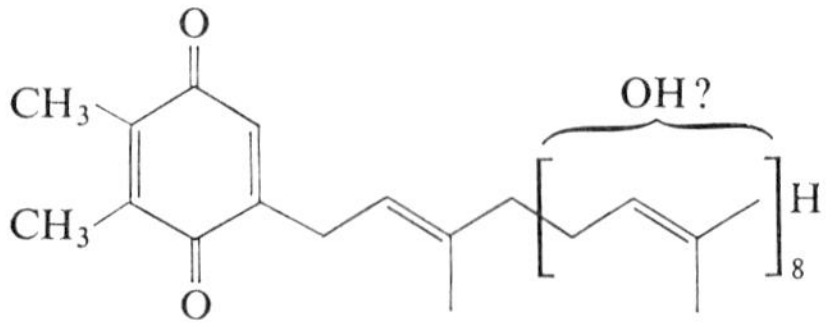

VII

violet absorption spectra of various tocopherylquinones and the oxidation product of plastochromanol (Whittle, Dunphy and Pennock, 1965), we propose that PQC′, the quinone with which most of these studies was concerned, and possibly PQD′ contain: (i) the PQ nucleus; (ii) a nonaprenyl side-chain the first isoprene unit of which is unmodified; and (iii) a hydroxyl in one of the remaining eight isoprenes (VII).

## Quinone Synthesis and Chloroplast Development

The effect on quinone levels of germinating maize or cultivating *E. gracilis* cells in the light or dark was first investigated, Table III. It was found that in the absence of light the maize shoot—the normal photosynthetic region of the maize seedling—contained reduced amounts of the chloroplastidic quinones vitamin $K_1$, PQ and $\alpha$-TQ, and no chlorophyll or $\beta$-carotene. The fall in quinone concentration was not uniform, $\alpha$-TQ showing the greatest decrease followed by PQ and vitamin $K_1$. Similar behaviour was observed in *E. gracilis*, although here all the chloroplastidic components virtually disappear under etiolating conditions. An intermediate stage can be induced by growing the organism in the light on a rich heterotrophic medium. In both maize and *Euglena*, ubiquinone levels were unaffected by the conditions of growth, which reflects the probable role of this quinone in terminal respiration.

Table III. The effect of different cultural conditions on quinone levels

| Tissue and growth conditions | /g dry wt of tissue | | | | |
|---|---|---|---|---|---|
| | Chlorophyll (mg) | Vitamin $K_1$ ($\mu$g) | PQ ($\mu$g) | $\alpha$-TQ ($\mu$g) | UQ ($\mu$g) |
| Maize shoots[1] | | | | | |
| (i) 6 days in light | 10 | 12 | 227 | 86 | 41 |
| (ii) 6 days in dark | 0 | 9 | 28 | 0 | 38 |
| *Euglena gracilis Z*[2,3] | | | | | |
| (i) autotrophic medium | 44 | +++ | 2260 | 432 | 213 |
| (ii) hetrotrophic+light | 13 | + | 171 | 39 | 120 |
| (iii) heterotrophic+dark | 0 | 0 | 17 | 0 | 200 |
| Tissue culture | | | | | |
| Paul's Scarlet Rose[4] (cambial tissue) | 0 | trace | 55 | ? | 43 |

[1] Barley showed a similar behaviour, although the differences were not as marked.
[2] (i) and (ii) 6 days old, (iii) 8 days old.
[3] Similar results were obtained by Fuller *et al.* (1961).
[4] Threlfall and Goodwin (1963).

Included in Table III, for comparison, are the results obtained for the analysis of non-photosynthetic cambial tissue of Paul's Scarlet Rose, grown in tissue culture (Threlfall and Goodwin, 1963). The presence of ubiquinone was not unexpected considering its function in respiration; the homologue found, $UQ_{10}$, is the form normally found in higher plants. Rather surprisingly, since this tissue is non-photosynthetic PQ was also found, although it was present in greatly reduced amounts compared to normal photosynthetic tissues. This observation is in agreement with similar work carried out on other non-photosynthetic tissues, i.e. maize roots (Lester and Crane, 1959) and the spadix of *Arum maculatum* (Hemming, Morton and Pennock, 1963), the significance of which is discussed shortly.

We went on to study the effect of light on dark grown photosynthetic tissues and some typical results are given in Table IV. The choice of excised shoots for this investigation, was governed by the fact that the specific terpenoid precursor mevalonic acid, cannot penetrate the root systems of seedlings (Goodwin, 1958), but can enter the vascular system of excised plants; essentially similar results are obtained when intact shoots are used, but the responses are more marked. In maize, barley and french bean 24-hr illumination led to the appearance of chlorophyll (and $\beta$-carotene) and an increase in the levels of the chloroplastidic quinones PQ and $\alpha$-TQ. The best response was obtained with maize

Table IV. The effect of 24 hr illumination on etiolated tissue

| Tissue | /100 shoots (*Euglena*/g dry wt) Chlorophyll (mg) | PQ ($\mu$g) | $\alpha$-TQ ($\mu$g) | $\alpha$-T ($\mu$g) | UQ ($\mu$g) |
|---|---|---|---|---|---|
| Maize shoots[1] | | | | | |
| Etiolated | 0 | 241 | 0 | 216 | 220 |
| Illuminated | 23 | 942 | 413 | 236 | 275 |
| Barley shoots[1] | | | | | |
| Etiolated | 0 | 87 | 0 | — | 22 |
| Illuminated | 18 | 180 | 20 | — | 32 |
| French bean leaves[2] | | | | | |
| Etiolated | 0 | 194 | 14 | — | — |
| Illuminated | 6·9 | 288 | 16 | — | — |
| *Euglena gracilis Z*[3] | | | | | |
| Etiolated | 0 | 17 | 0 | 156 | 234 |
| Illuminated 24 hr | 1·9 | 103 | 44 | 171 | 261 |
| 72 hr | 15 | 707 | 154 | 375 | 169 |

[1] Excised 5-day-old barley and 6-day-old maize shoots illuminated for 24 hr.
[2] Excised stems illuminated for 24 hr, only leaf taken for analysis.
[3] Seven-day-old heterotrophic dark grown cells, transferred to a "non-dividing" medium.

followed by barley and french bean. The increase in quinone levels of the latter plant were extremely poor and furthermore the tissue was unique in that the level of $\alpha$-TQ in the dark showed virtually no increase on illumination. An explanation for the poor response may be that the period of illumination (24 hr) was too short to produce effective chloroplast maturation. The results for the level of $\alpha$-T in maize, both before and after illumination, demonstrate that the $\alpha$-TQ synthesized is not at the immediate expense of the $\alpha$-T; since no fall in the level of this chromanol was observed. *E. gracilis* gave similar results to the other photosynthetic tissues, the response continuing up to 72 hr.

In agreement with previous work, ubiquinone again showed little significant variation.

From a knowledge of the development of the chloroplast in etiolated tissues and the effect of light on these structures these results led us to propose that in etiolated tissue (in this context dark-grown) the chloroplastidic quinones are associated with the plastid primordia—this is borne out by the fact that etiolated tissues of maize, barley and french bean where well-developed primordia are present, chloroplastidic quinones are found in significant amounts; whereas in etiolated *Euglena* where no structures readily recognizable as plastid primordia are found, vitamin $K_1$, PQ and $\alpha$-TQ are absent. The occurrence of some $\alpha$-T in etiolated cultures of *Euglena* would seem to be of importance and suggests that this chromanol is not wholly confined to the chloroplast. The presence of PQ in plant tissue cultures,

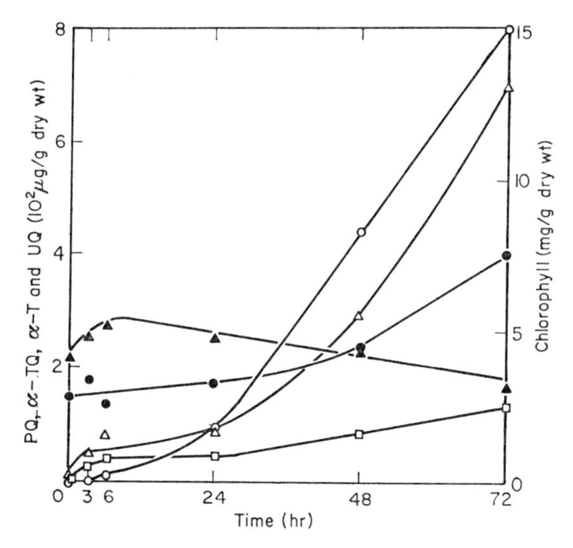


FIG. 1. Effect of light on the synthesis of quinones by etiolated non-dividing cells of *Euglena gracilis* strain Z. ○—○, chlorophyll; △—△, PQ; □—□, $\alpha$-TQ; ●—●, $\alpha$-T and ▲—▲, UQ.

maize roots and the spadix of *A. maculatum* is probably indicative of the occurrence of structures in these tissues which are either potentially capable, or, have lost their ability, of developing into chloroplasts.

The formation of the chloroplastidic quinones in relation to chloroplast development in illuminated etiolated tissues was examined in greater detail for maize and *Euglena*; although the investigation was primarily for determining the optimal conditions for exposure of the tissue, to radiochemical compounds, the results served as an adjunct to the studies just described.

Figure 1, shows the time course with etiolated non-dividing cells of *E. gracilis* strain Z. The graph shows that during the first 6 hr of illumination all the components measured increased in terms of absolute concentration; although this may be an apparent increase in concentration, owing to variations in the non-lipid components of the cells. At 24 hr all the chloroplast components (chlorophyll, PQ, α-T and α-TQ) are being actively synthesized and this trend continues up to the third day (72 hr); after this synthesis gradually stops. During this time UQ declines to the norm.

The results obtained for the study with maize are given in Fig. 2. On illumination of excised-etiolated shoots chlorophyll formation is apparent within 3 hr; if chlorophyll is taken as a measure of chloroplast development this means that functional chloroplasts are also present after this time. PQ and UQ levels fall slightly during this period owing to photodestruction and/or enzymic destruction brought about by excision of the shoots. After this PQ is rapidly synthesized, as is α-TQ and β-carotene in step with chlorophyll and chloroplast development; this synthesis proceeds rapidly up to 24 hr and then begins to fall away. In this system α-T showed no changes in concentration suggesting, if we

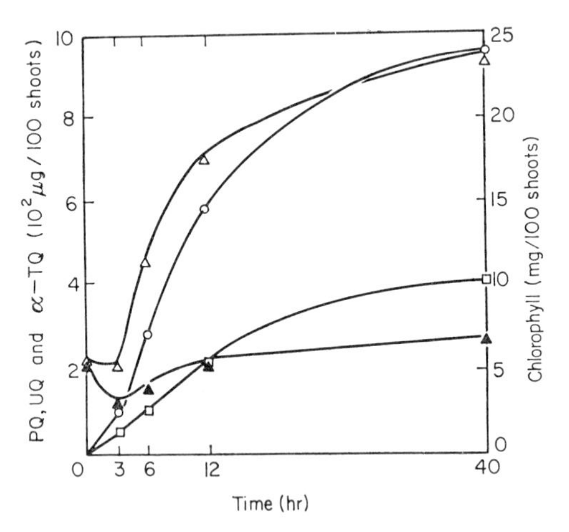


FIG. 2. Effect of light on the synthesis of quinones by excised etiolated 6-day-old maize shoots. ○—○, chlorophyll; △—△, PQ; □—□, α-TQ and ▲—▲, UQ.

accept that α-T is localized mainly in the chloroplast, and the consensus would seem to support this, that (i) α-TQ is not synthesized from α-T or (ii) any αT converted into α-TQ is rapidly replenished. The observation that the unesterified 3β-hydroxy sterols β-sitosterol and stigmasterol, showed no measurable incremental synthesis is important in relation to studies to be described shortly.

Thus, our investigations so far had shown that the chloroplastidic quinones (vitamin $K_1$, PQ and α-TQ) are apparently localized entirely in the chloroplast; that even when present in etiolated tissues they are probably associated with the plastid primordia and, on illumination of etiolated tissue chloroplastidic quinone synthesis parallels chloroplast formation. With these facts in mind we studied the biosynthesis of the PQ and UQ side-chains and the phytol side-chains, of vitamin $K_1$, α-TQ and α-T both in relation to each other and to β-carotene and the phytosterols, β-sito- and stigmasterol.

## Biosynthesis of Quinones

### TERPENOID SIDE-CHAIN

As a starting point for biosynthetic studies, we took mevalonic acid (MVA) to be the specific distal precursor of the isoprenoid side-chain and assumed that the quinone chains are elaborated, independently of the nucleus, by a pathway of sequential 5-carbon additions (Fig. 3). The alcohol pyrophosphate of requisite length would then condense with the appropriate nucleus or nuclear precursor, in the form of a quinol rather than a quinone (Lynen, 1961), to give either directly or after suitable methylations etc., the phytoquinol which would then be oxidized to the corresponding quinone. The region of "phytol" biosynthesis, i.e. Vitamin $K_1$, α-TQ and α-T is necessarily obscure, but synthesis of (say) vitamin $K_1$ could proceed via one of the two routes. Thus, the side-chain could be attached as geranylgeranyl pyrophosphate and then reduced, or altern-

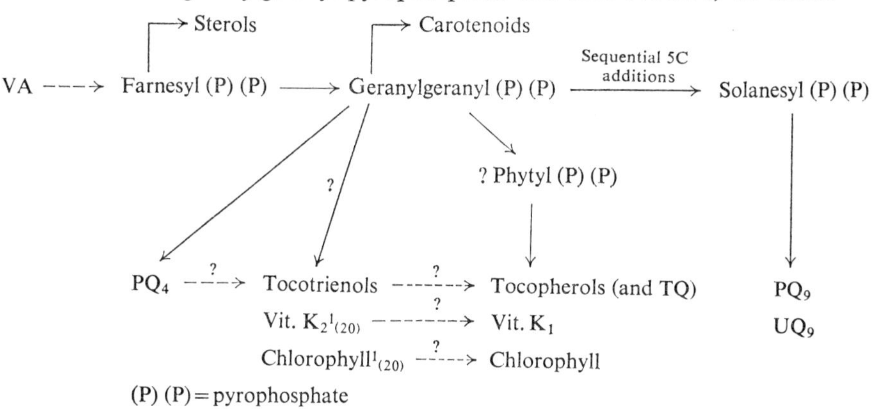


[1] No reports of these compounds occurring in plant tissues has yet appeared.

FIG. 3. Possible routes leading from MVA to the formation of the isoprenoid side-chains of quinones and chromanols found in plant tissues.

atively it could be reduced prior to attachment. In the first scheme $PQ_4$ (Eck and Trebst, 1963) would then become a possible precursor to the tocotrienol, γ-tocotrienol [see Pennock, Hemming and Kerr (1964), for nomenclature]. In view of the occurrence of the tocotrienols it would be reasonable to expect to find vitamin $K_{2(20)}$ and a chlorophyll with a geranylgeraniol side-chain in trace amounts.

The experimental system used routinely was greening-excised-etiolated maize shoots. The characteristics of this system are: (i) a large incremental synthesis of chlorophyll, β-carotene and the chloroplastidic quinones, PQ, α-TQ and possibly vitamin $K_1$, (ii) no net synthesis of the phytosterols (unesterified β-sito- and stigmasterol), α-T and UQ.

The approach adopted was that which has been used routinely in this department for the demonstration of intraplastidic synthesis of β-carotene (Treharne, Mercer and Goodwin, 1964) and phytol side-chain of chlorophyll (Mercer and

Goodwin, 1962) and involves comparing the incorporation of the non-specific substrate $^{14}CO_2$ and the specific distal terpenoid precursor [2-$^{14}$C]MVA into the various terpenoids under investigation. Then provided one of a given set of physiological factors prevails, in this instance probably membrane permeability (see below), it will be possible to differentiate between intra-and extraplastidic biosynthesis of the terpenoid molecules or terpenoid portions of molecules.

A. Common biosynthetic pathway

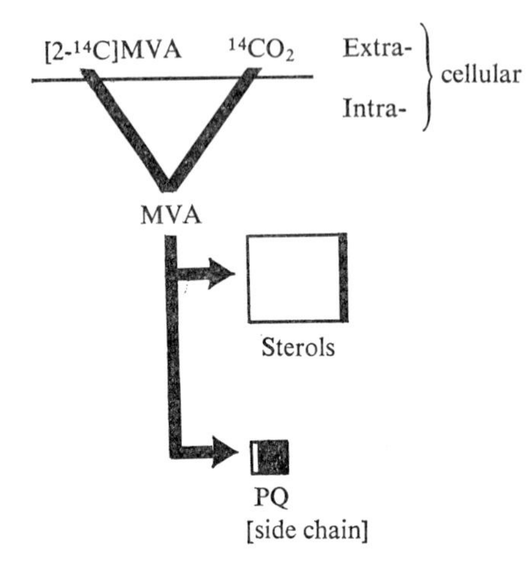


Expect
(a) Identical labelling pattern from $^{14}CO_2$ or [2-$^{14}$C]MVA.
(b) Specific activity of PQ [side-chain] much higher than the sterols due to the large incremental synthesis of the latter.

B. Two pathways differentiated by substrate availability

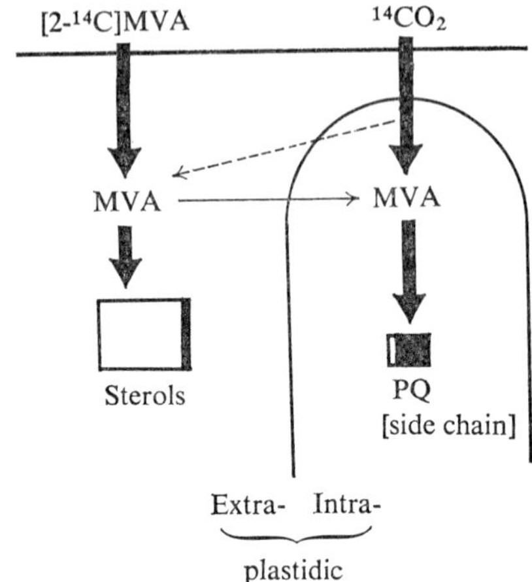


Pathways differentiated by either
(i) Relative availability of the substrate due to anatomy of the cell, or
(ii) impermeability of the chloroplast membrane to terpenoid precursors and intermediates.
Expect
Different labelling patterns from $^{14}CO_2$ and [2-$^{14}$C]MVA.

FIG. 4. Expected labelling patterns for the incorporation of [2-$^{14}$C]MVA and $^{14}CO_2$ into the intra- and extraplastidic terpenoids, dependent on substrate availability. The squares represent the pool sizes of sterol and PQ, in the maize shoot, at the end of the "greening up" period. The squares are subdivided into (i) black areas which represent the measurable incremental synthesis of the compound during the experiment and (ii) white areas which represent the original endogenous material.

The interpretation of the results obtained can best be appreciated by reference to Fig. 4.

In the diagram the squares represent the pool sizes of sterol and plastoquinone in the maize shoot at the end of the "greening up" period. The squares are further subdivided into (i) black areas which represent the measurable incremental synthesis of the compound during the experiment and (ii) white areas which represent the original endogenous material—the black leading edge to the sterol square is included as a reminder that even though no measurable increment in the level of this component occurs, nevertheless some newly synthesized material will appear as the result of turnover. The areas show that any [$^{14}$C]sterol formed will undergo a large dilution with non-radioactive material, PQ, on the other hand, will show hardly any dilution effect.

### *Common Biosynthetic Pathway*

Let us first assume that PQ and the sterols are biosynthesized at a common site. Then, since MVA and the early stages of the subsequent pathway are common to both, the relative labelling patterns from $^{14}CO_2$ and [2-$^{14}$C]MVA will be identical. In view of the larger pool-dilution of sterol the specific activity of the PQ should be much higher than the sterol.

### *Two Pathways Differentiated by Substrate Availability*

In this case we have two biosynthetic pathways, one extraplastidic for the formation of sterols (and the side-chain of UQ) and the other intraplastidic for the formation of PQ (β-carotene, α-TQ, vitamin $K_1$, and the phytol side-chain of chlorophyll). The pathways will, under our conditions, be experimentally detectable provided that the relative availability of one of the substrates, taking as an example [2-$^{14}$C]MVA, to the two sites of synthesis is affected by (i)

Table V. Maize: incorporation of [2-$^{14}$C]MVA and $^{14}CO_2$ into PQ and related terpenoids

Excised etiolated 6-day-old shoots exposed to [2-$^{14}$C]MVA or $^{14}CO_2$ for 24 hr with continuous illumination[1,2]

| Terpenoid | Level at the end of expt. mg/100 shoots | Increase in terpenoid level during expt. | [2-$^{14}$C]MVA (6 μc/100 shoots) d.$10^{-3}$/min/mg | $^{14}CO_2$ (200 μc/100 shoots) d.$10^{-3}$/min/mg |
|---|---|---|---|---|
| Chloroplastidic | | | | |
| Plastoquinone | 0·86 | 4-fold | 41 | 180 |
| β-Carotene | 0·68 | 10-fold | 47 | 236 |
| Extraplastidic | | | | |
| Ubiquinone | 0·26 | 2-fold | 456 | 25 |
| Sterol[3] | 25·0 | 0 | 96 | 4 |

[1] Six-day-old etiolated maize seedlings excised at the internode and either (i) exposed to an atmosphere containing $^{14}CO_2$, or (ii) the excised ends immersed in a solution of [2-$^{14}$C]-MVA (6 μc/20 ml of $H_2O$).

[2] These values are from the $^{14}CO_2$ experiment, the [2-$^{14}$C]MVA values were similar.

[3] Unesterified β-sito and stigmasterol.

the anatomy of the cell, such that the substrate is used up more rapidly at one site than the other, e.g. [2-$^{14}$C]MVA "mopped up" at some extraplastidic site before it can reach the chloroplast or (ii) the chloroplast membrane being impermeable or semi-permeable to [2-$^{14}$C]MVA. Then, even if early $^{14}CO_2$-fixation products were incorporated into extraplastidic-MVA as rapidly as intraplastidic-MVA, giving in effect the labelling pattern expected for a common biosynthetic pathway, two pathways will be detectable; since [2-$^{14}$C]MVA will be incorporated more readily into extraplastidic terpenoids and $^{14}CO_2$ into intraplastidic terpenoids, giving different labelling patterns for the two substrates. Rogers, Shah and Goodwin (1967) have recently demonstrated that condition (i), i.e. the relative impermeability of the chloroplast membrane to [2-$^{14}$C]MVA, is one of the governing factors in these experiments.

Before the hypothesis just proposed could be tested it was necessary to show that the various [$^{14}$C]-substrates are incorporated into the correct parts of the quinone molecules, i.e. [2-$^{14}$C]MVA into the solanesol side-chain and $^{14}CO_2$ uniformly throughout the molecule. Using an ozonolytic degradation procedure we have been able to demonstrate that in [$^{14}$C]PQ and UQ samples isolated from maize shoots which have been exposed to [2-$^{14}$C]MVA or $^{14}CO_2$, [2-$^{14}$C]MVA is incorporated specifically into the side-chain of PQ and UQ, whilst $^{14}CO_2$ is found uniformly distributed between the side-chain and the nucleus (Threlfall *et al.*, 1964), These results prove, as expected, that [2-$^{14}$C]-MVA can be regarded as the specific distal precursor to the isoprene side-chain of PQ and UQ in plant tissues.

Table VI. Maize: incorporation of [2-$^{14}$C]MVA and $^{14}CO_2$ into vitamin $K_1$ $\alpha$-tocopherol and $\alpha$-tocopherol quinone

Experimental conditions the same as the previous experiment

| Terpenoid | Level at the end of expt. mg/100 shoots | Increase in terpenoid level during expt. | [2-$^{14}$C]MVA (6 $\mu$c/100 shoots) d.$10^{-3}$/min/mg | $^{14}CO_2$ (263 $\mu$c/100 shoots) d.$10^{-3}$/min/mg |
|---|---|---|---|---|
| Chloroplastidic | | | | |
| Vitamin $K_1$ | 0·11 | 2-fold | 26 | 115 |
| $\alpha$-Tocopherol | 0·20 | 0 | 27[2] | 16 |
| $\alpha$-TQ | 0·44 | 22-fold | 4 | 17 |
| Reference cpds | | | | |
| $\beta$-Carotene | 0·30 | 10-fold | 37 | 210 |
| Sterols[1] | 17·0 | 0 | 102 | 4 |

[1] Unesterified $\beta$-sito and stigmasterol.
[2] Not completely pure (?).

The results for the incorporation of [2-$^{14}$C]MVA and $^{14}CO_2$ into the various terpenoids investigated are given in Tables V and VI. Table V summarizes a typical set of values for the incorporation of the two substrates into the solanesol side-chain containing quinones, PQ and UQ, using as reference compounds $\beta$-carotene and the unesterified 3$\beta$-hydroxysterols. Columns 2 and 3 give the increase and final levels of the terpenoid components examined. The results show quite clearly that two patterns of incorporation exist. Thus, [2-$^{14}$C]MVA is effectively incorporated into UQ and, bearing in mind the size of the sterol pool, the phytosterols. On the other hand the incorporation into

PQ and $\beta$-carotene, taking into consideration the large incremental synthesis of these compounds, is poor. $^{14}CO_2$ shows the reverse behaviour. These results then fit in with the predicted results for two sites of terpenoid synthesis within the photosynthetic plant cell differentiated by substrate availability.

The results for the incorporation of the two substrates into the phytol containing quinones vitamin $K_1$, $\alpha$-TQ and its cyclic isomer $\alpha$-T were not as clear cut, Table VI. The values show that vitamin $K_1$ behaved to the substrates in a manner similar to $\beta$-carotene and PQ; however $\alpha$-TQ and $\alpha$-T failed to fit into either pattern, even though a large incremental synthesis of $\alpha$-TQ had taken place. The explanation we offer at this time is that $\alpha$-T and its quinone are being formed from some late tocopherol precursor present in the dark grown shoots, possibly by a series of methylations, rather than a *de novo* synthesis of either the nucleus or side-chain. Furthermore, the $^{14}CO_2$ results would suggest that the chromanol and quinone are in a state of rapid equilibrium with each other.

These experiments demonstrate quite convincingly that the terpenoid portions of the vitamin $K_1$, and PQ molecules along with $\beta$-carotene are biosynthesized within the chloroplast, whilst the UQ side-chain along with the phytosterols are synthesized at some extraplastidic site. This separation of biosynthetic pathways is even more striking when it is remembered that the PQ and UQ homologues occurring in maize both have the same solanesol type side-chain.

## NUCLEUS

Several experiments have been carried out with the system just outlined in which maize shoots were incubated with either [methyl-$^{14}CT_3$] methionine or [U-$^{14}C$]*p*-hydroxybenzoic acid—a compound which has recently been shown to be an effective precursor of the ubiquinone nucleus in animals and micro-organisms (Parson and Rudney, 1964).

The results show, Table VII, that activity from the [methyl-$^{14}CT$]methionine was incorporated into PQ and the methylated 3$\beta$-hydroxysterols, $\beta$-sito- and stigmasterol. Furthermore, since no $^{14}C$ or T was found in the terpenoid reference compounds isolated, $\beta$-carotene and squalene, it follows that the methyl group from methionine has given rise to the quinone nuclear methyl group.[1] Here again the incorporation of label into PQ was much lower than expected which suggests that, in addition to synthesis of the PQ side-chain within the chloroplast, the nuclear methylations involved in the formation of this compound also take place within this organelle.

*p*-Hydroxybenzoic acid was incorporated into only one of the terpenoid compounds examined, ubiquinone (Table VIII). This would suggest that in plants, as well as in animals and micro-organisms (see above), this compound can give rise to the nuclear portion of the ubiquinone molecule. Its failure to be incorporated into the PQ-molecule does not exclude it as a precursor since low activity material was administered at a substrate level so that if the chloroplast

[1] It is perhaps worth mentioning that the results obtained for the retention of T by the nuclear methyl groups of PQ suggest that these methyls arise by the transfer of intact methyl groups from methionine.

membrane presents a barrier to this compound the incorporation of [$^{14}$C]-activity would be extremely low and escape detection. The experiment should be repeated using *p*-hydroxybenzoic of a higher specific activity.

Table VII. Maize: incorporation of [methyl-$^{14}$C/T]methionine and [U-$^{14}$C]*p*-hydroxybenzoic acid into plastoquinone and ubiquinone

| Terpenoid | Excised etiolated shoots exposed to the substrate[1] with continuous illumination for 24 hr | |
|---|---|---|
| | [methyl-$^{14}$C/T]methionine[2] (25 $\mu$c/140 shoots) d/min/mg | [U-$^{14}$C]*p*-hydroxybenzoic acid (2 mg, 837,000 d./min/mg) d./min/fraction |
| Plastoquinone | 4100 | 0 |
| Ubiquinone | Failed to purify | 6700 |
| Reference cpds | | |
| Squalene | 0 | 0 |
| $\beta$-Carotene | 0 | 0 |
| Sterols[3] | 15,040 | 0 |

[1] The shoots were exposed to the substrates by dipping the excised ends into an aqueous solution containing the radioactive material.
[2] Results given only in terms of $^{14}$C. However, T was present in both the plastoquinone and the sterols, but not the other reference compounds.
[3] Unesterified $\beta$-sito- and stigmasterol; one ethyl group derived from two methionine methyl groups present in each molecule.

## BIOSYNTHESIS OF SOLANESOL

In conclusion we should like to present briefly some results obtained with green tobacco plants. The reason for including them here is prompted by two reports, which taken together appear to contradict our basic tenet of two sites of terpenoid biosynthesis in the green plant cell. Reid (1961) reported that *Nicotiana affinis* incorporated [2-$^{14}$C]MVA to some 15% of the dose administered into the $C_{45}$ polysoprenoid alcohol solanesol (VIII). This observation

H[...]$_8$ ... $CH_2OH$

VIII

was of little significance in the present context until Stevenson, Hemming and Morton (1963) showed that the alcohol was located mainly in the chloroplast. In the light of this, the earlier observations of Reid's report could imply that MVA can effectively penetrate the chloroplast membrane and enter the biosynthetic sequence of solanesol and the other plastidic terpenoids.

We repeated Reid's experiment, Table VIII, and found that 40% of the available [$^{14}$C]-activity was incorporated into the lipid fraction; of this 17% was associated with the crude solanesol-containing fraction. On purification by thin-layer chromatography 0·018% of the original [$^{14}$C]-activity remained in the solanesol, the bulk of the activity ran just behind this compound. It is probable that in Reid's experiments, where large amounts of carrier material were

Table VIII. Tobacco: incorporation of [2-$^{14}$C]MVA into solanesol

Ten green seedlings (*N. tabacum*—11 weeks old) were excised and the cut ends dipped into a solution containing 30 $\mu$c [2-$^{14}$C]MVA. The seedlings were left in contact with the MVA for 24 hr with continuous illumination.

| Fraction | Incorporation of [2-$^{14}$C]MVA % administered[1] |
|---|---|
| Total lipid | 47 |
| Crude solanesol of alumina column | 17 |
| Solanesol after thin layer chromatography | 0·018 |

[1] Assuming only D-isomer utilized.

added, this highly active material would co-chromatograph and co-crystallize with the solanesol carrier and give the impression that solanesol was highly labelled. Our results show, as expected, that MVA is poorly incorporated into the plastidically located solanesol.

## Summary

The work described indicates that the chloroplasts of higher plants and some algae contain a basic quinone complement consisting of vitamin $K_1$, PQ and $\alpha$-TQ (and parent chromanol $\alpha$-T), which may be supplemented by the presence of PQC′, PQD′ and PQB (spinach and horsechestnut also contain shorter chain homologues of plastoquinone, i.e. $PQ_3$ (Misiti *et al.*, 1965) and $PQ_4$ (Eck and Trebst, 1963) respectively). It may be of systematic importance that all the monocotyledons examined contained only vitamin $K_1$, PQ and $\alpha$-TQ whereas the dicotyledons always contained in addition PQC′ and PQD′ and usually PQB. Characterization studies showed that PQC′ and PQD′ are almost certainly monohydroxy derivatives of PQ.

It has been demonstrated using maize and barley shoots, french bean leaves and *Euglena* cells that the concentration and synthesis of the chloroplastidic quinones is related to the degree of chloroplast development. Thus, although plastidic quinones could be detected in etiolated tissues of maize, barley and french bean the concentrations were lower than those found in normal green tissues, being entirely absent in dark grown *Euglena* cells. On illumination of etiolated maize shoots and *Euglena* cells it was found that quinone synthesis paralleled chloroplast development. Ubiquinone levels were relatively unaffected by the conditions of growth. These results led us to suggest that in etiolated tissues the plastidic quinones are entirely associated with the plastid primordia. The behaviour of α-T was not as clear cut, in maize no reduction in level was found in etiolated tissue—although the level of α-T+α-TQ was lower in etiolated than green tissue. In etiolated *Euglena* on the other hand, the level of α-T was much lower than that found in green cells and since the etiolated form of this organism contains no plastidic quinones, or structures recognizable as plastid primordia, its presence may be indicative of the association of this chromanol with structures other than the chloroplast.

Biosynthetic and degradation studies demonstrated that, in maize shoots, mevalonic acid can be regarded as the specific distal precursor of the PQ and UQ side-chains. The incorporation of this compound into vitamin $K_1$, α-TQ and α-T would indicate that MVA is also the precursor of the phytol side-chain. From a comparison of the relative incorporation of $^{14}CO_2$ and [2-$^{14}$C]MVA into the intra- and extraplastidic terpenoids evidence was obtained consistent with the tenet that the terpenoid portions of the plastidic quinones vitamin $K_1$ and PQ are biosynthesized within the confines of the chloroplast, the extraplastidic UQ and phytosterols being synthesized elsewhere within the cell. Experiments using [methyl-$^{14}$C]methionine and [U-$^{14}$C]*p*-hydroxybenzoic acid (if the latter compound should prove to be a PQ-ring precursor) provided preliminary indications that the PQ nucleus and methyl substituents may also be synthesized in the chloroplast. The results obtained for the incorporation of $^{14}CO_2$ and [2-$^{14}$C]MVA into α-T and α-TQ were not readily interpretable with regard to the site of synthesis of these compounds, but they did suggest that the maize experimental system contained large amounts of some early tocopherol precursor.

Although not reported here we have obtained evidence, using a similar experimental approach to that described for maize, for the chloroplastidic synthesis of the terpenoid portions of vitamin $K_1$, PQ, PQC′, α-T and α-TQ by tobacco seedlings (Griffiths, 1965). Using this tissue it was demonstrated that the earlier observations of Reid (1961) concerning the incorporation of [2-$^{14}$C]MVA into solanesol were erroneous and that the incorporation of radioactivity into the compound is consistent with its biosynthesis within the chloroplast.

The significance of our results in relation to the control of terpenoid biosynthesis during chloroplast development and germination is discussed elsewhere in this volume.

## Acknowledgements

The work from our laboratories reported here was supported in part by a grant to Professor T. W. Goodwin from the Scientific Research Council. One of us (W. T. G.) was in receipt of a Ministry of Agriculture Scholarship. We wish to thank Professor T. W. Goodwin for his continued interest and valuable suggestions, Drs. F. W. Hemming and J. F. Pennock, University of Liverpool (U.K.) and Dr. F. L. Crane, Purdue University (U.S.A.) for the free exchange of information, Dr. O. Isler, Hoffman La Roche & Co. Ltd. (Switzerland) for generously providing us with samples of plastoquinone, vitamin $K_1$ and ubiquinone and Mr. D. E. Lewis for valuable technical assistance.

## References

Eck, V. H. and Trebst, A. (1963). *Z. Naturf.* **18**, 446.
Fuller, R. C., Smillie, R. M., Rigopoulos, N. and Yount, V. (1961). *Archs Biochem. Biophys.* **95**, 197.
Goodwin, T. W. (1958). *Biochem. J.* **70**, 612.
Goodwin, T. W. and Mercer, E. I. (1963). *Biochem. Soc. Symp.* No. 24, p. 37.
Griffiths, W. T. (1965). Ph.D. Thesis (Univ. of Wales).
Griffiths, W. T., Threlfall, D. R. and Goodwin, T. W. (1964). *Biochem. J.* **90**, 40P.
Hemming, F. W., Morton, R. A. and Pennock, J. F. (1963). *Proc. R. Soc.* B **158**, 291.
Henninger, M. D. and Crane, F. L. (1963). *Biochemistry* **2**, 1168.
Henninger, M. D. and Crane, F. L. (1964). *Pl. Physiol.* **39**, 598.
Kegel, L. P., Henninger, M. D. and Crane, F. L. (1962). *Biochem. Biophys. Res. Commun.* **8**, 294.
Lester, R. L. and Crane, F. L. (1959). *J. biol. Chem.* **234**, 2169.
Lynen, F. (1961). *Fed. Proc.* **20**, 941.
Mercer, E. I. and Goodwin, T. W. (1962). *Biochem. J.* **85**, 13P.
McKenna, M., Henninger, M. and Crane, F. L. (1964). *Nature Lond.* **203**, 524.
Misiti, D., Moore, H. W. and Folkers, K. (1965). *J. Am. Chem. Soc.* **87**, 1402.
Parson, W. W. and Rudney, H. (1964). *Proc. natn. Acad. Sci., U.S.A.* **51**, 444.
Pennock, J. F., Hemming, F. W. and Kerr, J. D. (1964). *Biochem. Biophys. Res. Commun.* **17**, 542.
Reid, W. W. (1961). *Chem. & Ind.* 1489.
Rogers, L. J., Shah, S. P. J. and Goodwin, T. W. (1967). *In* "Biochemistry of Chloroplasts" Vol. II, (T. W. Goodwin, ed.) Academic Press, London and New York.
Stevenson, J., Hemming, F. W. and Morton, R. A. (1963). *Biochem. J.* **88**, 52.
Threlfall, D. R. and Goodwin, T. W. (1963). *Biochim. biophys. Acta* **78**, 532.
Threlfall, D. R. and Goodwin, T. W. (1964). *Biochem. J.* **90**, 40P.
Threlfall, D. R., Griffiths, W. T. and Goodwin, T. W. (1964). *Biochem. J.* **92**, 56P.
Threlfall, D. R., Griffiths, W. T. and Goodwin, T. W. (1965). *Biochim. biophys. Acta* **102**, 614.
Treharne, K. J., Mercer, E. I. and Goodwin, T. W. (1964). *Biochem. J.* **90**, 39P.
Whittle, K. J., Dunphy, P. J. and Pennock, J. F. (1965). *Biochem. J.* **96**, 17C.

## Discussion

*F. W. Hemming:* I would like to mention the results of some experiments carried out by Mr. R. Powls in our laboratory which have some bearing on Dr. Threlfall's paper and which to a large measure support his findings.

Mr. Powls also has been able to show that the incorporation of [2-$^{14}$C]mevalonic acid into the solanesol of tobacco leaves is very much lower than that into the sterols. In experiments of this sort one has to be very careful to remove traces of highly labelled $\beta$-amyrin. W. W. Reid (private communication) now agrees with this finding.

The metabolism of ring-labelled [$^{14}$C]*p*-hydroxybenzoic acid by the alga *Euglena gracilis* is very similar to that by the higher plant reported by Dr. Threlfall. This aromatic precursor was incorporated readily into the *Euglena* ubiquinone but not at all into plastoquinone. We do not know yet if the *p*-hydroxybenzoate was able to permeate the chloroplast membrane. Forty-eight hr after administering the *p*-hydroxybenzoate most of the radioactivity that had been incorporated by ubiquinone-9 had been transferred to rhodoquinone-9.

This is the first report of a rhodoquinone in an organism other than the photosynthetic bacterium *Rhodospirillum rubrum.* Since the level of this quinone in both dark-grown, heterotrophic cultures and in light-grown, autotrophic cultures of *Euglena gracilis* was similar to that of ubiquinone, it is clearly not concerned in photo-electron transport in this organism.

# Stereochemical and Evolutionary Aspects of Isopentenoid Biosynthesis

W. R. Nes, D. J. Baisted, E. Capstack, Jr.,
W. W. Newschwander and P. T. Russell

*The University of Mississippi, Oxford, Mississippi, U.S.A.*

In the course of work on isopentenoid biosynthesis in the germinating pea seed we have observed that marked changes occur during the time from initial absorption of water by the seed to emergence and development of the stem and green leaves. Although the evidence does not indicate that the observed changes occur in the chloroplasts, the fact that these changes do occur as the green parts of the plant develop seems to warrant a consideration of our findings here. We should like to summarize some published and unpublished information which we have obtained and to present some ideas which could conceivably lead to an explanation of our observations.

When dry pea seeds are allowed to absorb water containing [2-$^{14}$C]mevalonic acid (MVA) during several hours and then maintained wet for 5 days (stem and green leaves emerging after about the fourth day at room temperature),

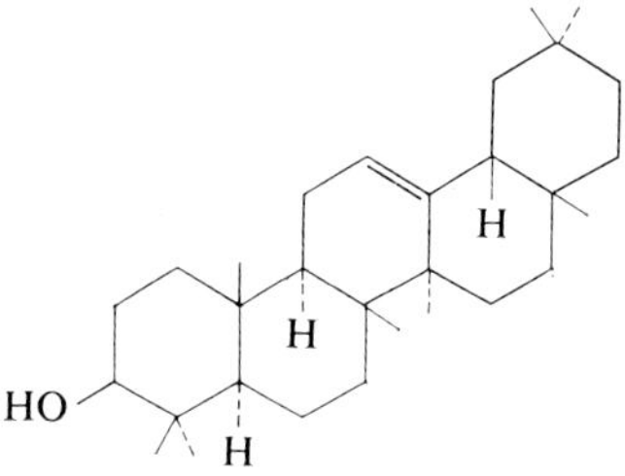

β-Amyrin

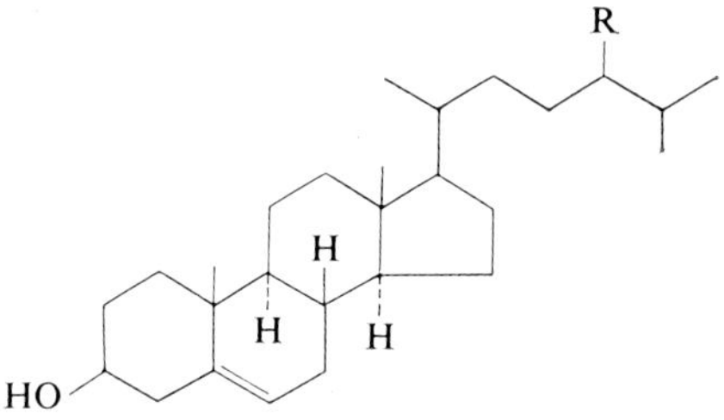

Campesterol
$R = CH_3$
β-Sitosterol
$R = C_2H_5$

nearly all the radioactivity in the lipid fraction is found in the pentacyclic triterpenoid β-amyrin (Baisted *et al.*, 1962). Moreover, nearly half of the radioactivity of one enantiomer of the MVA is incorporated into this substance, while only 2% is present as sterol (β-sitosterol and campesterol). Radioactive squalene is present in an amount equivalent to but 0·5%. On the other hand, when the incubation is carried out for only 1 day (during which time there is no emergence of stem or leaves), the yield of squalene is 38% (Capstack *et al.*,

(1962). This increased yield of squalene is at the expense of cyclized products, squalene now being the major metabolite. To determine whether this difference as a function of time is due to slow metabolism of the squalene or to changing metabolic capability, W. W. Newschwander and W. R. Nes (1962, unpublished) carried out incubations with MVA during 24 hr periods at various stages of seed development. Beyond 1 day the seed is first incubated with water, then presented with an MVA solution in a dry atmosphere. By transpiration the seed absorbs the solution. The amounts of products obtained show a strong dependence on time of germination. This is shown in Fig. 1. The most important feature is that the seed's metabolic capacity does change. The seed develops the capacity first to form squalene, then β-amyrin, then sterol. We have also shown

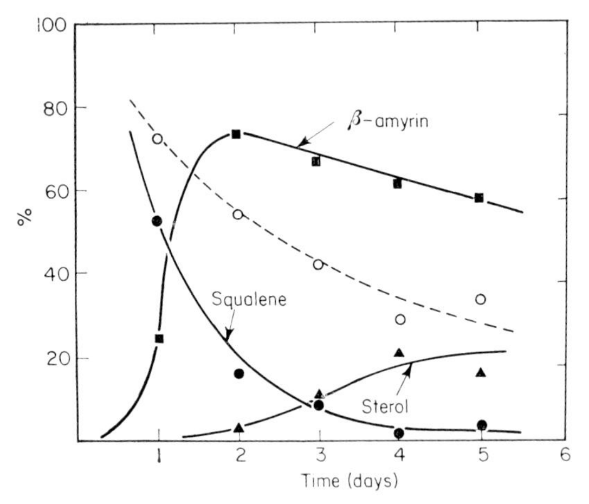


FIG. 1. The effect of time on the ability of germinating peas to biosynthesize squalene, β-amyrin, and sterol (campesterol and β-sitosterol) at 20°. The numbers on the abscissa are the total number of days the seeds had been germinated. At each time during the last 24 hr the experiment was carried out in the presence of [2-$^{14}$C]mevalonic acid. The ordinate for the solid lines is % of radioactivity extractable into organic solvents after saponification. The interrupted line represents the % of "available radioactivity" which was extractable into organic solvents after saponification. The "available radioactivity" is one-half of the total radioactivity of the D,L-mevalonic acid used.

(Capstack *et al.*, 1965) that with an uncentrifuged homogenate of 1-day-old peas squalene is cyclized to β-amyrin but not measurably to β-sitosterol, campesterol, or lanosterol. The ability to form squalene and β-amyrin is in the supernatant at 10,000 × ***g*** in 1-day-old peas (Capstack *et al.*, 1965). With 5-day-old peas (P. T. Russell and W. R. Nes, 1965, unpublished observations) an uncentrifuged homogenate yields squalene, but, contrary to the situation with 1-day peas, there is about an equal amount of material which chromatographs on alumina the way β-amyrin and sterol do (Table I). Although we have not yet completed identification of the metabolites, this information is so far in agreement with the *in vivo* observation that the ability to cyclize squalene develops with increasing time of germination. Since the question of the development of subcellular particles may well be involved here, we are also examining the cellular sites of biosynthesis. Based on the findings of Tchen and Bloch (1955) and others (Cornforth *et al.*, 1959) in mammalian liver, we expect that the squalene cyclizing ability will be microsomal.

On the assumption that the order in which squalene, pentacyclic triterpenoid, and sterol appear during germination is related to germination rather than specifically to the pea (and this of course remains to be established), we should like to offer a rationalization. This rationalization depends on certain speculations about evolution and is offered more as a guide to further thinking and experimentation than as any kind of firm explanation.

Table I. Relative amount of isopentenoids biosynthesized at different germination periods in pea homogenates[1]

| | Squalene[2] | Cyclized material[3] |
|---|---|---|
| 1-day old[4] | 55 | 12 |
| 5-day old[5] | 26 | 33 |

[1] Peas were germinated for the indicated period with water, then homogenized and the whole homogenate incubated with [2-$^{14}$C]MVA as described by Capstack *et al.* (1965). The figures given are the radioactivity in the metabolites × 100 divided by the radioactivity soluble in organic solvents after saponification.

[2] The radioactivity chromatographing as a hydrocarbon was taken to be squalene.

[3] The radioactivity chromatographing with $\beta$-amyrin and sterol was taken to be cyclized metabolites.

[4] From Capstack *et al.* (1965).

[5] From P. T. Russell and W. R. Nes (Unpublished observations).

Let us take $I_1$, $I_2$, $I_3$, etc. to refer to the monomeric, dimeric, trimeric, etc stages of isopentenoid polymerization and define an $I_n$–$I_n$-compound as one which has resulted like squalene ($I_3$–$I_3$), phytoene ($I_4$–$I_4$), and their metabolites from tail-to-tail coupling. Now, if we note that there is a greater variety of $I_2$-, $I_3$-, and $I_4$-compounds among plants than among animals and that among animals, especially mammals, $I_3$–$I_3$-compounds are restricted to squalene, lanosterol and the metabolites of the latter, it becomes possible to construct a "molecular tree" which seems to parallel the "biological tree". The "molecular tree" is shown in Fig. 2 with the main (vertical) trunk representing isopentenoid metabolism in mammals. As one proceeds down the evolutionary scale, more of the branches of the "molecular tree" are utilized and less of the uppermost portion of the main trunk. A quantitative and well documented example has to do with the metabolism of the sterol side-chain. Cholesterol is found (Johnson *et al.*, 1963) but found rarely in the plant kingdom where 24-alkylcholesterol is common (Fieser and Fieser, 1959). In lower animals so well investigated by Bergmann (1962) both cholesterol and 24-alkylcholesterol are found, and in higher animals from at least fish (Blondin *et al.*, 1966) to the well-known situation with man and other mammals only the unalkylated side-chain of cholestrol has been found.

One of the intriguing aspects of the "molecular tree" is that it proceeds upward from lesser to greater complexity as does the "biological tree". This means that increasing selection of alternatives has been made and obviously has its origin in increasing sophistication of enzymes. A striking example of this development of selectivity is found in the absolute configurations of isopentenoids.

Among the $I_2$-compounds it is common to find both enantiomeric configurations. This is true for the acyclic representatives such as citronellol and

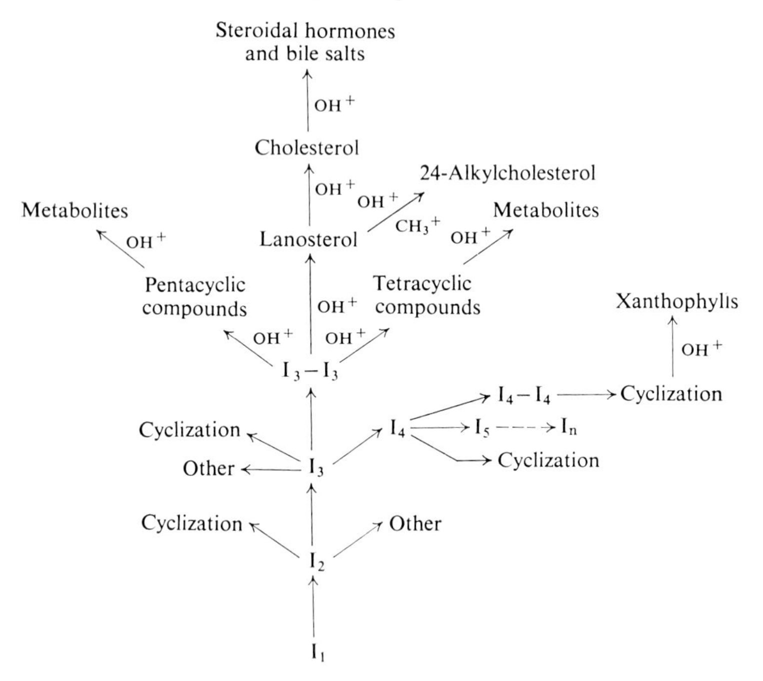


FIG. 2. "Molecular tree" of isopentenoid biosynthesis and metabolism. The main (vertical) trunk is the pathway followed by mammals. The term "I" represents the $C_5$-unit. "$OH^+$" represents hydroxylation by $NADPH/O_2$. "$CH_3^+$" represents the transfer of $C_1$- or $C_2$-groups to C-24 from methionine. The term "other" means metabolism other than cyclization, e.g., elimination, or hydrolysis followed by oxidation.

linalool which occur both as the D- and L-forms, for the monocyclic representatives such as limonene which occurs as the D-, L-, and D,L-forms, and for the bicyclic representatives such as $\alpha$-pinene which occurs in the D- and L-forms (Karrer, 1958).

At the $I_3$- and $I_4$-stages there is much more configurational selectivity in the known compounds. Nerolidol, the $I_3$-analog of the acyclic linalool, has so far been found only as the d-enantiomer (Karrer, 1958). Cadinene, which appears to be the most widely distributed of the bicyclic $I_3$-compounds, occurs predominantly as the l-form (Karrer, 1958). Among the monocyclic $I_3$-compounds usually only one enantiomer is known, although both forms have been found for bisabolol (Karrer, 1958). In addition, the known absolute configurations of the representatives of the selinene group (bicyclic $I_3$) are all the same e.g., α-selinene (Mills, 1952; Buchi *et al.*, 1959), β-eudesmol (Riniker *et al.*, 1954),

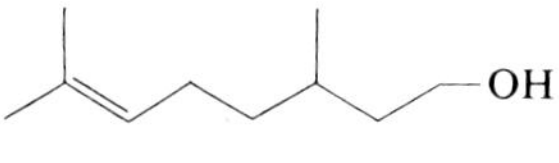

Citronellol 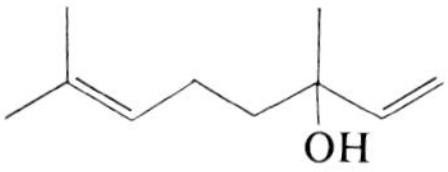 Linalool

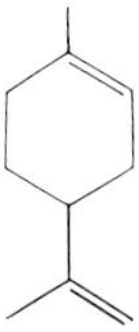

Limonene 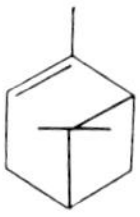 α-Pinene

carrisone (Djerassi *et al.*, 1956), and α-cyperone (Djerassi *et al.*, 1956), with the exception of maaliol (Buchi *et al.*, 1959). Furthermore, α-selinene has the same configuration as the eremophilone group (Zalkow *et al.*, 1959) when rearrangement in the latter series is taken into consideration.

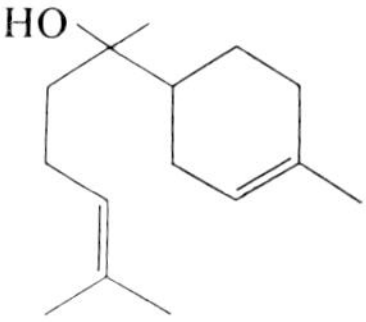

Bisabolol 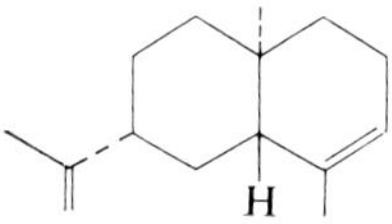 α-Selinene

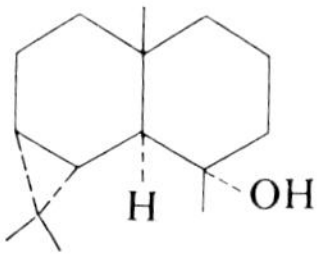

Maaliol 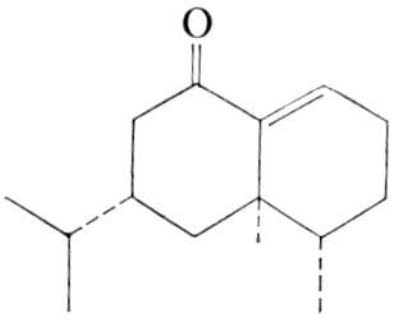 Eremophilone

Among the known $I_4$-compounds the acyclic representatives are few, but phytol appears to have a slight positive rotation, and 5,8-dimethyl-, 7,8-dimethyl-, and 5,7,8-trimethyltocol (the tocopherols) are of the same rotational series (Karrer, 1958). The cyclic $I_4$-compounds usually are found only as one enantiomer, but both absolute configurations of the *trans*-decalin-type of bicyclic system, such as labdanolic and eperuic acids (Djerassi and Marshall, 1957), are known among $I_4$-compounds and the same is true for the tricyclic perhydrophenanthrene system, such as darutigenol (Pudles *et al.*, 1959) and neoabietic acid (Barton, 1949; Klyne, 1953) and the tetracyclic system of the kaurene type. Kaurene itself has actually been found both in the d-form (Briggs *et al.*, 1950) and the l-form (Briggs and Cawley, 1948).

At the $I_3$-$I_3$-stage the situation is very different. Neither do any of the very

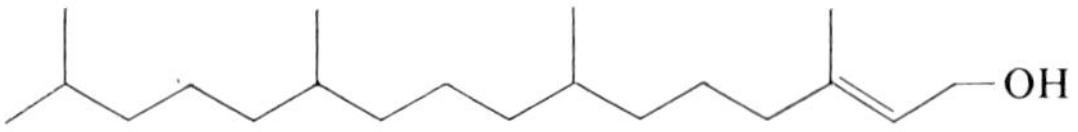

Phytol

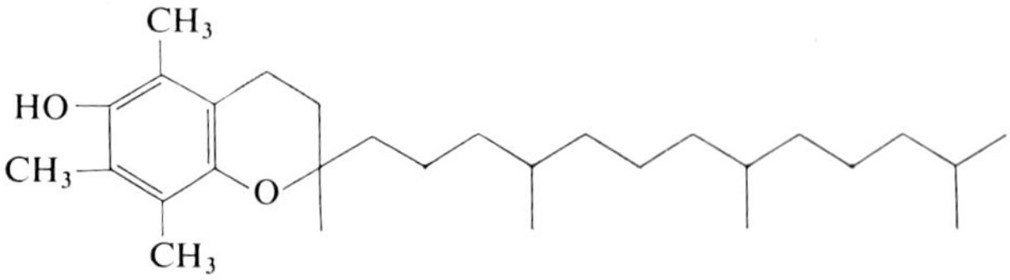

α-Tocopherol

many representatives occur in other but one form, but all of them are of the same absolute configurational series. This includes the tetracyclic steroids as well as the tetra- and pentacyclic triterpenoids.[1] Thus, there is an increasing stereoselectivity as the complexity of the isopentenoid increases from almost no selectivity with D-, L-, and D,L-forms found frequently among $I_2$-compounds,

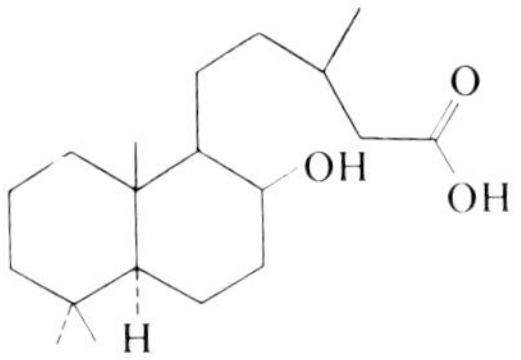

Labdanolic acid

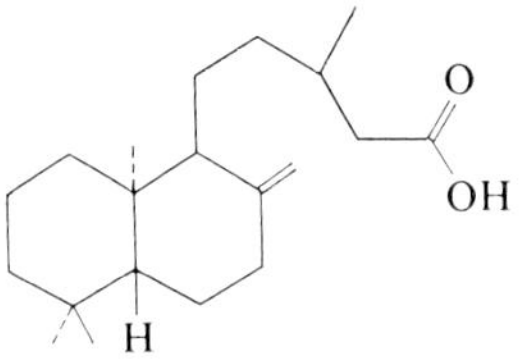

Eperuic acid

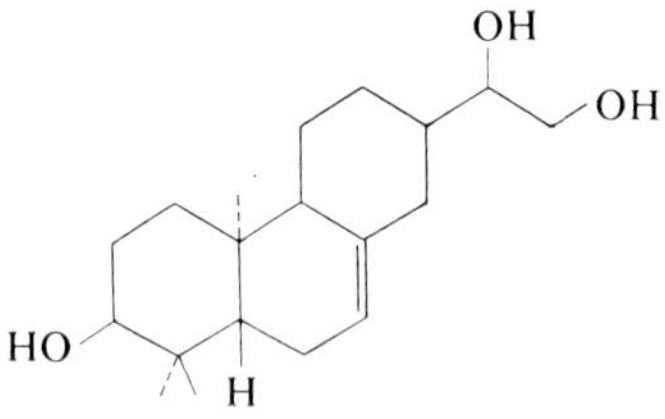

Darutigenol

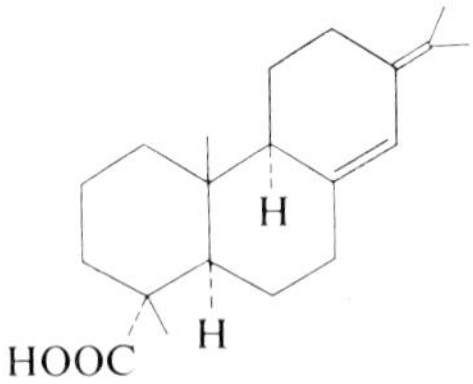

Neoabietic acid

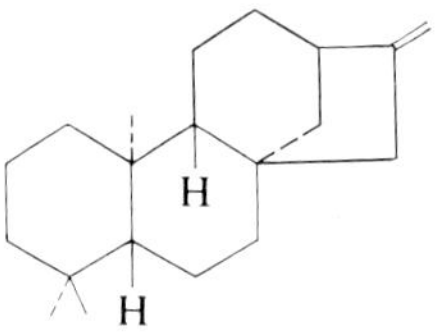

(−)-Kaurene

[1] For a detailed discussion of structure and stereochemistry of these compounds the books by Fieser and Fieser (1959), de Mayo (1959), and Ourisson *et al.* (1964) should be consulted.

through the $I_3$- and $I_4$-compounds where enantiomeric forms are less frequently encountered but where both configurational types are known in some cases, to complete specificity among $I_3$–$I_3$-compounds which occur so far as we know exclusively in one configurational series. Among $I_4$–$I_4$-compounds, it is also true that none has been reported in but one enantiomeric form (Karrer, 1958, and Karrer and Jucker, 1950).

Now, the order in which we find squalene, $\beta$-amyrin and sterol to appear during germination can be rationalized by assuming (a) that the organism is recapitulating its evolutionary history, (b) that hydroxylative cyclization is an "advanced" rather than "primitive" process and (c) that lanosterol is a less probable cyclization product of squalene than is $\beta$-amyrin. Thus, the pathway would proceed to squalene and stop until the organism had reached the stage of development corresponding in this respect to an evolutionary stage at which hydroxylation became possible. At this stage the product would not be lanosterol, but one of the other possibilities, such as $\beta$-amyrin. Only later at a stage corresponding to greater evolutionary sophistication would lanosterol be produced.

That hydroxylation is not primitive is strongly suggested by Bloch's (1962) careful considerations of oxygen in biosynthetic pathways. To what he has said we might add the fact that hydroxylative cyclization of isopentenoids appears to be more common at the $I_3$–$I_3$-level than at lower molecular levels of isopentenoid build-up, if the presently known occurrence of these compounds can be taken as a guide. Thus, iresin (Djerassi and Burstein, 1958; Rossmann and Lipscomb, 1958a,b) is a rare example of a cyclic $I_3$-compound which bears a hydroxyl group at the expected position for initiation of cyclization. Darutigenol (Pudles *et al.*, 1959) and andrographolide (Cava and Weinstein, 1959; Cava *et al.*, 1963; Chan *et al.*, 1963) represent the same situation among $I_4$-compounds.

The reasons for saying that lanosterol is less likely a product than $\beta$-amyrin are derived from the ingenious considerations of Woodward and Bloch (1953) and of Eschenmoser *et al.* (1955) and from the elegant experiments in the laboratories of Bloch (Tchen and Bloch, 1955; Maudgal *et al.*, 1958) and of Cornforth and Popják (Cornforth *et al.*, 1959 and 1965) which have led to our knowledge of how squalene folds and reacts to give lanosterol and the other products. For our present purposes, this work can be summarized and perhaps added to somewhat in the following way.

During its formation lanosterol must pass through a transition state which approximates the thermodynamically unfavorable (Johnson, 1953) *trans-syn-trans-anti-trans*-fusion of the four rings. This transition state corresponds to a folding in which ring B is a boat form with unfavorable non-bonded interactions. On the other hand, $\beta$-amyrin and other polycyclic compounds appear to arise by a transition state which approximates the stable *trans-anti-trans-anti-trans*-fusion. This is shown in the accompanying formulae. Transition state A has the *trans-syn*-arrangement of rings-A/B while transition state B has the *trans-anti*-arrangement. From these two types of transition state we can arrive at the various polycyclic $I_3$–$I_3$-structural types, assuming that (a) further intramolecular reactions will occur in a *trans*-fashion, i.e., the attacking and leaving atoms will be on opposite sides of the atom in question, that (b) the extent of ring closure will be determined by the conformation of the squalene adsorbed on the enzyme, and that (c) the extent of rearrangement will be determined by the position of a deprotonating (or hydroxide-donating) group on the enzyme.

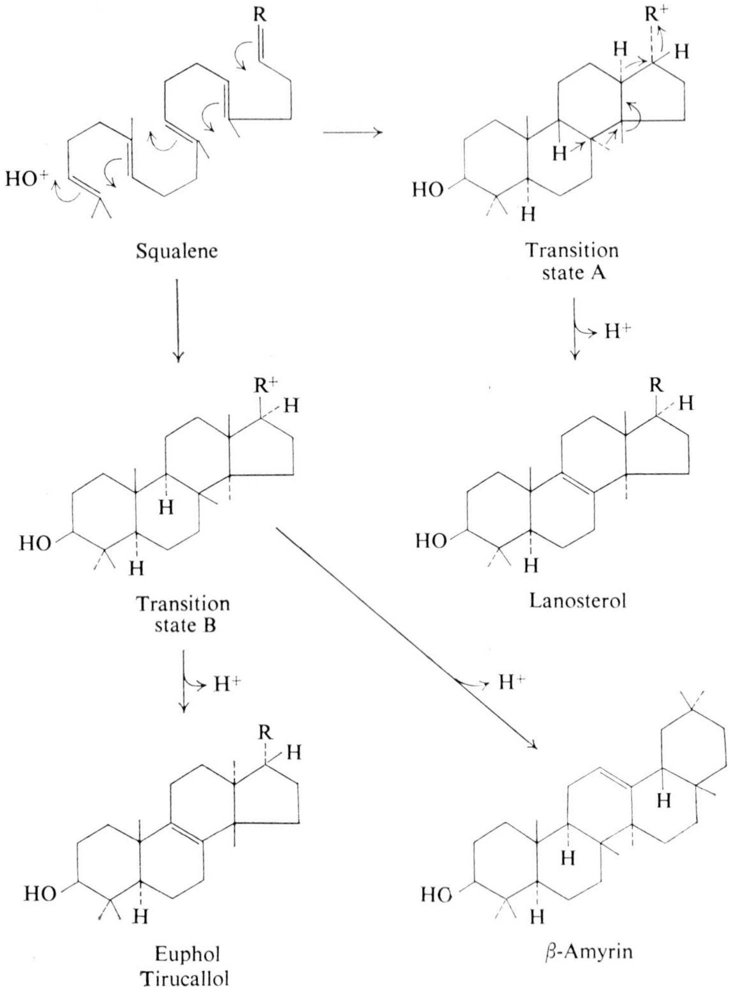

Squalene

Transition state A

Transition state B

Lanosterol

Euphol
Tirucallol

β-Amyrin

Thus, from transition state B by adsorption of squalene with the R-portion not arranged cyclically and with a deprotonating group arranged sterically at position-9, the euphol-tirucallol structure must result. With squalene adsorbed with the R-portion in the appropriate cyclic conformation and with a deprotonating agent at position-12, β-amyrin must result, or with the deprotonating group at position-15 taraxerol must result, etc.

From transition state A and with the same assumptions, the placing of a deprotonating group adjacent to position-9 leads to lanosterol. It is therefore apparent that lanosterol and its steroidal metabolites are set apart from the polycyclic $I_3$–$I_3$-compounds such as β-amyrin in that a less probable (less favorable thermodynamically) transition state is involved during biosynthesis. It seems possible, then, that this is why β-amyrin is formed before β-sitosterol in germinating peas.

There is a final aspect of our work which must be mentioned. Since we are unable to demonstrate sterol formation early in germination, how then does the endogenous sterol get there? Pea seeds contain about 67 mg of sterol per 100 g of seed which is actually more than we find for β-amyrin (13 mg/100 g) (Baisted *et al.*, 1962). It may be that the sterol is transferred from the parent plant. It has recently been reported that a hen passes cholesterol to its egg (Andrews *et al.*, 1965). Is it possible that embryonic tissue is in general, for stereochemical and evolutionary reasons, incapable of cyclizing squalene to lanosterol?

## References


Andrews, J. W., Wagstaff, R. K. and Edwards, H. M. (1965). *Fed. Proc.* **24**, 686.

Baisted, D. J., Capstack, E. and Nes, W. R. (1962). *Biochemistry* **1**, 537.

Barton, D. H. R. (1949). *Quart. Rev.* **3**, 36.

Bergmann, W. (1962). *In* "Comparative Biochemistry" (M. Florkin and H. S. Mason, eds.), p. 103, Vol. III, Part A, Academic Press, New York and London.

Bloch, K. (1962). *Fed. Proc*, **21**, 1058.

Blondin, G. A., Scott, J. L., Hummer, J. K., Kulkarni, B. D. and Nes, W. R. (1966). "Comparative Biochemistry and Physiology", **17**, 391.

Briggs, L. H. and Cawley, R. W. (1948). *J. chem. Soc.* 1888.

Briggs, L. H., Cawley, R. W., Loe, J. A. and Taylor, W. I. (1950). *J. chem. Soc.* 955.

Bruderer, H., Arigoni, D. and Jeger, O. (1956). *Helv. chim. Acta* **39**, 858.

Buchi, G., Wittenau, M. S. and White D. M. (1959). *J. Am. chem. Soc.* **81**, 1968.

Capstack, E., Baisted, D. J., Newschwander, W. W., Blondin, G., Rosin, N. L. and Nes, W. R. (1962). *Biochemistry* **1**, 1178.

Capstack, E., Rosin, N., Blondin, G. A. and Nes, W. R. (1965). *J. biol. Chem.* **240**, 3258.

Cava, M. P. and Weinstein, B. (1959). *Chem. & Ind.* 851.

Cava, M. P., Weinstein, B., Chan, W. R., Haynes, L. J. and Johnson, L. F. (1963). *Chem. & Ind.* 167.

Chan, W. R., Willis, C., Cava, M. P. and Stein, R. P. (1963). *Chem. & Ind.* 495.

Cornforth, J. W., Cornforth, R. H., Pelter, A., Horning, M. G. and Popják, G. (1959. *Tetrahedron* **5**, 311.

Cornforth, J. W., Cornforth, R. H., Donninger, C., Popják, G., Shimizu, Y., Ichii, S., Forchielli, E. and Caspi, E. (1965). *J. Am. chem. Soc.* **87**, 3224.

Djerassi, C., Riniker, R. and Riniker, B. (1956). *J. Am. chem. Soc.* **78**, 6362.

Djerassi, C. and Marshall, D. (1957). *Tetrahedron* **1**, 238.

Djerassi, C. and Burstein, S. (1958). *J. Am. chem. Soc.* **80**, 2593.

Eschenmoser, A., Ruzicka, L., Jeger, O. and Arigoni, D. (1955). *Helv. Chim. Acta* **38**, 1890.

Fieser, L. F. and Fieser, M. (1959). "Steroids." Reinhold Publishing Corp., New York.

Johnson, D. F., Bennett, R. D. and Heftmann, E. (1963). *Science, N.Y.* **140**, 198.

Johnson, W. S. (1953). *J. Am. Chem. Soc.* **75**, 1498.

Karrer, P. and Jucker, E. (1950). "Carotenoids." Elsevier Publishing Company, Inc., New York.

Karrer, W. (1958). "Konstitution und Vorkommen der Organisher Pflanzenstoffe." Berkhaüser Verlag, Basel.

Klyne, W. (1953). *J. chem. Soc.* 3072.

Maudgal, R. K., Tchen, T. T. and Bloch, K. (1958). *J. Am. chem. Soc.* **80**, 2589.

de Mayo, P. (1959). "The Higher Terpenoids." Interscience Publishers, Inc., New York.
Mills, J. A. (1952). *J. chem. Soc.* 4976.
Ourisson, G., Crabbe, P. and Rodig, O. (1964). "Tetracyclic Triterpenes." Holden-Day Inc., San Francisco, California.
Pudles, J., Diara, A. and Lederer, E. (1959). *Bull. Soc. chim. France*, 693.
Riniker, B., Kalvoda, J., Arigoni, D., Furst, A., Jeger, O., Gold, A. M. and Woodward, R. B. (1954). *J. Am. chem. Soc.* **76**, 313.
Rossmann, M. G. and Lipscomb, W. N. (1958a). *J. Am. chem. Soc.* **80**, 2592.
Rossmann, M. G. and Lipscomb, W. N. (1958b). *Tetrahedron* **4**, 275.
Tchen, T. T. and Bloch, K. (1955). *J. Am. chem. Soc.* **77**, 6085.
Woodward, R. B. and Bloch, K. (1953). *J. Am. chem. Soc.* **75**, 2023.
Zalkow, L. H., Markley, F. X. and Djerassi, C. (1959). *J. Am. chem. Soc.* **81**, 2914.

# The Intracellular Localization of Mevalonate Activating Enzymes: Its Importance in the Regulation of Terpenoid Biosynthesis

L. J. Rogers, S. P. J. Shah and T. W. Goodwin

*Department of Biochemistry and Agricultural Biochemistry, University College of Wales, Aberystwyth, Wales*

## Introduction

It has been observed that $^{14}CO_2$ is actively incorporated into $\beta$-carotene and phytol in illuminated maize seedlings whereas [2-$^{14}$C]mevalonic acid is insignificantly incorporated under similar conditions (Goodwin, 1958a,b; Mercer and Goodwin, 1962). Further studies showed that [2-$^{14}$C]mevalonic acid was not significantly incorporated into chloroplast terpenoids such as plastoquinone and tocopheryl quinone but was effectively incorporated into ubiquinone, sterols and related products (Mercer and Goodwin, 1963; Treharne *et al.*, 1964; Griffiths *et al.*, 1964).

These and earlier observations led Goodwin and Mercer (1963) to propose that the regulation of terpenoid biosynthesis in developing seedlings was achieved by a combination of enzyme segregation within the cell and specific impermeability of the intracellular membranes to certain compounds (see also Goodwin, 1965). In terpenoid biosynthesis both chloroplastidic and extra-chloroplastidic sites of synthesis are regarded as having a duplicated biosynthetic pathway by which acetyl-CoA is converted, via[1] MVA; MVA-5P; MVA-5PP; IPP; geranyl, farnesyl and geranylgeranyl pyrophosphates into solanesyl pyrophosphate and the $C_{50}$ isoprenoid pyrophosphate (Fig. 1). However, each "compartment" has certain enzymes specific to that site only the substrates of which may be those of the common biosynthetic pathway. For example, squalene synthetase would be extrachloroplastidic whereas phytoene synthetase would be chloroplastidic. Such a scheme would require that the chloroplast membrane be essentially impermeable to the intermediates of the biosynthetic pathway.

On the basis of this "compartmentalization" scheme mevalonate kinase [ATP: mevalonate 5-phosphotransferase], the enzyme acting on the first specific terpenoid precursor, should be found both inside and outside the chloroplast. Furthermore, the chloroplast membrane should be essentially impermeable to MVA. Experimental results supporting this conclusion have

[1] Abbreviations: MVA, mevalonic acid [3,5-dihydroxy-3-methylvaleric acid]; MVA-5P, mevalonic acid 5-phosphate; MVA-5PP, mevalonic acid 5-pyrophosphate; IPP, isopentenyl pyrophosphate.

recently been obtained. These studies have been briefly reported by Rogers *et al.* (1965), and are described in detail elsewhere (Rogers *et al.*, 1966).

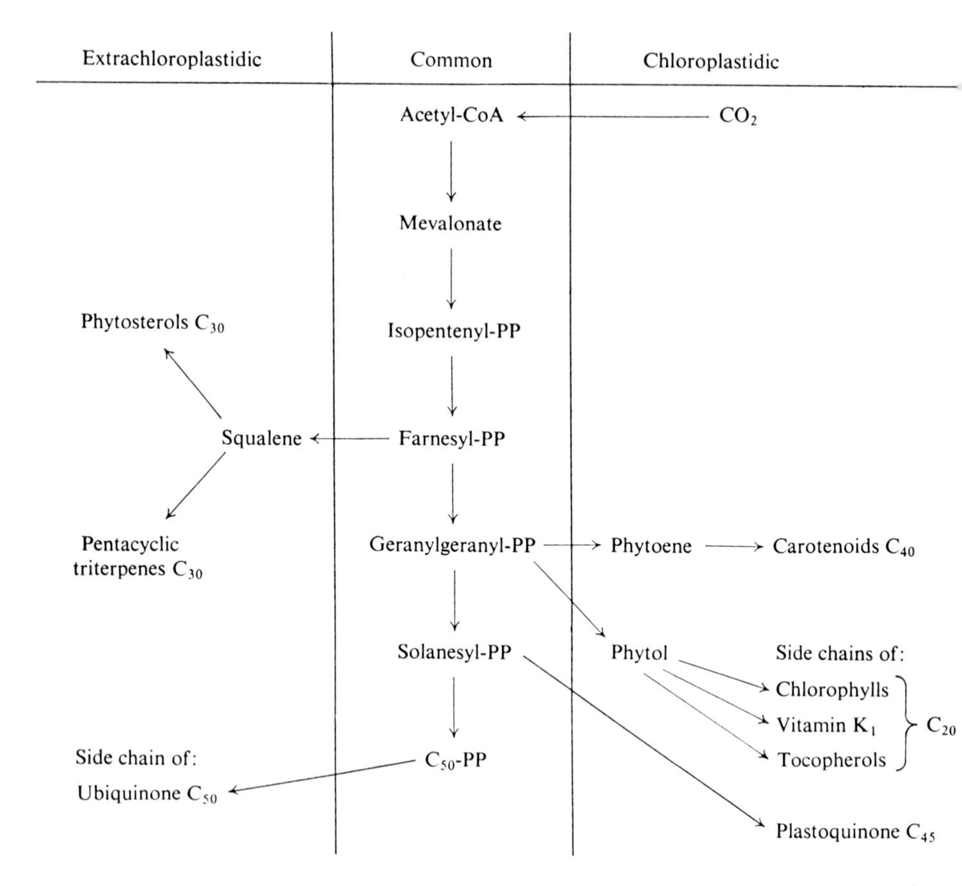


FIG. 1. Suggested scheme for compartmentalization of terpenoid biosynthesis in germinating seedlings.

## MATERIALS AND METHODS

French bean (*Phaseolus vulgaris*) plants were grown from seeds in a warm room for 6–8 days and then placed in the dark at the same temperature for a further 1–2 days. This treatment depletes the plants of starch; if it is omitted then starch grains in the chloroplast might be pulled through the bounding membranes of the chloroplast on subsequent centrifugation of chloroplast preparations.

### CHLOROPLAST ISOLATION

Density gradient techniques were employed in order to obtain intact chloroplasts as free as possible from contamination by, for example, nuclei, mitochondria, cytoplasm or the microsomal fraction of cells.

Two methods were used; that of Leech (1964), and a method based on Stocking (1959). The first method employs a discontinuous density gradient of

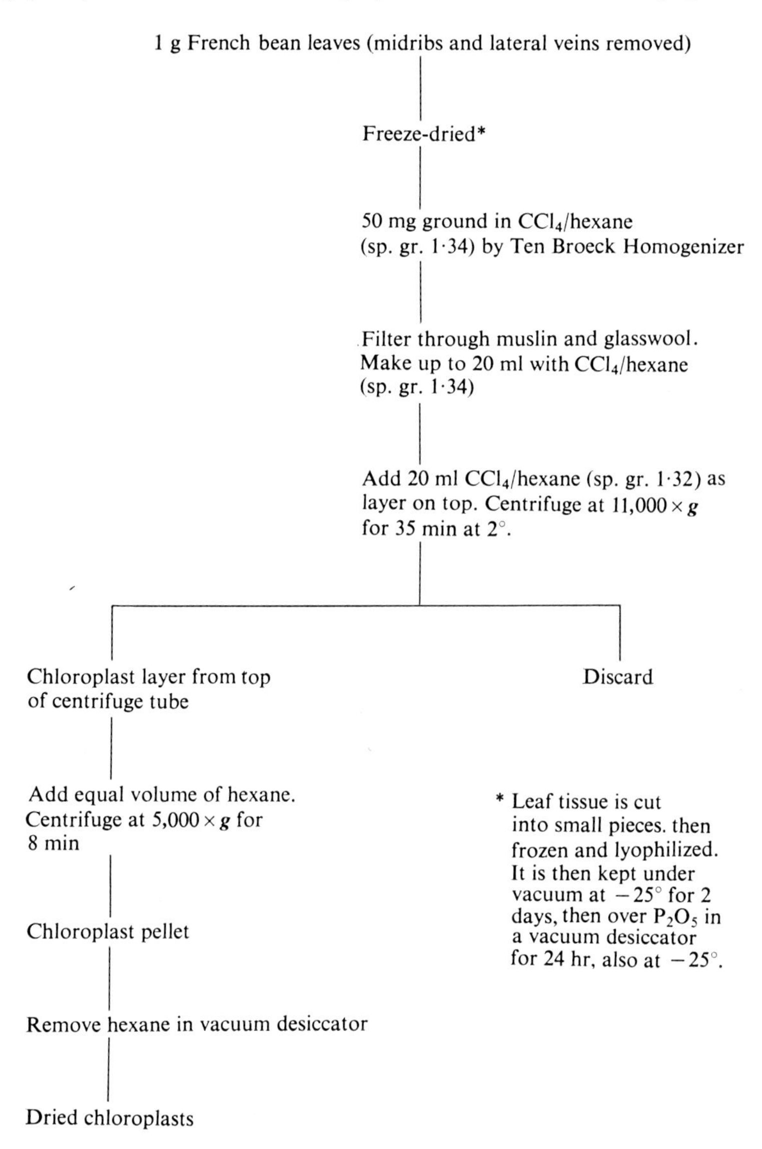


Scheme 1. Chloroplast preparation in non-aqueous media.

sucrose in phosphate buffer, while the second employs a discontinuous density gradient of carbon tetrachloride and hexane (Scheme 1).

The final chloroplast pellet in each case was dispersed immediately prior to use in 0·5–1·0 ml of 0·10 M phosphate buffer pH 7·4 containing 0·3 M sucrose.

### ACETONE-DRIED POWDERS

Acetone-dried powders of plant cell cultures of meristematic tissue of Paul's Scarlet Rose and of etiolated parts of variegated leaves of *Pelargonium kewensis* and *Hibiscus* sp. were prepared as described by Rogers *et al.* (1966).

### DETECTION OF ENZYME ACTIVITY

The assay method used was based on that of Markley and Smallman (1961), The substrate was DL-[2-$^{14}$C]MVA and the final incubation medium, pH 7·4. contained (μmole/tube): $KH_2PO_4$ 14·0; $MgSO_4$ 4·0; sucrose 120; L-cysteine 6·0; ATP 2·4; MVA 0·2, (23,100 disintegrations/sec). The protein present was of the order of 1–2 mg and the final volume per reaction tube was 0·4 ml. All incubations were performed at 36°.

To show whether the chloroplast membrane prevented access of MVA to the site of chloroplastidic mevalonate kinase intact chloroplasts were incubated with the assay medium while in other reaction tubes the chloroplasts were broken by ultrasonic treatment or by suspension of the chloroplasts in buffer not containing sucrose before their addition to the assay medium.

The reaction catalysed by mevalonate kinase is

$$\text{MVA}+\text{ATP} \xrightarrow{Mg^{2+}} \text{MVA-5P}+\text{ADP}$$

However, with crude enzyme preparations further reactions can occur:

$$\text{MVA-5P}+\text{ATP} \xrightarrow{Mg^{2+}} \text{MVA-5PP}+\text{ADP}$$

$$\text{MVA-5PP}+\text{ATP} \xrightarrow{Mg^{2+}} \text{IPP}+\text{ADP}+CO_2+P_i$$

catalysed by phosphomevalonate kinase (ATP: 5-phosphomevalonate phosphotransferase) and phosphomevalonate decarboxylase (ATP: 5-pyrophosphomevalonate carboxy-lyase) respectively. Therefore the method used to detect mevalonate kinase activity was to examine the reaction medium at the end of incubation for MVA-5P, MVA-5PP and IPP all of which might be expected to be present.

At the end of incubation the reaction was stopped by immersion of the tubes in a boiling water bath for 2 min. Precipitated protein was removed by centrifugation and samples of the incubation mixture (approx. 0·02 ml) were transferred to Whatman No. 1 chromatography paper and chromatographed at room temperature in n-butanol–formic acid–water (77:10:13, v/v); isobutyric acid–ammonia–water (66:3:30, v/v); t-butanol–formic acid–water (40:10:16, v/v) systems (Ohnoki *et al.*, 1962). The labelled compounds present were identified from the reported $R_f$ values of MVA, MVA-5P, MVA-5PP and IPP in these systems (Table I). Radioactive spots were located on the dried chromatograms by radioautography for 3–5 days. For quantitative assessment, 1·5 in wide strips of the chromatograms were scanned by a radiation detector connected to a chart recorder. From the area of the peaks on the recording chart the relative radioactivity of each spot was estimated. Finally, selected strips were cut into rectangles 1·5 in by 1·0 in and, after lightly spraying with N NaOH, were counted, following immersion in scintillation fluid, in a scintillation counter.

Table I. Reported $R_f$ values of MVA, MVA-5P, MVA-5PP and IPP in various solvent systems

| | n-butanol–formic acid–water (77:10:13, v/v) | isobutyric acid–ammonia–water (66:3:30 v/v) | t-butanol–formic acid–water (40:10:16, v/v) |
|---|---|---|---|
| MVA | 0·75 | 0·65–0·69 | 0·80–0·85 |
| MVA-5P | 0·12–0·18 | 0·35–0·43 | 0·53–0·65 |
| MVA-5PP | 0–0·03 | 0·20–0·27 | 0·29–0·35 |
| IPP | 0·30–0·32 | — | 0·54–0·62 |

## RESULTS

When intact chloroplasts prepared in the discontinuous sucrose density gradient were incubated for 3 hr with [2-$^{14}$C]MVA no mevalonate kinase activity was evident. However, under the same conditions chloroplasts ruptured by ultrasonic treatment or by osmotic shock converted MVA into MVA-5PP. In one experiment with a weakly active chloroplast preparation

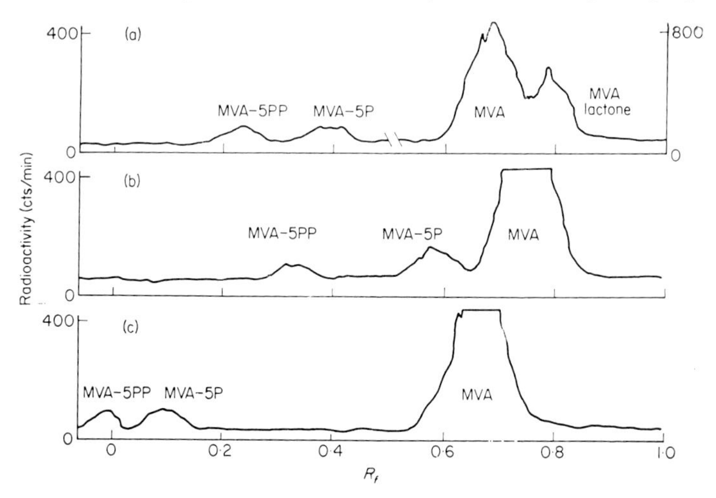


FIG. 2. Paper chromatographic separation of the reaction products obtained in the phosphorylation of MVA by chloroplasts isolated in aqueous media and ultrasonically treated; (a) in isobutyric–ammonia–water; (b) in t-butanol–formic acid–water; (c) in n-butanol–formic acid–water solvents.

both MVA-5P and MVA-5PP were detected (Fig. 2) but usually appreciable activity was found in MVA-5PP with much less in MVA-5P. The apparent percentage conversion into MVA-5PP appeared to be slightly less after 15 hr than after 3 hr.

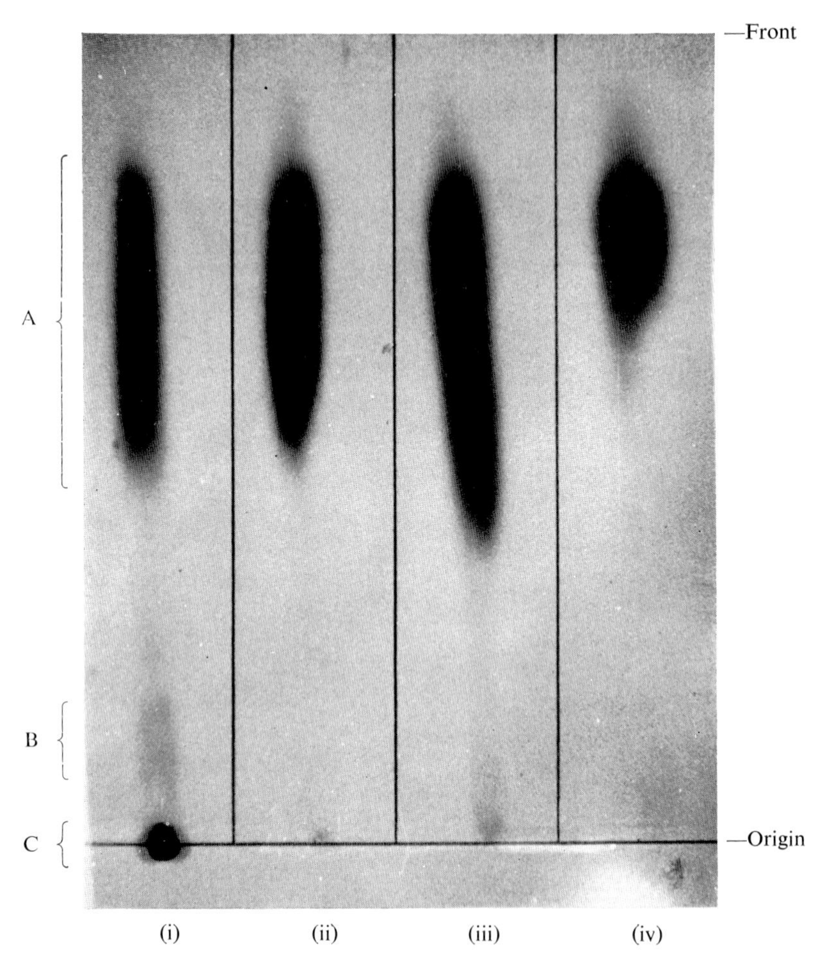


FIG. 3. Radioautograph of a descending chromatogram in n-butanol–formic acid–water (77:10:13, v/v) prepared from (a) the reaction mixture after incubation of ultrasonically treated "non-aqueous" chloroplasts with [2-$^{14}$C]MVA (i); (b) after incubation of intact chloroplasts with [2-$^{14}$C]MVA for 3 hr (ii) and 15 hr (iii); (c) after incubation of ultrasonically treated heat-inactivated chloroplasts with [2-$^{14}$C]MVA (iv). Experimental details are given in the text. The vertical lines indicate the strips scanned for radioactivity. A, position occupied by MVA; B, position occupied by MVA-5P; C, position occupied by MVA-5PP.

Similarly the intact chloroplasts isolated in carbon tetrachloride-hexane density gradients were found to be practically inactive in converting [2-$^{14}$C]-MVA into its phosphorylated derivatives even after 24 hr. However, ultrasonically treated chloroplasts were very active and on radioautography of the chromatogram developed with n-butanol–formic acid–water a radioactive spot very near the origin, corresponding to MVA-5PP, was clearly evident in a 3 hr incubate together with a very slight spot at an $R_f$ characteristic of MVA-5P (Fig. 3). In the case of intact chloroplasts only very slight activity was evident even when the chromatogram was substantially overloaded with radioactive material. This slight activity was observable on the radioautograph only; peaks corresponding to MVA-5P and MVA-5PP were not recorded when strips of the chromatogram were scanned for radioactivity even when the MVA peak on the chart was of considerable area. ATP was essential for conversion of

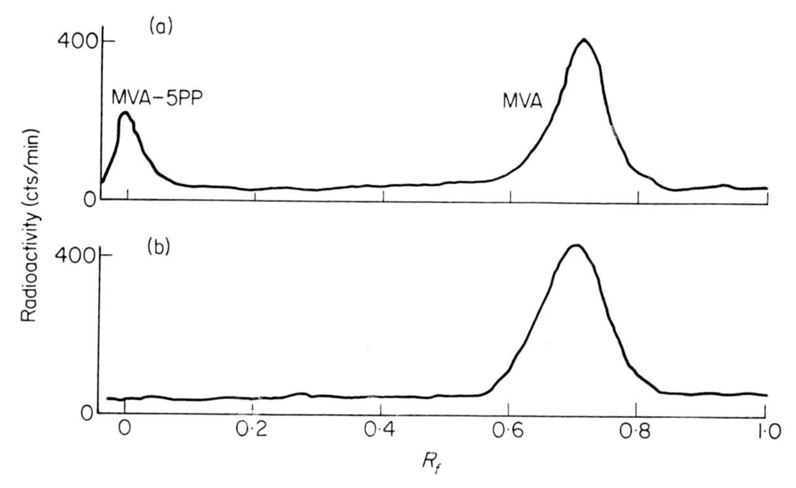


FIG. 4. Scan for radioactivity of paper chromatographic separation in n-butanol–formic acid–water of reaction products obtained in the phosphorylation of MVA by enzymes from chloroplasts isolated in non-aqueous media: (a) ultrasonically treated chloroplasts incubated with [2-$^{14}$C]MVA for 3 hr at 36° and pH 7·4; (b) intact chloroplasts incubated with [2-$^{14}$C]-MVA under the same conditions as (a). Similar traces to (b) were obtained with ultrasonically treated heat-inactivated chloroplasts, with ultrasonically treated chloroplasts if no ATP was present, or with [2-$^{14}$C]MVA alone.

[2-$^{14}$C]MVA by ruptured chloroplasts into the products observable by the methods used.

About 25% of added MVA was converted into MVA-5PP in the best "non-aqueous" chloroplast preparations (Fig. 4). This represents about 50% of the theoretical conversion since only one isomer of DL-MVA is biologically active. When counted in a scintillation counter the counts on areas of this chromatogram corresponding to the positions of radioactive spots were: MVA 187,000 disintegrations/min; MVA-5P 4440 disintegrations/min; and MVA-5PP 42,350 disintegrations/min.

Mevalonate-activating enzymes were also shown to be present in acetone-dried powders of cell cultures of Paul's Scarlet Rose, and in acetone-dried powders of white pigment-less tissue of the variegated leaves of *Pelargonium kewensis* and *Hibiscus* sp.

## Discussion

It has already been noted that one might expect to find varying amounts of MVA-5P, MVA-5PP and IPP in the reaction mixture at the end of incubation of crude enzyme preparations with [2-$^{14}$C]MVA. In nearly all cases, however, the only detectable product was MVA-5PP although MVA-5P was sometimes observed with weakly active chloroplast preparations; MVA-5P might have been in evidence more often if reaction mixtures from incubations of less than 2 hr had been examined. These findings were not entirely unexpected since work by Bloch *et al.* (1959) and by Henning *et al.* (1959) has shown that the reactions catalysed by mevalonate kinase and phosphomevalonate kinase proceed far towards formation of MVA-5PP in yeast autolysates provided sufficient ATP is present. Loomis and Battaile (1963) have also commented on the production from MVA of MVA-5PP as well as MVA-5P by crude enzyme extracts of pumpkin cotyledons.

No IPP was detected in any of the experiments; it may be that the equilibrium of the reaction in the *in vitro* systems studied is unfavourable for IPP formation or that any small amount of IPP formed is rapidly converted into further products not detected by our methods. Alternatively the fault may lie in the methods used to prepare the chloroplasts; an essential low molecular weight cofactor for the MVA-5PP into IPP conversion may have been lost from the "aqueously" prepared chloroplasts or the pyrophosphomevalonate decarboxylase inactivated during isolation of chloroplasts in "non-aqueous" media. Cofactor loss and enzyme inactivation are recognized hazards of these respective methods of chloroplast preparation.

The structural integrity of the chloroplasts prepared by the methods used is of interest. Leech (1964) has shown that chloroplasts prepared by the sucrose density gradient method have stroma material present and retain their bounding membranes. Those isolated in "non-aqueous" media, however, have lost the outer bounding membrane of the chloroplast and some of the chloroplast pigments and lipids (Stocking, 1959). However, chloroplasts so isolated are especially useful for enzymatic studies since the loss of soluble protein is far less than for chloroplasts prepared using "aqueous" media.

The interpretation of biochemical data obtained in terms of chloroplast function within the cell also depends on the purity of the chloroplast preparations with regard to other cellular components. The final purified chloroplast pellet in the sucrose density gradient method consists of over 90% intact chloroplasts, the rest of the pellet consists of almost equal proportions of chloroplast fragments and mitochondria (Leech, 1964). Since the preparation is washed before examination no extrachloroplastidic mevalonate activating enzymes nor any released from broken chloroplasts should be present. The total inactivity towards MVA of intact chloroplasts prepared by this method confirmed this supposition. It is worth noting, as Dr. Leech has already commented (p. 68, Vol. I), that though other "aqueous" density gradient methods of chloroplast preparation give purer preparations in terms of mitochondrial contamination the chloroplasts so isolated are invariably broken and almost devoid of stroma.

The only contaminants which appear to be present in "non-aqueous" chloroplast preparations are cytoplasmic proteins adsorbed to the chloroplast membrane (Stocking, 1959). Since the dry chloroplasts are not washed prior to incubation with [2-$^{14}$C]MVA it is possible that some mevalonate-activating enzymes released from structurally damaged chloroplasts would be present in unsonicated preparations. This would explain the very slight activity shown by these preparations (see Fig. 3). Our results therefore appear to confirm the reported structural integrity of chloroplasts prepared by this method.

These results showed that mevalonate kinase and phosphomevalonate kinase are present in chloroplasts and indicated that the chloroplast membrane is essentially impermeable to MVA. Other communications at this conference have shown that the chloroplast membrane is impermeable to acetyl-CoA, this also is a requirement of the compartmentalization scheme.

The presence of an extrachloroplastidic mevalonate kinase in addition to a chloroplastidic enzyme is difficult to show unequivocally. One cannot use cell homogenates from which chloroplasts have been merely spun off as the enzyme may be easily released from broken or damaged chloroplasts. The presence of extrachloroplastidic mevalonate-activating enzymes has been indicated, however, by demonstrating enzymic activity in acetone-dried powders of cultures of meristematic tissue of Paul's Scarlet Rose and in acetone-dried powders of the white pigmentless portions of variegated leaves of *Pelargonium* and *Hibiscus* sp. These tissues have no chloroplasts. Loomis and Battaile (1963) have also reported that mevalonate kinase occurs in the cytoplasm of cotyledons of pumpkin seedlings grown in the dark.

These findings support the view that a combination of segregation of enzymes and specific membrane permeability may be one of the primary ways by which the regulation of terpenoid biosynthesis in the developing seedling is achieved.

Similar compartmentalization of a biosynthetic pathway has been shown to exist by Kahn and Carell (1964) for the synthesis of porphyrins by isolated chloroplasts of *Euglena*, while Heber has also discussed (p. 71) the synthetic abilities of the chloroplast and the permeability of the chloroplast membrane to a number of photosynthetic products. Compartmentalization may therefore play a signficant role in the control of many other biosynthetic pathways in the cell.

## Acknowledgements

We thank Dr. K. J. Treharne of this department and Dr. R. M. Leech (Department of Botany, Imperial College, London) for helpful discussion of various aspects of this work. The S.R.C. provided substantial financial assistance to facilitate these studies.

## References

Bloch, K., Chaykin, S., Phillips, A. H. and de Waard, A. (1959). *J. biol. Chem.* **234**, 2595.
Goodwin, T. W. (1958a). *Biochem. J.* **68**, 26P.
Goodwin, T. W. (1958b). *Biochem. J.* **70**, 612.
Goodwin, T. W. and Mercer, E. I. (1963). *In* "The Control of Lipid Metabolism" (J. K. Grant, ed.) *Biochem. Soc. Symp.* No. 24, p. 37. Academic Press, London and New York.
Goodwin, T. W. (1965). *In* "Biosynthetic Pathways in Higher Plants" (J. B. Pridham and T. Swain, eds.). Academic Press, London and New York.

Griffiths, W. T., Threlfall, D. R. and Goodwin, T. W. (1964). *Biochem. J.* **90**, 40P.
Henning, U., Moslein, E. M. and Lynen, F. (1959). *Archs. Biochem. Biophys.* **83**, 259.
Kahn, J. S. and Carell, E. F. (1964). *Archs. Biochem. Biophys.* **108**, 1.
Leech, R. M. (1964). *Biochim. biophys. Acta* **79**, 637.
Loomis, W. D. and Battaile, J. (1963). *Biochim. biophys. Acta* **67**, 54.
Markley, K. and Smallman, E. (1961). *Biochim. biophys. Acta* **47**, 327.
Mercer, E. I. and Goodwin, T. W. (1962). *Biochem. J.* **85**, 13P.
Mercer, E. I. and Goodwin, T. W. (1963). *Biochem. J.* **88**, 46P.
Ohnoki, S., Suzue, G. and Tanaka, S. (1962). *J. Biochem. Tokyo* **52**, 423.
Rogers, L. J., Shah, S. P. J. and Goodwin, T. W. (1965). *Biochem. J.* **96**, 7P.
Rogers, L. J., Shah, S. P. J. and Goodwin, T. W. (1966). *Biochem. J.* **99**, 381.
Stocking, C. R. (1959). *Pl. Physiol.* **34**, 56.
Treharne, K. J., Mercer, E. I. and Goodwin, T. W. (1964). *Biochem. J.* **90**, 39P.

## Note Added in Proof

Existence of an extrachloroplastidic mevalonate kinase in addition to a chloroplastidic enzyme has been shown recently [Rogers, L. J., Shah, S. P. J., and Goodwin, T. W. (1966). *Biochem. J.* (In press)] by the fact that the two enzymes have different pH optima. The pH optimum of the chloroplast mevalonate kinase is at pH 7·5; while that of the extrachloroplastidic enzyme is at pH 5·5. Green leaves possess both enzymes and so do etiolated leaves, which shows the potentiality of the etiolated plastids for synthesis of terpenoid precursors.

# Part IV. Biosynthesis in Chloroplasts—Proteins and Nucleic Acids

# The Organization of Grana-Containing Chloroplasts in Relation to Location of Some Enzymatic Systems Concerned with Photosynthesis, Protein Synthesis, and Ribonucleic Acid Synthesis

S. G. Wildman

*Department of Botany and Plant Biochemistry, and the Molecular Biology Institute, University of California, Los Angeles, California, U.S.A.*

## Introduction

This essay is concerned with the grana-containing chloroplast found in the leaf cells of higher plants. In particular, I will consider the organization of these chloroplasts in relation to the location of certain enzymatic functions, placing greatest emphasis on those involved in protein synthesis. Compared to unicellular, photosynthetic organisms such as *Chlorella*, *Chlamydomonas*, and *Euglena* with all their attendant virtues, the higher plant offers one supreme advantage. The chloroplasts can be seen in living cells with magnificent clarity. Some of the chloroplasts in leaf mesophyll cells are as large, or larger, than an entire *Chlorella* cell. Chloroplasts in leaf cells can be observed with the phase microscope without any necessity for staining or other artifices which could alter the living condition. Therefore, I will first describe the appearance and behavior of chloroplasts as they are seen in the living condition. Following this, a consideration of what happens to their appearance when they are removed from the living cells will be made. Then, an analysis of the probable location of certain kinds of enzymatic activities in relation to the organization of the chloroplasts will be presented.

## Chloroplasts in Living Cells

A leaf cell can be rigorously defined as living when it exhibits vigorous protoplasmic streaming. Thus, whenever reference is made to *living* chloroplasts in the ensuing discussion, the observations were derived from cells which were clearly streaming. It may be noted in passing that a rigorous criterion for living chloroplasts is not always apparent in observations reported in previous literature. Undoubtedly, the previous confusion as to whether grana were an artifact produced by injury of cells, or a characteristic feature of chloroplasts in the living state, arose from failure to distinguish clearly between injured and living cells. In this connection, Brownian motion of organelles in injured and dying cells may have been confused with protoplasmic streaming. When a cell is injured and protoplasmic streaming is replaced by Brownian motion, the chloroplasts also markedly change in their appearance, but in a living cell, the conspicuous grana are the most prominent feature of every chloroplast brought into clear focus.

## THE BIPHASIC ORGANIZATION OF LIVING CHLOROPLASTS

Living, grana-containing chloroplasts are composed of two, microscopically distinct, materials: a *stationary component* where chlorophyll is located within the grana as revealed by fluorescence microscopy, and a *mobile phase*. Although micro-spectrophotometric analyses remain to be made, the absorption characteristic of the mobile phase, as judged by eye and interference filters, suggests that pigments other than chlorophylls could be present. There is a striking similarity in the biphasic appearance of grana-containing chloroplasts as seen in a variety of plant species. The living cells in Fig. 1 illustrate this fact. Without the species labels, classification of the plants by differences in the appearance of the chloroplasts would be very difficult. However, the appearance of chloroplasts may be greatly altered when plants are grown in media deficient in mineral elements. An excellent example of the effect on spinach chloroplast organization produced by Mn deficiency has been provided by Possingham *et al.* (1964).

The stationary component is remarkably static. In mesophyll cells of terrestrial plants (not so in *Elodea* and other plants that live in a submerged environment), chloroplasts gradually change their position within the cell, but the movement is imperceptible to the eye even with long periods of observation. In contrast, the mobile phase is a highly dynamic material, always in some kind of motion. Sometimes the mobile phase is very conspicuous; at other times, barely discernible. The mobile phase in one cell may be prominent while inconspicuous in an adjacent cell. An hour later, the situation could be reversed. There seems to be no obvious correlation between the extent of the mobile phase and the photosynthetic activity of the chloroplasts.

Figure 2 is a drawing of a model of a grana-containing chloroplast. The model is based on the extensive microscopic observations of living chloroplasts as well as of the behavior of isolated chloroplasts made by me with my collaborators at UCLA and in Australia (Spencer and Wildman, 1962; Spencer and Unt, 1965; Wildman *et al.*, 1962; Honda *et al.*, 1964; Hongladarom, 1964; Honda *et al.*, 1966). The concept seems to be consistent with both the light microscope observations as well as observations made on fixed and sectioned chloroplasts with the electron microscope. In constructing the model, the assumption is made that the continuous lamellar systems apparent in many electron micrographs are more representative of the condition least altered from the living state. The "fretworks" seen in some electron micrographs may have arisen from the more ordered system by swelling and disorganization during fixation, particularly where $KMnO_4$ has been employed. The orientation of the intergrana lamellae in relation to the grana lamellae has been left vague because of some ambiguity in the interpretation of electron micrographs.

As seen with the light microscope, the mobile phase obviously surrounds the stationary component and, in fact, often embraces more than one stationary component. It is surmised that the mobile phase also interpenetrates the intergrana lamellar regions. This surmise is based on the manner by which Fraction

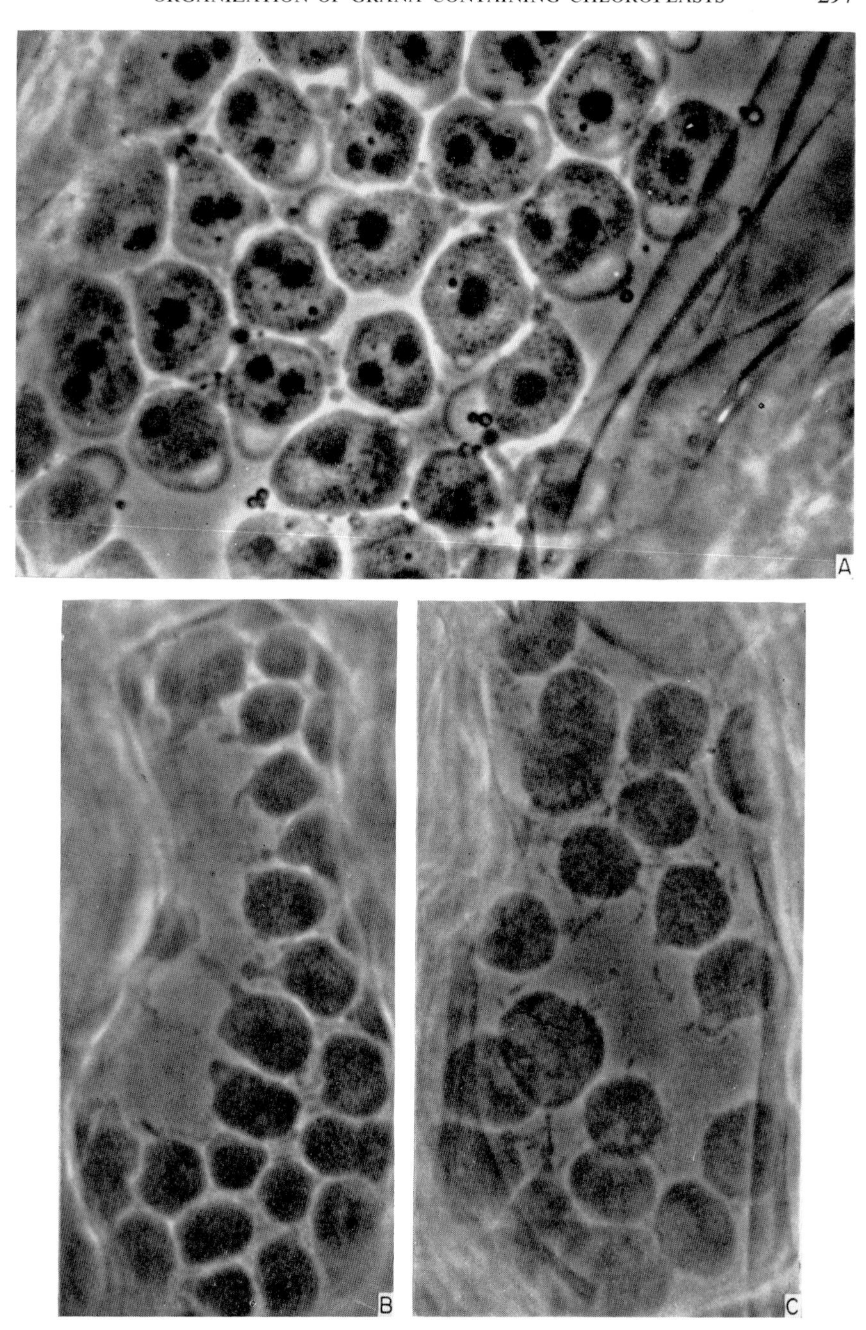


FIG. 1. The appearance of living chloroplasts displaying prominent grana and mobile phase. Living mesophyll cells from different kinds of leaves. A, *Nicotiana tabacum*; prominent starch grains in stationary components. B, *Spinacia oleracea*. C, *Beta vulgaris*. (Photomicrographs by courtesy of Dr. Shigeru I. Honda.)

I protein and ribosomes are released from isolated chloroplasts, and will be considered later. It is also based on the apparent continuity of the granular material found in regions between the intergrana lamellae and similar appearing granular material found to the exterior of the lamellar structure in electron micrographs. The mobile phase can be considered equivalent to stroma, but I prefer the term mobile phase because the words impart a more anthropomorphic sensation of the actual performance of a material which is in constant, ameboid-like motion.

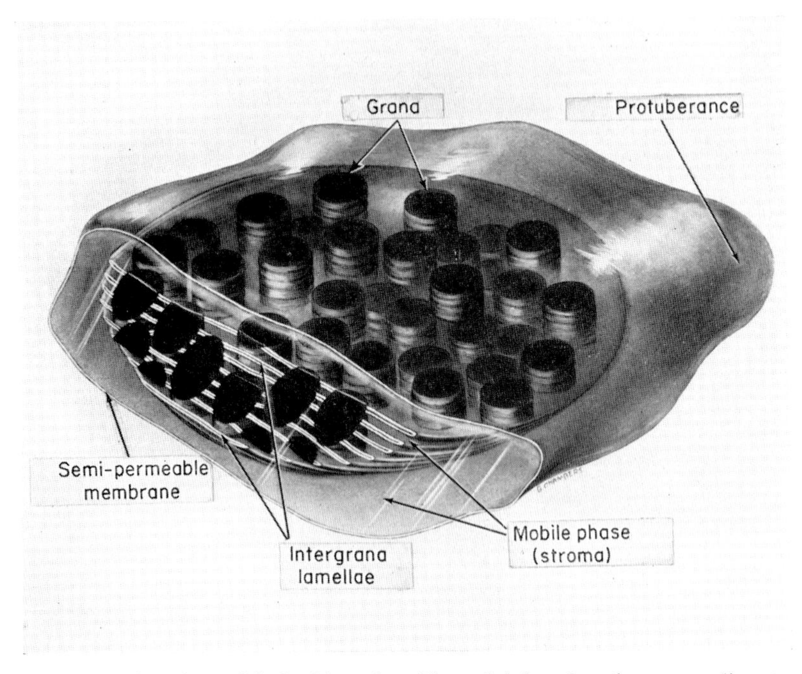


FIG. 2. Drawing of a model of a chloroplast. The scale is based on the average diameter of the grana (0·5 $\mu$). The chloroplast may have been drawn too thick; note how thin living chloroplasts can be in Fig. 1C where the grana of each chloroplast are in focus in the region where two chloroplasts overlap. Chloroplasts are so susceptible to swelling by fixatives that the correct dimensions are probably rarely seen in electron micrographs. (After Francki *et al.* 1965; reproduced with permission from *Biochemistry*.)

The outer surface of the mobile phase has the properties of a semipermeable membrane. Observing isolated chloroplasts with the phase microscope, Hongladarom and Honda (unpublished) were able to demonstrate the reversible swelling and contraction of the mobile phase without significant alteration in the structure of the stationary components using sucrose and mannitol as osmotic agents. Sodium chloride would not contract swollen chloroplasts, nor will it prevent both the mobile phase and stationary components from swelling except for short periods of time at ice-bath temperatures.

It appears that high concentrations of NaCl markedly alter the structure of the mobile phase membrane. The mobile phase membrane is different from the membrane surrounding nuclei. The neutral detergent, Triton X-100 completely and instantaneously solubilizes chloroplasts and mitochondria but does not attack the structure of nuclei nor spherosomes.

## MOBILE PHASE IN RELATION TO MITOCHONDRIA

The mobile phase may undulate slowly, or it may rapidly, or slowly, extend and withdraw protuberances. Protuberances may extend into the cytoplasm to a length several times larger than the diameters of the stationary components.

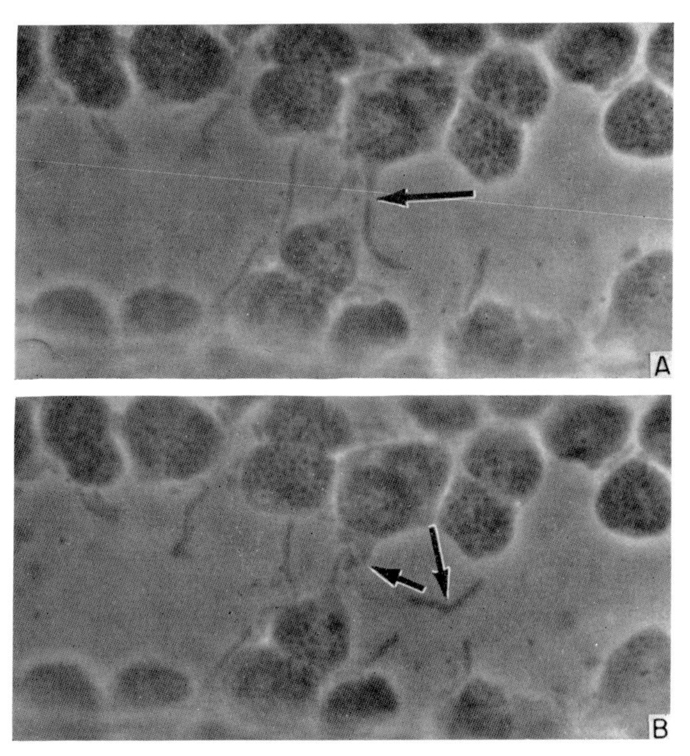


FIG. 3. Segmentation of a mobile phase protuberance. In A, the segment is still attached to the mobile phase by a fine thread. In B, the mobile phase has recoiled (left arrow) and the severed mitochondrion (right arrow) has begun to stream away. (After Wildman *et al.*, 1962; reproduced with permission from *Science, N.Y.*.) For sceptics: see "Organelles in Living Plant Cells", 16 mm, color, sound film, 26 min. Loaned by Extension Media Center, University of California Berkeley 94720.

The most dramatic activity of the mobile phase occurs when a protuberance segments into a free-streaming organelle that cannot be microscopically distinguished from other free-streaming mitochondria in the living cell. Figure 3 is an example of a segmentation of a mobile phase protuberance. The segmentation phenomenon need not be confused with a mitochondrion suddenly emerging from beneath a chloroplast. When a segmentation occurs, the process is immediately followed by the conspicuous recoil motion of the severed parts. Mitochondria also fuse with the mobile phase, first by the union of very fine threads, and then "melt" into the mobile phase and thereby lose their previous cytological identity.

Some of our descriptions of the segmentation phenomenon have been rather freely extrapolated by others to mean advocating chloroplasts as the origin of mitochondria. We disclaim reponsibility for this extrapolation because we prefer another view. The cells suitable for observation of the segmentation phenomenon are mature and contain large vacuoles. Mitotic activity ceased long before the cells had expanded to the state where they were suitable for observation. By this time, the chloroplasts were already formed. In our years of observation of these mature cells, we have yet to encounter a condition that might be reasonably interpreted as a chloroplast in the process of reproduction. How mitochondria and chloroplasts arise or reproduce before cells reach this stage of maturity has not been amenable to observation by light microscopy. When observations can be made, I suspect that there has already been formed what can be loosely called "mitochondrial substance". In the mature cells, the substance is now distributed in various forms which are interchangeable with each other. Part of the mitochondrial substance is in the form of recognizable mitochondria, whereas another part is in the form of the mobile phase of chloroplasts. Perhaps a relatively "steady state" condition is being observed where the total mitochondrial substance is in dynamic equilibrium with its parts. After mitotic formation of new cells, I would imagine that during further growth of the cells by expansion, mitochondria *per se* do not reproduce but rather that the total mitochondrial substance increases in amount and continues to be distributed in different forms. In this speculation, I am aware of the counter argument that "clean" chloroplasts in the isolated condition do not possess respiratory activities expected of mitochondria. My answer to this argument is simply that "clean" chloroplasts have had nearly all of their microscopically identifiable mobile phase removed in the process of "cleansing" them.

## POSSIBLE SIGNIFICANCE OF FLUID NATURE OF THE MOBILE PHASE

It seems to me that the ameboid movements and segmentation of the mobile phase may signify a novel method for preserving compartments wherein the photochemical machinery in the grana of the stationary components is separated physically from the enzymatic systems responsible for carbon assimilation and other metabolic activities. I imagine that the fluid nature of the mobile phase permits it to flow, or creep into the spaces between the intergrana lamellae and thereby make intimate contact with the stacks of grana lamellae. Energy derived from the interaction of light and chlorophyll could then be transferred to the enzymatic systems of the mobile phase at these areas of contact. Presumably, only a single membrane of about 30 Å would have to be surmounted in the transfer. The fluidity of the mobile phase would permit further enzymatic transformations of the energized compounds to occur at the same time as they were being physically removed from the region of photochemical activity, but still kept within the confines of the mobile phase.

The outer surface of the mobile phase appears to be a barrier that not only prevents passage of large molecules to and from the cytoplasm, but some small molecules as well. Apparently, the entire galaxy of enzymes ranging from those required for initial fixation of $CO_2$ to those finally involved in the deposition of starch molecules in starch grains are confined within the boundary of the mobile phase. Starch grains in the cells we look at are never found in the cytoplasm, only in the stationary components of chloroplasts.[1] Many enzymes involved in carbohydrate metabolism are found to be associated with chloroplasts (Stocking, 1956; Heber, 1962). Does this mean also that the substrates cannot readily pass from the mobile phase into the cytoplasm? In this connection, Goodwin (1965) provides evidence that mevalonic acid of the cytoplasm does not interchange with mevalonate of chloroplasts, with the strong implication that chloroplasts contain their own enzymatic systems for lipid metabolism separate and different from the lipid systems in the cytoplasm. As will be shown later, the chloroplasts also contain different systems for protein synthesis than those found in the cytoplasm, although amino acids appear to be able to penetrate through the surface of the mobile phase. The list of enzymatic systems apparently compartmentalized within chloroplasts could be extended but the question that remains in my mind is how photosynthetic energy, contained in compounds confined within the mobile phase, is made available for the energy-requiring reactions of the cytoplasm. Could it be that segmentation of mobile phase protuberances is a mechanism whereby the energized intermediates produced by photosynthesis are physically separated from the chloroplasts and then transported to other regions of the cell where the energy could be utilized for synthetic processes? Does fusion of mitochondria with the mobile phase represent a means for transporting materials into the chloroplast, or possibly the return of expended fuel cells for recharging?

Clearly, we have no reliable answers to such questions because the unsolved problem is how to design meaningful experiments. However, the information on how chloroplasts appear and behave in living cells provides a basis for judging what happens to their appearance when they are extracted from cells and subjected to various treatments for the purpose of understanding their behavior in the terms of biochemistry.

[1] The fact that starch grains are confined to the stationary components of chloroplasts may provide a clue to explain the final control exerted on the photosynthetic process. However naive, I am attracted by the notion that control occurs by the simple, physical distortion of the chloroplast resulting from starch grains produced by photosynthesis. A chloroplast, largely free of starch grains, is an extremely thin, nearly flat object in the living cell (Fig. 1C). As photosynthesis proceeds and new starch grains accumulate, apparently in between the intergrana lamellae, the chloroplast becomes more and more distorted in shape. With prolonged illumination, the stationary component may become so gorged with starch grains as to force the chloroplast into the shape of a football and almost a sphere. Such distortion in shape results in tilting the grana away from their previous position in the direct pathway of illumination. Thus, the change in position of the chlorophyll-containing grana with respect to light could be viewed as resulting in a diminished efficiency for light absorption by chlorophyll with a consequent reduction in the flow of electrons to the systems responsible for carbon assimilation.

## Isolated Chloroplasts

### APPEARANCE OF ISOLATED CHLOROPLASTS

When isolated by the conventional methods of biochemistry, phase microscopic examination shows most of the chloroplasts to bear little resemblance to their appearance in the living cell. However, by suitable adjustment of osmotic conditions and use of high molecular weight additivies such as Ficoll and Dextran, a majority of the chloroplasts in a cell-free extract can be preserved in a condition closely similar to the living state (Honda *et al.*, 1965). Figure 4 illustrates the appearance of isolated chloroplasts compared to their living condition. The total population of chloroplasts in a cell-free homogenate of leaves can be broadly grouped into two classes (Spencer and Wildman, 1962). Class I chloroplasts (Fig, 4a, b, c) still possess microscopically detectable mobile phase. The mobile phase is greatly diminished or missing from Class II chloroplasts (Fig. 4d, e). Even with the best preserved preparations, only about 70% of the chloroplasts in a cell-free homogenate are of the Class I variety. Moreover, only about one-half of the Class I chloroplasts will have the mobile phase present in a relaxed form where it is simple to identify with the microscope, particularly when protuberances are also preserved as in Fig. 4b. About 25% of the Class I chloroplasts have the appearance shown in Fig. 4a. These chloroplasts are folded, apparently via contraction of the mobile phase, and neither the grana nor the mobile phase can be resolved. However, folded chloroplasts can be caused to return to their native unfolded state by decreasing the concentration of osmotic agents in their environment. In the process of unfolding, the grana reappear and the mobile phase can be identified usually as a balloon enclosing the stationary component as illustrated by the Class I chloroplasts in Fig. 4c. In fact, about 5% of the Class I chloroplasts in a well-preserved homogenate will have the balloon-like appearance shown in Fig. 4c.

Of the Class II chloroplasts, constituting about 30% of the total population under the best of conditions, most of them will have the appearance shown in Fig. 4d. The grana are very prominent, but the edges of some of the chloroplasts are sufficiently irregular as to indicate that fragments of the stationary components have been broken off. Critical focus may often reveal small vesicles emerging from the surface of the stationary components. The mobile phase cannot be identified. The small proportion of remaining Class II chloroplasts show a marked disorganization and swelling of the stationary components (Fig. 4e).

### PHOTOSYNTHETIC ACTIVITIES ASSOCIATED WITH THE MOBILE PHASE

Spencer and Unt (1965) developed methods for separating Class I and Class II chloroplasts. They were then able to demonstrate that light-induced $CO_2$ fixation, non-cyclic photophosphorylation, and light-dependent, triphosphopyridine nucleotide reduction were all dependent upon the association of the mobile phase with Class I chloroplasts. The absence of the mobile phase in Class II chloroplasts resulted in a higher capacity for performing the Hill reaction, and this capacity was enhanced by first swelling the stationary components.

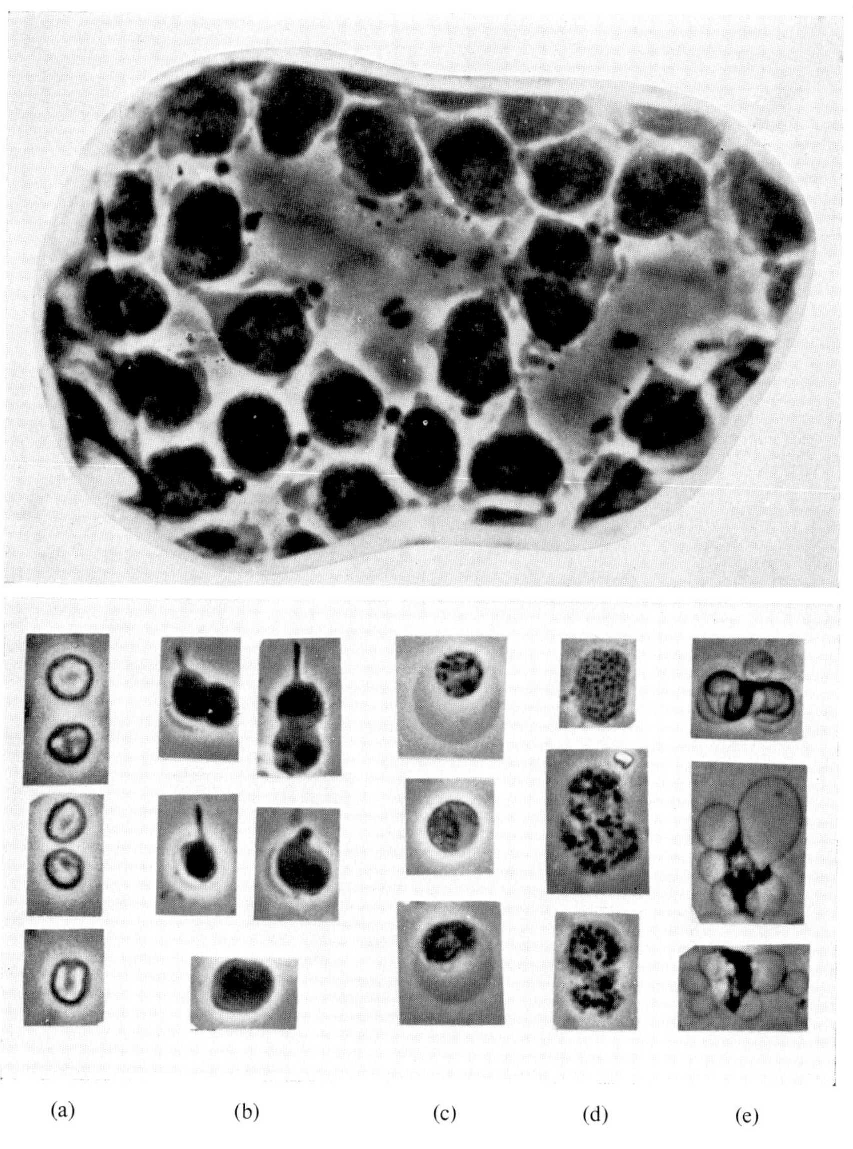


FIG. 4. Appearance of isolated chloroplasts compared to living chloroplasts. Class I chloroplasts which have retained the mobile phase are shown by *a*, *b*, and *c*; *d* and *e* are stationary components of Class II chloroplasts without mobile phase. In the living cell, the mobile phase is conspicuous on all chloroplasts. The smaller, ellipsoidal organelles are mitochondria. The tiny dots are spherosomes. (Photomicrographs by courtesy of Dr. S. I. Honda.)

That the enzymatic mechanism of $CO_2$ fixation should be associated with the presence of the mobile phase on Class I chloroplasts was not unexpected because Fraction I protein appears to be the principal macromolecular component of the mobile phase. Following on the isolation and characterization of carboxydismutase from leaves (Weissbach *et al.*, 1956; Jakoby *et al.*, 1956), the apparent identity of Fraction I protein and carboxydismutase has been shown by various workers (Dorner *et al.*, 1957; Van Noort and Wildman, 1964; Mendiola and Akazawa, 1964; Trown, 1965).

## MACROMOLECULAR COMPOSITION OF THE MOBILE PHASE

Lyttleton and T'so (1958) showed that when chloroplasts were protected from swelling during isolation by a gentle method, a large proportion of the Fraction I protein of leaves was contained within the chloroplasts, but could be readily solubilized by disrupting the chloroplasts. Park and Pon (1960) and

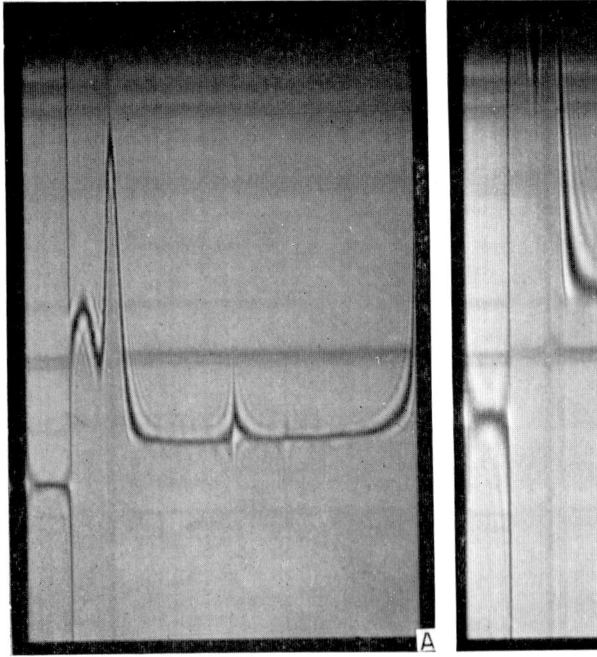


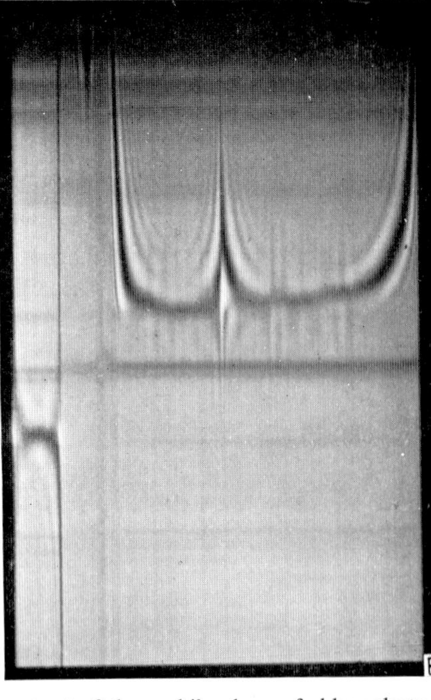


FIG. 5. Macromolecular composition of an extract of the mobile phase of chloroplasts resolved in the analytical ultracentrifuge. A, from left to right, 4 S, 18 S Fraction I Protein, and 70 S ribosome components, B, magnification of 70 S ribosome component.

Heber (1962) have shown carboxydismutase to be a chloroplast constituent which is also readily extracted from the chlorophyll-containing constituents of isolated chloroplasts. Lyttleton (1962) also has demonstrated that leaves contain two distinct species of ribosomes, and that one species appears to be in the chloroplasts. With the advent of the recognition of the mobile phase of chloroplasts, it has been possible to define the probable location of Fraction I protein and chloroplast ribosomes in relation to the biphasic structure of chloroplasts (Francki *et al.*, 1965; Boardman *et al.*, 1965, 1966).

The Schlieren diagrams presented in Fig. 5 were derived from examination of an extract of the mobile phase in the analytical centrifuge. The diagrams show that about 90% of the proteins in the mobile phase consist of a mixture of proteins with an average sedimentation constant of 4 S, and the 18 S Fraction I protein. About 10% of the proteins are ribosomes with a sedimentation con-

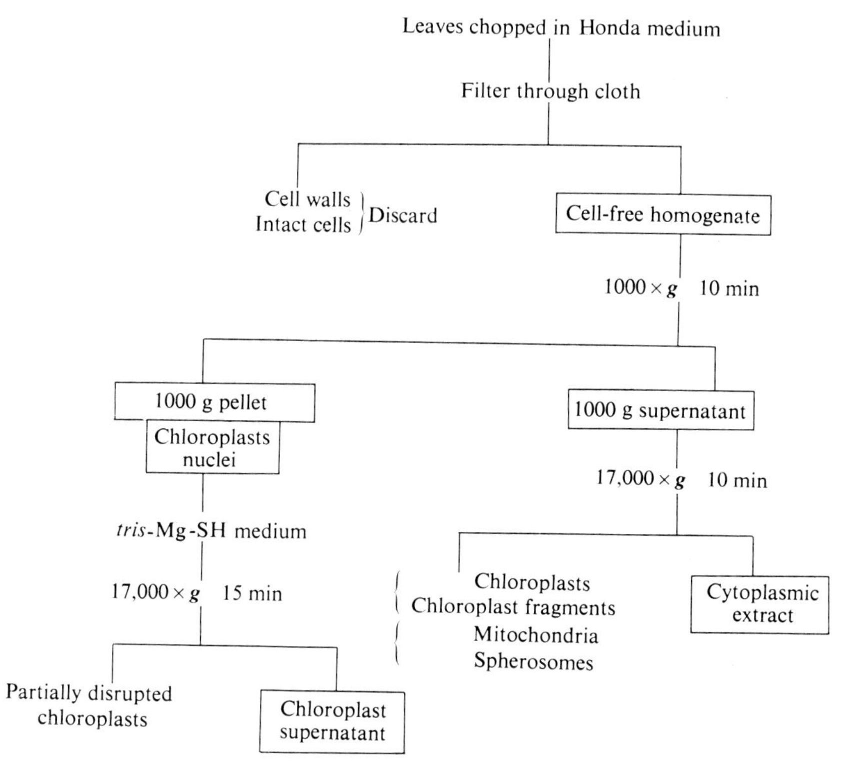


FIG. 6. Scheme for separating the mobile phase from isolated chloroplasts and for obtaining an extract of the cytoplasm of leaves. (After Boardman *et al.*, 1966; reproduced with permission from *J. mol. Biol.*)

stant of 70 S. Whereas the ratio of 4 S proteins to 18 S Fraction I protein is about 50:50 in this preparation, the ratio is dependent upon the size of the leaf from which the chloroplasts and the mobile phase are extracted. With very small, or very old leaves, the 4 S:18 S ratio could approach 80:20. The 50:50 ratio is achieved just about the time when the leaves have reached their maximum extension in length (Dorner *et al.*, 1957). The mobile phase extract shown in Fig. 5 was obtained by the fractionation procedure shown in Fig. 6.

The use of razor blades to cut the leaf cells for the release of their protoplasmic contents was adopted since microscopic examination of the organelles showed this method did the least fragmentation and disruption of the nuclei and chloroplasts. For experiments involving 10 g of leaf tissue, the method is not very laborious and the efficiency of extraction is reasonable, about 35 to 50% of the total leaf cells being extracted. Simultaneously with cutting the cells, the protoplasmic contents came into contact with Honda medium which has been found to preserve a majority of the chloroplasts in the Class I condition (Honda *et al.*, 1965). Application of low centrifugal force deposits most of the chloroplasts as a loosely packed pellet largely free of cytoplasmic constituents which remain in the supernatant. Resuspension of the pellet of chloroplasts in a medium of low osmotic strength causes the mobile phase to balloon around the stationary components, followed by disruption of the balloons and release of the mobile phase constituents. The chlorophyll-containing stationary components also swell to the extent that their starch grains are released, but the lamellar system remains sufficiently intact so that all the chlorophyll-containing material is deposited as a pellet after centrifugation at moderate force. The composition of the *chloroplast supernatant* is that shown in Fig. 5. Had the chloroplasts been resuspended in Honda medium, their microscopic appearance would not have changed appreciably nor would the supernatant solution have contained appreciable amounts of the macromolecular constituents shown in Fig. 5. Repeated washing of chloroplasts even in Honda medium, however, will successively remove portions of the mobile phase—many of the chloroplasts may fold as in Fig. 4a—which attests to the extreme lability of this material.

## COMPOSITION OF STATIONARY COMPONENTS

It is not within the province of this essay to examine the extensive literature on what might now be considered as related to the composition of the stationary components. Research on the composition of thylakoids, quantasomes, phosphatides and lipids, pigments, etc., would seem to be more likely involved with the stationary components than the mobile phase. However, I think that it might be possible to refine these analyses by devoting more attention to the selective removal of the mobile phase prior to analysis of the stationary components. I suspect that it should be possible to remove the mobile phase constituents by repeated washing of chloroplasts in a medium that would largely prevent swelling and disorganization of the stationary components. With this accomplished, I envisage the possibility of selective swelling, disruption, and separation of the intergrana lamellae from the grana lamellae. But I am convinced that all steps in the process deserve the most careful microscopic confirmation of what has happened. Perhaps this brief survey of the properties of isolated chloroplasts compared to living chloroplasts will awaken a greater desire to use the light microscope in research which purports to relate structure to function of organelles in higher plants. In recent years, the increasing reliance on the results of electron microscopy has fostered an attitude that tends to overlook the power of the light microscope for deciding on the degree of alteration in organelle structure produced by the treatment applied. My fond hope is that a combination of light and electron microscopy will produce more precise information than is now available. A critical assessment of just what happens to the structure of organelles when they are transformed from the living to the fixed condition preparatory to sectioning for electron microscopy could be accomplished by combining the critical use of these powerful tools. I have watched too many chloroplasts swell and go completely to pieces in the presence of fixatives such as $KMnO_4$ to have much confidence that what we see in electron micrographs of leaves and isolated chloroplasts is an accurate representation of the native structure of chloroplasts. With this complaint off my chest, I would now like to concentrate attention on amino acid incorporation and the mechanisms responsible for protein synthesis by isolated chloroplasts.

## AMINO ACID INCORPORATION INTO PROTEIN BY CELL-FREE EXTRACTS OF LEAVES

The ability of different organelle fractions obtained from leaves to incorporate labeled amino acids into protein was investigated by Stephenson *et al.* (1956). Their results pointed to the chloroplasts as being the most active fraction particularly when they were illuminated. However, no dependency on ATP, nor inhibition with RNase could be demonstrated. Sissakian *et al.* (1962) isolated chloroplast ribosomes which were capable of incorporating amino acids into protein, but these preparations showed little dependence on cofactors or sensitivity to RNase. In a later investigation (Sissakian *et al.*, 1965) a chloroplast ribosome preparation which incorporated amino acids would, after dialysis, reveal some ATP and other expected dependencies, but was surprisingly inhibited by actinomycin D while remaining insensitive to RNase. App and Jagendorf (1963) found results similar to those of Stephenson *et al.* (1956) in regard to lack of cofactor requirement for amino acid incorporation by isolated chloroplasts. By use of deoxycholate, ribosomes were extracted from chloroplasts which did reveal dependency on cofactors as well as being somewhat inhibited by RNase, but the level of incorporation on a ribosome basis was small. Biswas and Biswas (1965) were also unable to demonstrate ATP dependency for phenylalanine incorporation, only 50% inhibition with RNase, and no stimulation with Poly U until after the ribosomes were treated with RNase. In a later paper, App and Jagendorf (1964) have raised the important question as to what extent bacteria in the leaf preparation might have contributed to the incorporation that was measured. Whatever the explanation, the incorporating systems from leaves appeared to be different in kinetics, cofactor dependencies, and sensitivity to inhibitors when compared to the properties of such well characterized ribosomes as obtained, for example, from *E. coli* and red blood cells. More recently, a cell-free system that incorporates amino acids into protein and displays the kinetics and dependencies expected of ribosome systems such as those from *E. coli* and reticulocytes has been obtained from tobacco and spinach leaves (Spencer and Wildman, 1964; Spencer, 1965). These systems have been extensively characterized (Francki *et al.*, 1965; Boardman *et al.*, 1965; Boardman *et al.*, 1966) so that only the principal findings will be considered here.

The most active extracts for cell-free protein synthesis are obtained from leaves that are about one-third to midway through their total expansion in length. Extracts from older leaves are much less active. The chopping method of homogenization has been extensively employed although active preparations can be obtained by other gentle methods for extracting the protoplasmic contents of leaf cells.

Chloroplasts are far more active in cell-free protein synthesis than any other organelle fraction of a leaf homogenate. Nuclei are inactive. The activity of the chloroplasts is unaffected by light, is highly dependent upon ATP, is stimulated by the presence of a mixture of amino acids and also by trace amounts of GTP. Spencer (1965) had devised a medium whereby ATP produced by cyclic photophosphorylation can replace the need for exogenous ATP. The time course of incorporation of amino acids is similar to that of *E. coli* and red blood cell systems; there is no lag, the rate of incorporation is maximal for the first 10 min and then rapidly falls off with additional time until incorporation ceases after 30–45 min incubation at 25°. The chloroplast system is extremely sensitive to pancreatic ribonuclease, 25% inhibition being produced by $2 \times 10^{-3}$ $\mu$g/ml. The incorporation is strongly inhibited by puromycin and moderately inhibited by chloramphenicol. The incorporation of amino acids into protein is not affected by DNase, nor actinomycin D. Phenylalanine incorporation is specifically stimulated by the presence of Poly U.

## PROTEIN SYNTHESIS AND THE MOBILE PHASE

The enzymatic mechanisms responsible for protein synthesis by isolated chloroplasts appear to be located in the mobile phase. Eighty per cent of the protein synthesizing activity can be removed without loss in activity from the chlorophyll-containing stationary components as a *chloroplast supernatant* by repeated washing of the stationary components with a buffer of low osmotic strength. The fractionation method shown in Fig. 6, where no washes are employed, may result in about 50% of the activity in the chloroplasts being transferred to the *chloroplast supernatant*. The activity transferred into the *chloroplast supernatant*, the macromolecular composition of which was shown in Fig. 5, is entirely associated with the minor 70 S ribosome component. This result was obtained by resolving and isolating, the 4 S, 18 S, and 70 S, components in the *chloroplast supernatant* by density gradient centrifugation using linear sucrose columns. In the isolated condition, the 70 S chloroplast ribosomes become highly dependent upon added s-RNA and/or activating enzymes for maximum activity. These ingredients also appear to be located in the mobile phase of chloroplasts. Although Clark *et al.* (1964) were able to identify polyribosomes in leaf extracts prepared in polyvinyl sulphonate, the presence of this agent almost completely inhibits the incorporating activity of cell-free extracts (Boardman *et al.*, 1965). Thus, the question of whether the 70 S ribosomes within the chloroplast are in the form of polyribosomes lacks a satisfactory answer. In contrast to Lyttleton (1962) and Clark *et al.* (1964), Odintsova *et al.* (1964) reported that chloroplast ribosomes were of the 80 S variety and indistinguishable from cytoplasmic ribosomes. However, in a more recent communication from the same group (Sissakian *et al.*, 1965) the two kinds of ribosomes have been identified in pea seedlings.

## PHYSICAL AND MORPHOLOGICAL PROPERTIES OF CHLOROPLAST RIBOSOMES

The 70 S mobile phase ribosomes are remarkably similar in physical properties and structural morphology to the 70 S ribosomes isolated from bacteria and blue-green algae (Karlsson *et al.*, 1966). The corrected sedimentation coefficient of 69·9 S for chloroplast ribosomes (Boardman *et al.*, 1966) is close to the mean

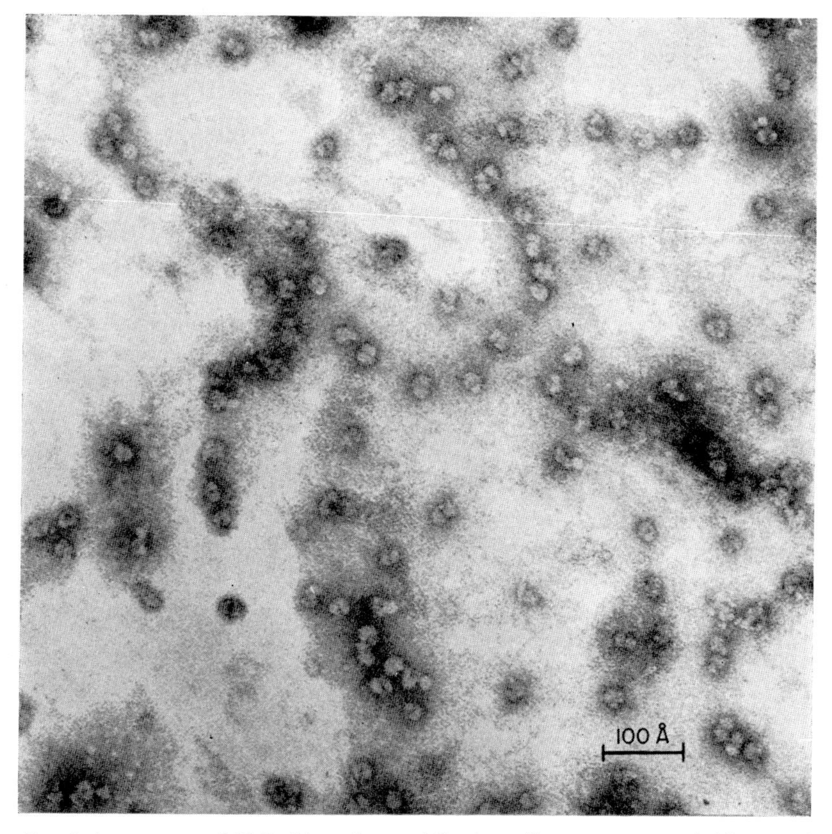


FIG. 7. Appearance of 70 S, chloroplast mobile phase ribosomes as revealed by negative staining and electron microscopy. They closely resemble *E. coli* ribosomes in clear outlines, dimensions, and asymmetrical clefts on many particles. (After Karlsson *et al.*, 1966; reproduced with permission from *J. mol. Biol.*)

value of 68·4 obtained by Taylor and Storck (1964) for twenty-five species of bacteria and two species of blue-green algae. The 70 S value is in excellent agreement with the value obtained by Clark *et al.* (1963, 1964) for chloroplast ribosomes from Chinese cabbage leaves, but is somewhat higher than the value reported by Lyttleton (1960,1962) for clover and spinach leaves, and Sissakian *et al.* (1965) for pea seedling chloroplast ribosomes.

A field of mobile phase, 70 S ribosomes subjected to negative staining and examination by electron microscopy is shown in Fig. 7. These particles strik-

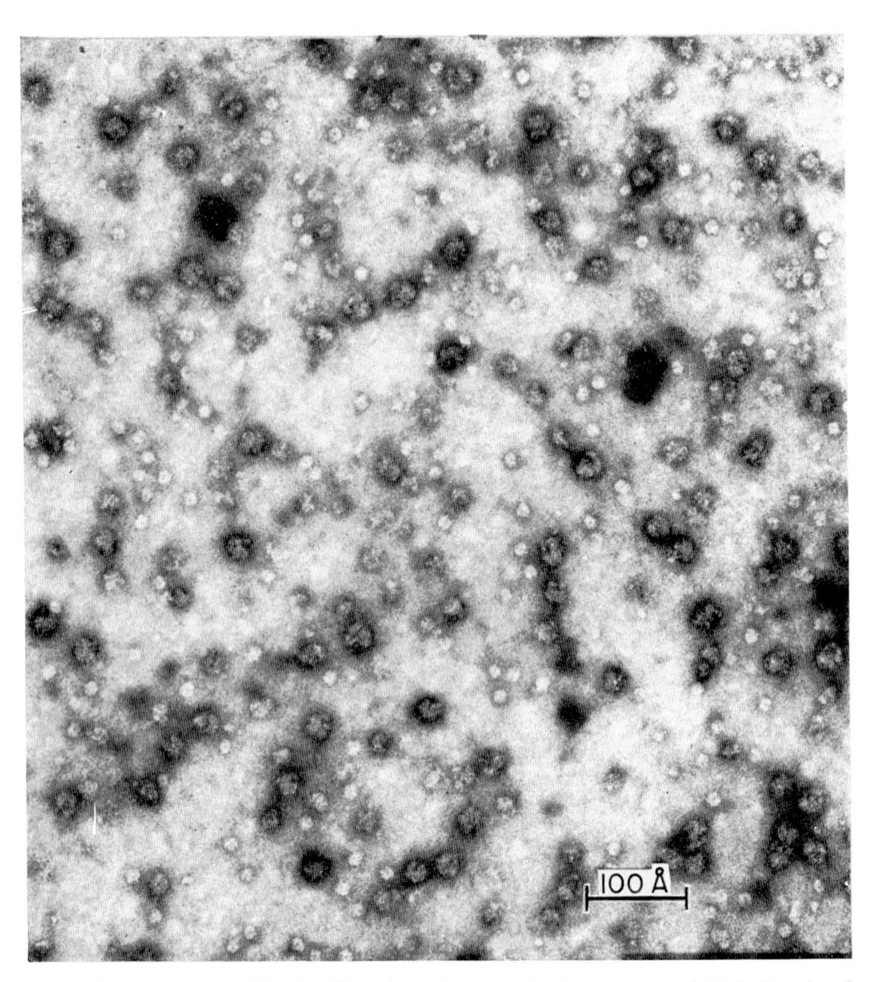


FIG. 8. Appearance of 70 S, chloroplast ribosomes in the presence of 18 S, Fraction I protein. Extract of the mobile phase of chloroplasts. The smaller Fraction I protein particles are more transparent than the ribosomes after negative staining. (After Karlsson *et al.*, 1966; reproduced with permission from *J. mol. Biol.*)

ingly resemble ribosomes isolated from *E. coli.* They have clear outlines and distinct clefts which are asymmetrical. Their average dimension is $214 \times 268 \pm 20$ Å with the cleft located about 140 Å from one end.

Chloroplast ribosomes are sharply different in appearance from the more compact 80 S ribosomes isolated from red blood cells and other mammalian sources. The 70 S ribosomes are also structurally distinguishable from the ribosomes isolated from pea seedlings (T'so *et al.*, 1958) and the 80 S ribosomes which are present in the cytoplasm of leaves and which will be considered later in this paper. In the isolated condition, the 70 S ribosomes are also readily distinguishable from Fraction I protein particles (average diameter, 110 Å) as shown by the electron micrograph in Fig. 8. However, the identification of the minute amount of 70 S ribosome particles in the presence of the large amounts of Fraction I protein particles might be more difficult in sections of chloroplasts observed by electron microscopy.

## STRUCTURE AND ACTIVITY OF CHLOROPLAST RIBOSOMES IN RELATION TO $Mg^{2+}$ AND PURIFICATION

When isolated from the chloroplasts, the 70 S ribosomes require concentrations of 15–20 mM $Mg^{2+}$ to preserve their structural integrity as well as for maximum catalysis of amino acid incorporation into protein. In lesser $Mg^{2+}$ concentrations, the 70 S ribosomes rapidly dissociate into 50 S and 35 S subunits. After dissociation, addition of $Mg^{2+}$ will cause the subunits to reform 70 S particles, but their capacity for amino acid incorporation will have largely disappeared. They resemble *E. coli* ribosomes in this respect also. When the 70 S ribosomes are confined within the chloroplasts by the boundary of the mobile phase, they are much less responsive to changes in $Mg^{2+}$ concentration in regard to incorporating activity. Chloroplasts in 3 mM $Mg^{2+}$ exhibit as much activity as produced by higher concentrations. In fact, high concentrations of $Mg^{2+}$ may hinder the release of 70 S ribosomes from the chloroplasts into the *chloroplast supernatant*.

The 70 S ribosomes lose up to 90% of their protein synthesizing activity when they are pelleted out of a *chloroplast supernatant* by high speed centrifugation. The mechanical act of pelleting the particles appears to be the primary cause for the loss in activity, although the physical properties of the particles are not noticeably changed. The dramatic loss of activity resulting from purification does not seem to result from nuclease activity, nor from a loss in s-RNA and messenger RNA. If the 70 S ribosomes are allowed to escape from the chloroplasts into the cytoplasm during the homogenization of leaf cells, they will also lose up to 90% of their activity. The inactivation appears to occur almost instantaneously with the release of the protoplasmic contents of the cut leaf cells and ensuing disorganization of the organelles. If the 70 S ribosomes present in a *chloroplast supernatant* are mixed with a *cytoplasmic extract*, inactivation does not occur.

In unpublished research, we have found that about 80% of the product of incorporation by the 70 S chloroplast ribosomes remains associated with the ribosomes after their resolution on a density gradient. The remaining 15–20% of the product appears in the 4–18 S position in the gradient. A shorter incubation period followed by a "chase" with [$^{12}$C]amino acid results in the transfer of 20–30% of the product into the soluble, 4–18 S region. Incubation followed by puromycin results in 50–60% of the product of incorporation being transferred into the 4–18 S region of soluble proteins. The solube product has been purified approximately forty-fold and at this stage is associated with the 18 S Fraction I protein component. Peptide mapping of the product in comparison to those obtained from Fraction I protein and the 4 S mixture of proteins may provide a clue as to what kind of protein the chloroplast ribosomes are making in the *in vitro* condition.

## PROTEIN SYNTHESIS BY CYTOPLASMIC EXTRACTS OF LEAVES

As shown by Lyttleton (1960; 1962) and confirmed by others as well as the research I am describing, plant leaves contain two species of ribosomes which are present in about equal proportions. This situation is illustrated by Fig. 9 which is a Schlieren diagram of a "total" extract of ribosomes from leaves

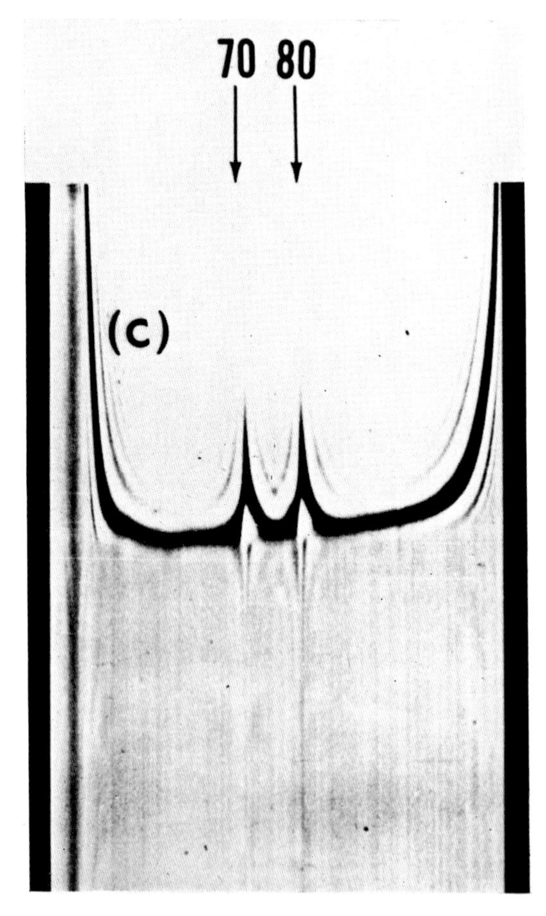


FIG. 9. Schlieren pattern similar to that discovered by Lyttleton (1962) of a "total" ribosome extract of leaves resolved in the analytical centrifuge. Areas occupied by the 70 and 80 S components are about equal. (After Boardman *et al.*, 1966; reproduced with permission from *J. mol. Biol.*)

examined in the analytical centrifuge. When a *cytoplasmic extract*, obtained according to Fig. 6 is analysed in the ultracentrifuge, Schlieren patterns such as shown in Fig. 10 are obtained. When compared to the macromolecular composition of a *chloroplast supernatant* (Fig. 5), the *cytoplasmic extract* is found to contain a larger ratio of 4 S to 18 S components, and about 80% of the ribosomes are of the 80 S variety. The small amount of Fraction I protein together with the slight amount of 70 S ribosomes in the *cytoplasmic extract*

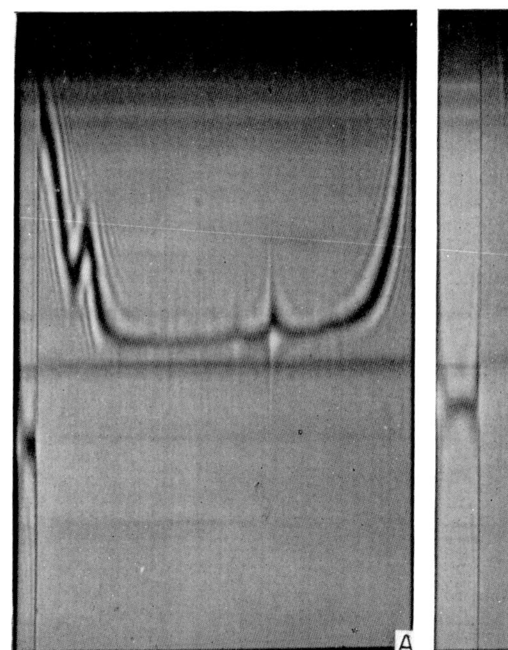


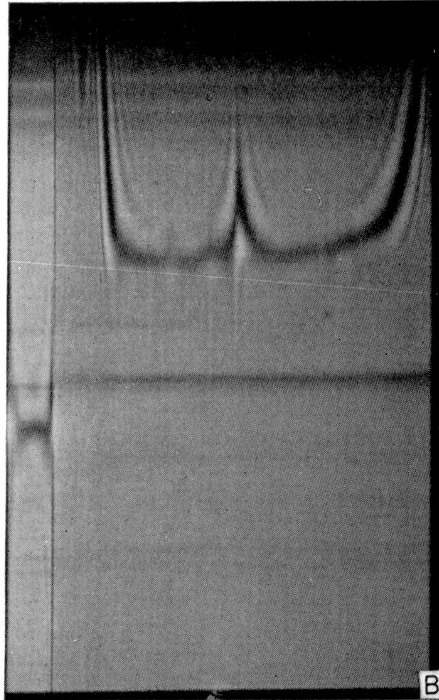


FIG. 10. Macromolecular composition of a *cytoplasmic extract* resolved by the analytical centrifuge. A, from left to right, 4 S, 18 S Fraction I protein, 70 S (minute),and 80 S ribosomes. B, magnification of 70 and 80 S components.

may have resulted from inadvertent fragmentation of chloroplasts during homogenization. Possibly, the presence of these components in the *cytoplasmic extract* is also the result of their having been removed from the chloroplasts by segmentation of mobile phase protuberances and consequently appearing in an organelle fraction different from the chloroplasts.

There is a very striking difference in the specific activity of the 80 S cytoplasmic ribosomes as compared to the 70 S chloroplast ribosomes. This difference is illustrated by the data in Table I, which show that the cytoplasmic ribosomes have only 1/10–1/20 the activity of the chloroplast ribosomes contained in a *chloroplast supernatant*. The 80 S ribosomes exhibit properties similar to the chloroplast ribosomes in regard to dependencies such as ATP, etc., and sensitivity to RNase. The 80 S ribosomes differ markedly in their $Mg^{2+}$ requirement for maximum incorporating activity. The cytoplasmic ribosomes require less than 5 mM $Mg^{2+}$ compared to 15 mM for the 70 S ribosomes. They also differ in their great resistance to dissociation in low $Mg^{2+}$; only after prolonged dialysis

Table I. An average picture of the concentration and incorporating activity of chloroplast and cytoplasmic ribosomes extracted from leaves (modified from Boardman *et al.*, 1966).

| | Chloroplast supernatant | Cytoplasmic extract |
|---|---|---|
| 70 S ribosome concentration, mg/ml | 0·11 | 0·03 |
| 80 S ribosome concentration, mg/ml | $<$0·01 | 0·13 |
| $^{14}C$ valine incorporation cpm/ml | 5320 | 272 |
| cpm/mg 70 S ribosomes | 48,000 | 1773 |
| cpm/mg 80 S ribosomes | $<$ 100 | 2080 |
| $\mu\mu$moles/mg ribosomes (70 S + 80 S) | 240 | 19 |

with the aid of EDTA to remove $Mg^{2+}$ will the 80 S ribosomes dissociate into 35 S and 56 S subunits. The 56 S subunit derived from dissociation of 80 S ribosomes can be resolved in the analytical centrifuge from the 50 S subunit derived from the 70 S ribosomes. The smaller 35 S subunits are not distinguished in the analytical centrifuge. Restoration of $Mg^{2+}$ does not result in the cytoplasmic ribosome subunits reconstituting into discrete 80 S particles, but rather they form into still larger aggregates. The morphology of the 80 S cytoplasmic ribosomes is shown by the electron micrograph in Fig. 11. When compared to the chloroplast ribosomes (Fig. 7), the cytoplasmic variety exhibit little evidence of medial clefts. The ribosomes have the gross appearance of acorns. A more or less spherical or ellipsoidal subunit seems to be associated with a larger cap-like unit. The average diameters of the 80 S ribosomes is not significantly different from 70 S ribosomes.

In contrast to the loss in activity experienced by 70 S mobile phase ribosomes when they are purified by pelleting, the 80 S ribosomes may double in activity by the same treatment. The reason for the increase in activity is as obscure as the reason for the loss in activity of 70 S ribosomes by the same procedure.

A similar picture of ribosome composition and location in relation to organelles is found in the extensive analyses of *Euglena* (*e.g.* Brawerman and Eisenstadt, 1964). We have also identified both species of ribosomes in extracts of *Chlorella* cells.

## SITES OF PROTEIN SYNTHESIS IN LIVING LEAF TISSUE

Isotopically labeled precursors of proteins have been extensively used for investigating the sites of protein synthesis in living leaf cells. In general, these experiments involve supplying whole plants, leaves, or portions of leaves with the precursor, and then homogenizing the tissue and separating the various organelle fractions prior to analysis for isotope contained in the proteins of the organelles. It would take us too far afield to examine and analyse this extensive literature. However, it is of interest to compare a few of these findings with living tissue with the results found for the cell-free protein synthesizing systems.

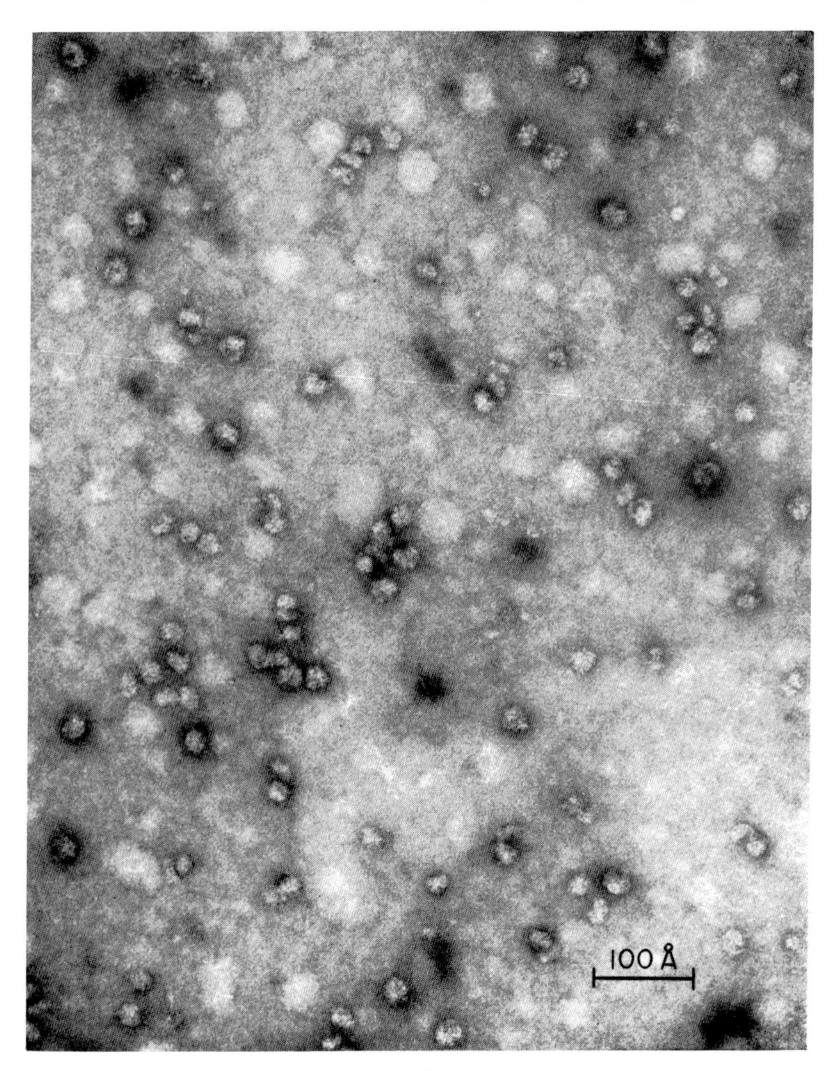


FIG. 11. Appearance of 80 S, cytoplasmic ribosomes after negative staining and electron microscopy. Many of the particles resemble acorns with a spherical subunit partly surrounded by a cap-like subunit. (After Karlsson *et al.*, 1966; reproduced with permission from *J. mol. Biol.*)

Stephenson *et al.* (1956) supplied leaf disks with [$^{14}$C]leucine and found a rapid uptake and incorporation of the amino acid into organelle fractions and soluble proteins. Fractionation of a homogenate, in the case of leaves also infected with tobacco mosaic virus, showed the highest specific activity in a pellet presumably enriched in cytoplasmic ribosomes. Somewhat lower values were found for a pellet enriched by mitochondria, and slightly lower specific activities were found in the chloroplast fraction.

Parthier (1963, 1964) supplied $^{14}CO_2$ to living leaf tissue for a brief period of photosynthesis. After a dark interval extending from 30 min to 24 hr wherein internal biosynthesis of [$^{14}$C]amino acids occurred, the tissue was homogenized and separated into organelle fractions which were analysed for radioactivity incorporated into proteins. The main conclusions were: (1) All organelle fractions were able to incorporate the photosynthetically formed amino acids into protein; (2) The proteins in the mitochondria fraction contained the highest specific activity, and the cytoplasmic fraction the lowest. The amino acids also appeared in the proteins of the chloroplasts, and those with the highest specific activity were structural proteins, the soluble proteins of the chloroplast having a lower activity.

In the field of chloroplastology, each investigator believes that his media and methods of homogenization and fractionation of organelles are the best available, and I am no exception. Consequently, with the collaboration of Dr. K. Jang we were curious to see how the distribution of amino acid incorporated into protein would appear after using our methods for organelle fractionation. Because of the great difference in specific activity of the 70 S and 80 S ribosomes in the cell-free systems, we hoped to see whether a difference could also be detected in the living cells. In unpublished experiments, [$^{14}$C]-valine was supplied to very small, thoroughly washed, pieces of leaf lamina which were also illuminated by 500 fc. of fluorescent light. Incorporation of the amino acid into protein was very rapid, 10 min of presentation of the amino acid to the living cells being sufficient to detect a significant incorporation into protein compared to zero time controls. The distribution of the amino acid incorporated into protein in relation to organelle fractions was investigated after 10 and 30 min presentation periods. The results were highly reproducible from one experiment to another. Less than 5% of the total amino acid incorporated into protein was found in the nuclei. The remaining 95% of the amino acid was distributed about equally between the chloroplasts and the *cytoplasmic extract* and the ratio did not change from 10 to 30 min. In the chloroplasts, less than 3% of the amino acid was incorporated into the structural proteins of the stationary components; 97% was extracted with the macromolecular constituents of the mobile phase. When the latter were resolved on sucrose columns by density gradient centrifugation, about 80% of the amino acid incorporated into protein was found in the 4–18 S region of the gradient and only 20% was associated with the 70 S ribosomes after a 10 min presentation time. When presentation was increased to 30 min, the total level of radioactivity associated with the ribosomes did not increase appreciably, but the amount in the 4–18 S soluble proteins increased more than ten-fold. Virtually the same pattern of labeling appeared in the *cytoplasmic extract*. The 80 S ribosomes became labeled to a low level and remained so with increasing time while the 4–18 S proteins became more radioactive as the time of presentation lengthened. Attempts to alter the pools of [$^{12}$C]amino acids in the living leaf cells and thereby label the ribosomes more intensely were unsuccessful and precluded the possibility of obtaining a reliable estimate of the rates of protein synthesis by the two species of ribosomes in the living condition. It seems clear, however, that the ribosomes do function in protein synthesis by living cells as anticipated, and that the soluble proteins of the mobile phase incorporate amino acids much earlier than the structural proteins of the stationary components.

## Ribonucleic Acid Synthesis by Cell-Free Extracts of Leaves

Homogenates of leaves have been shown to contain enzyme systems which catalyse the incorporation of radioactive triphosphate nucleotides into ribonucleic acid. The experimental results of Bandurski and Maheshwari (1962), Kirk (1964) and Semal *et al.* (1964) are in essential agreement in regard to the properties of the system and its gross location in relation to the organelles. The only significant incorporating systems in leaf extracts so far encountered are dependent upon DNA, and maximum synthesis requires the simultaneous presence of ATP, UTP, GTP, and CTP. In one case, the product of incorporation was shown to have the unequivocal properties of RNA (Moyer *et al.*, 1964). The systems are strongly inhibited by DNase, Actinomycin D, and RNase. The synthesized product is rendered acid-soluble by RNase, prolonged exposure to NaOH at room temperature, but is unaffected by DNase. Thus, the properties of the leaf system correspond to the DNA-directed, RNA polymerase synthesizing systems isolated from bacteria and other sources.

The problem of the location of the RNA polymerase system in relation to the organelles has acquired particular interest because of the findings that chloroplasts contain a DNA that is different in physical properties and composition from DNA extracted from leaf nuclei (Chun *et al.*, 1963; Kirk, 1963). Virtually all the polymerase activity of leaf extracts is associated with the organelle fraction which is composed mainly of nuclei and chloroplasts. The question being asked is whether the nuclear DNA, the chloroplast DNA, or both DNA's are functional in directing the synthesis of RNA in leaf extracts. The question has not been answered with the degree of confidence that I would like, but the weight of the experimental evidence points to the chloroplasts as mainly responsible for the cell-free synthesis of RNA. In some unpublished experiments, it appears that a more precise answer to the question is not easily forthcoming. Most of the polymerase activity in the chloroplast-nuclear fraction can be transferred into a *chloroplast supernatant*, and superficially it might appear that the polymerase is located in the mobile phase of chloroplasts. Unfortunately, the procedure for extracting the mobile phase also swells and disperses the contents of the nuclei and presumably they also appear in the *chloroplast supernatant*. This is not so much of a problem in the case of locating the protein-synthesizing systems of chloroplasts because the nuclei do not appear to participate in protein synthesis, either with the short time experiments with the living cells, nor in cell-free extracts. As the matter now stands, it seems that most of the activity can be tentatively associated with the mobile phase of chloroplasts, with the implication that a unique chloroplast DNA provides a template for synthesis of RNA.

ACKNOWLEDGEMENT

My interest in the chloroplast began with reading Chibnall's inspiring book, *Protein Metabolism in the Plant*. In the intervening 25 years, I have had many talented collaborators whose experimental work and ideas have greatly influenced my views —perhaps not enough. I hasten to add that they may not endorse the speculations I have presented. Those that have made their influence felt are: Drs. Andre Jagendorf, Morris Cohen, David Goodchild, Albert Kahn and Robert Dorner. More recently, Drs. Tasani Hongladarom and Shigeru Honda have shown me what can be seen with a phase microscope. Drs. Donald Spencer and N. K. Boardman educated me in biochemistry, at their own laboratory in Australia and together with Drs. R. I. B. Francki, J. Semal, Y. T. Kim and K. Jang produced a great deal of information on protein and RNA synthesis.

I am also grateful for the continued financial support received as research grants from the U.S. Atomic Energy Commission, the U.S. Public Health Service, and the U.S. National Science Foundation.

## References

App, A. A. and Jagendorf, A. T. (1963). *Biochim. biophys. Acta* **76**, 286.
App, A. A. and Jagendorf, A. T. (1964). *Pl. Physiol.* **39**, 772.
Bandurski, R. and Maheshwari, S. C. (1962). *Pl. Physiol.* **37**, 556.
Biswas, S. and Biswas, B. B. (1965). *Experientia* **21**, 251.
Boardman, N. K., Francki, R. I. B. and Wildman, S. G. (1965). *Biochemistry* **4**, 872.
Boardman, N. K., Francki, R. I. B. and Wildman, S. G. (1966). *J. mol. Biol.* **17**, 470.
Brawerman, G. and Eisenstadt, J. M. (1964). *J. mol. Biol.* **10**, 403.
Clark, M. F., Matthews, R. E. F. and Ralph, R. K. (1963). *Biochem. Biophys. Res. Commun.* **13**, 505.
Clark, M. F., Matthews, R. E. F. and Ralph, R. K. (1964). *Biochim. biophys. Acta* **91**, 289.
Chun, E. H. L., Vaughan, M. H. and Rich A. (1963). *J. mol. Biol.* **7**, 130.
Dorner, R. W., Kahn, A. and Wildman, S. G. (1957). *J. biol. Chem.* **229**, 945.
Francki, R. I. B., Boardman, N. K. and Wildman, S. G. (1965). *Biochemistry* **4**, 865.
Goodwin, T. W. ed. (1965). "Chemistry and Biochemistry of Plant Pigments", p. 143. Academic Press, London and New York.
Heber, U. (1962). *Nature, Lond.* **195**, 91.
Honda, S. I., Hongladarom, T. and Wildman, S. G. (1964). *In* "Primitive Motile Systems in Cell Biology" (R. D. Allan and N. Kamiya, eds.), p. 485. Academic Press, New York and London.
Honda, S. I., Hongladarom, T. and Laties, G. G. (1966). *J. exp. Bot.* **17**, 460.
Hongaldarom, T. (1964). Doctoral Dissertation, University of California, Los Angeles, U.S.A.
Jakoby, W. B., Brummond, D. O. and Ochoa, S. (1956). *J. biol. Chem.* **218**, 811.
Karlsson, U., Miller, A. and Boardman, N. K. (1966). *J. mol. Biol.* **17**, 487.
Kirk, J. T. O. (1963). *Biochim. biophys. Acta* **76**, 417.
Kirk, J. T. O. (1964). *Biochem. Biophys. Res. Commun.* **14**, 393.
Lyttleton, J. W. (1960). *Biochem. J.* **74**, 82.
Lyttleton, J. W. (1962). *Expl. Cell Res.* **26**, 312.
Lyttleton, J. W. and T'so, P. O. P. (1958). *Archs Biochem. Biophys.* **73**, 120.
Moyer, R., Smith, R., Semal, J. and Kim, Y. (1964). *Biochim. biophys. Acta* **91**, 217.
Mendiola, L. and Akazawa, T. (1964). *Biochemistry* **3**, 174.
Odintsova, M. S., Golubeva, E. V. and Sissakian, N. M. (1964). *Nature, Lond.* **204**, 1090.
Parthier, B. (1963). *Biochim. biophys. Acta* **72**, 505.
Parthier, B. (1964). *Z. Naturf.* **19**, 235.
Park, R. B. and Pon, N. G. (1960). *J. mol. Biol.* **3**, 1.
Possingham, J. V., Vesk, M. and Mercer, F. V. (1964). *J. Ultrastr. Res.* **11**, 68.

Semal, J., Spencer, D., Kim, Y. and Wildman, S. G. (1964). *Biochim. biophys. Acta* **91**, 205.
Sissakian, N. M., Filippovich, I. I. and Svetailo, E. N. (1962). *Doklady Akad. Nauk SSSR,* **147**, 488.
Sissakian, N. M., Filippovich, I. I. Svetailo, E. N. and Aliyev, K. A. (1965). *Biochim. biophys. Acta* **95**, 474.
Spencer, D. (1965). *Archs Biochem. Biophys.* **111**, 381.
Spencer, D. and Unt, H. (1965). *Aust. J. biol. Sci.* **18**, 197.
Spencer, D. and Wildman, S. G. (1962). *Aust. J. biol. Sci.* **15**, 599.
Spencer, D. and Wildman, S. G. (1964). *Biochemistry* **3**, 954.
Stephenson, M. L., Thimann, K. V. and Zamecnik, P. C. (1956). *Archs Biochem. Biophys.* **65**, 194.
Stocking, C. R. (1956). *Science, N.Y.* **123**, 1032.
Taylor, M. M. and Storck, R. (1964). *Proc. natn. Acad. Sci. U.S.A.* **52**, 958.
Trown, P. W. (1965). *Biochemistry* **4**, 908.
T'so, P. O. P., Bonner, J. and Vinograd, J. (1958). *Biochim. biophys. Acta* **30**, 570.
Van Noort, G. and Wildman, S. G. (1964). *Biochim. biophys. Acta* **90**, 309.
Weissbach, A., Horecker, B. L. and Hurwitz, J. (1956). *J. biol. Chem.* **218**, 795.
Wildman, S. G., Hongladarom, T. and Honda, S. I. (1962). *Science, N.Y.* **138**, 434.

# DNA Synthesis in Chloroplasts

A. Gibor[1]

*The Rockefeller University, New York, U.S.A.*

## Introduction

The genetic continuity of chloroplasts was indicated by analysis of cases of maternal inheritance, variegation and interspecies crosses in higher plants and by direct observations of multiplication of mutant plastids in lower plants. The extensive literature on this subject was reviewed by Granick in 1961.

More recent work reviewed by Gibor and Granick (1964) has shown that plastids contain DNA and RNA; such work suggests that the molecular basis for these genetic properties is similar to that established for nuclear genes. In general, the perpetuation of a genetic system requires the presence of the following components: 1. DNA; 2. RNA; 3. Enzymes for DNA synthesis; 4. Enzymes for RNA synthesis; 5. Enzymes for protein synthesis.

Evidence has now accumulated for the presence of most of these components in chloroplasts (Gibor, 1965).

Previous evidence that DNA was synthesized in the plastids themselves and not derived from the nucleus was indirect. It was based on the demonstration that the plastids of *Euglena* could be mutated by u.v. irradiation of the cytoplasm even though the nucleus was shielded from irradiation. Had the DNA of the plastids originated from the nucleus, then the unirradiated nucleus should have been capable of "curing" the irradiated cytoplasm. Since this did not occur it was inferred that the DNA of the chloroplasts was not derived from the nucleus.

Here we wish to present direct evidence that synthesis of DNA can occur in the chloroplasts in the absence of the nucleus. This evidence is based on the growing of enucleated *Acetabularia* cell fragments in the presence of $^{14}CO_2$ and subsequently isolating [$^{14}$C]-DNA from the chloroplasts of such cell fragments.

## Methods and Results

### cultivation and labeling of *Acetabularia*

*Acetabularia mediterranea* cultures, free from bacteria, were grown as previously described (Gibor and Izawa, 1963). Cells were harvested when they were about 2 cm long and before cap formation. At this stage of development the cells contained a single nucleus located in the basal portion of the cell. Enucleation was achieved by cutting off the basal third of the cells. The enucleated cells were collected into Petri dishes (5 cm. dia.). The collection and enucleation of the cells were carried out in a transfer room under a glass shield and precautions were taken to maintain sterile procedures throughout. To

[1] This work was supported by a grant from the U.S. National Science Foundation.

avoid possible cross contamination, the cells of different culture flasks were not pooled but each flask was treated separately.

To minimize the available $CO_2$ in the medium before exposure to $^{14}CO_2$, the culture medium was prepared as follows: 450 ml of sea water + 50 ml $H_2O$ were acidified to pH 3 with HCl then 10 mg $NaH_2PO_4$, 50 mg $KNO_3$ and 5 ml soil extract were added and the solution autoclaved in a flask fitted with a glass stopper. After cooling, sterile *tris* buffer (1·0 M, pH 8·2) was added to a final concentration of 0·02 M. About 8 ml of this sterile solution was dispensed

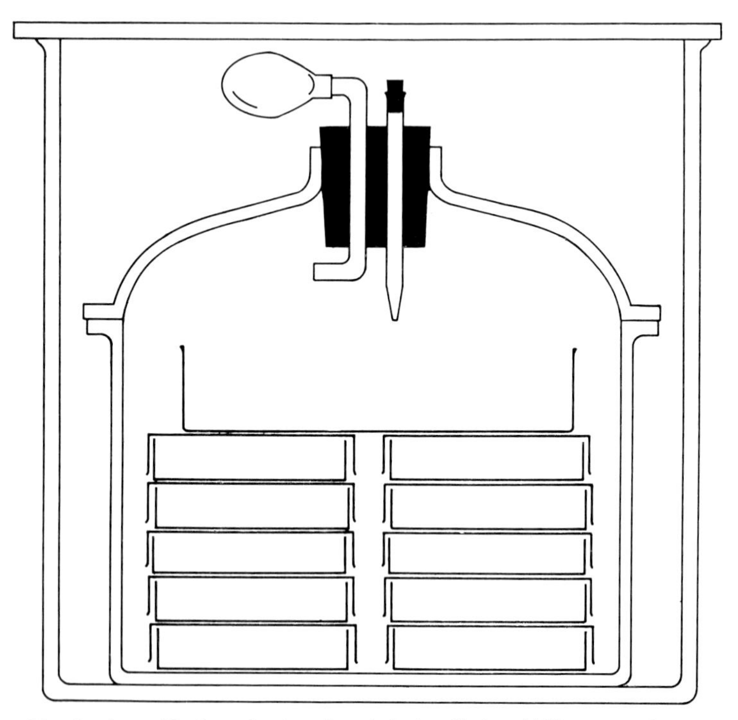

FIG. 1. Assembly for culturing *Acetabularia* cells in a $^{14}CO_2$ atmosphere.

into Petri dishes and the dishes were stored over solid KOH in a closed desiccator until needed.

Several thousand cells were collected into twelve dishes, and these were placed into a glass desiccator provided with a 2-hole rubber stopper. Exposure to $^{14}CO_2$ was accomplished by placing an open dish containing 4 mc $K_2{}^{14}CO_3$ (sp. act. 20 mc/$\mu$g) dissolved in 2 ml of 0·01 M KOH directly below the stopper. To one hole of the stopper was attached a rubber balloon to compensate for pressure changes inside the closed chamber. The other hole was plugged with an ampule stopper through which acid was injected into the open dish. The entire assembly was placed in an airtight glass chamber which contained KOH to trap any escaping $CO_2$ (Fig. 1). This culture assembly was then placed under continuous light provided by four G.E. cool-white "power-groove" lamps.

The light intensity reaching the inside of the desiccator was about 500 ft candles. The temperature inside the culture chamber varied between 20 and 25°.

After 9 days growth the experiment was terminated. The gas phase in the desiccator was flushed with room air, the unfixed radioactivity being trapped in KOH solution. The Petri dishes were removed and samples taken from each dish for microscopic examination. Only dishes in which no contaminating organisms were found were used for further analysis. The contents of the dishes were pooled and the cells washed with several changes of sterile sea water before fractionation.

## CHLOROPLAST ISOLATION

The washed cells were suspended in a solution composed of 0·4 M sucrose, 0·05 M KCl and 0·1 M *tris* buffer, pH 8·2. The cells were homogenized in this solution with a loose-fitting all-glass homogenizer. The homogenate was filtered through several layers of cheese cloth, then layered over a layer of 2·5 M sucrose in centrifuge tubes and centrifuged 30 min at 2000 ***g***. The green interphase on top of the heavy sucrose layer was collected, suspended in the buffered 0·4 M sucrose-KCl solution, and centrifuged at 2000 ***g*** for 10 min. The green chloroplast pellet so obtained was used for DNA isolation.

## DNA ISOLATION

The procedure for DNA isolation was essentially that described by Luck and Reich (1964) which in turn was an elaboration on the phenol procedure of Saito and Mirua (1963).

The chloroplasts were suspended in 5 v/v of a solution of 0·15 M NaCl, 0·1 M *tris* buffer, pH 8·6 and 1% sodium dodecylsulfate. The suspension was frozen and thawed once and then shaken for 10 min on a mechanical wrist shaker. An equal volume of phenol, pre-saturated with the same buffered salt solution, was added and the shaking continued for 30 min. The phases were separated and the phenol and interphase layers were re-extracted twice with 1 ml portions of 0·15 M NaCl. The combined aqueous phase was extracted three times with ether to remove all dissolved phenol. Traces of ether were removed by bubbling nitrogen through the solution. The aqueous solution was dialysed overnight against several liters of 0·15 M NaCl. The dialysed material was treated with RNase by adding 10 $\mu$g of enzyme per ml (Worthington, P-free enzyme was used. A solution of 1 mg/ml was prepared and boiled for 10 min before use) and *tris* buffer to 0·02 M, pH 7·8. The enzymic reaction was run at 30° for 3 hr. The entire solution was then returned to a dialysis bag and dialysed overnight. The dialysed material was concentrated by negative pressure dialysis to less than 1 ml. Then CsCl was added to a final concentration of 1·31 gm/ml, and *tris* buffer (pH 7·9) to a final concentration of 0·02 M. The final density of the CsCl solution was adjusted to 1·714 gm/ml. This solution was centrifuged in an SW-39 rotor of a Spinco model LII ultracentrifuge for 72 hr at 33,000 rev/min. The density gradient thus formed was fractionated by collecting consecutive drops from a pin-hole punctured in the bottom of the centrifuge tube. In all, thirty-two samples were collected. To each sample were added 50 $\mu$l of a *tris* buffer 0·1 M pH 7·8 and its u.v. absorption was measured on a Beckman D.U. spectrophotometer equipped with a microcuvette adapter. The optical density at 260 m$\mu$ and the difference in O.D. between 260 and 240 m$\mu$ of the different samples are presented in Fig. 2.

Samples 17–21 contain u.v. absorbing material with absorption characteristics of DNA.

For measurements of radioactivity in the different samples aliquots were counted directly in a scintillation counter. The distribution of radioactivity is plotted in Fig. 3. A peak of radioactivity was found to correspond to the peak

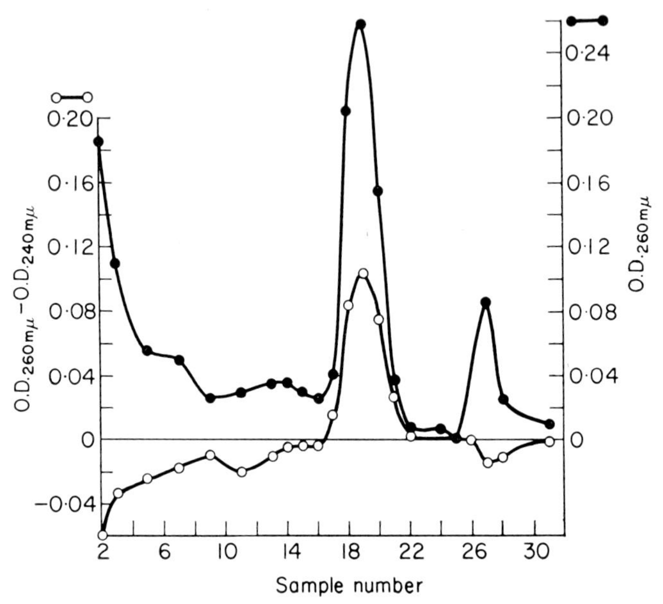


FIG. 2. U.v.-absorption of fractions from a DNA preparation which was fractionated in a CsCl density gradient.

of u.v. absorption in samples 17–21. Samples with peak radioactivities were pooled as follows: 1–10, 18–20, and 25–28. The pooled samples were dialysed and the u.v. absorption curves for the dialysed pooled samples measured (Fig. 4). Only sample 18–20 had an absorption curve typical of nucleic acids.

To characterize further the DNA of sample 18–20, it was analysed by a density gradient centrifugation in an analytical ultracentrifuge as described by Schildkraut *et al.* (1962). We are grateful to Drs. D. Luck and E. Reich for performing this analysis for us. The micro-densitometer tracing of a u.v. photograph of the centrifuged CsCl gradient is reproduced in Fig. 5. The DNA sample appeared as uniform band with a buoyant density of 1·695 gm./ml. This buoyant density is identical with the density of the satellite DNA obtained from *Chlamydomonas* and *Chlorella* (Chun *et al.*, 1963); the satellite DNA of these organisms was thought to be derived from their chloroplasts, since DNA

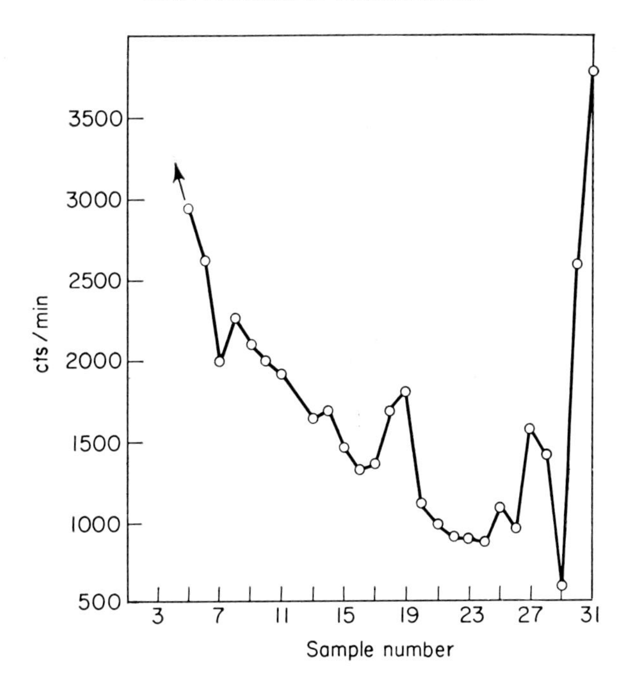


FIG. 3. Distribution of radioactivity in fractions of the same CsCl gradient as in Fig. 2.

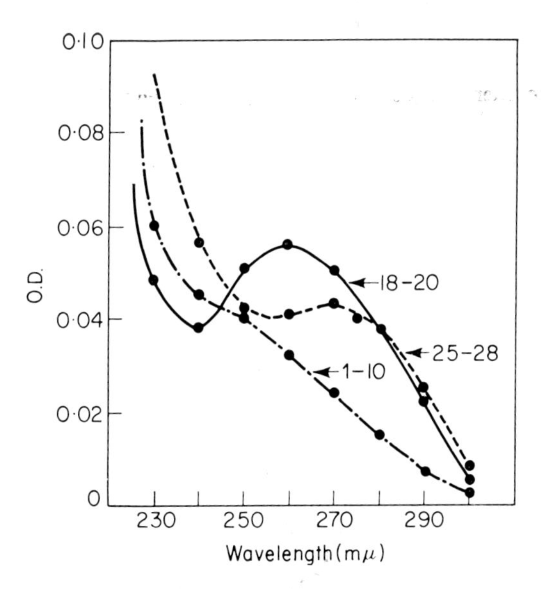


FIG. 4. U.v.-absorption spectra of pooled fractions from the same CsCl gradient as in Fig. 2. The numbers over the curves identify the fractions of the gradient which were pooled together.

extracted from isolated chloroplasts of these organisms were proportionately richer in the satellite-type DNA.

To verify that the radioactivity found in samples 18–20 was due to labeling of the DNA molecules, the DNA was hydrolysed enzymatically, chromatographed on paper, and the distribution of activity on the paper localized by radioautography, The following procedure was followed: the volume of the sample was reduced by negative pressure dialysis to 0·2 ml. An enzyme mixture was prepared containing 1·6 mg DNase I, 1 mg snake venom phosphodiesterase, 1 mg *E. coli* phosphatase, and 10 mg $MgCl_2$, all dissolved in 1 ml of 0·005 M *tris* buffer pH 8·4. The enzyme solution (50 $\mu$l) was added to the DNA sample and the mixture incubated at 37° for 4 hr. Then 10 $\mu$g of each of the deoxynucleosides were added to the reaction mixture. The solution was boiled

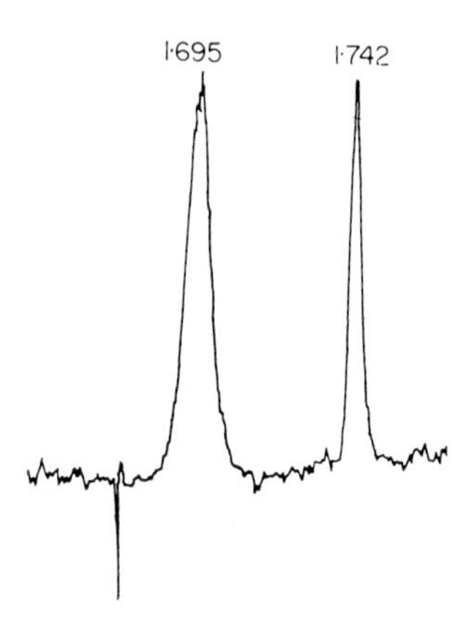


FIG. 5. Densitometer tracing of u.v.-absorbing photograph of the DNA of fractions 18–20 from the CsCl gradient of Fig. 2. The reference DNA has a density of 1·742 g/ml, the symmetrical band of the DNA of the chloroplasts has a density of 1·695 g/ml.

for several minutes and centrifuged. The precipitate was washed with water containing again 10 $\mu$g of each of the deoxynucleosides. The combined supernatant solution was applied to a strip of Whatman No. 1 paper for chromatography. The ascending chromatogram was developed with n-butanol-containing 1% ammonia. The u.v. absorbing spots were marked after drying the paper and the paper was then left in contact with X-ray film for 1 week. A photograph of a developed X-ray film superimposed over the paper chromatograph is presented in Fig. 6. It is apparent from this figure that deoxynucleosides, especially deoxyadenosine and deoxyguanosine were radioactive. Most of the radioactivity remained however at the origin; this probably represented the non-hydrolysed core-DNA. To determine whether the bases themselves were also labeled all the radioactive spots were eluted from the paper and the nucleosides were hydrolysed with 88% formic acid at 175° in a sealed tube for 30 min. The hydrolysate was chromatographed again in the same solvent system. Radioactive spots which corresponded to adenine and guanine were readily seen in these chromatograms.

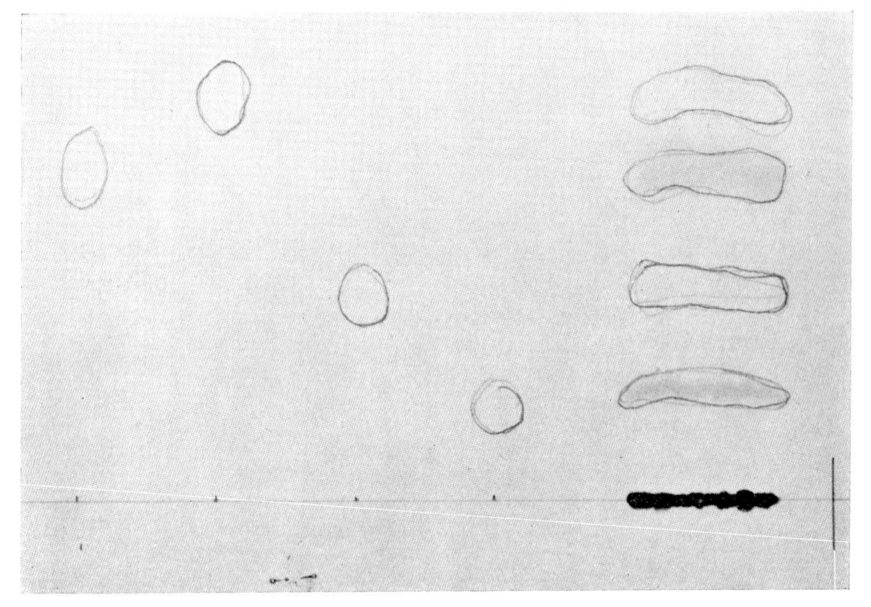

FIG. 6. Radioautograph of the chromatogram on which the deoxynucleosides of the hydrolysed DNA were separated. Reference spots from left to right are of deoxyadenosine, thymidine, deoxycytosine and deoxyguanosine.

## DISCUSSION

It is apparent that enzyme systems for the synthesis of deoxynucleosides and nucleotides and their polymerization to DNA are present in the cytoplasm of *Acetabularia* cells. Whether all these enzyme systems are localized within the chloroplasts themselves is not yet established.

Also unanswered is the intriguing question whether these enzymes are different from those responsible for nuclear DNA synthesis. It is possible that the enzymes are coded for by the DNA of the chloroplasts but it is also possible that the enzymes are synthesized via a nuclear message.

We are also intrigued by the apparent higher labeling of the purine deoxynucleosides of the enucleated cells. It is well established from many genetic studies on higher plants that the nucleus possesses many genes which control the development of chloroplasts. Thus despite their possession of a separate genetic apparatus, the chloroplasts are only semi-autonomous structures of the cell. The regulatory mechanisms by which the nucleus controls the development and multiplication of the chloroplasts is the next exciting problem before us.

## SUMMARY

Bacteria-free *Acetabularia* cells from which the nuclei were removed were grown in $^{14}CO_2$, then their chloroplasts were separated and DNA was isolated and purified from these chloroplasts. The purified DNA contained radioactive carbon. The radioactivity in this DNA was due to the labeling of the deoxy-nucleosides, especially deoxyadenosine and deoxyguanosine. It is concluded that synthesis of deoxynucleosides, their phosphorylation and polymerization can proceed in the absence of the nucleus.

## References

Chun, E. H. L., Vaughan, M. H. and Rich, A. (1963). *J. mol. Biol.* **7**, 130.
Granick, S. (1961). *In* "The Cell" (J. Brachet and A. E. Mirsky, eds.) Vol. II, p. 490. Academic Press, New York and London.
Gibor, A. (1965). *Am. Nat.* **99**, 229.
Gibor, A. and Granick, S. (1964). *Science, N.Y.* **145**, 890.
Gibor, A. and Izawa, M. (1963). *Proc. natn. Acad. Sci. U.S.A.* **50**, 1164.
Luck, D. J. and Reich, E. (1964). *Proc. natn. Acad. Sci. U.S.A.* **52**, 931.
Saito, H. and Mirua, K. (1963). *Biochim. biophys. Acta* **72**, 619.
Schildkraut, C. L., Marmur, J. and Doty, P. (1962). *J. mol. Biol.* **4**, 430.

## Note Added in Proof

Recent experiments indicated that all four deoxynucleosides are synthesized in the absence of the nucleus.

# Chloroplasts and Virus-RNA Synthesis

J. M. Bové, Colette Bové, Marie-José Rondot and G. Morel

*Service de Biochimie, IFAC, Station de Physiologie Végétale, Centre National de Recherches Agronomiques, Versailles, France*

## Introduction

When an infectious virus-RNA molecule penetrates a cell, it behaves as a messenger-RNA coding for the synthesis of the virus-protein as well as for the synthesis of one, or possibly two, enzymes necessary for the replication of the virus-RNA itself (Weissmann, 1965). According to this view, synthesis of the virus should be independent of the genetic material of the cell, namely the DNA in the nucleus. Indeed, Sänger and Knight (1963) have shown *in vivo*, with tobacco leaves, that TMV-RNA is still synthesized when the synthesis of cellular RNA is almost completely inhibited by actinomycin D. In the uninfected cell on the contrary, the synthesis of normal RNA occurs on a DNA-template.

We have investigated this problem by trying to isolate, from healthy and TYMV (turnip yellow mosaic virus)-infected chinese cabbage leaves, a subcellular preparation, active in the synthesis of RNA *in vitro*. We have found that a preparation containing essentially whole and broken chloroplasts, obtained by differential centrifugation, followed by a sucrose density-gradient centrifugation, incorporates $UM^{32}P$ from $UT^{32}P$ into an acid-insoluble product in the presence of ATP, CTP and GTP. This incorporation is inhibited up to 95% by DNase or actinomycin D when the preparation is obtained from *healthy* leaves, whereas active incorporation continues, in the presence of either one of these substances, when *infected* leaves are used to isolate the subcellular preparation. However, the purity of this preparation remains to be investigated further and will be considered in the discussion.

## Material and Methods

### Plant material

Healthy and TYMV-infected chinese cabbage plants (*Brassica chinensis*, var. Wong Bok) were grown in the greenhouse (Goffeau and Bové, 1964). Very young systemically infected leaves of the centre of the rosette, and corresponding leaves of healthy plants were used.

### Preparation of active fractions

Healthy or TYMV-infected leaves (10–20 g) were washed in ice-cold water and ground for 15 sec in a Virtis-45 homogenizer at full speed with twice their weight of ice-cold grinding buffer (*tris* [*tris* (hydroxymethyl) amino methane]-HCl pH 7·4, 0·05 M; $MgCl_2$, 0·008 M; β-mercaptoethylamine, 0·009 M). In some experiments, the grinding buffer contained 0·5 M sucrose. The homogenate was gently squeezed through four layers of cheese-cloth, discarding the solid material that starts to come through the cheese-cloth at the end of the squeezing action. The filtrate (approximately 35 ml for 15 g of leaves) was centrifuged for 5 min at 100 × ***g*** at 2°. The small pellet (0 P 100) was discarded. Centrifugation of the supernatant at 10,000 × ***g*** for 15 min yielded a pellet (100 $P_1$ 10,000) that was resuspended in 20 ml of pH 9·0 buffer (*tris*-HCl pH 9·0, 0·05 M; $MgCl_2$, 0·008 M; β-mercaptoethylamine, 0·009 M) and centrifuged again at 10,000 × ***g*** for 15 min. In some experiments, the pH 9·0 buffer contained 0·5 M sucrose. The washed pellet (100 $P_2$ 10,000) was resuspended in pH 9·0 buffer (2 ml of buffer per 12 ml of initial filtrate). The 100 $P_2$ 10,000 preparation contained most of the activity and was used either as such or after sucrose density-gradient centrifugation (Kirk, 1963) in the following way. Four ml of 35% sucrose were layered on top of 12 ml of 45% sucrose already layered on top of 8 ml of 60% sucrose in the bottom of a tube fitting the SW 25 rotor of the Spinco centrifuge. Three ml of 100 $P_2$ 10,000, resuspended in pH 9·0 buffer (containing 0·5 M sucrose or not), were placed on top of the sucrose density-gradient and centrifuged for 60 min at 24,000 rev/min. in a Spinco model L centrifuge. Two clearly defined green zones and a small pellet were obtained. The upper zone (I) at the boundary of the 35–45% sucrose contained essentially broken chloroplasts as seen under the light microscope. The lower zone (III) at the 45–60% sucrose boundary, revealed whole chloroplasts. The pellet (P) contained starch and some chloroplasts, but no tests have been done to identify the nuclei in this pellet.

Zones I and III were collected with a syringe and a hypodermic needle (Kirk, 1963), diluted to twice their volume with pH 9·0 buffer and centrifuged for 15 min at 20,000 × ***g***. Pellets I and III, as well as pellet P, were resuspended with pH 9·0 buffer.

### INCORPORATION TEST

The complete reaction mixture contained: *tris*-HCl pH 9·0, 30 μmole; $MgCl_2$, 2 μmole; β-mercaptoethylamine, 2 μmole; $PO_4 K_x Y_y$ pH 8·0, 2 μmole; ATP, 0·5 μmole; CTP, 0·5 μmole; UTP-$^{32}$P, 0·02 to 0·12 μmole (500,000 to 1,000,000 cts/min per reaction mixture); PEP, 5 μmole; PEP-kinase, 20 μg; cell-free preparation, 0·20 ml containing 10 μmole *tris*-HCl pH 9·0, 1·6 μmole $MgCl_2$ and 1·8 μmole β-mercaptoethylamine. The total volume varied between 0·35 ml and 0·41 ml according to the experiment.

The reaction was carried out at 30° or 37° and allowed to proceed for 15 or 30 min, depending on the experiment. A zero-time control was always run. The reaction was stopped with 2·5 ml of 15% TCA (trichloroacetic acid) followed by the addition of 0·5 ml of a 0·1 M solution of potassium orthophosphate and potassium pyrophosphate. The precipitate was recovered by centrifugation, washed twice with 10% TCA-containing pyrophosphate and orthophosphate at a concentration of 0·1 M, resuspended in 0·5 ml of 50%

ethanol and plated for radioactivity determination. A FD-1 Tracerlab gas-flow counter with a Mylar window was used.

The cts/min/0·2 ml indicated in the results, represent the values obtained after deduction of the cts/min/0·2 ml of the zero-time control.

### PRODUCTS

ATP, CTP, GTP, UTP, ADP, CDP, GDP, PEP, PEP-kinase, DNase (DN-C, crystalline, from bovine pancreas) and RNase (Type 1 A, 5 × crystallized, protease free, from bovine pancreas), were Sigma products. Actinomycin D was a gift of Merck Sharp & Dohme Research Lab., UTP-$^{32}$P, labelled in the $\alpha$-phosphate, was obtained by preparing UM$^{32}$P by the method of Tener (1961) and by phosphorylating the UM$^{32}$P with an enzyme purified from *Escherichia coli* (Canellakis *et al.*, 1960). UM$^{32}$P and UT$^{32}$P were isolated from the respective reaction mixtures by column chromatography on Dowex-1 formate (Canellakis *et al.*, 1960). The final UT$^{32}$P was used as the free acid.

### ANALYTICAL DETERMINATION

Proteins were determined by the method of Lowry *et al.* (1951) and chlorophyll by the method of Arnon (1949).

## RESULTS

The general properties of the system used in this work will be described extensively elsewhere. Only relevant properties will be documented here.

### GENERAL REQUIREMENTS OF THE INCORPORATION SYSTEM

Table I shows that most of the activity was localized in the fraction that comes down between 100 and 10,000 × ***g*** (100 $P_1$ 10,000) and that washing of this fraction affected the activity only slightly (100 $P_2$ 10,000).

Table I. Location of activity. Effect of washing 100 P 10,000 (*V*) (E 101)

| | Cts/min/0·2 ml | Total cts/min |
|---|---|---|
| 0 P 100 (*V*) | 1809 | 8750 |
| 100 $P_1$ 10,000 (*V*) | 1557 | 37,275 |
| 100 $P_2$ 10,000 (*V*) | 1407 | 33,725 |
| S 10,000 (*V*) | 40 | 4800 |

Experimental conditions: 30 min incubation at 37° with complete reaction mixture. Zero-time controls: between 58 and 66 cts/min/0·2 ml. The complete reaction mixture (total volume: 0·35 ml) contained 0·08 $\mu$mole UT$^{32}$P (1,035,000 cts/min). Subcellular fractions were prepared from 13 g of virus-infected leaves.

Table II shows that all four nucleotide triphosphates were required for full activity, and Table III that only nucleotide triphosphates, but not nucleotide diphosphates, were used as substrates. Thus, no polynucleotide phosphorylase activity was present, as further illustrated by the fact that the presence of inorganic phosphate did not inhibit the reaction: on the contrary, a stimulation, as yet unexplained, was observed (Table III). It will be shown elsewhere that the incorporation reaction was time and temperature-dependent, and proportional to the amount of cell-free preparation added. Mg and Mn ions stimulated the incorporation. Finally, the optimum pH of the reaction was around 8·5, whereas grinding of the leaves had to be done at pH 7·0.

Table II. Requirement for all four nucleotide triphosphates (E 104)

| Reaction mixture | 100 $P_2$ 10,000 (*H*) cts/min/0·2 ml | 100 $P_2$ 10,000 (*V*) cts/min/0·2 ml |
|---|---|---|
| Complete | 1456 (100%) | 2019 (100%) |
| −ATP | 447 (31%) | 608 (30%) |
| −CTP | 951 (65%) | 839 (42%) |
| −GTP | 198 (14%) | 536 (27%) |
| −ATP−CTP−GTP | 139 (10%) | 256 (13%) |

Experimental conditions: 30 min incubation at 37°. Zero-time control: 82 cts/min/0·2 ml of 100 $P_2$ 10,000 (*H*); 73 cts/min/0·2 ml of 100 $P_2$ 10,000 (*V*). Both preparations contained 4·2 mg dry weight/0·2 ml. 100 $P_2$ 10,000 (*H*) contained 1·8 mg chlorophyll/ml; 100 $P_2$ 10,000 (*V*) contained 0·8 mg chlorophyll/ml. The complete reaction mixture contained 0·10 $\mu$mole of $UT^{32}P$ (952,000 cts/min) in 0·41 ml.

Table III. Requirement for nucleotide triphosphates. Absence of polynucleotide-phosphorylase activity (E 111)

| NTP ($\mu$moles) | NDP ($\mu$moles) | PEP+ PEPkinase | $PO_4^{-3}$ ($\mu$moles) | Cts/min/ 0·2 ml | % |
|---|---|---|---|---|---|
| 0·1[1] | − | + | − | 761 | 100 |
| 0·1 | − | − | − | 372 | 49 |
| − | − | + | − | 92 | 12 |
| − | − | − | − | 80 | 10 |
| 0·5[1] | − | + | − | 1218 | 160 |
| − | 0·1[2] | − | − | 147 | 19 |
| − | 0·1 | + | − | 562 | 74 |
| 0·1 | − | + | 2 | 1441 | 190 |

Experimental conditions: 15 min incubation at 37°. Zero-time control: 38 cts/min/0·2 ml. 100 $P_2$ 10,000 (*V*) contained 1·5 mg Protein/0·2 ml and was prepared from 15 g of virus-infected leaves. All reaction mixtures contained 0·03 $\mu$mole $UT^{32}P$ (340,000 cts/min) in 0·35 ml.

[1] 0·1 or 0·5 $\mu$mole of ATP, CTP and GTP.
[2] 0·1 $\mu$mole of ADP, CDP and GDP.

All these properties applied both to the preparation obtained from the healthy leaves [100 $P_2$ 10,000 *H* (healthy)] and to the one from the virus-infected leaves [100 $P_2$ 10,000 *V* (virus-infected)].

## EFFECT OF DNASE, RNASE AND ACTINOMYCIN D ON THE INCORPORATION REACTION

Whereas the preceding properties applied to both preparations, the effect of DNase, RNase and actinomycin D on the incorporation reaction revealed a fundamental difference between the incorporation obtained with 100 $P_2$ 10,000 *H* and that shown by 100 $P_2$ 10,000 *V*.

Table IV illustrates, for several experiments, that in the presence of DNase, the incorporation obtained with 100 $P_2$ 10,000 *H* was decreased to a level as low as 10% whereas, under the same condition, the incorporation achieved by the preparation from the virus-infected leaves (100 $P_2$ 10,000 *V*) remained at a level as high as 50%.

Figure 1 extends these data for various concentrations of DNase and Fig. 2 illustrates that, as expected, the same results were obtained when actinomycin was used instead of DNase.

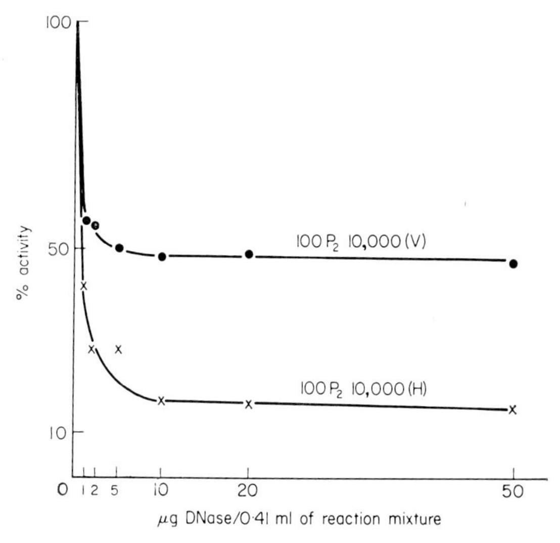


FIG. 1. Effect of DNase on RNA synthesis (DNase is present during incubation). Experimental conditions: 15 min incubation at 37°. The complete reaction mixture contained 0·06 μmole $UT^{32}P$ (640,000 cts/min). With 100 $P_2$ 10,000 *H*, 100% = 684 cts/min/0·2 ml or 456 cts/min/mg protein. With 100 $P_2$ 10,000 *V*, 100% = 1350 cts/min/0·2 ml or 794 cts/min/mg protein.

Finally, Fig. 3 shows that, in the presence of RNase, practically no incorporation occurred with 100 $P_2$ 10,000 *H*, whereas noticeable incorporation (20%) could still be found with 100 $P_2$ 10,000 *V*.

Table IV. Effect of DNase on RNA synthesis

| Exp. No. | Time of reaction (min) | DNase ($\mu$g) | 100 $P_2$ 10,000 ($H$) | | | | 100 $P_2$ 10,000 ($V$) | | | |
|---|---|---|---|---|---|---|---|---|---|---|
| | | | − DNase | | + DNase | | − DNase | | + DNase | |
| | | | cts/min/0·2 ml | % | cts/min/0·2 ml | % | cts/min/0·2 ml | % | cts/min/0·2 ml | % |
| 103 | 30 | 17 | 2630 | 100 | 247 | 9 | 2857 | 100 | 1442 | 50 |
| 104 | 30 | 17 | 1456 | 100 | 98 | 7 | 2019 | 100 | 900 | 45 |
| 106 | 60 | 17 | | | | | 2754 | 100 | 1247 | 45 |
| 108 | 30 | 17 | | | | | 1600 | 100 | 914 | 57 |
| 113 | 15 | 10 | 684 | 100 | 114 | 17 | 1350 | 100 | 638 | 48 |
| 115 | 20 | 10 | 928 | 100 | 105 | 11 | 1848 | 100 | 1065 | 57 |
| 116 | 20 | 10 | | | | | 1927 | 100 | 1131 | 58 |
| 117 | 20 | 10 | | | | | 2124 | 100 | 1185 | 56 |
| 118 | 20 | 10 | | | | | 802 | 100 | 410 | 51 |
| 120 | 20 | 10 | | | | | 1920 | 100 | 805 | 42 |
| 121 | 20 | 10 | 523 | 100 | 56 | 11 | 1105 | 100 | 489 | 44 |
| 122 | 20 | 10 | 356 | 100 | 48 | 13 | 815 | 100 | 423 | 53 |
| 123 | 20 | 10 | | | | | 1390 | 100 | 917 | 66 |
| 124 | 30 | 10 | | | | | 2049 | 100 | 1348 | 66 |
| | | | | | Average | 11 | | | | 53 |

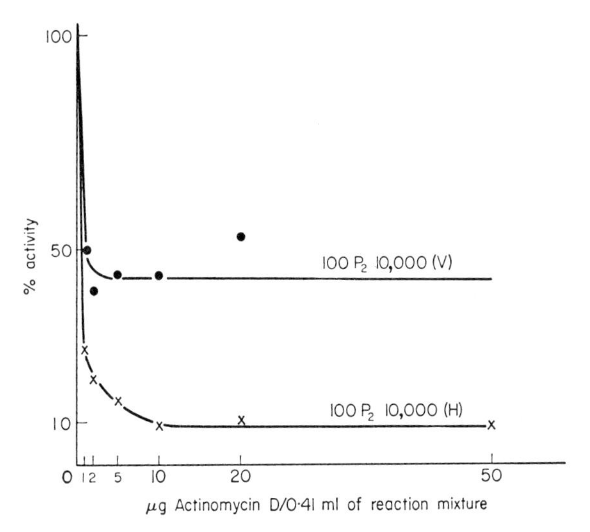


FIG. 2. Effect of actinomycin D on RNA synthesis (E 113) (Actinomycin D is present during incubation). Experimental conditions: Same as for Fig. 1.

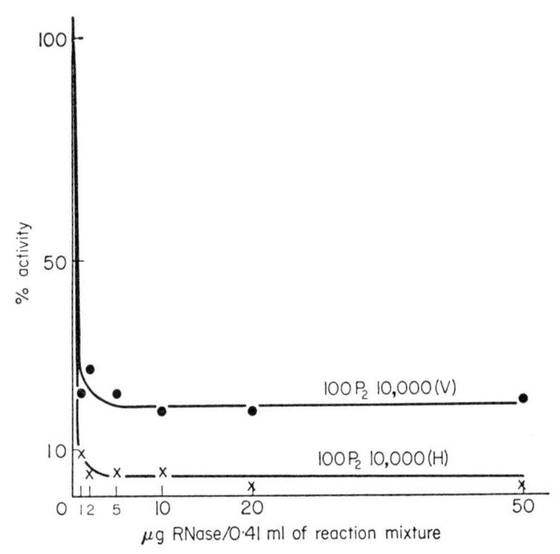


FIG. 3. Effect of RNase on RNA synthesis (RNase is present during incubation). Experimental conditions: Same as for Fig. 1.

## EFFECT OF RNASE ON THE SYNTHETIZED PRODUCT

Figure 4 shows that the product synthetized by 100 $P_2$ 10,000 *H* was entirely hydrolysed by RNase, whereas with 100 $P_2$ 10,000 *V* 87% of the material synthetized in the presence of DNase remained resistant to RNase.

It will be documented elsewhere that this RNase-resistant material becomes susceptible to RNase, after it has been heated several minutes at 108°.

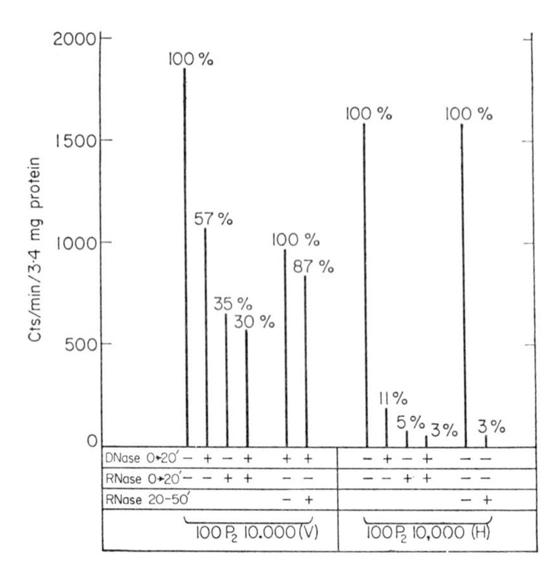


FIG. 4. Effect of RNase on synthesized product. Experimental conditions: 20 min incubation at 30°. The complete reaction mixture (0·40 ml) contained 0·12 μmole $UT^{32}P$ (770,000 cts/min). DNase and RNase are used at the concentration of 10 μg/0·4 ml reaction mixture where indicated. When present, DNase was added at time zero, and RNase either at time zero or at time 20 min as indicated. At time 20 min 1·0 μmole of "cold" UTP was added to the reaction mixtures that were further incubated for 30 more min.

## FRACTIONATION OF ACTIVITY BY SUCROSE DENSITY-GRADIENT CENTRIFUGATION

Centrifugation of 100 $P_2$ 10,000 *H* or 100 $P_2$ 10,000 *V* on a sucrose density gradient yielded an upper dark green zone (Zone I) containing practically only broken chloroplasts as far as could be seen in the light microscope, a lower dark green zone (Zone III) in which mainly whole chloroplasts but some broken chloroplasts could also be seen, and a small pellet in which only starch grains and some chloroplasts have so far been identified.

Table V illustrates the pattern of incorporation given by the two preparations, 100 $P_2$ 10,000 *H* and 100 $P_2$ 10,000 *V*, and by Zone I, Zone III and Pellet P obtained by centrifuging these preparations on a sucrose density-gradient. It can be seen that the only preparations that retained more than 50% of their activity in the presence of DNase were Zone I and III obtained from 100 $P_2$ 10,000 *V*. The activity of Zone I from 100 $P_2$ 10,000 *V*, weak but significant, was completely resistant to DNase, whereas Zone I from 100 $P_2$ 10,000 *H* was completely inactive in the presence or in the absence of DNase. Zone III was much more active and showed the same 50% resistance to DNase than the original 100 $P_2$ 10,000 *V* preparation. Finally, the pellet from 100 $P_2$ 10,000 *V*

Table V. Fractionation of activity by sucrose density-gradient centrifugation

| | | 100 $P_2$ 10,000 (*H*) | | | | | | 100 $P_2$ 10,000 (*V*) | | | | | |
|---|---|---|---|---|---|---|---|---|---|---|---|---|---|
| | | −DNase | | | +DNase | | | −DNase | | | +DNase | | |
| Fraction | Exp. No. | Cts/min/ 0·2 ml | Total cts/min | % | Cts/min/ 0·2 ml | Total cts/min | % | Cts/min/ 0·2 ml | Total cts/min | % | Cts/min/ 0·2 ml | Total cts/min | % |
| 100 $P_2$ 10,000 | 120 | | | | | | | 1558 | 15,580 | 100 | 667 | 6670 | 43 |
| 100 $P_2$ 10,000[1] | 122 | 356 | 5350 | 100 | 48 | 720 | 13 | 815 | 12,200 | 100 | 423 | 6345 | 53 |
| I | 120 | | | | | | | 112 | 560 | 100 | 96 | 480 | 86 |
| I | 122 | 0 | 0 | | 0 | 0 | | 202 | 860 | 100 | 198 | 840 | 98 |
| III | 120 | | | | | | | 930 | 4650 | 100 | 463 | 2315 | 50 |
| III | 122 | 253 | 1200 | 100 | 44 | 209 | 17 | 381 | 1810 | 100 | 187 | 890 | 49 |
| P | 120 | | | | | | | 817 | 4085 | 100 | 99 | 495 | 12 |
| P | 122 | 380 | 1425 | 100 | 70 | 262 | 18 | 889 | 3560 | 100 | 164 | 656 | 18 |

Experimental conditions: 30 min incubation at 30°. Zero time control: 40 cts/min/0·2 ml of 100 $P_2$ 10,000 (*H*) (E 122); 35 cts/min/0·2 ml of 100 $P_2$ 10,000 (*V*) (E 122); 57 cts/min/0·2 ml of 100 $P_2$ 10,000 (*V*) (E 120).

[1] In experiment 122, the grinding buffer, and the pH 9·0 buffer used to wash 100 $P_1$ 10,000, contained 0·5 M sucrose.

showed the same marked inhibition by DNase as the pellet from the 100 $P_2$ 10,000 *H*.

## Discussion

The product that is synthesized by 100 $P_2$ 10,000 *H* is most likely RNA, since all four nucleotide triphosphates are required for full activity and since it is highly susceptible to RNase. Because its synthesis is inhibited by DNase and actinomycin D, the template must be DNA. Thus, 100 $P_2$ 10,000 *H* contains DNA-dependent RNA-polymerase activity.

In the case of 100 $P_2$ 10,000 *V*, two products can be detected: the first showing all the properties of the RNA synthesized by a DNA-dependent RNA-polymerase (susceptibility to RNase and DNase), and a second product, the synthesis of which is not inhibited by DNase or actinomycin D, and which is almost entirely (87%) resistant to RNase unless it has been heated at 108°C. These properties are in agreement with the idea that this material could possess the double-stranded RNA-structure characteristic of the replicative form of virus-RNA (Burdon *et al.*, 1964; Ralph *et al.*, 1965; Shipp and Haselkorn, 1964).

Experiments are under way to characterize this material further.

Thus, it can be concluded that the 100 $P_2$ 10,000 *V* preparation contains a DNA-dependent RNA-polymerase-system also present in the normal leaves, plus an enzyme-system involved in the synthesis of a DNA-independent RNA. This second enzyme-system does not function in the preparation from healthy leaves (100 $P_2$ 10,000 *H*). The nature of this enzyme-system will not be discussed here in the context of the recent findings on the replication of bacteriophage-RNA (Weissman, 1965).

The results of this work support those obtained *in vivo* by Sänger and Knight (1963) with TMV-infected tobacco leaves, but they are not in agreement with those of Semal *et al.* (1964) who were unable to find a difference in sensitivity to DNase or actinomycin between the RNA-synthesis systems in cell-free extracts of healthy and TMV-infected tobacco leaves.

Most of the RNA-synthesizing activity is associated with the 100 $P_2$ 10,000 fraction, no matter whether it is prepared from normal or TYMV-infected leaves. The association of this activity with an insoluble fraction (mixture of chloroplasts and nuclei, essentially) is in agreement with the results of Semal *et al.* (1964). However, Astier-Manifacier and Cornuet (1965) have used the supernatant of a 15,000 **g** centrifugation in their investigations.

The sucrose density-gradient centrifugation experiments have made it possible to subdivide the 100 $P_2$ 10,000 preparations further into a fraction containing essentially broken chloroplasts (Zone I), a fraction containing mainly whole chloroplasts together with broken chloroplasts (Zone III) and a Pellet (P). Zone I and Zone III could also contain some mitochondria.

The pellet P is very active on a protein basis, and the chloroplasts that it still contains are too few to account for the RNA-synthesizing activity. This activity is markedly inhibited by DNase. It is thus very likely that the pellet represents the nuclear fraction, although no experiments have yet been made to stain the nuclei for microscopic examination. However, Kirk (1963) whose density-gradient centrifugation technique has been used here, has established the presence of nuclei in such a pellet.

The whole chloroplast-fraction from 100 $P_2$ 10,000 *V* contains most of the DNase-resistant RNA-synthesizing activity. Although this fraction is very rich in whole chloroplasts, it remains still impossible at the present time to associate the DNase-resistant RNA-synthesizing activity definitely with the chloroplasts. The whole chloroplast fraction might still be contaminated by some nuclear material which would be responsible for the activity. However, this nuclear material would be of a different type than that present in the pellet, since the RNA-synthesizing system in the pellet is much more inhibited by DNase (90%) than the one in the whole chloroplast fraction (50%). The same reasoning applies with even more force to the broken chloroplast fraction, where no inhibition by DNase could be found.

It is possible that during the grinding of the leaves the nuclei are broken. Most of the DNA-part of the nuclei could be centrifuged down in the pellet whereas the RNA-rich nucleoli, with more or less DNA attached to them, could be found in the chloroplast zones. The enzyme system responsible for the synthesis of DNase-resistant RNA might then be associated with the nucleolus. Reddi (1964) and Schlegel (1965) have pointed towards the role of nuclei in TMV-synthesis.

With respect to the DNA-dependent RNA-synthesizing activity present in the various fractions from both healthy and TYMV-infected leaves, it is also difficult, at the present time, to associate this activity with either the chloroplasts, the nuclei, or both. It is however relevant to note that Kirk (1964) has presented evidence for the occurrence of a DNA-dependent RNA polymerase in chloroplasts.

## References


Arnon, D. I. (1949). *Pl. Physiol.* **24**, 1.

Astier-Manifacier, S. and Cornuet, P. (1965). *Biochem. biophys. Res. Commun.* **18**, 283.

Burdon, R. H., Billeter, M. A., Weissmann, C., Warner, R. C., Ochoa, S. and Knight, C. A. (1964). *Proc. natn. Acad. Sci. U.S.A.* **52**, 768.

Canellakis, E. S., Gottesman, M. E. and Kammen, H. O. (1960). *Biochim. biophys. Acta* **39**, 82.

Goffeau, A. and Bové, J. M. (1964). *Physiol. Végét.* **2**, 75.

Kirk, J. T. O. (1963). *Biochim. biophys. Acta* **76**, 417.

Kirk, J. T. O. (1964). *Biochem. biophys. Res. Commun.* **16**, 233.

Lowry, O. H., Rosebraigh, N. J., Farr, A. L. and Randall, R. J. (1951). *J. biol. Chem.* **193**, 265.

Ralph, R. K., Matthews, R. E. F., Matus, A. I. and Mandel, H. G. (1965). *J. mol. Biol.* **11**, 202.

Reddi, K. K. (1964). *Proc. natn. Acad. Sci. U.S.A.* **52**, 397.

Sänger, H. L. and Knight, C. A. (1963). *Biochem. biophys. Res. Commun.* **13**, 455.

Schlegel, D. E. (1965). Conference on Plant Viruses, Symposium I, Wageningen, 5–9 July, 1965.

Semal, J., Spencer, D., Kim, Y. T., Moyer, R. H., Smith, R. A. and Wildman, S. G. (1964). *Virology* **24**, 155.

Shipp, W. and Haselkorn, R. (1964). *Proc. natn. Acad. Sci. U.S.A.* **52**, 401.

Tener, G. M. (1961). *J. Am. chem. Soc.* **83**, 159.

Weissmann, C. (1965). Conference on Plant Viruses, Symposium V, Wageningen, 5–9 July, 1965.

# Protein Synthesis in Chloroplasts and Chloroplast Ribosomes

Jerome M. Eisenstadt
*Yale University School of Medicine, New Haven, Connecticut, U.S.A.*

## Introduction

In recent years evidence has begun to accumulate suggesting that chloroplasts can be considered as semi-autonomous units with physiological activities different from their cellular environment. Isolated chloroplasts from algae and higher plants have been shown to contain both RNA (Brawerman, 1963) and DNA (Chun *et al.*, 1963; Sager and Ishida, 1963; Ray and Hanawalt, 1964; Brawerman and Eisenstadt, 1964a) with physical and chemical properties quite distinct from those found in the nucleus and cytoplasm.

In many instances growth in the absence of light results in colorless cells. Exposure of such cells to light produces a rapid adaptive response involving the synthesis of chlorophyll, various structural proteins of the chloroplasts, and the specific enzymes required for photosynthesis (Brawerman *et al.*, 1960). Chloroplast synthesis has been shown to be reduced by various culture conditions (App and Jagendorf, 1963) and is preferentially inhibited by inhibitors of protein synthesis (Pogo and Pogo, 1965). Exposure of cultures of *Euglena gracilis* to elevated temperatures, u.v. irradiation, or streptomycin under conditions having little effect on cell proliferation, results in irreversible bleaching of the cells (Pringsheim and Pringsheim, 1952; Provasoli *et al.*, 1948). These cells apparently have lost the system responsible for the self-replication of the chloroplasts. Micro-beam irradiation of the nucleus or the cytoplasm alone indicates that the irreversible mutation to bleached cells occurs only when the plastids are exposed (Gibor and Granick, 1962). This may be correlated with the observation that bleached cells no longer contain the chloroplast-associated DNA, and it implies that the DNA of the plastids did not originate from the nucleus.

The evidence cited suggests that chloroplasts possess intrinsic genetic and biochemical capabilities. The recent finding that chloroplasts contain ribosomes which, by the criteria of size, RNA content, and biochemical properties, are sharply different from cytoplasmic ribosomes has lent additional support to the concept that chloroplasts possess an independent protein synthesizing system (Eisenstadt and Brawerman, 1964a).

This paper will describe the characteristics of the protein synthesizing system isolated from the chloroplasts of *Euglena gracilis* and compare such a system to that found in the cytoplasm. The properties exhibited by these two systems will be discussed in an attempt to correlate their physiological processes with *in vitro* protein synthesis.

## The Protein Synthesizing System Isolated from Chloroplasts of *E. gracilis*

### Isolation of Chloroplasts

Crude chloroplast preparations isolated from cells of *E. gracilis* are heavily contaminated with cytoplasmic ribosomes and nuclear DNA. Nucleotide composition of the RNA of colorless and green cells indicated that a species of RNA rich in adenylic and uridylic acids is associated with the appearance of chloroplasts (Brawerman and Chargaff, 1959). The purification process, therefore, utilizes the composition of chloroplast RNA as a criterion of the efficiency of the removal of cytoplasmic contaminants. Chloroplast isolation is essentially a three-step process after the cells are broken in a protective medium containing 10% sucrose (Eisenstadt and Brawerman, 1964a). First, a low-speed centrifugation sediments the chloroplasts leaving the bulk of the cytoplasmic ribosomes and soluble materials necessary for protein synthesis in the supernatant fluid. The low speed pellet is resuspended and allowed to stand for 10 min in the cold. A light fluffy material aggregates and is removed by filtration through gauze. The chloroplasts are concentrated from the filtrate by centrifugation and mixed with a solution of sucrose having a density slightly higher than the chloroplasts. Centrifugation of this mixture results in the chloroplasts floating to the top as a dark green pellicle. Contaminating cytoplasmic material still present at this stage is to be found in the pellet. The RNA content of the intact chloroplasts purified in this manner is identical with that of the purified chloroplast ribosomes. The chloroplast purification procedure results in a substantial lessening of the RNA and the protein content of the preparation. This procedure can work only on cells grown under autotrophic conditions.

### Isolation of Chloroplast Ribosomes

The purified chloroplasts are lysed with sodium deoxycholate followed by centrifugation at 23,000 × ***g*** (Fig. 1). This treatment releases the chloroplast ribosomes. Most of the pigment and the protein of the chloroplasts are found in P-23. The chloroplast-associated DNA remains attached to the structure and is also found in this pellet. The 23,000 × ***g*** supernatant fluid is centrifuged at high speed to sediment the ribosomes. Most of the RNA associated with the chloroplasts is found in P-100. The pellet is resuspended in a small volume of buffer and centrifuged at low speed to remove any aggregated materials. Chloroplast ribosomes are found in both the pellet and the supernatant fluid of the final low-speed centrifugation. Ribosomes prepared in this way have an RNA content with a nucleotide composition different from the cytoplasmic ribosomal RNA (Brawerman and Eisenstadt, 1964b). The ratio of RNA to protein of 0·52–0·71 in chloroplast ribosomes may indicate contamination by non-ribosomal materials. Phenol-extracted RNA from purified chloroplast ribosomes shows two components with sedimentation values of approximately 19 S and 14 S. By contrast, cytoplasmic ribosomes show a single RNA species having a value of 19 S. The nucleotide composition of each of the 19 S peaks is distinctly different.

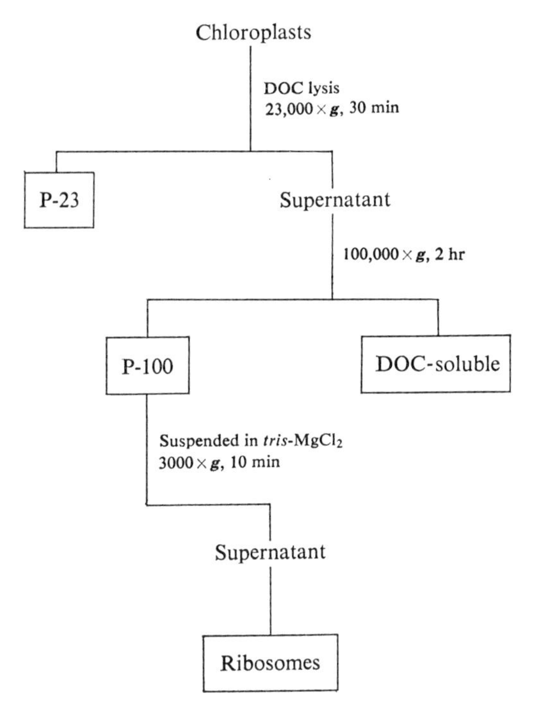


FIG. 1. Isolation of ribosomes from chloroplasts. [Data reproduced from Eisenstadt and Brawerman (1964a)].

## CHARACTERISTICS OF THE *in vitro* AMINO ACID INCORPORATION BY CHLOROPLAST PREPARATIONS

Purified chloroplasts are active in incorporating amino acids into hot, acid-insoluble material (Eisenstadt and Brawerman, 1963; 1964a). The cell-free reaction requires an energy-generating system, is dependent on additional transfer RNA, and the soluble enzymes found in the high-speed supernatant fluid (Eisenstadt and Brawerman, 1964b) (Table I). The rate of incorporation slows down considerably after 15 min, and little, if any, activity remains after 30 min. The chloroplasts are but slightly stimulated by the addition of *Euglena* template RNA, although polyuridylic acid produces a marked stimulation of phenylalanine incorporation. Puromycin and ribonuclease strongly inhibit the incorporating activity. Illumination of the chloroplasts also inhibits the incorporating activity, although these chloroplast preparations apparently are capable of photophosphorylation (Eisenstadt and Brawerman, 1964a). The requirements for transfer RNA and soluble supernatant materials suggest that these chloroplasts may have a damaged permeability barrier. Examination by light microscopy does not indicate any gross damage. Chloroplasts appear similar in size and shape to those seen within the living cell.

Chloroplasts exhibit a small but reproducible inhibition of incorporation

Table I. Characteristics of incorporation of amino acids into *Euglena* chloroplasts

| Conditions | Amino acid incorporated leucine | phenylalanine |
|---|---|---|
| Complete system | 430 | 192 |
| Minus S-RNA | 52 | 127 |
| Minus ATP system | 86 | |
| Plus puromycin (110 μg/ml) | 65 | |
| Plus RNase (50 μg/ml) | 6 | |
| Plus template RNA (100 μg/ml) | 580 | |
| Plus poly U (120 μg/ml) | 403 | 397 |

Values expressed as μμmoles amino acid incorporated/mg chloroplast RNA. [Data reproduced from Eisenstadt and Brawerman (1964a)].

in the presence of actinomycin (Table II). The addition of the other three nucleoside triphosphates shows a weak stimulation of amino acid incorporation which is abolished in the presence of actinomycin. A slight inhibition, similar to that occurring as a result of actinomycin addition, is also produced in the presence of deoxyribonuclease. These results suggest that some DNA-dependent RNA synthesis may occur in the chloroplast preparations. It is not yet known whether the RNA polymerase or the DNA primer is intrinsic to the chloroplasts.

Table II. Effect of ribonucleoside triphosphate and actinomycin on incorporation of leucine into the protein of *Euglena* chloroplasts

| Conditions | No actinomycin | Actinomycin added |
|---|---|---|
| Complete system | 472 | 384 |
| Plus UTP, GTP, CTP | 523 | 410 |
| Plus DNase | 370 | |

Additions per ml of reaction mixture: DNase, 20 μg/ml; UTP, GTP, CTP each added at a concentration of 0·8 μmole/ml. Amount of chloroplasts in incubation mixtures equivalent to 36 μg/ml of chloroplast RNA. Values expressed as μμmoles amino acid incorporated/mg of chloroplast RNA. [Data reproduced from Eisenstadt and Brawerman (1964a)].

Chloroplast RNA polymerase activity has been established in the broad bean (Kirk, 1964). Isolated chloroplasts from *Acetabularia* also exhibit amino acid incorporation into proteins (Goffeau and Brachet, 1965). The incorporation is insensitive to the addition of streptomycin, deoxyribonuclease, and, in contrast to *Euglena* chloroplasts, is insensitive to ribonuclease. Actinomycin completely inhibits incorporation of amino acids, possibly indicating that, in this case, synthesis of proteins is dependent on chloroplast DNA. The apparent insensitivity to the nucleases suggests that these chloroplasts are impermeable to macromolecules.

## CHARACTERISTICS OF AMINO ACID INCORPORATION USING CHLOROPLAST RIBOSOMES

Ribosomes isolated from chloroplasts exhibit an amino acid incorporating activity appreciably lower than that of the isolated chloroplasts or cytoplasmic ribosomes. The chloroplasts and cytoplasmic ribosomes have similar specific incorporation activities (Tables I and III). Because of the excessively large protein to RNA ratio in the chloroplasts, and since most of the RNA associated with chloroplasts sediments with the ribosomes, incorporation activities are

Table III. Incorporation of leucine into *Euglena* ribosomes

| Conditions | Chloroplast ribosomes | Cytoplasmic ribosomes |
|---|---|---|
| Complete system | 43 | 425 |
| Minus supernatant, S-RNA | 2 | 8 |
| Plus ribosomal RNA (215 μg/ml) | 66 | |
| Plus template RNA (100 μg/ml) | 212 | 423 |
| Plus puromycin (40 μg/ml) | 58 | |
| Plus actinomycin (56 μg/ml) | 209 | |

Values expressed as μμmoles leucine incorporated/mg ribosomal RNA. [Data reproduced from Eisenstadt and Brawerman (1964a)].

expressed in terms of RNA content. Addition of template RNA isolated from *Euglena* results in a marked stimulation of the chloroplast ribosomes. The stimulation is inhibited by puromycin while actinomycin has little effect. Addition of template RNA to the cytoplasmic ribosomes or isolated chloroplasts also produces little effect. Treatment of cytoplasmic ribosomes with deoxycholate in a manner similar to that used to isolate chloroplast ribosomes does not make them less refractory to template RNA. A slight stimulation of cytoplasmic ribosomes by template RNA can be demonstrated only when ribosomes and supernatant fractions are pre-incubated (Eisenstadt and Brawerman, 1964a). Stimulation of chloroplast ribosomes is a specific effect because the addition of ribosomal RNA, not stimulatory in an *Escherichia coli* test system, is relatively inactive. The fact that chloroplast ribosomes require template RNA, as cytoplasmic ribosomes do not, may indicate a basic physiological difference with respect to the binding of messenger RNA. Isolated chloroplasts are unaffected by the addition of messenger RNA. This suggests the isolated chloroplasts contain sufficient messenger RNA for maximal activity. The lack of effect is probably not due to permeability problems since the macromolecules in the supernatant fraction are necessary for the activity of the chloroplasts.

The deoxycholate treatment used to prepare chloroplast ribosomes depletes the particles of messenger RNA. This may mean that the messenger RNA is rather loosely bound to the ribosomes. Such an interpretation is substantiated by the isolation from the deoxycholate-soluble fraction of RNA with a specific template activity three to four times that of RNA isolated from the ribosomes. The cytoplasmic ribosomes have a strong affinity for template RNA which is not dissociated by deoxycholate treatment.

Sissakian *et al.* (1965) have isolated from pea seedlings chloroplast ribosomes active in amino acid incorporation. The incorporation is inhibited by chloramphenicol, puromycin, and actinomycin, but is not significantly inhibited by ribonuclease.

## THE EFFECT OF CHLORAMPHENICOL

The biochemical differences between the cytoplasmic and chloroplast ribosomes can be further illustrated by examining their responses to chloramphenicol (Table IV). Cytoplasmic ribosomes are unaffected by concentrations of chloramphenicol greatly in excess of those which completely block protein synthesis in bacteria. The higher concentrations produce a marked inhibition of the chloroplasts. The chloroplast ribosomes also show a definite sensitivity to chloramphenicol. This could be correlated with the presence of template

Table IV. Effect of chloramphenicol on incorporation of leucine by the chloroplast and cytoplasmic cell-free systems

| Chloramphenicol μg/ml | Chloroplasts | Ribosomes from chloroplasts | | | Ribosomes from cytoplasm |
|---|---|---|---|---|---|
| 60 | 99 | 99 | 98 | 99 | 99 |
| 120 | 84 | 74 | 82 | 88 | 103 |
| 210 | 55 | 61 | 78 | 87 | 94 |
| 360 | | 59 | 68 | 74 | 94 |
| Template RNA added μg/ml | 0 | 0 | 40 | 120 | 0 |

Values expressed as percentage of controls without chloramphenicol. Values with template RNA are percentages of controls in presence of some concentration of template RNA. Cytoplasmic ribosomes were treated with DOC. [Data reproduced from Eisenstadt and Brawerman (1964a)].

RNA. Higher levels of template RNA decrease the effect of the antibiotic. Chloramphenicol seems to affect the interaction between the ribosome and the messenger RNA. If this were the mechanism by which chloramphenicol inhibits protein synthesis, ribosomes which bond template RNA weakly would be affected more readily by the antibiotic (Weisberger *et al.*, 1963). The apparent resistance of cytoplasmic ribosomes to chloramphenicol could be related to their high affinity for template RNA as previously discussed. Chloramphenicol preferentially inhibits *Euglena* chloroplast synthesis *in vivo* and confirms the results obtained in the cell-free experiments.

Cytoplasmic ribosomes have some striking similarities to those derived from mammalian systems. They are refractory to deoxycholate treatment, insensitive to chloramphenicol, and are very poorly stimulated by template RNA. By contrast, chloroplast ribosomes are similar to those obtained from bacterial cells. The latter types can easily be depleted of template RNA and, therefore, can be highly stimulated by the addition of such RNA. Both chloroplast and bacterial ribosomes are also sensitive to chloramphenicol.

Such similar physiological properties may be related in that both bacterial and chloroplast replications are extremely sensitive to environmental conditions. Bacterial cells respond to changing conditions by exceedingly rapid and sensitive control mechanisms of protein synthesis. Chloroplasts respond to light in a like manner.

## SEDIMENTATION CHARACTERISTICS OF ACTIVE RIBOSOMES

Zone centrifugation of purified cytoplasmic ribosomes yields a single peak with a sedimentation value of approximately 70 S. Centrifugation of the reaction mixture after amino acid incorporation indicates that the major radioactivity peak coincides with the 70 S ribosomal peak (Fig. 2). Two heavier, minor radioactive components are also present.

Chloroplast ribosomes sediment more slowly than the cytoplasmic ribosomes. The major radioactivity peak from the incubation mixture also has a sedimentation peak lower than 70 S.

Sedimentation over a longer period of time resolves the ribosomal component into a 60 S peak which contains a substantial portion of the incorporated radioactivity. The smaller size of the chloroplast ribosome has been confirmed by electron microscopy of thin sections of *Euglena* cells (S. Dales, unpublished). A 62 S chloroplast ribosome component has been isolated from pea seedlings (Sissakian *et al.*, 1965).

## SIGNIFICANCE OF AN INDEPENDENT PROTEIN-SYNTHESIZING SYSTEM IN CHLOROPLASTS

The above results indicate that the isolated chloroplasts of *Euglena gracilis* are capable of incorporating amino acids into proteins and that this protein-synthesizing machinery is different from that found in the cytoplasm. The kinetics of incorporation, as well as the requirement for an energy source, eliminate the possibility that unbroken cells contribute to the observed activity. The association of specific proteins with the plastid structure raises the question of the physiological mechanism responsible for the localization of these proteins. The existence in chloroplasts of distinct species of DNA, RNA, and ribosomes strongly suggests that plastids may be considered as semi-autonomous units possessing all the components essential for the specification of the amino acid sequence in protein. Other mechanisms could be responsible for the localization of chloroplast proteins at the site of synthesis in the structure.

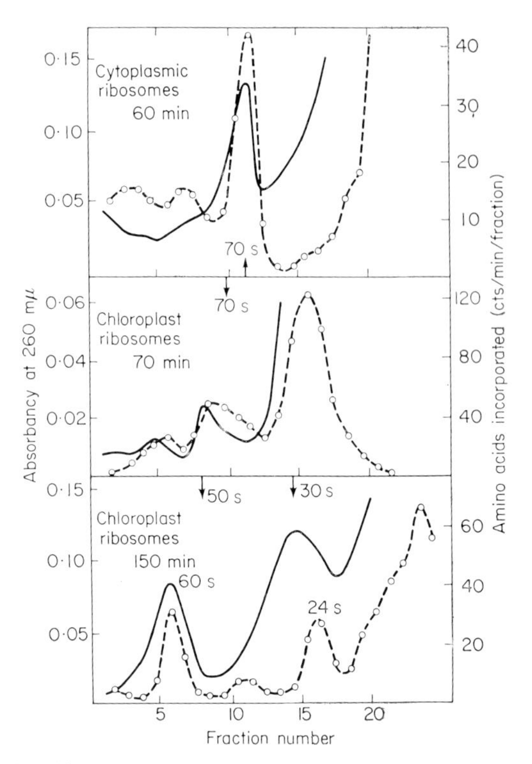


FIG. 2. Zone centrifugation of incubation mixtures with cytoplasmic and chloroplast ribosomes. Duration of centrifugation is indicated in each figure. Solid lines represent absorbancy at 260 mμ and dashed lines hot acid-insoluble radioactivity. 70 S, 50 S, and 30 S positions obtained by using *Escherichia coli* ribosomes as a standard. [Data derived from Eisenstadt and Brawerman (1964a).]

If the chloroplast ribosomes determine the amino acid sequence in proteins, only chloroplast proteins would be synthesized. The possibility has been eliminated by the observation that chloroplast ribosomes from *Euglena* can synthesize the coat protein of $f_2$ coliphage when incubated with $f_2$ RNA (Schwartz *et al.*, 1965). The messenger RNA, therefore, determines the amino acid sequence of the protein and establishes that the ribosomes do not play an active role in such a determination.

The active selection of messenger RNA by the ribosomes could also insure the localization of plastid proteins within the structure. By this mechanism only chloroplast-specific messenger RNA would be recognized and utilized for protein synthesis by the ribosomes. Comparison of the relative abilities of *E. coli* and *Euglena* chloroplast ribosomes to be stimulated by RNA's isolated from several sources failed to indicate any specificity. Neither of the ribosomal preparations could ascertain differences among the various RNA's, including the RNA from the homologous organism.

A simple working hypothesis would be that cytoplasmic messenger RNA is prevented from entering the plastids by a permeability barrier. Each plastid contains enough DNA for the specification of chloroplast proteins. Messenger RNA would be synthesized within the chloroplast and utilized by the protein synthesizing machinery in the formation of proteins. This is confirmed by the observation that treatments which result in permanent bleaching of *Euglena gracilis* cultures can be correlated with the loss of chloroplast DNA from the cells (Ray and Hanawalt, 1965). As suggested by Gibor and Granick (1964), nuclear control of chloroplast development could take place by either activators or inhibitors acting on the chloroplast genome.

## References

App, A. A. and Jagendorf, A. T. (1963). *J. Protozool.* **10**, 340.
Brawerman, G. (1963). *Biochim. biophys. Acta* **72**, 317.
Brawerman, G. and Chargaff, E. (1959). *Biochim. biophys. Acta* **31**, 172.
Brawerman, G. and Eisenstadt, J. M. (1964a). *Biochim. biophys. Acta* **91**, 477.
Brawerman, G. and Eisenstadt, J. M. (1964b). *J. mol. Biol.* **10**, 403.
Brawerman, G., Konigsberg, N. and Chargaff, E. (1960). *Biochim. biophys. Acta* **43**, 364.
Chun, E. H. L., Vaughan, N. H., Jr. and Rich, A. (1963). *J. mol. Biol.* **7**, 130.
Eisenstadt, J. M. and Brawerman, G. (1963). *Biochim. biophys. Acta* **73**, 319.
Eisenstadt, J. M. and Brawerman, G. (1964a). *J. mol. Biol.* **10**, 392.
Eisenstadt, J. M. and Brawerman, G. (1964b). *Biochim. biophys. Acta* **80**, 493.
Gibor, A. and Granick, S. (1962). *J. biophys. biochem. Cytol.* **15**, 599.
Gibor, A. and Granick, S. (1964). *Science, N.Y.* **145**, 890.
Goffeau, A. and Brachet, J. (1965). *Biochim. biophys. Acta*, **95**, 302.
Kirk, J. T. O. (1964). *Biochem. biophys. Res. Commun.* **14**, 393.
Pogo, B. G. T. and Pogo, A. O. (1965). *J. Protozool.* **12**. 96.
Pringsheim, E. G. and Pringsheim, O. (1952). *New Phytolog.* **51**, 65.
Provasoli, L., Hutner, S. and Shatz, A. (1948). *Proc. Soc. exp. Biol. Med.* **69**, 279.
Ray, D. S. and Hanawalt, P. C. (1964). *J. mol. Biol.* **9**, 812.
Ray, D. S. and Hanawalt, P. C. (1965). *J. mol. Biol.* **11**, 760.
Sager, R. and Ishida, M. R. (1963). *Proc. natn. Acad. Sci. U.S.A.* **50**, 725.
Schwartz, J. H., Eisenstadt, J. M., Brawerman, G. and Zinder, N. D. (1965). *Proc. natn. Acad. Sci. U.S.A.* **53**, 195.
Sissakian, N. M., Filippovich, I. I., Svetailo, E. N. and Aliyev, K. A. (1965). *Biochim. biophys. Acta* **95**, 474.
Weisberger, A. D., Armentrout, S. and Wolfe, S. (1963). *Proc. natn. Acad. Sci. U.S.A.* **50**, 86.

# A Light Effect on Amino Acid and Nucleotide Incorporating Systems Derived from Higher Plants

Rusty J. Mans

*Botany Department, University of Maryland,*
*College Park, Maryland, U.S.A.*

## Introduction

Sundry morphological, physiological and biochemical events are initiated with irradiation of etiolated plant tissues (Butler, 1964). The onset of a series of rapid responses of dark grown tissue to light, present the experimentalist with an excellent probe to investigate many regulatory modes of biosynthetic activity (Price *et al.*, 1964; Gordon, 1964). In our investigations on the mechanism of protein biosynthesis in cell-free systems from maize, we, too, have observed a "light effect". A preliminary report of this observation has been published (Williams and Novelli, 1964).

We have observed an enhancement in the ability of ribosomes isolated from irradiated corn seedlings to incorporate amino acids as compared to the activity of ribosomes prepared from etiolated seedlings. This phenomenon reflects, we think, a fundamental metabolic response of the tissue to light; i.e. a light-induced stimulus to protein biosynthesis. The changes in amino acid incorporating activity are correlated with changes in the RNA composition of the tissue. This reinforces our supposition that the response to light which we are measuring is intermeshed with the systems regulating the biosynthetic activities of the plant, particularly nucleic acid synthesis.

It is my purpose to describe in some detail the experimental system in which the phenomenon is observed; to elaborate on the nature of the light response as measured in the cell-free system; and finally, to indicate our approach to understanding the light-induced stimulus to protein synthesis via RNA synthesis.

## Amino Acid Incorporating System

Implicit in our interpretation of the light effect as an enhancement of protein synthetic activity, is the assumption that the amino acid incorporation we detect is a measure of protein synthesis. This assumption requires experimental support, especially when the amino acid incorporating systems are derived from higher plants (see App and Jagendorf, 1964; Lett *et al.*, 1963).

The cell-free amino acid incorporating system derived from corn seedlings (Mans and Novelli, 1961a) is in many respects analogous to the classical cell-free system from rat liver (see Zamecnik, 1960). Aside from general considerations regarding the unity of nature (Baldwin, 1940) the striking similarity between the two systems can be accounted for in two ways. (1) Both systems are obtained from organized tissues composed of nucleated and highly differentiated cells. (2) The experimental approach; i.e. the isolation and characterization of the active components, was patterned after those employed with the liver system.

## ACTIVE COMPONENTS

### *Isolation*

A resume of techniques employed in the preparation of active cellular components from corn seedlings (Mans and Novelli, 1964a) is presented in Fig. 1. Essentially two modifications have been introduced. The first involves germination of the hybrid waxy maize seed in perforated plastic pails (see Huang and Bonner, 1962) instead of on trays. After inhibition the seed is germinated for 64 hr under a shower at 23°. Considerably more material can be germinated and uniformity in shoot length is retained as compared with germination on trays. The second modification involves harvesting. The shoots are broken off at the first internode and immediately frozen in liquid nitrogen rather than held in ice-cold medium. A powder is prepared by grinding 10 to 100 g of frozen shoots under liquid nitrogen. The powder is thawed in a medium containing bentonite, and homogenized in a loose fitting glass homogenizer or, alternatively, passed through a French Pressure Cell. Water-washed microsomal and deoxycholate-washed ribosomal particles and the high speed supernatant as well as polysomes (Wettstein *et al.*, 1963) can be prepared by this technique. The components are two- to five-fold more active and much larger quantities of plant tissue can be processed via the liquid nitrogen freezing technique.

### *Characterization*

The particulate components are characterized as microsomes, ribosomes, or polysomes on the basis of the method of isolation. The ribosomes isolated by treatment of the 100,000 $\times$ ***g*** pellet with deoxycholate consist almost exclusively of ribonucleoprotein particles with sedimentation constants of 71 S and 95 S. The 71 S component predominates. The polysome preparations show the characteristic distribution upon sucrose density gradient centrifugation (see Fig. 6). Reference to the water-washed particles as microsomes (Fig. 1) may be a misnomer. Examination of a preparation (highly active in amino acid incorporation) in the analytical centrifuge reveals a population of ribonucleoprotein particles with discrete sedimentation values of 96 S, 121 S, 132 S, 220 S, 266 S and 305 S (see Fig. 2). The major component is 132 S, and were it a polysome, corresponds to a trimer of 70 S monomers (see Wettstein *et al.*, 1963). Electron micrographs of the same preparation show aggregates of two to ten particles. The physical characteristics and high level of activity of the water-washed particles suggest that they are polysomes.

The high speed supernatant component is replete with the enzymic activities required for the formation of the aminoacyl RNAs and the transfer of the amino acid residue to the particulate components (Mans *et al.*, 1964). The level of soluble RNA in the supernatant, although low (0·1 to 0·3 mg/ml), is sufficient to support maximum levels of amino acid incorporation obtained with the corn system. The occurrence of a DNA-dependent RNA polymerase

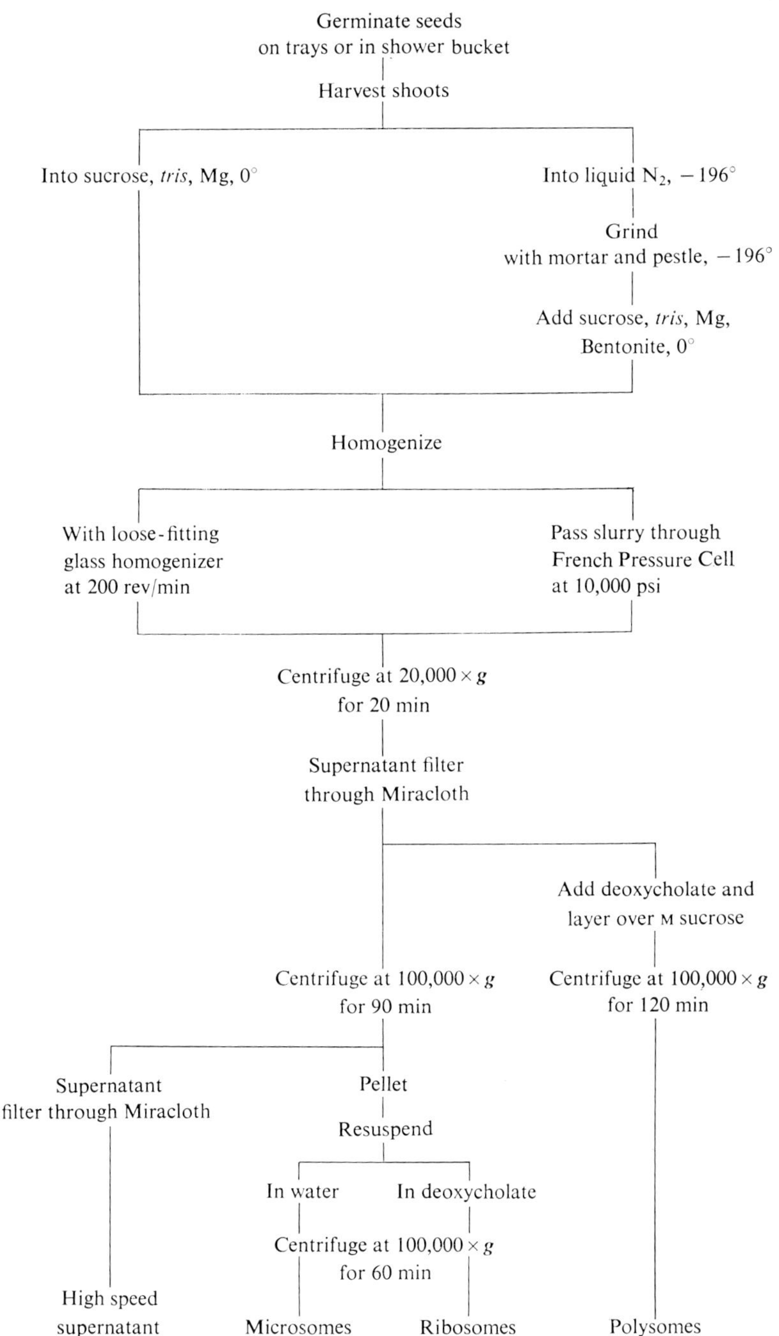


FIG. 1. Resume of procedures for the isolation from corn seedlings of components active in amino acid and ribonucleotide incorporation.

in the soluble component (Mans and Novelli, 1964b) will be discussed later. Preparation of the supernatant component using the French Pressure Cell (see

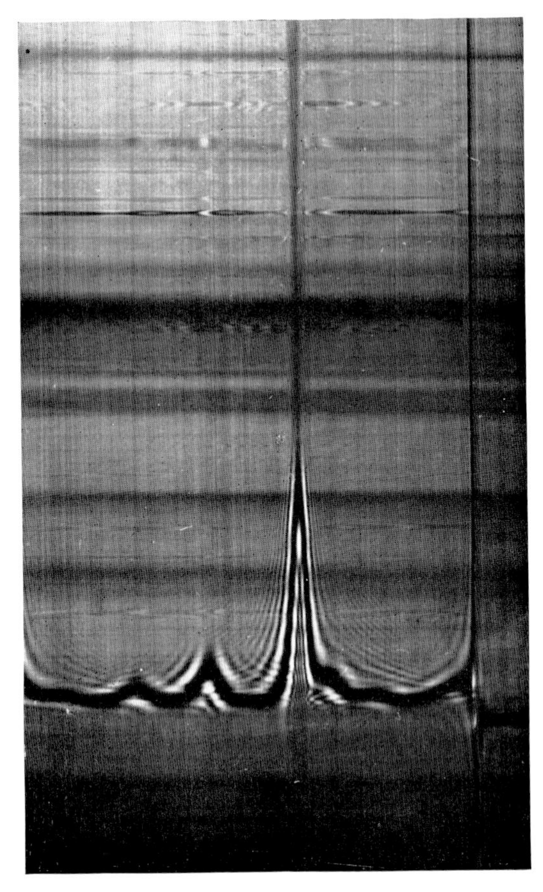

FIG. 2. Corn "microsomes". Water washed particles (8 mg protein/ml) were centrifuged at 24,630 rev/min in distilled water at 4° in a model E Spinco ultracentrifuge equipped with Schlieren optics. The photograph was taken at a 65° bar angle 12 min after attaining speed. The meniscus is to the right. The major component is 132s.

Fig. 1) insures that the bulk of the DNA-dependent polymerase activity is recovered in the soluble fraction.

## ASSAY

The small amount of amino acid fixed in the presence of the relatively gross amounts of protein fractions required for activity, precludes the use of chemical methods for the detection of newly formed protein. Therefore, the extreme sensitivity afforded by isotope detection devices and the availability of [$^{14}$C]-

amino acids at high specific activities are utilized in the detection of amino acid incorporated into new protein.

In Fig. 3 are presented the three phases of the assay used to follow the fate of a small portion of the [$^{14}$C]amino acid added to the cell-free incorporating system from corn seedlings. Radioactivity is detected by the filter paper disk method of Mans and Novelli (1961b). Thc extensive extraction procedure performed on each sample also removes lipid-soluble materials including: lipids, phospholipids, sterols, and pigments. We infer that the conversion of trichloroacetic acid-soluble radioactivity into hot trichloroacetic acid-insoluble radioactivity is a measure of protein biosynthesis (Phase I and II). The inference is drawn from the following experimental findings: (1) The components and

| | | | |
|---|---|---|---|
| PHASE I | [$^{14}$C]-Amino acid<br>ATP<br>SRNA<br>(Soluble in cold TCA) | Aminoacyl synthetase<br>$Mg^{++}$ at 30°<br>→ | [$^{14}$C]-Aminoacyl-RNA<br>P~P<br>AMP<br>(Insoluble in cold TCA) |
| PHASE II | [$^{14}$C]-Aminoacyl-RNA<br>GTP<br>Particulate<br>(Soluble in hot TCA) | Aminoacyl transferases<br>$Mg^{++}$ at 30°<br>→ | [$^{14}$C]-Particulate<br>$P_i$<br>GDP<br>SRNA<br>(Insoluble in hot TCA) |
| PHASE III | [$^{14}$C]-Particulate<br>(Sedimented at 100,000 × *g*) | EDTA at 0° centrifuge<br>→ | [$^{14}$C]-Supernatant particulate<br>(Not sedimented at 100,000 × *g*) |

FIG. 3. Fate of radioactivity during incorporation assay.

factors required for amino acid incorporation, and the intermediate reactions occurring and the intermediates formed are analogous to those found in other cell-free systems demonstrated to have synthesized a specific peptide (see Schweet *et al.*, 1958; Nirenberg and Matthaei, 1961), and (2) the [$^{14}$C]amino acid can be re-isolated from a specific acid-insoluble polypeptide.

### *Requirements and characteristics*

The data presented in Table I summarize the requirements for leucine incorporation by corn ribosomes. The level of incorporation (830 $\mu\mu$moles [$^{14}$C]leucine/mg ribosomal protein) is calculated from the specific activity of the added [$^{14}$C]leucine. The endogenous leucine contribution is neglected in the calculation and, therefore, this represents the lower limit of the total amount of leucine incorporated into acid-insoluble protein.

The endergonic nature of the incorporation is reflected in the requirement for an energy generator in the presence of catalytic amounts of ATP and GTP (see line 2, Table I). The requirement for an exogenous source of chemical energy is definitely demonstrated by the data presented in Fig. 4. Curve A represents the incorporation of phenylalanine with time in a complete reaction mixture. Omission of the phosphoenolpyruvate, pyruvic kinase, nucleotide

Table I. Requirements for leucine incorporation into maize particle protein

| System[1] | Specific activity[2] (counts/min/mg) | ($\mu\mu$moles/mg) |
|---|---|---|
| Complete | 10,010 | 830 |
| Complete less pyruvic kinase and PEP | 684 | 50 |
| Complete less $MgCl_2$ | 352 | 30 |
| Complete less GTP | 4880 | 400 |
| Complete less dialysed supernatant | 344 | 30 |
| Complete less deoxycholate particles | 132 | 10 |
| Complete plus RNase (200 $\mu$g)[3] | 314 | 30 |
| Complete plus chloramphenicol (200 $\mu$g) | 1947 | 160 |
| Complete plus 18 amino acids | 7467 | 600 |

[1] The complete system was: 0·1 M *tris* buffer (pH 7·6), 1 mM ATP, 0·3 mM GTP, 10 mM $MgCl_2$, 0·1 M phosphoenolpyruvate, 50 $\mu$g pyruvic kinase, 0·2 mM L-[$^{14}$C]leucine (specific activity 11 mc/mmole) and contained 0·37 mg ribosomes and 0·17 mg dialysed supernatant protein. The reaction mixture was incubated at 37° in a total volume of 0·5 ml and aliquots removed at 0, 10, 20, and 30 min.

[2] The specific activity is reported as total [$^{14}$C]leucine incorporated in 45 min/mg particle protein.

[3] The system is inhibited 97% in the presence of 0·2 $\mu$g pancreatic ribonuclease. From Mans and Novelli (1964).

regenerating system (Curve B) from the mixture incubated at 30°, results in no incorporation of phenylalanine during the first 20 min. Subsequent addition of the energy generator triggers incorporation. The level of incorporation obtained when the generator is added 20 min after incubation of the system at 30° (Curve B), is considerably less than that obtained when the generator is present initially (Curve A). A second addition of generator to this reaction mixture (Curve C, Fig. 4) or to the reaction mixture containing generator initially (Curve D, Fig. 4) results in no further increase in phenylalanine incorporation. Therefore, the energy generator, once added, is in excess throughout the incubation period. The reduced level of incorporation seen in Curve B reflects partial denaturation of the particulate component upon incubation at 30° for 20 min. These data, together with the insensitivity of leucine incorporation to streptomycin and penicillin (see Mans and Novelli, 1964a) preclude the possibility that the amino acid incorporation is other than endergonic or the result of microbial contamination.

The virtual lack of incorporation in the absence of the high speed supernatant component (line 5, Table I) indicates the requirements for the soluble enzymic activities and soluble ribonucleic acid (see Mans *et al.*, 1964). Also consistent with the involvement of soluble ribonucleic acid in the amino acid incorporation is the inhibition of the system by puromycin (Table II). At $5 \times 10^{-4}$ M puromycin, the maize incorporating system is 75% inhibited and almost completely inhibited at $5 \times 10^{-3}$ M. The level of inhibition by puromycin

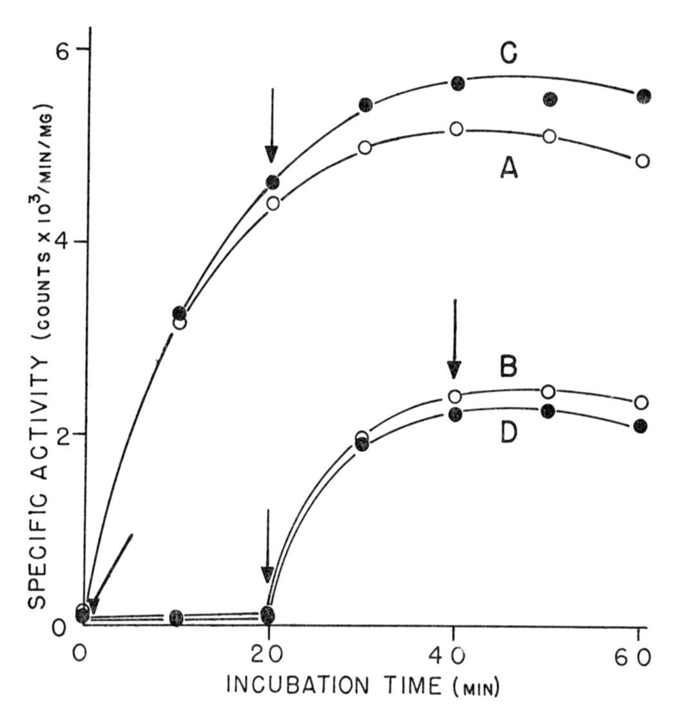


FIG. 4. Energy generator required for phenylalanine incorporation. The reaction mixtures were as described in Table I as modified in Table II. However, the phosphoenolpyruvate, pyruvic kinase, ribonucleotide generating system was added after incubation at 30°. The generator was added (arrows) to mixture in Curve A at zero time, Curve B after 20 min incubation, Curve C at zero time, and after 20 min incubation, Curve D after 20 min and after 40 min incubation. Appropriate additions of buffer were added to mixtures not receiving generator. The data are corrected for volume changes introduced during the 60 min incubation period.

Table II. Inhibition of phenylalanine incorporation by puromycin[1]

| Puromycin added ($\mu$moles) | Phenylalanine incorporated (cts/min/mg) | Inhibition (%) |
|---|---|---|
| 0 | 6040 | |
| 0·025 | 3690 | 39 |
| 0·25 | 1460 | 76 |
| 2·50 | 225 | 96 |

[1] The reaction mixture described in Table I was modified as follows: 0·2 mg of polyurdylic acid was added, 0·009 mM L-[$^{14}$C]phenylalanine (specific activity 225 mc/mmole) was substituted for L-[$^{14}$C]leucine and 2 mg microsomal and 0·4 mg supernatant protein were added. Puromycin at the indicated concentrations was added prior to incubation of the mixture at 30° for 30 min.

is precisely that found in other cell-free systems capable of polypeptide synthesis (Darken, 1964).

The inhibition of leucine incorporation by addition of a mixture of eighteen other amino acids (last line, Table I) is inconsistent with the known requirements for polypeptide synthesis. When ten of the amino acids constituting the mixture are added individually to the incorporating system as radioactive precursors, each is incorporated (Table III). It can also be seen that the level

Table III. Incorporation of several labeled amino acids into maize protein[1]

| L-Amino acid | Added[2] (total μmoles) | Incorporated[3] (total mμmoles) | Ratio (incorp./added $\times 10^{-3}$ |
|---|---|---|---|
| Glutamic | 0·23 | 1·2 | 5·2 |
| Histidine | 0·31 | 0·075 | 0·24 |
| Leucine | 0·044 | 0·21 | 5·2 |
| Lysine | 0·17 | 0·26 | 1·5 |
| Methionine | 0·19 | 0·20 | 1·0 |
| Phenylalanine | 0·11 | 0·34 | 3·2 |
| Proline | 0·19 | 0·16 | 0·086 |
| Serine | 0·97 | 0·52 | 0·54 |
| Threonine | 0·28 | 0·057 | 0·20 |
| Valine | 0·16 | 0·096 | 0·60 |

[1] The reaction mixture described in Table I was modified as follows: 1·9 mg microsomal protein and 0·44 mg supernatant protein were added. The specific activities of the added [$^{14}$C]amino acids in mc/mmole were: DL-glu. 3·1, L-his. 6·2, L-leu. 25, L-lys. 4·7, L-met. 4·8, L-phe. 10, L-pro. 5, DL-ser. 6, L-thr. 17, and L-val. 12.

[2] The amino acid composition of the supernatant component was determined on the amino acid analyser and the amount of amino acid added represents the sum of that added as [$^{14}$C] amino acid plus that in the supernatant component.

[3] The amount of amino acid incorporated was determined from the specific activity of the [$^{14}$C]amino acid incorporated after 30 min. incubation at 37°. The specific activity of the amino acid was calculated from the specific activity of the added [$^{14}$C]amino acid and the dilution by the supernatant contribution of the particular unlabeled amino acid.

of incorporation of a given amino acid is determined by factors other than its concentration in the reaction mixture. Some of the endogenous amino acids are removed by rapid passage of a freshly prepared supernatant component over G-50 Sephadex, previously equilibrated with the homogenizing medium. In the presence of the Sephadex-treated supernatant, a two-fold stimulation in leucine incorporation now can be elicited by addition of the amino acid mixture (Table IV).

### *Product*

Essentially all the radioactivity incorporated by the corn system remains associated with the particle component throughout the incubation period, thus hampering attempts to isolate and characterize the product. More than

Table IV. Stimulation of leucine incorporation by added amino acids[1]

| Supernatant added | Specific activity (cts/min/mg) | (mμmoles/mg) |
|---|---|---|
| Untreated | 4840 | 0·26 |
| Untreated plus amino acid[2] | 4665 | 0·25 |
| Sephadex[3] treated | 1250 | 0·07 |
| Sephadex treated plus amino acids | 3020 | 0·16 |

[1] The reaction mixture described in Table I was modified as follows: 0·68 mg microsomal protein and 0·25 mg of the indicated supernatant protein were added.

[2] The mixture contained nineteen amino acids exclusive of leucine (Mans and Novelli, 1961a).

[3] Passage of the high speed supernatant over G-50 Sephadex is described in the text.

half of the radioactivity can be released into the supernatant component by dissociation of the ribosomes in 0·01 M ethylenediaminetetraacetic acid and 0·5% deoxycholate (see Phase III of Fig. 1). However, much of the ribosomal protein is also released. If it is assumed that a polypeptide of twenty amino acids is synthesized and that 5% of the polypeptide is leucine, then only 2 μg of new protein is synthesized and associated with each mg of ribosomal protein in the incubation mixture. Such a 0·2% change is beyond the sensitivity of the colorimetric protein determination of Lowry *et al.* (1951).

Prompted by the expository experiments of Nirenberg and Matthaei (1961) this experimental impasse has been partly circumvented. The maize incorporating system can be stimulated to synthesize a specific polypeptide composed of a single amino acid. As seen in Table V addition of the oligonucleotide,

Table V. Specificity of polyuridylic acid stimulation of phenylalanine incorporation[1]

| Labeled amino acid added | Incorporation[2] without poly u. (cts/min/mg) | with poly u. (cts/min/mg) |
|---|---|---|
| Leucine | 154 | 140 |
| Lysine[3] | 158 | 120 |
| Phenylalanine | 197 | 2225 |
| Proline | 113 | 144 |

[1] The reaction mixture described in Table I was modified as follows: 5 mg water washed particles and 0·26 mg supernatant protein were added and 0·2 mg polyuridylic acid added where indicated. The [$^{14}$C]amino acids were added in the following concentrations and specific activities (mc/mmole): 0·003 mM L-leucine (150), 0·09 mM L-lysine (10·9), 0·009 mM L-phenylalanine (225) and 0·2 mM L-proline (5).

[2] Incorporation is reported as total [$^{14}$C]amino acid incorporated in 30 min/mg particle protein at 30°.

[3] Product was precipitated with 5% trichloroacetic acid and 0·25% tungstic acid at pH 2.

polyuridylic acid, to the corn system specifically stimulates the incorporation of phenylalanine. The incorporation of leucine, lysine and proline are unaffected by the presence of polyuridylic acid. Again, more than 90% of the radioactivity incorporated in the presence of polyuridylic acid remains associated with the particle protein component. However, the newly synthesized polypeptide can be separated from the protein components added to the reaction mixture. The radioactivity incorporated in the presence of polyuridylic acid behaves as authentic poly-L-phenylalanine (Nirenberg and Matthaei, 1961). The radioactivity is insoluble in hot, trichloroacetic acid, ether–ethanol and 80% formic acid. The formic acid-insoluble radioactivity is soluble in 32% HBr in glacial acetic acid. The HBr-extracted material was hydrolysed in 12 N HCl for 48 hr at 121° and 15 psi, in the presence of carrier poly-L-phenylalanine. Phenylalanine is the only radioactive, ninhydrin-positive spot recovered on paper chromatography of the hydrolysate in n-butanol, acetic acid, water solvent system. The maize amino acid incorporating system, therefore, can catalyse the formation of a specific polypeptide in response to an added oligonucleotide; albeit, the peptide synthesized is not naturally occurring.

Synthesis of the specific peptide together with the cofactor requirements for incorporation, the effects of various protein synthesis inhibitors and the chemical and physical characteristics of the supernatant and particle components support the inference that the amino acid incorporation catalysed by isolated components is a measure of the protein synthetic capacity of the maize tissue.

## Light Effect

The light effect on the maize amino acid incorporating system was discovered quite inadvertently. In working with components derived from etiolated maize seedlings it was noted that young seedlings, 2 and 3 days old, are the most active (see Fig. 5). Etiolated tissue was used to minimize the interference from chlorophyll and other light-induced pigments not completely extracted from protein samples prepared for liquid scintillation counting. As seedlings mature, 5 and 6 days old, components isolated from them are essentially inactive; however, at about 7 to 8 days after germination, greater activity was detected in components from these and older seedlings than from the 5- and 6-day-old seedlings. The older seedlings also exhibited slight greening. It was postulated that the "return" in amino acid incorporating activity reflects the synthesis of new enzymic proteins accompanying a metabolic transition of the seedling; i.e. a change from a heterotrophic existence on the endosperm to an autotrophic metabolism as a green plant.

### POLYSOME INVOLVEMENT

Dr. Dure reinvestigated the changes in amino acid incorporating activity of components isolated from maturing corn seedlings. He was able to repeat the earlier observation regarding the rapid decline in activity (see Fig. 5). He did not, however, observe the "return" of incorporating activity in components isolated from the 7- to 9-day-old seedlings. During the period between the initial time course study and Dr. Dure's experiments, the techniques employed in the germination and subsequent incubation of the seedlings were grossly modified. In the earlier experiments the seedlings received intermittent exposure to white light upon removal of samples and watering during the germination period. We deduced that the intermittent light exposure induced the "return" of incorporating activity and greening in the older maize seedlings.

Dr. Dure and Dr. Williams then undertook a careful investigation of the

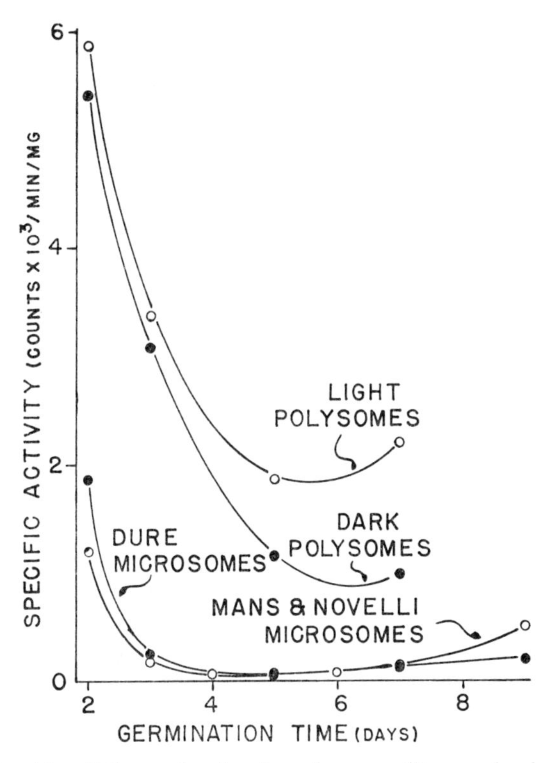


FIG. 5. Effects of irradiation and maturation of corn seedlings on leucine incorporating activity of isolated particle components. Experimental details are presented in the text. All data are normalized to specific activity 25 mc/mmole of the added [$^{14}$C]leucine.

effects of irradiation on etiolated seedlings of various ages. Three hours prior to harvesting, one half batch of seedlings of a given age was exposed to low intensity, incandescent light. Greening of the shoots was perceptible during the exposure period. Two preparations were then prepared from each batch of seedlings of a given age; one etiolated and the other irradiated. The shoots were broken away from the hypocotyl at the first internode and the high speed supernatant and polysome components prepared by the liquid nitrogen procedure described in Fig. 1. Each of the polysome preparations was assayed for leucine-incorporating activity in the presence of the optimum level of high speed supernatant derived from etiolated 2-day-old corn shoots. The salient results of this experiment are presented in Fig. 5. The particles prepared from irradiated seedlings are consistently more active than those prepared from the etiolated tissue. The light effect is manifest in the particle component of the incorporating system, and not in the soluble component. In fact, the high speed supernatant prepared from 2-day-old, etiolated seedlings is more active than any of the other seedling supernatants tested and is, therefore, used routinely in the assay of various particle preparations.

The light effect is greater in older seedlings, 5 to 7 days old, than in younger seedlings, although there is a decline in the specific activity of all particle components with maturation of the seedlings. The increased activity of the components from older irradiated shoots may represent a greater retention of the activity present in the younger shoots as compared to the more rapid loss in activity in the older etiolated shoots, rather than a "return" in activity. Con-

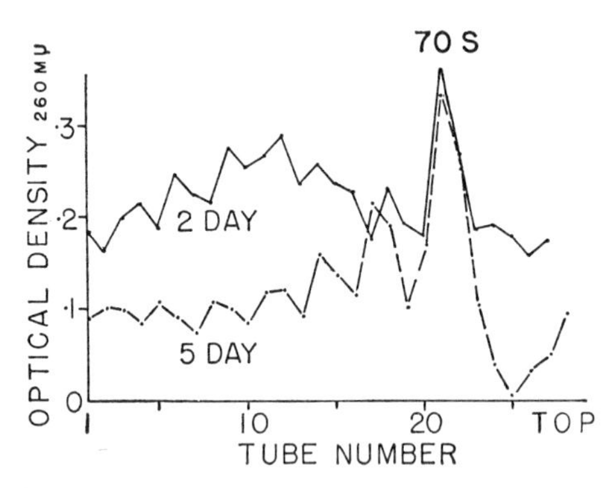


FIG. 6. Sedimentation patterns of corn polysomes by sucrose density gradient centrifugation. Maize shoots were homogenized as described in the text and polysomes isolated and detected by the procedures of Wettstein *et al.* (1963).

sideration of changes induced in RNA metabolism, discussed below, would seem to make this possibility unlikely.

Note in Fig. 5 that the Wettstein *et al.* (1963) procedure yielded particles consistently more active than the microsomes prepared from seedlings at all ages tested. The increase in incorporating activity correlates with an increase in polysome level in the particle preparations. A comparison of the polysome profile in 2-day-old, etiolated *vs* 5-day-old, irradiated seedlings is shown in Fig. 6. Particles from the 2-day-old seedlings are three times as active as the comparable preparation from 5-day-old, irradiated, seedlings. A rigorous comparison of particles prepared from irradiated *vs* etiolated seedlings at a given age shows the level of heavier aggregates consistently higher in the irradiated tissue.

The light effect on amino acid incorporation activity is not restricted to corn seedlings. The particulate component has been isolated from the primary leaves of irradiated and etiolated lima beans and soy beans. When assayed for leucine incorporation in the presence of the maize high speed supernatant, particles from the irradiated bean leaves are almost three times as active as particles from the etiolated tissue (Table VI). The effect is clearly manifested in the particle component and is not peculiar to corn but can also be demonstrated even more dramatically in dicotyledonous plants.

Table VI. Effect of light treatment on [$^{14}$C]leucine incorporation by ribosomes subsequently prepared from dark-grown plants[1]

| Particle source | Cts/min incorporated/mg protein | | Stimulation (%) |
|---|---|---|---|
| | Untreated | Treated | |
| Maize shoot | 2085 | 3710 | 70 |
| | 1877 | 3193 | 70 |
| | 2359 | 4196 | 78 |
| | 1893 | 2882 | 52 |
| Bean leaf | 2992 | 7780 | 160 |
| | 2742 | 8220 | 199 |
| Soybean leaf | 4470 | 12,130 | 172 |

[1] The data are expressed as cts/min of [$^{14}$C]leucine incorporated into the acid-insoluble product per mg ribosomal protein. The reaction mixture contained, in a final volume of 0·5 ml: 0·05 mg pyruvic kinase, crude maize supernatant (0·5 to 1·0 mg protein), 0·1 ml ribosomes (0·5 to 1·0 mg protein), and the following in $\mu$moles: *tris* buffer, pH 7·6, 50; ATP, 0·5; GTP, 0·15; $MgCl_2$, 5·0; phosphoenol pyruvate, 6·4; KCl, 8·0; [$^{14}$C]leucine (spec. act. 49·2), 0·004. The reaction was initiated by adding ribosomes and was assayed after 30 min at 37°. From Williams and Novelli (1964).

## EFFECTS ON RNA

The increase in amino acid incorporating activity after irradiation of dark-grown seedlings requires approximately 3 hr for realization of the maximum response. Data bearing on this point, originally presented in a figure by Williams and Novelli (1964) are presented in Table VII. Five-day-old, etiolated seedlings

Table VII. Development of light-stimulated [$^{14}$C]leucine incorporation by ribosomes isolated after treatment[1]

| Incubation after light exposure (hr) | Light stimulated incorporation (% increase of dark control) |
|---|---|
| 1 | 6 |
| 2 | 6, 7 |
| 2·5 | 21, 23, 38 |
| 3 | 51, 61, 70, 78 |
| 11 | 64, 51 |

[1] Data are expressed as cts/min incorporated into the acid-insoluble product per mg ribosomal protein. Assay conditions are described in Table VI. From Williams and Novelli (1964).

were irradiated for 30 min and then reincubated in the dark for the times indicated. Polysomes were then prepared from the irradiated tissue as well as the control, dark-grown seedlings and both preparations were assayed for leucine incorporating activity. The data clearly indicate, (1) a 3 hr lag occurs prior to maximum stimulation of incorporation, and (2) once the particle component is activated, it retains the increased activity for many hours.

The lag period suggests that the change in the amino acid-incorporating ability of the particles requires the formation or accumulation of a particle component not present in etiolated shoots. The increase in the level of polysomes in irradiated seedlings implicates ribonucleic acid as the newly formed component responsible for the increased incorporating activity. If these inferences about RNA synthesis are valid, then it should be possible to demonstrate a stimulation in RNA synthesis by irradiation of dark-grown seedlings.

Dr. Williams has been able to demonstrate (Williams, 1965) a stimulation of the incorporation of [$^{32}$P] into ribonucleic acid by a brief exposure of dark-grown seedlings to white light. Furthermore, the stimulation of incorporation of $^{32}$P precedes or accompanies the stimulation of leucine-incorporating activity in the particle component. The labeled ribonucleic acid isolated from the irradiated seedlings has an altered base ratio as well as a higher specific radioactivity when compared with ribonucleic acid isolated from the non-irradiated seedlings.

## DNA-DEPENDENT RNA POLYMERASE

The changes in ribonucleic acid metabolism coincident with an increase in the level of polysomes in irradiated tissue are consistent with the notion of induced synthesis of a specific messenger RNA in the irradiated tissue (Brenner *et al.*, 1961; Gros *et al.*, 1961; Jacob and Monod, 1961; Spiegelman and Hayashi, 1963). We assume that the persistence of the induced activity for at least 11 hr (Table VII) either represents a continual messenger RNA synthesis in the absence of further light induction or indicates that the newly formed RNA in relatively slower dividing corn tissue is more stable than that formed in rapidly dividing bacterial cells. If the light effect on ribonucleic acid metabolism in the corn shoot entails synthesis of messenger RNA, then the DNA-dependent RNA polymerase, as first described by Weiss (1960), should be active in the corn shoots. Furthermore, the level of RNA polymerase activity should be greater in irradiated than dark-grown tissue.

### ISOLATION

We have been able to isolate a DNA-dependent RNA polymerase from corn shoots (Mans and Novelli, 1964b). Unlike the enzyme preparations from rat liver (Weiss, 1960) and peas (Huang *et al.*, 1960) the bulk of the activity is dissociated from the insoluble DNA rich component of a tissue homogenate (Fig. 7). Both the 20,000 × *g* pellet and supernatant fractions of an homogenate were assayed for the incorporation of radioactively labeled ATP into acid insoluble material. Calf thymus deoxyribonucleic acid was added as the exogenous template. The preparations were assayed with and without the addition of the three other ribonucleotides required for RNA synthesis. Passage of the tissue through a French Pressure Cell (see Fig. 1) insures that 90% of the total XTP-stimulated, ATP-incorporating activity is recovered in the $20,000 \times g$ supernatant.

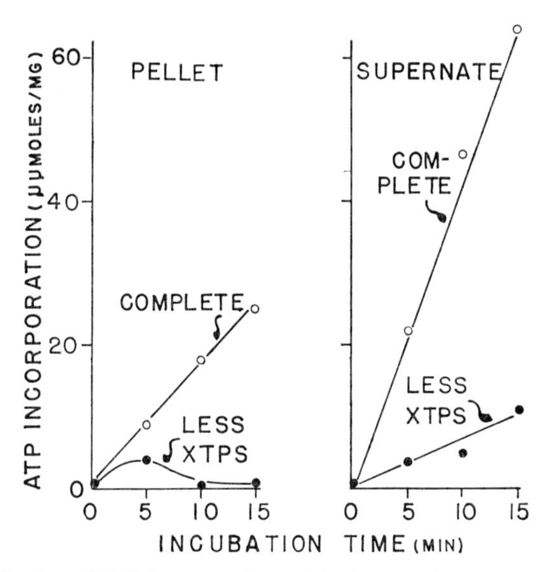


FIG. 7. Distribution of ATP-incorporating activity in centrifugal fractions of an homogenate of maize shoots. The homogenate was prepared by the liquid nitrogen procedure through the $20,000 \times g$ centrifugation as outlined in Fig. 1. Both the pellet and supernatant fractions were assayed with and without the other ribonucleotides (XTP'S) as described in Table VIII. $NH_4Cl$ was omitted from the reaction mixtures.

## IDENTIFICATION

The soluble ATP-incorporating activity is identified as a DNA-dependent RNA polymerase. The substrate, template and cofactor requirements for the incorporation of the four ribonucleotides, the effect of highly specific inhibitors on the ribonucleotide incorporation, and the characterization of the radioactive product as an oligonucleotide, are the criteria used in identification of the enzymic activity in the supernatant component.

The substrate and cofactor requirements for ATP incorporation are presented in Table VIII. Essentially no ATP is incorporated upon omission of the three other ribonucleoside triphosphates from the reaction mixture. Deletion of any one of the ribonucleoside triphosphates reduces ATP incorporation by at least 75%. Magnesium acetate can be replaced with $MnCl_2$ (0·005mM) in stimulating ATP incorporation. Of real significance is the eight-fold stimulation elicited by the addition of calf thymus deoxribonucleic acid to the crude enzyme preparation. Occurrence of the ATP-incorporating activity in the supernatant component fortuitously afforded the opportunity of establishing

Table VIII. Requirements for DNA-stimulated ATP incorporation

| System[1] | Incorporation[2] ($\mu\mu$moles ATP/mg protein) |
|---|---|
| Complete | 410 |
| Complete less CTP | 104 |
| Complete less GTP | 70 |
| Complete less UTP | 106 |
| Complete less CTP, GTP, UTP | 10 |
| Complete less magnesium acetate | 127 |
| Complete less DNA | 48 |

[1] The complete system was 0·1 M *tris* buffer (pH 8·0), 0·05 M 2-mercaptoethanol, 0·1 M $NH_4Cl$, 2·5 mM CTP, 2·5 mM GTP, 2·5 mM UTP, 0·21 mM [8-$^{14}$C]-ATP (9·1 mc/mmole), 0·025 M magnesium acetate and contained 110 $\mu$g calf thymus DNA and 1 mg 20,000 × ***g*** supernatant protein. The reaction mixture was incubated at 30° in a total volume of 0·4 ml and aliquots were removed at 0, 5, 10, 15 and 20 min.

[2] The specific activity is calculated from the total cts/min incorporated into acid-insoluble material in 20 min and the specific activity of the added ATP. From Mans and Novelli (1964).

this essential criterion for identification of a DNA-dependent activity in a crude preparation.

The two inhibitors specific for DNA, deoxyribonuclease and actinomycin D, depress ATP or UTP incorporation to a level essentially that obtained in the absence of added DNA (Table IX). The presence of ribonuclease in the incubation mixture prevents accumulation of radioactive product and addition of ribonuclease subsequent to incubation solubilizes the radioactive product. The incorporation is also inhibited by pyrophosphate and refractory to inorganic phosphate.

Table IX. Inhibitors of nucleotide incorporation[1]

| System | Incorporation ($\mu\mu$moles/mg) | |
|---|---|---|
| | ($^{14}$C) ATP | ($^{14}$C) UTP |
| Complete | 382 | 166 |
| Complete plus RNase | 22 (94) | 6 (96) |
| Complete plus DNase | 36 (91) | 9 (95) |
| Complete plus actinomycin D | 49 (87) | 27 (84) |
| Complete less DNA | 59 (85) | 14 (92) |

[1] The reaction mixture described in Table VIII was modified as follows: 0·024 mM [2-$^{14}$C]UTP (11·3 mc/mmole) and 2·5 mM ATP were substituted for UTP and radioactive ATP where indicated. Inhibitors were added at zero time and amounts added per reaction mixture were: 25 $\mu$g pancreatic ribonuclease, 25 $\mu$g spleen deoxyribonuclease I, and 1 $\mu$g actinomycin D. The numbers in parenthesis indicate percent inhibition of the complete system after 20 min incubation.

The radioactive product formed from labeled ribonucleotides behaves as an oligonucleotide. The radioactive product is non-dialysable, insoluble in acid or ethanol and is solubilized upon treatment with alkali or ribonuclease. The product formed from either [$\alpha^{32}$P]-ATP or [$\alpha^{32}$P]-GTP upon alkaline hydrolysis yields all four $^{32}$P labeled 2′,3′-ribonucleotides. See Table X. The base ratios of the product formed by the crude enzyme are not, however, equal or complementary to the base ratio in the primer DNA. As the crude enzyme is purified away from competing nucleotide incorporating activities, nucleases, and other nucleic acids, the base composition of the product more nearly equals that of the added DNA. ATP, GTP, and UTP, when added as radioactive precursors in the presence of the other three unlabeled ribonucleoside triphosphates, are all incorporated into the product (see Tables IX and X).

Table X. Analysis of alkaline hydrolysate of radioactive product[1]

| Labeled precursor | Isolated 2′,3′-ribonucleotide[2] | | | | $\frac{A+U}{G+C}$ |
|---|---|---|---|---|---|
| | AMP | UMP | GMP | CMP | |
| [$\alpha^{32}$P]-ATP | 23·1 | 21·3 | 23·9 | 32·1 | 0·80 |
| [$\alpha^{32}$P]-GTP | 22·0 | 35·7 | 24·3 | 17·9 | 1·36 |

[1] The reaction mixture described in Table VIII was modified as follows: 0·4 mM [$\alpha^{32}$P]-ATP (35 mc/mmole) or 0·3 mM [$\alpha^{32}$P]-GTP (34 mc/mmole), 84 $\mu$g calf thymus DNA and 2 mg high speed supernatant protein were incubated in a total volume of 2 ml at 30° for 40 min. The radioactive product was isolated by the phenol procedure, hydrolysed in 0·3 M KOH for 18 hr at 37° and the 2′,3′-ribonucleotides separated by paper electrophoresis as described by Weiss (1960).

[2] Values are expressed as moles % calculated from the total radioactivity recovered after electrophoresis.

## Concluding Remarks

A direct comparison of the level of the DNA-dependent RNA polymerase activity in irradiated and etiolated maize shoots is imminent. At least two responses might be anticipated. Firstly, an increase in the level of polymerase activity in the irradiated tissue owing to an increase in the amount of enzyme protein present. Secondly, no difference in the DNA-dependent polymerase activity in the two preparations. Dr. Stout has conducted two preliminary experiments with irradiated and etiolated maize shoots and found the latter result, i.e. no difference These assays were conducted using calf thymus DNA as the required template rather than the exogenous DNA in the 20,000 × ***g*** pellet of the maize homogenate. We suspect that the changes in ribonucleic acid metabolism induced by irradiation of etiolated tissue might result from changes in the template activity of the "Cellular" DNA (see Allfrey and Mirsky, 1963; Bonner *et al.*, 1963). The crude RNA polymerase from maize seedlings would seem to be an excellent test system for detecting a light-induced change in the template activity of the endogenous deoxyribonucleic acid.

As a participant in a study institute on the biochemistry of chloroplasts, I feel obliged at least to mention these organelles. The light effect on the maize amino acid incorporating system may result in the organization of the chloroplast and/or subsequent synthetic activity of the newly formed organelle. Dr. Williams, again in preliminary studies, has found that red light can be substituted for white light to induce the changes in ribosomes reported here. Furthermore, the ribosomal changes are correlated with the synthesis of chlorophyll by the irradiated shoot. Albino mutants of maize fail to show either response. These results suggest that a comparison be made between the ribosomes and ribonucleic acids isolated from the cytoplasm and those isolated from the chloroplasts of irradiated corn shoots as has been accomplished in *Euglena gracilis* (Eisenstadt and Brawerman, 1964; Brawerman and Eisenstadt, 1964). Such studies could help to resolve whether the increase in amino acid incorporating activity induced in the ribosomal component upon irradiation of etiolated maize shoots, precedes chloroplast development and is thus required for it, or whether it reflects the synthetic activity of the completed organelle, or both.

## Acknowledgements

The initial stimulus to utilize the cell-free approach with tissues of higher plants in the resolution of the mechanisms of regulation of biosynthetic processes; the techniques of isolation of active amino acid incorporating systems, and the engendering in the author of the necessary fortitude to persist with an elusive enzyme isolation (RNA polymerase) are among Dr. G. David Novelli's many contributions to the work presented here. Essentially all but the most recent experiments[1] were conducted in his laboratory and often in collaboration with many people then associated with the Biology Division of the Oak Ridge National Laboratory.[2]

Some of the experiments described in this paper were supported by a contract with the United States Atomic Energy Commission (No. AT (30)-173536).

[1] Dr. Ernest Stout of the Botany Department, University of Maryland, College Park, Maryland.

[2] Dr. Leon Dure of the Department of Biochemistry, University of Georgia, Athens. Dr. Gene R. Williams of the Department of Biology, Indiana University, Bloomington. Dr. Oscar Miller and Dr. Peter Pfuderer of the Biology Division of the Oak Ridge National Laboratory, Oak Ridge, Tennessee.

## References

Allfrey, V. G. and Mirsky, A. E. (1963). *Cold Spr. Harb. Symp. quant. Biol.* **28**, 247.

App, A. A. and Jagendorf, A. T. (1964). *Pl. Physiol.* **39**, 772.

Baldwin, E. (1940). "An Introduction to Comparative Biochemistry". Cambridge University Press, London.

Bonner, J., Huang, R. C. and Gilden, R. V. (1963). *Proc. natn. Acad. Sci. U.S.A.* **50**, 893.

Brawerman, G. and Eisenstadt, J. M. (1964). *J. mol. Biol.* **10**, 403.

Brenner, S., Jacob, F. and Meselson, M. (1961). *Nature, Lond.* **190**, 576.

Butler, W. L. (1964). *Quart. Rev. Biol.* **39**, 1.

Darken, M. A. (1964). *Pharmac. Rev.* **16**, 223.

Eisenstadt, J. M. and Brawerman, G. (1964). *J. mol. Biol.* **10**, 392.

Gordon, S. A. (1964). *Quart. Rev. Biol.* **39**, 19.

Gros, F., Hiatt, H., Gilbert, W., Kurland, C. G., Risebrough, R. W. and Watson, J. D. (1961). *Nature, Lond.* **190**, 581.
Huang, R. C. and Bonner, J. (1962). *Proc. natn. Acad. Sci. U.S.A.* **48**, 1216.
Huang, R. C., Maheshwari, N. and Bonner, J. (1960). *Biochem. biophys. Res. Commun.* **3**, 689.
Jacob, F. and Monod, J. (1961). *J. mol. Biol.* **3**, 318.
Lett, J. T., Takahashi, W. N. and Birnstiel, M. (1963). *Biochim. biophys. Acta* **76**, 105.
Lowry, O. H., Rosebrough, N. J., Farr, A. L. and Randall, R. J. (1951). *J. biol. Chem.* **193**, 265.
Mans, R. J. and Novelli, G. D. (1961a). *Biochim. biophys. Acta* **50**, 287.
Mans, R. J. and Novelli, G. D. (1961b). *Archs Biochem. Biophys.* **94**, 48.
Mans, R. J. and Novelli, G. D. (1964a). *Biochim. biophys. Acta* **80**, 127.
Mans, R. J. and Novelli, G. D. (1964b). *Biochim. biophys. Acta* **91**, 186.
Mans, R. J., Purcell, C. M. and Novelli, G. D. (1964). *J. biol. Chem.* **239**, 1762.
Nirenberg, M. W. and Matthaei, J. H. (1961). *Proc. natn. Acad. Sci. U.S.A.* **47**, 1588.
Price, L., Mitrakos, K. and Klein, W. H. (1964). *Quart. Rev. Biol.* **39**, 11.
Schweet, R., Lamfrom, H. and Allen, E. (1958). *Proc. natn. Acad. Sci. U.S.A.* **44**, 1029.
Spiegelman, S. and Hayashi, M. (1963). *Cold Spr. Harb. Symp. quant. Biol.* **28**, 161.
Weiss, S. B. (1960). *Proc. natn. Acad. Sci. U.S.A.* **46**, 1020.
Wettstein, F. O., Staehelin, T. and Noll, H. (1963). *Nature, Lond.* **197**, 430.
Williams, G. R. (1965). *Pl. Physiol.* Supp. 40, 1.
Williams, G. R. and Novelli, G. D. (1964). *Biochim. biophys. Res. Commun.* **17**, 23.
Zamecnik, P. C. (1960). *Harvey Lect.* **54**, 256.

# Part V. Biosynthesis in Chloroplasts—Pigments

# The Heme[1] and Chlorophyll Biosynthetic Chain

S. Granick[2]

*The Rockefeller University, New York, U.S.A.*

## Introduction

The biosynthetic chain of the tetrapyrroles contains two end products, heme and chlorophyll. These are the two major pigments of protoplasm. Both pigments are derived from the same intermediate, protoporphyrin. When iron is inserted into protoporphyrin, then the iron protoporphyrin or heme molecule is made (Fig. 1). When magnesium is inserted into protoporphyrin, magnesium protoporphyrin is made which is converted in a series of further steps into chlorophyll (Fig. 1). A number of recent comprehensive reviews on heme biosynthesis and chlorophyll biosynthesis are available to which the reader is referred for a fuller discussion of various topics and a more complete bibliography (Lascelles, 1964; Granick and Levere, 1964; Falk, 1963; Bogorad, 1960; Smith and French, 1963; Granick and Mauzerall, 1961).

In this review we have attempted to survey the steps in the heme-chlorophyll biosynthetic chain in order to provide a basis for discussions of control mechanisms and some evolutionary aspects of this chain.

Heme and chlorophyll (Fig. 1) are related not only because they are two end products of the same biosynthetic chain. They are also related in function. Heme serves as the catalyst for respiration (i.e., in the form of heme enzymes and cytochromes) to release the energy stored in organic bonds to be used for useful work. Chlorophyll serves as a catalyst to convert the energy of sunlight into the stored chemical energy of organic bonds. The basic energetics of protoplasm are thus catalysed by these two pigments derived from the same biosynthetic chain.

When we examine the structures of heme and chlorophyll and attempt to relate them to their functions, a simple relation is evident. The redox activities of heme which are responsible for its catalytic functions reside in the property of the iron atom; the properties of the porphyrin surrounding the iron atom merely modify the properties of the iron atom. On the other hand, the porphyrin or chlorophyll molecules are dyestuffs that have properties suitable for photochemical reactions; these dyes absorb light and fluoresce intensely in the visible region of the spectrum. A number of biological pigments are known which can act photochemically, e.g., derivatives of flavins or reduced carotenoids. The ability of the tetrapyrrole to do two jobs, i.e., to carry out both redox and photochemical reactions gave it such an evolutionary advantage that it easily pre-empted the biological stage for these functions (Granick, 1965a).

[1] "Heme" is used throughout this article; it is indexed as "haem". Ed.

[2] This work has been supported in part by U.S. Public Health Service Grant No. GM 04922

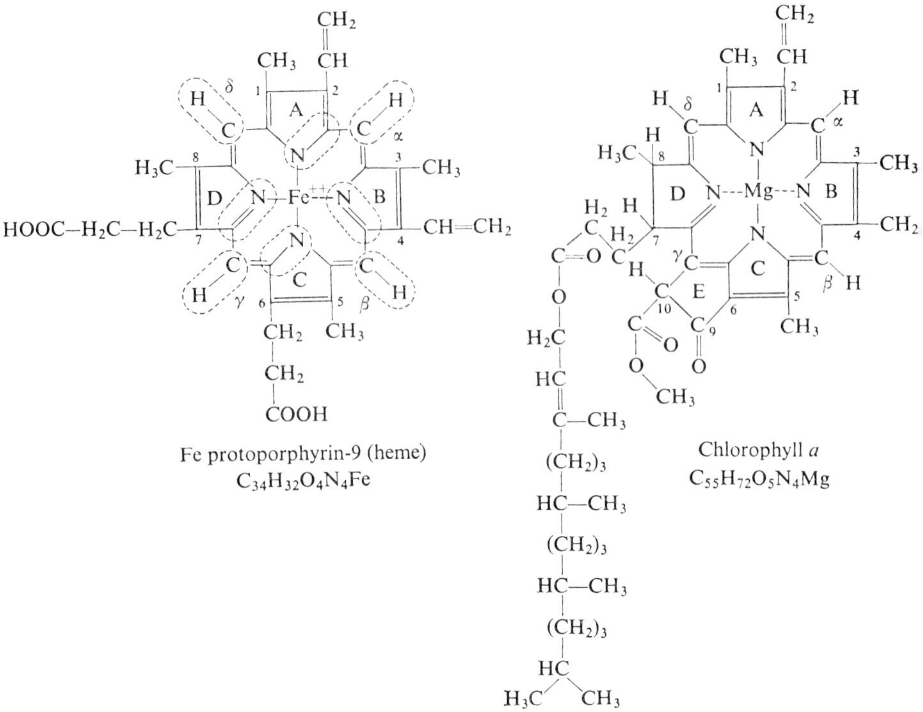

Fe protoporphyrin-9 (heme)
$C_{34}H_{32}O_4N_4Fe$

Chlorophyll *a*
$C_{55}H_{72}O_5N_4Mg$

FIG. 1. Numbering system for heme and chlorophyll. In heme the circled atoms are derived from eight glycine molecules and the other atoms are derived from succinyl-CoA. The variations in the side chains of related compounds are as follows, for:

Heme: *Spirographis* hemoglobin: 2 is –CHO. Cytochrome *c*: 2 and 4 are

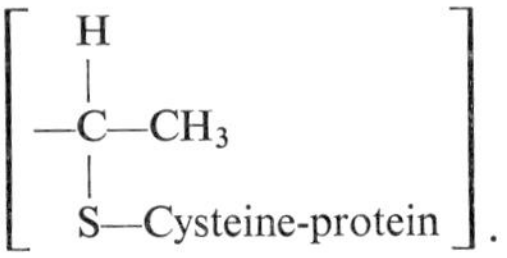

$$\left[ -\overset{\displaystyle H}{\underset{\displaystyle S\text{—Cysteine-protein}}{\overset{|}{\underset{|}{C}}}}\text{—}CH_3 \right].$$

Cytochrome oxidase: 8 is –CHO and 2 is –CHOH—CO—$CH_2R$(?). Cytochrome $\alpha_2$ of *E. coli*: one of the pyrrole rings may have two extra H atoms; this "chlorin" type corresponds to protoporphyrin in side chains.

Porphyrins: Deuteroporphyrin: 2 and 4 are –H. Pyroporphyrin: 6 is H and 2 and 4 are ethyl. Phylloporphyrin is pyroporphyrin and at $\gamma$-C an H is replaced by –$CH_3$.

Chlorophylls: Chlorophyll *b*: 3 is –CHO. Chlorophyll *d*: 2 is –CHO. Bacteriochlorophyll: at 3 and 4 add H atoms, 2 is –$COCH_3$. Pheophytin: magnesium is removed and replaced by two H atoms on the two pyrrole nitrogens. Pheophorbide: both magnesium and phytol are removed. Protochlorophyll: the two H atoms at 7 and 8 are removed and a double bond is added. Protochlorophyllide or magnesium vinyl pheoporphyrin $\alpha_5$: the phytol is removed from protochlorophyll. Vinyl pheoporphyrin $\alpha_5$: the phytol and magnesium are removed from protochlorophyll.

## ORGANELLES AND TETRAPYRROLE SYNTHESIS

Two kinds of organelles are concerned with tetrapyrrole synthesis. One kind is the mitochondrion present in plant and animal cells. The mitochondrion of the animal cells makes heme. The mitochondrion of the plant cells has yet to be investigated in this regard. For want of specific information we shall review the work on heme synthesis in the animal mitochondrion and assume tentatively that the plant mitochondrion has similar properties. The second kind of organelle that is concerned with tetrapyrrole synthesis is the plastid. The plastid makes chlorophyll. Unlike the mitochondria it has the ability to form Mg complexes of tetrapyrroles.

The first intermediate in tetrapyrrole biosynthesis is ALA (δ-aminolevulinic acid, Fig. 2). It is not yet known whether this synthesis is exclusively the function of mitochondria or can also be accomplished by plastids. Nor is it yet known whether heme is only formed by mitochondria or can also be formed by plastids. It seems likely that both organelles will be found to synthesize their own ALA and also heme. Because both organelles contain their own DNA "chromosomes" (Gibor and Granick, 1964), the question arises whether these DNA's code for the enzymes of the tetrapyrrole chain that are contained in them, and whether the enzymes are different in composition. If the enzyme, e.g., of function A, in both the mitochondrion and plastid of the same cell were found to be identical in amino acid composition then coding by the DNA of the nucleus might be suspected, otherwise coding by the DNA's of the organelles might be considered.

It should be possible to study heme biosynthesis in mitochondria of plants almost uncontaminated with plastids. For example, roots contain proplastids which are few in number and undifferentiated as compared to the mitochondria. Similarly, in plants like *Euglena* when grown in the dark the mitochondria are profuse and occupy most of the volume ordinarily occupied by the chloroplasts of *Euglena* when grown in the light; so that comparison of enzymes of heme biosynthesis of dark-grown and light-grown *Euglena* should give useful information. In addition, colorless mutant *Euglena* strains could also be used for such a study.

Important studies of chlorophyll biosynthesis have been made on photosynthetic bacteria and blue-green algae. These procaryotes lack double membranes that separate photosynthetic, oxidative respiratory, and nucleic synthetic functions from one another. Double membrane lamellae and vesicles that contain the plastid pigments and carry out photosynthesis have been identified in them (Cohen-Bazire, 1964). It will be important to examine whether in these cells oxidative respiratory functions also are carried out independently by small vesicles containing organizations of enzymes similar to those found on mitochondrial cristae. In the case of the procaryotes there is no evidence for any DNA other than that of the central body. It may be that in procaryotes the tetrapyrrole enzymes, at least to the step of protoporphyrin synthesis, are identical both for heme and chlorophyll synthesis. A comparison of synthesis to the heme step, between aerobic dark-grown and anaerobic dark-grown *Rhodopseudomonas spheroides*, would be interesting; for under aerobic conditions oxidative metabolism predominates and under anaerobic conditions bacteriochlorophyll-containing vesicles are formed which can carry out photosynthesis (Lascelles, 1964).

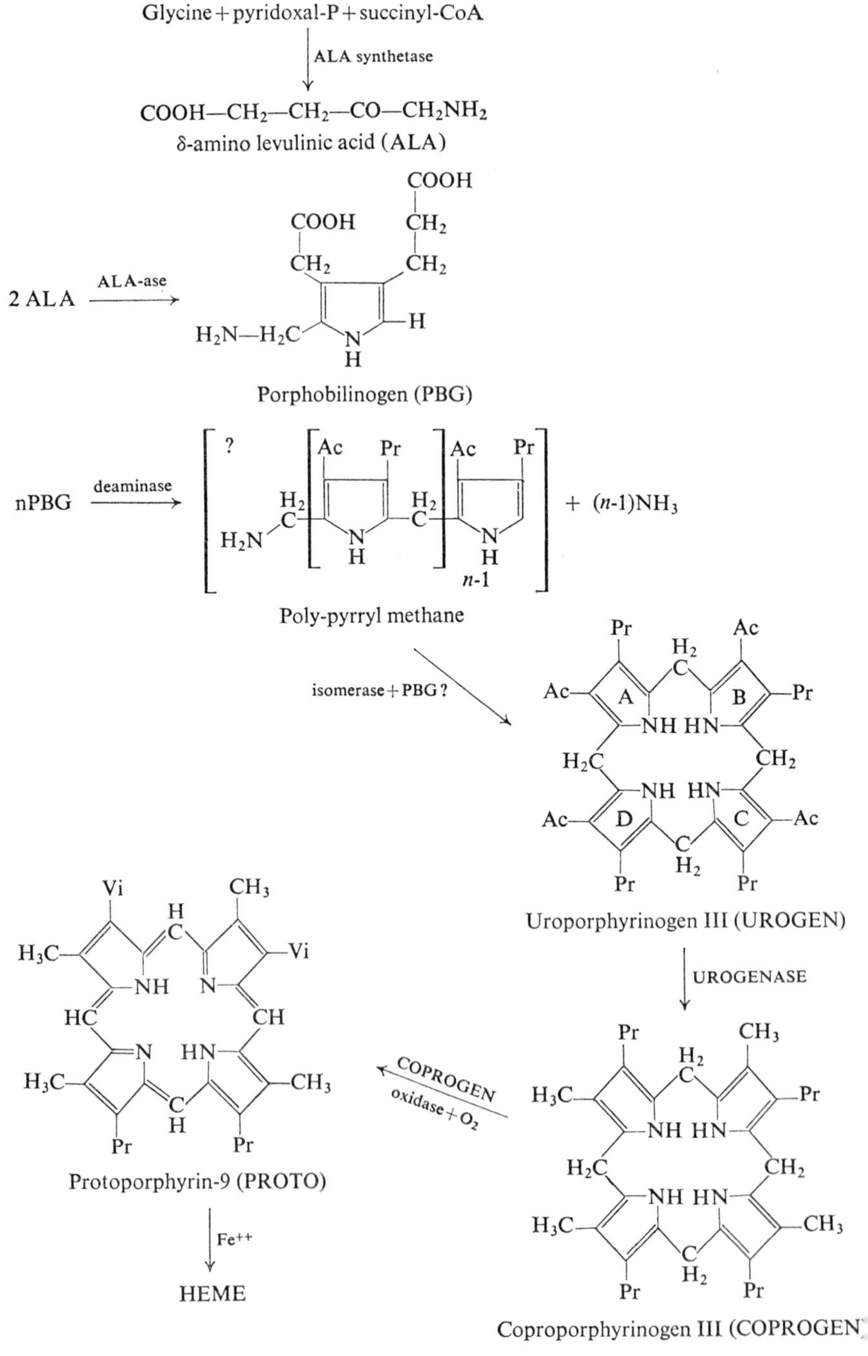

Glycine + pyridoxal-P + succinyl-CoA

↓ ALA synthetase

COOH—$CH_2$—$CH_2$—CO—$CH_2NH_2$

δ-amino levulinic acid (ALA)

2 ALA —ALA-ase→

Porphobilinogen (PBG)

nPBG —deaminase→ [Poly-pyrryl methane] + (n-1)$NH_3$

Poly-pyrryl methane

isomerase + PBG ?

Uroporphyrinogen III (UROGEN)

↓ UROGENASE

Coproporphyrinogen III (COPROGEN)

COPROGEN oxidase + $O_2$

Protoporphyrin-9 (PROTO)

↓ $Fe^{++}$

HEME

FIG. 2. Biosynthetic chain to heme.

## SUMMARY OF THE BIOSYNTHETIC CHAIN TO HEME

The steps in the synthesis of HEME (ferrous or ferric protoporphyrin) are presented in Fig. 2. This synthesis is beautiful in its essential simplicity. It is as if a molecular engineer had set up the synthesis on an assembly line. First, two ubiquitous small molecules—glycine and succinate—become attached by one machine (i.e. an enzyme) to form a short chain molecule, δ-aminolevulinic acid (ALA). Next, two ALA molecules are placed side by side on another machine and tied together to form a small, ring-shaped molecule, a monopyrrole, porphobilinogen (PBG). Then, on the next machine, four PBG molecules are tied together in succession to form a large ring, a colorless tetrapyrrole, called uroporphyrinogen (UROGEN). At the next machine, this molecule is trimmed by cutting off successively four carboxyl groups from four acetic side chains that stick out from the edges of the molecule, thus forming coproporphyrinogen (COPROGEN). At the next machine, oxidations with $O_2$ occur on two of the side chains and protoporphyrinogen (PROTOGEN) is formed. As the colorless molecule issues from the end of the assembly line it is given a paint job by being autoxidized with $O_2$, and out pops a shiny, red protoporphyrin molecule. Finally, into the center of this molecule is added the engine, the iron atom, to form the molecule HEME or iron protoporphyrin.

Only the first reaction requires energy, i.e., for the activation of succinate (as succinyl-CoA) to combine with glycine to form ALA. All the other reactions, except possibly the conversion of PBG to UROGEN, are downhill or irreversible. They are strongly favored thermodynamically because they involve the formation of resonating pyrrole and porphyrin rings, decarboxylations, and oxidations of propionic acid groups. $O_2$ is required in two places. One is in the oxidation of citric acid via the cycle, to supply succinyl-CoA. The other is in the oxidation of COPROGEN to PROTOGEN. It is interesting to note that both these enzymic oxidations occur in mitochondria.

## DISTRIBUTION OF ENZYMES OF THE BIOSYNTHETIC CHAIN TO HEME

All aerobic cells, bacterial as well as plant and animal, have the ability to synthesize heme. Heme synthesis has frequently been studied in the avian erythrocyte or rabbit reticulocyte cells. When such cells are hemolysed, a number of the enzymes of the heme biosynthetic chain are found to be soluble and can be separated from each other by zone electrophoresis on starch (Granick and Mauzerall, 1958a). The insoluble enzymes are presumably localized in the mitochondria or particles derived from mitochondria.

Figure 3 shows this interesting distribution. The first enzyme, ALA-synthetase, is in the mitochondria; the second, third and fourth enzymes which convert ALA into PBG into UROGEN III and COPROGEN III are soluble and reside in the cytoplasm. The last two which convert COPROGEN into PROTOGEN and PROTO into HEME are insoluble and are localized in the mitochondria (Sano and Granick, 1961). Evidence for this localization will be discussed under the heading for each enzyme.

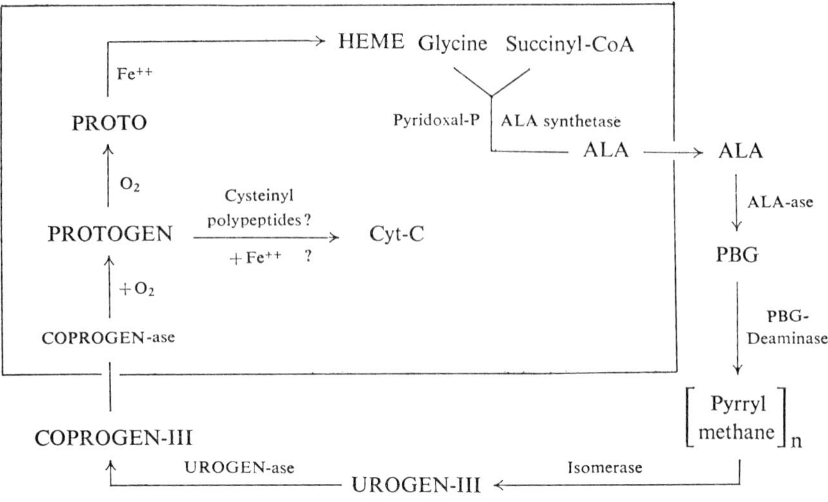


FIG. 3. Distribution of the enzymes of HEME biosynthesis in the liver cell. The first enzyme ALA-synthetase (ALA-ase), and the last two enzymes, coproporphyrinogen oxidase (COPROGEN-ase) and iron chelatase, are localized in the mitochondria and the other enzymes, which are soluble, are localized in the surrounding cytoplasm.

## SYNTHESIS OF δ-AMINOLEVULINIC ACID

The now classic labeling experiments of Shemin, Rittenberg and coworkers (Shemin, 1957) led to the recognition that ALA was the first product of the biosynthetic chain of heme. ALA is formed from glycine and succinyl-CoA according to reaction I. A similar reaction (II) which forms aminoacetone in liver mitochondria is carried out by a separate enzyme that condenses glycine with acetyl-CoA (Urata and Granick, 1963). Studies by Schulman and Richert

$$\text{Succinyl-CoA} + \text{Glycine} \xrightarrow[\text{Pyridoxal-P}]{\delta\text{-ALA-synthetase}} \delta\text{-Aminolevulinic acid} + CO_2 + \text{CoA} \quad \text{(I)}$$

$$\text{Acetyl-CoA} + \text{Glycine} \xrightarrow[\text{Pyridoxal-P}]{\text{AA-synthetase}} \text{Aminoacetone} + CO_2 + \text{CoA} \quad \text{(II)}$$

(1957) on vitamin $B_6$- and pantothenic acid-deficient ducklings showed that pyridoxal phosphate and coenzyme A are required for this reaction. Similarly, Lascelles (1957) showed that porphyrin synthesis in *Tetrahymena vorax*, a protozoan, requires both of these vitamins. Weintrobe (1950) found that pigs deficient in vitamin $B_6$ produced small, pale, red cells deficient in heme and very low in free protoporphyrin. The low heme content is due to the low activity of ALA-synthetase which requires pyridoxal phosphate as its prosthetic group. The smallness of the cells indicates a low protein content which may occur because pyridoxalphosphate is also required for transamination reactions to supply some amino acids that are used for protein synthesis.

The synthesis of ALA in mammalian cells, e.g. the young red blood cell, requires the cooperation of several enzyme systems that are localized in the mitochondria. The requirement for a citric acid cycle has been shown by tracer studies with acetate and succinate, and by inhibition studies with malonate, *trans*-aconitate and fluoracetate and arsenite. The requirement for an electron transfer system from the citric acid cycle to $O_2$ has been shown by inhibition studies with anaerobiosis and CO. The requirement for oxidative phosphorylation has been shown by dinitrophenol inhibition of ALA synthesis; dinitrophenol may also inhibit ALA-synthetase (Falk, 1963; Granick and Mauzerall, 1958a).

ALA-synthetase is found in the insoluble fraction of red cell hemolysates obtained from chickens that have been treated with phenylhydrazine. This insoluble or particle fraction consists of nuclei, mitochondria and cell membranes. Arsenite poisons the ability of these particles to make succinyl-CoA via the citric acid cycle which is localized exclusively in mitochondria. ALA-synthetase is also found in mitochondria obtained from guinea pigs with chemically induced porphyria. Thus, both ALA-synthetase and the mechanism for making its substrate, succinyl-CoA ,are present in mitochondria. Neuberger and Turner (1963) have reported that $\gamma,\delta$-dioxovalerate can be transaminated to ALA by an L-alanine aminotransferase. This enzyme is dependent on pyridoxal phosphate and requires free thiol groups for activity. It has been partly purified from *R. spheroides*. Its significance for ALA synthesis *in vivo* is not known.

## SUCCINYL-COA SYNTHESIS

Neuberger's laboratory was the first to demonstrate that succinyl-CoA is one of the substrates of ALA-synthetase (Gibson *et al.*, 1958). The particles from the chicken red cells form ALA when incubated in air with glycine, $\alpha$-KG, CoA, $Mg^{2+}$ and pyridoxal phosphate. The intact particles are, however, impermeable to succinyl-CoA. Freeze-drying renders them permeable to this compound. The disrupted particles form ALA from glycine and added or generated succinyl-CoA. Gibson *et al.* also showed that the disrupted particles, in addition to ALA-synthetase, contain $\alpha$-ketoglutarate dehydrogenase and glutamic dehydrogenase.

In order to demonstrate the ALA-synthetase in the disrupted particles, Gibson *et al.* generated succinyl-CoA from $\alpha$-KG using an oxidase from pig heart, as shown in the following series of reactions (III). Glutamic dehydrogenase and ammonia were added to regenerate the $NAD^+$.

(a) $\alpha$-KG + TPP + $NAD^+$ + CoA + Lipoic oxidase (pig heart) $\rightarrow$
Succinyl-CoA + $CO_2$ + NADH + $H^+$.

(b) NADH + $H^+$ + $NH_3$ + $\alpha$-KG + Glutamic dehydrogenase $\rightarrow$ $NAD^+$ + Glutamic acid.

(c) Succinyl-CoA + Glycine + Pyridoxal-P + ALA-synthetase (disrupted particles from chicken reticulocytes) $\rightarrow$ ALA + $CO_2$ + CoA.

(III)

Succinyl-CoA may be formed in several ways. In erythrocytes, the main pathway appears to be via the enzyme complex α-KG-oxidase and the citric acid cycle. It is usually assumed that α-KG-oxidase is localized only in the mitochondria. This complex has a molecular weight of about two million and contains five enzyme activities. Massey (1960) succeeded in obtaining a soluble complex, mol wt 260,000, which contained TPP (thiamine pyrophosphate), lipoic acid and FAD. The steps in the oxidation of α-KG are given in the series of reactions IV.

(a) α-KG ($Km = 1{\cdot}3 \times 10^{-5}$ M) + TPP → Succinyl-TPP + $CO_2$

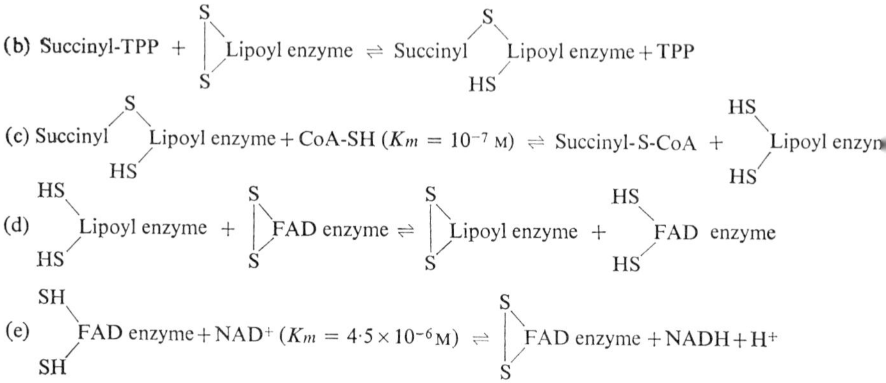

(b) Succinyl-TPP + (S–S)Lipoyl enzyme ⇌ Succinyl-S(HS)Lipoyl enzyme + TPP

(c) Succinyl-S(HS)Lipoyl enzyme + CoA-SH ($Km = 10^{-7}$ M) ⇌ Succinyl-S-CoA + (HS)(HS)Lipoyl enzyme

(d) (HS)(HS)Lipoyl enzyme + (S–S)FAD enzyme ⇌ (S–S)Lipoyl enzyme + (HS)(HS)FAD enzyme

(e) (SH)(SH)FAD enzyme + $NAD^+$ ($Km = 4{\cdot}5 \times 10^{-6}$ M) ⇌ (S–S)FAD enzyme + NADH + $H^+$

(IV)

The succinyl-S CoA reacts with a nucleoside diphosphate to form a nucleoside triphosphate as shown in reaction V.

$$\text{Succinyl-S-CoA} + \text{NuDP} + P_i \underset{\longleftarrow}{\xrightarrow{\text{succinyl-CoA synthetase}}} \text{Succinate} + \text{NuTP} + \text{CoASH}$$

(V)

The reversibility of reaction V was suggested by the studies of Shemin and Kumin (1952). They blocked succinate oxidation in intact duck erythrocytes with 0·02 M malonate so that very little α-KG would be generated from the citric acid cycle. When they now added methylene-labeled succinate the label was incorporated into newly formed heme. From this experiment it was inferred that about 30–50% of the succinyl-CoA could be derived directly from added succinate.

Reactions IV and V also occur in liver and have been studied in liver mitochondria obtained from chemically porphyric animals that have a relatively high ALA-synthetase activity (Granick and Urata, 1963). To distinguish between the extent of the forward and reverse reactions the mitochondria were poisoned with arsenite. With α-KG and glycine as substrates it was estimated that 85% of the succinyl-CoA which was used for ALA synthesis was derived from the oxidation of α-KG, and 15% from succinate by the reverse of reaction V. When succinate and glycine were substrates under the same conditions, 35–60% of the succinyl-CoA was derived from succinate.

ATP, when added to hemolysed chicken erythrocytes, enhanced ALA synthesis three-fold only when succinate was substrate, a result that also suggests the physiological importance of the reverse of reaction V.

The properties of succinyl-CoA synthetase have been reviewed by Hager (1962). The enzyme is specific for succinate. The equilibrium constant is 3·7 and the free energy 750 cal and, therefore, the reaction is thermodynamically easily reversible. For the kidney enzyme the $K_m$ for succinate is $5 \times 10^{-3}$ M. The animal synthetase requires either GTP or ITP whereas the enzyme from spinach requires ATP.

Burnham (1963) has obtained a purified succinyl-CoA synthetase from *Rhodopseudomonas spheroides* which should be useful to provide continuous generation of succinyl-CoA in further studies of ALA-synthetase.

A third type of reaction for succinyl-CoA synthesis is the one from propionyl-CoA. The propionyl-CoA reacts with $CO_2$ and a biotin enzyme to form methyl malonyl-CoA. The latter is converted by a soluble enzyme containing $B_{12}$ into succinyl-CoA. These reactions provide a pathway for the conversion of propionic acid into a compound of the citric acid cycle. This method of succinyl-CoA formation appears to be especially important in sheep and other ruminants which use the propionic acid produced in the rumen as an energy source. The $B_{12}$ reaction has not been shown in erythrocytes. The absence of $B_{12}$ from plants eliminates this mechanism for succinyl-CoA synthesis from consideration in plants.

A fourth method of succinyl-CoA synthesis is via a CoA transferase from acetoacetyl-CoA to succinate. This enzyme has low activity in liver.

The available data indicate that mitochondria of animal cells and presumably of leaf cells (Zelitch, 1964) are the principal site of ALA synthesis because succinyl-CoA is formed via $\alpha$-KG-oxidase, a mitochondrial enzyme complex, and because ALA-synthetase is present in mitochondria. In erythrocytes, the presence of mitochondria correlates well with the ability to synthesize heme.

As yet no evidence has been obtained that chloroplasts can make succinyl-CoA.

## PROPERTIES OF ALA-SYNTHETASE

ALA-synthetase of erythrocyte particles or liver mitochondria is inhibited by a number of substances such as cysteine, CoASH and L-penicillamine. These probably react with the aldehyde group of the pyridoxal phosphate coenzyme on the enzyme to form a thiazolidine ring, or with HCN to form a cyanohydrin (Gibson *et al.*, 1958). Aminomalonate is also a good inhibitor ($K_i = 2 \times 10^{-5}$ M) (Gibson *et al.*, 1961).

ALA-synthetase from the purple photosynthetic bacterium *R. spheroides* has been obtained in a soluble form by Kikuchi *et al.* (1958). Shemin (1962) has described the preparation of the enzyme. The enzyme occurs in the chromatophores. The ALA-synthetase has been localized in the solubilized fraction which precipitates with 30% ammonium sulfate. The mechanism of action proposed for ALA-synthetase (Fig. 4) is the condensation of pyridoxalphosphate with glycine to form a Schiff-base on the enzyme $E_n$, thus forming a stabilized carbanion with loss of a proton. This is followed by condensation in which the acyl-C atom of succinyl-CoA acts as an electron acceptor. The decarboxylation occurs simultaneously or shortly thereafter. Enzyme activity is maximal at pH 6·9. The $K_m$ for pyridoxal phosphate is $5 \times 10^{-6}$ M, for succinyl-CoA it is $2{\cdot}2 \times 10^{-5}$ M and for glycine $3 \times 10^{-3}$ M. Inhibitors of this enzyme material reveal that SH groups are required for activity. The enzyme is inhibited

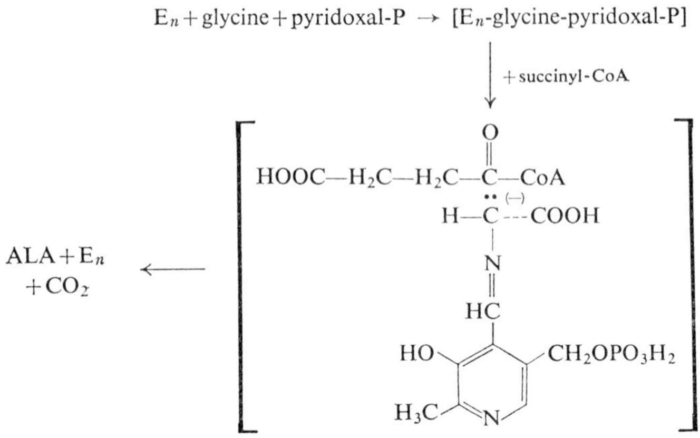

$E_n$ + glycine + pyridoxal-P → [$E_n$-glycine-pyridoxal-P]

↓ + succinyl-CoA

[ HOOC—$H_2$C—$H_2$C—C(=O)—CoA ; H—C$^{(-)}$---COOH ; N ; HC ; HO, $CH_2OPO_3H_2$, $H_3$C, N (pyridine ring) ]

ALA + $E_n$ + $CO_2$ ←

FIG. 4. Mechanism of ALA-synthesis.

50% by ALA itself at $5 \times 10^{-2}$ M. α-KG is a competitive inhibitor of glycine utilization which may explain the fact that in erythrocytes, concentrations of α-KG above 0·001 M inhibit ALA synthesis.

In *R. spheroides* Shemin *et al.* (1963) have reported growth inhibition in ALA (0·16 mM) both under anaerobic conditions under light, and under aerobic conditions in the dark. Growth occurred when the ALA decreased to one-fourth of the original value. The inhibition was therefore reversible. The authors suggest that ALA may be acting as a repressor of ALA-synthetase.

Iron-deficient bird erythrocytes have a diminished rate of ALA synthesis (Vogel *et al.*, 1960). When such erythrocytes are incubated for 30 min with $Fe^{2+}$ the rate of ALA synthesis increases two to three times. Similarly, the iron chelating agent *o*-phenanthroline ($1 \times 10^{-3}$ M) decreases, by half, the PROTO synthesis of chicken erythrocytes (Gibson *et al.*, 1958). Neither hemolysates nor cell particles from iron deficient cells respond to the addition of iron. It is possible that in iron deficiency non-heme enzymes of the citric acid cycle, e.g. aconitase, and flavin-Fe enzymes (Granick, 1958), e.g. succinic dehydrogenase, may be diminished as well as the heme enzymes, e.g. the cytochromes of the oxidative pathway. This would lead to a diminution in the rate of ALA synthesis. To test this hypothesis direct measurements of low ALA-synthetase activity will be required.

In plants, iron deficiency leads to an inability to synthesize chlorophyll. Marsh *et al.* (1963) showed that the lack of iron in young leaves of cow pea resulted in a marked decrease in the ability of illuminated leaf discs to incorporate labeled citrate, α-KG or succinate into chlorophyll. In contrast, the ability to convert ALA into protoporphyrin was not impaired. These investigators also found that ferredoxin is also diminished in iron deficiency. Ferredoxin is an iron (non-heme) enzyme which functions during photosynthesis to reduce $NADP^+$. A decrease in this enzyme could prevent the energy of photosynthesis from being available for chloroplast synthesis. Thus in higher plants chlorosis due to iron deficiency is suggested to be due to a decreased supply of succinyl-CoA for ALA-synthesis, as well as to the inability of photosynthesis to supply energy for chloroplast synthesis.

In photosynthetic and non-photosynthetic bacteria iron deficiency is accompanied by the production of excessive amounts of COPRO III. In these organisms Lascelles (1964) has presented evidence for the hypothesis that heme regulates the formation of ALA. Heme inhibits ALA-synthetase activity by allosteric inhibition. Also the synthesis of ALA-synthetase is repressed by heme. In iron deficiency, when heme is low, more ALA-synthetase is formed, and the ALA-synthetase that is present is not inhibited. Because ALA-synthetase is the limiting enzyme, all other enzymes of the chain being normally in excess, an increase in ALA-synthetase activity leads to an increased synthesis of ALA and thus to an increase in porphyrin formation. The control of the ALA-synthetase level will be discussed further on pp. 394 and 404.

### ALA-DETERMINATION

The synthesis of ALA is best followed by its reaction with acetylacetone to form a pyrrole and the determination of the pyrrole colorimetrically with Ehrlich's reagent (*p*-dimethylaminobenzaldehyde in acid) (Mauzerall and Granick, 1956). Methods for the separation and determination of ALA, PBG and aminoacetone have been developed (Urata and Granick, 1963). A non-chromatographic method for the determination of ALA and aminoacetone is now available (Granick, 1966).

## Conversion of δ-Aminolevulinic Acid into Porphobilinogen by ALA-Dehydrase (PBG-Synthetase)

ALA and the class of α-aminoketo compounds to which it belongs are, in general, stable in acid but rapidly and reversibly self-condense to dihydropyrazines in alkaline solution (pH 8) as shown in Fig. 5. The dihydropyrazines autoxidize in air to form aromatic pyrazines.

ALA-dehydrase has been partially purified from ox liver (Gibson *et al.*, 1955) rabbit reticulocytes (Granick and Mauzerall, 1958a), and from *R. spheroides* (Shemin, 1962). The $K_m$ for these enzymes is, respectively, $1{\cdot}4 \times 10^{-4}$ M, $5 \times 10^{-4}$ M and $3 \times 10^{-4}$ M. The optimal pH is about 6·7 for the animal enzymes and 8·6 for the *R. spheroides* enzyme. The enzyme requires activation by SH compounds such as glutathione or cysteine. Metals inhibit roughly in the order of the solubility products of their sulfides. The enzyme will only accept ALA as substrate. Inhibitor studies (Granick and Mauzerall, 1958a) suggest that a carbonyl group $\gamma$ to an ionized carboxyl group, but not an amino group, is required for binding of the substrate to the enzyme. Kinetic studies suggest the following mechanism indicated in Fig. 5. Both molecules of ALA are

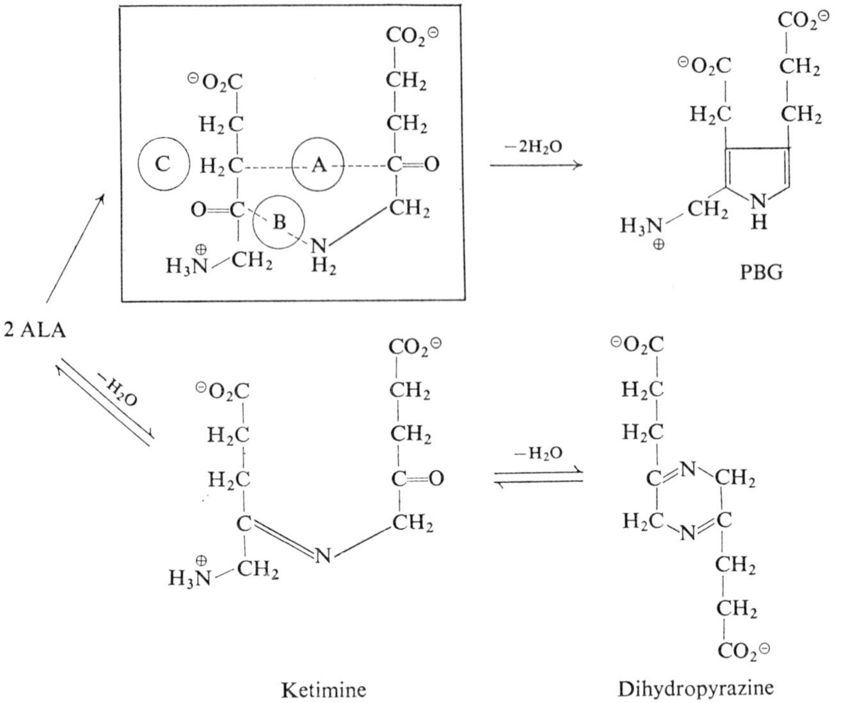

2 ALA

$-2H_2O$ PBG

$-H_2O$

$-H_2O$

Ketimine Dihydropyrazine

FIG. 5. The condensation of two ALA molecules to PBG by ALA-dehydrase is shown. Also shown is the condensation of two ALA molecules non-enzymically to a ketimine and thence to a dihydropyrazine.

bound by the enzyme,the first at least ten times more firmly than the second. The binding of the second ALA may involve spontaneous formation of the ketimine at *B*. The presence of a metal at *C* (as Lewis acid) would favor the formation of an enolate ion, but no metal was found. The aldol condensation requires that an enolate ion attack the carbonyl of the adjacent ALA molecule at *A*. The condensation at *A* with resultant hydrogen shift would then result in pyrrole formation. The enzyme is inhibited by EDTA "uncompetitively". Both the liver and reticulocyte enzymes are inhibited at low concentrations ($K_i = 10^{-5}$ M). The enzyme from chicken erythrocytes is only partly inhibited at $10^{-5}$ M EDTA and the inhibition is not increased even at $10^{-2}$ M (Granick and Mauzerall, 1958a). Possibly, a change in configuration of the enzyme or removal of a metal or both are involved in this peculiar effect. The *R. spheroides* enzyme, on the other hand, is not inhibited by EDTA nor by αα-dipyrridyl.

## Conversion of Porphobilinogen into Uroporphyrinogen III

This reaction requires two enzymes, as was first recognized by their differential susceptibility to heat. When an enzyme preparation is incubated with

PBG, UROGEN III is formed. If the preparation is heated to 60° for 15 min, only UROGEN I is formed. Preparations from *Chlorella* cells (Bogorad, 1960), erythrocytes and *R. spheroides* (Heath and Hoare, 1959) behave in the same way.

The four possible isomers of uroporphyrinogen are shown in Fig. 6. Only isomers I and III are ever encountered in nature. A stepwise condensation of four PBG molecules with the elimination of $NH_3$ at each step would give rise to isomer I. However, for isomer III to be formed one of the pyrrole residues must

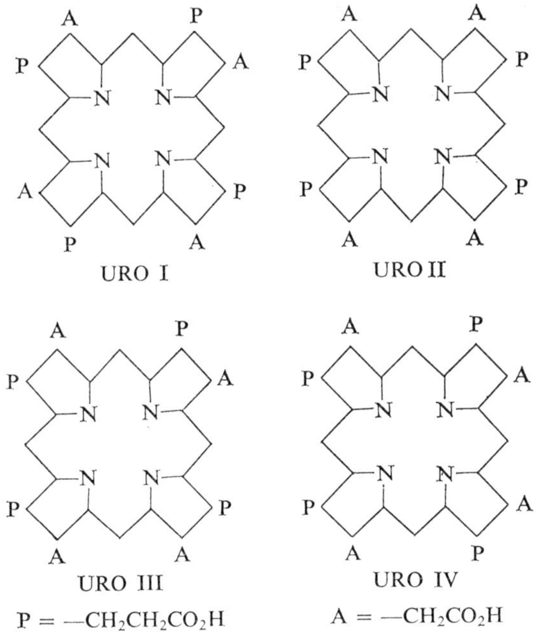

URO I

URO II

URO III

URO IV

P = $—CH_2CH_2CO_2H$ A = $—CH_2CO_2H$

FIG. 6. Schematic representation of the four isomers of uroporphyrin. Isomer III differs from I in that one pyrrole ring (in the lower left-hand corner of the porphyrin ring) has been "flipped over".

be "flipped over" during the condensation. The many synthetic mechanisms that have been offered in explanation have been discussed in detail by Bogorad (1960). Here we mention only the ingenious hypothesis of Mathewson and Corwin (1961). They suggest that PBG condenses to a protonated tetramer (Fig. 7, No. 3) which retains the $\alpha$-hydrogen atoms, thus allowing for greater flexibility of the molecule. The tetramer contains three pyrrolenine rings and may cyclize in one of three ways. It may cyclize directly to form UROGEN **I,** with the splitting out of $NH_3$. Or, it may cyclize, then open and recyclize to form either UROGEN III or a corrin ring precursor of $B_{12}$. In the formation of UROGEN III the $-CH_2NH_3^+$ group of one pyrrole "d" attacks not at the

$\alpha$ position of pyrrole "a" containing the H atom (as for UROGEN I synthesis) but rather at the $\alpha$position which contains a $-CH_2-$ group.

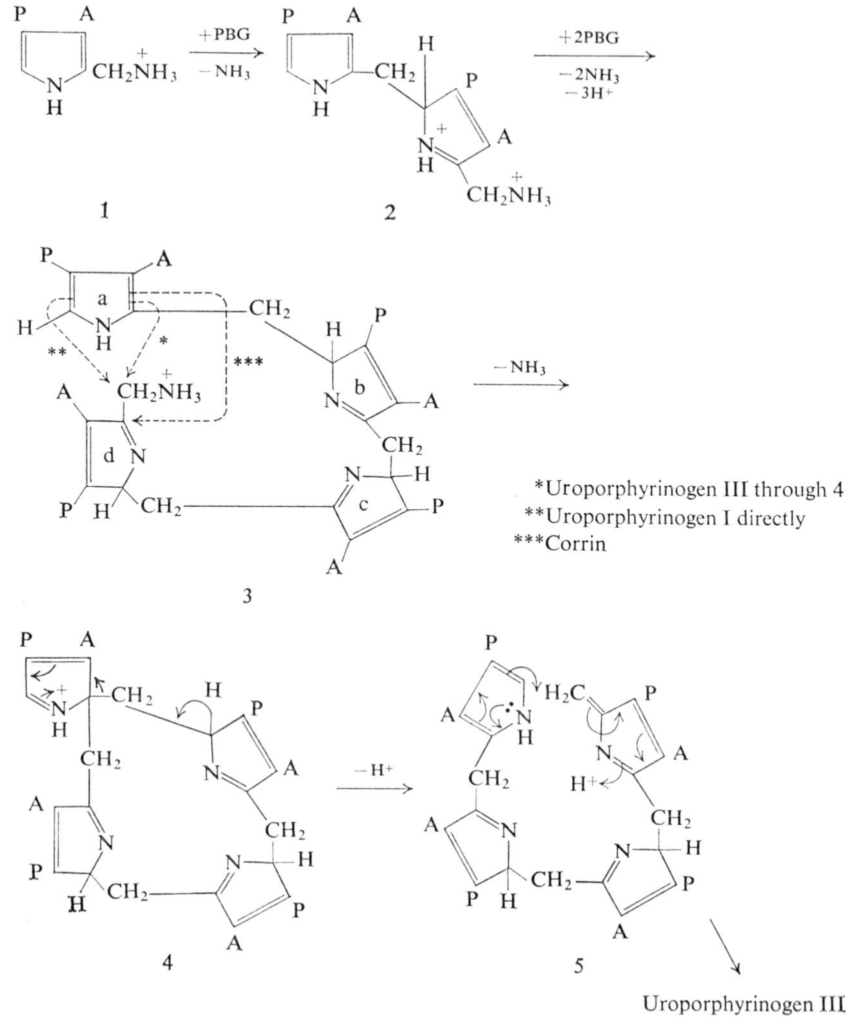

FIG. 7. Mechanism of Mathewson and Corwin (1961) for the synthesis of uroporphyrinogen isomers I and III and the corrin ring of vitamin $B_{12}$.

Bray and Shemin (1963) have found that the main skeleton of the corrin structure of $B_{12}$ is derived from the condensation of eight ALA molecules in a manner similar to porphyrin biosynthesis. The six "extra" methyl groups are derived from the methyl group of methionine probably via adenosyl methionine. Bonnet (1963) has reviewed the chemistry of $B_{12}$.

Bogorad (1960) has investigated the two enzyme activities required for the synthesis of UROGEN III in plant material. Using an acetone-dried powder of spinach he obtained an extract which, when heated to 60°, had only uroporphyrinogen I synthetase activity (i.e. PBG deaminase). This enzyme converts 4 PBG quantitatively under anaerobic conditions into UROGEN I as indicated by the reaction

$$4\ \text{PBG} \rightarrow 1\ \text{UROGEN I} + 4\ NH_3$$

The $K_m$ at pH 8·2, 37° is $7 \times 10^{-5}$ M. The enzyme requires SH groups for activity. It will act only on PBG. It will not act on isoporphobilinogen (in which the acetic and propionic acid groups are exchanged) nor on opsopyrrole-dicarboxylic acid (PBG in which H is substituted for $CH_2NH_3^+$). The latter two compounds act as competitive inhibitors. Dipyrrylmethanes with two $CH_2NH_3^+$ side chains, or with none, are not used by the enzyme.

Intermediates accumulate when PBG is condensed enzymatically in the presence of $NH_2OH$ or $NH_3^+$. The second enzyme, uroporphyrinogen III-cosynthetase (PBG isomerase) was isolated by Bogorad (1962) from an aqueous extract of wheat germ. This enzyme does not act directly on PBG. However, UROGEN III is formed when both enzymes are incubated together with PBG. The overall $K_m$ at pH 8·2, 37° is $1 \times 10^{-4}$ M. Hydroxylamine is a powerful inhibitor of UROGEN III formation. UROGEN I is not converted into UROGEN III by the enzyme. The lack of inhibition of sulfite or dithionite ions suggests that the condensations occur at the methane (i.e. fully reduced) not methene level (Mauzerall and Granick, 1958). Probably PBG is condensed by the first enzyme to a trimer, and the second enzyme acts with another PBG to form UROGEN III. Although two enzyme activities must be present, electrophoresis of hemolysates of red cells from three different species indicates that only one zone contains activity to convert PBG into UROGEN III (Granick and Mauzerall, 1958a). There is no zone with activity to form only UROGEN I. This suggests that both enzymes may be associated in some kind of complex in animal cells whereas in plant cells the complex is not present or is readily dissociated.

## Oxidation States of Uroporphyrinogen and the Photochemical Properties of Uroporphyrin

The porphyrinogens are colorless tetrapyrrylmethanes containing six more H atoms than the porphyrins (Fig. 8). Fischer had early reported their occurrence in biological material and obtained COPROGEN from a congenital porphyria patient. Watson, Schwartz and coworkers reported that COPROGEN was present in the urine of porphyria patients. Suspicions arose that these porphyrinogens were the actual intermediates in porphyrin biosynthesis when it was observed that the porphyrins URO and COPRO could not be used for heme or chlorophyll synthesis by cells. The structure of PBG gave further indirect evidence for this view since the condensation of this pyrrole produces UROGEN. In 1955 Bogorad obtained COPRO by the incubation of a *Chlorella* preparation with UROGEN; and Neve, Labbe and

Aldrich (1956) found that the addition of UROGEN to hemolysed red cells increased the incorporation of Fe into heme.

The porphyrinogens of URO and COPRO are most readily prepared by shaking the porphyrin with sodium amalgam in the dark in the absence of air. Traces of the colored porphyrins markedly sensitize the autoxidation of the

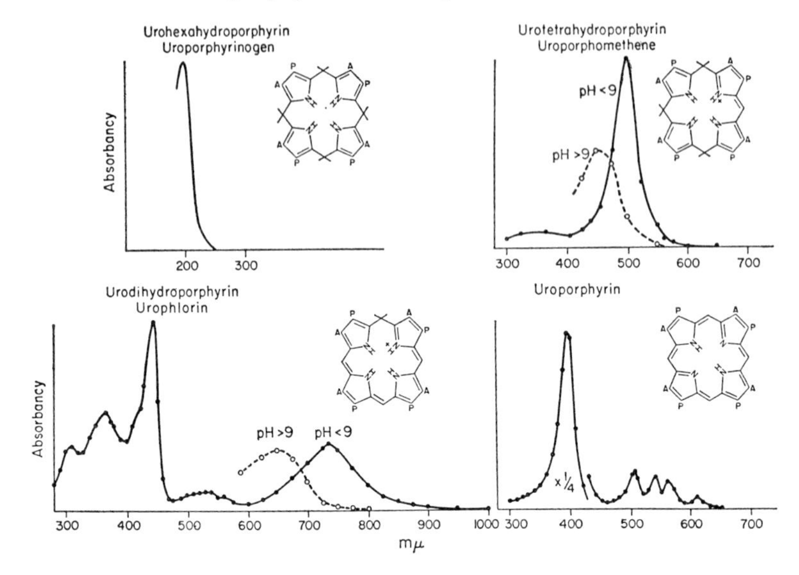


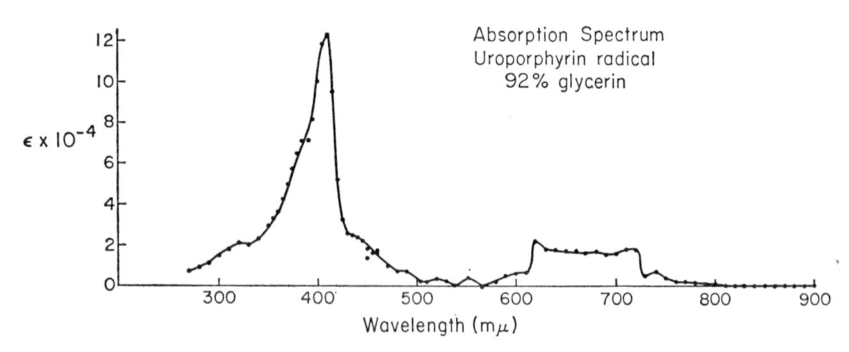


FIG. 8. Formulae and absorption spectra of the different oxidized states of uroporphyrinogen.

porphyrinogens in air. The spectra of the porphyrinogen (hexahydroporphyrin), the tetrahydroporphyrin (with a strong absorption band at 500 mμ), the dihydroporphyrin or phlorin (with a band at 735 mμ) and the porphyrin itself are shown in Fig. 8 and were obtained by Mauzerall (1962). The mechanism of autocondensation of PBG in acid and neutral solutions to form various UROGEN isomers was also delineated by Mauzerall (1960a,b). Recently the free radical of uroporphyrin has been obtained by Mauzerall and Feher (1965). This probably represents uroporphyrin with an added electron. It has a broad absorption band in the visible region from 620 to 730 m$\mu$ including an extension to about 800 m$\mu$.

Uroporphyrin is of especial interest photochemically. It must have been the first compound of the tetrapyrrole chain to serve in protoplasm in a photochemical capacity. Because of the outstanding investigations of Mauzerall, some of the remarkable photochemical properties of uroporphyrin have become known. When uroporphyrin is illuminated it will react with EDTA, one of the mild reductants, and pull out an an electron from EDTA. Thus a radical of uroporphyrin may be formed (Fig. 8). Two of these radicals can disproportionate to form a molecule of dihydroporphyrin and a molecule of porphyrin (Fig. 9). Thus by the action of light, uroporphyrin can be excited to accept and store up to four electrons. If these electrons are on a potential level

$$\text{Porphyrin} + \text{EDTA} \xrightarrow{h\nu} [\text{Porphyrin radical}^{\bullet}] + \text{EDTA oxidized}$$

$$2\ [\text{Porphyrin radical}^{\bullet}] \rightleftharpoons \text{Dihydroporphyrin} + \text{Porphyrin}$$

$$2\ \text{Dihydroporphyrin} \rightleftharpoons \text{Tetrahydroporphyrin} + \text{Porphyrin}$$

FIG. 9. Photochemical reduction of porphyrin to tetrahydroporphyrin.

of the hydrogen electrode then with the correct coupling enzymes, $H_2$ may be generated. This reaction would be analogous to the overall reaction that occurs anaerobically in purple photosynthetic bacteria; when these bacteria are provided with a reductant such as sodium thiosulfate they generate $H_2$ in the presence of light.

At present we do not know whether the light energy of this uroporphyrin reaction is used merely as energy of activation, that is, to catalyse electron transfer, or whether the energy of light can be stored in the form of chemical energy. Even if light acting on URO merely catalysed electron transfer from organic compounds, and $H_2$ was generated, this might have served as a useful mechanism in evolution. It could have brought about the removal of $H_2$ from the highly reduced "organic soup" that has been postulated by Urey as a starting condition for life on earth. The $H_2$ would have escaped into the atmosphere and eventually into outer space thus leaving the lithosphere less reducing (Granick, 1965a).

The isomers of URO found in nature belong to the I and III series but it is only the III isomer which is represented in the structure of heme, the chlorophylls and vitamin $B_{12}$. Why was isomer III selected out of the four possible isomers? When PBG is heated in acid solution all four isomers are obtained. When PBG is heated in neutral or alkaline solution only isomers I and III are obtained (Cookson and Rimington, 1954; Mauzerall, 1960). One may conjecture that at an early time in evolution the biosynthetic chain ended with the enzyme that formed PBG in a slightly basic environment such as sea-water. PBG might spontaneously form small amounts of UROGEN isomers I and III. If these compounds represented an advantage to the cell then an enzyme would be developed to do the job more efficiently. This leads to the larger question whether there is a general tendency in evolution to select an enzyme to carry out a step that may take place spontaneously in neutral or alkaline rather than in acid solution. In addition, is there selection for compounds that are more stable in neutral or alkaline rather than in acid solution, e.g. DNA *vs* RNA? These selections would follow from an evolution in a somewhat basic environment like that of sea-water.

## Conversion of Uroporphyrinogen into Coproporphyrinogen by Uroporphyrinogen-Decarboxylase

An enzyme preparation that converts UROGEN to COPROGEN was obtained from rabbit reticulocytes by zone electrophoresis on starch. The enzyme has a high affinity for substrate ($K_m$ $5 \times 10^{-6}$ M) and a pH optimum of 6·8 (Mauzerall and Granick, 1958). It acts only on UROGEN to decarboxylate the 4 acetic side chains to $-CH_3$. URO is not acted on. Sulfite or dithionite ions which only complex tightly with di- or tetrahydroporphyrins at methene bridge C atoms do not inhibit the enzymatic decarboxylation. The decarboxylation proceeds stepwise but the intermediate 7 to 5 carboxyl-containing porphyrinogens, although they can be detected, are converted rapidly into the 4 carboxyl-containing COPROGEN. The enzyme decarboxylates all four isomers of UROGEN in the order III > IV > II > I. Isomer III reacts at twice the rate of isomer I. The enzyme from human red cells acts on III 7·5 times as fast as on I (Cornford, 1964).

The decarboxylase is assumed to be a single enzyme since electrophoresis of the red cell hemolysates of three different species yields in each instance only one zone with the enzymic activity. It is inferred from the lack of isomer specificity that the enzyme decarboxylates the side chains at random and thus a large number of intermediates may exist, all of which, on continued decarboxylation, yield the appropriate isomer of COPROGEN.

## Conversion of Coproporphyrinogen into Protoporphyrinogen by Coproporphyrinogen Oxidase

The formation of PROTO from COPROGEN has been observed in preparations from chicken erythrocytes (Granick and Mauzerall, 1958b), *Euglena* and beef liver mitochondria (Sano and Granick, 1961). In contrast to the non-specific nature of UROGEN decarboxylase for isomer types, the oxidase enzyme is highly specific. It attacks COPROGEN III, not I. With a *Euglena* preparation neither isomer I nor II is a substrate. Isomer IV undergoes oxidative decarboxylation only one-tenth as fast as isomer III. Porra and Falk (1964) have also reported a slight activity with isomer IV. COPRO III is not a substrate.

This high specificity for COPROGEN III is the main reason for the ubiquitous distribution of PROTO of series III and the absence of PROTO of series I. However, two other enzymes of the biosynthetic chain of heme have discriminated in favor of isomer III. One is the UROGEN-cosynthetase, the action of which results in the formation of over 99·9% of UROGEN III and only traces of isomer I, as judged by the amounts of isomer I found in normal urine.

UROGEN-decarboxylase also favors isomer III, acting two to three times as fast on III as on I.

The coproporphyrinogen oxidase has a greater activity in liver and bone marrow than in other tissues (Sano and Granick, 1961). It has a $K_m$ of about $2 \times 10^{-5}$ M and pH optimum of 7·7. Only two of the four propionic acid groups are oxidatively decarboxylated. Specifically, those at side chain positions 2 and 4 form vinyl groups (Fig. 1). During the reaction, an intermediate that has one vinyl and three propionic acid groups appears and then disappears. Deuteroporphyrinogen IX-4-propionic acid can also be converted into the 4-monovinyl derivative (Porra and Falk, 1964). The enzyme does not act on the porphyrinogens of hematoporphyrin (Bogorad and Marks, 1960), 2,4-diacetyldeuteroporphyrin, or *trans*-2,4-diacrylicdeuteroporphyrin (Granick and Sano, 1961). These observations eliminate from consideration steps similar to fatty acid oxidation. Preliminary experiments in tritiated water suggest that the oxidation and decarboxylation occur simultaneously. A tentative hypothesis is that the oxidation of the propionic acid proceeds by hydride ion removal, with $CO_2$ leaving simultaneously. Cyanide does not inhibit the enzyme. No substitute for $O_2$ as oxidant has been found that would oxidize the propionic acid groups yet not oxidize the porphyrinogen as well. Even $H_2O_2$ cannot replace $O_2$ as oxidant with this enzyme. The impure enzyme preparation from guinea-pig liver contains a flavin. In anaerobic photosynthetic bacteria an oxidant other than $O_2$ must be postulated to explain heme and chlorophyll formation.

In *Euglena* the rate of synthesis of chlorophyll can be made linearly dependent on the iron content of the cells. Carell and Price (1965) have found that in iron-depleted *Euglena* coprogenase is not a limiting enzyme.

The protoporphyrinogen which is formed from COPROGEN is autoxidized with $O_2$; possibly this autoxidation is catalysed by an enzyme (Sano and Granick, 1961; Porra and Falk, 1964).

## Methods for Analysis of Porphyrinogens

Methods for the preparation and determination of porphyrinogen have been described by Mauzerall and Granick (1958). The porphyrinogens are best determined by oxidation to the porphyrins with $I_2$ at neutral pH, avoiding excess $I_2$. However, protoporphyrinogen must be photo-oxidized or autoxidized because it is easily destroyed by $I_2$ (Sano and Granick, 1961). The number of carboxyl groups on a porphyrin is readily determined by paper chromatography with a lutidine-$NH_3$-$H_2O$ system (Falk, 1961). The specific isomers of COPRO are best analysed by paper chromatography with the same system (Mauzerall, 1960b). To determine the isomers of URO, they must first be decarboxylated to the respective COPRO isomers (Edmundson and Schwartz, 1953).

## Insertion of Ferrous Iron into Protoporphyrin
## The Iron Chelatase Enzyme

Attempts to isolate a ferrous iron chelating enzyme have been hampered by the relative ease of the non-enzymic incorporation of iron into porphyrins.

Conditions which enhance the chelation of porphyrins with iron by enzymes also enhance the non-enzymic chelation. These conditions are: the maintenance of iron in the ferrous form by anaerobiosis or by reducing agents such as ascorbic acid, cysteine or glutathione; the avoidance of buffers such as those containing phosphate or amines that complex with iron; the solubilization of highly insoluble or colloidal PROTO by means of various detergents so that PROTO may become available in a mono-molecularly dispersed form.

The enzyme catalysing the chelation of iron to form heme has been reported to be present in liver mitochondria and in particles from hemolysed chicken or duck erythrocytes. The enzyme seems to be localized in mitochondria. However, the nuclei of avian erythrocytes have not yet been eliminated as another site of this enzyme.

Labbe and Hubbard (1961) studied the chelatase enzyme from rat liver mitochondria and from duck erythrocytes. Tween 20 was employed as the solubilizing agent in their system. They suggested that SH groups are needed by the enzyme to bind the porphyrin. Both $Fe^{2+}$ and $Co^{2+}$ can be inserted into the porphyrin by the enzyme; other metals are inhibitory. Dicarboxylic porphyrins but not tetracarboxylic porphyrins are used by the enzyme. The rate of iron incorporation by the enzyme is more than ten times the non-enzymic rate. They considered that the enzyme is more selective with respect to the kind of metal than to the porphyrin because the enzyme in the cell is presented with only one porphyrin, PROTO, but needs to select $Fe^{2+}$ from among a number of other metal ions.

Oyama *et al.* (1961) obtained a solubilized preparation from duck erythrocytes using sodium cholate; their preparation incorporated ferrous, zinc, and cobalt ions into PROTO in a ratio of 100:10:2.

Porra and Jones (1963) isolated a chelatase enzyme from pig liver mitochondria. Their test system contained a detergent to render the porphyrin more soluble. GSH was also added and $O_2$ removed to maintain the iron in the ferrous state. SH groups were required to keep the iron reduced, not to activate the enzyme. Only the dicarboxylic porphyrins like proto-, meso-, deutero-, and hemato- were substrates but not tetracarboxylic porphyrins like copro-. The pH curve of the reaction had two maxima of activity, suggesting the presence of two enzymes. At pH 7·8 and 37° their enzyme preparation exchanged 1·25% of the iron between protoheme and labeled ferrous citrate in the presence of dithionite in 1 hr.

Neve (1961) prepared a Tween 40 extract from chicken erythrocyte particles and fractionated it with ammonium sulfate. Lead acetate ($10^{-5}$ M) inhibited the chelatase activity of this preparation by 65%. Schwartz *et al.* (1961) found that in the presence of globin the incorporation of iron into PROTO by their preparation of enzyme was increased three-fold, but Yoneyama *et al.* (1963) found globin inhibitory.

## The Hemes of Heme-proteins and the Detection of Heme

All the hemes of heme-containing proteins are derivatives of the type III isomer. [A summary of their properties are presented in Granick and Mauzerall

(1961), Falk (1963), see also Vol. 8 of *The Enzymes* (Boyer, Lardy and Myrbäck, eds.) (1963).] The proteins include cytochrome *b* and catalase which, like hemoglobin, contain iron protoporphyrin as prosthetic group. Cytoxidase has a modified iron porphyrin in which side chain position 2 is [$-CHOH-CH_2-[CH_2.CH_2.CH(CH_3).CH_2]_3H$] in place of a vinyl group, and position 8 has a $-CHO$ group in place of a $-CH_3$. These hemes can be readily removed from their respective proteins by organic solvents.

Cytochrome *c*, on the other hand, contains a heme which is bound through each vinyl side chain to a cysteine residue forming a thioether bond at the 2 and the 4 positions (Fig. 1). This heme, because it is covalently linked to the protein, cannot be removed with organic solvents. The mechanism assumed for cytochrome *c* formation is an enzymatic reaction of apocytochrome *c* with iron protoporphyrin through the vinyl groups. Recently, a different mechanism has been suggested based on the fact that SH compounds react with vinyl groups of reduced porphyrins. Sano and Tanaka (1964) have shown that apocytochrome *c* reacted readily with reduced protoporphyrin to form thioether bonds. The iron was then inserted into the ring to produce a molecule resembling cytochrome *c*. Perhaps cytochrome *c* is formed enzymatically in cells by a similar mechanism.

Plant microsomes contain a cytochrome $b_3$ ($E_0' = +0{\cdot}04$), and chloroplast grana contain a cytochrome $b_6$ ($E_0' = -0{\cdot}06$) and a cytochrome *f* ($E_0' = +0{\cdot}36$) that resembles cytochrome *c*. Cytochrome *f* has a mol wt of $10^5$ and in parsley is present in a molar ratio of 1 : 400 chlorophylls.

Porphyrins, dihydroporphyrins and metalloporphyrins have an intense absorption band in the 400 m$\mu$ region (the Soret band), with an $E_m > 10^5$. Ferrous porphyrin in aqueous pyridine has a characteristic "hemochromogen" absorption in which form it can be determined quantitatively (Paul *et al.*, 1953). Iron porphyrin or heme may be detected by its peroxidase activity with the benzidine-$H_2O_2$ test. Heme may also be converted into porphyrin by removal of iron from the molecule and the porphyrin can then be detected by its fluorescence. The iron may be removed by ultraviolet light in 2N perchloric acid in the presence of SH groups; this makes possible a fluorescence histological technique for the detection of heme in tissues (Granick and Levere, 1965).

## Relation of Iron to Porphyrin and Chlorophyll Synthesis

Studies of the requirement of iron for the growth of various organisms have revealed two major effects. One is the relatively high requirement of iron for the greening of plants. The other is the excessive production of porphyrins by certain photosynthetic and non-photosynthetic bacteria when the iron in the medium is low.

In general it appears that under conditions of limited iron, the iron is coordinated most tightly first with porphyrins to form the heme that is required for heme enzymes. As the iron is increased, looser complexes of iron are formed, e.g. the Fe-flavin complexes of succinic dehydrogenase, xanthine oxidase, NAD-cytochrome *c* reductase, etc. With still more iron available, the iron sulfide complexes contained in ferredoxin may be formed. And with still more iron, the basic ferric hydroxide-phosphate form of iron contained in ferritin may be formed to serve as a storage form of iron; or the complex form of iron in transferrin may be formed to serve for iron-transport in the blood stream and as a temporary storage form in egg white. At relatively high iron concentrations iron may also be used as a cofactor for enzymes such as aconitase. Thus, depending on the level of iron depletion studied and the organism, it may be expected that different effects of iron deficiency will be observed.

In photosynthetic bacteria such as *Rhodopseudomonas spheroides* and a number of other non-photosynthetic bacteria under growth conditions with limited iron, but not when iron is ample, it is found that porphyrins are excreted into the medium. Lascelles (1964) has presented the reasonable and well documented hypothesis that the porphyrin excretion is caused by a lack of heme. Heme decreases ALA-synthetase activity both by acting as a repressor at the DNA level, and as an inhibitor of ALA-synthetase itself.

In plants the chloroplasts contain 50–80% of the total iron of the leaf. The molar ratio of Fe to chlorophyll is 1:4 to 1:10 in the leaves of most plants. The chloroplasts contain cytochromes $b_6$ and $f$ and ferredoxin. The leaves also contain ferritin. When the supply of iron to the leaf is limited, Marsh *et al.* (1963) suggest that ALA synthesis is limited; this might be a result of a limited synthesis of succinyl-CoA if iron is required for enzymes of the citric acid cycle. Tests for heme as a repressor or inhibitor of ALA-synthetase in leaves have not yet been made; these may require drastic iron-depletion techniques. Studies of the conversion of ALA into porphyrins, including MgVP, indicate that chelators of iron, e.g. $\alpha\alpha$-dipyridyl or pyridyl-2-aldoxime, enhance the porphyrin yield. From this fact it is surmised that iron enzymes are not involved directly in most of the biosynthetic chain of chlorophyll (Granick, 1961a). If iron is deficient in the leaf, ferredoxin might be limited and under such conditions photosynthesis would be limited; therefore the synthesis of developing chloroplasts would be depressed. In iron-deficient plants the protein and the lamellae of the chloroplasts are fewer in number (Bogorad *et al.*, 1959). Because there is a close coupling between chlorophyll formation and the formation of the lipoprotein lamellae of the chloroplasts, it has not been possible to determine which is affected first by the lack of iron.

In higher animals, iron deficiency is most apparent as an anemia. The red blood cells although not decreased in number are small and have a decreased content of hemoglobin. The synthesis of globin is controlled by the availability of heme (Levere and Granick, 1965). Free PROTO may be ten times as high in the anemic cells as it is in normal. Possibly the lack of heme as repressor has resulted in an overproduction of ALA and therefore of PROTO. In iron-deficient duck blood, ALA-synthetase activity is depressed; the cells, but not cell fragments, when incubated with iron for an hour or longer acquire an increased ability to convert glycine into PROTO (Vogel *et al.*, 1960). It has yet to be determined whether the iron is involved in the synthesis of ALA-synthetase or in some other limiting condition.

## Bile Pigments

The formation of Fe PROTO appears to be the end of the line for Fe chelate porphyrins as far as it concerns the biosynthetic chain which we are considering. No naturally occurring Fe pigments are known that are related to the Mg series. The phycobilin pigments may represent a decomposition of heme (Bogorad, 1965a). This is an assumption based on the mechanisms proposed for bile pigment formation in animals. Here the $\alpha$-methene carbon of heme is oxidized and eliminated as CO, thus bringing about the opening of the porphyrin ring (Granick and Mauzerall, 1961). Studies on *Cyanidium* by Bogorad (1965a) have shown that these cells in the dark can form porphyrins and bile pigments when supplied with ALA; however, the capacity to make ALA for bile pigment and chlorophyll formation requires light. The phycobilins have been reviewed recently by Ó hEocha (1965). The phycobilins are not only of interest as accessory plastid pigments that function to absorb and transmit light energy to chlorophyll, but the important growth regulator, phytochrome, also appears to contain a bile pigment that is activated by light forming a reversible $P_{650} \leftrightarrow P_{730}$ system (Hendricks and Borthwick, 1965).

## Steps from Protoporphyrin to Chlorophyll—General Aspects

The postulated biosynthetic steps from PROTO to chlorophyll are not yet supported by enzyme studies except in a few instances. However, a number of Mg derivatives of the porphyrins have been found which may be arranged into a reasonable sequence (Fig. 10).

Studies with *Chlorella* mutants first suggested this sequence to chlorophyll (Granick, 1948). One mutant accumulated PROTO in granular masses in cells but lacked chlorophyll. Another mutant formed MgPROTO. From another mutant evidence was obtained for MgPROTO monomethyl ester (Granick, 1961b); here the propionic acid at side chain position 6 (Fig. 1) is presumably the one that is esterified. From another mutant Mg vinyl pheoporphyrin $a_5$ (MgVP), i.e. protochlorophyllide, was obtained.

Results with plant materials treated with ALA also lend support to this formulation of the pathway. They suggest that all the enzymes from ALA to MgVP are already present in etiolated leaves in non-limiting amounts, except ALA-synthetase. When etiolated tobacco leaves in the dark were fed ALA via the petioles, Duranton *et al.* (1958) found that traces of MgVP were formed containing about 5% of the original ALA supplied. Studies by Granick (1961a,b), with etiolated barley and bean leaves fed ALA in the dark, have shown that about 20% of the ALA was converted into MgVP. In this case the leaves turned pale green because of the high content of this compound and they fluoresced intensely. When $\alpha\alpha$-dipyridyl or, better still, pyridyl-2-aldoxime was added together with ALA, the conversion of ALA into the porphyrins was about 40%. In addition to MgVP, there also were formed PROTO, MgPROTO monoester and trace amounts of MgPROTO. No increase in carotenoids or chlorophyll was observed. By the technique of fluorescence microscopy,

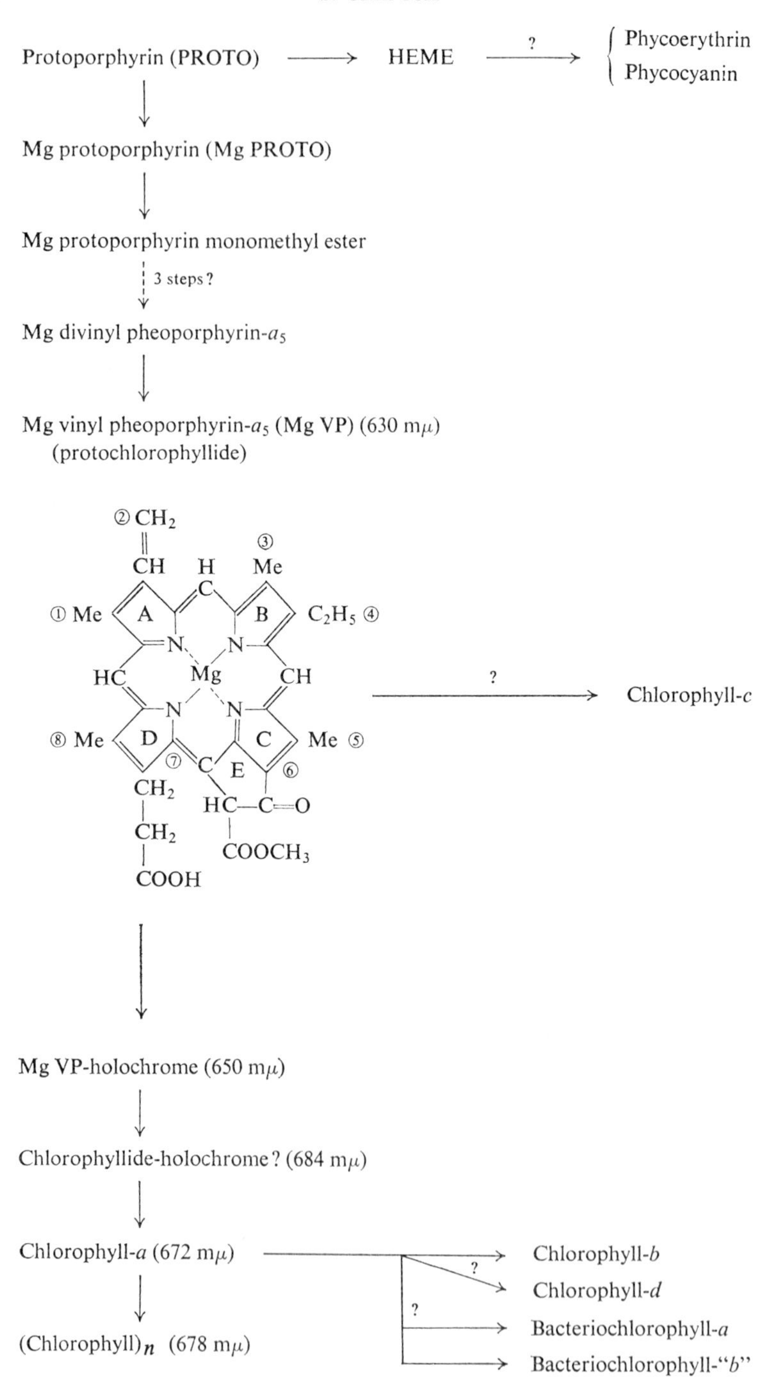


FIG. 10. Biosynthetic chain from protoporphyrin to bacteriochlorophyll.

coupled with an analysis of the wavelengths of the emission bands, it was possible to identify PROTO, MgPROTO and MgVP in the proplastids and also to recognize MgVP fluorescence in the prolamellar body of the proplastid.

Are the enzymes of the earlier steps also present in plastids? Carrel and Kahn (1964) have isolated *Euglena* chloroplasts and observed that they convert ALA into PROTO. Thus, one may assume that all the enzymes that convert ALA into chlorophyll are present in the plastids. It remains to be established whether ALA itself is formed in the plastids especially since no evidence is yet at hand that the plastids can make succinyl-CoA.

A consideration of the various pigments of the biosynthetic chain is interesting from an evolutionary viewpoint. A useful step in the biosynthetic chain was reached only when UROGEN and its oxidation products were formed. Uroporphyrin is an intensely absorbing, fluorescent and photoactive dye which can carry out photochemical reactions. Uroporphyrin can also form a tight chelate complex with iron which can serve as a redox catalyst.

Why did this biosynthetic chain not stop with the formation of URO? It is evident from the 10 or more additional steps to chlorophyll that URO was just one step in an evolutionary process to seek out the compounds and mechanisms that would function most efficiently for redox and photosynthetic reactions. In general these elaborations proceeded from water-soluble compounds to more lipid-soluble compounds which then became organized in lipoprotein membranes.

Once URO had been formed at an early stage in evolution it would have been the end product of this biosynthetic chain. As the chain evolved, modifying the URO to COPRO to PROTO, etc., these would successively become end products of the chain and be explored for various uses. Certain of these intermediates are known to take part in photochemical reactions today. For example, the plant protein-bile pigments phycocyanin and phycoerythrin which are found in red and blue-green algae, are possibly derived from iron protoporphyrin. These compounds are combined with protein to form intensely fluorescent pigments. At present these pigments serve as antennae to catch the light and transfer it to chlorophyll *a* 674 m$\mu$. Another protein which contains a bile pigment is phytochrome 650. It undergoes a photochemical change to phytochrome 730; in the latter form it triggers a biochemical reaction that leads to important physiological growth changes in plants. Near the apex of the present-day biosynthetic chain is the greenish pigment protochlorophyllide. This pigment in the presence of light and an unknown reductant is converted into chlorophyllide. In this photochemical reaction pyrrole ring IV is reduced by the addition of 2 H atoms (Fig. 1). There is no evidence that this photoreduction is a reversible reaction at present. However, it may have been the basis for a photosynthetic mechanism in the past. In the brown algae is a pigment chlorophyll *c* which is related to protochlorophyllide in structure and serves, like the phycobilins, as an accessory pigment to transmit energy from sunlight to chl *a*. 674.

Why did plants go on to vary the porphyrin structure until they arrived at chlorophyll? The answers must lie in the chemistry and photochemistry of these pigments of which we know so little.

## Magnesium Protoporphyrin

This compound was first found in a *Chlorella* mutant (Granick, 1948) and in etiolated leaves fed ALA.

The formation of MgPROTO probably occurs by insertion of Mg into the ring at the oxidation level of protoporphyrin. Undoubtedly an enzyme and some form of activated Mg is required. Although Fe, Cu and Zn can enter the porphyrin ring spontaneously under the right conditions, Mg does not. Organo-chemically the insertion of Mg can be accomplished with a decomposed Grignard under anhydrous conditions but even this reaction goes slowly (Granick, 1948). A Mg $\alpha\alpha$-dipyridyl reagent has also been used for this purpose (Wei and Corwin, 1962).

## Mg Protoporphyrin Monomethylester

This compound has been found in a *Chlorella* mutant and in etiolated plants treated with ALA (Granick, 1961b). It has been identified in culture filtrates of *R. spheroides* (Jones, 1963) where it is increased on a low-iron medium at the same time that bacteriochlorophyll is decreased. In *R. capsulata* it is accumulated in a relatively large amount when glycine, succinate and methionine are in the growth medium (Cooper, 1963).

Tait and Gibson (1961) have studied the methylating enzyme from *R. spheroides*. It is firmly bound to the chromatophore containing fraction of the cell. The active methyl used is S-adenosyl methionine. Tracer studies show a dilution factor of only three, using the methyl labelled compound (Gibson *et al.*, 1963). MgPROTO is esterified fifteen times more readily than is PROTO supporting the idea that Mg is coordinated before esterification. Other metal porphyrins besides MgPROTO are also esterified. S-adenosyl ethionine inhibits

$$\text{MgPROTO} + \text{S-adenosyl methionine} \rightarrow \text{MgPROTO monomethyl ester} + \text{S-Adenosyl homocysteine}$$

the enzyme completely and accounts for the fact that relatively low concentrations of ethionine ($10^{-4}$ M), added to the growth medium of *R. spheroides*, can inhibit bacteriochlorophyll formation by whole cells although there is little inhibition of protein synthesis and growth. This block in esterification leads to the excretion of COPRO into the medium. On this basis it is assumed that a repressor or inhibitor of ALA synthesis may require a methylation step for its formation. Perhaps this is an intermediate Mg porphyrin after the methylation step. Leslie and Sistrom (1964) have also found that *R. spheroides* deprived of methionine or a sulphur source excretes COPRO and does not form bacteriochlorophyll, but the fact that protein synthesis is also limited makes this experiment difficult to intrepret.

## Mg Divinyl Pheoporphyrin-$a_5$

Nothing is known about the conversion of the propionic acid methyl ester side chain into the cyclopentanone ring. Three steps may be required. Possibly two are needed for the oxidation of the $\beta$-carbon atom to a keto group; and in addition, a cyclization step between the $\alpha$-carbon atom and the $\gamma$C-methene bridge atom of the porphyrin ring which may consist of the removal of two H atoms.

Stanier and Smith (1959) and Griffiths (1962) described protochlorophyll-like compounds from mutants of *R. spheroides* which had spectra whose bands were shifted about 4 mμ to the long wavelength side. This result suggested that an additional vinyl group was present. Jones (1963) then found that when normal *R. spheroides* was grown in a medium containing 25 mM 8-hydroxyquinoline a number of pigment intermediates accumulated in the cells and the medium, among them being the Mg divinyl pheoporphyrin-$a_5$. Vinyl pheoporphyrin-$a_5$ in dioxane has the following band maxima in the visible: 638, 587, 567, 524 mμ compared to the divinyl compound: 644, 592, 569, 528 mμ. The effective inhibitor may be the Cu complex of 8-hydroxyquinoline.

## Mg Vinyl Pheoporphyrin-$a_5$ (MgVP) i.e. Protochlorophyllide, and Chlorophyll *c*

The normal intermediate in chlorophyll biosynthesis is MgVP. MgVP was first isolated from the paper-thin greenish seed coats of cucurbits. These seed coats contain MgVP as the phytylated derivative (Fischer and Bohn, 1958), i.e. as protochlorophyll. In 6-day-old etiolated barley leaves grown in the dark the compound is largely in the form of MgVP, but in 17-day-old leaves in the dark a more hydrophobic MgVP is also present to the extent of about one-fourth (Sironval *et al.*, 1965). According to Fischer and Rüdiger (1959) the hydrophobic alcohol in protochlorophyll is not phytol. In the etiolated bean leaf about 20% of the MgVP bears a hydrophobic alcohol (Smith, 1960).

When ALA is fed to etiolated barley or bean leaves in the dark sufficient MgVP accumulates to make the leaves pale green. Spectroscopic studies of these leaves have revealed three forms of MgVP. One has a band maximum at 631 mμ and is readily bleached by light. Another has a maximum at 650 mμ (the MgVP-holochrome of Smith) which is reduced at pyrrole ring IV to chlorophyllide, by brief illumination. In holochrome material that was isolated from bean leaves absorption maximum was about 640 mμ, and was converted by light to a 673 mμ absorbing form (Smith, 1960; Boardman, 1962). Evidence for the existence of a reducing component is suggested by the experiments of Robbeln (1956) who found a mutant of *Arabidopsis*, containing MgVP. This could be photoreduced only when the leaf homogenate was mixed with a soluble unpigmented fraction isolated from normal etiolated leaves of *Arabidopsis*. The reducing group is not a free thiol or a reduced pyridine nucleotide (Boardman, 1962).

Etiolated barley leaves fed ALA in the dark, in 24 hr form about ten times their normal content of MgVP. The MgVP is found to be localized in the proplastids. This means that the leaves, probably the etiolated proplastids themselves, contain all the enzymes for adequate conversion of ALA into MgVP. When such treated leaves are exposed to light of low intensity, insufficient to result in photosensitized bleaching, MgVP diminishes only slightly, whereas chlorophyll is formed abundantly, yet not more than in controls which have not been fed ALA. The excessive MgVP (631 mμ) in the plastids therefore is not used for chlorophyll formation, nor does it inhibit chlorophyll formation, i.e. MgVP does not itself serve as a feedback control. Because the excessive MgVP is not converted into chlorophyllide in the light, either the holochrome or its reducing coenzyme may be limiting. Furthermore, since light results in chlorophyll synthesis independent of the large amount of MgVP present, this result suggests that light enhances not only chlorophyllide formation but also ALA formation. Perhaps chlorophyllide formation is tightly coupled with ALA formation (Granick, 1961c, 1963).

The MgVP holochrome of Smith has its absorption maximum shifted approx. 20 m$\mu$ to the red as compared to the MgVP not on the holochrome. This shift suggests that the MgVP on the lipoprotein is associated with some resonating group. When light is absorbed by the pigment two H atoms are introduced *trans* to each other at the double bond of the D ring (Fig. 1). Our knowledge of this protoreduction is mainly due to the outstanding work of J. H. C. Smith (1960). The quantum yield is about 0·6 which suggests that probably one light quantum is sufficient for reduction and that this is a photosensitized reduction. The excited MgVP becomes reduced probably by some component on the protein as is suggested by the fact that the reduction occurs to the extent of 60% at $-70°$ but is inhibited at $-195°$. Boardman (1962) has isolated from etiolated bean leaves a particle, mol wt 600,000, 100 Å diam. containing one MgVP. After irradiation to convert MgVP into chlorophyllide, the MgVP is renewed even in the dark in 10–15 min. Thus, only a limited amount of MgVP together with its protein and reductant are present in the leaf at any one time. It is not known whether the holochrome protein serves as an enzyme or is converted into a structural component of the developing chloroplast. The action of light on the 650 m$\mu$ MgVP form results in the formation of chlorophyllide with an absorption maximum at 685 m$\mu$. (Fig. 10) However, the isolated holochrome has a maximum at 637 m$\mu$ which on illumination is shifted to the 673 m$\mu$ band of chlorophyllide. This difference between the maxima in solution and in the leaf has yet to be explained.

Chlorophyll *c* is found in brown algae and appears to be related to MgVP in structure. Jeffrey (1963) obtained crystals of chlorophyll *c* and on the basis of Mg content calculated a mol wt of 1050. The HCl number of the pigment is 12. Therefore if the mol wt is correct some large substituents, relatively non-hydrophobic, must be attached to the molecule. In ether the absorption band maxima are at 628, 580 and 442 nm.

## Chlorophyllide to Chlorophyll

After chlorophyllide has been formed in the etiolated leaf by a brief exposure to light the enzymic esterification with phytol to form chlorophyll may proceed in the dark to be completed within 10–60 min (Wolf and Price, 1957; Godnev and Akulovich, 1960). By repeated illumination with light flashes of adequate intensity as short as a millisecond, with 10–15 min dark periods interposed, a considerable amount of chlorophyll can be accumulated (Madsen, 1963). During the dark time after a flash of light, progressive changes in the absorption maxima are observed. The following tentative interpretation may be made of these changes: chlorophyllide when formed and while still attached to the holochrome has a max. at 685 mμ or somewhat higher (Shibata, 1957). When the chlorophyllide separates from the holochrome its absorption max. becomes 673 mμ. After about 10 min at 20° chlorophyllide 673 mμ becomes esterified with phytol to form chlorophyll 673 mμ which is in the monomolecular form. Concomitant with esterification the chlorophyll may become organized into the lamellar lipoprotein membrane where carotenoids are present, so that light energy may now be transferred to a limited extent from carotenoid to chlorophyll (Butler, 1960). The chlorophyllide 685 and 673 forms are both very labile to photobleaching (Goedheer, 1961). The 673 form usually becomes photostable within 15 min, possibly indicative of its becoming phytylated to chlorophyll, and then associated with and protected by the carotenoids. Shylk and Nikolayeva (1962) have found by $^{14}CO_2$ labeling that chlorophyll 673 is more readily extracted by light petroleum containing small amounts of alcohol than is older, more firmly bound chlorophyll. On the basis of high chlorophyll fluorescence at first, and a gradual decrease in fluorescence, Goedheer (1961) suggested that new chlorophyll molecules did not add together to form units of high pigment density, but rather that as more chlorophyll molecules were formed the distances between them gradually decreased during greening. In the grana of the mature leaf the absorption max. is at 678 mμ and the fluorescence is decreased, suggestive of an interaction with neighboring chlorophyll molecules and a more efficient transfer of energy to an energy sink. Goedheer has shown that chlorophyll 673 mμ is capable of photosynthesis provided very high light intensities are used.

The mechanism of phytylation is probably via a phytol pyrophosphate rather than by the hydrolytic enzyme, chlorophyllase. The synthesis of phytol and carotenoids has been reviewed recently by Goodwin (1965). Corn mutants which lack phytylating ability do not show an appreciable shift of the spectrum to 673 mμ (Smith *et al.*, 1959). In mature leaves the phytol is equivalent to the chlorophyll within 5% (Fischer and Bohn, 1958). Etiolated barley leaves do not have a reserve of phytol. When they are exposed to light the production of phytol parallels that of chlorophyll. This indicates an intimate control mechanism. However, in young pine needles the phytol may be 80% greater than the chlorophyll but in the mature leaf the ratio is close to one. In dying leaves the phytol portion of the chlorophyll disappears more slowly than the chlorophyllide portion.

The conversion of chlorophyllide into chlorophyll in the etiolated barley leaf exposed to a 3-min illumination and then placed in the dark is not complete according to Shylk and Nikolayeva (1962) even on prolonged darkening at 22°. They did not observe the presence of chlorophyllide *b*. There is some evidence for a turnover of phytol which may depend on the presence of chlorophyllase. In spinach and sugar beet leaves chlorophyllase activity is high but not in barley seedlings. It is low in dark grown pea seedlings but increases to a maximum in 48 hr. In general, the activity is usually lower in monocots than in dicots. Holden (1960) has isolated a water-soluble chlorophyllase from sugar beet leaves and studied its specificity. In aqueous acetone chlorophyllase hydrolyses chlorophylls *a* and *b*, pheophytins, purpurins, bacteriochlorophyll and

Chlorobium chlorophyll, but not protochlorophyll or compounds lacking the carbomethoxy group on C-10 and the hydrogens on C-7 and C-8. Sironval *et al.* (1965) have suggested that chlorophyllides may play a role in photosynthesis and have identified the 680 m$\mu$ *in vivo* band of the mature leaf as chlorophyllide *a*, but both the identification and hypothesis appear tenuous.

## Chlorophyll *a*

The organic synthesis of chlorophyll *a* has been achieved by Woodward and coworkers (1960) in a masterful series of steps which include the ingenious photochemical insertion of the 2H atoms into the pyrrole ring at C-7 and C-8 in the *trans* positions, confirming the analytical findings of Ficken *et al.* (1956).

The turnover of chlorophyll *a* appears to be negligible in mature leaves of monocots as in lily, oats, philodendron, and tradescantia. However, in mature leaves of red clover, petunia, geranium, spruce and Boston fern, appreciable amounts of $^{14}C$ appear to be incorporated into the dihydroporphyrin ring (Perkin and Roberts, 1963). $^{28}MgCl_2$ fed to excised tobacco shoots of mature leaves does not exchange with the Mg of chlorophyll, and Aronoff (1963) concludes that there is no turnover or synthesis of chlorophyll in these leaves. Wickliff and Aronoff (1963) observed no labeling of the chlorophyllide portion of chlorophyll in mature soy bean leaves, although there was some labeling of phytol.

The course of chlorophyll synthesis in the leaves parallels the differentiation of the proplastid to the chloroplast and has been reviewed by Granick (1963) and by Smith and French (1963). Several light reactions are involved which we briefly summarize. The first is the transformation of protochlorophyllide to chlorophyllide; this requires very brief (a few seconds) exposure to light. The second light reaction causes the prolamellar body of the proplastid to fall apart into tiny vesicles.

The third light reaction causes the vesicles to aggregate into tubules and requires light of moderate intensity; it can occur at 3°. The fourth light reaction requires higher light intensity and is temperature-dependent; it causes the tubules to develop into primary double membrane layers or disks. This latter change requires ATP (it is inhibited by dinitrophenol) and enzymes, for it goes only at higher temperatures (e.g. 20°). Gunning (1965) has calculated that the membrane areas of the tubules in the prolamellar body of *Avena*, when converted into primary layers, are sufficient to form about ten disks, each 2$\mu$ in diameter. Embedded in the prolamellar body are ribosomes which are now released to serve in protein synthesis.

In addition to these light effects there is an early photo-conversion of phytochrome P650 into P730 which appears to trigger protein synthesis, and control chlorophyll synthesis and leaf expansion. Once started by light, the effects of this change can go on in the dark; proteins and enzymes will be formed. After 2 hr, when the leaf is exposed to bright light the lag period for chlorophyll synthesis has disappeared. Now chlorophyll can be synthesized at a rapid steady level ten times that during the lag period. The continuous light of increased intensity now serves for photosynthesis and continued protein synthesis to bring about an increase in the number of disks and their fusion to form grana. At the same time stroma proteins are synthesized that fix $CO_2$ by the pentose cycle (Smillie, 1963).

Metabolic inhibitors that block protein synthesis decrease or block the greening of the chloroplast. Margulies (1964) observed that treatment of etiolated bean leaves in light with chloramphenicol (4 mg/ml) results in a 50% decrease in the synthesis of chloroplast protein. The Hill reaction does not develop nor does photosynthetic phosphorylation. Ribulose diphosphate carboxylase activity does not increase but glyceraldehyde 3-phosphate dehydrogenase does.

In wheat seedlings, developing in the light, changes in carotenoids also go on (Wolf, 1963). As compared to dark grown seedlings of the same age, the carotenoids double, lutein and lutein epoxide double, a nine-fold increase in β carotene takes place and the ratio of lutein to β carotene of 2:1 is attained.

Studies of French on derivative spectra of intact leaves have revealed three chlorophyll *a* maxima: at 673, 683 and 695 mμ which have fluorescence maxima respectively at 686, 696 and 717 mμ. Smith and French (1963) have summarized various studies on attempts to fractionate the disk membranes into units that differ in pigments and photosynthetic activities. Only the work of Park and Biggins (1964) will be mentioned. These investigators isolated quantasomes from the disk membranes. The quantasome is considered to have a mol wt of two million, to contain 230 chlorophyll molecules and to be made up of four or more subunits.

## CHLOROPHYLL *b*

Chlorophyll *b* has a formyl group on side chain position-3 in place of a methyl group as in chlorophyll *a*. Smith and French (1963) as well as Bogorad (1965), have summarized the various hypotheses on its origin. Because no protochlorophyllide *b* has been found, it is considered likely that chlorophyll *b* arises either from chlorophyllide *a* or chlorophyll *a*. No plants have been found which contain chlorophyll *b* but not *a*. However, brown algae and a *Chlorella* mutant contain only *a* and not *b*. These facts support the idea of the conversion of chlorophyll *a* into *b*. The investigations of Shylk and Nikolayeva (1962) indicate that chlorophyll *b* can be formed from *a* in the dark. Pulse labeling of barley seedlings with $^{14}CO_2$ resulted in the early labeling of *a*, and the later labeling of *b*. The labeling activity of *b* was lower than of *a* as if older (i.e. unlabeled *a*) was being preferentially converted into *b*. Because these investigations found no chlorophyllide *b* they suggest that chlorophyll *a* was the precursor of chlorophyll *b*. An interesting *Chlorella* mutant has been found by Allen (1959) which forms only chlorophyll *a* in the dark but forms both chlorophyll *a* and *b* in the light.

## BACTERIOCHLOROPHYLL

Bacteriochlorophyll is the functional primary photosynthetic pigment in the red sulfur bacteria (Thiorhodaceae) and the non-sulfur purple bacteria (Athiorhodaceae). This compound differs from chlorophyll *a* in that it contains a reduced pyrrole ring II as well as IV. Both H atoms of each reduced pyrrole ring are *trans* (Golden *et al.*, 1958). In addition, an acetyl group is present in side chain position -2. Evidence that bacteriochlorophyll is synthesized via chlorophyllide *a* is the presence of pheophorbide *a* in a mutant of *R. spheroides* treated with 8-hydroxy quinoline (Jones, 1963).

Bacteriochlorophyll attached to its protein has been isolated from alkaline extracts of the green bacterium *Chloropseudomonas ethylicum* apparently in its native state by Olson *et al.* (1964). The protein–chlorophyll complex contains twenty bacteriochlorophyll molecules and has a mol wt of $1{\cdot}2 \times 10^5$. The protein contains equal amounts of basic and acidic amino acids. This is the first pure holochrome that has ever been isolated. A bacteriochlorophyll *b* has been reported in a species of *Rhodopseudomonas* (Eimhjellen, 1963). No determination of its structure has been made. Its spectrum in organic solvents has maxima at 368, 407 and 794 m$\mu$ as compared to bacteriochlorophyll *b* maxima at 358, 390 and 771 m$\mu$.

## Other Chlorophylls

Chlorophyll *d* is a constituent of red algae. It contains a formyl group in side chain position -2 in place of the vinyl of chlorophyll *a*. Holt (1961) has isolated it from *Gigartina papillata* and confirmed the structure.

Chlorobium chlorophylls or bacterioviridins are the major chlorophylls in the green sulfur bacteria; in addition, bacteriochlorophyll is also present but only one-twentieth the concentration of the other chlorophylls. Jones (1963) isolated Chlorobium-like pheophorbides, lacking phytol that gave a phase test. However, the studies of Holt (1965) on two Chlorobium chlorophylls with maxima at 650 and 660 nm in organic solvents indicate that they lack the $C_{10}$-carbmethoxy group and at C-2 have a hydroxyethyl group in place of a vinyl group. In place of phytol they have farnesol (*trans-trans*) (Rapaport and Hamlow, 1961).

Reports of other less defined chlorophylls are noted in the reviews by Smith and French (1963), and Holt (1965).

## Chlorophyll Methods

Methods for the identification of structural components of porphyrins and chlorophylls have been reviewed by Holt (1965) and Smith and Benitez (1955).

Methods for the isolation, purification and spectral properties of the chlorophylls have been reviewed by Holden (1965), Perkins and Roberts (1962), Smith and Benitez (1955) and Granick and Mauzerall (1961).

## Control Mechanisms

The obvious place to control heme and chlorophyll biosynthesis is at the first step of the biosynthetic chain where ALA is formed. Studies of animal and bacterial cells suggest that, at least for heme biosynthesis, a control at the first step is present. Control of chlorophyll biosynthesis is more complex and will be discussed separately.

Because animal cells of higher organisms are highly specialized, contain mitochondria but no plastids, and live in a fixed environment, the analysis of the control of heme biosynthesis in animal cells appears to be easier to understand. As an example, we shall first consider the control of heme biosynthesis in a liver cell. In this cell all of the enzymes of the biosynthetic chain to heme are at non-limiting activities except the first enzyme, ALA-synthetase. This enzyme in liver cells is normally repressed by heme (Granick, 1965). ALA-synthetase is present at a level of activity normally so low that it is difficult to detect; yet its action must be sufficient to permit the synthesis of heme for all the cytochromes, catalase, etc., of the liver cell. By treatment of the cell with certain "porphyria-inducing" chemicals, which appear to compete with heme for the repressor-protein, more ALA synthetase will be formed. When this occurs, more ALA will be formed and the rate of porphyrin and heme synthesis by the liver cell may even become as great as the rate in a developing erythrocyte. In the case of the liver cell there is no evidence that heme inhibits the activity of ALA-synthetase, i.e. there is no allosteric inhibition. Although the kidney cells (of a chick embryo) contain all the enzymes in non-limiting amounts that can convert ALA to heme, their repressor mechanism is not affected by the chemicals that affect the liver cells in the same embryo. Thus the liver cells may have a special repressor mechanism.

As another example of control over heme we may consider the events in the differentiation of red blood cells (of chick blastoderm). The colorless proerythroblast is the early cell. It contains all the enzymes at non-limiting activity that convert ALA to heme. It also contains the mRNA and ribosomes ready to synthesize globin. During the development of the proerythroblast the repression of the synthesis of ALA-synthetase must be lifted. Then excess ALA is made, heme is made, and in the presence of heme, globin can be made. In this case the synthesis of hemoglobin is controlled by a repressor of ALA-synthetase presumably acting at the DNA level (Levere and Granick, 1965). Here, too, the repressor mechanism may differ from that in the liver or kidney.

The effect of limiting iron supply (see p. 393) also fits in well with an interpretation of heme, controlling ALA synthesis. Under conditions of limiting iron in non-photosynthetic as well as in photosynthetic bacteria, porphyrins, mainly COPRO, are formed in large amounts. Lascelles (1964) has marshalled the evidence to support the hypothesis that a repressor mechanism is involved. The ALA-synthetase of *R. spheroides* is inhibited by the presence of heme, both by a repressor mechanism and by a direct allosteric inhibition. In these bacteria, none of the enzymes to convert ALA into heme is limiting. With limiting iron in the medium, insufficient heme is made to repress or inhibit ALA-synthetase. Much ALA is thus made and COPRO is formed in amounts equivalent to ten times the amount of bacteriochlorophyll that would be formed normally under non-limiting iron conditions. The reason that COPRO rather than PROTO or heme is increased may be because, under stress conditions of excessive ALA, the enzyme that converts COPROGEN into PROTOGEN may become limiting.

Thus, in all of the above cases of animal and of bacterial cells the control of heme biosynthesis appears to be by heme-repression of the synthesis of ALA-synthetase, or inhibition of its activity.

The repressor mechanism probably acts at the DNA level to control the synthesis of mRNA which will code for ALA-synthetase. In animal cells it is not known whether this DNA is localized within the mitochondrion where ALA synthetase is localized, or whether this DNA is nuclear. In the case of bacterial cells it is presumably the DNA of the single chromosome which contains the structural gene for ALA-synthetase.

In some organisms like yeast which can be converted from an anaerobic to an aerobic metabolism by aeration of the medium, complications are added. Not only must one consider the mechanism for control of heme biosynthesis, but one must also consider the mechanism which controls the differentiation of the promitochondrion to mitochondrion. In this differentiation cristae are formed together with enzymes for the citric acid cycle, the electron transport chain and oxidative phosphorylation. Possibly the DNA of the mitochondrion is involved in this differentiation process (Gibor and Granick, 1964). In bacterial species such as *Haemophilus*, the levels of various cytochromes as well as the kinds of cytochromes may change dramatically, depending on the $O_2$ concentration of the medium (White and Granick, 1963); these cells may become red-brown because of the large amount of cytochromes that are formed. The control mechanisms related to $O_2$ in such bacteria are not understood.

The control of chlorophyll or bacteriochlorophyll biosynthesis is similarly complicated because it appears to involve a number of overlapping systems that partake in the differentiation of the proplastid to the chloroplast. Let us first consider the plastids of leaf cells of higher plants. These require light for greening. Concerning the chlorophyll biosynthetic chain, it is not yet known whether the proplastids can make their own ALA or have to be supplied with this compound by the mitochondria. It is probable that all the enzymes that convert ALA into protochlorophyllide are present in the proplastids. At any one time, the amount of protochlorophyllide present in a leaf is only $10^{-2}$ to $10^{-3}$ that of the chlorophyll of a mature leaf. To be converted into chlorophyllide 685 by light, the protochlorophyllide must be attached to a special protein on which must reside a reducing group perhaps in the form of a coenzyme.

When the complex is briefly irradiated, another small aliquot of protochlorophyllide is synthesized from ALA during the next 15 min. Then no further synthesis occurs. The stop in the synthesis probably signifies that no further ALA is made. Therefore on re-exposure to a brief irradiation, not only is protochlorophyllide 650 converted to chlorophyllide 685 but repression or inhibition of ALA-synthetase must be lifted temporarily to permit another small batch of protochlorophyllide to be made. Studies with ALA show that protochlorophyllide is not the inhibitor or repressor of ALA-synthetase (Granick, 1961).

In the conversion of the proplastid into chloroplast in the presence of light other new and different enzymes and proteins must be made that come to function in photosynthesis. Some evidence suggests that the protein synthesis of certain enzymes is under phytochrome control. When protein synthesis is blocked by metabolic inhibitors greening is prevented (Hudock *et al.*, 1964;

Bogorad, 1965); the inability to make more ALA, as well as the inability to make enzymes and structural lipoprotein lamellae, may all contribute to the prevention of greening.

In the case of some algae like *Chlorella* or *Chlamydomonas*, or the primary leaves of conifer seedlings, greening and synthesis of chlorophylls *a* and *b* can proceed in the dark. No light reaction is necessary for the conversion of MgVP into chlorophyllide. The reduction of MgVP can be done enzymatically in the dark. It is possible that these plants have two mechanisms for reduction of MgVP, one for accomplishing the reduction in the dark; and a more effective one that they can use in the light, as in the higher plants. This possibility is suggested by the fact that there are yellow mutants of *Chlorella* and of *Chlamydomonas* that have lost the ability to form chlorophyll in the dark and as in higher plants are yellow in the dark. However, these mutants can form chlorophyll when grown in the light and can carry on photosynthesis. Perhaps the key step in the control of chlorophyll biosynthesis is the one connected with the reduction of MgVP. In algae that can develop chlorophyll in the dark, the rate of reduction of MgVP may govern chlorophyll and chloroplast synthesis. In algae that require light, the control, as in higher plants, is at the MgVP reduction step. The reduction step may in some way be related to the temporary removal of a block in ALA-synthetase.

Levine and Smillie (1962) have shown that the yellow *Chlamydomonas* mutant segregates in a Mendelian fashion. Therefore the gene that governs the reduction of MgVP in the dark and which is mutated may be a nuclear gene and not a plastid gene.

In the case of the photosynthetic bacteria, such as *R. spheroides* which has been studied extensively (Lascelles, 1964), the problem of the control of chlorophyll synthesis is even more complicated because not only is there overlapping with the controls of differentiation of photosynthetic organelles, but also with the controls of respiratory systems. In the absence of $O_2$, photosynthetic vesicles (chromatophores) develop. In the presence of $O_2$ the photosynthetic vesicles break down and other structures and enzymes develop that bring about an oxidative respiration. The exposure of these photosynthesizing bacteria to $O_2$ inhibits bacteriochlorophyll synthesis promptly. High light intensity is equivalent to the presence of $O_2$. Under light plus anaerobic conditions ALA-synthetase and ALA-dehydrase activities increase; these increases can be prevented by inhibitors of protein synthesis. A unifying concept of all these controls might be that they are intimately related to some one redox mechanism that senses the presence of $O_2$ or an oxidant.

Lascelles (1964) has discussed the $O_2$-sensitive differentiation process in bacteria and higher organisms. The reader is referred to her excellent summaries of this intriguing subject.

The frequency with which redox control mechanisms that govern differentiation appear in biology suggests the possibility that there may be redox mechanisms associated with the activation or repression of DNA operons.

## References


Allen, M. B. (1959). *Brookhaven Symp. Biol.* **11**, 339.
Aronoff, S. (1963). *Pl. Physiol.* **38**, 628.
Boardman, N. K. (1962). *Biochim. biophys. Acta* **64**, 63, 279.
Bogorad, L. (1955). *Science, N.Y.* **121**, 878.
Bogorad, L. (1960). *In* "Comparative Biochemistry of Photoreactive Systems" (M. B. Allen, ed.), p. 227. Academic Press, New York and London.
Bogorad, L. (1962). *In* "Methods in Enzymology" (S. P. Colowick and N. O. Kaplan, eds.) Vol. 5, p. 885. Academic Press, New York and London.
Bogorad, L. (1965a). *Rec. Chem. Prog.* **26**, 1.
Bogorad, L. (1965b). *In* "Chemistry and Biochemistry of Plant Pigments" (T. W. Goodwin, ed.), p. 29. Academic Press, London and New York.
Bogorad, L. and Marks, S. G. (1960). *J. biol. Chem.* **235**, 2127.
Bogorad, L., Peres, G., Swift, H. and McIlrath, W. J. (1959). *Brookhaven Symp. Biol.* **11**, 132.
Bonnet, R. (1963). *Chem. Rev.* **63**, 573.
Bray, R. C. and Shemin, D. (1963). *J. biol. Chem.* **238**, 1501.
Burnham, B. F. (1963). *Acta chem. scand.* **17**, 123.
Butler, W. L. (1960). *Biophys. biochem. Res. Commun.* **2**, 419.
Carell, E. F. and Kahn, J. S. (1964). *Archs. Biochem. Biophys.* **108**, 1.
Carell, E. F. and Price, C. A. (1965). *Pl. Physiol.* **40**, 1.
Cohen-Bazire, G. (1964). *J. biophys. biochem. Cytol.* **22**, 207.
Cookson, G. H. and Rimington, C. (1954). *Biochem. J.* **57**, 476.
Cooper, R. (1963). *Biochem. J.* **89**, 100.
Cornford, P. (1964). *Biochem. J.* **91**, 64.
Duranton, J., Galmiche, J. M. and Roux, E. (1958). *C. r. hebd. séanc. Acad. Sci. Paris* **246**, 992.
Edmundson, P. R. and Schwartz, S. (1953). *J. biol. Chem.* **205**, 605.
Eimhjellen, K. E., Aasumdrud, O. and Jensen, A. (1963). *Biochem. biophys. Res. Commun.* **10**, 232.
Falk, J. E. (1961). *J. Chromatog.* **5**, 277.
Falk, J. E. (1963). *In* "Comprehensive Biochemistry" (M. Florkin and E. H. Stotz, eds,). Vol. 9, p. 3. Elsevier, New York.
Ficken, G. E., Johns, R. B. and Linstead, R. P. (1956). *J. chem. Soc.* **78**, 2272.
Fischer, F. G. and Bohn, H. (1958). *Ann. Chem* **611**, 224.
Fischer, F. G. and Rüdiger, W. (1959). *Ann. Chem.* **627**, 35.
Gibor, A. and Granick, S. (1964). *Science, N.Y.* **145**, 890.
Gibson, K. D., Neuberger, A. and Scott, J. J. (1955). *Biochem. J.* **61**, 618.
Gibson, K. D., Laver, W. G., and Neuberger, A. (1958). *Biochem. J.* **70**, 71.
Gibson, K. D., Matthew, W., Neuberger, A. and Tait, G. H. (1961). *Nature, Lond.* **192**, 204.
Gibson, K. D., Neuberger, A. and Tait, G. H. (1963). *Biochem. J.* **88**, 325.
Golden, J. H., Linstead, R. P. and Whitham, G. H. (1958). *J. chem. Soc.* 1725.
Godnev, T. N. and Akulovich, N. K. (1960). *Doklady Akad. Nauk. SSSR* **134**, 710. Eng. Transl.
Goedheer, J. C. (1961). *Biochim. biophys. Acta* **51**, 494.
Goodwin, T. W. (1965). *In* "Chemistry and Biochemistry of Plant Pigments" (T. W. Goodwin, ed.), p. 143. Academic Press, London and New York.
Granick, S. (1948). *Harvey Lect.* **44**, 220.
Granick, S. (1958). *In* "Trace Elements" (C. A. Lamb, O. G. Bentley and J. M. Beattie, eds), p. 365. Academic Press, New York and London).
Granick, S. (1961a). *Pl. Physiol.* **36**, XLVII.
Granick, S. (1961b). *J. biol. Chem.* **236**, 1168.
Granick, S. (1961c). *Proc. 5th Int. Congr. Biochem. Moscow.* **6**, 176.
Granick, S. (1963). *In* "Cytodifferential and Macromolecular Synthesis" (M. Locke, ed.) p.144, Academic Press, New York and London.
Granick, S. (1965). *In* "Evolving Genes and Proteins" (V. Bryson and H. J. Vogel, eds.), p. 67. Academic Press, New York and London.
Granick, S. (1966). *J. biol. Chem.* **241**, 1359.

Granick, S. and Levere, R. D. (1964). *Prog. Hematol.* **4**, 1. Grune and Stratton. New York.
Granick, S. and Levere, R. D. (1965). *J. biophys. biochem. Cytol.* **27**, 167.
Granick, S. and Mauzerall, D. (1958a). *J. biol. Chem.* **222**, 1119.
Granick, S. and Mauzerall, D. (1958b). *Ann. N.Y. Acad. Sci.* **75**, 115.
Granick, S. and Mauzerall, D. (1958c). *Fed. Proc.* **17**, 233.
Granick, S. and Mauzerall, D. (1961). *In* "Chemical Pathways of Metabolism" (D. Greenberg, ed.), Vol. 2, p. 526. Academic Press, New York and London.
Granick, S. and Sano, S. (1961). *Fed. Proc.* **20**, 376.
Granick, S. and Urata, G. (1963). *J. biol. Chem.* **238**, 821.
Griffiths, M. (1962). *J. gen. Microbiol.* **27**, 427.
Gunning, B. E. S. (1965). *Protoplasma* **60**, 111.
Hager, L. P. (1962). *In* "The Enzymes" (P. D. Boyer, H. Lardy and K. Meyrböck, eds.), Vol. 6, p. 387. Academic Press, New York and London.
Heath, H. and Hoare, D. S. (1959). *Biochem. J.* **72**, 14.
Hendricks, S. B. and Borthwick, H. A. (1965). *In* "Chemistry and Biochemistry of Plant Pigments" (T. W. Goodwin, ed.), p. 405. Academic Press, London and New York.
Holden, M. (1961). *Biochem. J.* **78**, 359.
Holden, M. (1965). *In* "Chemistry and Biochemistry of Plant Pigments" (T. W. Goodwin, ed.), p. 462. Academic Press, London and New York.
Holt, A. S. (1961). *Can. J. Bot.* **39**, 327.
Holt, A. S. (1965). *In* "Chemistry and Biochemistry of Plant Pigments" (T. W. Goodwin, ed.), p. 3. Academic Press, London and New York.
Hudock, G. A., McLeod, G. C., Moravkova-Kiely, J. and Levine, R. P. (1964). *Pl. Physiol.* **39**, 898.
Jeffrey, S. W. (1963). *Biochem. J.* **86**, 313.
Jones, O. T. G. (1963a). *Biochem. J.* **89**, 182.
Jones, O. T. G. (1963b). *Biochem. J.* **88**, 335.
Kikuchi, G., Kumin, A., Talmage, P. and Shemin, D. (1958). *J. biol. Chem.* **233**, 1214.
Labbe, R. F. and Hubbard, N. (1961). *Biochim. biophys. Acta* **52**, 130.
Lascelles, J. (1957). *Biochem. J.* **66**, 65.
Lascelles, J. (1964). *In* "Tetrapyrrole Biosynthesis and Its Regulation." 132 pp. W. A. Benjamin, Inc., New York.
Leslie, T. G. and Sistrom, W. R. (1964). *Biochim. biophys. Acta* **86**, 250.
Levere, R. D. and Granick, S. (1965). *Proc. natn. Acad. Sci. U.S.A.* **54**, 134.
Levine, R. P. and Smillie, R. M. (1962). *Proc. natn. Acad. Sci. U.S.A.* **48**, 417.
Madsen, A. (1963). *Photochem. Photobiol.* **2**, 93.
Margulies, M. M. (1964). *Pl. Physiol.* **39**, 579
Margulies, M. M. (1965). *Pl. Physiol.* **40**, 57.
Marsh, H. V., Evans, H. J. and Matrone, G. (1963). *Pl. Physiol.* **38**, 632.
Massey, V. (1960). *Biochim. biophys. Acta* **38**, 447.
Mathewson, J. H. and Corwin, A. H. (1961). *J. Am. chem. Soc.* **83**, 135.
Mauzerall, D. (1960a). *J. Am. chem. Soc.* **82**, 2605.
Mauzerall, D. (1960b). *J. Am. chem. Soc.* **82**, 2601.
Mauzerall, D. (1962). *J. Am. chem. Soc.* **84**, 2437.
Mauzerall, D. and Granick, S. (1956). *J. biol. Chem.* **219**, 435.
Mauzerall, D. and Granick, S. (1958). *J. biol. Chem.* **232**, 1141.
Mauzerall, D. and Feher, H. (1965). *Biochim. biophys. Acta* **79**, 430.
Neuberger, A. and Turner, J. M. (1963). *Biochim. biophys. Acta* **67**, 342.
Neve, R. A., Labbe, R. F. and Aldrich, R. A. (1956). *J. Am. chem. Soc.* **78**, 691.
Neve, R. A. (1961). *In* "Hematin Enzymes" (J. E. Falk, R. Lemberg and R. K. Morton, eds.), Vol. 2, p. 207. Pergamon Press, Oxford.
Ó hEocha, C. (1965). *In* "Chemistry and Biochemistry of Plant Pigments" (T. W. Goodwin, ed.), p. 175. Academic Press, London and New York.
Olson, J. M., Graham, J. and Latham, G. (1964). *Int. Congr. Biochem.* **6**(10), 784.
Oyama, H., Suzita, M. Y., Yoneyama, Y. and Yoshikaya, H. (1961). *Biochim. biophys. Acta* **47**, 413.
Perkins, H. J. and Roberts, D. W. A. (1962). *Biochim. biophys. Acta* **58**, 486.

Perkins, H. J. and Roberts, D. W. A. (1963). *Can. J. Bot.* **41**, 221.
Park, R. B. and Biggins, J. (1964). *Science, N.Y.* **144**, 1009.
Paul, K. G., Theorell, H. and Akeson, A. (1953). *Acta chem. scand.* **7**, 1284.
Porra, R. J. and Jones, O. T. G. (1963). *Biochem. J.* **87**, 186.
Porra, R. J. and Falk, J. F. (1964). *Biochem. J.* **90**, 69.
Rapaport, H. and Hamlow, H. P. (1961). *Biochem. biophys. Res. Commun.* **6**, 134.
Robbeln, G. (1956). *Planta* **47**, 532.
Sano, S. and Granick, S. (1961). *J. biol. Chem.* **236**, 1173.
Sano, S. and Tanaka, K. (1964). *J. biol. Chem.* **239**, PC 3109.
Schulman, M. P. and Richert, D. A. (1957). *J. biol. Chem.* **226**, 181.
Schwartz, H. C., Goudsmit, R., Hill, R. I., Cartwright, S. F. and Wintrobe, M. M. (1961). *J. clin. Invest.* **40**, 188.
Shemin, D. (1957). *Ergeb. Physiol.* **49**, 299.
Shemin, D. (1962). *In* "Methods in Enzymology" (S. P. Colowick and N. O. Kaplan, eds.), Vol. 5, p. 883. Academic Press, New York and London.
Shemin, D. and Kumin, S. (1952). *J. biol. Chem.* **198**, 827.
Shemin, D., Kikuchi, G. and Abramsky, T. (1963). *In* "Les Maladies du Metabolisme des Porphyrines", p. 173. Presse Universitaires de France, Paris.
Shibata, K. (1957). *J. Biochem. Tokyo* **44**, 147.
Shylk, A. and Nikolayeva, G. A. (1962). *In* "La Photosynthese Colloques", Internationaux du Centre National de la Recherche Scientifique, No. 119, Gif-sur-Yvette et Saclay. July 23–27.
Sironval, C., Michel-Wolwertz, M. R. and Madsen, A. (1965). *Biochem. biophys. Acta* **94**, 344.
Sistrom, W. R., Griffiths, M. and Stanier, R. Y. (1956). *J. cell. comp. Physiol.* **48**, 459.
Smillie, R. M. (1963). *Can. J. Bot.* **41**, 123.
Smith, J. H. C. (1960). *In* "Comparative Biochemistry of Photoreactive Systems" (M. B. Allen, ed.), p. 257. Academic Press, New York and London.
Smith, J. H. C. and Benitez, A. (1955). *In* "Moderne Methode der Pflanzenanalyse" (K. Paech and M. V. Tracy, eds.), Vol. 4, p. 142. Springer, Berlin.
Smith, J. H. C., Durham, L. J. and Wurster, C. F. (1959). *Pl. Physiol.* **34**, 340.
Smith, J. H. C. and French, S. (1963). *Annu. Rev. Pl. Physiol.* **14**, 181.
Stanier, R. Y. and Smith, J. H. C. (1960). *Biochim. biophys. Acta* **41**, 478.
Tait, G. H. and Gibson, K. D. (1961). *Biochim. biophys. Acta* **52**, 614.
Urata, G. and Granick, S. (1963). *J. biol. Chem.* **238**, 811.
Virgin, H. I. (1961). *Physiol. Plantarum* **14**, 439.
Vogel, W., Richert, D. A., Pixley, B. O. and Schulman, M. P. (1960). *J. biol. Chem.* **235**, 1769.
Wei, P. and Corwin, A. H. (1962). *J. Org. Chem.* **27**, 3344.
White, D. C. and Granick, S. (1963). *J. Bacteriol.* **85**, 842.
Weintrobe, M. M. (1950). *Harvey Lect.* **45**, 87.
Wickliff, J. L. and Aronoff, S. (1963). *In* "Studies on Microalgae and Photosynthetic Bacteria" p. 441. Univ. Tokyo Press, Tokyo.
Woodward, R. B. (1960). *J. Am. chem. Soc.* **82**, 3800.
Wolf, F. T. (1963). *Pl. Physiol.* **38**, 649.
Wolf, J. B. and Price, L. (1957). *Archs Biochem. Biophys.* **72**, 293.
Yoneyama, Y., Okyama, H., Sugita, A. and Yoshikawa, H. (1963). *Biochem. biophys. Acta* **74**, 635.
Zelitch, I. (1964). *Annu. Rev. Pl. Physiol.* **15**, 121.

# Adaptation of *Rhodopseudomonas spheroides* from Aerobic to Semi-anaerobic Conditions

A. Gorchein, A. Neuberger and G. H. Tait

*Medical Research Council Research Group, Department of Chemical Pathology, St Mary's Hospital Medical School, London, England*

It has been known for some time that certain groups of photosynthetic bacteria can be grown on fairly simple media in the dark using reactions of the tricarboxylic acid cycle, and gaseous oxygen as ultimate hydrogen acceptor. With *Rhodopseudomonas spheroides*—an organism with which most of our work has been done—the pigments required for photophosphorylation are absent in cells grown in the dark, especially if the oxygen content of the medium is high; alternatively, if such pigments are present at all, they are found under such conditions in minute quantities only. Similarly aerobically grown cells have no, or at least very few chromatophores, particles which are characteristic of light grown organisms. These structures which can be separated by differential ultracentrifugation (Schachman *et al.*, 1952) used to be considered as separate or discrete entities, like mitochondria or chloroplasts. However, there is now an impressive body of evidence indicating that chromatophores are vesicles derived from the cell membrane and it has been suggested that they are formed by invagination of the cytoplasmic membrane, extending more and more into the cytoplasm as the pigment content of the cell increases (Cohen-Bazire and Kunisawa, 1963; Drews and Giesbrecht, 1963). Chromatophores are therefore not considered to be structurally or functionally independent subcellular units. On the other hand, chromatophores of the same size as those obtained by ultracentrifugation and separation by density gradients from cell-free extracts, can be seen in electron micrographs of thin sections of whole cells (e.g. Gibson, 1965) and this means that they are morphological entities, whatever their origin is.

The present paper is concerned with a more detailed study of the changes which occur when *Rhodopseudomonas spheroides* is transferred from aerobic conditions in the dark to semi-anaerobic conditions in the light. It is important to emphasize that aerobic conditions in most of our work mean that the cultures were shaken in an atmosphere of pure oxygen, not air as used, for example, by Lascelles or Kikuchi. Light means illumination of the order of 200 foot candles, which means moderate light intensity.

The adaptation from an aerobic to an anaerobic environment involves profound changes in the composition of the organism and some of these have already been studied and analysed in a classical paper by Cohen-Bazire *et al.* (1957). There is a very large increase of synthesis of bacteriochlorophyll and of carotenoids such as spheroidene and the contents of anaerobically grown cells, with respect to these substances is high. The induced pigment synthesis is associated with the formation of new protein as shown by Lascelles (1959), Bull and Lascelles (1963), who demonstrated that dietary deficiency of nitrogen and the presence of inhibitors of protein synthesis in the medium slowed down or prevented formation of new bacteriochlorophyll. Higuchi *et al.* (1965) have extended these findings and also produced evidence to show that synthesis of

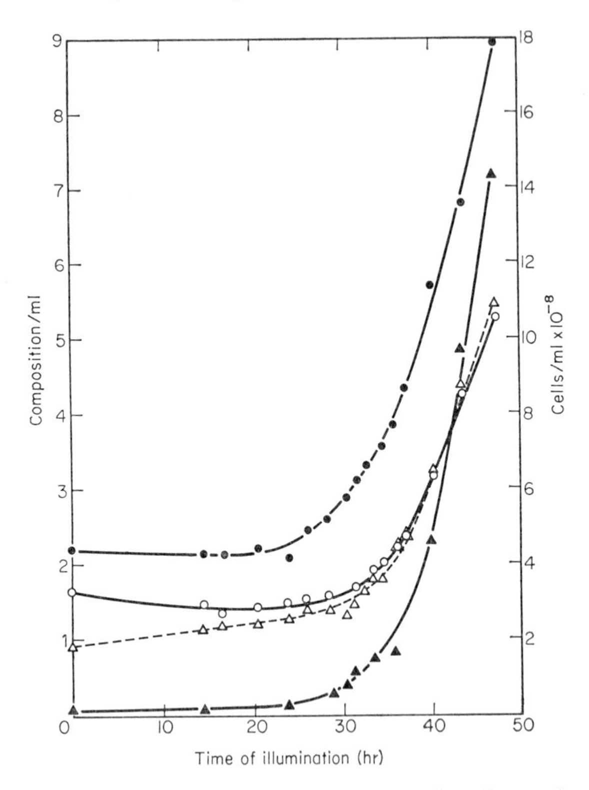


FIG. 1. Changes in composition of *Rhodopseudomonas spheroides* on adaptation. At 0 hr the culture growing under oxygen in the dark was diluted ×2 with medium *S*, the flow of oxygen was stopped and the culture was illuminated. ● Cells /ml of culture. For convenient graphical presentation the contents of constituents were multiplied as follows: ▲ bacteriochlorophyll (μmmoles/ml of culture), by 1; ○ dry wt (μg/ml of culture), by 0·01; △ protein (μg/ml of culture), by 0·02.

deoxyribonucleic acid is needed for the induced formation of photosynthetic pigments. Lascelles (1959) also showed that the level of ALA synthetase increased four- or five-fold during adaptation to low oxygen pressure. In a recent paper Lascelles and Szilagyi (1965) showed that pigmented cells of *R. spheroides* contained approximately 60% more lipid-phosphorus/mg protein than non-pigmented and they suggested that this indicated that pigment formation may be associated with the formation of new membrane material.

In our experiments, we started with cells grown in oxygen in the dark and then transferred to new medium $S$; the flow of oxygen was stopped at time $O$ and the culture was illuminated. Cell numbers were counted in the Coulter counter, dry weight was estimated either by turbidity measurement or more reliably by direct weighing, whilst protein content was obtained with the Lowry method. Cell numbers remained constant for 18–25 hr and then increased rapidly (Fig. 1). This lag period varied somewhat in different experi-

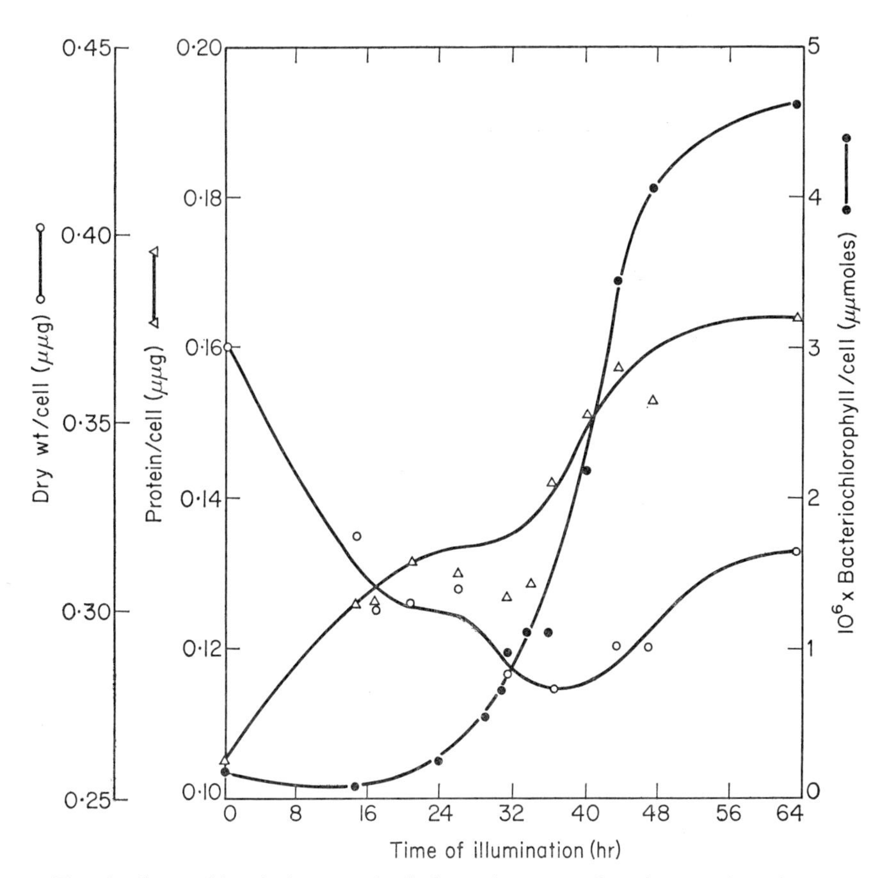


FIG. 2. Compositional changes of *Rhodopseudomonas spheroides* on adaptation. The amount of each component per cell was calculated from the data in the previous figure.

ments, but was never less than 15 hr. During this lag period there was a definite loss of dry weight which was first noticed from turbidity measurements, but was confirmed by direct weighing of washed cell suspensions. On the other hand protein content increased markedly. Both changes took place whilst the cell number remained unchanged. The increase of bacteriochlorophyll coincided with the increase of cell numbers, i.e. it did not begin until the end of the lag period. Figure 2 shows clearly that there is a loss of dry weight per cell from about $0{\cdot}36 \times 10^{-12}$ g to $0{\cdot}27 \times 10^{-12}$ g which is followed by a slight increase to about $0{\cdot}30 \times 10^{-12}$ g. Protein content increases steadily during this period from $0{\cdot}105 \times 10^{-12}$ to about $0{\cdot}15 \times 10^{-12}$ g. Lipid phosphorus increases faster than dry weight and the increase precedes in time the increase in cell number, as shown in Fig. 3. In this particular experiment the poly-$\beta$-hydroxybutyrate which had accumulated during aerobic growth fell by 50% during the latent period when cell numbers remained stationary. In most other experiments poly-$\beta$-hydroxybutyrate remained fairly constant or decreased to a lesser extent. However the very marked loss in total carbohydrate, measured in terms of glucose after hydrolysis by N HCl at 100° for 1 hr, was always observed

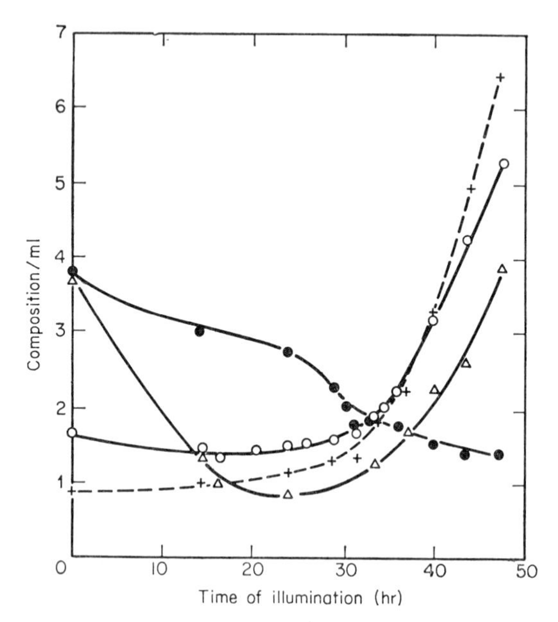


FIG. 3. Changes in composition of *Rhodopseudomonas spheroides* on adaptation. At 0 hr the culture growing under oxygen in the dark was diluted ×2 with medium *S*, the flow of oxygen was stopped and the culture was illuminated. For convenient graphical representation, the contents of constituents (expressed as $\mu$g/ml of culture) were multiplied as follows: (○——○) dry weight, by 0·01; (△——△) total reducing sugar, by 0·1; (●——●) poly-$\beta$-hydroxybutyrate, by 0·25; (+-----+) lipid phosphorus, by 2·5.

during this latent period. Table I gives a comparison of the composition of the aerobic and anaerobic cells. It is again obvious that the anaerobic cell has a dry weight which is about 20% lower than the aerobic cell. On the other hand the protein content of the latter is just below 30%, whilst the anaerobic cell has a protein content of 50%; in other words the latter has 50% more protein than the former. The difference is to a large extent caused by a doubling of the particulate protein, i.e. the protein which can be sedimented at 105,000 ***g*** for 1 hr; the soluble protein is increased by only 20%. There are no very marked differences in the total nucleic acid or in DNA. It is not quite certain whether the increase of about 10% in DNA found with anaerobic as compared with aerobic cells is significant. Again, the lipid phosphorus which is a measure of the phospholipid content is about 85% higher for anaerobic cells than for the aerobic cell. This difference is somewhat greater than the value found by Lascelles and Szilagyi (1965) who carried out this comparison on the basis of protein content which however is different for the cell in the two conditions. The question now remains as to why the dry weight of the anaerobic is smaller than that of the aerobic cell when the latter has a lower protein content. We are not certain that the Lowry method is completely reliable if used under these conditions. However the other methods available, such as estimation of total

Table I. Composition of *Rhodopseudomonas spheroides* grown under different conditions

| Constituent | Aerobic cells ($\mu\mu$g/cell) | Anaerobic cells ($\mu\mu$g/cell) |
|---|---|---|
| Dry wt | 0·37 | 0·29 |
| Protein | | |
| Total | 0·103 | 0·154 |
| "Particulate" | 0·036 | 0·073 |
| "Particulate" (as % of total protein) | 35 | 47·5 |
| Nucleic acid | 0·024 | 0·026 |
| DNA | 0·011 | 0·012 |
| Lipid phosphorus | 0·00078 | 0·00145 |
| Poly-$\beta$-hydroxybutyrate | 0·034 | 0·003 |
| Carbohydrate (as total reducing sugar) | 0·085 | 0·022 |

Organisms were grown in Medium *S* of Lascelles under the following conditions: 1. In the dark, gassing with oxygen; 2. Semi-anaerobically in the light (approx. 200 ft candles). Cells were disrupted in an ultrasonic disintegrator for 7 min at 20 Kc and centrifuged at 105,000 × *g* for 1 hr, giving a "particulate" and a "supernatant" fraction.

nitrogen or assay of nitrogen precipitated by trichloroacetic acid, have more serious or obvious drawbacks and even estimation of individual amino acids with an Auto-analyser will include amino acids derived from the cell wall material. We have therefore provisionally accepted the results of the Lowry method as giving at least approximate and valid comparative values. The major part of the difference in dry weight is accounted for by the much higher content in carbohydrate (which is probably mainly polyglucose) and poly-$\beta$-hydroxybutyrate of the aerobic cell than of the anaerobic cell. A smaller part of the loss in dry weight which occurs when *R. spheroides* adapts itself to anaerobic conditions may be accounted for by loss of other cytoplasmic components which have not been estimated or by a loss in cell wall substances. On the other hand, the increase in protein and phospholipid may be explained by the formation of chromatophores. The composition of this fraction, as determined by ourselves, agrees with the figures given by others and is shown in Table II.

Thus the dry weight of chromatophores consists of just over 60 % protein, 20 % phospholipid and about 7 % bacteriochlorophyll.

The finding that aerobic cells have a greater dry weight and thus presumably a greater mass than anaerobic cells is also supported by measuring the size distribution of these types of organisms with the aid of the Coulter counter. It can be seen that aerobic cells are significantly bigger than anaerobic cells (Fig. 4).

We have pointed out above that bacteriochlorophyll formation begins when cell numbers start to increase. This may indicate that pigment formation is linked to cell division; such an idea is also supported by experiments of Higuchi *et al.* (1965) which show that substances which are considered specific inhibitors of DNA synthesis interfere with induced pigment formation. How-

Table II. The composition of purified chromatophores from *Rhodopseudomonas spheroides*

| Component | mg/100 mg dry wt |
|---|---|
| Protein | 61–63 |
| Total lipid (including pigments) | 26·7 |
| Carbohydrate | 4·0 |
| Nucleic Acid | 0·10 |
| Total nitrogen | 10·20 |
| Total phosphorus | 1·03 |
| Lipid phosphorus | 0·78 |
| Bacteriochlorophyll | 6–9 |
| Yellow carotenoid | 1·5 |
| Red carotenoid | 0·5 |

Organisms were disrupted in the French pressure cell. Chromatophores were isolated by differential centrifuging and purified on a sucrose density gradient.

ever our data suggest, but do not prove, that bacteriochlorophyll formation precedes by a short interval an increase in cell number. It is also possible that cell division can only occur when a certain value of bacteriochlorophyll per cell has been reached. More work on this problem is necessary, but our data are compatible with the hypothesis that the aerobically grown cell itself can acquire the capacity to form pigments, without first undergoing a division.

Figure 5 shows changes in the levels of various enzymes during the adaptation process. The zinc protoporphyrin chelatase, which is a particulate enzyme also probably responsible for the insertion of iron into porphyrin, remains essentially constant during adaptation. A second enzyme which however is soluble, the threonine dehydrogenase forming aminoacetone from threonine, also remains constant. However the activities of two enzymes specifically concerned with bacteriochlorophyll synthesis increase very markedly. The ALA synthetase increases thirty-fold, whilst the level of the methylating enzyme which catalyses the esterification of one of the two carboxyl groups of magnesium protoporphyrin increases up to about forty- to fifty-fold. The other interesting feature shown in Fig. 5 is that the increase starts almost immediately when the oxygen supply is withdrawn. In fact, with the methylating enzyme, the greatest rate of increase occurs during the first 8 hr, long before any pigment

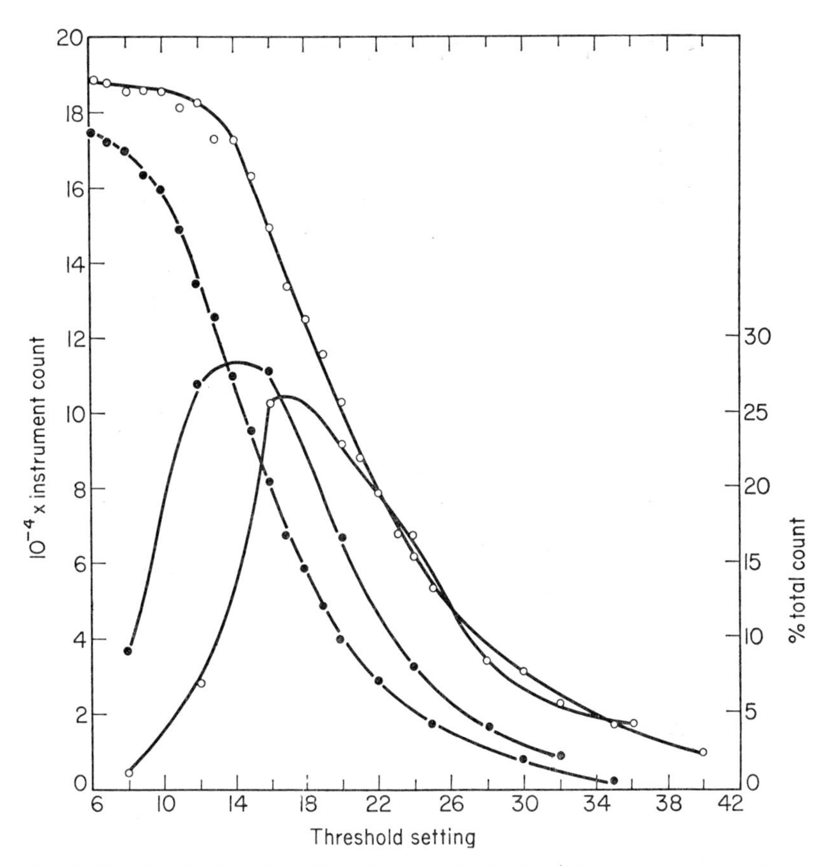


FIG. 4. Size distribution of aerobic and anaerobic *Rhodopseudomonas spheroides*. Successive counts at increasing threshold intervals were performed on a Coulter counter model A fitted with a 30 $\mu$ aperture. Values were corrected for background and coincidence. Relative cell size distribution curves were obtained from the differences in counts between the successive numbered threshold intervals. ○——○ Aerobic cells. ●——● Anaerobic cells.

formation occurs or before cell numbers begin to increase. It thus appears that an increase in protein and phospholipid and also an increase in the level of the two enzymes specifically concerned with pigment formation precedes by several hours the appearance of bacteriochlorophyll and even more markedly cell division under anaerobic conditions. The latent period is thus the time during which the cell produces its new equipment necessary for life under anaerobic conditions, where light energy is available.

As already mentioned some of our figures, for example the level of ALA synthetase under aerobic conditions, differ from those of Lascelles who observed on adaptation only an increase of four- to five-fold. Table III shows how this discrepancy can be resolved. We have grown *R. spheroides* in oxygen and in air (both in the dark), and anaerobically in the light and have then compared

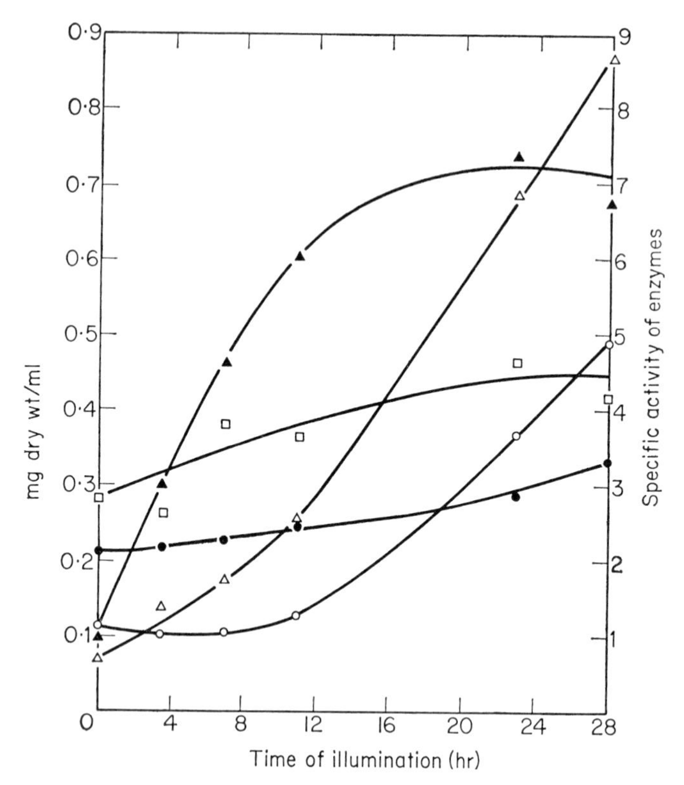


FIG. 5. Changes of enzyme activities in *Rhodopseudomonas spheroides* on adaptation. At 0 hr the culture growing under oxygen in the dark was diluted ×2 with medium *S*, the flow of oxygen was stopped, and the culture was illuminated. ○ Dry wt per ml of culture. For convenient graphical representation, specific activities of enzymes ($\mu$mmoles/mg of particulate or supernatant protein) were multiplied as follows: △ ALA synthetase (supernatant), by 0·05; ▲ magnesium protoporphyrin methylating enzyme (particulate), by 1; □ zinc protoporphyrin chelatase (particulate) by 0·25; ● threonine dehydrogenase (supernatant), by 0·05.

the composition of the organisms and enzyme levels. It will be seen that the low level of bacteriochlorophyll doubles in going from oxygen to air but this increase is still small compared with that produced by anaerobiosis and light. The levels of two enzymes which increase so markedly during adaptation show an intermediate value if the organisms are grown in air. Thus ALA synthetase increases about ten-fold in going from oxygen to air and only three-fold in changing from the latter to anaerobic conditions. With the methylating

Table III. Composition of *Rhodopseudomonas spheroides* grown under different conditions

| Constituent | Dark | | Light |
|---|---|---|---|
| | Oxygen | Air | Semi-anaerobic |
| Bacteriochlorophyll (μmmoles/mg dry wt) | 0·25 | 0·52 | 24·5 |
| Total protein (% of dry weight) | 21, 26 | 48 | 42, 51 |
| Particulate protein (% of total protein) | 32, 37 | 38 | 44, 53 |
| ALA synthetase (μmmoles product/mg supernatant protein/hr) | 6, 9 | 72 | 189, 230 |
| Threonine dehydrogenase (μmmoles product/mg supernatant protein/hr.) | 0·26, 0·43 | —* | 0·30, 0·39 |
| Magnesium protoporphyrin methylating enzyme (μmmoles product/mg particulate protein/2 hr) | 0·2, 0·3 | 13 | 10, 14 |
| Zinc protoporphyrin chelatase (μmmoles product/mg particulate protein/hr) | 12·7 | 15·4 | 13·2 |
| Succinic dehydrogenase (μmmoles INT reduced/mg particulate protein/30 min) | 0·06, 0·18 | 0·39 | 0·87, 0·93 |

*R. spheroides* were grown in medium S of Lascelles (1956) under the following conditions: 1. By bubbling oxygen in the dark; 2. By bubbling air in the dark; 3. Semi-anaerobically in the light (approx. 200 ft candles). Organisms were passed twice through the French pressure cell and centrifuged at 105,000 × ***g*** for 1 hr to give a particulate fraction and a supernatant fraction. Where two values are given they refer to two independent experiments.

* Not measured.

enzyme the activities in air and in semi-anaerobic conditions are practically identical. It is now accepted that it is mainly the oxygen content of the medium which determines adaptation and it is thus perhaps not surprising that a reduction of oxygen tension should bring about changes associated with or necessary for an anaerobic existence in light. Finally, it is of interest that the succinic dehydrogenase level should be also greatly increased in changing from oxygen to air and from air to anaerobic conditions.

## References

Bull, M. J. and Lascelles, J. (1963). *Biochem. J.* **87**, 15.
Cohen-Bazire, G., Sistrom, W. R. and Stanier, R. Y. (1957). *J. cell. comp. Physiol.* **49**, 25.
Cohen-Bazire, G. and Kunisawa, R. (1963). *J. Cell Biol.* **16**, 401.
Drews, G. and Giesbrecht, P. (1963). *Zbl. Bakt.* (1. Abt. Orig.) **190**, 508.
Gibson, K. D. (1965). *J. Bact.* **90**, 1059.
Higuchi, M., Goto, K., Fujimoto, M., Namiki, O. and Kikuchi, G. (1965). *Biochim. biophys. Acta* **95**, 94.
Lascelles, J. (1956). *Biochem. J.* **62**, 78.
Lascelles, J. (1959). *Biochem. J.* **72**, 508.
Lascelles, J. and Szilagyi (1965). *J. gen. Microbiol.* **38**, 55.
Schachman, H. K., Pardee, A. B. and Stanier, R. Y. (1952). *Archs Biochem. Biophys.* **38**, 245.

# Production of Bile Pigments from δ-Aminolevulinic Acid by *Cyanidium caldarium*

Robert Troxler and Lawrence Bogorad[1]
*Department of Botany, University of Chicago, Chicago, Illinois, U.S.A.*

Information regarding biliprotein biogenesis in cell-free systems is entirely lacking. Our present understanding of biliprotein formation is based on indirect evidence obtained from studies on the action spectrum for phycocyanin formation in a chlorophyll-less *Cyanidium caldarium* mutant (Nichols and Bogorad, 1962) and of the effect of pre-illumination with red (600–700 mμ) or green (500–550 mμ) light on the formation of phycocyanin and phycoerythrin precursors, respectively, in *Tolypothrix tenuis* (Fujita and Hattori, 1963).

Label from [$^{14}$C]-δ-aminolevulinic acid (ALA) is incorporated into phycocyanin by illuminated *C. caldarium* cells (L. Bogorad, unpublished data). This provides indirect evidence for the existence of a porphyrin or heme precursor of the phycocyanin chromophore in this alga (Nichols and Bogorad, 1962). Biliproteins are produced only by illuminated cells of some *C. caldarium* strains and are spectrally undetectable in dark-grown cells. Rabbit antiserum to the phycocyanin protein does not react with soluble protein extracts of dark-grown cells of several biliprotein-containing *Cyanidium* strains (L. Bogorad and R. Troxler, unpublished observations). This suggests that both the protein and bile-pigment components of phycocyanin in this alga are produced *de novo* during "greening" of illuminated cells previously grown in darkness.

In an attempt to determine whether protein synthesis is necessary for biliprotein accumulation to occur, suspensions of dark-grown *Cyanidium* cells were exposed to chloramphenicol ($10^{-3}$ M), ethionine ($10^{-3}$ M), or *p*-fluorophenylalanine ($10^{-3}$ M) or deprived of inorganic nitrogen during illumination. Both chlorophyll *a* and phycocyanin failed to accumulate when nitrogen was omitted from the suspending medium or when these inhibitors were incubated with the cells during illumination (Table I). Pigment synthesis was stopped even after accumulation had begun by adding chloramphenicol ($10^{-3}$ M) or by resuspending cells in fresh nutrient medium without nitrogen. After deprivation of nitrogen or incubation with inhibitors for 48 hr, the cells were collected by centrifugation and resuspended in fresh "complete" nutrient medium in order to see whether the algae were living. Pigment synthesis resumed during illumination for another 48 hr although the total chlorophyll *a* and phycocyanin produced was less than in cell suspensions illuminated for the same time but never exposed to inhibitors or deprived of nitrogen.

In view of the incorporation of label from [$^{14}$C]-ALA into phycocyanin by

[1] Research Career Awardee of the National Institute of General Medical Sciences, U.S.P.H.S.

illuminated cells of *C. caldarium*, the failure of the alga to produce phycobiliproteins in darkness, and the likely origin of the phycocyanin chromophore from a metalloporphyrin (e.g. heme) or a metalloporphyrin–protein complex (e.g., hemoprotein), the fate of ALA administered to cells which had been maintained in darkness or in the light for 24 hr was studied.

Table I. Effect of inhibitors and nitrogen starvation on phycocyanin and chlorophyll *a* formation in *C. caldarium*, strain III-D-2

| Conditions[1] | Pigment: Chlorophyll *a* µmole/flask × $10^{-4}$ | Pigment: Phycocyanin µmole/flask × $10^{-5}$ |
|---|---|---|
| 1. Un-illuminated Cells | 6·1 | 0 |
| 2. Illuminated Cells | 277·0 | 1·4 |
| a. +Ethionine, $10^{-3}$ M | 5·3 | 0 |
| b. +*p*-Fluorophenylalanine, $10^{-3}$ M | 6·1 | 0 |
| c. +Chloramphenicol, $10^{-3}$ M | 7·4 | 0 |
| d. −nitrogen[2] | 6·8 | 0 |

[1] Each treatment consisted of 10 ml of cell suspension. The cell density was $5 \times 10^8$ cells/ml. Inhibitors were pre-incubated with the cells for 6 hr prior to illumination. Illumination time was 48 hr.

[2] NaCl was substituted for $NH_4Cl$ in the nutrient medium.

During 24 hr incubations in darkness with ALA ($7{\cdot}0 \times 10^{-3}$ M), both dark-grown and previously illuminated cells excreted porphobilinogen (PBG), porphyrins (primarily coproporphyrin III and uroporphyrin III), and a blue phycobilin (bile-pigment) into the suspending medium (Troxler and Bogorad, 1965). Cells which had been illuminated prior to administration of ALA formed two to three times more pyrrolic pigment than un-illuminated algae (Table II).

Table II. ALA utilization by variously treated III-D-2 cell suspensions

| Conditions[1] | Porphyrins[2] µmole/ml × $10^3$ | Bile-Pigment[3] µmole/ml × $10^3$ |
|---|---|---|
| 1. Dark-Grown Cells | 5·60 | 16·30 |
| 2. Pre-illuminated Cells | 14·32 | 38·30 |
| a. +Ethionine, $10^{-3}$ M | 0·40 | 0·30 |
| b. +Chloramphenicol, $10^{-3}$ M | 0·42 | 0·40 |

[1] Cells were grown heterotrophically in darkness for 10 days. Some cells were illuminated for 24 hr before incubation with ALA in darkness for an additional 24 hr.

[2] Porphyrins estimated by the method of Shemin (1962).

[3] The bile-pigment yields were determined by measuring the optical density of the pigment in chloroform.

Cells which had been pre-incubated with chloramphenicol or ethionine (both at $10^{-3}$ M) failed to produce porphyrins or phycobilin during treatment with ALA (Table II). (After incubation with ALA and inhibitors in darkness, the cells produced chlorophyll *a* and phycocyanin when washed with water and

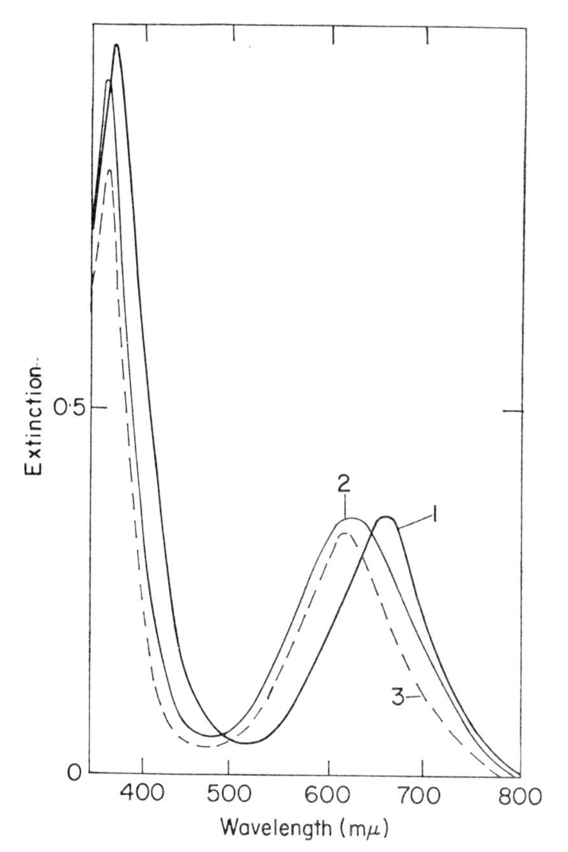


FIG. 1. Absorption spectra of bile-pigments in chloroform. 1. Biliverdin. 2. Phycobilin excreted into the suspending medium by *Cyanidium caldarium* during 24-hr incubations with ALA ($7{\cdot}0 \times 10^{-3}$ M). 3. Phycobilin produced by refluxing *Cyanidium* phycocyanin for 2–3 hr in 90% methanol containing 1% L-ascorbic acid.

resuspended in fresh nutrient medium without ALA or inhibitors present. Obviously some of the cells survived the treatment with the inhibitors.)

After 24-hr incubations with ALA the bile pigment was isolated by centrifuging the cells from the suspending medium (pH 2·0) and transferring the phycobilin into chloroform by three successive extractions. The spectral properties of the *C. caldarium* bile-pigment are shown in Table III and Fig. 1. In chloroform and in methanol, the maximum absorption in the visible portion of the spectrum of the algal phycobilin was observed at 610–625 mμ (Table III). For comparison, the spectral properties of biliverdin prepared by the method of Gray *et al.* (1961), and of the phycobilin produced by refluxing purified *Cyanidium* phycocyanin in 90 % methanol containing 1 % L-ascorbate (method of C. O'hEocha and P. O'Carra, personal communication), are shown (Table III and Fig. 1). The phycobilin obtained from the culture medium of ALA-treated cells, and of the pigment prepared by the methanol-ascorbate extraction (i.e. absorption maxima at 367 and 610–625 mμ in chloroform) are spectrally similar (Fig. 1).

Table III. Data on the spectral properties of biliverdin,[1] phycocyanobilin, and the blue phycobilin produced by *C. caldarium*, strain III-D-2, after incubation with ALA ($7{\cdot}0 \times 10^{-3}$ M) in darkness for 24 hr.

| Pigment | Solvent | Max (mμ) 1 | Max (mμ) 2 | $E_2/E_1$ |
|---|---|---|---|---|
| Biliverdin | Chloroform | 645–655 | 378–380 | 3·20 |
| | Methanol | 645–655 | 378–380 | 3·10 |
| | Methanol-HCl 5 % | 665–675 | 375 | 2·16 |
| Phycocyanobilin (Methanol-ascorbate) | Chloroform | 612 | 366 | |
| Phycobilin (ALA) | Chloroform | 610–625 | 367–372 | 2·70 |
| | Methanol | 610–625 | 367–375 | 3·25 |
| | Methanol-HCl 5 % | 680–690 | 370 | 1·70 |

[1] Biliverdin prepared by the method of Gray *et al.* (1961).

The *Cyanidium* bile-pigment produced from ALA did not fluoresce under ultraviolet light (3600 Å) in neutral chloroform or in methanol. The pigment developed an intense red fluorescence, however, when the zinc complex was prepared. The absorption maximum of the zinc complex (in methanol) in the visible portion of the spectrum was shifted from 610–625 mμ to 675–690 mμ (Table III). In agreement with the finding of Gray *et al.* (1961) the development of red fluorescence did not accompany a similar spectral shift in the visible absorption maximum of the biliverdin zinc complex. The formation of a red-fluorescing zinc complex suggested a violin-type structure for the algal tetrapyrrole (Gray 1953). In contradiction to the evidence for its presumptive violin-type structure, the phycobilin produced by the ALA-treated algae reacted similarly to biliverdin in the Gmelin reaction, suggesting the pigment may be a bilitriene (Lemberg and Legge, 1949).

## Conclusions

The inhibition of phycocyanin accumulation by chloramphenicol in illuminated *C. caldarium* cells suggests that protein synthesis is needed for pigment formation to occur. This presumes the effect of the antibiotic in the algal system is similar to its effect *in vitro* on bacterial ribosomes (Kucan and Lipmann, 1964) and on reticulocyte ribosomes (Weisberger and Wolfe, 1964). Chloramphenicol also inhibits biliprotein synthesis in *Tolypothrix tenuis* (Fujita and Hattori, 1960). The effects of nitrogen starvation and of amino acid analogs on phycocyanin synthesis in *C. caldarium* further supports the contention that proteins must be made for "greening" of this alga.

Porphyrin and phycobilin synthesis from ALA was also blocked by treating cells of *C. caldarium* with chloramphenicol or ethionine. These data suggest that: (1) a synthesis of proteins (e.g. enzymes) is necessary for bile-pigment production to occur, in agreement with the observations on inhibition of phycocyanin synthesis in illuminated cells by chloramphenicol, ethionine, and *p*-fluorophenylalanine, (2) the concentration of enzymes for ALA utilization and phycobilin synthesis in the strains tested is very low, perhaps because the rate of turnover of porphyrin- and bile-pigment-making enzymes is very high, and (3) in view of the inability of cells treated with inhibitors to utilize ALA, it is likely that terminal steps in the synthesis of the phycobilin from ALA by *Cyanidium* are enzymatic rather than occurring by non-enzymatic reactions analogous to the coupled oxidation of hemoglobin with ascorbate in air.

Although the *C. caldarium* phycobilin was produced during incubation with ALA in darkness, synthesis of the phycocyanin protein and chromophore (e.g. phycocyanobilin) normally requires light. Phycobilin formation from ALA as described in this report, however, is stimulated by but does not *require* any pre-illumination of the cells (Table II). On the basis of these observations, it is concluded that: (1) light is probably required for synthesis of ALA or its precursors, (2) ALA utilization is stimulated by light as the yield of bile-pigment is generally two times greater in pre-illuminated cells than in dark-grown cells, and (3) since ALA utilization does occur in non-illuminated cells, bile-pigment synthesis (i.e. phycocyanin chromophore production) in this system does not occur via photochemical reactions [for instance, similar to the photo-oxidation of the aetioporphyrin sodium complex into aetiobiliverdin (Lemberg and Legge, 1949)].

## Acknowledgement

This work was supported in part by grants from the National Science Foundation, the National Institute of Arthritis and Metabolic Diseases of the U.S.P.H.S., and the Abbott Memorial Fund of the University of Chicago.

The authors wish to express their appreciation to Dr. H. W. Siegelman for this helpful advice on the extraction of phycocyanin with methanol-ascorbate.

## References

Fujita, Y. and Hattori, A. (1960). *Pl. Cell Physiol.* **1**, 281.
Fujita, Y. and Hattori, A. (1963). *In* "Microalgae and Photosynthetic Bacteria" (Japanese Society of Plant Physiologists, ed.) p. 431. University of Tokyo Press, Tokyo.
Gray, C. H. (1953). "The Bile Pigments," Methuen and Co. Ltd., New York.
Gray, C. H., Lichtarowicz-Kulczycka, A., Nicholson, D. C. and Petryka, Z. (1961). *J. chem. Soc.* **440**, 2264.

Kucan, Z. and Lipmann, F. (1964). *J. biol. Chem.* **239**, 516.
Lemberg, R. and Legge, J. W. (1949). "Haematin Compounds and Bile Pigments."
Nichols, K. and Bogorad, L. (1962). *Bot. Gaz.* **124**, 85.
Shemin, D. (1962). *In* "Methods in Enzymology" (S. P. Colowick and N. O. Kaplan, eds.) p. 883. Academic Press, New York and London.
Troxler, R. F. and Bogorad, L. (1965). *Pl. Physiol.* **40**, xxix. Abstract.
Weisberger, A. S. and Wolfe, S. (1964). *Fed. Proc.* **23**, 976.

# The Use of Fully Deuterated Pigments to Study Their Function[1]

R. C. Dougherty[2], H. H. Strain and J. J. Katz
*Argonne National Laboratory, Argonne, Illinois, U.S.A.*

The availability of fully deuterated micro-organisms (Katz, 1965), and the unequivocal assignment of the proton magnetic resonance spectra for the chlorophylls (Closs *et al.*, 1963; Dougherty *et al.*, 1966) and other biologically important compounds provide a new approach to the general problems of biogenesis and reactivity in micro-organisms. Intact and physiologically competent organisms can now be used to obtain detailed information about the formation and function of normal metabolites.

Massive or complete deuterium substitution has been successful with a large number of organisms, of which the following species are representative: the green and blue-green algae *Chlorella vulgaris, Scenedesmus obliquus, Nostoc commune, Phormidium luridium;* photosynthetic and non-photosynthetic bacteria, including *Rhodospirillum rubrum* and *Escherichia coli*; the fungi *Claviceps purpurea* and *Saccharomyces cerevisiae*. The protozoan *Euglena gracilis* has most recently been cultured in fully deuterated medium, and is the most complex organism so far obtained in fully deuterated form. Higher plants and animals appear to tolerate only moderate deuterium levels, but complete deuteration is not a pre-requisite for useful studies of plant biosynthesis.

The magnetic properties of hydrogen and deuterium nuclei are very different, and this makes possible a novel methodological approach to *in vivo* biochemical studies. The fully deuterated compounds obtained from deuterio-organisms are transparent in ordinary high resolution proton magnetic resonance spectrometers. However, any hydrogen present in the compound can be readily detected, and both the location of the hydrogen in the molecule and the relative amount present can be determined (Jackman, 1959). Thus the combination of deuterated organisms and proton magnetic resonance spectroscopy provides a way to follow the path of hydrogen in biological systems. This procedure employs physiologically competent organisms, instead of mutants that are not truly viable, and does not require extensive degradation of the natural products.

Hydrogen can be introduced into a deuterated organism in several ways: (a) the organism can be grown in a mixture of $H_2O$ and $D_2O$; (b) the organism growing in $D_2O$ can be transferred to $H_2O$; (c) the organism can be grown in $D_2O$ on carbon-hydrogen substrates; (d) the organism can be grown in $H_2O$

[1] Based on work performed under the auspices of the U.S. Atomic Energy Commission.

[2] Resident Research Associate 1963–1965; present address, Department of Chemistry, Ohio State University, Columbus, Ohio, 43210.

on carbon-deuterium substrates. (c) and (d) constitute a "mirror image" experiment, in which the isotopic composition of the biosynthesized compounds are related in a unique way. To complete the experiment, a particular compound is then isolated from the organism and examined by proton magnetic resonance. Each of these procedures relates to a different aspect of the metabolic activities of the organism.

Procedure (b) has been used to study possible cyclic hydrogen transport between water and chlorophyll during photosynthesis (Katz *et al.*, 1964). Algae growing in $D_2O$ are transferred to $H_2O$, allowed to photosynthesize, and the chlorophyll subsequently extracted. The chlorophylls are found by proton magnetic resonance spectroscopy to contain no hydrogen at the δ, 7, or 8 positions. This result strongly suggests that exchangeable or labile hydrogen in chlorophyll (Dougherty *et al.*, 1965) is not directly involved in photosynthesis.

Studies that employ isotopically mixed media, i.e. $D_2O$–$H_2O$ mixtures (procedure a) make it possible to study kinetic isotope effects on the composition of the hydrogen-deuterium containing natural products. In order to analyse accurately these effects, however, information about the exchange behavior of the intermediates in the milieu of the cell is required. We have, therefore, examined the exchange behavior of bacteriochlorophyll intermediates by the following experiment.

*Rhodospirillum rubrum* was grown in $H_2O$ with succinic acid-$d_4$ as the only reduced carbon source. The chlorophyll was extracted and purified by conventional techniques. Since the NMR spectra of bacteriochlorophyll and its derivatives have been assigned (Dougherty *et al.*, 1966), the assessment of relative hydrogen abundance at each position in the molecule from the isotope experiment merely involved accurate integration of the spectra, taken under high resolution with deuterium decoupling.

The resonances due to hydrogen at the 4 positions (see Fig. 1) suggest that the strongly predominant (>95%) species at that locus in bacteriochlorophyll ($H_2O$ succinic acid-d) was:

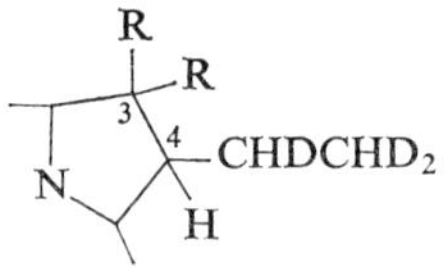

R
R
3
4
CHDCHD$_2$
N
H

The protons at position 2′ in the molecule appeared in the deuterium decoupled NMR spectrometer as a series of three lines with relative intensities 1:1·4:2·0. These are due to the three species $-COCH_3$, $-COCH_2D$ and $-COCHD_2$. The relative intensities of the lines suggested that the (approximate) relative abundance of these three species was 10, 23, and 66% respectively. From these data we expect that the 2′ resonance should have a relative area of 1·4 (compared to a methine proton as 1), which is in agreement with the integrated intensity of the 2′ methyl peak. The methyl group at position 4 thus has a different isotopic composition from the methyl at position 2. If both of these methyl groups originated in the same way, they should have the same isotopic composition.

The isotopic composition of the methyl group at position 4 may be explained by assuming that an original deuterium propionic acid side chain at that position underwent an oxidative decarboxylation with proton abstraction to yield a per-deuterio vinyl group[1]. Direct reduction of this group with a reducing agent of the same isotopic composition as that of the medium would be expected to yield the observed product. If the acetyl group at position 2 had a common origin with the ethyl group at position 4, the hydration and oxidation necessary for the conversion of vinyl into acetyl would have had to involve a substantial amount of hydrogen exchange with the hydrogen of the medium. This amount of exchange, $>20\%$, is considerably more than would ordinarily be expected for the hydration and oxidation of an aliphatic double bond. There is reason to believe, however, that a cation center adjacent to a porphyrin nucleus should show unusual stability, and the presumed intermediates in the hydration and oxidation reactions might be sufficiently stable to exchange hydrogen on the forming methyl group with the water in the cell.

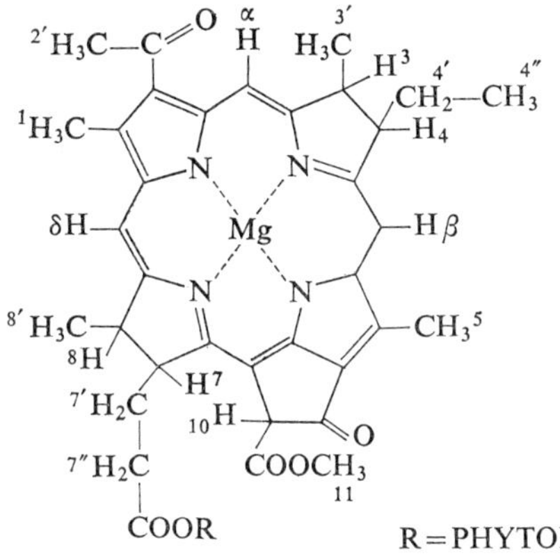

FIG. 1. Structure and NMR numbering system for bacteriochlorophyll.

An alternative explanation for the differences in isotopic composition of the methyl groups at positions 2′ and 4″ in bacteriochlorophyll is that these groups do not have a common origin. This raises the possibility that protoporphyrin IX is not necessarily implicated as the intermediate in the normal biogenesis of bacteriochlorophyll.

A set of "mirror image" isotope experiments (c and d above) have provided some further insight into bacteriochlorophyll biogenesis. Bacteriochlorophyll was prepared as above from cultures of *R. rubrum* that had been grown in $D_2O$ with succinate-$H_4$ and in $H_2O$ with succinic acid-$D_4$ as the only reduced carbon source. Under these conditions the isotopic composition of the two bacteriochlorophylls should be "mirror images" for those biosynthetic

[1] This proposal is not in conflict with Granick's suggestion for the origin of the vinyl group by hydride abstraction followed by decarboxylation. We tend to favour oxidative decarboxylation from the point of view of energetics (see pp. 373-410).

reactions that did not involve uncatalysed hydrogen exchange of the intermediates with the medium. In fact, the two chlorophylls obtained in this way showed substantial differences from the theoretical "mirror image" composition. These differences could not be rationalized entirely on the basis of isotope effects on the *de novo* synthesis of bacteriochlorophyll in the deuterio-organisms. It would thus appear that the biosynthetic pathway to bacteriochlorophyll is dependent upon the isotopic composition of the growth medium, and suggests that more than one biosynthetic pathway for the synthesis of bacteriochlorophyll is available to the organism.

Further work is in progress, which should elucidate these points.

ACKNOWLEDGEMENT

The authors enjoyed the able collaboration of Dr. H. L. Crespi, Mr. Walter Svec and Miss Gail Norman in these studies.

## References

Closs, G. L., Katz, J. J., Pennington, F. C., Thomas, M. L. and Strain, H. H. (1963). *J. Am. chem. Soc.* **85**, 3809.
Dougherty, R. C., Strain, H. H. and Katz, J. J. (1965). *J. Am. chem. Soc.* **87**, 104.
Dougherty, R. C., Strain, H. H. and Katz, J. J. (1966). To be published.
Jackman, L. M. (1959). "Applications of Nuclear Magnetic Resonance Spectroscopy in Organic Chemistry." Pergamon Press, New York.
Katz, J. J. (1965). "Thirty-ninth Annual Priestley Lectures," The Pennsylvania State University, University Park, Pennsylvania.
Katz, J. J., Dougherty, R. C., Svec, W. A. and Strain, H. H. (1964). *J. Am. chem. Soc.* **86**, 4220.

# Oxygen Reactions of Xanthophylls[1]

H. V. Donohue, T. O. M. Nakayama and C. O. Chichester
*Department of Food Science and Technology,*
*University of California, Davis, California, U.S.A.*

In photosynthetic tissue, such as leaves of the higher plants, there is almost universally the occurrence of both oxygenated carotenoids and carotenes. The fading of leaves which occurs in the fall was shown by the extensive studies of Strain (1938, 1959) to signify a decrease in the carotene level, accompanied by a net increase in the xanthophylls. On the basis of these analyses it was proposed that the carotenes are converted by oxygenation to the xanthophylls, although in senescence the xanthophylls were not assigned a specific function. As the concentration of xanthophylls in photosynthetic tissue is normally quite high, it was however proposed that they serve a physiological function.

The photosynthetic tissue always contains, in addition to the normally colored carotenoids, an amount of the carotene precursors (or non-colored C-40 polyenes). The analysis of the tissue indicates, however, that the chloroplasts normally contain the colored carotenoids and the xanthophylls, while the cytoplasmic matrix contains primarily the non-colored (or highly saturated) polyenes. Hence the suggestion that the saturated C-40 polyenes do not play a role in photosynthesis.

Since there is a net transfer of oxygen to carotenes to form xanthophylls, it was postulated at quite an early date that xanthophylls could serve as subsidiary oxygen-carrying systems, and in fact Blass, Anderson and Calvin in 1959 suggested that within the chloroplasts xanthophylls carried out this function. In addition to the hydroxylated carotenes it was also noted that carotenoid epoxides were metabolically active, and these were suggested as intermediates, or products of reactions between oxygen and the hydroxylated carotenoids. Cholnoky *et al.* (1956, 1958), working with peppers, found that the concentrations of β-carotene and zeaxanthin were dynamic elements of the chromoplasts, and that in addition to this there appeared to be a direct relation between the epoxide pigments and β-carotene and zeaxanthin. He postulated, on studying the kinetics of the rise and fall of pigment concentrations in ripening peppers, that both β-carotene and zeaxanthin could be converted into epoxides, and these in turn could be reduced to α-carotene and the dihydroxy derivative of α-carotene, lutein, or to the original pigments.

It is interesting to note in this respect that lutein has always been a problem in the carotenoid field, since the major carotene of leaves (β-carotene) has the β-ionone configuration, yet under most circumstances the major xanthophyll is lutein, which possesses an α-ionone configuration in one end of its chain.

[1] This work was supported in part by NIH Grant GM 08869.

Cholnoky's proposal would explain this situation, and also the presence of large amounts of individual pigments at different times, and relate them to the physiological activities of the plant. What was lacking in many of these experiments was direct evidence that carotenes could be converted into xanthophylls. Earlier, Liaaen and Sørensen (1956) had observed the conversion of violaxanthin into zeaxanthin, but they interpreted the phenomenon as a post-mortem change. Claes (1959) and Claes and Nakayama (1959), using a mutant of *Chlorella* (5/520), first gave substantial proof to this possibility. Growing the *Chlorella* mutant in the dark aerobically, they found only the accumulation of carotene percursors, or highly saturated carotenoids, such as phytoene, phytofluene, etc. When the mutants were exposed to dim red light anaerobically, the colored carotenoids, α- and β-carotene, and in addition a carotene labeled "carotene X", were formed. If these mutants were then returned to the dark and an aerobic atmosphere, there was a net increase or formation of xanthophylls, which was accompanied by a decrease in the carotenes. The establishment of a stoichiometric relationship between the two classes of compounds thus constituted the first substantial proof that the carotenes in a photosynthetic organism could be interconverted on a mole per mole basis to their oxygenated forms. It also suggested the direct participation of molecular oxygen in the conversion.

A study of these interconversions in higher plants was described by Yamamoto *et al.* (1962). Normally, if harvested in the evening, bean leaves contain a considerable amount of violaxanthin, antheraxanthin, and zeaxanthin. When these leaves are exposed to light in a nitrogen atmosphere, a radical decrease in violaxanthin is observed. Concomitant with this, antheraxanthin first increases and then decreases, and the zeaxanthin content increases in direct proportion to the decrease in violaxanthin and antheraxanthin. On this basis it was postulated that violaxanthin was converted into zeaxanthin through the intermediate mono-epoxide, antheraxanthin, and thus the reduction occurred under illumination in two steps. If the leaves were then placed in oxygen and held in the dark, the zeaxanthin decreased, while equivalent increases in antheraxanthin and violaxanthin were observed. The stoichiometry would suggest that zeaxanthin was reconverted to the epoxides. Similar leaves that were not transferred to the dark continued losing violaxanthin, and a continued increase in zeaxanthin was noted.

At about the same time Krinsky (1962) found that *Euglena gracilis* when exposed to the light showed a decrease in antheraxanthin. Since *Euglena* does not contain the di-epoxide violaxanthin, the increase in zeaxanthin could be accounted for solely by the decrease in antheraxanthin. These experiments were performed under light and anaerobic conditions, and confirm in general the results in bean leaves.

On the basis of the interconversions in photosynthetic tissues and suggestions relating the pigments to oxygen transports, the origin of the oxygen in the pigments was investigated by Yamamoto *et al.* (1962a, b, c). If pigments which were kinetically related to a light reaction were acting as oxygen carriers, one should be able to observe the incorporation of oxygen into various oxygenated pigments, and if indeed these derived their oxygen from water photosynthetically, there should be substantial incorporation of water-derived oxygen into hydroxyl as well as epoxy groups. The incorporation of oxygen from water enriched in $H_2O^{18}$ into epoxide pigments in chloroplast systems undergoing a Hill reaction was investigated. No enrichment was detectable in violaxanthin or neoxanthin; a very low incorporation, however, was noted in lutein, but this was not dependent upon the Hill reaction. In other experiments *Chlorella vulgaris* was grown heterotrophically in the dark in the presence of $H_2O^{18}$. When grown under such conditions *Chlorella* contains a substantial amount of lutein. Upon isolation and analysis for $O^{18}$ content it was found that there was no significant labeling in the lutein. Consequently it could be postulated that the oxygen incorporated in the lutein did not come from the water, and it was thus unlikely that the lutein would normally serve as a transfer agent for water-derived oxygen. The antheraxanthin was also analysed in this experiment, but as the quantity was extremely small, the results were indefinite. The complementary experiment was then performed. *Chlorella vulgaris* was grown in the presence of $O_2{}^{18}$, and again the pigments were isolated and analysed. In this case lutein was found to be labeled in sufficient quantity to account for the incorporation of oxygen from $O_2{}^{18}$ into both hydroxyl groups. Antheraxanthin (produced in only very small quantities) was also analysed. Since it contains one epoxy and two hydroxyl groups and had in this experiment more apparent label than the lutein, but not enough to account for the three oxygens, some doubt remained as to the origin of the epoxide oxygen. On the basis of this work a biosynthetic pathway was postulated wherein the light reaction was one in which de-epoxidation of a xanthophyll di-epoxide occurred through a two-step route, the removal of a single oxygen at a time. Since the configuration of the $\alpha$ and $\beta$ compounds are different, it was assumed that the hydroxylation of the $\alpha$ compound also proceeded in two steps, since in most instances one can find both the monohydroxy derivative of $\alpha$ and $\beta$-carotenes. Thus $\beta$-cryptoxanthin and $\alpha$-cryptoxanthin were postulated as intermediates between $\beta$-carotene and zeaxanthin and $\alpha$-carotene and lutein, aerobic oxygen supplying both oxygens in these cases. It was also postulated that zeaxanthin was epoxidized in two steps, first being converted to the mono-epoxy compound antheraxanthin, and then to the di-epoxy compound violaxanthin. Lutein could also be epoxidized in the same manner, the epoxidation in both pathways occurring in the dark.

Krinsky (1964) found that in broken *Euglena gracilis* chloroplasts the conversion from antheraxanthin to zeaxanthin proceeded in either light or dark, provided the co-factors FMN and NADPH were added. He also demonstrated that the conversion of zeaxanthin into antheraxanthin in broken chloroplast preparations was non-enzymatic and dependent only upon the presence of oxygen and light.

A similar non-enzymatic production of epoxide carotenoids from both carotenes and hydroxylated carotenes has also been demonstrated in methanol using a photosensitizer and high intensity light as the activating agents (K. Hasegawa, personal communication).

Because of the ambiguous status of antheraxanthin in both leaves and in other photosynthetic tissues a more careful investigation of antheraxanthin was warranted. A series of experiments was performed utilizing bean leaves, which accumulate a considerable amount of antheraxanthin as a source of pigments (Yamamoto and Chichester, 1965). In order to obtain sufficient antheraxanthin for a mass spectroscopic analysis it was necessary to perform the experiment several times. Leaves were exposed first anaerobically to light and then placed in an enriched $O_2^{18}$ atmosphere in the dark. The antheraxanthin was isolated and crystallized. It was found that the newly-formed antheraxanthin incorporated oxygen from the atmosphere and thus the epoxidation in the intact leaf appeared to proceed by the incorporation of oxygen from the atmosphere. It would appear that most of the oxygen conversions of the carotenes are oxygen-dependent and derive their oxygen from the air, rather than from

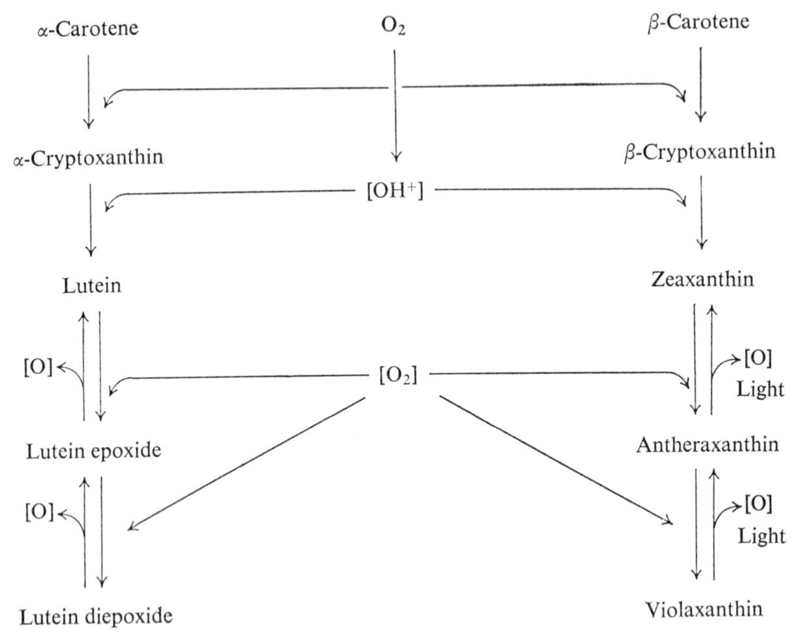


FIG. 1. A proposed scheme for the formation of the xanthophylls and epoxy pigments.

photosynthetically derived oxygen. This conversion, however, does not take place in boiled, or otherwise inactivated, leaves.

With the same methods it was shown that the carboxyl group of the pigment torularhodin derived one oxygen of the carboxyl group from molecular oxygen (Simpson *et al.*, 1964).

On the basis of these results the scheme shown in Fig. 1 is postulated. Direct experimental proof is available for conversions of the β-compounds; the α-based compounds require additional investigation. A possible alternate route to the α-ionone-derived hydroxylated compounds could be through the di-epoxy pigment violaxanthin, which could be converted through an intermediate in three steps to lutein. This is shown in Fig. 2. The structure of the intermediate is unknown,[1] although Curl (1965) has recently published the

[1] Curl (1965) suggested a structure for neoxanthin identical to that of the intermediate in Fig. 2. Recent evidence, however, casts some doubt upon this structure for neoxanthin.

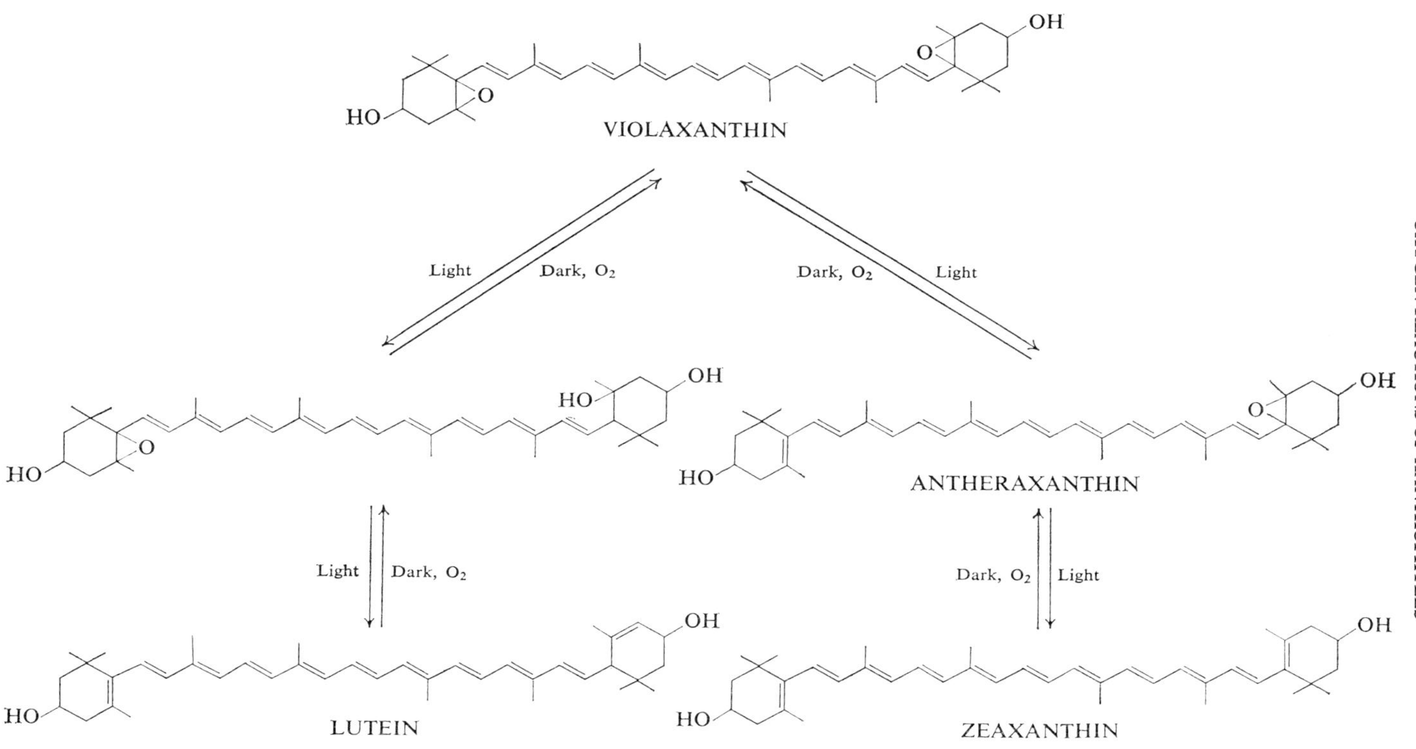


FIG. 2. Possible route to the $\alpha$-ionone derived hydroxylated compounds.

structure shown. This pathway could conceivably account for the production of lutein from zeaxanthin or violaxanthin, despite the fact that in most preparations neoxanthin is relatively inactive metabolically compared to zeaxanthin and antheraxanthin. The synthesis of $\alpha$-carotene still presents some problems, however.

Most of the experiments described were performed over a fairly long-term period, i.e., intervals between measurements ranging from $\frac{1}{2}$ hr to 1 hr or longer. They did not necessarily reveal any short time constant reactions of the individual pigments. If these pigments are connected with photosynthesis there should be changes occurring in the xanthophylls over comparatively short

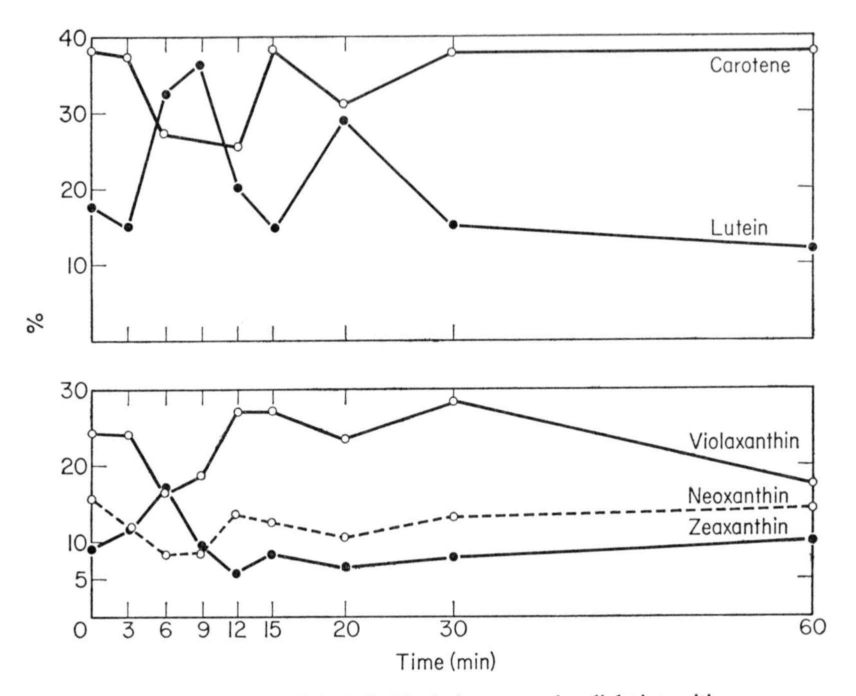


FIG. 3. Changes of the individual pigments at low light intensities.

periods of time. In order to determine such changes, New Zealand spinach leaves were selected as the tissue to be investigated, and samplings were made at intervals as short as $1\frac{1}{2}$ min after the initiation of illumination. Spinach leaves were usually obtained in the evening, stored overnight in the dark in an oxygen atmosphere, and the experiments initiated the next morning. When the leaves were illuminated it was found that there was a substantial initial drop of violaxanthin, usually occurring in the first 2 to 3 min, followed by a transitory rise in epoxys, i.e., neoxanthin and violaxanthin, followed finally by a decrease in the concentration for the next 30 to 60 min. The overall result, of course, was the same as that shown previously, in that there is a total decrease in the first hour of the epoxide concentration. What was unexpected, however, was the transitory increase in the zeaxanthin, following inversely the concentration of violaxanthin. As the violaxanthin decreased, zeaxanthin increased, and the trend was then reversed. As zeaxanthin decreased, the concentration of violaxanthin, neoxanthin, and lutein increased to a maximum, and fell again as zeaxanthin increased. In leaves, the oscillatory behavior of the different pigment concentrations could be reproduced routinely as to time intervals, although the relative concentrations varied somewhat. Figure 3 is representative of the changes of the individual pigments under low light intensities; the

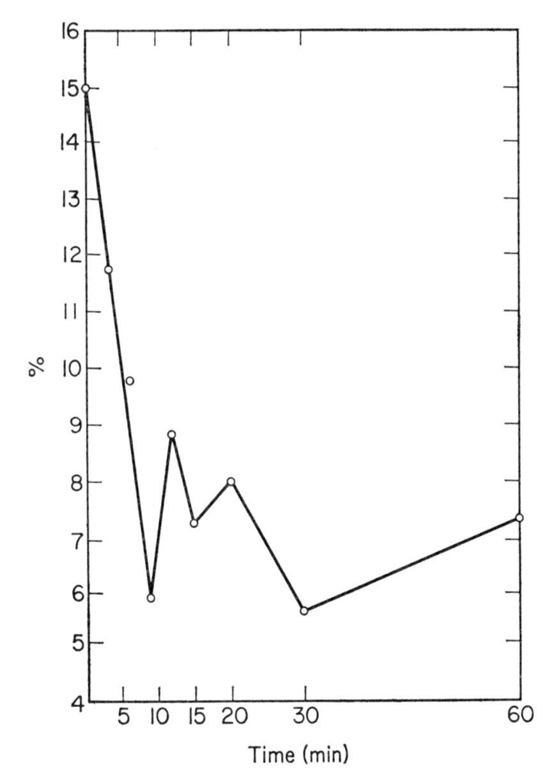


FIG. 4. Changes of epoxide concentration in whole leaves under higher light intensities.

concentrations shown are relative and not absolute. Figure 4 shows the changes of epoxide concentration under higher light intensities.

If the levels of illumination were varied, quantitative differences were noted, although qualitatively the oscillations still took place. At very low levels of illumination there was a comparatively small loss in epoxy pigments initially, followed by increases in concentration to values on occasion higher than the initial concentration. But after a period of 60 min the concentration of epoxides was essentially equivalent to or slightly lower than the initial starting concentration. In contrast, if the experiment was performed at high levels of illumination (in this case 2000 to 10,000 foot candles), the initial decrease was extremely high, and the oscillating behavior was generally downward, i.e., the maximum of the epoxys after 10 min was in all cases below the initial concentration, and the relative increase above the minimum was small. After a 60-min period of illumination the concentrations of the epoxides were significantly below the initial concentrations. These oscillations, however, occurred in approximately the same time period.

An attempt was made to reproduce the same conversions in isolated chloroplasts. The chloroplasts were obtained by homogenizing the New Zealand spinach leaves in a cold 0·5 M sucrose 0·1 M phosphate buffer, pH 7·4, for 2 min. The extract was then pressed through cheese cloth and centrifuged at 200 × ***g*** for 8 min to remove the debris. The supernatant was then centrifuged at 1000 × ***g*** for 10 min and the supernatant discarded. The chloroplasts were washed in the

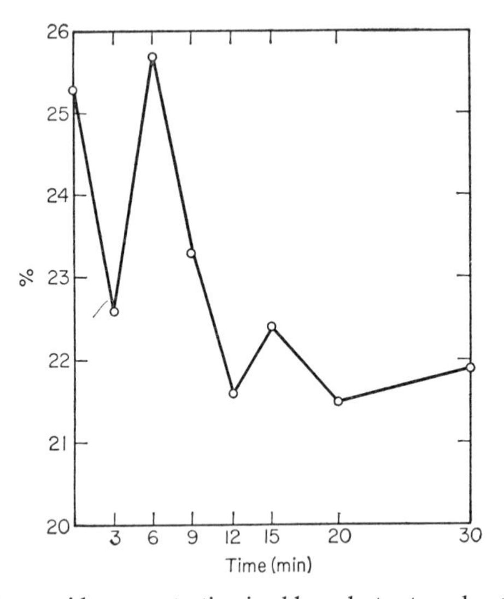


FIG. 5. Changes in epoxide concentration in chloroplasts at moderate light intensities.

sucrose-phosphate buffer and again centrifuged at 1000 × ***g*** for 10 min. The washed chloroplasts were then gently resuspended. The preparation was tested for photosynthetic activity by noting that it would carry out a Hill reaction. When such a preparation was exposed to light, the same changes in pigment concentration were observed as found in whole leaves. Figure 5 shows the results of a typical experiment utilizing the chloroplast preparation. In this case a rapid method was used to determine the epoxy concentration which gave total change rather than the change in individual pigments (Yamamoto *et al.*, 1965). The results were confirmed on occasion by chromatography. It is interesting to note that the epoxy oscillations found in the chloroplasts are the same as noted in the whole tissue, although the absolute values do vary from experiment to experiment. The variation of epoxide concentration was within a narrow range representing a change of between 14 and 18% of the total epoxides. Analysis of variance, however, showed that effects noted were real, and replication indicated that the changes were reproducible. The same experiment was performed under anaerobic conditions and approximately the same fluctuations in epoxide concentration were obtained. If the chloroplasts are broken, utilizing the French press, the system is inactive. Thus the work with the chloroplasts would confirm the fact that in an active photosynthetic tissue oxygen was not required for the de-epoxidation reaction of the xanthophyll pigment.

In some further experiments the effect of different wavelength illumination was investigated. Chloroplasts were isolated conventionally, and exposed to approximately 500 foot candles of light passed through gelatin filters. In one experiment the effect of illumination above 550 m$\mu$ was contrasted to that below 550 m$\mu$. In the experiment using blue light no changes were noted in the concentration of epoxys, while controls under white light gave the normal response. The red light group gave essentially the same response as those exposed to white light. Although the carotenoids absorb below 550 m$\mu$, photosynthetic activity (initiated by light above 550 m$\mu$) is required under our experimental conditions for de-epoxidation to take place.

In other experiments the effect of several inhibitors of photosynthesis was investigated. In presence of CMU the light reaction of the conversion of the violaxanthin into zeaxanthin was inhibited. Krinsky (1964) had shown earlier that FMN and NADP were required co-factors in sonicated chloroplasts for the conversion of antheraxanthin into zeaxanthin, so that in experiments which were recently completed the effect of PCMB was tested. It was found that the oscillating behavior of pigment concentrations in the chloroplasts was inhibited at concentrations of $1 \times 10^{-3}$ M.

These experiments, and some others which are in progress, would support the speculation that the xanthophylls in most photosynthetic tissue are indeed involved in the photosynthetic mechanism. The fact that in whole tissue, as well as in isolated chloroplasts, the xanthophylls appear to exhibit cyclic behavior with rapid changes at the onset of photosynthesis would implicate them in the dynamic system. As it can be seen, the conversion of the epoxys to the xanthophylls is a net electron-accepting reaction, and the reverse, of course, acts as an electron donor reaction. While the pigments do not seem to play a direct role in oxygen transport, they may very well serve as an electron buffering system, being able to accept electrons from the photosynthetic mechanism when these are in excess, and able to return them to the system when they can be accepted. The reducing potentials of these compounds have not been established, and therefore the level at which they might function in the system cannot readily be established. The experiments described here, and others reported in the literature, would now substantiate their cyclic behavior, as well as suggest their function in the photosynthetic system.

## References

Blass, U., Anderson, J. M. and Calvin, M. (1959). *Pl. Physiol.* **34**, 329.
Cholnoky, L., Gyorgyfy, C., Nagy, E. and Panczel, M. (1956). *Nature, Lond.*, **178**, 410.
Cholnoky, L., Szabolcs, J. and Nagy, E. (1958). *Liebigs. Ann. Chem.* **515**, 207.
Claes, H. (1959). *Z. Naturf.* **14b**, 4.

Claes, H. and Nakayama, T. O. M. (1959). *Nature, Lond.* **183**, 1053.
Curl, A. L. (1965). *J. Fd. Sci.* **30**, 426.
Krinsky, N. I. (1962). *Fed. Proc.* **21**, 92.
Krinsky, N. (1964). *Biochim. biophys. Acta* **88**, 487.
Liaaen, S. and Sørensen, N. A. (1956). *In* "Second International Seaweed Symposium" (T. Braarud and N. A. Sørensen, eds.) p. 25. Pergamon Press.
Simpson, K. L., Nakayama, T. O. M. and Chichester, C. O. (1964). *Biochem. J.* **92**, 508.
Strain, H. H. (1938). *Leaf Xanthophylls.* Carnegie Inst. Wash. Publ. No. 490.
Strain, H. H. (1959). *Chloroplasts Pigments and Chromatographic Analysis*, p. 1. Penn. State Univ. Press, Univ. Park, Penn.
Yamamoto, H. Y., Chichester, C. O. and Nakayama, T. O. M. (1962). *Archs Biochem. Biophys.* **96**, 645.
Yamamoto, H. Y., Chichester, C. O. and Nakayama, T. O. M. (1962). *Archs Biochem. Biophys.* **97**, 168.
Yamamoto, H. Y., Chichester, C. O. and Nakayama, T. O. M. (1962). *Photochem. Photobiol.* **1**, 53.
Yamamoto, H. Y. and Chichester, C. O. (1965). *Biochim. biophys. Acta* **109**, 303.
Yamamoto, H. Y., Go, G. and Chang, J. L. (1965). *Analyt. Biochem.*, **12**, 344.

## Discussion

*S. Liaaen Jensen:* Many years ago we [S. Liaaen and N. A. Sørensen, in *Proc. 2nd Int. Seaweed Symp.* Trondheim (1955) p. 25] were able to demonstrate the transformation of violaxanthin to zeaxanthin in the brown algae *Fucus vesiculosus* under so-called postmortem conditions. We later found that this reaction also occurs in other brown seaweeds such as *Fucus serratus*, *Fucus inflatus*, *Pelvetia canaliculata* and *Ascophyllum nodosum.*

After the report by Krinsky [*Fed. Proc.* **21** (1962) 92] we examined the effect of light on this reaction in *Fucus vesiculosus* and found that under anaerobic conditions no de-epoxidation occurred and that light indeed was required for this reaction.

# Action Spectrum of Light-dependent Carotenoid Synthesis in *Chlorella vulgaris*

Hedwig Claes

*Max-Planck-Institut für Biologie, Abt. Melchers, Tübingen, Germany*

Stimulation of carotenoid synthesis in chloroplasts by irradiation has been described by various authors. Besides changes within the mixture of the different polyenes, more carotenoids are synthesized in light than in the dark. Especially observations of Bandursky (1949) indicate that the increase of the yellow pigments in light is not necessarily linked to the formation of end products of photosynthesis.

A mutant of *Chlorella vulgaris* was used to study the action spectrum of light-dependent carotenoid synthesis. Strain 520 is distinguished from the wild type by the fact that it synthesizes the normal bicyclic carotenoids, which are found in chloroplasts of higher plants, only in the light. It differs from etiolated plants by accumulating acyclic $C_{40}$-polyenes in the dark. The amount of chlorophylls synthesized in the dark is only a fraction of the normal chlorophyll concentration of *Chlorella*.

The following sequence of reactions was found in strain 520 (Claes, 1959):

$$\left.\begin{array}{l}\text{Phytoene}\\ \text{Phytofluene}\\ \zeta\text{-Carotene}\\ \text{Proneurosporene}\\ \text{Prolycopene}\end{array}\right\} \xrightarrow[N_2]{\text{light}} \begin{array}{l}\alpha\text{-Carotene}\\ \beta\text{-Carotene}\\ (\beta\text{-Zeacarotene})\end{array} \xrightarrow[O_2]{\text{dark}} \text{Xanthophylls}$$

Probably in neurosporene and lycopene ring closure is initiated by proton attack (Olson and Knizley, 1962; Porter and Anderson, 1962). Both polyenes are synthesized by strain 520 in the dark. Light may influence but is not essential for xanthophyll synthesis. Therefore it must be assumed that, at least in *Chlorella*, light takes part in cyclization of the acyclic carotene precursors.

The action spectrum of light-induced carotene synthesis can be studied in strain 520 in the absence of oxygen. By use of anaerobic irradiation formation of carotenes is isolated from xanthophyll synthesis, from photo-oxidative destructions of carotenes, and from reactions leading from low molecular weight precursors to the immediate precursors of cyclic carotenes.

For the experiments described below, strain 520 was grown in the dark and harvested in dim light. During the irradiation at +25°, nitrogen was streaming through the cuvette, which contained the cells in mineral solution. α-Carotene,

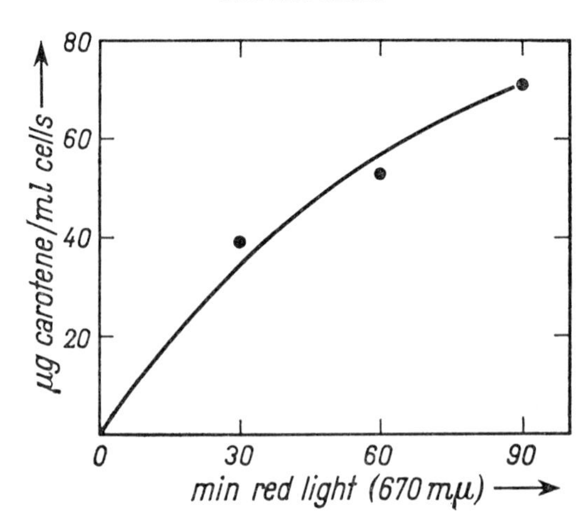


FIG. 1. Time course of carotene synthesis by *Chlorella* mutant 520 in red light (670 mμ).

β-carotene and β-zeacarotene synthesized in light were isolated chromatographically. Narrow spectral regions were isolated with interference filters (Schott u.Gen. Mainz, Germany).

The time course of synthesis of cyclic carotenes in red light of 670 mμ is shown in Fig. 1. Carotene formation starts immediately at the beginning of

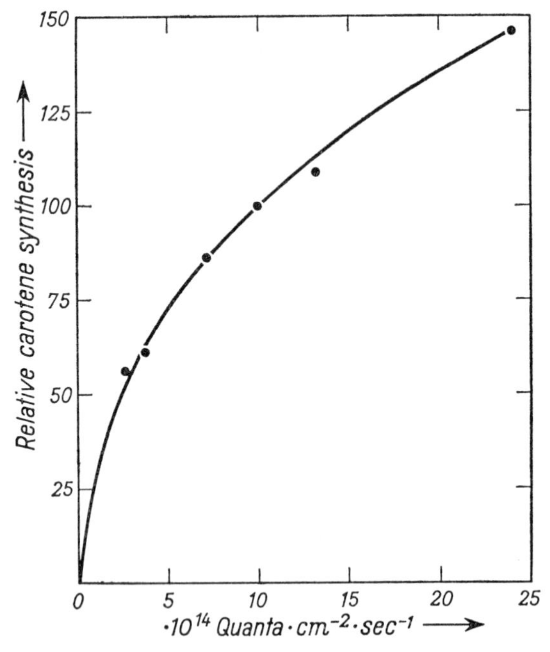


FIG. 2. Influence of light intensity on carotene synthesis by *Chlorella* mutant 520 during 90 min irradiation with red light (670 mμ).

anaerobic irradiation. When the light is turned off, no further increase of cyclic polyenes can be observed. The influence of light intensity at 670 m$\mu$ on carotene synthesis is illustrated in Fig. 2. With the possible exception of the lowest range of intensities, formation of cyclic polyenes does not increase linearly with light intensity.

To measure the effect of light of different wavelength on carotene synthesis, the cells were irradiated for 90 min with light intensities below saturation values (cf. Fig. 2). In red and green light $10 \times 10^{14}$ quanta/cm$^2$ sec and in blue light $4 \times 10^{14}$ quanta/cm$^2$ sec were measured in the centre of the light beam and at the bottom of the cuvette containing the cells. To reduce the influence of variability of algal cultures, one sample of each culture was irradiated at 670 m$\mu$ as a control. Carotene synthesis at this wavelength was designed as 100%.

The results are demonstrated in Fig. 3. The shape of the curve is certainly

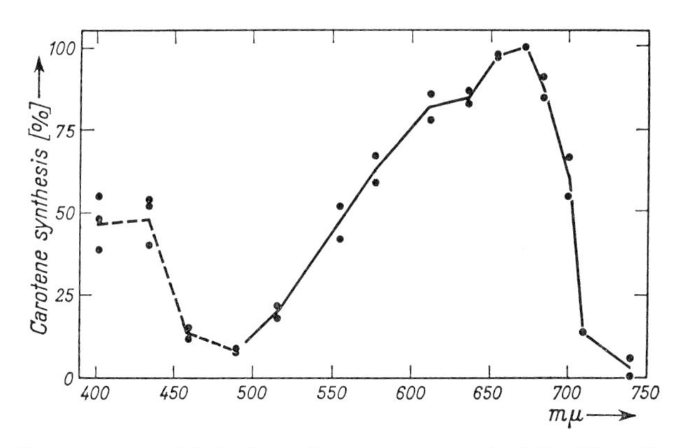


FIG. 3. Action spectrum of light-dependent carotene synthesis by *Chlorella* mutant 520. 90 min irradiation at wavelength indicated.

influenced by the absence of a linear relationship between light intensity and carotene synthesis. The slopes would be steeper if light intensities necessary to produce the same effect could have been plotted. Nevertheless it must be concluded that red light near 670 m$\mu$ is most efficient in promoting ring-closure. Experiments still in progress show that the maximum near 670 m$\mu$ becomes more pronounced, if the background transmission of the interference filters is reduced. While carotene synthesis is low in green light, another maximum is indicated in the blue part of the spectrum.

The pigments with an absorption in red and green light similar to the action spectrum in Fig. 3 are the chlorophylls or phytochrome. However, reversibility of the effect of red light by far-red light, a characteristic property of the phyto-chrome system, could not be observed. The effect of red light (670 m$\mu$) could not be reduced by simultaneous irradiation of the cells with dark red light (727 m$\mu$) of equal intensity. Therefore it must be concluded that the chloro-phylls are involved. The absence from Fig. 3 of a pronounced peak in blue light corresponding to the absorption of chlorophylls in this part of the spectrum might be explained by a screening effect of carotenes. Further experiments are necessary to find out, if the results obtained with *Chlorella* can be extended to chloroplasts of higher plants.

## References

Bandursky, R. S. (1949). *Bot. Gaz.* **111**, 95.
Claes, H. (1959). *Z. Naturf.* **14**b, 4.
Olson, J. A. and Knizley, H. (1962). *Archs Biochem. Biophys.* **97**, 138.
Porter, J. W. and Anderson, D. G. (1962). *Archs Biochem. Biophys.* **97**, 520.

# Studies on the Metabolism of the Assimilatory Pigments in Cotyledons of Four Species of Pine Seedlings Grown in Darkness and in Light

S. Więckowski[1] and T. W. Goodwin

*Department of Biochemistry and Agricultural Biochemistry, University College of Wales, Aberystwyth, Wales*

## Introduction

For several years intensive investigations on the metabolism of terpenoids in plant cells have been carried out in our department (for review see: Goodwin, 1965, 1967). All these experiments show that there are two distinct sites of terpenoid biosynthesiswi thin the plant cell: one in the chloroplast in which "photosynthetic terpenoids" (carotenoids, phytol, plastoquinone, vitamin $K_1$, tocopherols) are formed, and one in the cytoplasm where such terpenoids as phytosterols, pentacyclic triterpenes and ubiquinone are synthesized. Much evidence has been obtained in confirmation of this picture, but the main support has come from the comparison of the incorporation of [$2^{14}$C]mevalonic acid ([2-$^{14}$C]MVA) and $^{14}CO_2$ into various terpenoids; after illumination of etiolated plants [$2^{14}$C]MVA is incorporated primarily into terpenoids formed outside the chloroplast, although considerable amounts of carotenoids and chlorophylls are being synthesized at the same time. Under the same conditions $^{14}CO_2$ is much better incorporated into photosynthetic terpenoids (Goodwin, 1958; Goodwin and Mercer, 1963; Threlfall *et al.*, 1964; Treharne *et al.*, 1964). The view that the chloroplast membrane is impermeable either way to MVA, which is an important aspect of the hypothesis proposed to explain this differential synthesis (Goodwin and Mercer, 1963), has recently been experimentally demonstrated. It has been shown that mevalonic kinase in isolated chloroplasts of bean leaves cleaves MVA only when the chloroplasts are destroyed by ultrasonics or osmotic pressure (Rogers *et al.*, 1965).

All these experiments have been carried out on angiosperms (mono- and dicotyledons). The studies described in this paper are concerned with similar experiments on gymnosperms. From this point of view seedlings of many gymnosperms are interesting because they are able to synthesize chlorophyll in darkness. We felt that chloroplasts may not be fully developed in darkness and that their membranes might have different properties especially concerning permeability to MVA.

Before experiments were undertaken with [2-$^{14}$C]MVA and $^{14}CO_2$, the concentration of chlorophylls and carotenoids were determined in pine seedlings grown in darkness and in light.

[1] Present address: Laboratory of Plant Physiology, Jagellonean University, Cracow (Poland).

Table I. Concentration of plant pigments in cotyledons of pine seedlings grown in darkness and in light. Seedlings 6 days after germination. Light intensity about 300 lux, temperature 29°

| | | Concentration (mg/g fresh wt) | | | | Ratios | | |
|---|---|---|---|---|---|---|---|---|
| | | Chlorophyll *a* | Chlorophyll *b* | Carotenes | Xanthophylls | Chl. *a* / Chl. *b* | Xanth. / Carot. | Chlorophylls / Carotenoids |
| *P. jeffrei* | dark | 0·416 | 0·135 | 0·0313 | 0·0615 | 3·09 | 1·96 | 5·93 |
| | light | 0·856 | 0·342 | 0·0485 | 0·0870 | 2·50 | 1·79 | 8·84 |
| *P. silvestris* | dark | 0·129 | 0·0192 | 0·0113 | 0·0298 | 6·72 | 2·63 | 3·61 |
| | light | 0·562 | 0·187 | 0·0370 | 0·0507 | 3·00 | 1·37 | 8·54 |
| *P. radiata* | dark | 0·238 | 0·074 | 0·0163 | 0·0601 | 3·20 | 3·69 | 4·09 |
| | light | 0·789 | 0·333 | 0·0502 | 0·0863 | 2·37 | 1·72 | 8·22 |
| *P. contorta* | dark | 0·136 | 0·0346 | 0·0144 | 0·0269 | 3·93 | 1·87 | 4·14 |
| | light | 0·617 | 0·281 | 0·0408 | 0·0715 | 2·20 | 1·75 | 8·00 |

## Results

### COMPARISON OF THE ASSIMILATORY PIGMENTS IN PINE SEEDLINGS GROWN IN DARKNESS AND IN LIGHT

The biosynthesis of carotenoids in angiosperms is stimulated by light (Bandurski, 1949; Blaauw-Jansen *et al.*, 1950). Etiolated plants contain only small amounts of these pigments with xanthophylls predominating (Kay and Phinney, 1956; Goodwin, 1958; Goodwin and Phagpolngarm, 1960) and no chlorophylls but traces of protochlorophyll. Our experiments (Goodwin and Więckowski, 1966a) show that seedlings of four species of *Pinus* grown in darkness are able not only to synthesize chlorophylls but also to synthesize greater amounts of carotenoids than do etiolated angiosperms. Typical results are presented in Table I. It shows that seedlings cultivated in darkness contain about four times less chlorophyll that those cultivated in continuous illumination. The ratio of chlorophyll *a* to chlorophyll *b* is greater in darkness. Table I also shows that the carotenoid content is about 2–3 times greater in light. The xanthophyll: carotene ratio does not vary in light and in darkness in *P. jeffrei* and *P. contorta*, but is marked by different in *P. radiata* and *P. silvestris*. The last column in Table I indicates that the ratio of chlorophylls to carotenoids is significantly greater in light; that means that in darkness there are relatively more carotenoids.

It is known that gymnosperms generally contain more α-carotene than other plants (Wierzchowski *et al.*, 1962). It was interesting to compare the α-carotene content of pine seedlings grown in darkness and in light. The results obtained are summarized in Table II. In light there are 2–3 times as much β-carotene as

Table II. Concentration of β and α-carotene in *Pinus* seedlings growing in darkness and in light. Seedlings 6 days after germination. Light intensity about 300 lux, temperature 29°

| | | β-Carotene (mg/g fresh weight) | α-Carotene (mg/g fresh weight) | Ratio $\frac{\beta}{\alpha}$ |
|---|---|---|---|---|
| *P. jeffrei* | dark | 0·0240 | 0·0073 | 3·29 |
| | light | 0·0302 | 0·0182 | 1·66 |
| *P. silvestris* | dark | 0·0092 | 0·0022 | 4·19 |
| | light | 0·0292 | 0·0079 | 3·70 |
| *P. radiata* | dark | 0·0137 | 0·0026 | 5·27 |
| | light | 0·0368 | 0·0135 | 2·73 |
| *P. contorta* | dark | 0·0122 | 0·0022 | 5·55 |
| | light | 0·0300 | 0·0108 | 2·78 |

α-carotene; in darkness the relative amount of β-carotene to α-carotene increases to 3–5. Thus pine seedlings grown in the light contain relatively more α-carotene than those grown in the dark.

We suppose that the chloroplasts of pine seedlings grown in darkness are at a more advanced stage of morphological development than the plastids in etiolated angiosperms. Moreover, we feel that the structural changes of plastids which take place after illumination of etiolated plants (see Virgin *et al.*, 1963) may occur to a certain extent in pine seedlings in darkness. However, the different ratio of β-carotene to α-carotene in light and darkness indicates that light affects the biosynthesis of carotenes not only through the structural changes of plastids which it induces.

## INCORPORATION OF DL[2-$^{14}$C]MVA LACTONE INTO β-CAROTENE AND INTO PHYTOL SIDE CHAIN OF CHLOROPHYLLS

The amount of radioactivity in the total lipid fraction per 100 plants (Table III) is dependent on the species, being considerably greater in *P. jeffrei* than in the other species examined. In many cases plants growing in light incorporated [2-$^{14}$C]MVA to a greater extent than plants growing in the dark. The specific radioactivities of β-carotene and phytol are also dependent on species. We found the greatest specific radioactivities of β-carotene and phytol in the cotyledons of *P. silvestris* and the lowest in *P. jeffrei* and *P. contorta*. If we compare the percentage of label incorporated into β-carotene and phytol we cannot establish any consistent species differences: the incorporation into β-carotene varies from 0·1 to 0·4 % and into phytol from 0·5 to 3 %. Differences due to the growth conditions or age of seedlings are also not very marked.

Using the same methods for introducting the label into plants, we obtained (Goodwin and Więckowski, 1966b) a similar picture for the incorporation of [2-$^{14}$C]MVA into β-carotene and phytol after illumination of etiolated bean plants. In comparison with the radioactivity of the whole lipid fraction the same relative amounts of label were also incorporated into β-carotene and phytol.

If we introduced the same activity of [2-$^{14}$C]MVA into the plant not through the leaves but through the stems we found a much smaller uptake of radioactivity. The specific radioactivities of β-carotene and phytol were about ten times smaller, but if we compare the percentage of radioactivity incorporated into β-carotene and phytol with the radioactivity of the whole lipid fraction, we do not find any important differences between the various methods of introducing MVA into the plant cell.

## INCORPORATION OF $^{14}CO_2$ INTO β-CAROTENE AND THE PHYTOL SIDE CHAIN OF CHLOROPHYLL IN PINE SEEDLINGS

In *P. jeffrei* $^{14}CO_2$ is incorporated to a very much greater extent into β-carotene and the phytol side chain of chlorophyll than the [2-$^{14}$C]MVA. After introduction of $^{14}CO_2$ 1·5 % of assimilated radioactivity is incorporated into β-carotene and about 7 % into phytol, compared with about 0·2 and 1·0 %, respectively when [2-$^{14}$C]MVA is the substrate. These differences are even greater if we compare the radioactivity incorporated into β-carotene and phytol not with whole lipid fraction but with the sterol fraction (Table IV). After

Table III. Total radioactivity of lipid fraction, specific radioactivity of $\beta$-carotene and the phytol side chain of chlorophyll in pine seedlings grown in darkness (d.) and in light (l.) (6 hr in the presence of [2-$^{14}$C] MVA — 8 $\mu$c per about 400 mg of fresh wt). Growth conditions as indicated in Table I.

| | | Two days after germination | | | | | Six days after germination | | | | |
|---|---|---|---|---|---|---|---|---|---|---|---|
| | | Total radioact. of lipid fraction per 100 plants cts/min × $10^6$ | $\beta$-Carotene | | Phytol | | Total radioact. of lipid fraction per 100 plants cts/min × $10^6$ | $\beta$-Carotene | | Phytol | |
| | | | Specific radioact. cts/min/mg × $10^5$ | % of total radioact. | Specific radioact. cts/min/mg × $10^5$ | % of total | | Specific radioact. cts/min/mg × $10^5$ | % of total | Specific radioact. cts/min/mg × $10^5$ | % of total |
| *P. jeffrei* | d. | 42·2 | 2·48 | 0·07 | 4·14 | 0·81 | 54·2 | 3·26 | 0·14 | 3·83 | 0·81 |
| | l. | 44·3 | 5·24 | 0·21 | 4·22 | 1·61 | | | | | |
| *P. silvestris* | d. | 5·33 | 17·8 | 0·24 | 33·8 | 1·84 | 6·10 | 10·95 | 0·19 | 9·66 | 0·62 |
| | l. | 7·76 | 17·2 | 0·13 | 28·2 | 2·43 | | | | | |
| *P. radiata* | d. | 10·22 | 6·61 | 0·38 | 7·55 | 1·58 | 3·70 | 0·32 | 0·07 | 0·68 | 0·69 |
| | l. | 15·4 | 6·21 | 0·32 | 9·76 | 3·07 | | | | | |
| *P. contorta* | d. | 3·37 | 5·00 | 0·13 | 5·35 | 0·50 | 2·78 | 2·43 | 0·05 | 7·30 | 0·58 |
| | l. | 3·14 | 11·96 | 0·29 | 6·71 | 0·89 | | | | | |

introduction of [2-$^{14}$C]MVA only 0·2–0·3% of the label resides in the β-carotene, whilst after the introduction of $^{14}CO_2$ the percentage is about 7. This difference is independent of the previous history of the seedlings, that is whether they were grown in darkness or light before being exposed for 6 hr in the light to the labelled substrate.

Table IV. Incorporation of $^{14}CO_2$ and [2-$^{14}$C]MVA into sterols and β-carotene in cotyledons of seedlings of *Pinus jeffrei* (activity/ten plants). Seedlings germinated for 3 days in either darkness or light and then transferred to light for 6 hr in presence of labelled substrate. Light intensity about 600 lux Temperature 29°

| | "Crude" sterols cts/min × $10^4$ | β-Carotene cts/min × $10^3$ | % of total counts of sterols in β-carotene |
|---|---|---|---|
| $^{14}CO_2$ (60 μc) | | | |
| light light | 27·4 | 20·5 | 7·47 |
| dark light | 8·45 | 5·74 | 6·78 |
| [2-$^{14}$C]MVA (8 μc) | | | |
| light light | 351 | 9·50 | 0·27 |
| dark light | 465 | 10·7 | 0·23 |

## General Remarks

The results presented above show that the specific radioactivity of β-carotene and the phytol side chain of chlorophyll after introduction of [2-$^{14}$C]MVA depends very markedly on the manner of introduction of the label into the plants. However this is only a quantitative difference because the percentage of label incorporated into β-carotene and phytol is the same in two methods examined. About the same percentage of [2-$^{14}$C]MVA is also incorporated into β-carotene in some micro-organisms, e.g. *Phycomyces blakesleeanus* (Braithwaite and Goodwin, 1960).

All the results show that only a very small per cent of assimilated [2-$^{14}$C]MVA is incorporated into β-carotene and phytol side chain of chlorophyll and that this is independent of the species of pine used and is the same in beans. It seems that the distribution of [2-$^{14}$C]MVA into various terpenoids is similar in all plant organisms and it is independent of the manner of introduction of this label into the plant. On the other hand the distribution of radioactivity from $^{14}CO_2$ in various terpenoids is different than from [2-$^{14}$C]MVA.

These results tend to confirm the hypothesis that there exists two sites of terpenoid biosynthesis in plant cells, and that the relative impermeability of chloroplast membrane to some metabolites, for example MVA, may play an important part in the regulation of synthesis at these two sites. The difficulty with which exogenous MVA penetrates into plant cells, and its low incorporation into β-carotene in some micro-organisms which have no plastids also confirm the hypothesis (Goodwin and Mercer, 1963) that other plasmatic membranes present a barrier to the easy migration of MVA. In this respect the pine seedlings grown in darkness behave as some other etiolated plants behave when illuminated in the presence of [2-$^{14}$C]MVA.

## References

Bandurski, R. S. (1949). *Bot. Gaz.* **111**, 95.
Blaauw-Jansen, G., Komen, J. and Thomas, J. B. (1950). *Biochim. biophys. Acta* **5**, 179.
Braithwaite, G. D. and Goodwin, T. W. (1960). *Biochem. J.* **76**, 5.
Goodwin, T. W. (1958). *Biochem. J.* **70**, 503.
Goodwin, T. W. (1965). *In* "Chemistry and Biochemistry of Plant Pigments", (T. W. Goodwin, ed.), p. 143. Academic Press, London and New York.
Goodwin, T. W. (1967). *In* "Biochemistry of Chloroplasts" Vol. II, Academic Press, London and New York.
Goodwin, T. W. and Mercer, E. I. (1963). In "The Control of Lipid Metabolism" (J. K. Grand, ed.) p. 37. Academic Press, New York and London
Goodwin, T. W. and Phagpolngarm, S. (1960). *Biochem. J.* **76**, 197.
Goodwin, T. W. and Więckowski, S. (1966a). *Phytochem.* (in press).
Goodwin, T. W. and Więckowski, S. (1966b). (in preparation).
Kay, R. E. and Phinney, B. (1956). *Pl. Physiol.* **31**, 226.
Rogers, L. J., Shah, S. P. J. and Goodwin, T. W. (1965). *Biochem. J.* **96**, 7p.
Threlfall, D. R., Griffiths, T. W. and Goodwin, T. W. (1964). *Biochem. J.* **92**, 56p.
Treharne, K. J., Mercer, E. J. and Goodwin, T. W. (1964). *Biochem. J.* **90**, 39p.
Virgin, I. H., Kahn, A. and Wettstein, D. von (1963). *Photochem. Photobiol.* **2**, 83.
Wierzchowski, Z., Leonowicz, A., Sapiecha, K. and Sykut, A. (1962). *Roczn. Nauk. Roln.* **81B**, 87.

# An Action Spectrum for Vesicle Dispersal in Bean Plastids

Knud W. Henningsen

*Institute of Genetics, University of Copenhagen, Denmark*

The formation of the lamellar system in plastids of dark-grown leaves proceeds through three light-dependent steps (von Wettstein and Kahn, 1960; Eriksson *et al.*, 1961; Virgin *et al.*, 1963). The structural arrangements formed by these three processes are given in the diagram of Fig. 1. The tubes in the prolamellar body constitute a specific arrangement of the membrane material in proplastids of dark-grown leaves. They function as the starting material in the development of the chloroplast lamellar system. In the process of tube transformation the tubes are converted into vesicles simultaneously with the photochemical reduction of protochlorophyll to chlorophyll *a*, and protochlorophyllide to chlorophyllide *a*. In a second reaction, called vesicle dispersal, the vesicles of the prolamellar body move into regularly spaced primary layers, which form concentric scales more or less parallel to the surface of the plastid. From earlier experiments (Eriksson *et al.*, 1961; Virgin *et al.*, 1963: A. Kahn and D. von Wettstein, unpublished observations) with white light it was concluded that this reaction has a considerably higher energy requirement than the tube transformation, that the rate of vesicle dispersal is intensity-dependent, and that reciprocity holds between 75 and 7500 ft candles. With low light intensity dispersal is completed in a few hours, whereas with sufficiently high intensity the arrangement of the vesicles into layers is achieved in a few minutes. The third light-dependent step consists in the fusion of the vesicles into discs, as well as multiplication and aggregation of the discs into grana. Grana formation is directly correlated with the rapid phase of chlorophyll synthesis (Virgin *et al.*, 1963; A. Kahn and D. von Wettstein, unpublished observations).

The present report presents an action spectrum for the vesicle dispersal in plastids of 16–18 days old dark-grown primary leaves of *Phaseolus vulgaris* L. (variety Alabaster). Monochromatic light was obtained from an interference filter monochromator using a Xenon lamp (Osram XBO 450) as light source. The Schott interference filters (DEPIL and UV-PIL) employed have a half-bandwidth of about 9 m$\mu$. Energy measurements were made with a Hilger-Schwartz thermopile (FT 17) connected to a KIPP lightspot galvanometer (AL 1). After irradiation the detached leaves were fixed immediately in $KMnO_4$ and processed for electron microscopy. Operations were carried out in dim green safe light. The amount of vesicle dispersal was determined in micrographs of ultra-thin sections of the leaf pieces by counting the number of plastid sections containing undispersed prolamellar bodies. To randomize the analysis serial sections were not used and counts were made on as many different parts of the leaves as possible. Each point on the curves of Fig. 2 and 3 represents counts from 300 to 2000 plastid sections. In the dark controls 60% of the plastid sections contain prolamellar bodies (range 50–70%). Fifty per cent vesicle dispersal is defined as the stage, when 30% of the plastid sections contain undispersed prolamellar bodies.

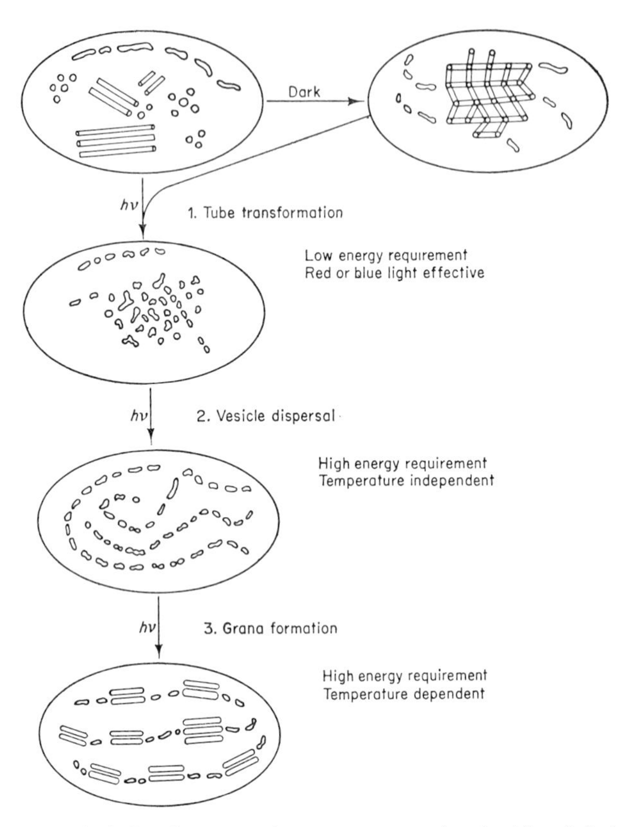


FIG. 1. The light-induced structural rearrangements in plastids of dark-grown bean leaves. [After Virgin, Kahn and von Wettstein (1963).]

Using monochromatic light at a wavelength of 450 m$\mu$ a pronounced dose-rate effect was found for the vesicle dispersal reaction (Fig. 2). With dose rates of less than 1000 erg/cm$^2$ sec 50% vesicle dispersal is obtained after irradiation with $2{\cdot}5 \times 10^{16}$ quanta/cm$^2$, whereas $8 \times 10^{17}$ quanta/cm$^2$ are required when a dose-rate of 7500 erg/cm$^2$ sec is employed. For the determinations of the energy necessary for 50% vesicle disperal at the various wavelengths irradiation was carried out with the lower dose rates. In addition the irradiation times were kept in the order of 20 min to allow possible dark reactions involved, to be completed. In some of the preliminary experiments, where the leaves after

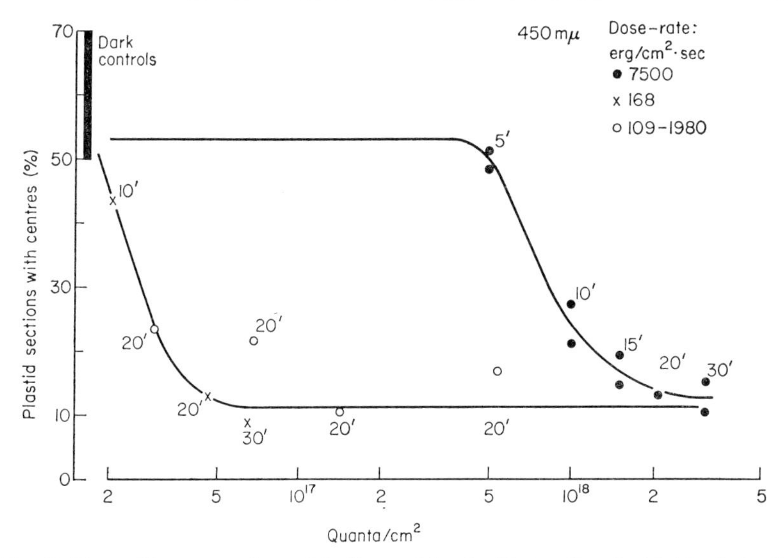


FIG. 2. Dose-effect curves for vesicle dispersal in plastids irradiated with blue light (450 mμ) using different dose-rates.

irradiation were stored for up to several hours in the dark before fixation, an indication of such a dark process was found at room temperature. The final values of dispersion were, however, even in these cases reached within 30 min after irradiation.

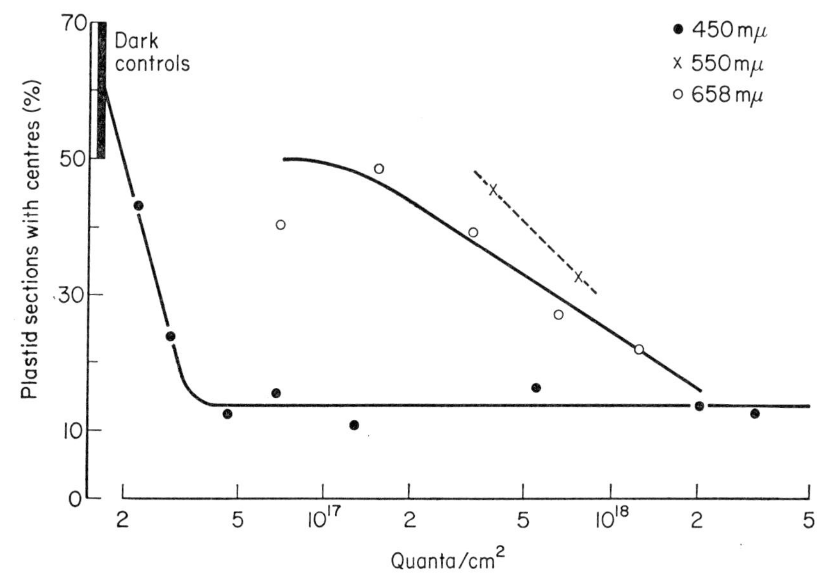


FIG. 3. Dose-effect curves for vesicle dispersal in plastids irradiated with light of different wavelength.

The dose-effect curves of Fig. 3 are representative for the experiments made to find the 50% dispersal energies at various wavelengths. The effectiveness ($1/Q$) for 50% dispersal of the wavelengths studied is plotted in Fig. 4. The action spectrum shows a sharp maximum at 450 mμ with a minor peak at 402 mμ. The wavelengths around 660 mμ and 550 mμ seem to be slightly more effective than the other regions tested.

The action spectrum for vesicle dispersal is clearly different from that for

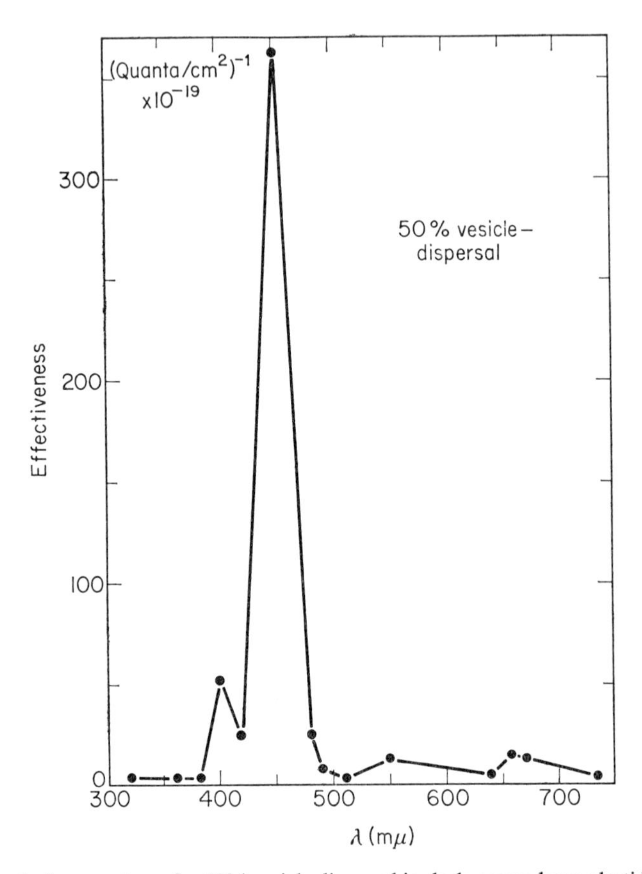


FIG. 4. Action spectrum for 50% vesicle dispersal in dark-grown bean plastids.

tube-transformation (Virgin *et al.*, 1963) only blue light being highly effective. The low effectiveness of red light excludes chlorophyll *a* and phytochrome as the pigment responsible for the absorption of the energy driving the vesicle dispersal. Of the carotenoids absorbing in the blue region, β-carotene and xanthophylls seem also excluded since mutants of *Helianthus annuus* lacking β-carotene and xanthophylls can carry out the vesicle dispersal (Walles, 1965). The low effectiveness of the region 300–380 mμ is not compatible with the absorption spectrum of riboflavin. Further studies are necessary to identify the pigment responsible for the energy absorption in the vesicle dispersal process.

## Summary

An action spectrum for the second light-induced reaction in the formation of the lamellar system in chloroplasts is presented. Only blue light is highly effective in this process of vesicle dispersal, whereby the vesicles of the pro-lamellar body are arranged into regularily spaced primary layers parallel to the plastid envelope.

## Acknowledgements

This work has been supported financially by grants to Professor D. von Wettstein from the U.S. Public Health Service, National Institutes of Health (Gm-10819), the Carlsberg Foundation and the Danish National Science Research Council.

## References

Eriksson, G., Kahn, A., Walles, B. and Wettstein, D. von (1961). *Ber. dtsch. Bot. Ges.* **LXXIV**, 221.
Virgin, H. I., Kahn, A. and Wettstein, D. von (1963). *Photochem. Photobiol.* **2**, 83.
Walles, B. (1965). *Hereditas* **53**, 247.
Wettstein, D. von and Kahn, A. (1960). *Proc. Eur. Reg. Conf. Electron Microsc.* **2**, 1051.

# VI. Biochemistry of Photosynthetic Phosphorylation

# Photosynthetic Phosphorylation: Facts and Concepts

Daniel I. Arnon[1]
*Department of Cell Physiology, University of California, Berkeley, California, U.S.A.*

## Introduction

Photosynthesis may be broadly defined as the utilization of solar energy by plants and photosynthetic bacteria for the synthesis of organic carbon compounds. In green plants, the known endergonic enzyme reactions concerned with the actual synthesis of organic carbon compounds are driven by ATP and a reductant at the level of reduced nicotinamide adenine dinucleotides. Thus, what distinguishes carbon assimilation in green plants from carbon assimilation in nonphotosynthetic cells is the manner in which they generate ATP and a strong reductant. Nonphotosynthetic cells form them at the expense of chemical energy whereas plant cells form them at the expense of radiant energy by the process of photosynthetic phosphorylation.

Photosynthetic phosphorylation (photophosphorylation) is subdivided into cyclic photophosphorylation, a reaction which produces only ATP, and noncyclic photophosphorylation, in which the production of ATP is stoichiometrically coupled with electron transfer that results in oxygen evolution and, indirectly, in the reduction of $NADP^+$. The discovery and characterization of these partial reactions of photosynthesis came not from investigations with intact cells but with isolated chloroplasts. It will be useful, therefore, to begin our discussion of photophosphorylation with a discussion of the overall photosynthetic capacity of isolated chloroplasts.

No comprehensive survey of the literature will be attempted here since several such reviews are available elsewhere (Jagendorf, 1962; San Pietro and Black, 1965; and Vernon and Avron, 1965). This report will be mainly concerned with the work of our laboratory.

## Photosynthetic Capacity of Isolated Chloroplasts

Biochemical research on photosynthesis by isolated chloroplasts rests on the premise that in photosynthesis, as was the case earlier in fermentation and respiration, the elucidation of the constituent reactions and their mechanisms would most likely come when the process is reconstructed outside the intact cell. Since, within photosynthetic cells, chloroplasts contain all the photosynthetic pigments and were observed, almost a century ago (Sachs, 1887; Engelmann, 1883, 1888), to produce starch and oxygen on illumination, it was

[1] The research from the author's laboratory has been supported, in part, by the National Institutes of Health, the Office of Naval Research and the Charles F. Kettering Foundation.

thought for many years that photosynthesis in green plants begins and ends in chloroplasts.

It is often difficult for the student of photosynthesis today to realize that this view was never supported by critical experimental evidence and was largely abandoned after Hill (1939, 1951) showed that isolated chloroplasts could evolve oxygen but could not assimilate carbon dioxide. In this reaction, which became known as the Hill reaction, isolated chloroplasts evolved oxygen when carbon dioxide was replaced by ferric oxalate or, as found later by Warburg (1949), by other nonphysiological electron acceptors, benzoquinone and ferricyanide. The ability of isolated chloroplasts to assimilate $CO_2$ was reinvestigated when the very sensitive $^{14}CO_2$ technique became available, but here again the results were essentially negative. Such $^{14}CO_2$ fixation as was observed was limited in scope. Thus, Fager (1952, 1954) found no fixation of $^{14}CO_2$ by chloroplasts but by a protein preparation ("enzyme") from spinach leaves. The fixation of $CO_2$ by the "enzyme" was enhanced in the presence of the illuminated chloroplast preparation but did not proceed beyond phosphoglycerate. There was no evidence for a reductive assimilation of $CO_2$ to the level of carbohydrate.

Without proof that isolated chloroplasts were the site of total photo-assimilation of carbon dioxide, photosynthesis in the early 1950's came to be regarded, like fermentation in the days of Pasteur, as a process that cannot be separated from the structural and functional complexity of whole cells.[1] Nevertheless, the possibility remained that the observed restricted photosynthetic capacity of isolated chloroplasts was merely a consequence of inappropriate experimental methods that were used in different laboratories, including our own. This proved to be the case. By changing our experimental methods, we found in 1954 that isolated spinach chloroplasts, unaided by other cellular particles or enzyme systems, reduced $CO_2$ to the level of carbohydrates, including starch, with a simultaneous evolution of oxygen at physiological temperatures and with no energy supply except visible light (Arnon *et al.*, 1954; Allen *et al.*, 1955). By using the new experimental methods, or modifications thereof, the conversion of $^{14}CO_2$ into phosphorylated sugars and starch by isolated chloroplasts was confirmed and extended in other laboratories (Gibbs and Cynkin, 1958; Tolbert, 1958; Gibbs and Calo, 1959; Smillie and Fuller, 1959; Smillie and Krotkov, 1959; compare also Ueda, 1949; Irmak, 1955; Thomas *et al.*, 1957). Thus, reproducible biochemical evidence has finally documented the frequently asserted, but never before proved, thesis that chloroplasts are the cytoplasmic structures in which the complete photosynthetic process takes place.

[1] To illustrate: In 1953, Rabinowitch wrote that "the task of separating it [photosynthesis] from other life processes in the cell and analysing it into its essential chemical reactions has proved to be more difficult than was anticipated. The photosynthetic process, like certain other groups of reactions in living cells, seems to be bound to the structure of the cell; it cannot be repeated outside that structure." In a review in 1954, Lumry *et al.* summarized the many investigations with isolated chloroplasts as pointing to the conclusion that the chloroplast was "a system much simpler than that required for photosynthesis", and was the site of only "the light-absorbing and water-splitting reactions of the overall photosynthetic process".

## CHLOROPLASTS AND THE EARLY REACTIONS OF PHOTOSYNTHESIS

Once the complete photosynthetic capacity of isolated chloroplasts was experimentally established, it was possible to concentrate on them rather than on whole cells in the search for those photochemical reactions which generate the first chemically defined, energy-rich products that are formed *prior* to, and are essential for, the conversion of $CO_2$ into organic compounds. The advantages of isolated chloroplasts for investigations of this aspect of photosynthesis are substantial. Chloroplasts cannot respire—they lack the terminal respiration enzyme, cytochrome oxidase (Hill, 1956; James and Das, 1957; Lundegårdh, 1961). This feature insures that the early products of photosynthesis in chloroplasts would not be confused with intermediate products (including ATP) of respiration—a possibility that cannot be excluded with certainty in intact cells and which, therefore, has been the subject of much controversy in research on photosynthesis. Furthermore, isolated and fragmented chloroplasts, unlike whole cells, do not have permeability barriers to the entry of such key intermediates as ADP, NADP and, as will be shown later, small protein molecules (ferredoxin). Thus, working with isolated chloroplasts, it is possible to supply these normally catalytic substances in substrate amounts and thereby follow, on a fairly large scale, their conversion, under the influence of light, into energy-rich products. This feature proved to be a great experimental advantage in the discovery of photosynthetic phosphorylation.

## CYCLIC PHOTOPHOSPHORYLATION

The first experiments with the sensitive $^{32}P$ technique to test the ability of isolated chloroplasts to form, on illumination, ATP gave negative results (Aronoff and Calvin, 1948). The most plausible model for ATP formation in photosynthesis became one that envisaged a collaboration between chloroplasts and mitochondria. Chloroplasts would, in that scheme, reduce NAD photochemically and mitochondria would reoxidize it with oxygen and form ATP via oxidative phosphorylation (Vishniac and Ochoa, 1952). This model posed a serious physiological problem. Photosynthesis in saturating light can proceed at a rate almost thirty times greater than the rate of respiration. It was difficult to see, therefore, how the respiratory mechanisms of mitochondria could cope with the ATP requirement in photosynthesis.

In 1954, Arnon, Allen and Whatley discovered a light-induced ATP formation by isolated spinach chloroplasts unaided by mitochondria. This process, which they named photosynthetic phosphorylation (photophosphorylation) to distinguish it from the respiratory (oxidative) phosphorylation by mitochondria, was independent of $CO_2$ assimilation. ATP was formed under conditions when no $CO_2$ was supplied to the reaction mixture and the reaction vessels contained KOH in the center well. The possibility cannot be excluded that, even under these conditions, residual $CO_2$ may have had a catalytic function on the photochemical reaction, as observed by Warburg (1949) and Stern and Vennesland (1960) for the photoproduction of oxygen by chloroplasts. What can be excluded is that substrate amounts of carbon compound(s) were first synthesized in the light from exogenous $CO_2$ and were then used as electron donors for the formation of ATP.

Several unique features distinguished this type of photophosphorylation from ATP formation in fermentation (substrate level phosphorylation) or in oxidative phosphorylation (Arnon *et al.*, 1954a, b): (1) ATP was formed only in chlorophyll-containing structures and was independent of any other organelles or enzyme systems; (2) no oxygen was consumed or produced; (3) no energy-rich chemical substrate was consumed, the only source of energy being that of the absorbed photons; (4) ATP formation was not accompanied by a net change in any external electron donor or acceptor (Eq. 1).

$$n \cdot \mathrm{ADP} + n \cdot \mathrm{P_i} \xrightarrow{h\upsilon} n \cdot \mathrm{ATP} \qquad (1)$$

The discovery of photosynthetic phosphorylation was confirmed and extended to other photosynthetic organisms. Photosynthetic phosphorylation in cell-free preparations of photosynthetic bacteria was observed by Frenkel (1954) and later by Williams (1956), Geller and Gregory (1956), Kamen and Newton (1957), and Anderson and Fuller (1958); in algae by Thomas *et al.* (1955) and Petrack and Lipmann (1961); and in isolated chloroplasts by Avron and Jagendorf (1957b), Wessels (1957) and many others. It was thus established that in all photosynthetic cells a major phosphorylating site was always associated with the chlorophyll pigments and could supply, independently of respiration or fermentation, the ATP needed in photosynthesis.

Soon after the demonstration of photosynthetic phosphorylation in isolated chloroplasts, attempts were made to compare its rate with that of $CO_2$ assimilation by illuminated whole cells. Since, as with most newly discovered cell-free reactions, the rates of photosynthetic phosphorylation were rather low, there was little inclination at first to accord this process quantitative importance as a mechanism for converting light into chemical energy (Rabinowitch, 1957).

With further improvement in experimental methods, we obtained rates of photosynthetic phosphorylation up to 170 times higher (Allen *et al.*, 1958) than those originally described (Arnon *et al.*, 1954) and even these high rates were exceeded by Jagendorf and Avron (1958). The improved rates of photosynthetic phosphorylation were equal to, or greater than, the maximum known rates of carbon assimilation in intact leaves. It appeared, therefore, that isolated chloroplasts retain, without substantial loss, the enzymatic apparatus for photosynthetic phosphorylation—a conclusion which is in harmony with evidence that the phosphorylating system is tightly bound in the water-insoluble grana portion of the chloroplasts.

When photophosphorylation was first extended from chloroplasts to photosynthetic bacteria, a question arose whether these two processes were fundamentally similar in not requiring a chemical substrate. Frenkel (1954) found that photophosphorylation in a cell-free preparation from *Rhodospirillum rubrum* became dependent on a substrate ($\alpha$-ketoglutarate) when the chlorophyll-containing particles were washed. However, in later experiments, Frenkel (1956) and other investigators (Kamen and Newton, 1957; Anderson and Fuller, 1958; Geller and Lipmann, 1960) found that the role of $\alpha$-ketoglutarate and other organic acids in the bacterial system was catalytic and regulatory and not that of a substrate. When this basic point was clarified, the fundamental similarity of cyclic photophosphorylation in chloroplasts and in bacterial systems was no longer in doubt.

Once the main facts of photophosphorylation were firmly established, the next objective was to explain its mechanism. All known cellular phosphorylations occur at the expense of free energy liberated during electron transport from a high-energy electron donor to an electron acceptor, but there was no direct evidence for this in photophosphorylation. After early attempts to link photophosphorylation with photochemical splitting of water (Arnon, 1955), we postulated that ATP formation was coupled to a special type of electron flow that is induced by light but is hidden in the structure of the chloroplast. The hypothesis was (Arnon, 1959, 1961) that a chlorophyll molecule, on absorbing a quantum of light, becomes excited and promotes an electron to an outer orbital with a higher energy level. This high-energy electron is then transferred to an adjacent electron acceptor molecule with a strongly electronegative oxidation-reduction potential. The transfer of an electron from excited chlorophyll to an adjacent electron acceptor molecule, present in chloroplasts, is the energy conversion step proper. It transforms a flow of photons into a flow of electrons; that is, it constitutes a mechanism for generating a strong reductant at the expense of the excitation energy of chlorophyll. Once the strong reductant is formed, the subsequent electron transfer steps are exergonic. In subsequent reactions within the chloroplast, an electron is transferred, without any additional input of radiant energy, to a second electron acceptor with a more electropositive oxidation-reduction potential, from the second to the third and so on.

The number of these exergonic electron transfer steps in this type of photophosphorylation is still under investigation. It is envisaged that such an electron "bucket brigade" liberates free energy that is used to form one or more ATP's from ADP and orthophosphate. In the end, the electron originally emitted by the excited chlorophyll molecule returns to the electron-deficient chlorophyll molecule and the quantum absorption process is repeated. Because of the envisaged cyclic pathway traversed by the emitted electron, we named the process *cyclic photophosphorylation* (Arnon, 1959).

A puzzling feature of cyclic photophosphorylation in chloroplasts, a feature which distinguished it from cyclic photophosphorylation in bacterial systems, was a dependence on an added catalyst or electron carrier—a function fulfilled by many different substances of a physiological or nonphysiological character. An example of the former is menadione (Arnon *et al.*, 1955) and of the latter, phenazine methosulfate (Jagendorf and Avron, 1958). However, as discussed later, recent evidence points to ferredoxin, an iron-containing protein native to chloroplasts, as being the endogenous catalyst of cyclic photophosphorylation in chloroplasts.

## Noncyclic Photophosphorylation

A second type of photophosphorylation was discovered in 1957 (Arnon *et al.*, 1958), which provided the first direct experimental evidence for a coupling between light-induced electron transport and the synthesis of ATP. Here, in contrast to cyclic photophosphorylation, ATP formation was stoichiometrically coupled with a light-driven transfer of electrons from water to NADP (or to a nonphysiological electron acceptor such as ferricyanide) and a concomitant evolution of oxygen.

In extending the electron flow concept to this reaction, we envisaged that a chlorophyll molecule excited by a captured photon transfers an electron to NADP (or to ferricyanide). Electrons thus removed from chlorophyll are replaced by electrons from water ($OH^-$ at pH 7) with a resultant evolution of oxygen. In this manner, light induces an electron flow from $OH^-$ to $NADP^-$ and a coupled phosphorylation. Because of the unidirectional or noncyclic nature of this electron flow, we have named this process *noncyclic photophosphorylation* (Arnon, 1959).

More recent evidence has established that illuminated chloroplasts do not react directly with NADP but react with ferredoxin (Tagawa and Arnon, 1962). As will be discussed later, the capture of photons by chloroplasts induces a noncyclic electron flow to ferredoxin with a coupled phosphorylation and a concomitant evolution of oxygen. Reduced ferredoxin in turn reduces $NADP^+$ by a mechanism that is independent of light.

Ferredoxin has thus emerged as a key substance in cyclic and noncyclic photophosphorylation. Since the properties of ferredoxin and its role in the energy conversion process of photosynthesis have only recently been recognized, the pertinent evidence will now be reviewed in some detail.

## Ferredoxins in Bacteria and Green Plants

Prior to 1961, there was no evidence to challenge the view that chloroplasts, and only chloroplasts, contain a protein factor or an enzyme that catalyses the photochemical reduction of $NADP^+$. But in that year K. Tagawa and M. Nozaki (unpublished data) and Losada *et al.* (1961) isolated in this laboratory a "pyridine nucleotide reductase" from an organism devoid of chloroplasts, the photosynthetic bacterium, *Chromatium*. The bacterial protein was able to replace the native chloroplast protein in mediating the photoreduction of $NADP^+$ and the evolution of oxygen by chloroplasts, although *Chromatium* cells, from which this protein was isolated, are incapable of evolving oxygen in light. This finding indicated that proteins similar to those functioning in the $NADP^+$-reducing apparatus of chloroplasts were also present in photosynthetic bacteria devoid of chloroplasts, but the full significance of this observation was understood a year later when Tagawa and Arnon (1962) obtained the same effect with a protein, ferredoxin, from a nonphotosynthetic organism.

Ferredoxin is the name given by Mortenson *et al.* (1962) to a protein containing iron which is neither a heme protein nor a flavin protein. Mortenson *et al.* (1962) isolated this protein from *Clostridium pasteurianum*, a nonphotosynthetic anaerobic bacterium which normally lives in the soil without any exposure to light. In this, and in other nonphotosynthetic, obligately anaerobic bacteria where ferredoxin was later found, it appeared to function as a link between the enzyme hydrogenase and different electron donors and acceptors (Mortenson *et al.*, 1962; Buchanan *et al.*, 1963). Thus, the distribution of ferredoxin seemed likely to be limited to those obligately anaerobic, nonphotosynthetic bacteria that contain an active hydrogenase system. There was nothing to indicate that ferredoxin was in any way linked with photosynthesis.

It soon became clear, however, that ferredoxin-like proteins are present in all photosynthetic cells and play a key role in the energy transfer mechanisms of photosynthesis. In fact, Tagawa and Arnon (1962) recognized that, between 1952 and 1960, proteins which we now call ferredoxins had been isolated from chloroplasts of several species of green plants and had been assigned various functions under such different names as "methaemoglobin-reducing factor" (Davenport *et al.*, 1952), "TPN-reducing factor" (Arnon *et al.*, 1957), "photosynthetic pyridine nucleotide reductase" (PPNR) (San Pietro and Lang, 1958), and the "haem-reducing factor" (Davenport and Hill, 1960). All these terms are now known to be synonymous and have been replaced by the term ferredoxin. It also became apparent that the "red enzyme" isolated in 1962 in Warburg's laboratory is analogous to ferredoxin (Warburg, 1963; Gewitz and Voelker, 1962).

## Definition of Ferredoxin

Tagawa and Arnon (1962) crystallized ferredoxin from the nonphotosynthetic bacterium *C. pasteurianum* (Fig. 1) and found that it was also able to replace the native chloroplast protein in the photoreduction of NADP (Fig. 2). The same investigation led also to other findings: (a) the chloroplast protein, like ferredoxin of *C. pasteurianum*, contained iron and was reversibly oxidized and reduced with characteristic changes in its absorption spectrum [the presence of iron in the chloroplast protein ("PPNR") was also independently observed by Fry and San Pietro (1962), Horio and Yamashita (1962), Katoh and Takamiya (1962), and Gewitz and Voelker (1962)]; (b) crystalline ferredoxin from *C. pasteurianum* has a remarkably low oxidation-reduction potential ($E_0' = -417$ mV, at pH 7·55), close to the potential of hydrogen gas and about 100 mV more electronegative than the oxidation-reduction potential of pyridine nucleotides; and (c) the oxidation-reduction potential of the spinach chloroplast protein was also strongly electronegative ($E_0' = -432$ mV, at pH 7·55).

These similarities led Tagawa and Arnon (1962) to extend the name ferredoxin to the chloroplast protein and to other iron-containing proteins of photosynthetic cells and anaerobic bacteria that have an oxidation-reduction potential close to that of hydrogen gas and are, at least in part, functionally interchangeable in the photoreduction of $NADP^+$ by isolated chloroplasts. In the new terminology, the family of ferredoxins would include those non-heme, non-flavin proteins that transfer to appropriate enzyme systems some of the most "reducing" electrons in cellular metabolism—electrons released by the photochemical apparatus of photosynthesis or by the $H_2$-hydrogenase system. Ability to catalyse the photoreduction of $NADP^+$ by washed chloroplasts was included provisionally in the definition of ferredoxins because, in the experience of this laboratory, all the ferredoxins that were tested so far exhibit this

FIG. 1. Photomicrograph of recrystallized ferredoxin from *C. pasteurianum* (Tagawa and Arnon, 1962).

property. By contrast, the replaceability of different ferredoxins in other enzymic reactions is less consistent.

It is to be noted that this provisional definition allows for dissimilarities of some properties among different ferredoxins. For example, the absorption spectra of ferredoxins from bacterial cells, whether photosynthetic or non-photosynthetic, resemble each other but differ significantly from the type of spectrum common to ferredoxins from algae and from chloroplasts of higher plants. In fact, we will distinguish, on the basis of spectral characteristics, two types of ferredoxins: the bacterial type and the chloroplast type.

A definitive characterization of ferredoxins as a group of electron carriers must await the isolation of a common prosthetic group in ferredoxins of

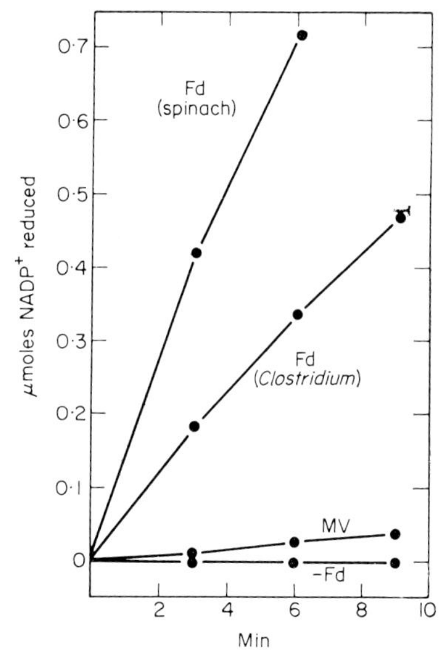


FIG. 2. Substitution of bacterial ferredoxin (from *C. pasteurianum*) for spinach ferredoxin (Fd) in the $NADP^+$ reduction of illuminated spinach chloroplasts. MV=methyl viologen (Tagawa and Arnon, 1962).

different species. Pending the isolation of a common prosthetic group, it seems useful to retain the tentative definition of ferredoxins as iron-containing proteins which function as electron carriers on the "hydrogen side" of pyridine nucleotides. This definition stresses the present distinction between ferredoxins and all the heme or non-heme iron proteins (including flavo-proteins) with more electropositive oxidation-reduction potentials that serve as electron carriers on the "oxygen side" of pyridine nucleotides.

## SPECTRAL CHARACTERISTICS AND OXIDATION-REDUCTION POTENTIALS

Unlike cytochromes, which exhibit well-defined absorption peaks in the reduced state, ferredoxins have distinctive absorption peaks in the oxidized state. On reduction, these absorption peaks of ferredoxins disappear.

The first ferredoxin to be crystallized, that of *C. pasteurianum*, exhibited in its oxidized state a distinctive spectrum with peaks in the visible and ultraviolet at 390, 300 and 280 mμ (Tagawa and Arnon, 1962). The crystalline preparation gave an absorption ratio of 390 mμ/280 mμ=0·79. These spectral characteristics of ferredoxin from *C. pasteurianum* were confirmed and extended by Buchanan *et al.* (1963) and Lovenberg *et al.* (1963) to ferredoxins of other species of *Clostridium*, which they prepared in crystalline form.

Figure 3 shows that the absorption spectrum of ferredoxin of the photosynthetic bacteria *Chromatium* closely resembles that of ferredoxin from the nonphotosynthetic *Clostridium* species. In the oxidized state, the absorption

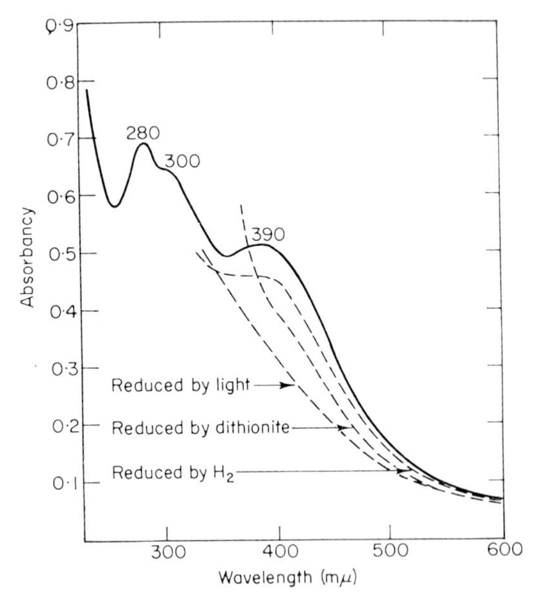


FIG. 3. Absorption spectra and reduction of *Chromatium* ferredoxin by $H_2$, sodium dithionite and by illuminated spinach chloroplasts; —— oxidized form; ------ reduced form (Bachofen and Arnon, 1966).

spectrum of *Chromatium* ferredoxin exhibits a flat peak at 385 mμ, a shoulder at 300 mμ and a peak at 280 mμ. In our purest preparation, the ratio of optical density, 385 mμ/280 mμ was 0·74 (Bachofen and Arnon, 1966).

As shown in Fig. 3, *Chromatium* ferredoxin was reduced by three methods: (a) $H_2$ gas in the dark, in the presence of a hydrogenase preparation from *C. pasteurianum*; (b) sodium dithionite, in the dark; and (c) photochemically, using a heated preparation of spinach chloroplasts and reduced dichlorophenol indophenol as the electron donor.

Complete reduction of *Chromatium* ferredoxin was obtained only photochemically. Taking the reduction in this system as 100%, the reduction by the $H_2$-hydrogenase system is only 24%. The reduction by the dithionite system was intermediate between the two. From these data the oxidation-reduction potential of *Chromatium* (at pH 7) was calculated (Bachofen and Arnon, 1966) to be about −490 mV, i.e. considerably more electronegative than that of the spinach chloroplast ferredoxin or of the *Clostridium* ferredoxin. It remains to be seen whether the ferredoxins from other photosynthetic bacteria will also prove to be as strongly reducing as *Chromatium* ferredoxin.

Figure 4 shows that the absorption spectrum of ferredoxin (oxidized state) from the blue-green alga *Nostoc* is of the chloroplast type (A. Mitsui and D. I. Arnon, 1963, unpublished data). It resembles closely the absorption spectrum of ferredoxin from spinach chloroplasts and is different from the absorption spectrum of bacterial ferredoxins. The absorption peaks of *Nostoc*

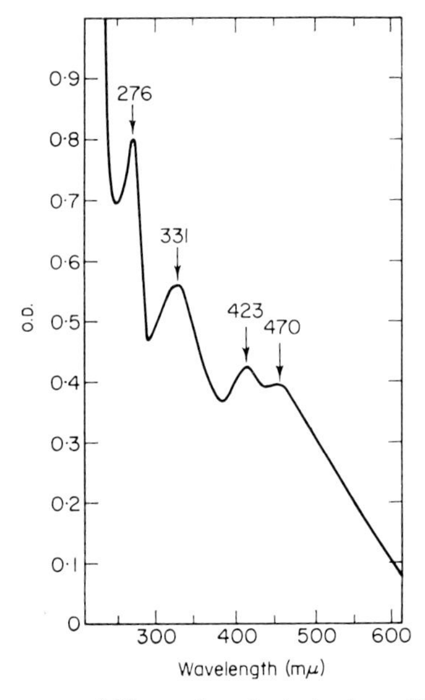


FIG. 4. Absorption spectrum of *Nostoc* ferredoxin in the oxidized state (A. Mitsui and D. I. Arnon, 1963, unpublished data).

ferredoxin in the visible and in the ultraviolet are 470, 423, 331 and 276 mμ, as compared with 463, 420, 325 and 274 mμ for spinach ferredoxin. Our purest preparation of *Nostoc* ferredoxin had a ratio of optical density, 423 mμ/276 mμ, of 0·57 (A. Mitsui and D. I. Arnon, 1963, unpublished data).

The similarity between the ferredoxin of *Nostoc* and spinach is interesting from an evolutionary point of view. Blue-green algae are considered to be the most primitive algae, not too distant on the evolutionary scale from photosynthetic bacteria. They reproduce vegetatively and, like photosynthetic bacteria, do not have their photosynthetic pigments localized in chloroplasts but distributed throughout the outer part of the cell. However, unlike photosynthetic bacteria, the photosynthesis of blue-green algae is accompanied by the evolution of oxygen. It is an interesting question whether the occurrence of the chloroplast type rather than of the bacterial type of ferredoxin in blue-green algae is related to the type of the photosynthetic pigment system and oxygen evolution that distinguish algal photosynthesis from bacterial photosynthesis.

A preliminary determination of the oxidation-reduction potential of *Nostoc* ferredoxin gave a value of $E_0' = -405$ mV, at pH 7·55 (Mitsui and Arnon, unpublished).

Since ferredoxin and not NADP is the terminal electron acceptor in the photochemical reactions of chloroplasts, the experimentally established reducing potential that is generated by chloroplasts in the course of photophosphorylation is extended by over 100 mV. It should be pointed out that the emphasis here is not on a theoretical reducing potential that can be generated by chloroplasts but on the experimental isolation and characterization of a reductant, native to photosynthetic cells, that is formed by the photochemical act of photosynthesis. It is theoretically possible for illuminated chloroplasts to generate stronger reductants than reduced ferredoxin: one einstein of red light ($\lambda = 663$ m$\mu$) is equivalent to 43 kcal or to 1·87 eV. However, all such possibilities must remain speculative without evidence that photosynthetic cells contain reductants stronger than ferredoxin.

## Some Chemical Properties of Ferredoxin

Apart from iron, ferredoxins of chloroplasts and bacteria are noted for containing "labile sulfide," i.e. an inorganic sulfide group which is equimolar with iron. This was first observed in spinach ferredoxin by Fry and San Pietro (1962) and independently in Warburg's laboratory in the "red enzyme" or ferredoxin of *Chlorella* (Warburg, 1963; Gewitz and Voelker, 1962). Buchanan *et al.* (1963) found inorganic sulfide in bacterial ferredoxins. The inorganic sulfur in ferredoxin is liberated as hydrogen sulfide upon acidification. Both iron and inorganic sulfide are loosely bound to the protein and the removal of

Table I. Iron analysis of spinach ferredoxin (Tagawa, Chain and Arnon, 1963)

| Ferredoxin used (mg) | Fe found ($\mu$g) | % Fe | Minimum mol wt |
|---|---|---|---|
| 1·0 | 8·6 | 0·86 | 6490 |
| 2·0 | 17·9 | 0·90 | 6200 |
| 3·7 | 33·3 | 0·89 | 6260 |
| 7·4 | 64·4 | 0·87 | 6420 |

one is accompanied by the removal of the other. Upon the loss of iron or labile sulfur, ferredoxin loses its spectral characteristics and also its biochemical activity.

Ferredoxins are small proteins. The bacterial ferredoxin, first estimated to have a molecular weight of around 12,000 (Tagawa and Arnon, 1962), is now

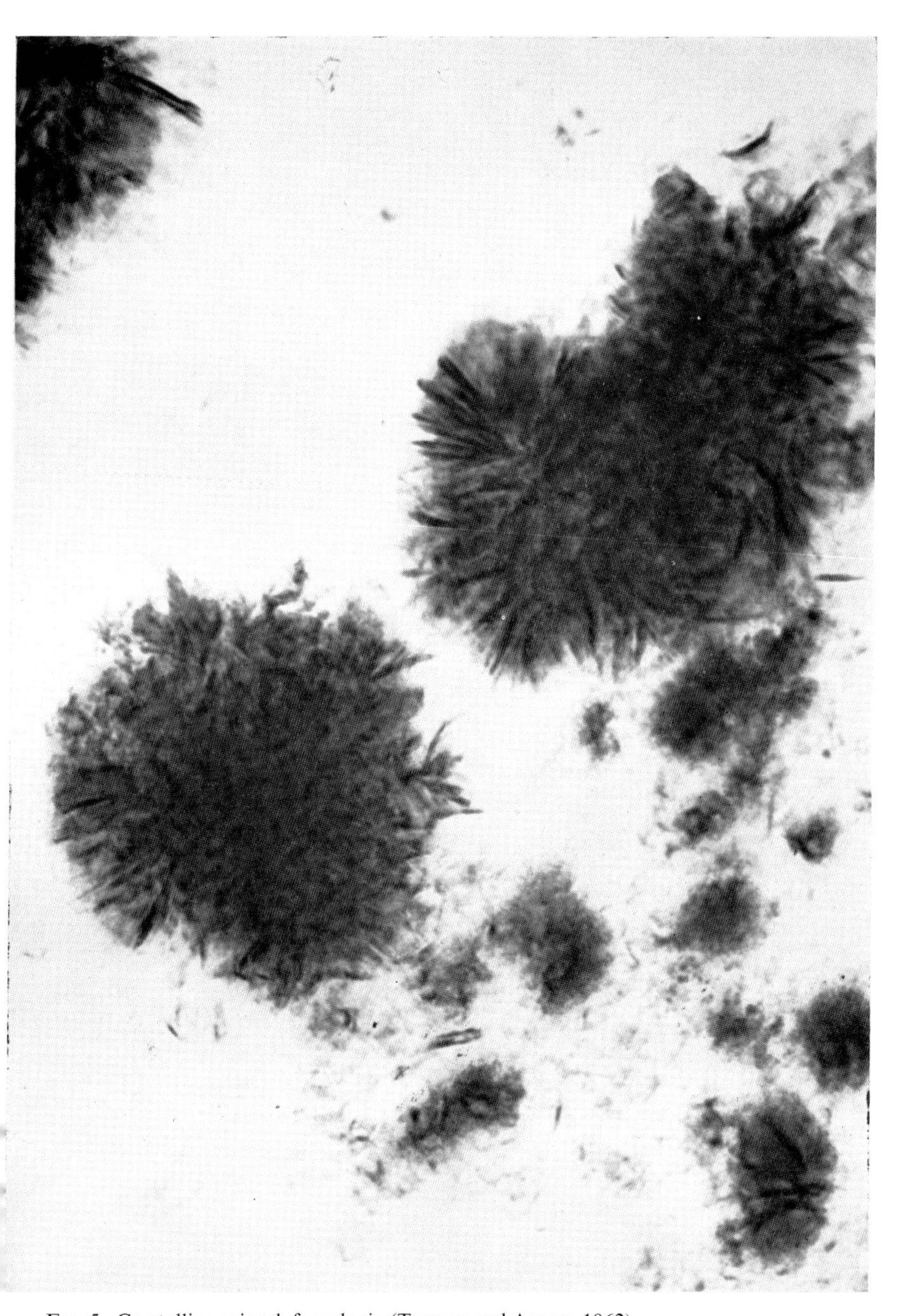

FIG. 5. Crystalline spinach ferredoxin (Tagawa and Arnon, 1962).

FIG. 6. Crystalline ferredoxin from *Nostoc* (Mitsui and Arnon, unpublished).

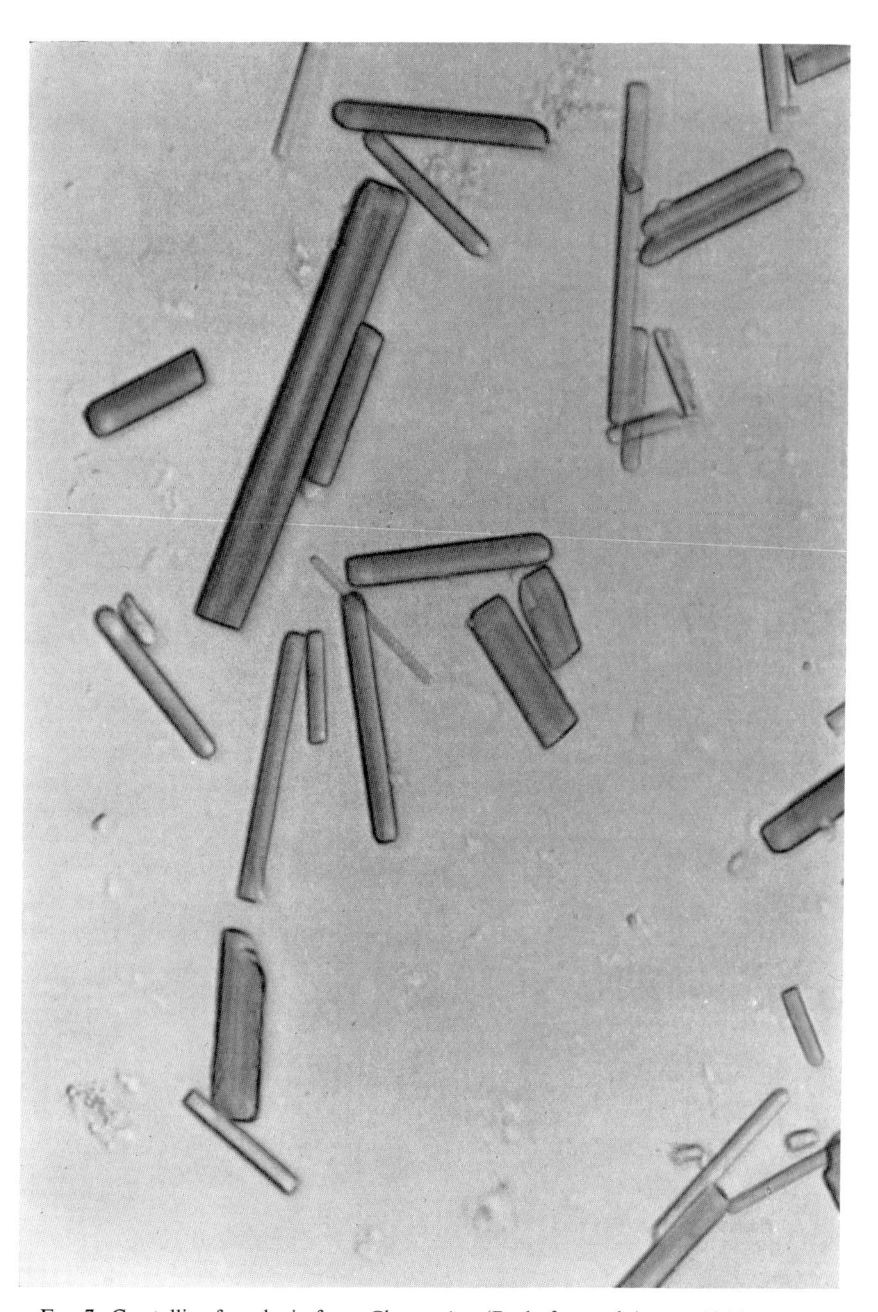

FIG. 7. Crystalline ferredoxin from *Chromatium* (Bachofen, and Arnon, 1966).

known to have a molecular weight of around 6000 (Lovenberg *et al.*, 1963). The chloroplast ferredoxin is estimated to have a molecular weight of approximately 13,000 (Whatley *et al.*, 1963). The iron content of bacterial and chloroplast ferredoxin varies. Thus the chloroplast ferredoxin of spinach has, on a molar basis, two atoms of iron per molecule (Table I) whereas the bacterial ferredoxin of *Chromatium* has three and that of *Clostridium* has seven. A summary of some chemical properties of several ferredoxins is given in Table II.

Table II. Some properties of bacterial and spinach ferredoxins

| | *C. pasteurianum* | *Chromatium* | Spinach |
|---|---|---|---|
| Iron (atoms/molecule protein) | 7 | 3 | 2 |
| Inorganic sulfide (moles/mole protein) | 7 | 3 | 2 |
| Molecular weight (approx.) | 6000 | 6000 | 13,000 |
| Redox potential (mV at pH 7·55) | −417 | −490 (approx.) | −432 |

## Ferredoxins and NADP+ Reduction

To test the effectiveness of different ferredoxins in catalysing the photoreduction of $NADP^+$, we have crystallized several ferredoxins from organisms other than *C. pasteurianum*. Crystalline ferredoxin from spinach chloroplasts is shown in Fig. 5 (Tagawa and Arnon, unpublished). Figure 6 shows crystalline ferredoxin from the blue-green alga *Nostoc* (Mitsui and Arnon, unpublished) and Fig. 7 shows crystalline ferredoxin from the photosynthetic bacterium *Chromatium* (Bachofen and Arnon, 1966). Despite the diversity of source, all these ferredoxins were effective as substitutes for the native spinach ferredoxin in catalysing the reduction of NADP by illuminated spinach chloroplasts. Figure 8 shows the replaceability of the spinach chloroplast ferredoxin by varying amounts of crystalline *Chromatium* ferredoxin (Bachofen and Arnon, 1966)—an effect which, as previously mentioned, was already observed with crude *Chromatium* ferredoxin when it was still called "pyridine nucleotide reductase" (Losada *et al.*, 1961). Figure 9 shows the substitution of *Nostoc* ferredoxin for spinach ferredoxin (Mitsui and Arnon, unpublished).

## Stoichiometry of Ferredoxin Oxidation-Reduction

Normally the role of ferredoxin in $NADP^+$ reduction is catalytic. The photoreduced ferredoxin is promptly reoxidized by $NADP^+$ and the accumulation of NADPH is measured spectrophotometrically by the increase in absorption at 340 m$\mu$. However, in the absence of $NADP^+$ "substrate" amounts of ferredoxin were quantitatively reduced by an illuminated chloroplast preparation (Fig. 10). The progressive photoreduction of ferredoxin by

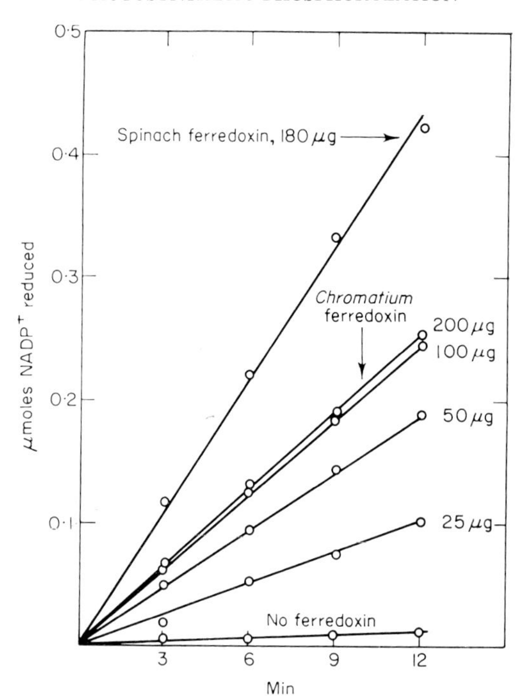


FIG. 8. Substitution of *Chromatium* ferredoxin for spinach ferredoxin in the reduction of $NADP^+$ by illuminated spinach chloroplasts (Bachofen and Arnon, 1966).

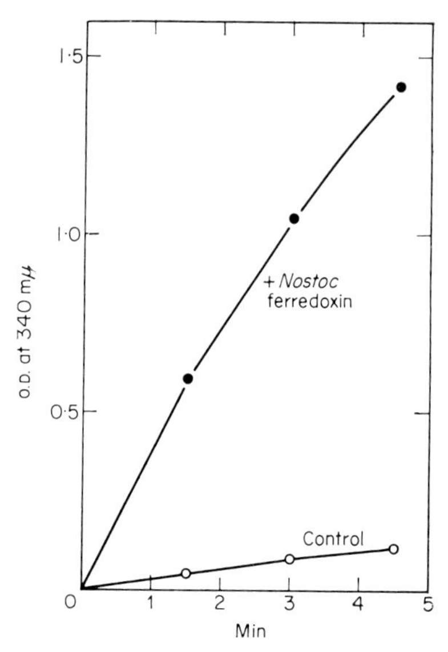


FIG. 9. Substitution of *Nostoc* ferredoxin for spinach ferredoxin in the reduction of $NADP^+$ by illuminated spinach chloroplasts (Mitsui and Arnon, unpublished).

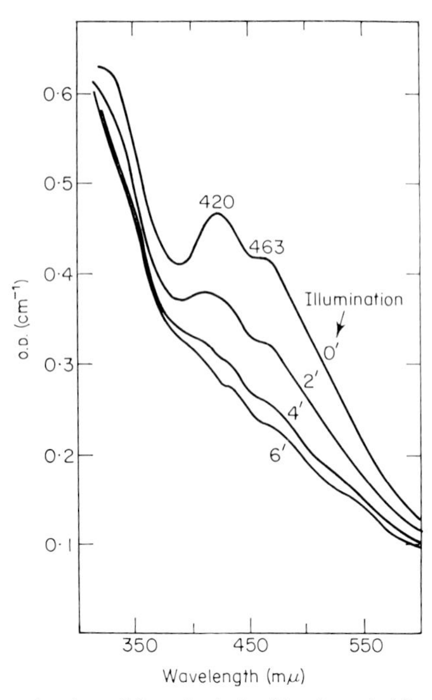


FIG. 10. Progressive reduction of ferredoxin by illuminated chloroplasts (Whatley *et al.*, 1963).

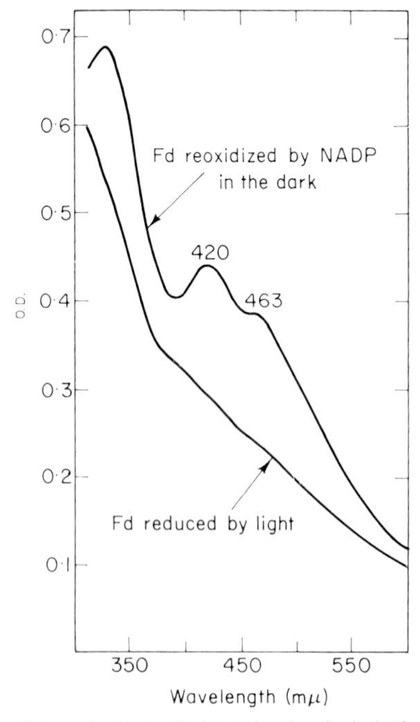


FIG. 11. Reoxidation of ferredoxin by NADP in the dark (Whatley *et al.*, 1963).

illuminated chloroplasts was followed by measuring its spectral changes in a recording spectrophotometer (Whatley *et al.*, 1963).

The ferredoxin reduced by light was stable under the strictly anaerobic conditions of the experiments and was not spontaneously reoxidized in a subsequent dark period. However, the reduced ferredoxin became quickly and completely reoxidized when $NADP^+$ was added (Fig. 11). The results, which showed that 1 mole of $NADP^+$ reoxidized 2 moles of ferredoxin (Table III), established that the reduction and oxidation of 1 molecule of ferredoxin involves a transfer of one electron (Whatley *et al.*, 1963). This stoichiometry between ferredoxin and $NADP^+$ was confirmed by Horio and San Pietro (1964).

Table III. Stoichiometry of photoreduction of spinach ferredoxin and its subsequent reoxidation by NADP in the dark (Whatley *et al.*, 1963)

| | $\mu$moles |
|---|---|
| Ferredoxin (Fd) photoreduced | 0·102 |
| Fd reoxidized by NADP in the dark | 0·106 |
| NADP reduced | 0·047 |
| Ratio Fd:NADP | 2·17 |

## Mechanism of $NADP^+$ Reduction by Chloroplasts

As already mentioned, the elucidation of the nature of ferredoxin as an electron carrier led to the resolution of the mechanism of $NADP^+$ reduction by chloroplasts into three steps: (a) a photochemical reduction of ferredoxin; (b) reoxidation of ferredoxin by a flavoprotein enzyme, ferredoxin-$NADP^+$ reductase; and (c) reoxidation of the reduced ferredoxin-$NADP^+$ reductase by $NADP^+$ (Shin and Arnon, 1965).

Crystalline ferredoxin-$NADP^+$ reductase isolated from spinach chloroplasts is shown in Fig. 12 (Shin *et al.*, 1963). Its absorption spectrum shows a typical flavoprotein absorption spectrum with peaks at 275, 385 and 456 m$\mu$ and minima at 321 and 410 m$\mu$ (Fig. 13). Figure 14 shows ferredoxin-$NADP^+$ reductase after it was first reduced ($E_{red}$) by reduced ferredoxin and then reoxidized by the addition of $NADP^+$ ($E_{red}$ + NADP). Thus, the oxidation-reduction of the flavin component of ferredoxin-$NADP^+$ reductase was shown to be an intermediate step in the transfer of electrons from reduced ferredoxin to $NADP^+$. This mechanism of NADP (and NAD) reduction is shown diagrammatically in Fig. 15.

Ferredoxin-$NADP^+$ reductase catalysed directly the reduction of either $NADP^+$ or $NAD^+$ but its affinity for $NADP^+$ was very much greater (Shin and Arnon, 1965). The Michaelis constant for $NAD^+$ was found to be $3{\cdot}75 \times 10^{-3}$ M, which was about 400 times greater than the $K_m$ found for $NADP^+$ ($9{\cdot}78 \times 10^{-6}$ M). The great difference between the affinities for ferredoxin-$NADP^+$ reductase for $NADP^+$ and $NAD^+$, the approximately equal concentrations of $NAD^+$ and $NADP^+$ in the cell (Anderson and

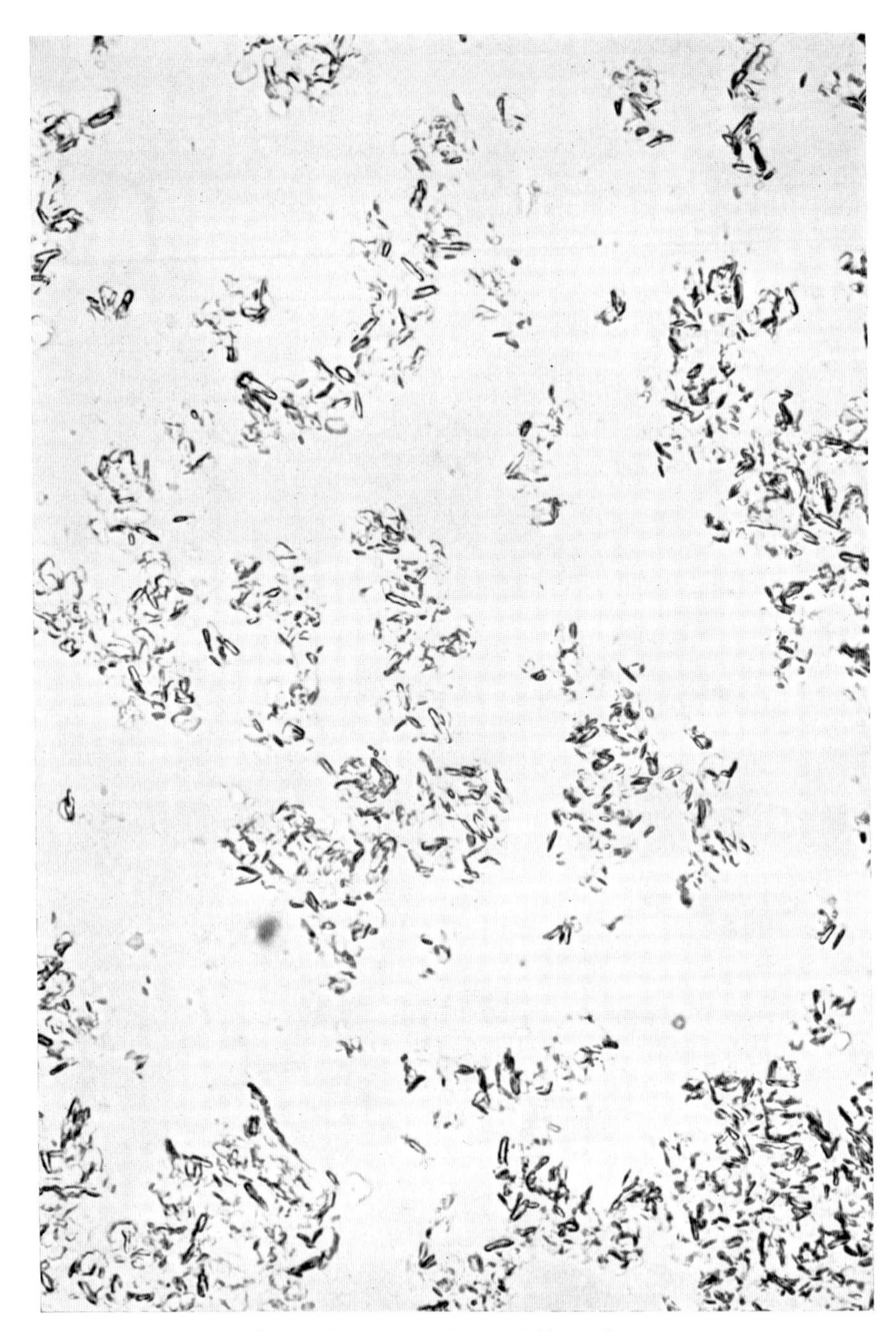

FIG. 12. Crystalline ferredoxin–NADP reductase (Shin *et al.*, 1963).

Vennesland, 1954), and the competition between $NAD^+$ and $NADP^+$ (Shin and Arnon, 1965) account for the apparent specificity of the pure enzyme for $NADP^+$.

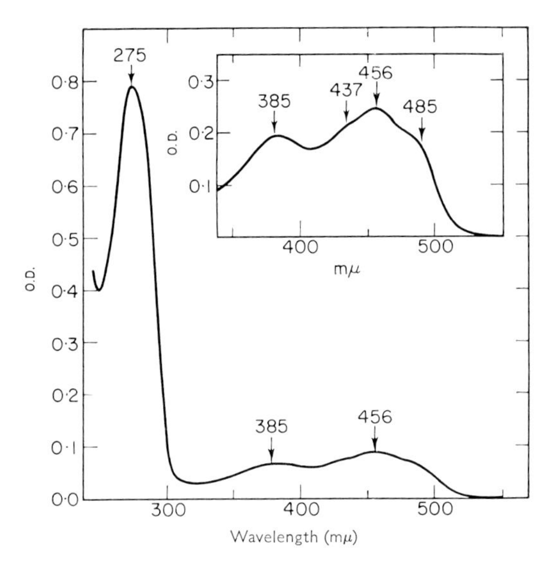


FIG. 13. Absorption spectra of ferredoxin–NADP reductase (Shin *et al.*, 1963).

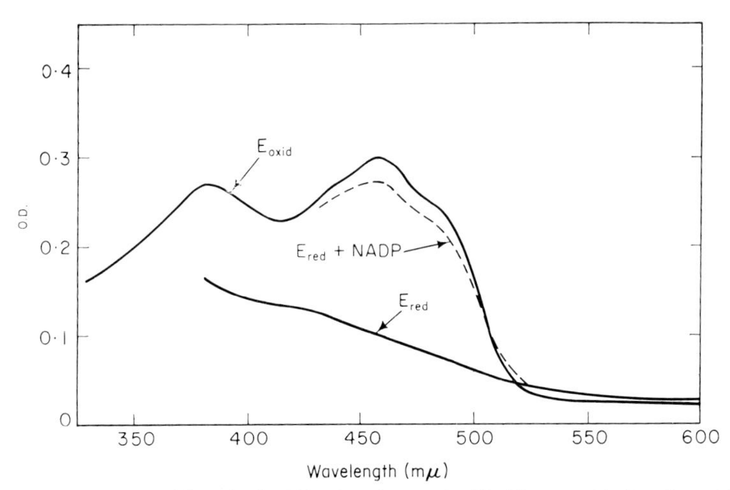


FIG. 14. Reduction of ferredoxin–NADP reductase by $H_2$ ($E_{oxid} \rightarrow E_{red}$) and reoxidation of the reduced enzyme ($E_{red}$) by NADP (Shin and Arnon, 1965).

The reduction of $NAD^+$ and $NADP^+$ by the reduced ferredoxin-$NADP^+$ reductase is reversible. Figure 16 shows that both NADPH and NADH reduce the oxidized ferredoxin-$NADP^+$ reductase. The absorption spectrum of the reduced enzyme shows the disappearance of the characteristic flavoprotein peaks (Shin and Arnon, 1965). This reversibility of the action of the enzyme accounts for its apparent secondary function as diaphorase (Avron and Jagendorf, 1957a) and as a transhydrogenase (Keister *et al.*, 1960, 1962)

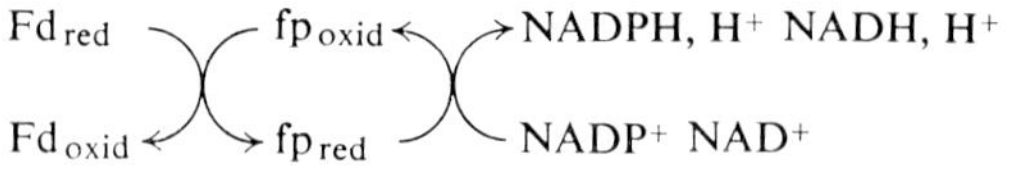

$Fd_{red}$ $fp_{oxid}$ $NADPH, H^+$ $NADH, H^+$

$Fd_{oxid}$ $fp_{red}$ $NADP^+$ $NAD^+$

FIG. 15. Reduction of nicotinamide adenine dinucleotides by chloroplasts. *Fd* denotes ferredoxin; *fp* denotes the flavoprotein, ferredoxin–NADP reductase (Shin and Arnon, 1965).

which had been reported before its primary function as a $NADP^+$ reductase was recognized. The reported specificity of its diaphorase and transhydrogenase action for NADPH can now be explained by the low affinity of the enzyme for NADH.

The experiments which have shown that reduced ferredoxin does not transfer

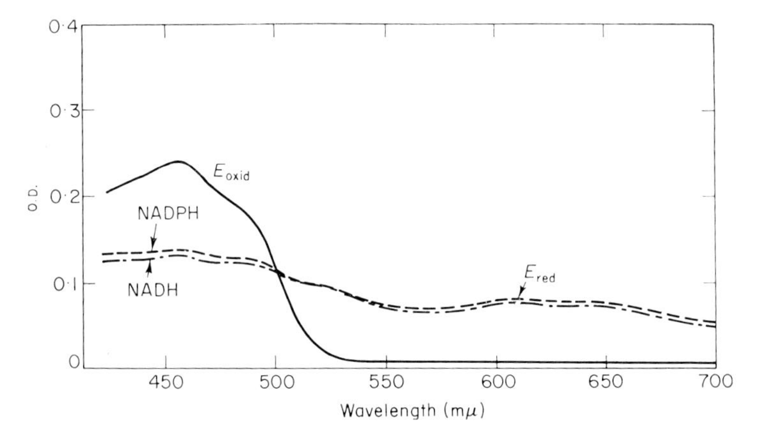


FIG. 16. Reduction of ferredoxin–NADP reductase by reduced NAD and NADP (Shin and Arnon, 1965).

electrons directly to $NADP^+$ have yielded no evidence for the existence of a "bound" NADP (Keister *et al.*, 1962). Our evidence indicates that the electron transfer in the reduction of NADP by chloroplasts is not by a transhydrogenation reaction from a reduced "bound" NADP but by direct reduction of free $NADP^+$ by reduced ferredoxin-NADP reductase. This is discussed in greater detail elsewhere (Shin and Arnon, 1965).

Under physiological conditions (solid lines in the scheme below), the electron flow in the reduction of pyridine nucleotides by chloroplasts can now be summarized as follows:

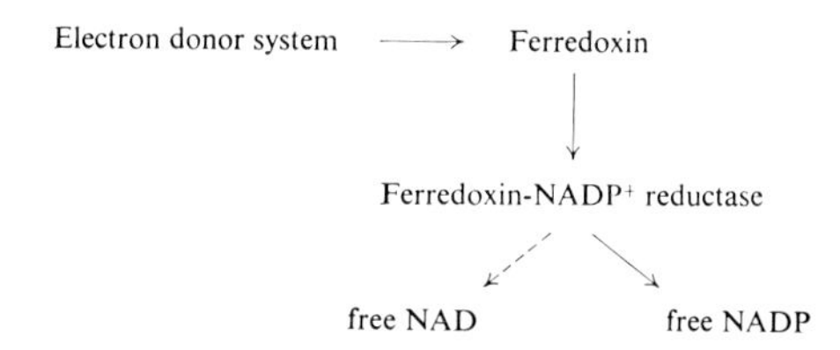


## Photoreduction of Ferredoxin Coupled with Photoproduction of Oxygen

The key role assigned to ferredoxin in the photosynthetic electron transport is subject to a rigid test. The evolution of oxygen by chloroplasts is uniquely dependent on light, and it occurs only in the presence of a proper electron acceptor. Thus, photoproduction of oxygen by chloroplasts should accompany the photoreduction of ferredoxin. Such direct demonstration, however, was technically difficult because reduced ferredoxin is readily oxidized by oxygen.

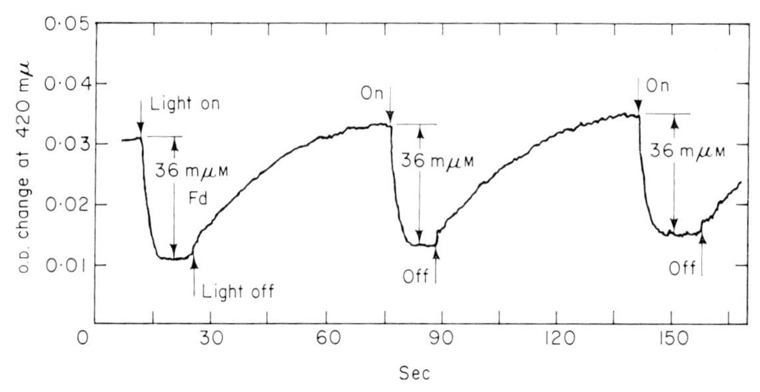


Fig. 17. Quantitative photoreduction of ferredoxin (Fd) by spinach chloroplasts and its reoxidation in the dark (Arnon *et al.*, 1964).

However, when the rapid back reaction between reduced ferredoxin and evolved oxygen was impeded, the stoichiometry between the photoreduction of ferredoxin and photoproduction of oxygen became measurable.

The techniques used involved measuring oxygen evolution polarographically and determining the photoreduction of ferredoxin by the decrease in optical density at 420 mμ (Arnon *et al.*, 1964). Traces of oxygen were rigidly excluded prior to turning on the light.

Figure 17 shows the photoreduction of added ferredoxin by chloroplasts under these experimental conditions. On turning off the light, ferredoxin was reoxidized. The amount of ferredoxin reduced in the light was equal to the amount of ferredoxin reoxidized in the dark. The sequence of photoreduction followed by dark reoxidation was reproducible at least three consecutive times. The added ferredoxin was completely photoreduced.

The photoproduction of oxygen, when ferredoxin was the terminal electron acceptor, is shown in Fig. 18. No oxygen was evolved without addition of ferredoxin. On turning off the light, the oxygen evolved during the preceding illumination period was consumed. The successive evolution and consumption of oxygen parallels the photoreduction and reoxidation of ferredoxin shown in Fig. 17. The sequence of oxygen production in the light and consumption in the dark was also reproducible several times in succession. Of special interest

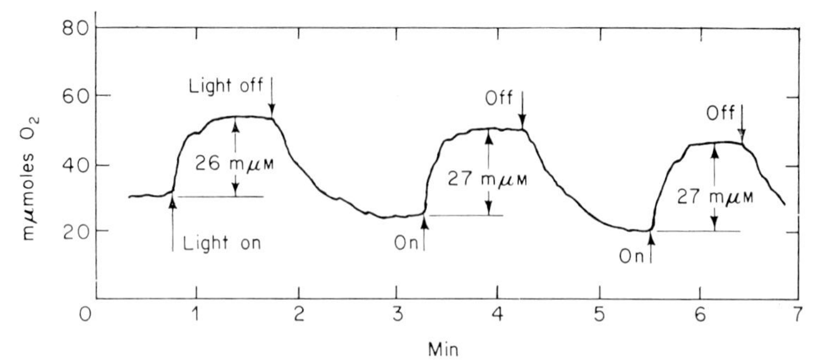


FIG. 18. Ferredoxin-linked oxygen evolution in the light and oxygen consumption in the dark (Arnon *et al.*, 1964).

is the stoichiometry between the ferredoxin added and oxygen produced. Table IV shows that the stoichiometry between ferredoxin added and oxygen produced was 4 to 1 and remained the same with different amounts of added ferredoxin. This again substantiates the conclusion that the photoreduction of ferredoxin involves a transfer of one electron.

Table IV. Stoichiometry between oxygen evolution and photoreduction of ferredoxin by isolated chloroplasts (mμmoles) (Arnon *et al.*, 1964)

| Exp. | Test no. | Ferredoxin added | $O_2$ produced | Ferredoxin added: $O_2$ produced |
|---|---|---|---|---|
| A | 1 | 110 | 26 | 4·2 |
| | 2 | 110 | 27 | 4·1 |
| | 3 | 110 | 27 | 4·1 |
| B | 1 | 128 | 32 | 4·0 |
| | 2 | 128 | 33 | 3·9 |
| | 3 | 128 | 32 | 4·0 |
| C | 1 | 154 | 37 | 4·2 |
| | 2 | 154 | 42 | 3·7 |
| | 3 | 154 | 39 | 3·9 |

## NONCYCLIC PHOTOPHOSPHORYLATION WITH FERREDOXIN

The stoichiometric evolution of oxygen, coupled with the photoreduction of ferredoxin, was also accompanied by a stoichiometric ATP formation

(Arnon *et al.*, 1964). As shown in Table V, the amount of ATP formed was proportional to the amount of ferredoxin added in a molar ratio of approximately 1 ATP to 2 ferredoxins (P:2e = 1). This ratio is consistent with the other evidence that the oxidation-reduction of ferredoxin involves a transfer of one electron. Thus, noncyclic photophosphorylation can now be summarized by equation 2.

$$4Fd_{ox} + 2ADP + 2P_i + 2H_2O \xrightarrow{h\upsilon} 4Fd_{red} + 2ATP + O_2 + 4H^+ \quad (2)$$

In experiments with isolated chloroplasts, it is usually more convenient to measure noncyclic photophosphorylation by using catalytic amounts of ferredoxin and stoichiometric amounts of NADP, which, unlike chloroplast ferredoxin, is commercially available and relatively stable to oxygen. However, this is merely an operational convenience which must not obscure the key role of ferredoxin in this type of photophosphorylation.

Table V. Stoichiometry of noncyclic photophosphorylation with ferredoxin (μmoles/ml) (Arnon *et al.*, 1964)

| Fd → <br> Min. | 0 | 0·20 | 0·40 | 0·60 |
|---|---|---|---|---|
| 2·5 | 0·00 ATP | 0·13 ATP | 0·13 ATP | 0·08 ATP |
| 5 | 0·00 ,, | 0·13 ,, | 0·21 ,, | 0·16 ,, |
| 10 | 0·00 ,, | 0·12 ,, | 0·21 ,, | 0·27 ,, |
| 15 | 0·00 ,, | 0·11 ,, | 0·21 ,, | 0·26 ,, |
| Fd:ATP[1] | — | 1·5 | 1·9 | 2·2 |

[1] Based on the highest values of ATP for each concentration of ferredoxin.

## Cyclic Photophosphorylation with Ferredoxin

As already mentioned, apart from noncyclic photophosphorylation, ferredoxin was also found to catalyse cyclic photophosphorylation. Evidence for a ferredoxin-catalysed cyclic photophosphorylation which proceeds anaerobically without the addition of other cofactors was obtained after the experimental conditions for this type of photophosphorylation had been established (Tagawa *et al.*, 1963a). These conditions include use of an effective inhibitor of electron flow from $OH^-$ which results in oxygen evolution.

Table VI shows the effect of added ferredoxin on anaerobic cyclic photophosphorylation in the absence of other catalysts of photophosphorylation. The dependence of cyclic photophosphorylation on the presence of *p*-chlorophenyldimethyl urea (CMU) is shown in Table VII. The addition of CMU increased the rate of photophosphorylation about seven-fold, but excessively high concentrations of CMU were inhibitory.

It thus became clear that ferredoxin-catalysed cyclic photophosphorylation and noncyclic photophosphorylation are mutually exclusive. Cyclic photophosphorylation catalysed by ferredoxin can be unmasked only when noncyclic photophosphorylation is stopped.

Another way to demonstrate this mutually exclusive relation between cyclic and noncyclic photophosphorylation is to use monochromatic light above

Table VI. Effect of ferredoxin on cyclic photophosphorylation in an atmosphere of argon (Tagawa *et al.*, 1963a)

| Ferredoxin added (μg) | ATP formed | |
|---|---|---|
| | cts/min | μmoles |
| 0 | 23 | 0·00 |
| 30 | 146 | 0·02 |
| 75 | 410 | 0·07 |
| 150 | 789 | 0·14 |
| 300 | 2109 | 0·37 |
| 750 | 7404 | 1·29 |
| 1500 | 10,816 | 1·88 |

700 mμ (Tagawa *et al.*, 1963a). Chloroplasts illuminated in this region of far-red light cannot produce oxygen; that is, they cannot remove electrons from water but can still carry on cyclic photophosphorylation catalysed by ferredoxin. Here no inhibitor of photoproduction of oxygen is necessary since the far-red light serves as a physical equivalent of a chemical inhibitor: it allows cyclic

Table VII. Effect of *p*-chlorophenyldimethyl urea (CMU) on ferredoxin-catalysed cyclic photophosphorylation in an atmosphere of argon (Tagawa *et al.*, 1963a)

| CMU added (μg) | ATP formed (μmoles) |
|---|---|
| None | 0·21 |
| 0·1 | 0·21 |
| 1·0 | 0·53 |
| 10 | 1·57 |

photophosphorylation to proceed and makes photoproduction of oxygen impossible (Tagawa *et al.*, 1963b).

Cyclic and noncyclic photophosphorylation are also sharply distinguished by their differential sensitivity to several inhibitors. Low concentrations of antimycin A, 2,4-dinitrophenol and desaspidin, a phlorobutyrophenone derivative, inhibit cyclic but do not inhibit noncyclic photophosphorylation. Figure 19 illustrates a differential sensitivity to inhibition by antimycin A of cyclic photophosphorylation catalysed by ferredoxin, menadione or phenazine methosulfate; it also illustrates that, under the very low light intensity under which these experiments were carried out, ferredoxin is a much more effective catalyst for the conversion of radiant energy into pyrophosphate bond energy than either menadione or phenazine methosulfate. This result is also consistent with the physiological nature which we ascribe to the ferredoxin-catalysed cyclic photophosphorylation.

Heretofore, a notable distinction between oxidative phosphorylation by mitochondria and photosynthetic phosphorylation by chloroplasts was the insensitivity of the latter to low concentrations ($5 \times 10^{-5}$ M) of 2,4-dinitrophenol (see review, Losada and Arnon, 1963). It was interesting to find, therefore, that the ferredoxin-dependent cyclic photophosphorylation is

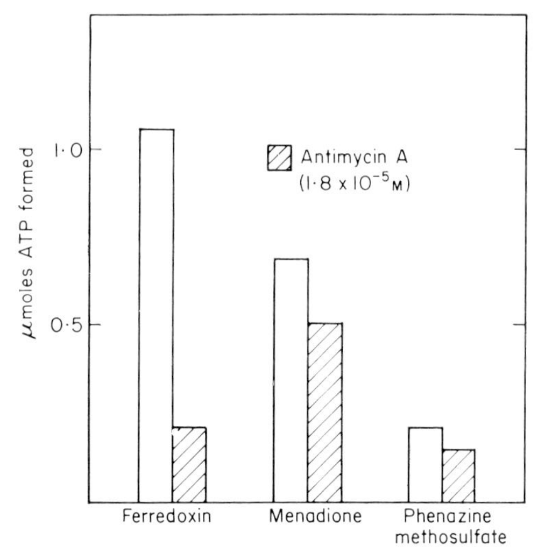


FIG. 19. Differential effect of antimycin A on cyclic photophosphorylation catalysed by ferredoxin, menadione or phenazine methosulfate. Illumination: monochromatic light 714 mμ (Arnon *et al.*, 1964).

distinct from cyclic photophosphorylation catalysed by other cofactors in being strongly inhibited by low concentrations of 2,4-dinitrophenol (Tagawa *et al.*, 1963b). Fifty per cent inhibition was observed at $3 \times 10^{-5}$ M dinitrophenol concentration (Fig. 20).

Baltscheffsky and de Kiewiet (1964) introduced to the study of photosynthetic phosphorylation a new inhibitor, a phlorobutyrophenone derivative, desaspidin, used in medicine as an anthelmintic agent and found by Runeberg (1963) to act as a powerful uncoupler of oxidative phosphorylation. The remarkable property of desaspidin is that at very low concentrations (*ca.* $10^{-7}$ M) it inhibits cyclic photophosphorylation catalysed by phenazine methosulfate, menadione and dichlorophenol indophenol, whereas a similar degree of inhibition of noncyclic photophosphorylation requires desaspidin at about a 100 times greater concentration. In the experiments of Baltscheffsky and de Kiewiet (1964) and Gromet-Elhanan and Arnon (1965), sensitivity to inhibition by low concentrations of desaspidin gave a new and unambiguous criterion by which cyclic photophosphorylation was distinguished from the

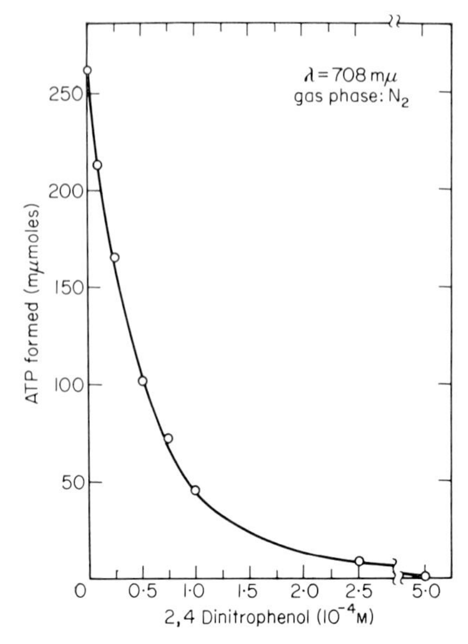


FIG. 20. 2,4-Dinitrophenol inhibition of ferredoxin-catalysed cyclic photophosphorylation (Tagawa *et al.*, 1963b).

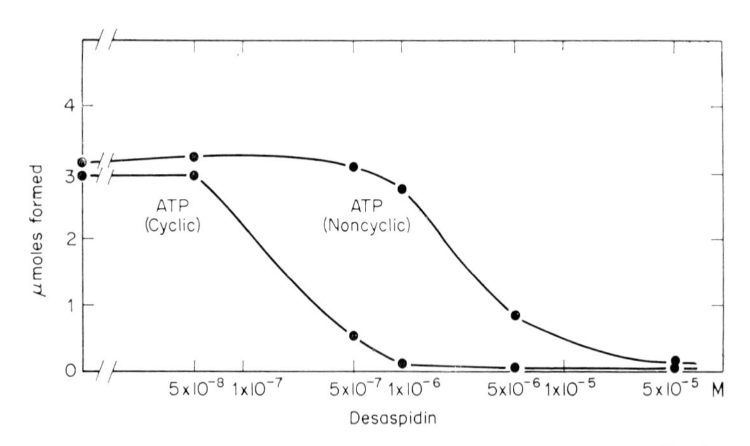


FIG. 21. Effect of desaspidin on ferredoxin-catalysed cyclic and noncyclic photophosphorylation (Arnon *et al.*, 1965b).

noncyclic phosphorylation that is associated with the transfer of electrons from $OH^-$ to $NADP^+$ (via ferredoxin) or to such nonphysiological agents as ferricyanide. Figure 21 shows that $5\times10^{-7}$ M to $10^{-6}$ M desaspidin inhibited almost completely ferredoxin-catalysed cyclic photophosphorylation but

$5 \times 10^{-6}$ M to $5 \times 10^{-5}$ M desaspidin was required to inhibit noncyclic photophosphorylation to the same degree.

An interesting effect of desaspidin was observed on ferredoxin-catalysed photophosphorylation in the presence of air. We have previously reported that this type of photophosphorylation is of the pseudocyclic type (Arnon *et al.*, 1961), a special case of noncyclic photophosphorylation in which electrons are transferred from $OH^-$ to ferredoxin and thence to molecular oxygen. This noncyclic electron flow from $OH^-$ to $O_2$ via ferredoxin gives the appearance of a cyclic electron flow since manometric measurements give no indication that an electron donor and acceptor are being consumed concomitantly with ATP formation. In pseudocyclic photophosphorylation the consumption of oxygen at the terminal end of the electronic pathway is balanced by the production of oxygen at the site of electron donation by $OH^-$ (Arnon *et al.*, 1961).

Table VIII shows that, in the presence of air, ferredoxin-catalysed phosphorylation was 83% inhibited by desaspidin ($5 \times 10^{-7}$ M) at the lowest concentration of ferredoxin (0·2 mg/1·5 ml). However, ATP formation became progressively more resistant to desaspidin as the concentration of ferredoxin increased (only 7% inhibition at 3·0 mg ferredoxin/1·5 ml). Conversely, CMU, the well-known inhibitor of oxygen evolution and hence of pseudocyclic phosphorylation by chloroplasts, gave 11% inhibition at the lowest concentration of ferredoxin and 65% inhibition at the highest concentration of ferredoxin (Table VIII).

Table VIII. Effect of desaspidin on ferredoxin-catalysed phosphorylation in air (Arnon *et al.*, 1965b)

| Ferredoxin added (mg) | ATP formed (μmoles) | | |
|---|---|---|---|
| | No inhibitor | Desaspidin | CMU |
| 0·2 | 1·12 | 0·19 | 1·0 |
| 1·0 | 2·38 | 1·06 | 1·32 |
| 2·0 | 3·13 | 2·25 | 1·41 |
| 3·0 | 3·43 | 3·17 | 1·20 |

These results indicate that, in the presence of air, ferredoxin catalysed a mixed type of photophosphorylation. At the lowest concentration of ferredoxin, the phosphorylation was predominantly of the cyclic type, whereas at the highest concentration of ferredoxin used, the phosphorylation was predominantly of the noncyclic type (more specifically, of the pseudocyclic type).

## Noncyclic Photophosphorylation of the Bacterial Type

Desaspidin proved to be specially useful in clarifying the nature of a phosphorylation in chloroplasts that is coupled with an artificial noncyclic electron flow system, i.e. a system in which $OH^-$, the natural electron donor for noncyclic photophosphorylation in chloroplasts, is replaced by a reduced indophenol dye (Vernon and Zaugg, 1960). The background of this problem is as follows.

Losada *et al.* (1961) found that this artificial noncyclic electron flow in chloroplasts is coupled with a phosphorylation. They characterized it as a "bacterial type" of noncyclic photophosphorylation because photosynthetic bacteria cannot use $OH^-$ as an electron donor but exhibit a noncyclic electron flow from other, less oxidized, electron donors (e.g., succinate) to pyridine nucleotide (Frenkel, 1958; Vernon and Ash, 1959). Losada *et al.* (1961) also arranged experimental conditions in such a manner that they obtained oxygen evolution coupled to ferricyanide reduction but without an accompanying phosphorylation. They interpreted these findings as a separation of noncyclic

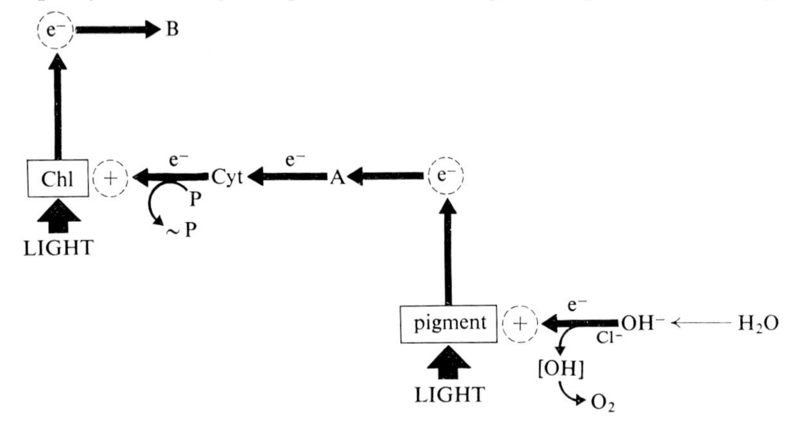


Non-cyclic photophosphorylation of green plant type

FIG. 22. The 1961 scheme for noncyclic photophosphorylation in chloroplasts. One light reaction (right) was envisaged as responsible for oxygen evolution and another light reaction (left) for the reduction of $NADP^+$ (B) and the concomitant phosphorylation "A" was assumed to be a quinone or cytochrome that joined the two light reactions.

photophosphorylation by chloroplasts into two component light reactions: (a) a photo-oxidation of water ($OH^-$) which results in oxygen evolution but not in phosphorylation and supplies electrons, at an intermediate reducing potential, for (b) a second photochemical reaction, the bacterial type of noncyclic phosphorylation, in which the reduction of NADP is coupled with photophosphorylation (Fig. 22).

This interpretation seemed to receive additional support when Nozaki *et al.* (1961) found a similar noncyclic photophosphorylation in particles of photosynthetic bacteria. Thus, the site of noncyclic photophosphorylation in plants was considered to be analogous to that in bacterial noncyclic photophosphorylation. In accordance with earlier formulations (Arnon, 1959, 1961) the same site was also shared by cyclic photophosphorylation.

The validity of these conclusions became open to some question after Trebst and Eck (1961), Gromet-Elhanan and Avron (1963), Keister (1963), Wessels

(1964) and Avron (1964) found that dichlorophenol indophenol in a reduced form can catalyse cyclic photophosphorylation. The matter remained complicated because, under the conditions of the experiments of Losada *et al.* (1961) with chloroplasts, and of Nozaki *et al.* (1961, 1963) with bacterial particles (strict anaerobicity and a great excess of ascorbate), dichlorophenol indophenol failed to catalyse any phosphorylation unless a noncyclic electron flow from reduced dye to pyridine nucleotide was established and maintained.

The matter was finally resolved by the recent experiments with desaspidin which, at a low concentration, strongly inhibited the bacterial type of noncyclic photophosphorylation but not the noncyclic photophosphorylation associated with a flow of electrons from $OH^-$ to $NADP^+$ (Baltscheffsky and de Kiewiet, 1964; Gromet-Elhanan and Arnon, 1965). Thus, contrary to our old interpretations, ATP formation in the bacterial type of noncyclic photophosphorylation appears to occur at a "cyclic" site. It became necessary, therefore, to abandon the previous hypothesis (Losada *et al.*, 1961) and its later elaborations (Tagawa *et al.*, 1963c; Arnon *et al.*, 1964) which envisaged a phosphorylation site common to cyclic and noncyclic photophosphorylation. Our new hypothesis of cyclic and noncyclic photophosphorylation fits the new results and is not in conflict with the earlier findings on the relation between electron flow and photophosphorylation.

## Mechanisms of Cyclic and Noncyclic Photophosphorylation

Our new hypothesis envisages that in chloroplasts the cyclic electron transport chain and its coupled phosphorylations are distinct from noncyclic photophosphorylation. This conclusion is based on the fact that ferredoxin-catalysed cyclic photophosphorylation will occur only under conditions when noncyclic photophosphorylation is excluded. We consider cyclic photophosphorylation in chloroplasts as including, under physiological conditions, phosphorylations that are coupled with a flow of electrons from excited chlorophyll to ferredoxin, and then from reduced ferredoxin to cytochromes $b_6$ and $f$ and back to chlorophyll (Fig. 23). We include the chloroplast cytochromes (Hill, 1954; Davenport, 1952) $b_6$ and $f$ as electron carriers in cyclic but not in noncyclic photophosphorylation although, so far as we are aware, there is so far no direct evidence for the involvement of chloroplast cytochromes in either type of photophosphorylation. Indirect evidence comes from inhibition of ferredoxin-catalysed cyclic photophosphorylation by antimycin A and 2,4-dinitrophenol, two well-known inhibitors of oxidative phosphorylation where the direct participation of cytochromes is well documented. Antimycin A inhibition of oxidative phosphorylation is considered to be indicative of the participation of cytochrome $b$ in electron transport (Chance and Williams, 1956; Racker, 1961). It is likely that participation of cytochrome $b_6$ accounts for the sensitivity of ferredoxin-catalysed cyclic photophosphorylation to antimycin A. As for cytochrome $f$, its joint participation in an electron transport chain with cytochrome $b_6$ of chloroplasts is considered likely by analogy with oxidative phosphorylation in mitochondria.

The span between the redox potential of cytochromes $b_6$ and $f$ ($-0{\cdot}06$ V and

0·365 V, respectively) is large enough to accommodate a phosphorylation. Likewise, the span between the redox potentials of ferredoxin (−0·43 V) and cytochrome $b_6$ is large enough to accommodate at least one phosphorylation in this segment of the cyclic chain. We specify tentatively two phosphorylation sites in this cyclic electron transport chain but this does not exclude the possibility of additional phosphorylation sites.

As discussed elsewhere (Tagawa *et al.*, 1963a; Arnon, 1965), we consider that in chloroplasts the ferredoxin-catalysed cyclic photophosphorylation is the physiological one. However, experimentally, cyclic photophosphorylation proceeds readily without ferredoxin when catalysed by one of several dyes or

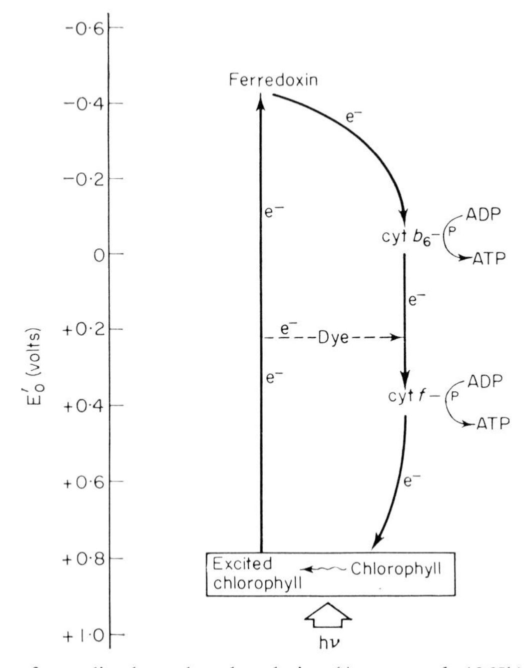


FIG. 23. Scheme for cyclic photophosphorylation (Arnon *et al.*, 1965b).

other artificial cofactors. Since such cyclic photophosphorylations are resistant to inhibition by antimycin A, it seems reasonable to conclude that they bypass the cytochrome $b_6$ site (see dotted line in Fig. 23).

Our present concept of noncyclic photophosphorylation in plants is shown in Fig. 24. The phosphorylation is envisaged as being coupled to the oxidation of $OH^-$, a coupling that would account for the consistent stoichiometry, P/2e = 1, between oxygen evolution and ATP formation. We consider that an electron from $OH^-$ is transferred via chlorophyll to ferredoxin in a single light reaction. In isolated chloroplasts, ferredoxin may be replaced by non-physiological electron acceptors (Hill reagents) with an attendant drop in the light-generated reducing potential. Figure 24 (dotted lines) illustrates this for ferricyanide and benzoquinone (BQ).

The "chlorophyll" in Figs. 23 and 24 represents the complex of chlorophyll *a* and *b* pigments in their various forms and includes those accessory pigments which are involved in light absorption by chloroplasts (see reviews, Smith and French, 1963; Vernon and Avron, 1965). We assume that excitation energy is transferred among the chloroplast pigments with the great efficiency that is well documented in algal cells (see review, Duysens, 1964).

In laboratory experiments with monochromatic light, it has been established that cyclic photophosphorylation and the previously discussed noncyclic photophosphorylation of the bacterial type can occur at wavelengths longer

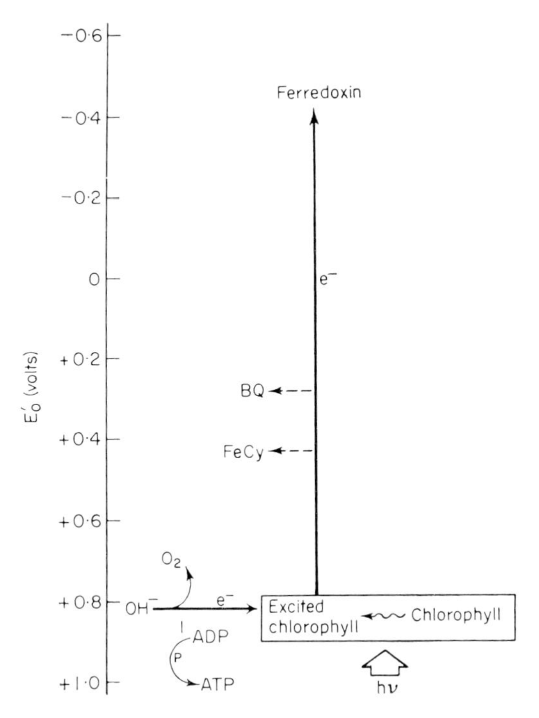


FIG. 24. Scheme for noncyclic photophosphorylation of the plant type (Arnon *et al.*, 1965b).

than 700 mμ that will not support noncyclic photophosphorylation of the plant type. Thus, it appears that noncyclic photophosphorylation in chloroplasts requires the shorter wavelengths of light and is perhaps most efficient at wavelengths not longer than those at which photosynthesis in intact cells occurs with maximum efficiency. The energy balance given in Table IX shows that 1 quantum of light at 680 mμ—the longest wavelength that would still support maximum efficiency of plant photosynthesis (Duysens, 1964)—contains sufficient energy for the transfer of an electron from $OH^-$ to ferredoxin and the coupled phosphorylation.

The electron pathway from $OH^-$ to chlorophyll is the least understood part in the mechanism of photosynthesis and its elucidation must be left to future research. Only a few of the cofactors and catalysts involved therein are now known: chloride ions (Warburg, 1949; Bové *et al.*, 1963), manganese (Pirson, 1937; Kessler, 1955) and plastoquinone (Bishop, 1959; Krogmann, 1961; Arnon and Horton, 1963) [plastoquinone is also involved in cyclic photophosphorylation in chloroplasts (Krogmann, 1961; Whatley and Horton, 1963)]. The nature of the linkage between electron flow from $OH^-$ and phosphorylation remains obscure but there is good reason to believe that the two are closely linked since the rate of electron flow from $OH^-$ is greatly increased by the concurrent phosphorylation (Arnon *et al.*, 1958; Arnon *et al.*, 1959; Avron *et al.*, 1958; Davenport, 1960).

Table IX. Energy balance of noncyclic photophosphorylation of the plant type

Energy input of 1 Einstein, $\lambda = 680$ m$\mu$, is 42 kcal

Ferredoxin$_{ox}$ $\rightleftharpoons$ Ferredoxin$_{red}$ + e$^-$; $E_0' = -0{\cdot}43$ V (pH 7)

$\frac{1}{2}H_2O = \frac{1}{4}O_2 + H^+ + e^-$; $E_0' = +0{\cdot}82$ V (pH 7)

Potential span between $O_2$ and ferredoxin$_{red}$ is $1{\cdot}25$ eV $= 1{\cdot}25 \times 23{\cdot}06 = 28{\cdot}8$ kcal

Energy requirement (per 1 electron) to form 1 ATP: 10 kcal/2 = 5 kcal

Excess of energy input over output: $42 - (28{\cdot}8 + 5) = 8{\cdot}2$ kcal

The flow of electrons from $OH^-$ to ferredoxin and the resultant oxygen evolution are easily susceptible to damage when chloroplasts are removed from intact cells. The noncyclic electron flow of the bacterial type is much more stable but, again, it is also uncertain whether it can give a true measure of the efficiency of electron transport in the intact cell. It is not surprising, therefore, that a requirement of one quantum per electron transferred to ferredoxin (or NADP) has not been obtained in investigations with isolated chloroplasts (Sauer and Biggins, 1965).

How can ferredoxin participate in both cyclic and noncyclic photophosphorylation? We have already stressed the fact that in isolated chloroplasts cyclic photophosphorylation occurs only when noncyclic photophosphorylation (coupled to oxygen evolution) is stopped. The regulatory mechanism(s) in the cell for switching from noncyclic to cyclic photophosphorylation is unknown, but one possibility has experimental support: the availability of NADP in the oxidized form (Tagawa *et al.*, 1963a). As long as oxidized NADP is available, we envisage that electrons will flow from water to ferredoxin and thence (via ferredoxin-NADP reductase) to NADP. However, when $CO_2$ assimilation temporarily ceases for lack of ATP, NADP accumulates in the reduced state and electrons from reduced ferredoxin begin to "cycle" within the chloroplasts, giving rise to cyclic photophosphorylation. The additional ATP thus generated would re-establish $CO_2$ assimilation, which, in turn, would restore NADPH to its oxidized form and thereby re-establish noncyclic photophosphorylation.

## Cyclic and Noncyclic Photophosphorylation as the Two Light Reactions of Photosynthesis

Our hypothesis identifies cyclic and noncyclic photophosphorylation as two complementary and parallel pathways of energy conversion which jointly generate the assimilatory power required for carbon assimilation in green plants. Based on their response to monochromatic light, cyclic photophosphorylation would be the "long wavelength" light reaction and noncyclic photophosphorylation the "short wavelength" light reaction. The overall efficiency of photosynthesis would thus depend on the efficient functioning of both these parallel photochemical processes.

This hypothesis explains some of the observations that are now used to support the concept of two light reactions working in series (see reviews by Smith and French, 1963; Duysens, 1964; Vernon and Avron, 1965). Thus, in our scheme "red drop" would mean that at the longer wavelengths of light noncyclic photophosphorylation becomes limiting, thereby reducing the availability of NADPH and decreasing the overall efficiency of the process. "Enhancement" would mean that the addition of light of short wavelength to light of long wavelength restores noncyclic photophosphorylation, removes the shortage of NADPH and thereby makes for a much more efficient utilization of the ATP produced by cyclic photophosphorylation. Thus, a supplement of short wavelength radiation would result in a synergistic rather than in an additive increase in the overall efficiency of photosynthesis.

Among the other observations that would appear to fit the present hypothesis is the oxidation of chloroplast cytochromes in longer wavelengths of light and their reduction in shorter wavelengths (see reviews cited above). Our hypothesis envisages that at the shorter wavelengths ferredoxin is reduced by electrons from water and reduced ferredoxin in turn reduces the chloroplast cytochromes. They would remain in the reduced state as long as noncyclic photophosphorylation is in operation. Cytochromes would become oxidized during cyclic photophosphorylation, which, under laboratory conditions, is established either by long-wavelength illumination or by short-wavelength illumination in the presence of inhibitors of oxygen evolution. Further experiments are now in progress to test this hypothesis.

We have not discussed so far whether cyclic photophosphorylation is essential for photosynthesis in plants; that is, whether the ATP that it supplies to supplement that produced by noncyclic photophosphorylation is required for carbon assimilation. Two moles of reduced ferredoxin are required to give 1 mole of NADPH. Thus, noncyclic photophosphorylation gives rise to NADPH and ATP in a ratio of 1:1 (Eq. 2). Were this ratio adequate for carbon assimilation there would be no need for cyclic photophosphorylation. However, the carbon reduction cycle appears to have a requirement of 2 NADPH and 3 ATP per 1 mole of $CO_2$ assimilated to the level of glucose (Bassham and Calvin, 1957). Even if this requirement were reduced to 2 ATP and 2 NADPH, additional ATP would still be necessary to form starch, the main product of photosynthesis in leaves, because ATP is expended in the formation of ADP-glucose from which the glucosyl residue is transferred to a starch primer (Ghosh and Preiss, 1965). Moreover, as was pointed out elsewhere (Arnon *et al.*, 1958), cyclic photophosphorylation may be an important mechanism for providing the large supplies of ATP that are required for protein synthesis and for other endergonic processes in the cell.

## Quenching of Chloroplast Fluorescence by Cyclic and Noncyclic Photophosphorylation

Fluorescence of chloroplasts represents that portion of absorbed radiant energy which is not converted into chemical energy (or heat) but is re-emitted as radiation. If our thesis is correct that photosynthetic phosphorylation constitutes the prime energy conversion process of photosynthesis, then it should act as a quencher of chloroplast fluorescence.

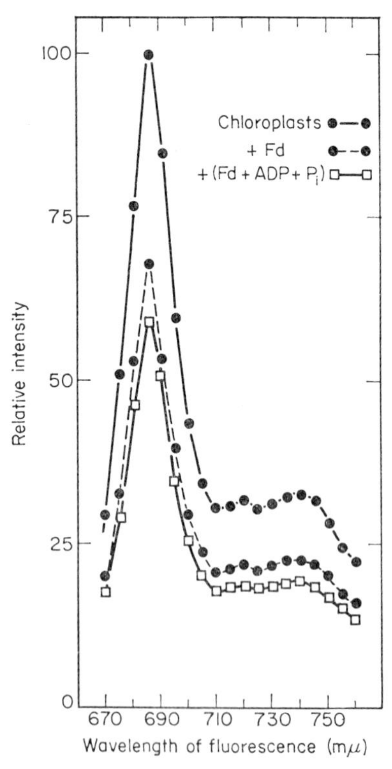


FIG. 25. Quenching of chloroplast fluorescence by ferredoxin, ADP and inorganic phosphate (Arnon *et al.*, 1965a).

Figure 25 shows a marked quenching of chloroplast fluorescence by the addition of ferredoxin (Arnon *et al.*, 1965). This result is consistent with our hypothesis that ferredoxin establishes a cyclic electron flow: the degradation of the energy of a molecule excited by photon capture can occur by electron transfer to an appropriate electron acceptor molecule. An additional quenching of chloroplast fluorescence was observed upon adding ADP and inorganic phosphate (Fig. 25)—an observation which supports the idea that the cyclic electron flow catalysed by ferredoxin is accelerated by concomitant phosphorylation.

Quenching of chloroplast fluorescence by noncyclic photophosphorylation is shown in Fig. 26. The addition of ferredoxin and NADP gave a pronounced

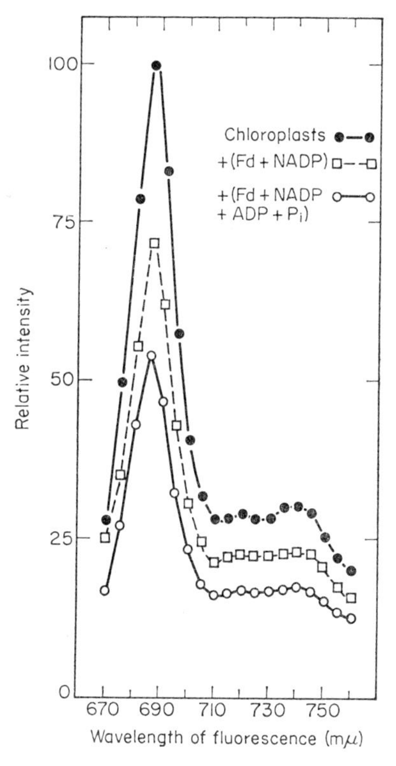


FIG. 26. Quenching of chloroplast fluorescence by NADP, ferredoxin, ADP and inorganic phosphate (Arnon *et al.*, 1965a).

quenching effect. The addition of ADP and inorganic phosphate to the ferredoxin-NADP system gave a marked additional quenching effect. These results are consistent with earlier findings that the rate of noncyclic electron flow is markedly increased by a concomitant phosphorylation.

The quenching of chloroplast fluorescence by cyclic and noncyclic photophosphorylation supports their characterization as the primary photochemical reactions in photosynthesis (Arnon, 1959; 1961). The energy of captured photons may be dissipated as fluorescence or may generate an electron flow which yields the chemical energy stored in the pyrophosphate bonds of ATP and in the reducing potential of ferredoxin.

## References

Allen, M. B., Arnon, D. I., Capindale, J. B., Whatley, F. R. and Durham, L. J. (1955). *J. Am. chem. Soc.* **77**, 4149.
Allen, M. B., Whatley, F. R. and Arnon, D. I. (1958). *Biochim. biophys. Acta* **27**, 16.
Anderson, D. G. and Vennesland, B. (1954). *J. biol. Chem.* **207**, 613.
Anderson, I. C. and Fuller, R. C. (1958). *Archs Biochem. Biophys.* **76**, 168.
Arnon, D. I. (1955). *Science, N.Y.* **122**, 9.
Arnon, D. I. (1959). *Nature, Lond.* **184**, 10.
Arnon, D. I. (1961). *In* "Light and Life" (W. D. McElroy and B. Glass, eds.), p. 489. Johns Hopkins Press, Baltimore, Maryland.
Arnon, D. I. (1965). *Science, N.Y.* **149**, 1460.
Arnon, D. I. and Horton, A. A. (1963). *Acta chem. scand.* **17**, S135.
Arnon, D. I., Allen, M. B. and Whatley, F. R. (1954a). *Nature, Lond.* **174**, 394.
Arnon, D. I., Whatley, F. R. and Allen, M. B. (1954b). *J. Am. chem. Soc.* **76**, 6324.
Arnon, D. I., Whatley, F. R. ane Allen, M. B. (1955). *Biochim. biophys. Acta* **16**, 607.
Arnon, D. I., Whatley, F. R. and Allen, M. B. (1957). *Nature, Lond.* **180**, 182, 1325.
Arnon, D. I., Whatley, F. R. and Allen, M. B. (1958). *Science, N.Y.* **127**, 1026.
Arnon, D. I., Whatley, F. R. and Allen, M. B. (1959). *Biochim. biophys. Acta* **32**, 47.
Arnon, D. I., Losada, M., Whatley, F. R., Tsujimoto, H. Y., Hall, D. O. and Horton, A. A. (1961). *Proc. natn. Acad. Sci. U.S.A.* **47**, 1314.
Arnon, D. I., Tsujimoto, H. Y. and McSwain, B. D. (1964). *Proc. natn. Acad. Sci. U.S.A.* **51**, 1274.
Arnon, D. I., Tsujimoto, H. Y. and McSwain, B. D. (1965a). *Proc. natn. Acad. Sci. U.S.A.* **54**, 927.
Arnon, D. I., Tsujimoto, H. Y. and McSwain, B. D. (1965b). *Nature, Lond.* **207**, 1367.
Aronoff, S. and Calvin, M. (1948). *Pl. Physiol.* **23**, 351.
Avron, M. (1964). *Biochem. biophys. Res. Commun.* **17**, 430.
Avron, M. and Jagendorf, A. T. (1956). *Archs Biochem. Biophys.* **65**, 475.
Avron, M. and Jagendorf, A. T. (1957a). *Archs Biochem. Biophys.* **72**, 17.
Avron, M. and Jagendorf, A. T. (1957b). *Nature, Lond.* **179**, 428.
Avron, M., Krogmann, D. W. and Jagendorf, A. T. (1958). *Biochim. biophys. Acta* **30**, 144.
Bachofen, R. and Arnon, D. I. (1966). *Biochim. biophys. Acta* **120**, 259.
Baltscheffsky, H. and de Kiewiet, D. Y. (1964). *Acta chem. scand.* **18**, 2406.
Bassham, J. A. and Calvin, M. (1957). "The Path of Carbon in Photosynthesis." Prentice Hall, Englewood Cliffs, New Jersey.
Bishop, N. I. (1959). *Proc. natn. Acad. Sci. U.S.A.* **45**, 1696.
Bové, J. M., Bové, C., Whatley, F. R. and Arnon, D. I. (1963). *Z. Naturf.* **18**b, 683.
Buchanan, B. B., Lovenberg, W. and Rabinowitz, J. C. (1963). *Proc. natn. Acad. Sci. U.S.A.* **49**, 345.
Chance, B. and Williams, G. R. (1956). *Adv. Enzymol.* **17**, 65.
Davenport, H. E. (1952). *Nature, Lond.* **170**, 1112.
Davenport, H. E. (1960). *Biochem. J.* **77**, 471.
Davenport, H. E. and Hill, R. (1960). *Biochem. J.* **74**, 493.
Davenport, H. E., Hill, R. and Whatley, F. R. (1952). *Proc. R. Soc.* B**139**, 346.
Duysens, L. N. M. (1964). *Prog. Biophys.* **14**, 1.
Engelmann, Th. W. (1883). *Arch ges. Physiol.* (*Pflüger's*) **30**, 95.
Engelmann, Th. W. (1888). *Bot. Z.* **46**, 661 ff.
Fager, E. W. (1952). *Archs Biochem. Biophys.* **41**, 383.
Fager, E. W. (1954). *Biochem. J.* **57**, 264.
Frenkel, A. W. (1954). *J. Am. chem. Soc.* **76**, 5568.
Frenkel, A. W. (1956). *J. biol. Chem.* **222**, 823.
Frenkel, A. W. (1958). *Brookhaven Symp. Biol.* **11**, 276.
Fry, K. T. and San Pietro, A. (1962). *Biochem. biophys. Res. Commun.* **9**, 218.
Geller, D. M. and Gregory, J. S. (1956). *Fed. Proc.* **15**, 260.
Geller, D. and Lipmann, F. (1960). *J. biol. Chem.* **235**, 2478.
Gewitz, H. S. and Voelker, W. (1962). *Z. Phys. Chem.* **330**, 124.
Ghosh, H. P. and Preiss, J. (1965). *Biochemistry* **4**, 1354.

Gibbs, M. and Calo, N. (1959). *Pl. Physiol.* **34**, 318.
Gibbs, M. and Cynkin, M. A. (1958). *Nature, Lond.* **182**, 1241.
Gromet-Elhanan, Z. and Arnon, D. I. (1965). *Pl. Physiol.* **40**, 1060.
Gromet-Elhanan, Z. and Avron, M. (1963). *Biochem. biophys. Res. Commun.* **10**, 215.
Hill, R. (1939). *Proc. R. Soc.* B**127**, 192.
Hill, R. (1951). *Symp. Soc. exp. Biol.* **5**, 223.
Hill, R. (1954). *Nature, Lond.* **174**, 501.
Hill, R. (1956). *Proc. 3rd Int. Congr. Biochem., Brussels*, 1955, p. 225. Academic Press, New York and London.
Horio, T. and San Pietro, A. (1964). *Proc. natn. Acad. Sci. U.S.A.* **51**, 1226.
Horio, T. and Yamashita, T. (1962). *Biochem. biophys. Res. Commun.* **9**, 142.
Irmak, L. R. (1955). *Rev. Fac. Sci. Univ. Istanbul* B**20**, 237.
Jagendorf, A. T. (1962). *In* "Survey of Biological Progress" (B. Glass, ed.), Vol. 4, p. 181. Academic Press, New York and London.
Jagendorf, A. T. and Avron, M. (1958). *J. biol. Chem.* **231**, 277.
James, W. O. and Das, V. S. R. (1957). *New Phytol.* **56**, 325.
Kamen, M. and Newton, J. W. (1957). *Biochim. biophys. Acta* **25**, 462.
Katoh, S. and Takamiya, A. (1962). *biochem. biophys. Res. Commun.* **8**, 310.
Keister, D. L. (1963). *J. biol. Chem.* **238**, PC2590.
Keister, D. L., San Pietro, A. and Stolzenbach, F. E. (1960). *J. biol. Chem.* **235**, 2989.
Keister, D. L., San Pietro, A. and Stolzenbach, F. E. (1962). *Archs Biochem. Biophys.* **98**, 235.
Kessler, E. (1955). *Archs Biochem. Biophys.* **59**, 527.
Krogmann, D. W. (1961). *Biochem. biophys. Res. Commun.* **4**, 275.
Losada, M. and Arnon, D. I. (1963). *In* "Metabolic Inhibitors" (R. M. Hochster and J. H. Quastel, eds.), Vol. 2, p. 559. Academic Press, New York and London.
Losada, M., Whatley, F. R. and Arnon, D. I. (1961). *Nature, Lond.* **190**, 606.
Lovenberg, W., Buchanan, B. B. and Rabinowitz, J. C. (1963). *J. biol. Chem.* **238**, 3899.
Lumry, R., Spikes, J. D. and Eyring, H. (1954). *Annu. Rev. Pl. Physiol.* **5**, 271.
Lundegårdh, H. (1961). *Nature, Lond.* **192**, 243.
Mortenson, L. E., Valentine, R. C. and Carnahan, J. E. (1962). *Biochem. biophys. Res. Commun.* **7**, 448.
Nozaki, M., Tagawa, K. and Arnon, D. I. (1961). *Proc. natn. Acad. Sci. U.S.A.* **47**, 1334.
Nozaki, M., Tagawa, K. and Arnon, D. I. (1963). *In* "Bacterial Photosynthesis" (H. Gest, A. San Pietro and L. P. Vernon, eds.), p. 175. Antioch Press, Yellow Springs, Ohio.
Petrack, B. and Lipmann, F. (1961). *In* "Light and Life" (W. D. McElroy and B. Glass, eds.), p. 621. Johns Hopkins Press, Baltimore.
Pirson, A. (1937). *Z. Bot.* **31**, 193.
Rabinowitch, E. I. (1953). *Sci. Am.* Nov. 1953, p. 80.
Rabinowitch, E. (1957). *In* "Research in Photosynthesis" (H. Gaffron, ed.), p. 345. Interscience Publishers, New York.
Racker, E. (1961). *Adv. Enzymol.* **23**, 323.
Runeberg, L. (1963). "Commentationes Biologicae XXVI", 7 Diss, 69 pp. Societas Scientiarum Fennica, Helsinki, Finland.
Sachs, J. (1887). "Lectures on the Physiology of Plants", p. 299. Clarendon Press, Oxford.
San Pietro, A. and Black, C. C. (1965). *Annu. Rev. Pl. Physiol.* **16**, 155.
San Pietro, A. and Lang, H. M. (1958). *J. biol. Chem.* **231**, 211.
Sauer, K. and Biggins, J. (1965). *Biochim. biophys. Acta* **102**, 55.
Shin, M. and Arnon, D. I. (1965). *J. biol. Chem.* **240**, 1405.
Shin, M., Tagawa, K. and Arnon, D. I. (1963). *Biochem. Z.* **338**, 84.
Smillie, R. M. and Fuller, R. C. (1959). *Pl. Physiol.* **34**, 651.
Smillie, R. M. and Krotkov, G. (1959). *Can. J. Bot.* **37**, 1217.
Smith, J. H. C. and French, C. S. (1963). *Annu. Rev. Pl. Physiol.* **14**, 181.
Stern, B. K. and Vennesland, B. (1960). *J. biol. Chem.* **235**, PC51.
Tagawa, K. and Arnon, D. I. (1962). *Nature, Lond.* **195**, 537.
Tagawa, K., Tsujimoto, H. Y. and Arnon, D. I. (1963a). *Proc. natn. Acad. Sci. U.S.A.* **49**, 567.
Tagawa, K., Tsujimoto, H. Y. and Arnon, D. I. (1963b). *Proc. natn. Acad. Sci. U.S.A.* **50**, 544.
Tagawa, K., Tsujimoto, H. Y. and Arnon, D. I. (1963c). *Nature, Lond.* **199**, 1247.

Thomas, J. B., Haans, A. J. M. and Van der Leun, A. A. (1957). *Biochim. biophys. Acta* **25**, 453.
Tolbert, N. E. (1958). *Brookhaven Symp. Biol.* **11**, 271.
Trebst, A. and Eck, H. (1961). *Z. Naturf.* **16**b, 455.
Ueda, R. (1949). *Bot. Mag.* **62**, 731.
Vernon, L. P. and Ash, O. K. (1959). *J. biol. Chem.* **234**, 1878.
Vernon, L. P. and Avron, M. (1965). *Annu. Rev. Biochem.* **34**, 269.
Vernon, L. P. and Zaugg, W. S. (1960). *J. biol. Chem.* **235**, 2728.
Vishniac, W. and Ochoa, S. (1952). *J. biol. Chem.* **198**, 501.
Warburg, O. (1949). "Heavy Metal Prosthetic Groups and Enzyme Action", p. 213. Clarendon Press, Oxford.
Warburg, O. (1963). Remarks in discussion of a paper by D. I. Arnon in "La Photosynthese" p. 540. Colloq. Intern. Centre Natl. Rech. Sci. Paris, No. 119.
Warburg, O., Krippahl, G., Gewitz, H. S. and Volker, W. (1959). *Z. Naturf.* **14**b, 712.
Wessels, J. S. C. (1957). *Biochim. biophys. Acta* **25**, 97.
Wessels, J. S. C. (1964). *Biochim. biophys. Acta* **79**, 640.
Whatley, F. R. and Horton, A. A. (1963). *Acta chem. scand.* **17**, S140.
Whatley, F. R., Tagawa, K. and Arnon, D. I. (1963). *Proc. natn. Acad. Sci. U.S.A.* **49**, 266.
Williams, A. M. (1956). *Biochim. biophys. Acta* **19**, 570.

## Discussion

*Z. Sestak:* Why don't your schemes of photophosphorylation fit in with the two-pigment system?

*D. I. Arnon:* I do not question the participation of different pigment systems in the photochemistry of photosynthesis but our data lead me to question the current hypothesis (which we had ourselves supported in the past) that noncyclic photophosphorylation requires the collaboration of two separate photochemical reactions that are joined by intervening dark reactions. Our new hypothesis explains our data better and is also consistent with other observations that are often cited in support of the former hypothesis: "red drop" and enhancement effects.

*Z. Sestak:* In one of your review articles you did mention that cyclic photophosphorylation goes on in the plants also after closing the stomata, when very limited or no gas exchange appears. Do you have some experimental evidence for this effect and is it consistent with your new observations?

*D. I. Arnon:* All our work has been with isolated chloroplasts and hence we have had no occasion to measure the closing or opening of stomata. Our initial suggestion (Arnon *et al.*, 1958) that cyclic photophosphorylation could proceed when stomata were closed was based on the fact that it does not involve an exchange of gases. This suggestion is consistent with, and is fortified by, our recent observations that cyclic photophosphorylation in isolated chloroplasts occurs at wavelengths longer than 700 m$\mu$ which do not support oxygen evolution and is independent of noncyclic photophosphorylation. Furthermore, other investigators who worked with intact cells found that cyclic phosphorylation can proceed under conditions which prevent oxygen evolution (for example, inhibition by DCMU).

*W. W. Hildreth:* Will you please compare your proposed electron transport scheme with the one-center theory of J. Franck and J. Rosenberg, *J. Theoret. Biol.* **7**, 276 (1964).

*D. I. Arnon:* Such comparison is difficult because the scheme of Franck and Rosenberg involves such hypothetical entities as "X, primary oxidant" and "Y", which they define as, "as yet unidentified electron transport enzyme built into the reaction center complex". In some points in which the scheme of Franck and Rosenberg is specific, it is at variance with our hypothesis; for example, they include cytochrome *f* in the oxygen evolution reaction and they assign no function to cytochrome $b_6$. However, their proposal that there is only one reaction center in the photochemical reactions of chloroplasts is not in conflict with our hypothesis and with the key role that we assign to ferredoxin in both cyclic and noncyclic photophosphorylation.

*H. Baltscheffsky:* Your new scheme for cyclic photophosphorylation in the ferredoxin system included both cytochromes $b_6$ and *f* as well as two coupling sites in the cyclic electron transport chain, one at the $b_6$ level and another at the *f* level. Is your inclusion of cytochrome $b_6$ and of two coupling sites in this cyclic chain purely a hypothesis or do you have any direct experimental evidence on this point?

*D. I. Arnon:* Our strongest evidence, so far, for the participation of both cytochromes $b_6$ and *f* in the ferredoxin-catalysed cyclic photophosphorylation is the inhibition of this type of photophosphorylation by antimycin A, which in mitochondria inhibits electron flow between cytochrome *b* and *c* (cytochrome *f* of chloroplasts corresponds to cytochrome *c* of mitochondria). Moreover, when ferredoxin is replaced by the nonphysiological catalyst, phenazine methosulfate, the resultant cyclic photophosphorylation is resistant to inhibition by antimycin A. These observations suggest that ferredoxin-catalysed cyclic photophosphorylation has an additional cytochrome $b_6$-linked phosphorylation site which is bypassed by phenazine methosulfate.

*H. Baltscheffsky:* The answer you gave to my first question was a good introduction to the second. You used about two orders of magnitude higher concentrations of antimycin A to inhibit cyclic photophosphorylation in chloroplasts than those required to inhibit photophosphorylation in bacterial chromatophores and oxidative phosphorylation in animal mitochondria. Although this may be just a reflection of solubility or distribution differences, I was a little worried about the conclusions you drew from the inhibition data. There may be a relief though, if other inhibitors, which act in a similar way as antimycin A on the other systems, also are active in your ferredoxin-system. The question is: have you tried SN5949 or HOQNO (or NOQNO) and were they active or not?

*D. I. Arnon:* I am not sure that these differences in the concentration of antimycin A are not merely a reflection of different experimental conditions. In our experiments, antimycin A at a concentration of $1{\cdot}8 \times 10^{-5}$ M gave an inhibition of about 81% of the ferredoxin-catalysed cyclic photophosphorylation (Fig. 19). We have not investigated in detail the effects of antimycin A concentration because we were impressed by the striking difference between such drastic inhibition of the ferredoxin-catalysed cyclic photophosphorylation by low concentrations of antimycin A and the absence of such inhibition in cyclic photophosphorylation catalysed by phenazine methosulfate.

We have not as yet tried the other inhibitors which you mention.

*A. Trebst:* How do you picture the site of function of plastoquinone and plastocyanin in your new scheme?

*D. I. Arnon:* We are now investigating the place of plastocyanin and plastoquinone in our scheme and will report on this in due course.

*H. Gest:* I would like to comment on one point raised by Dr. Arnon. On the basis of our own work with bacterial chromatophore preparations and of published reports by others, Bose and I concluded that the phenomenon of "noncyclic photophosphorylation" does not exist in bacterial photosynthesis. In other words, we believe that there are significant differences between bacterial and green plant photosyntheses. As I understood your remarks, Dr. Arnon, you have now abandoned your previous contention that noncyclic photophosphorylation occurs in bacteria. Is that correct?

*D. I. Arnon:* No, that is incorrect. My paper here does not invalidate the work on noncyclic photophosphorylation of the bacterial type but, on the contrary, it clarifies it.

We give an extensive discussion of the data and of the arguments of Bose and Gest in the article by Nozaki *et al.* (1963) which is cited in my manuscript. Here, I would like to add that some of the apparent confusion that surrounds bacterial noncyclic photophosphorylation seems to center, in part, on nomenclature. As one who must bear the responsibility for introducing it (Arnon, 1959), I would like to clarify it a bit further at this opportunity. By noncyclic electron flow I mean a light-driven, "uphill", unidirectional electron transfer against the thermodynamic gradient. One example is a light-driven transfer of an electron from water ($OH^-$) to ferredoxin and thence to $NADP^+$. This type of noncyclic electron transfer is found only in plants. Another example is a light-driven transfer of an electron from reduced dichlorophenol indophenol to pyridine nucleotides. This type has been demonstrated in bacterial chromatophores and also, under special experimental conditions, in isolated chloroplasts. So far as I know, no one disputes the existence of such noncyclic electron flow either in chloroplasts or in chromatophores.

Turning now to noncyclic photophosphorylation, I mean by this term ATP formation by either chloroplasts or chromatophores that is coupled to their respective type of noncyclic electron flow and proceeds under conditions that make cyclic photophosphorylation impossible. When such noncyclic phosphorylation is coupled with a noncyclic electron transport in which water is the electron donor, we speak of it as noncyclic photophosphorylation of the plant type. When such noncyclic phosphorylation is coupled with a noncyclic electron transport in which the electron donor is not water but a substance of a much more positive redox potential (e.g., reduced dichlorophenol indophenol, succinate), we speak of it as noncyclic photophosphorylation of the bacterial type.

Thus, when we speak of noncyclic photophosphorylation of the bacterial type, we emphasize rather than obscure the difference between photosynthesis in plants and bacteria. We emphasize the fact that only plants can use water as an electron donor for noncyclic photophosphorylation. It is here that we depart from the concept of van Niel that a primary photolysis of water is a common photochemical act in the photosynthesis of green plants and photosynthetic bacteria (C. B. van Niel (1941). *Adv. Enzymol.* **1**, 324).

*O. Kandler:* I appreciate the type of scheme suggested by Prof. Arnon. Some data obtained from experiments with whole cells fit more easily in an alternative than in a consecutive system, demanding both light reactions in stoichiometric amounts.

We now have found, that the anaerobic photo-assimilation of glucose proceeds at an almost undiminished rate even at 710 m$\mu$ (if light saturation is provided), whereas $CO_2$ assimilation drops strongly. This cyclic photophosphorylation is light-saturated at the same very low light intensities as the nitrite reduction shown by Kessler, or the decrease in organic phosphate upon illumination shown by Wintermans and ourselves.

The photo-assimilation of glucose is only sensitive to higher concentrations of DCMU. The table of Prof. Arnon, too, showed a 20–30% inhibition of cyclic phosphorylation by DCMU at high concentrations.

Regarding the remark of Dr. Baltscheffsky, I should like to mention that in our experiments antimycin inhibited the oxidative glucose assimilation to the same extent as the photo-assimilation, which suggests a common cofactor.

It is remarkable that antimycin does not essentially inhibit $CO_2$ assimilation in *Chlorella*, thus showing the opposite effect of DCMU. The fact, that $CO_2$ assimilation may be separated from cyclic phosphorylation *in vivo* may be more readily explained by a system of two parallel light reactions which may be described in the following way:

Reaction II requires a wavelength $< 680$ m$\mu$ and leads to the splitting of water, the reduction of Fd and $CO_2$ and the formation of a certain amount of ATP according to the scheme proposed by Arnon for higher plants.

Reaction I which is saturated at very low light intensities utilizes in addition to short-wave light also light $> 680$ m$\mu$ (e.g. 710 m$\mu$) but does not lead to the splitting of water since the redox potential of the pigment involved (P 700?) is less positive than the one of reaction II. Only a cyclic flow of electrons probably via Fd and cytochromes which is coupled to phosphorylation, is induced. This system can be converted to the one proposed by Arnon for the bacterial type of photosynthesis, where electron donors like succinate, ascorbate, etc., are supplied, thus allowing the utilization of the reductant (Fd) for NADP and $CO_2$ reduction.

While the light reaction II takes care of the requirements of the $CO_2$ reducing cycle, light reaction I provides the relatively small amount of ATP which is lost from the cycle by the withdrawal of phosphorylated intermediates like sugar phosphates or PGA for the formation of polysaccharides or amino acids and other products.

On the basis of such a parallel scheme the Emerson enhancement effect may be explained as a sparing effect: If only light $< 680$ m$\mu$ is supplied, the quanta have to be shared between both light reactions, but upon the addition of light $> 680$ m$\mu$ the latter supplies light reaction I and all the quanta of the short-wave light are available for light reaction II, thus leading to an increased $O_2$ evolution. Such a view explains the finding of Bannister *et al.* and Govindjee, that the enhancement is much earlier light-saturated than $CO_2$ fixation, which is almost unexplainable on the basis of the consecutive scheme.

# Some Factors Affecting the Onset of Cyclic Photophosphorylation

Bruce R. Grant[1] and F. R. Whatley[2]
*Department of Cell Physiology, University of California, Berkeley, California, U.S.A.*

Cyclic photophosphorylation was first observed in chloroplasts from spinach (Arnon, Allen and Whatley, 1954) and in chromatophores from *Rhodospirillum rubrum* (Frenkel, 1954). It has been identified in several species of higher plants, algae, and photosynthetic bacteria since that time. In cyclic photophosphorylation light absorbed by the pigment system is considered (Arnon, 1959, 1961) to drive a cyclic flow of electrons through a number of catalytic intermediates, the process leading to a coupled formation of ATP. In the earlier experiments with chloroplasts it was found necessary to add one of a number of catalytic cofactors to obtain high rates of ATP formation (Allen *et al.*, 1958), but at no time was the addition of a substrate found necessary. By contrast Frenkel (1954) found it necessary at first to add α-ketoglutarate to the chromatophore system, and because of this, some slight confusion occurred as to whether or not the two processes in chloroplasts and chromatophores were closely similar. This confusion vanished when Frenkel showed subsequently (1956) that very small amounts of α-ketoglutarate, insufficient to act as a substrate, were effective in the chromatophore system and that α-ketoglutarate could be replaced by a small amount of NADPH. The idea then arose that the chromatophore system may need to be brought to a suitable redox state before effective cyclic electron flow could take place. It is, of course, clear that in any cyclic system, in which a catalytic intermediate is alternately oxidized and reduced, part of the intermediate must be oxidized at all times (to accept electrons from the component which precedes it in the cycle) and part reduced (to donate electrons to the component which follows it).

The question of the balancing of the redox state in the chromatophores of *Rhodospirillum rubrum* to bring them into a condition in which cyclic photophosphorylation could take place was further taken up by Horio and Kamen (1962). They found that washed chromatophores gave minimal rates of photophosphorylation which could be restored to a high rate by the addition of soluble factors present in the washings. Horio and Kamen (1962) point out that the soluble factors may (1) "restore concentrations of endogenous cofactors to magnitudes originally present in the chromatophores", (2) "create

[1] Present address: C.S.I.R.O., Fisheries and Oceanography Division, Cronulla, N.S.W., Australia.
[2] Present address: Botany Department, King's College, 68 Half Moon Lane, London, S.E.24.

an optimal reducing environment in which components of the electron transport system assume reduction levels best suited for optimal coupling to the phosphorylation process"—in a poising effect, and for (3) "bypass 'portions' of the electron transport system which are rate-limiting for photophosphorylation." Let us look principally at their evidence supporting (2) as a significant factor.

The addition under anaerobic conditions of ascorbate alone in the range between $10^{-3}$ M and $10^{-2}$ M reactivated the photophosphorylation by the washed chromatophores to rates similar to those obtained by adding the washings. When amounts of ascorbate above or below the range indicated were added, the rates of photophosphorylation decreased, i.e. a definite optimal ascorbate concentration was observed. Cytochrome $c_2$ is almost completely reduced, and *Rhodospirillum* haem-protein largely reduced in the presence of these concentrations of ascorbate. These two proteins appear to be important components of the electron transport chain of chromatophores. Horio and Kamen concluded that the oxidation-reduction potential within the chromatophore system when maximal photophosphorylation is taking place is about 0·0 volts.

These results are consistent with the idea that ascorbate acts as a redox balancing reagent, an idea further supported by the observation that under aerobic conditions the rate of photophosphorylation was markedly depressed at lower ascorbate concentrations but attained the same maximum rates as that obtained with an optimal concentration of ascorbate anaerobically on increasing the ascorbate concentration about tenfold. At even higher ascorbate concentrations in air the rate of photophosphorylation was diminished. Horio and Kamen also discussed the significance of cytochrome $c_2$ and the *Rhodospirillum* haem protein factors as important natural cofactors in the system, and pointed out that their data, in addition, indicates that phenazine methosulphate (PMS), which, when added to washed chromatophores, stimulated the rate of photophosphorylation to that obtained with ascorbate, and may function as a redox balancing reagent.

Nozaki, Tagawa and Arnon (1963) observed that fresh washed chromatophores from *Rhodospirillum rubrum*, which had been isolated under anaerobic conditions, did not require any addition, but could sustain a high rate of cyclic photophosphorylation alone when illuminated under anaerobic conditions. On ageing of the particles the rate of photophosphorylation decreased, but could be restored to the original rate by addition of ascorbate or other reductant. Similar results indicating the effect of various reducing substances on the redox balance of the chromatophore system were obtained by Bose and Gest (1963). An excess of ascorbate was inhibitory. Low concentrations of reduced pyridine nucleotide were effective in restoring the original rate to aged chromatophores, although an excess was inhibitory. Incubation with hydrogen gas in the presence of hydrogenase also restored the phosphorylating activity, but an excessive hydrogen treatment proved to be inhibitory. The addition of air to these systems depressed the photophosphorylation in a manner consistent with a redox balancing action for these substances. The restored photophosphorylation was antimycin-sensitive, like the original. The original rate of photophosphorylation was also restored to aged particles by the addition of phenazine methosulphate, but in this case the reaction was not antimycin sensitive. On the basis of the antimycin sensitivity, M. Nozaki. K. Tagawa and D. I. Arnon (unpublished observations, 1963) inclined to the view that the systems stimulated by ascorbate and the other reductants are as "physiological" as the fresh chromatophore system, and are subjected to redox balancing adjustments by ageing, air and reductants. On the contrary. the PMS-stimulated photophosphorylation was regarded as "artificial" and most probably represents not principally a redox balancing action but rather a bypass of the rate-limiting antimycin sensitive site.

By contrast, in chloroplasts the redox balancing has in the past appeared less significant than the bypass effect. Most workers have found little evidence for such effects, the cofactors apparently acting simply as bypass agents to re-establish cyclic electron flow. However, a few examples have been observed where the state of oxidation/reduction of an added cofactor (which may indirectly represent the state of an endogenous component with which it reacts) was important in establishing the onset of cyclic photophosphorylation.

Thus Jagendorf and Avron (1959) and Tsujimoto, Hall and Arnon (1961) observed that with sub-optimal concentrations of various cofactors there was a large stimulation in the presence of oxygen. While the participation of oxygen in a "pseudocyclic" electron flow could account satisfactorily for some of these observations it was not valid for the case of phenazine methosulphate. Whatley (1963) confirmed and extended the observations of Jagendorf and Avron (1959) that the rate of ATP formation at lower concentrations of phenazine methosulphate was much less under argon than in air. This is interpreted (Whatley, 1963) to mean that, under argon, electrons flowing from photoreaction B over-reduce the intermediates of the electron transport chain; and this slows down or prevents the return of electrons from photoreaction A via phenazine methosulphate to the chain. [Compare with Fig. 1 for ferredoxin-catalysed cyclic photophosphorylation—phenazine methosulphate occupies the place of ferredoxin, and oxygen plays a part equivalent to that of $NADP^+$.] Oxygen could counterbalance the over-reduction by oxidizing some portion of the intermediates and so bring about a regulation of the electron flow. Smaller amounts of oxygen produced a proportionately smaller response, although the action of oxygen was clearly catalytic. The fact that very small amounts of ferricyanide stimulated the cyclic photophosphorylation with phenazine methosulphate in a similar fashion emphasizes the idea that redox reagents other than oxygen can regulate ("poise") the system by partly oxidizing the intermediates. It appeared likely that even phenazine methosulphate itself in larger amounts could bring about the redox regulation of the electron transport chain. When photoreaction B was prevented by the addition of the specific inhibitor, CMU (*p*-chlorophenyl-dimethyl urea), the flow of electrons from water to the electron transport chain was stopped, and no over-reduction could occur. There was thus no need for oxygen as a regulator under argon in the presence of CMU. On the other hand, in the presence of air some contribution of electrons from water by photoreaction B appeared to be necessary in order to maintain a suitable redox balance. The addition of CMU caused a large inhibition of cyclic photophosphorylation in air, probably because oxygen over-oxidized the intermediates of the electron transport chain, thus preventing a supply of electrons to photoreaction A.

In the case of phenazine methosulphate the "poising" of the phosphorylating system appears connected with a situation in which over-reduction of the cofactor itself is the main danger. Other cases are known in which the problem is to reduce the cofactor enough to make it function in cyclic photophosphorylation. For example, Trebst and Eck (1961) found that under anaerobic conditions menadione (vitamin $K_4$) catalysed cyclic photophosphorylation, which was unexpectedly inhibited by CMU. If, however, the menadione-containing system was first illuminated briefly to reduce the vitamin $K_3$ prior to adding CMU no inhibition was found. If the reduced analogue, vitamin $K_5$, was substituted no inhibition by CMU under anaerobic conditions was found. This indicated that the cofactor must be adequately reduced before it would catalyse cyclic electron flow, a view confirmed by the observation that the addition of reduced pyridine nucleotide, itself unable to catalyse cyclic photophosphorylation, to the anaerobic menadione system rendered the photophosphorylation insensitive to CMU.

Similar results were obtained with the dye 2:6-dichlorophenolindophenol, which was shown to be a cofactor of cyclic photophosphorylation in the presence of CMU only if it was first reduced.

Recently Tagawa, Tsujimoto and Arnon (1963) found that illuminated chloroplast fragments can carry out an anerobic cyclic photophosphorylation catalysed by a naturally occurring chloroplast protein, ferredoxin. This reaction shows a number of characteristics which indicate the importance of an initial reduction of the cofactor before the onset of cyclic photophosphorylation becomes possible. In the reaction described by Tagawa *et al.* ferredoxin catalyses a cyclic electron flow which differs in an important fashion from the electron flow occurring when ferredoxin becomes the terminal acceptor in a non-cyclic photophosphorylation. It may be useful to summarize the steps involved in these electron flow systems.

The noncyclic electron flow appears to involve the participation of a sequence of two photoreactions, A and B. In photoreaction B electrons are donated in the light by water ($OH^-$) to a chloroplast pigment system, in which chlorophyll *b* predominates, and are accepted by a moderately reducing intermediate, believed to be plastoquinone. Oxygen is released as a result of the photo-oxidation of the water. In photoreaction A electrons are donated in the light from cytochrome *f* to a chloroplast pigment system, in which chlorophyll *a* predominates, and are accepted by ferredoxin. The reduced ferredoxin can subsequently donate its electrons to other acceptors in secondary dark reactions, for example, to $NADP^+$, and finally to $CO_2$. Photoreactions A and B are connected by a thermochemical bridge, the passage of electrons from reduced plastoquinone to cytochrome *f* being coupled with the production of ATP. These relationships are expressed in Fig. 1. Experimentally, photoreaction B can be replaced by an ascorbate/dichlorophenolindophenol dye couple which donates electrons chemically to the thermochemical bridge. In this modified noncyclic photophosphorylation only photoreaction A is involved, and the system is analogous to the *noncyclic* photophosphorylation of the photosynthetic bacteria.

In contrast to the combined participation of photoreactions A and B in the normal noncyclic electron flow, the electron flow in cyclic photophosphorylation requires only photoreaction A as an energy-converting reaction; photoreaction B is not required. This is analogous to the cyclic electron flow in photosynthetic bacteria.

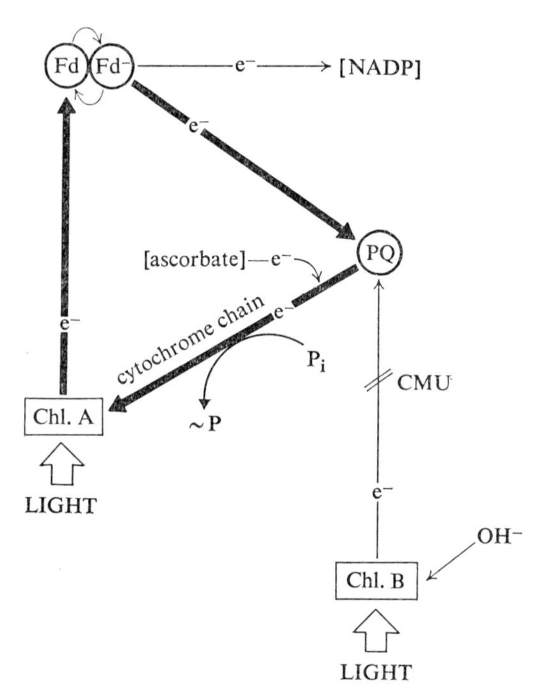


FIG. 1. A scheme to represent cyclic photosynthetic phosphorylation catalysed by ferredoxin. The cyclic electron flow is envisaged as being "primed" by electrons supplied by slow functioning of photoreaction B (inhibited by CMU). The ascorbate/dye couple can function as an alternative "priming" electron donor. $NADP^+$ can remove "priming" electrons by accepting them from $Fd^-$. Variations in the rate of accumulation of the electrons are reflected in variations in the time lag which preceeds the onset of cyclic electron flow. Ferredoxin is represented as two interacting "pools" of oxidized (Fd) and reduced ($Fd^-$) respectively. For details see the text.

The mechanisms for the cyclic electron flow catalysed by ferredoxin which Tagawa *et al.* have proposed may be summarized as follows (compare with Fig. 1). Electrons released in the light by photoreaction A are accepted by ferredoxin, which passes them on to the next component of the cyclic pathway, cytochrome $b_6$, as is suggested by the antimycin A sensitivity of the ferredoxin-catalyzed system. In turn, the reduced cytochrome $b_6$ directly or indirectly reduces cytochrome *f*. The reoxidation of cytochrome *f* by chlorophyll in photoreaction A completes the cycle, the operation of which brings about the production of ATP coupled with the passage of electrons along the cytochrome chain. In this formulation there is no need for photoreaction B to take place; and, in fact, its continued operation might well completely reduce the components of the electron flow pathway, thus rendering further cyclic electron flow impossible.

Experimentally, the cyclic photophosphorylation catalysed by ferredoxin was found to proceed (1) under strictly anaerobic conditions, (2) in the presence of the inhibitor CMU (*p*-chlorophenyldimethyl urea) which was used to suppress photoreaction B, (3) in the presence of a relatively large amount of

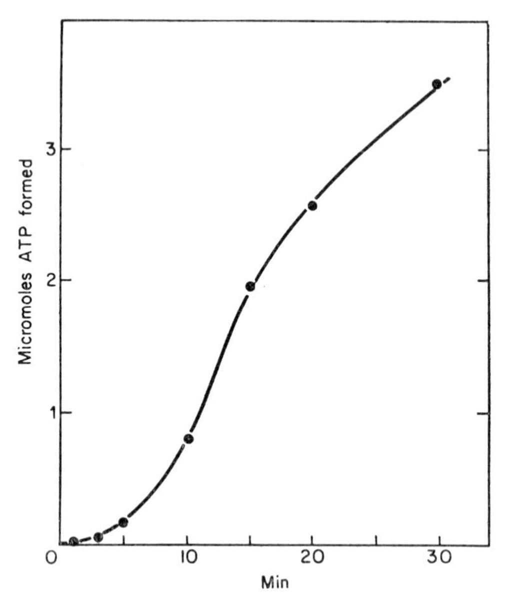


FIG. 2. Onset of ferredoxin-catalysed cyclic photophosphorylation. The reaction mixture contained, per 3 ml, chloroplasts ($C_{IS}$) equivalent to 0·2 mg chlorophyll, 1·5 mg purified ferredoxin and the following in micromoles: *tris* buffer pH 8·0, 100; ADP, 10; $MgCl_2$, 10; $K_2HPO_4$, 10; CMU, 0·06 (12 $\mu$g). The experiments were carried out as described under EXPERIMENTAL METHODS. The results are calculated as micromoles ATP formed per 3 ml reaction mixture to permit ease of comparison with the earlier experiments of Tagawa *et al.* (1963).

ferredoxin as a catalyst (up to 0·1 $\mu$moles) and (4) at a moderate light intensity, not exceeding 1000 f.c. These "optimal conditions" were determined in experiments in which cyclic phosphorylation was allowed to proceed for a definite time, after which the extent of the reaction was determined. In the present investigation the time course of the ATP formation under conditions similar to those above were studied and a consistent lag of several minutes (5 min or more) was observed before any ATP formation began. Following the lag the cyclic phosphorylation proceeded at a rapid rate for 15–20 min and finally slowed to a continuing but slower rate (see Fig. 2) after about 20 min illumination.

An attempt to interpret this unexpected progress curve has suggested (a) that the ferredoxin in the system must be first partly reduced before it can react at a significant rate with the component of the cyclic electron flow adjacent to it (perhaps cytochrome $b_6$), and (b) that the electrons needed for this initial "priming" reduction are supplied from water ($OH^-$) by the slow continuation of photoreaction B under the particular experimental conditions employed—this external source of "priming" electrons is needed because there are insufficient endogenous electrons in the systems.

The results reported in this paper show the effect of varying several of the experimental conditions used in Fig. 2. The results show striking effects on the lag in ATP formation which support the correctness of the theory proposed to account for this lag.

## Experimental Methods

Chloroplasts ($C_{IS}$) were prepared from spinach leaves as described by Whatley and Arnon (1963). To determine the progress curves for ATP formation under various conditions, complete reaction mixtures were prepared in sufficient amounts to permit a suitable number of points to be taken. One ml aliquots of the reaction mixture were placed in Warburg manometer flasks. The flasks were transferred to a 15° water bath and gassed for 3 min in the dark with pure argon. The reaction mixture was then illuminated with incandescent light (filtered through a red, Jena G2, filter) at 1000 foot candles, or at the light intensity specified. Vessels were removed from the light at selected intervals and 0·5 ml aliquots were taken. $AT^{32}P$ in the aliquots was measured by the method of Hagihara and Lardy (1960).

Photoreduction of ferredoxin was measured by illuminating the reaction mixture in a 1 cm cuvette with red light of approximately 1000 foot candles (passed through a Kodak Wratten filter, No. 29) and reading the optical density at 420 m$\mu$ in a Zeiss spectrophotometer connected to a recorder to give a continuous record of the photoreduction. The interfering actinic light was screened out by a blue filter (Kodak Wratten filter, No. 47B). The amount of ferredoxin in the reaction mixture was checked at the end of each experiment by the addition of $Na_2S_2O_4$ to obtain complete reduction.

## Results

### Effect of Light Intensity

When the light intensity was varied over a ten-fold range, but all other experimental conditions were kept the same as in Fig. 2, the results summarized in Fig. 3 were obtained. As the light intensity increased the duration of the lag prior to the onset of ATP formation diminished. However, once the "priming reaction" (the initial photoreduction of ferredoxin) was completed, the rate of phosphorylation was not much affected by varying the light intensity. The slopes of the rapidly rising portions of the time course curves at the three intensities employed were very similar, from which fact it may be concluded that the operation of photoreaction A readily supplies a saturating flow of electrons to the cyclic electron flow pathway even at the lowest light intensity used.

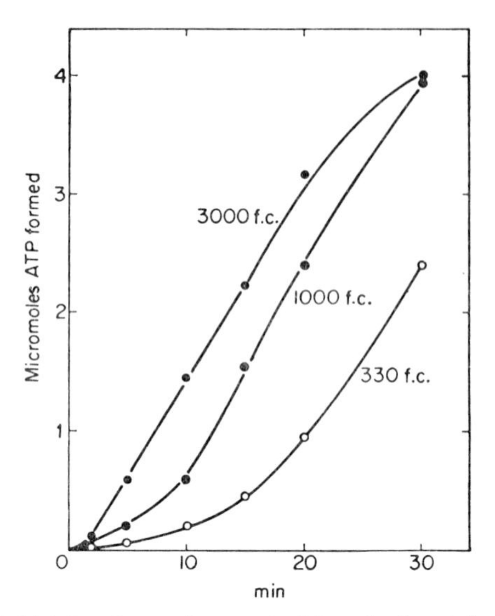


Fig. 3. Effect of light intensity on the onset of ferredoxin-catalysed cyclic photophosphorylation. Experimental details as described in legend to Fig. 2, except that the light intensity (incandescent filament source) was varied between 330 foot candles and 3000 foot candles.

## EFFECT OF CHLOROPLAST CONCENTRATION

When the amount of chloroplast material added to the reaction mixture was varied but all other experimental conditions were maintained as in Fig. 2, the results summarized in Fig. 4 were obtained. As the chloroplast concentrations increased, the lag became shorter. However, the rate of ATP formation, when once established, was greater as the chloroplast concentration was increased, presumably since the insoluble catalytic components of the cyclic electron flow were now being added in larger amounts.

## EFFECT OF CMU CONCENTRATION

When the concentration of CMU in the reaction mixture was varied while the other conditions were maintained as in Fig. 2, the results summarized in Figs. 5 and 6 were obtained. Two ranges of CMU concentration were employed, above (Fig. 5) and below (Fig. 6) an arbitrary "*standard*" of 15 $\mu$g per vessel. The experiments showed that an increase of the CMU above the amount used in the "*standard*" increased the duration of the lag before ATP formation began. In the presence of 120 $\mu$g CMU the lag was extended to more than the 30 min of illumination used in this experiment and the cyclic photophosphorylation seemed to be completely "inhibited". When the CMU was decreased

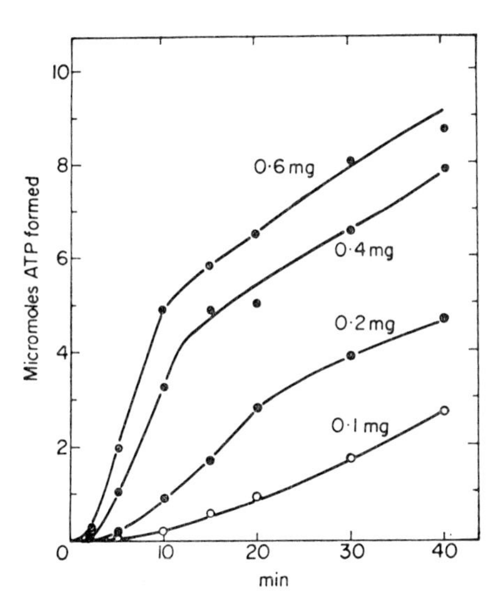


FIG. 4. Effect of chloroplast concentration on the onset of ferredoxin-catalysed cyclic photophosphorylation. Experimental conditions as described in legend to Fig. 2, except that the chlorophyll content of the chloroplast suspension added was varied between 0·1 mg and 0·6 mg per 3 ml reaction mixture.

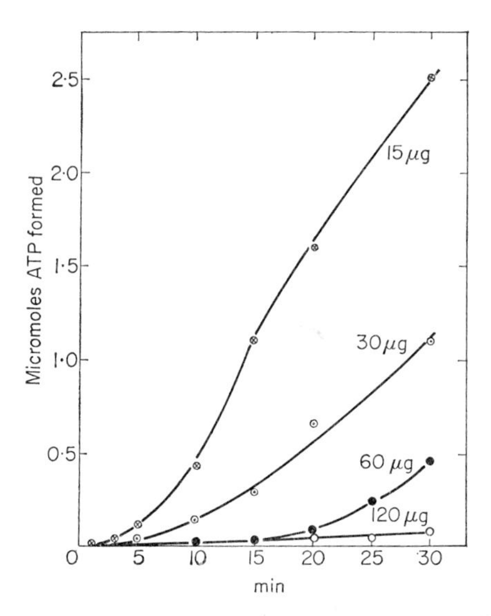


FIG. 5. Effect of CMU concentration on ferredoxin-catalysed cyclic photophosphorylation. Experimental conditions as described in legend to Fig. 2 except that the CMU concentration was varied as indicated on the graph.

below the "*standard*" the lag was sharply decreased. However, the rate of ATP formation was now markedly depressed after proceeding for a relatively short time at the maximum rate. This phenomenon will be discussed later.

CMU was shown by Jagendorf and Avron (1959) to be an inhibitor of PMS-catalysed cyclic photophosphorylation at a concentration higher than that required to inhibit the Hill reaction with ferricyanide almost completely. This inhibition was observed in a system with washed chloroplast fragments; on the readdition of a protein component isolated from the washings a reversal of the CMU inhibition of cyclic photophosphorylation took place. In the experiments of Figs. 5 and 6, unwashed chloroplast fragments were used, and an inhibition of the cyclic photophosphorylation system itself by CMU would

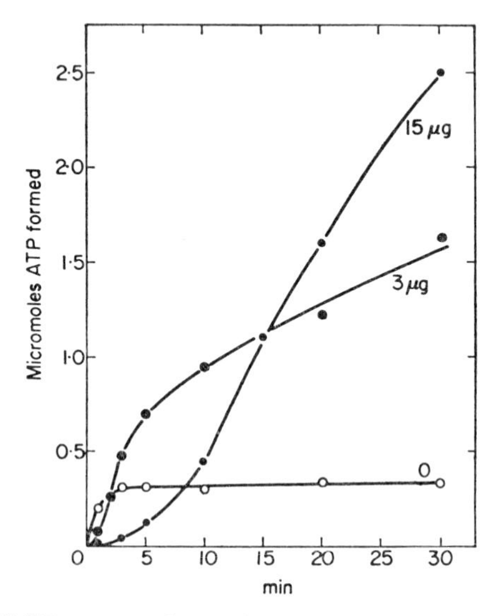


FIG. 6. Effect of CMU concentration on ferredoxin-catalysed cyclic photophosphorylation. Experimental conditions as described in legend to Fig. 2 except that the CMU concentration was varied as indicated on the graph.

therefore be unlikely. The apparent inhibition by the largest amount of CMU (120 μg) would therefore seem most probably due to the extension of the lag beyond the total time of the experiment. In view of the results summarized in Fig. 4, we considered it likely that increasing the amount of chloroplast material (which decreases the duration of the lag) would counteract the effect of CMU (which increases the duration of the lag). An experiment was therefore carried out in which 120 μg CMU was present in all vessels but in one set the amount of chloroplast material was kept the same as in Fig. 5, and in the second set it was increased three-fold. The results are shown in Fig. 7. A simple increase in chloroplast concentration was sufficient to decrease the lag time to less than 10 min, and subsequently cyclic photophosphorylation took place at a rapid rate. It can therefore be concluded that, under the experimental conditions of Figs. 5 and 7, CMU decreases the amount of cyclic photophosphorylation not primarily by inhibiting the cyclic electron flow itself but rather by greatly increasing the duration of the lag.

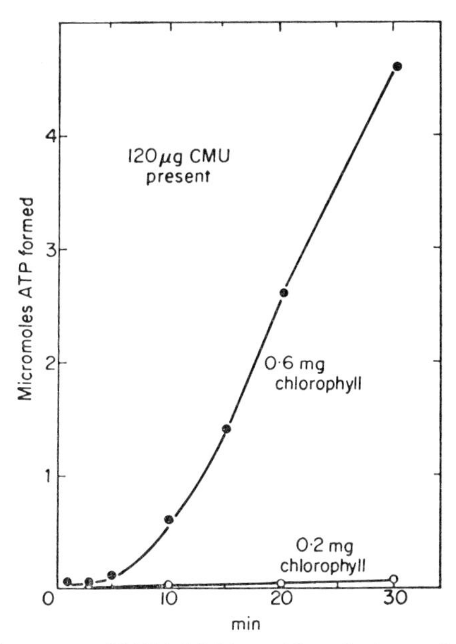


FIG. 7. Reversal of apparent CMU inhibition of ferredoxin-catalysed cyclic photophosphorylation. The reaction mixture contained, per 3 ml, chloroplasts ($C_{IS}$), equivalent to 0·2 mg or 0·6 mg chlorophyll, as noted, 1·5 mg purified ferredoxin, and the following in μmoles: *tris* buffer, pH 8·0, 100; ADP, 10; $MgCl_2$, 10; $K_2H^{32}PO_4$, 10; CMU, 0·67 (120 μg): illumination at 1000 foot candles. Other conditions as described in the section on EXPERIMENTAL METHODS.

## EFFECT OF SMALL AMOUNTS OF $NADP^+$

Reduced ferredoxin has a very high affinity for $NADP^+$ and is rapidly and completely reoxidized in the presence of low concentrations of $NADP^+$ (Whatley, Tagawa and Arnon, 1963). The reaction requires the presence of the enzyme, ferredoxin-$NADP^+$ reductase (Shin and Arnon, 1964), which is a normal constituent of chloroplasts. In the presence of $NADP^+$, therefore, reduced ferredoxin should not accumulate, and the duration of the lag should be extended for as long as there is any $NADP^+$ remaining. The initial rate of the "priming reaction" (the initial photoreduction of ferredoxin) is very slow—the total amount of ferredoxin added to the system is only 0·1 μmole (see also Fig. 10). An amount of $NADP^+$ equivalent to the ferredoxin would therefore be expected to prevent the accumulation of reduced ferredoxin long enough at least to double the duration of the lag. The results shown in Fig. 8 indicate that this is so. The addition of 0·05 μmoles $NADP^+$ to the system markedly increased the lag, while very obviously not affecting the rate of ATP formation once the lag had been overcome.

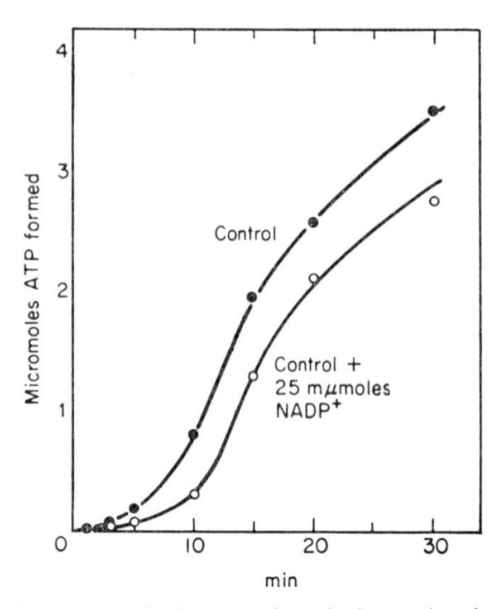


FIG. 8. Effect of small amounts of $NADP^+$ on ferredoxin-catalysed cyclic photophosphorylation. Experimental details for the "control" curve exactly as described in legend to Fig. 2. Twenty-five mμmoles $NADP^+$ were added per ml reaction mixture for the "control+ $NADP^+$" curve.

## EFFECT OF ASCORBATE/DYE COUPLE

In the preceding experiments the conditions which have been varied have all, according to the theory briefly proposed above to explain the lag in the onset of ATP formation, caused a change in the rate of operation of photoreaction B. But it has already been shown experimentally that photoreaction B can be replaced by the ascorbate/dye couple in a number of reactions (Losada, Whatley and Arnon, 1961; Vernon and Zaugg, 1960). In the system described in Fig. 2, in which there was a 5 min lag, photoreaction B was largely inhibited by CMU. On adding ascorbate/dye to this system there was a complete elimination of the lag, as far as could be experimentally determined, and ATP formation started at once. The results of an experiment in which the ascorbate/dye couple was added are summarized in Fig. 9. It will be noted from the figure that after a short time the rate of formation in the "ascorbate/dye system" decreased and came to a stop after 10 min, whereas the "*control*" system continued to produce ATP until the total extent of phosphorylation in the "*control*" in which there was a considerable lag, was finally greater than that with the "ascorbate/dye system", where there was no lag.

In a parallel experiment, shown in Fig. 10, the rate of photoreduction of ferredoxin in "*control*" and "*ascorbate/dye*" systems similar to those in Fig. 9 was determined. As is clearly indicated, the complete photoreduction of 0·1 μmole ferredoxin was accomplished in 30 sec in the "ascorbate/dye" system but only after 10 min in the "control", in which photoreaction B was largely suppressed with CMU.

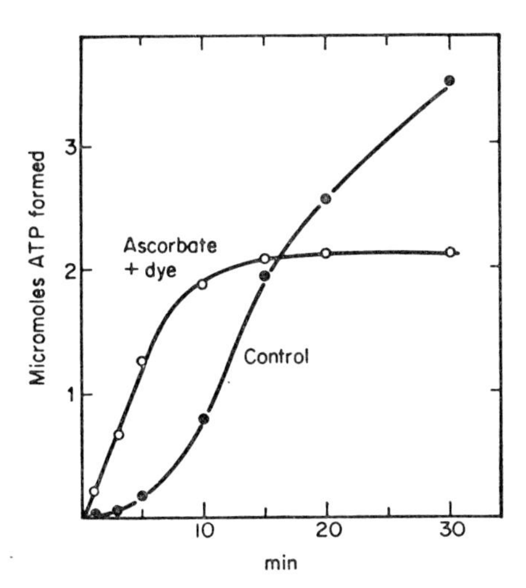


FIG. 9. Effect of ascorbate/dye couple on the onset of ferredoxin-catalysed cyclic photophosphorylation. Experimental details for the "control" curve exactly as described in legend to Fig. 2. The "ascorbate/dye system" received in addition 20 $\mu$moles sodium ascorbate and 0·2 $\mu$moles dichlorophenolindophenol.

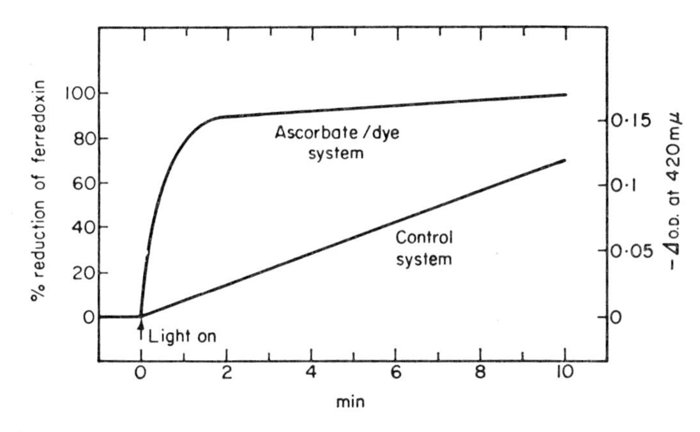


FIG. 10. Photoreduction of ferredoxin in a cyclic photophosphorylation system. The reaction mixture labelled "control system" contained, in a final volume of 3 ml, chloroplasts ($P_{IS}$) equivalent to 0·1 mg chlorophyll, 1·5 mg purified ferredoxin (0·1 $\mu$mole), and the following in micromoles: *tris* buffer, pH 8·0, 50; ADP, 5; $K_2HPO_3$, 5; $MgCl_2$, 5; and CMU. 0·06 (12 $\mu$g). In this system the electrons used to reduce ferredoxin originate in water. In addition to the components listed above the ascorbate/dye system received 20 $\mu$moles ascorbate+0·2 $\mu$mole dichlorophenolindophenol. In this system ascorbate is the principal electron donor. The reaction mixture was placed in a 1 cm cuvette fitted as a Thunberg tube, and was made anaerobic by four repeated cycles of evacuation and filling with purified argon Other details are noted in the section on EXPERIMENTAL METHODS.

## Discussion

The evidence shows clearly that under specified experimental conditions there is a time lag before the onset of ferredoxin-catalysed cyclic photophosphorylation. By altering the conditions, the duration of the lag could be influenced in a manner consistent with the proposal that the ferredoxin added must first be partly reduced before it can react at significant rate with cytochrome $b_6$, or another unidentified component adjacent to it in the electron flow pathway. It is suggested that until the ferredoxin is sufficiently reduced there will be no measurable electron flow by the cyclic pathway and hence no ATP formation. The lag would thus correspond to the time taken for the ferredoxin to become sufficiently reduced. It is not implied that the redox state of ferredoxin is necessarily the critical factor determining the onset of cyclic photophosphorylation; other components of the electron flow pathway which may be in equilibrium with externally added ferredoxin could well be the significant limiting factors.

According to the proposed cyclic electron flow pathway suggested earlier (Arnon, 1961), which does not require any external supply of electrons to continue, it would seem likely that the operation of photoreaction A alone would reduce the ferredoxin sufficiently to enable it to donate electrons to cytochrome $b_6$ to complete the cyclic electron flow. However, the amount of ferredoxin experimentally added, although catalytic in quantity, is 10–100 times larger than the amounts of the other components of the electron flow pathway, which are added to the reaction mixture as insoluble components of the chloroplasts. The insoluble components appear to be unable to photoreduce the ferredoxin sufficiently for cyclic electron flow to begin and the ferredoxin acts as a "redox buffer" against the onset of the cyclic photophosphorylation. In order for a degree of ferredoxin reduction to take place which is sufficient to allow cyclic electron flow, a source of electrons external to the cyclic electron flow pathway must be available. This source may be water, donating electrons through photoreactions B and A, or the ascorbate/dye couple, donating electrons through photoreaction A alone. If it is to be water, then the rate of operation of photoreaction B must be small (e.g. by adding CMU), or the limited noncyclic electron flow needed to photoreduce the ferredoxin will be exceeded and the system will become over-reduced. The initial photoreduction of ferredoxin by a noncyclic electron flow thus has the character of a "priming reaction". Once the ferredoxin is sufficiently reduced, the cyclic photophosphorylation obtains all its energy exclusively from photoreaction A and requires no continued input of electrons from an external source, such as the ascorbate/dye couple, or from water via photoreaction B. The actual amount of electron flow which has to take place in the priming reaction is small (less than 0·1 $\mu$moles of electrons!) and the priming reaction itself can proceed at a slow rate which is insignificant by comparison with the rate of electron flow established in the cyclic photophosphorylation system.

In the time courses observed a sequence of three phases in the progress of ATP formation can be distinguished: (1) an initial portion, the lag, in which the significant reaction is thought to be the initial reduction of ferredoxin by a noncyclic electron flow, (2) a phase of rapid ATP formation which occurs when the cyclic electron flow has become established following Phase 1, and (3) a terminal portion, in which the rate of ATP formation becomes adjusted to a new lower rate.

The duration of Phase 1, the lag, could be experimentally shortened by any treatment which permitted a more rapid initial photoreduction of ferredoxin. An increase in the light intensity (Fig. 3), an increase in the chloroplast concentration (Fig. 4) or a decrease in the CMU concentration (Figs. 5 and 6), all shortened the lag, in agreement with the idea that all these experimental treatments increased the rate at which photoreaction B operated to supply electrons for the initial priming reduction of ferredoxin. Alternatively, the addition of the ascorbate/dye couple could effectively substitute for photoreaction B, and reduce the duration of the lag to a very short time. Phase 1, the lag, could be lengthened by any treatment which slowed down the initial photoreduction of ferredoxin. A decrease in the light intensity (Fig. 3), a decrease in the chloroplast concentration (Fig. 4) or an increase in the CMU concentration (Figs. 5 and 6), all lengthened the lag, presumably by decreasing the rate at which photoreaction B was operating. Alternatively, the addition of a small amount of $NADP^+$ prevented the accumulation of reduced ferredoxin (even though the ferredoxin was being continuously reduced in the system) by causing its re-oxidation as long as any $NADP^+$ remained oxidized. The onset of cyclic photophosphorylation was consequently delayed by the addition of $NADP^+$, an observation quite consistent with the suggestion that the ferredoxin must be partly reduced before it can catalyse cyclic electron flow.

The duration of Phase 2 was variable and was related to the duration of Phase 1. However, the maximum *rate* of ATP formation was always approximately the same (compare especially the slopes in Figs. 3, 5, 8 and 9) when the chlorophyll and ferredoxin concentrations were kept constant. This indicates that Phase 2 represents the activity of the ferredoxin-catalysed cyclic photophosphorylation system itself under optimal conditions. The maximum rate of ATP formation attained in Phase 2 was not affected by the duration of the lag, Phase 1.

When the lag was experimentally made very short, cyclic photophosphorylation ceased a short time after it had started. This was observed either in the absence of CMU (Fig. 6) or when the ascorbate/dye couple was used in place of photoreaction B in the "priming reaction". It is probable that the rapid flow of electrons to ferredoxin, which at first quickly establishes a suitable redox state in the system (and thus virtually abolishes the lag), soon brings about an over-reduction of the components of the system, causing the cyclic electron flow and, hence, ATP formation to cease.

In these two cases Phase 3 of the time course is thus represented by a complete cessation of ATP formation, which supervenes after a rather short time.

When the initial photoreduction of ferredoxin was experimentally made slower (corresponding to lags of intermediate duration) the duration of Phase 2 was longer, and when Phase 3 was established the formation of ATP continued for some time at a rate markedly slower than in Phase 2. These observations (see especially Figs. 2 and 6) suggest that when the rate of photoreaction B was limited by the use of CMU the tendency for it to over-reduce the system became correspondingly less. The ferredoxin thus retained a suitable degree of reduction for a longer time, thus allowing Phase 2 to occupy an extended period of time. Phase 3 of the time course under these conditions may thus correspond to a secondary equilibrium position of the redox balance in which ferredoxin and the other components of the cyclic electron flow chain are reduced a little beyond the optimum redox state for cyclic electron flow and ATP formation.

In summary we may comment that chromatophores of *Rhodospirillum rubrum* show very clearly the profound effects of redox buffers. Ascorbate shows an optimal concentration at which it stimulates cyclic photophosphorylation, the chromatophores becoming "over-reduced" at higher concentrations. Other reductants behave similarly. The bypassing of rate-limiting steps by the addition of cofactors such as phenazine methosulphate is a less important though still dramatic way of stimulating ATP formation by chromatophores.

By contrast the redox state of a number of cofactors of cyclic photophosphorylation by isolated chloroplasts has been shown to be significant for the onset of electron flow only under closely specified experimental conditions. We have noted the need to prevent over-reduction of phenazine methosulphate and the requirement for vitamin $K_3$ (but not its reduced analogue, vitamin $K_5$) and for dichlorophenolindophenol to be largely in the reduced state before they can bring about cyclic electron flow. To these we can now add the evidence about the need for an initial "priming" reduction for ferredoxin.

However, chloroplasts do not normally show the profound effects due to added redox regulators which are observed with chromatophores. The most striking effects of adding substances to chloroplasts are due to the participation of these substances as cofactors in bypass reactions which circumvent rate-limiting steps in the electron flow pathway as it occurs in the isolated chloroplast. The experiments with ferredoxin are of particular interest since they concern the redox state of a natural protein component participating in a reaction which is sensitive to antimycin and perhaps of physiological importance.

## References

Allen, M. B., Whatley, F. R. and Arnon, D. I. (1958). *Biochim. biophys. Acta* **27**, 16.
Arnon, D. I. (1959). *Nature, Lond.* **184**, 10.
Arnon, D. I. (1961). *Bull. Torrey Bot. Club* **88**, 215.
Arnon, D. I., Allen, M. B. and Whatley, F. R. (1954). *Nature, Lond.* **174**, 394.
Bose, S. K. and Gest, H. (1963). *Proc. natn. Acad. Sci. U.S.A.* **49**, 337.
Frenkel, A. W. (1954). *J. Am. chem. Soc.* **76**, 5568.
Frenkel, A. W. (1956). *J. biol. Chem.* **222**, 823.
Hagihara, B. and Lardy, H. A. (1960). *J. biol. Chem.* **235**, 889.
Horio, T. and Kamen, M. D. (1962). *Biochemistry* **1**, 144.
Jagendorf, A. T. and Avron, M. (1959). *Archs Biochem. Biophys.* **80**, 246.
Losada, M., Whatley, F. R. and Arnon, D. I. (1961). *Nature, Lond.* **190**, 606.
Shin, M. Tagawa, K. and Arnon, D. I. (1963). *Biochem. Z.* **338**, 84.
Tagawa, K., Tsujimoto, H. Y. and Arnon, D. I. (1963). *Proc. natn. Acad. Sci. U.S.A.* **49**, 567.
Trebst, A. and Eck, H. (1961). *Z. Naturf.* **16b**, 455.

Tsujimoto, H. Y., Hall, D. O. and Arnon, D. I. (1961). *In* "Light and Life" (W. D. McElroy and B. Glass, eds.), p. 489. Johns Hopkins Press.
Vernon, L. P. and Zaugg, W. S. (1960). *J. biol. Chem.* **235**, 2728.
Whatley, F. R. and Arnon, D. I. (1963). *In* "Methods of Enzymology" (S. P. Colowick and N. O. Kaplan, eds.), Vol. VI, p. 308. Academic Press, New York and London.
Whatley, F. R. (1963). *In* "Photosynthetic Mechanisms in Green Plants", p. 243. Publication 1145, National Academy of Sciences—National Research Council.
Whatley, F. R., Tagawa, K. and Arnon, D. I. (1963). *Proc. natn. Acad. Sci. U.S.A.* **49**, 266.

# Nicotinamide Adenine Dinucleotide Phosphate-Cytochrome *f* Reductase of Chloroplasts

Giorgio Forti and Giuliana Zanetti
*Laboratorio di Fisiologia Vegetale, Istituto di Scienze Botaniche, Milano, Italy*

## Identity of Cytochrome *f* Reductase with Ferredoxin–NADP Reductase, Diaphorase and Transhydrogenase

The flavoprotein of chloroplasts discovered by Avron and Jagendorf (1956) as a NADPH-diaphorase has subsequently been found to have a number of other enzymic activities. Keister *et al.* (1960) have shown that the flavoprotein is endowed with transhydrogenase activity, and have demonstrated that it is an absolute requirement for $NADP^+$ photoreduction by chloroplasts (San Pietro, 1963; San Pietro and Keister, 1962). More recently, Shin and his collaborators (Shin *et al.*, 1963; Shin and Arnon, 1965) have crystallized the flavoprotein and shown that the transhydrogenase, diaphorase and ferredoxin-NADP reductase activities are associated with the same flavoprotein. These authors emphasize the fact that the NADP reductase activity is the most important physiological function of the enzyme in photosynthetic electron transport (Shin and Arnon, 1965).

It was demonstrated in this laboratory (Forti *et al.*, 1963) that the chloroplasts have a high NADPH cytochrome *f* reductase activity and that this activity is purified together with the transhydrogenase of Keister *et al.* (1960). It was also shown that the ratio between cytochrome *f* reductase, diaphorase and transhydrogenase activities is approximately the same in washed chloroplasts as in the purified flavoprotein (Forti *et al.*, 1963). The association of these three activities and of ferredoxin-NADP reductase with the flavoprotein enzyme crystallized according to Shin *et al.* (1963) has now been demonstrated. Table I shows that a rabbit antibody against transhydrogenase (kindly supplied by Dr. A. San Pietro) inhibits to the same extent the diaphorase, transhydrogenase and NADPH-cytochrome *f* reductase activity of the crystalline enzyme, and $NADP^+$ photoreduction by chloroplasts supplied with excess ferredoxin. An antibody prepared against the crystalline enzyme gave the same results.

Table II shows the activities of the crystalline enzyme. It can be seen that oxidized cytochrome *f* is a very efficient electron acceptor, the turnover number of the reaction being from 4000 to 5000. It would be interesting to compare this figure with the turnover number of the ferredoxin-$NADP^+$ reductase activity; unfortunately, this is not yet known.

The measurement of the flavoprotein enzyme catalysis of electron transfer from reduced ferredoxin to cytochrome *f* would be of interest. However, such

Table I. Inhibition of crystalline spinach flavoprotein activities by a rabbit antibody

| Additions | | Activities | | | |
|---|---|---|---|---|---|
| Enzyme ($\mu$l) | Antibody (ml) | Diaphorase activity (m$\mu$ equivalents/min) | Cytochrome *f* reductase (m$\mu$ equivalents/min) | Transhydrogenase (m$\mu$ moles/min) | NADP+ photoreduction (m$\mu$ moles/min) |
| 4 | 0·00 | 88 | 2·00 | 0·48 | 35·5 |
| 4 | 0·05 | 42 | 1·06 | 0·26 | 14·1 |

Conditions: the activities were measured at 20°, in the presence of *tris* buffer 0·05 M, pH 8. Diaphorase, cyt. *f* reductase and transhydrogenase activities were measured as previously described (Forti *et al.*, 1963). Cyt.$f^{ox}$ was $2 \times 10^{-5}$ M. NADP+ photoreduction was measured in the presence of *Tetrargonia* chloroplasts containing 32 $\mu$g of chlorophyll, and containing NADPH-diaphorase activity corresponding to the reduction of 86 m$\mu$moles of ferricyanide per minute. Purified spinach ferredoxin was added. The flavoprotein enzyme was crystallized according to Shin *et al.* (1963). Cytochrome *f* was purified from parsley as described elsewhere (Forti *et al.*, 1965a).

Table II. Activities of the crystalline chloroplast flavoprotein

| Electron donor | Electron acceptor | Specific[3] activity | Turnover number | Reference |
|---|---|---|---|---|
| NADPH 0·5 mM | Ferricyanide 0·8 mM | 280 | 18,000 | Zanetti and Forti, 1965 |
| | Ferricytochrome[1] *f* 0·3 mM | 45 | 4000–5000 | Zanetti and Forti, 1965 |
| | Ferricytochrome[2] c 1·5 mg/ml | 0·00 | — | Zanetti and Forti, 1965 |
| | NAD+ 1 mM | 2·55 | ? | Zanetti and Forti, 1965 |
| | $O_2$ (air) | 0·28 | ? | Zanetti and Forti, 1965 |
| | DCPIP 0·05 mM | 91 | ? | Zanetti and Forti, 1965 |
| NADH 1 mM | Ferricytochrome *f* 0·3 mM | 0·000 | — | Forti *et al.* (1963) |
| Reduced ferredoxin 54 $\mu$g/ml | NADP+ 0·2 mM | *ca* 340 | ? | Calculated from Fig. 2 of Shin and Arnon, 1965 |

[1] Parsley cytochrome *f*, purified according to Forti *et al.* (1965).
[2] Horse heart cytochrome *c*.
[3] Microequivalents × min$^{-1}$ mg protein$^{-1}$.

activity cannot be easily measured, because ferredoxin even in the oxidized state reduces ferricytochrome *f*. This is shown in Table III.

It seems most probable that this reaction is due to "labile" –SH groups of ferredoxin (see Fry and San Pietro, 1963). It can be seen that the decrease of

Table III. Reduction of cytochrome *f* by ferredoxin

| Addition | | | |
|---|---|---|---|
| Ferredoxin[1] (mμ moles) | Cyt. $f^{ox}$ haematin (mμmoles) | $-\Delta$ O.D. 465 mμ | % Reduction of cyt. *f* |
| 12 | 7 | 0·063 | 100 |
| 12 | 14 | 0·114 | 100 |
| 48 | 7 | 0·059 | 100 |

Conditions: *tris* buffer 0·05 M, pH 8—Final volume 1 ml—Temp. 25°.

[1] Ferredoxin purified according to San Pietro (1963). A molecular weight of 18,000 is assumed (Appella and San Pietro, 1961). The reduction of cytochrome *f* was measured as the 555 mμ absorption change (Davenport and Hill, 1952; Forti *et al.*, 1965).

absorbance at 465 mμ, due to the disappearance of "labile –SH" of ferredoxin (Fry and San Pietro, 1963), is stoicheiometrically related to the cytochrome *f* reduced. On the other hand, it is well known that ferricytochrome *f* is rapidly reduced, at neutral or slightly alkaline pH, by sulphydryl compounds.

## Properties of the Flavoprotein Enzyme

### ABSORPTION SPECTRUM

The chloroplast flavoprotein has a typical flavin absorption spectrum, with peaks at 275 mμ, 385 mμ, 456–460 mμ and a shoulder at 485 mμ (Shin *et al.*, 1963; Zanetti and Forti, 1965). On reduction with dithionite in excess or with the system $H_2$-hydrogenase-ferredoxin the spectrum of a fully reduced FAD-flavoprotein appears (Shin and Arnon, 1965). However, when the enzyme is reduced anaerobically with excess NADPH full reduction is not achieved, and a large absorption band above 530 mμ appears (Shin and Arnon, 1965). If the enzyme is reduced (anaerobically) by NADPH in the presence of 4 M urea, the absorption at 520–550 mμ appears initially, accompanied by a small bleaching of the 460 mμ peak and the 485 mμ shoulder; subsequently, the long-wavelength band disappears, and further bleaching at the flavin peak (460 and 485 mμ) occurs, until the typical spectrum of a fully reduced flavoprotein is obtained with excess NADPH (Zanetti and Forti, 1965). The appearance of an absorption band at 530–550 mμ upon partial reduction, and its disappearance when excess reductant (in this case, NADPH) is added is suggestive of the formation of a semiquinone form of FAD (Massey and Gibson, 1963; Masters *et al.*, 1965). The formation of the semiquinone $FADH^0$ is considerably faster than the further reduction to the level of $FADH_2$ (Zanetti and Forti, 1965);

this might be indicative of an important role of the semiquinone state in the catalytic activity.

## INACTIVATION BY NADPH AND BY UREA

It was shown (Zanetti and Forti, 1965) that the diaphorase activity of the enzyme, but not the cytochrome *f* reductase activity, is inactivated if the enzyme is preincubated a few minutes with NADPH in large molar excess over

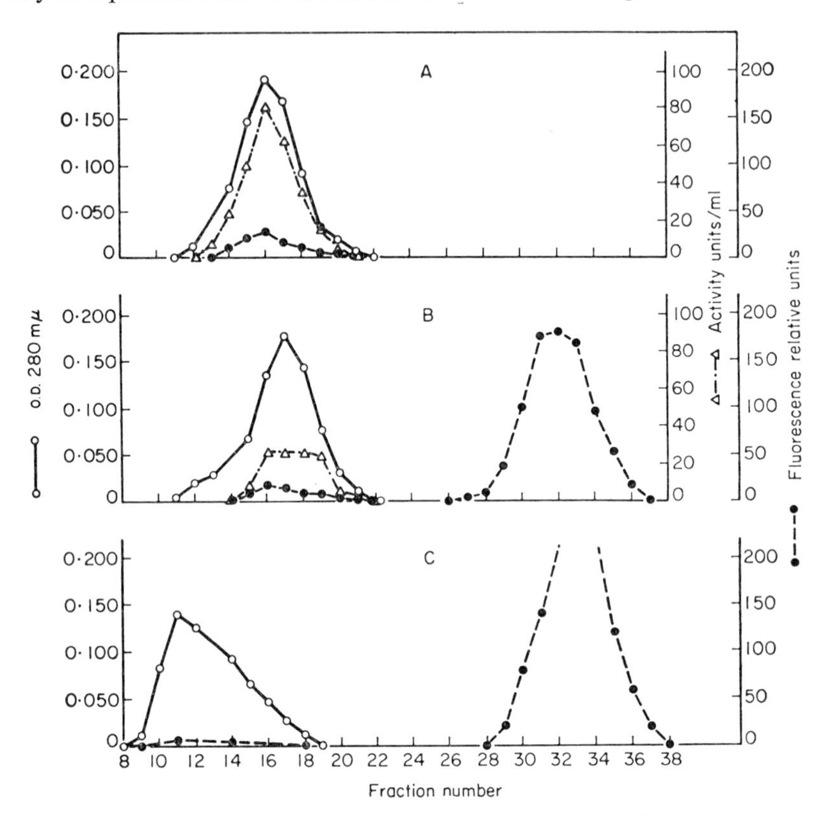


FIG. 1. Inactivation of the enzyme and splitting of FAD from the afoenzyme upon reduction with NADPH in the presence of 4 M urea. Conditions: 640 μg of enzyme were incubated at 25°, for 2 hr, in a final volume of 0·4 ml in *tris* buffer 0·05 M, pH 8. Other additions: A, none; B: NADPH 0·5 mM, urea 4 M, atmosphere: air—C: NADPH 0·5 mM, urea 4M, atmosphere: $N_2$. Zt the end of the preincubation, the mixtures were applied on a column (0·6 cm × 80 cm) of Sephadex G-100, equilibrated with *tris* buffer, 0·05 M ,pH 8. Elution with the same buffer. FAD fluorescence was measured in a spectrophotofluorimeter as the light emitted at 535 mμ upon excitement with 450 mμ light.

the enzyme. Such inactivation is completely prevented and also reversed by $NADP^+$ in a one-to-one ratio with NADPH, and is inversely proportional to the enzyme concentration. However, if the pretreatment with NADPH is performed in the presence of urea (4 M), a concentration which by itself gives very little inhibition, the inactivation by NADPH becomes complete and irreversible (Zanetti and Forti, 1965). Under these conditions, as was discussed above, the enzyme is fully reduced by NADPH; however, even if the enzyme is reoxidized, the activity is not recovered to any extent (Zanetti and Forti, 1965). These observations lead to the conclusion that enzyme-bound FAD in the oxidized but not in the reduced state, and hydrogen bonds (urea-sensitive bonds) are operative in preserving the structural integrity of the enzyme required for activity. Treatment of the enzyme with NADPH and 4 M urea does indeed split FAD from the enzyme, as shown in Fig. 1.

The denatured apoenzyme cannot be reactivated by addition of FAD. Figure 1A shows the chromatography of native enzyme on a Sephadex G-100 dextran column. It can be seen that the enzyme is very weakly fluorescent but all the FAD fluorescence observable is associated with the protein. In Fig. 1B,

Table IV. Inhibition by N-ethylmaleimide of the diaphorase activity

| Additions during preincubation | Preincubation time | | |
|---|---|---|---|
| | 4 min | 30 min (% Inactivation) | 70 min |
| NEM[1] 3 mM | 6 | 27 | 68 |
| NADPH 0·5 mM | 53 | 56 | 58 |
| NADPH 0·5 mM+NEM 3 mM | 95 | 100 | — |
| Urea 4 M | 5 | 9 | — |
| Urea 4 M+NEM 3 mM | 19 | 35 | — |

Conditions: preincubation was performed at 30°, in 1 ml spectrophotometer cuvettes containing 2·8 $\mu$g of enzyme, *tris* buffer 0·05 M, pH 8, EDTA 0·5 mM and the other additions as indicated. At the end of preincubation, ferricyanide and NADPH were added to start the diaphorase reaction, at the final concentration of 0·8 mM and 0·5 mM, respectively.

[1] = N-ethylmaleimide.

the enzyme incubated in air with NADPH and 4 M urea is chromatographed on the same column. It can be seen that about 67% of the activity is lost and a considerable amount of FAD (detected by its fluorescence) is separated from the protein. In Fig. 1C the chromatography, on the same column, of the enzyme treated for 2 hr anaerobically with NADPH (0·5 mM) and 4 M urea is shown. The protein, completely devoid of any enzymatic activity, is eluted with a smaller volume of effluent buffer as compared to the native enzyme, and all the FAD is eluted in fractions 29 to 38 with a peak at fraction 33. A sample of authentic FAD, applied to the same column, was eluted in the same fractions. The nature of the binding of FAD to the apoenzyme is an interesting problem for future investigation. However, some indication of the involvement of a sulphur atom is provided by the studies on activity inhibition by N-ethylmaleimide (NEM), the well-known sulphydryl reagent. As shown in Table IV, inactivation by NEM occurs slowly in the absence of NADPH, but is immediate if the enzyme is treated with NADPH and NEM simultaneously.

Also urea accelerates the inhibitory action of NEM, but the effect is much slower than the one of NADPH. These observations are favourable to the hypothesis that the reduction by NADPH of a disulphide and/or a sulphur atom of the protein possibly bound to FAD is required to permit the binding of N-ethylmaleimide to the sulphydryl, which results in enzyme inactivation.

## Conclusion

The demonstration of the identity of NADPH-cytochrome *f* reductase with ferredoxin-$NADP^+$ reductase of chloroplasts, provided here and elsewhere (Zanetti and Forti, 1965) emphasizes the major role of this flavoprotein enzyme in photosynthetic electron transport. The physiological role of the NADPH-cytochrome *f* reductase is still a subject of speculation; however, the high turnover number of this activity is favourable to a major role of it in photosynthetic electron transport.

The enzyme is an FAD flavoprotein with a molecular weight of between 40,000 and 45,000, as determined by gel filtration and FAD content (Zanetti and Forti, 1965). An interesting property of the enzyme is that it is completely and irreversibly inactivated upon reduction by NADPH in the presence of a relatively low concentration of urea, which by itself is without effect (Zanetti and Forti, 1965). It has been demonstrated here that this type of inactivation is accompanied by the splitting of FAD from the apoenzyme (see Fig. 1). The denatured apoenzyme, however, cannot be reactivated by the addition of FAD, and the alteration of the structure of the protein is also indicated by its elution pattern from a Sephadex G-100 dextran column, different from the one of the native enzyme. The importance of one or more protein-bound sulphur atom or sulphide group for activity is shown by the fact that NADPH reduction makes –SH group(s) available to N-ethylmaleimide binding, resulting in complete inactivation of the enzyme (Table IV). Whether protein sulphur is *directly* involved in the binding of FAD to the apoenzyme is still a problem open to investigation.

## Acknowledgement

This work was supported by an equipment grant from Charles F. Kettering Foundation, which is gratefully acknowledged.

## References

Appella, E. and San Pietro, A. (1961). *Biochem. biophys. Res. Commun.* **6**, 349.
Avron, M. and Jagendorf, A. T. (1956). *Archs Biochem. Biophys.* **65**, 475.
Davenport, H. E. and Hill, R. (1952). *Proc. R. Soc.* **B139**, 327.
Forti, G., Bertolè, M. L. and Parisi, B. (1963). *In* "Photosynthetic Mechanisms of Green Plants", p. 284. Natl. Acad. Sci., Natl. Res. Council publication 1145. Washington, D.C.
Forti, G., Bertolè, M. L. and Zanetti, G. (1965). *Biochim. biophys. Acta* **109**, 33.
Fry, K. T. and San Pietro, A. (1963). *In* "Photosynthetic Mechanisms of Green Plants", p. 252. Natl. Acad. Sci., Natl. Res. Council publication 1145, Washington, D.C.
Keister, L. D., San Pietro, A. and Stolzenbach, F. E. (1960). *J. biol. Chem.* **235**, 2989.
Massey, V. and Gibson, Q. H. (1963). *In* "Proceedings of the Fifth International Congress of Biochemistry", Vol. V, p. 157. Pergamon Press.

Masters, B. S. S., Kamin, H., Gibson, Q. H. and Williams, C. H., Jr. (1965). *J. Biol. Chem.* **240**, 921.
San Pietro, A. (1963). *Ann. N.Y. Acad. Sci.* **103**, 1093.
San Pietro, A. (1963). *In* "Methods in Enzymology" (S. P. Colowick and N. O. Kaplan, eds.), Vol. VI, p. 439. Academic Press, New York and London.
San Pietro, A. and Keister, D. L. (1962). *Archs Biochem. Biophys.* **98**, 235.
Shin, M., Tagawa, K. and Arnon, D. I. (1963). *Biochem. Z.* **38**, 84.
Shin, M. and Arnon, D. I. (1965). *J. biol. Chem.* **240**, 1405.
Zanetti, G. and Forti, G. (1966). *J. biol. Chem.* **241**, 279.

# Plastocyanin as Cofactor of Photosynthetic NADP+ Reduction in Digitonin-Treated Chloroplasts[1]

Achim Trebst and Erich Elstner

*Pflanzenphysiologisches Institut der Universität Göttingen, Abt. Biochemie der Pflanzen, Göttingen, Germany*

From studies with inhibitors (Green *et al.*, 1939; Arnon, 1950; Trebst and Eck, 1963; Trebst, 1963) and deficient cells (Spencer and Possingham, 1960; Bishop, 1964) it seems to be clearly established that copper is essential for photosynthesis. A copper-containing enzyme of chloroplasts of higher plants participating in the electron transport chain of photosynthesis is very probably identical with plastocyanin, which was first isolated by Katoh (1960) (Katoh and Takamiya, 1961; Katoh *et al.*, 1961). The identification of the site of function of plastocyanin as electron carrier between the two light reactions of photosynthesis and close to the donor site for light reaction 1 resulted from a number of different investigations: from studies of the site of action of the copper-chelating agent salicylaldoxime (Trebst *et al.*, 1963), of the dependence of cytochrome *c* photo-oxidation in digitonin-fragmented chloroplasts (Katoh and Takamiya, 1963a, b) on plastocyanin, of the stimulation of $NADP^+$ reduction in *Brassica* chloroplasts by plastocyanin (Katoh and Takamiya, 1963a, b, 1965), and of light-induced spectroscopic changes of plastocyanin in intact cells (de Kouchkovsky and Fork, 1964; Fork and Urbach, 1965) as well as in detergent-treated chloroplasts (Kok, 1963; Kok *et al.*, 1964; Kok and Rurainski, 1965).

Digitonin fragmentation of spinach chloroplasts seemed to be particularly suitable to obtain a plastocyanin-free system for demonstrating the requirement of plastocyanin for photosynthetic $NADP^+$ reduction. A relatively simple procedure which consists of dissolving chloroplast fragments in digitonin and then centrifugation at $200{,}000 \times g$ is now reported. It leads to highly active particles which on illumination reduce $NADP^+$ if ferredoxin, ferredoxin-$NADP^+$-reductase, a suitable electron donor system, and plastocyanin are supplied.

## Results and Discussion

Nieman and Vennesland (1959) found that after digitonin treatment broken chloroplasts, in contrast to untreated broken chloroplasts, are able to photo-oxidize reduced cytochrome *c*. By further purification by ethanol precipitation a particular fraction which required the supernatant in order to photo-oxidize

[1] The following abbreviations are used: DCPIP = dichlorphenolindophenol; DAD = diaminodurol = 2,3,5,6-tetramethyl-*p*-phenylenediamine.

cytochrome *c* was obtained (Nieman *et al.*, 1959). Katoh and Takamiya (1963a, b) could later identify the factor in the supernatant as plastocyanin. Further spectroscopic studies of this system by Kok (1963) and Kok *et al.* (1964), Kok and Rurainski (1965) indicated directly that reduced plastocyanin is photo-oxidized and donates electrons to pigment system 1 of photosynthesis. [Kok *et al.* (1964) found that cytochrome *f* also stimulates cytochrome *c* photo-oxidation in detergent-treated chloroplasts]. Since pigment system 1 of photosynthesis normally reduces $NADP^+$ via ferredoxin and ferredoxin-$NADP^+$-reductase, digitonin-fragmented chloroplasts should be able to photoreduce $NADP^+$ at the expense of reduced cytochrome *c*. Whatley (1963) was able to show that this is the case. According to these results $NADP^+$ reduction in

Table I. Stimulation of photosynthetic $NADP^+$ reduction in digitonin-fragmented chloroplasts by the supernatant[1]

| Additions | μMoles NADPH formed | | |
|---|---|---|---|
| | Particles | Particles + supernatant | Particles + supernatant + $10^{-2}$ M salicylaldoxime |
| — | 0·13 | 0·18 | 0·1 |
| + 10 μmoles ascorbate | 0·37 | 1·25 | 0·15 |
| + 10 μmoles ascorbate + 0·2 μmoles DCPIP | 1·8 | 0·6 | 1·0 |
| + 10 μmoles ascorbate + 0·2 μmoles DAD | 0·9 | 2·8 | 0·25 |

[1] Experimental conditions: Broken chloroplasts of spinach, prepared according to Whatley *et al.* (1959), with a chlorophyll content of 10 mg were suspended in 10 ml 1% digitonin solution and centrifuged after 1 hr at 0° at 18,000 × *g* for 30 min. The green supernatant, i.e. digitonin-fragmented chloroplasts according to Nieman and Vennesland (1959) (Nieman *et al.*, 1959), were centrifuged at 200,000 × *g* for 1 hr at 5°. The pellets are designated as particles. Particles with 0·4 mg chlorophyll were illuminated with 35,000 lux at 15° under $N_2$ in: 80 μmoles *tris* buffer pH = 8·0, 5 μmoles $MgCl_2$, 6 μmoles $NADP^+$, saturating amounts of ferredoxin and ferredoxin–NADP–reductase, purified according to Losada and Arnon (1964) and the indicated addition in 3 ml. After 15 min light NADPH formation was measured at 340 mμ.

digitonin-treated chloroplasts should require plastocyanin. We have recently found (Trebst and Pistorius, 1965b) that the photoreduction of $NADP^+$ and quinone-stimulated photo-oxidation of ascorbate is, indeed, dependent on plastocyanin in particles obtained by the method of Nieman and Vennesland (1959). The procedure using ethanol precipitation, however, damaged the system to a certain extent so that the rates of $NADP^+$ reduction were only $\frac{1}{5}$–$\frac{1}{4}$ of the original rates (Trebst and Pistorius, 1965b). Furthermore DAD, which is a very effective electron donor in intact chloroplasts (Trebst and Pistorius, 1965a), was ineffective and the electron donor system DCPIP/ascorbate was not stimulated by the supernatant. By the simple procedure of centrifugation instead of ethanol precipitation, which also removes all solubilized enzymes, we have now obtained a preparation of particles which (if properly supplemented) showed almost the same rate of $NADP^+$ reduction as the initial preparation of broken chloroplasts. Furthermore these particles have the advantage of being stable for at least a week if stored at 4° in 0·05 M *tris* buffer pH = 8·0 (with 6·5 mg chlorophyll/ml).

As Table I indicates, such particles are not able to reduce $NADP^+$ on illumination in the presence of ascorbate even if ferredoxin and ferredoxin-$NADP^+$-reductase are added. Addition of DCPIP together with ascorbate leads to some NADPH formation whereas DAD is ineffective. However, the $NADP^+$ reduction is stimulated by the addition of the supernatant particularly in the DAD system, but the DCPIP system is inhibited. Salicylaldoxime

Table II. Stimulation of photosynthetic $NADP^+$ reduction by plastocyanin and inhibition of the DCPIP system by an ammonium sulphate fraction of the supernatant in particles after digitonin treatment of chloroplasts[1]

| Additions | μMoles NADPH formed | | |
|---|---|---|---|
| | Washed particles | Washed particles + plastocyanin | Washed particles + plastocyanin + fraction 2 |
| — | 0·18 | 0·15 | 0·29 |
| + 10 μmoles ascorbate | 0·29 | 1·25 | 0·55 |
| + 10 μmoles ascorbate + 0.2 μmoles DCPIP | 1·55 | 3·35 | 0·3 |
| + 10 μmoles ascorbate + 0·2 μmoles DCPIP + $1{\cdot}2 \times 10^{-3}$ M KCN | | 3·05 | 2·65 |
| + 10 μmoles ascorbate + 0·2 μmoles DAD | 0·57 | 2·4 | 2·45 |

[1] Experimental procedure as in Table I, except that the particles were washed once. The supernatant was precipitated by 80% acetone. The precipitate was fractionated by ammonium sulphate. The fraction from 0–40% saturation was discarded, the fraction from 40–70% is designated "fraction 2" and the blue fraction from 70–100% was used to obtain plastocyanin after further purification according to Katoh *et al.* (1961). Using an extinction coefficient of $9{\cdot}8 \times 10^3$ [according to Katoh and Takamiya (1964)], about 0·002–0·005 μmoles plastocyanin were added in this and in experiments recorded in the following tables.

abolishes the stimulatory effect of the supernatant in the DAD system and partly restores the activity of the DCPIP system inhibited by the supernatant.

This unexpected behaviour of the DCPIP system was clarified by recombination experiments using washed particles, acetone and ammonium sulphate fractions of the supernatant and purified plastocyanin obtained from the ammonium sulphate fraction of 70–100% saturation from the supernatant according to Katoh *et al.* (1961).

As Table II shows, $NADP^+$ reduction by all three electron-donor systems (ascorbate alone, DCPIP/ascorbate and DAD/ascorbate) are now greatly stimulated by the addition of plastocyanin. The rates in the DCPIP system are now definitely higher than those in the DAD system. Addition of "fraction 2" from the supernatant (40–70% ammonium sulphate saturation) completely abolishes $NADP^+$ reduction in the DCPIP system but has no effect on the DAD system. The inhibitory effect of this fraction of the supernatant is prevented if the experiment is run in the presence of $10^{-3}$ M KCN. This explains why the supernatant does not show any stimulation of the DCPIP system. It contains an inhibitor which is removed by ammonium sulphate fractionation and which is inhibited by low concentrations of KCN. Close inspection of the inhibitory ammonium sulphate fraction revealed that it contains the originally latent polyphenoloxidase which is liberated by detergent treatment of chloroplasts, as was shown some years ago (Mayer and Friend, 1960; Trebst and Wagner, 1962). The polyphenol oxidase is able to oxidize reduced DCPIP but not DAD. This activity invalidates the NADPH estimation, i.e. measurement of the extinction at 340 mμ after the experiment, since the polypheno-oxidase

Table III. Requirements for photosynthetic $NADP^+$ reduction in particles after digitonin treatment of chloroplasts[1]

| | μMoles NADPH formed |
|---|---|
| Complete | 2·9 |
| DAD omitted | 1·1 |
| Ascorbate omitted | 0·1 |
| Plastocyanin omitted | 0·45 |
| Ferredoxin omitted | 0·18 |
| Ferredoxin–$NADP^+$ reductase omitted | 1·9 |

[1] The complete system contained in 3 ml: 80 μmoles *tris* buffer pH=8·0, 5 μmoles $MgCl_2$, 6 μmoles $NADP^+$, 10 μmoles ascorbate, 0·2 μmoles DAD, about 0·005 μmoles plastocyanin, saturating amounts of ferredoxin and ferredoxin–$NADP^+$–reductase and washed particles with 0·4 mg chlorophyll. Illumination for 15 min at 15° under $N_2$.

quickly oxidizes the accumulated NADPH via DCPIP when the illuminated vessel is opened and exposed to air.

This photosynthetic $NADP^+$ reduction in digitonin-fragmented chloroplasts depends not only on plastocyanin but also on ferredoxin and the ferredoxin-$NADP^+$-reductase, as summarized in Table III. The plastocyanin still contains some of the reductase since omission of the reductase sample reduces the rate only to about 40%. The water photo-oxidation system and the complex of light reaction 2 does not function any more as no $NADP^+$ is formed without the artificial electron-donor ascorbate. Katoh and Takamiya (1963a) have reported that plastocyanin even restores the coupling of light reaction 2 to light reaction 1. This is apparently not possible in our preparation.

It is interesting that DCPIP but not DAD partly replaces plastocyanin as indicated by the result that the washed particles have some activity when plastocyanin is left out if DCPIP is present (Tables II and IV). This seems to indicate that DCPIP is able to donate electrons directly into pigment system 1 [as Witt *et al.* (1963) have shown already by spectroscopic experiments]. This plastocyanin-independent $NADP^+$ reduction in the DCPIP system is not inhibited by a high concentration of KCN (Table IV). Formation of NADPH in the DCPIP system is reduced to the value obtained without plastocyanin, but disappears completely in the DAD system. The effect of salicylaldoxime is not so clearcut but shows the same tendency (Table IV, see also Table I). The KCN-insensitivity seems to support the assumption that DCPIP is able to react directly with a site in the electron transport chain after plastocyanin, though with a lower rate.

Table IV. Inhibition of plastocyanin-dependent $NADP^+$ reduction by KCN and salicylaldoxime in particles after digitonin treatment of chloroplasts[1]

| Additions to 10 μmoles ascorbate | μMoles NADPH formed | |
|---|---|---|
| | − Plastocyanin | + Plastocyanin |
| — | 0·2 | |
| 0·2 μmoles DCPIP | 1·8 | 4·4 |
| 0·2 μmoles DCPIP + $10^{-2}$ M KCN | 1·5 | 1·3 |
| 0·2 μmoles DCPIP + $5 \times 10^{-3}$ M KCN | 1·6 | 3·3 |
| 0·2 μmoles DCPIP + $2 \times 10^{-2}$ M salicylaldoxime | 0·8 | 1·3 |
| 0·2 μmoles DAD | 0·5 | 3·0 |
| 0·2 μmoles DAD + $10^{-2}$ M KCN | | 0·6 |
| 0·2 μmoles DAD + $5 \times 10^{-3}$ M KCN | | 1·0 |
| 0·2 μmoles DAD + $10^{-2}$ M salicylaldoxime | | 0·23 |

[1] Experimental conditions as in Table III.

Boardman and Anderson (1964) have recently also obtained a particulate fraction after ultracentrifugation (at $144{,}000 \times g$) of digitonin-treated chloroplasts. These particles reduced $NADP^+$ at the expense of TCPIP/ascorbate when a crude acetone fraction of a spinach leaf extract was added. However, they reported later (Anderson *et al.*, 1964) that the $144{,}000 \times g$ fraction was enriched in copper (presumably plastocyanin) which seems to be in complete contradiction to our results.

## Summary

Ultracentrifugation of digitonin-treated broken chloroplasts removes the digitonin, solubilized enzymes, cofactors and inhibitors and leads to a particulate fraction. These particles are able to photoreduce $NADP^+$ at good rates if plastocyanin, ferredoxin, ferredoxin-$NADP^+$-reductase and an artificial electron-donor system are added. The particles are stable at 4° for at least a week and constitute a sensitive test system for plastocyanin. As an electron donor system ascorbate may be used but DCPIP/ascorbate or DAD/ascorbate give higher rates. The DCPIP system donates electrons also without plastocyanin but at a reduced rate. The plastocyanin-stimulation of the photo-reduction of $NADP^+$ is inhibited by high concentrations of KCN and of salicylaldoxime. The dependence of photosynthetic $NADP^+$ reduction on plastocyanin supports the conclusion, particularly of Katoh and Takamiya (1963a, b, 1965), Kok (1963; Kok *et al.*, 1954) and Fork and Urbach (1965), as to the essential role and site of function of plastocyanin in photosynthesis.

## Acknowledgements

The Research reported in this document has been sponsored by the Air Force Cambridge Research Laboratories under Contract Nr. AF 61 (052)-716 through the European Office of Aerospace Research (OAR), United States Air Force and by Deutsche Forschungsgemeinschaft.

## References

Anderson, J. M., Boardman, N. K. and David, D. J. (1964). *Biochem. biophys. Res. Commun.* **17**, 685.
Arnon, D. I. (1950). *In* "A Symposium on Copper Metabolism" (W. D. McElroy and B. Glass, eds.). Johns Hopkins Press, Baltimore, Marlyand.
Bishop, N. I. (1964). *Nature, Lond.* **204**, 401.
Boardman, N. K. and Anderson, J. M. (1964). *Nature, Lond.* **203**, 166.
Fork, D. C. and Urbach, W. (1965). *Proc. natn. Acad. Sci. U.S.A.* **53**, 1207.
Green, L. F., McCarthy, J. F. and King, C. G. (1939). *J. biol. Chem.* **128**, 447.
Katoh, S. (1960). *Nature, Lond.* **186**, 533.
Katoh, S. and Takamiya, A. (1961). *Nature, Lond.* **189**, 665.
Katoh, S. and Takamiya, A. (1963a). *In* "Photosynthetic Mechanisms of Green Plants", National Academy of Sciences—National Research Council, Washington, D.C.
Katoh, S. and Takamiya, A. (1963b). *Pl. Cell Phys.* **4**, 335.
Katoh, S. and Takamiya, A. (1964). *J. Biochem.* (*Japan*) **55**, 378.
Katoh, S. and Takamiya, A. (1965). *Biochim. Biophys. Acta* **99**, 156.
Katoh, S., Suga, I., Shiratori, I. and Takamiya, A. (1961). *Archs. Biochem. Biophys.* **94**, 136.
Kok, B. (1963). *In* "Photosynthetic Mechanisms of Green Plants", National Academy of Sciences—National Research Council, Washington, D.C.
Kok, B. and Rurainski, H. J. (1965). *Biochim. biophys. Acta* **94**, 588.
Kok, B., Rurainski, H. J. and Harmon, E. A. (1964). *Pl. Physiol.* **39**, 513.
de Kouchkovsky, J. and Fork, D. C. (1964). *Proc. natn. Acad. Sci.* **52**, 232.
Losada, M. and Arnon, D. I. (1964). *In* "Modern Methods of Plant Analysis" (H. F. Linskens, B. D. Sanwal, and M. V. Tracey, eds.), Bd. VII. Springer Verlag, Berlin.
Mayer, A. M. and Friend, J. (1960). *Nature, Lond.* **185**, 464.
Nieman, R. H. and Vennesland, B. (1959). *Pl. Physiol.* **34**, 255.
Nieman, R. H., Nakamura, H. and Vennesland, B. (1959). *Pl. Physiol.* **34**, 262.
Spencer, D. and Possingham, J. V. (1960). *Aust. J. biol. Sci.* **13**, 441.
Trebst, A. (1963). *Z. Naturf.* **18**b, 817.
Trebst, A. and Eck, H. (1963). *Z. Naturf.* **18**b, 105.
Trebst, A. and Pistorius, E. (1965a). *Z. Naturf.* **20**b, 143.
Trebst, A. and Pistorius, E. (1965b). *In* "Beiträge zur Biochemie und Physiologie von Naturstoffen", Festschrift zum 65. Geburtstag v. K. Mothes, VEB Fischer Verlag, Jena.
Trebst, A. and Wagner, S. (1962). *Z. Naturf.* **17**b, 396.
Trebst, A., Eck, H. and Wagner, S. (1963). *In* "Photosynthetic Mechanisms of Green Plants", National Academy of Sciences—National Research Council, Washington, D.C.
Whatley, F. R. (1963). *In* "Photosynthetic Mechanisms of Green Plants", National Academy of Sciences—National Research Council, Washington, D.C.
Whatley, F. R., Allen, M. B. and Arnon, D. I. (1959). *Biochim. Biophys. Acta* **32**, 32.
Witt, H. T., Müller, A. and Rumberg, B. (1963). *Nature, Lond.* **197**, 987.

# Oxidation-Reduction Reactions of Endogenous Plastoquinone in Chloroplasts and Digitonin-Fractionated Chloroplasts

J. Friend, R. Olsson and E. R. Redfearn

*Department of Botany, University of Hull,*
*and*
*Department of Biochemistry, University of Leicester, England*

Evidence for the involvement of endogenous plastoquinone in photosynthetic electron transport reactions has been based on two different types of experiment, viz. (i) direct spectrophotometric measurements in the ultraviolet region of chloroplasts and of algae (Klingenberg *et al.*, 1962; Amesz, 1964), and (ii) estimation of the oxidation-reduction level of plastoquinone after its extraction from an appropriately treated chloroplast suspension (Crane *et al.*, 1960; Redfearn and Friend, 1961).

Experiments of this last type have shown that plastoquinone may be either photoreduced or photo-oxidized in isolated sugar beet chloroplasts (Redfearn and Friend, 1962; Friend and Redfearn, 1963). The present paper describes similar experiments on digitonin particles prepared from isolated chloroplasts.

## Micro-method for the Determination of the Oxidation-Reduction Level of Endogenous Plastoquinone

The method used for the determination of the oxidation-reduction level of endogenous plastoquinone may be outlined as follows:

Chloroplast suspensions are extracted with methanol-containing pyrogallol and light petroleum in a stoppered test-tube. After shaking, the pigments and the plastoquinone are found in the upper light petroleum layer. A second extraction with light petroleum completes the extraction of plastoquinone.The combined light petroleum layers are then washed several times with aqueous methanol (100 ml methanol plus 5 ml water) to remove chlorophylls which otherwise interfere with spectrophotometric measurements in the ultraviolet region. The final yellow, light petroleum layer is taken to dryness *in vacuo*, the residue dissolved in a known volume of ethanol and its absorption spectrum plotted. After addition of one crystal of sodium borohydride the absorption spectrum is redetermined. The difference between the two spectra corresponds to the difference spectrum of plastoquinone minus plastoquinol, and the amount of plastoquinone can be calculated from measurements of the change in absorbance at 255 m$\mu$. An example of the type of spectra obtained is shown in Fig. 1. Such a determination gives merely the amount of oxidized plastoquinone in a given sample of chloroplast suspension. To determine the total level of oxidized plastoquinone in the chloroplast suspension, 2,6-dichlorophenolindophenol is added to another aliquot of the same suspension which is then left in the dark for at least 5 min. This oxidizes any plastoquinol to the quinone form. The plastoquinone is then extracted and estimated as just described. Comparison of the ratio of plastoquinone found in untreated chloroplasts and chloroplasts treated with dichlorophenolindophenol gives the percentage of oxidized plastoquinone in the chloroplast suspension.

Thin-layer chromatography of the light petroleum solution from which the

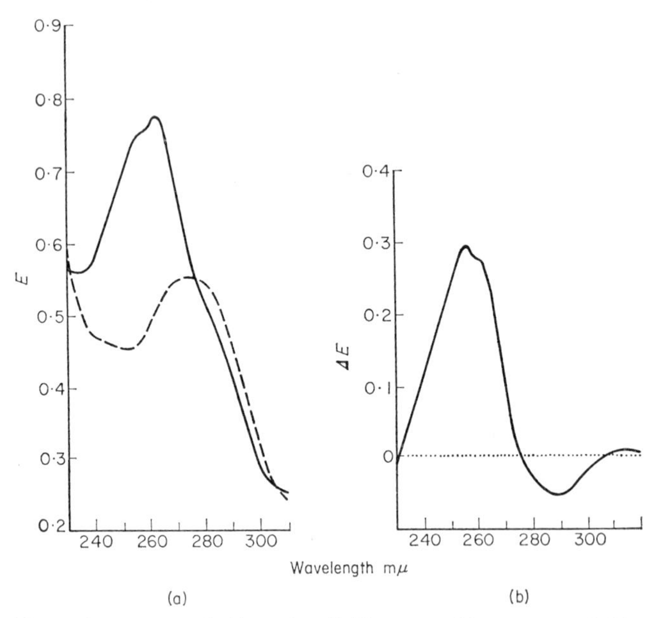


FIG. 1. Absorption spectra of chloroplast lipid extract. (a) spectrum of chloroplast lipid extract before (——) and after (------) reduction with $NaBH_4$; (b) difference spectrum (oxidized *minus* reduced) of chloroplast lipid extract.

chlorophyll had been removed with aqueous methanol showed that there was only one compound present which would react with leuco-methylene blue and which had a difference spectrum corresponding with that of plastoquinone/plastoquinol. This compound had the same $R_f$ values in several solvent systems as a sample of synthetic plastoquinone-9 and it was concluded that our method when applied to sugar beet chloroplasts measured only plastoquinone-9. Furthermore, comparison of the value for total plastoquinone measured by this method gave excellent agreement with a value obtained after elution of the plastoquinone-9 band from a thin-layer chromatogram of an acetone-extract of chloroplasts which had been treated with dichlorophenolindophenol.

Although there are traces of two other plastoquinone-like compounds in thin-layer chromatograms of acetone extracts of sugar beet chloroplasts, and also traces of α-tocopherol quinone, these compounds are present in such small amounts that they are not detectable in thin-layer chromatograms of the light-petroleum solutions used in the micro-determination.

Although this method for the determination of the oxidation-reduction level of plastoquinone works well with isolated chloroplasts suspended in buffered sucrose solutions, it was found to be unsatisfactory for measurements on chloroplasts which had been treated with digitonin since this compound interferes with the extraction of the pigments into light petroleum. A modification to overcome this difficulty has therefore been made. The chloroplast suspension is pipetted into 4 volumes of cold acetone (−20°) and shaken with light petroleum. The digitonin is precipitated by the acetone and the pigments pass into the upper layer. A second extraction with light petroleum completes the extraction of all the pigments. The light petroleum layer is then washed carefully with water to remove the bulk of the acetone, and chlorophyll is then removed by treatment with aqueous methanol as in the previous method.

## EFFECT OF LIGHT

Crane, Ehrlich and Kegel (1960) found an 80% reduction of endogenous plastoquinone after illumination of spinach chloroplasts. We found (Redfearn and Friend, 1961; Friend and Redfearn, 1963) that in sugar beet chloroplasts there was a drop in the oxidation level of plastoquinone of about 20–30% after illumination of sugar beet chloroplasts. The quantitative difference between these results is probably a reflection of the different plant materials from which the chloroplasts were isolated, although it should be noted that Crane *et al.* added trichloroacetic acid to their chloroplasts before extraction with isooctane.

It was also found (Friend and Redfearn, 1962; 1963) that photoreduction of plastoquinone was inhibited by DCMU; however DCMU did not inhibit the photoreduction of plastoquinone using the ascorbate-indophenol couple. In these experiments plastoquinone behaved similarly to added cofactors such as NADP, FMN and menadione. A photo-oxidation of plastoquinone occurred in the presence of either ammonia or methylamine hydrochloride, both of which uncouple photophosphorylation and stimulate the Hill reaction (Krogmann *et al.*, 1959; Good, 1960). It was suggested that these compounds could possibly act by stimulating electron flow at a point after the site of action of plastoquinone, and thus account for the observed photo-oxidation of the quinone. The increased photoreduction of endogenous plastoquinone caused by the addition of either KCN or $NaN_3$ was explained by suggesting that these compounds inhibited the photo-oxidation of the quinol by a haemoprotein or other metallo-enzyme. Thus it was suggested that plastoquinone probably acted between the two light reactions.

Direct spectrophotometric measurements of Amesz (1964) on the blue-green alga *Anacystis nidulans* shows that a compound which had a difference spectrum corresponding to that of plastoquinone was reduced by light absorbed by system 2 and oxidized by system 1, and was therefore situated between the two light reactions.

Some preliminary studies of our own using light of 650 m$\mu$ (system 2) and light of 730 m$\mu$ (system 1) showed that plastoquinone photoreduction was more effective in the short wavelength light than in white light and that in the

Table I. The effect of light on the oxidation level of plastoquinone in sugar beet chloroplasts

| | % of total of plastoquinone in oxidized form |
|---|---|
| Dark | 85·7 |
| White light | 77·6 |
| 650 m$\mu$ light | 72·8 |
| 730 m$\mu$ light | 86·6 |

Incubation time, 5 min.
The 730 m$\mu$ light was 4·2 times more intense than the 650 m$\mu$ light.
Each filter transmitted 80% of the incident light at the appropriate wavelength, i.e. 650 m$\mu$ or 730 m$\mu$, and cut off steeply at 680 m$\mu$.

long wavelength light there was a photo-oxidation of the quinone (Table I). These results thus confirmed, by analysis of extracted plastoquinone, the *in vivo* spectrophotometric observations of Amesz (1964).

## DIGITONIN FRACTIONS OF CHLOROPLASTS

Boardman and Anderson (1964) have reported that after incubation of spinach chloroplasts with digitonin a range of particles which sediment at centrifugal speeds ranging from 1000 to 144,000 × **g** could be obtained which had different proportions of chlorophylls *a* and *b*. The "lower speed" particles showed high rates of photoreduction of 2,6-dichlorophenolindophenol and $NADP^+$ but the "higher speed" particles were inactive. However, the "higher speed" particles showed much higher rates of photoreduction of $NADP^+$ using the ascorbate-indophenol couple as electron donor. It was suggested by Boardman and Anderson that possibly the two photochemical systems had been separated by digitonin treatment, the smaller particles, which had a higher ratio of chlorophyll *a* to chlorophyll *b*, belonging to system 1 and larger particles, which had a lower ratio of chlorophylls *a* and *b*, belonging to system 2.

We have carried out a series of similar fractionations using sugar beet chloroplasts. In general, a pattern of results qualitatively similar to those of Boardman and Anderson have been obtained although the quantitative picture is not the same. The ratio of chlorophylls *a* and *b* increases from the 1000 × **g** to the 144,000 × **g** fraction but it is never as high as that reported for the spinach particles. Morevoer, the ratios show a wide variation between preparations, although the 144,000 × **g** particles and the supernatant always have higher $C_a/C_b$ ratios than the starting chloroplasts (Table II). In Table II the distribution of plastoquinone and of the molar ratios of plastoquinone to total chlorophyll is also recorded. Plastoquinone is present in all the fractions and the supernatant; the amount in the latter varies from about 20 % to nearly 50 % of the plastoquinone in the original chloroplasts. This may well represent solubilization of the osmiophilic globules which are reported to contain about 50 % of the total plastoquinone of the chloroplast (Bailey and Whyborn, 1963). The plastoquinone/total chlorophyll ratio is highest in the supernatant; in some cases a ratio cannot be obtained because the amount of chlorophyll is too low to be measured in 80 % acetone. The plastoquinone/chlorophyll ratio for the 144,000 × *g* fraction is similar to that for the whole chloroplast but the ratios for the other fractions are much lower (Table II).

Table II. Distribution of plastoquinone in digitonin fractions of sugar beet chloroplasts

| | Molar ratios | | | |
|---|---|---|---|---|
| | Experiment A | | Experiment B | |
| | $C_a/C_b$ | Plastoquinone / Total chlorophylls | $C_a/C_b$ | Plastoquinone / Total chlorophylls |
| Intact chloroplasts | 2·87 | 0·071 | 2·44 | 0·067 |
| 1000 × *g* | 2·99 | 0·030 | 2·51 | 0·052 |
| 10,000 × *g* | 2·87 | 0·030 | 2·46 | 0·043 |
| 50,000 × *g* | 5·16 | 0·039 | 3·44 | 0·076 |
| 144,000 × *g* | 7·78 | 0·079 | 3·68 | 0·24 |
| Supernatant | 3·65 | 1·03 | — | — |

Percentage of plastoquinone in supernatant: 46·9, 19·8.

Three photochemical activities of the chloroplasts and of digitonin fragments have been compared: (i) photoreduction of 2,6-dichlorophenolindophenol (DCPIP); (ii) photoreduction of $NADP^+$ with water as electron donor; (iii) photoreduction of $NADP^+$ with the ascorbate-indophenol couple as electron donor in the presence of DCMU.

The pattern of activities in the fractions (Table III) is similar to that found by Boardman and Anderson; the photoreduction of DCPIP and $NADP^+$ is virtually confined to the 1000 × *g* and 10,000 × *g* fractions. These two activities in the 50,000 × *g* fraction are very low, but the 50,000 × *g*–144,000 × *g* fractions show a higher photoreduction of $NADP^+$ in the presence of the ascorbate-indophenol couple.

If we accept Boardman's hypothesis then we have not achieved full separation of the two photochemical systems. The 144,000 × *g* fractions would appear to contain no active system 2 but the other fractions would appear to have both photochemical systems operative.

In those fractions which show a photoreduction of DCPIP there is also a photoreduction of plastoquinone, whereas in the 144,000 × *g* fraction there is a photo-oxidation of plastoquinone (Table IV).

Thus there appears to be some correlation between the presence of an active

Table III. Distribution of photochemical activities in digitonin fractions of chloroplasts

<table>
<tr><th></th><th>A</th><th>B</th><th>C</th><th>A</th><th>B</th><th>C</th></tr>
<tr><td>Chloroplasts</td><td>182·3</td><td>32·8</td><td>25·5</td><td>193·0</td><td>31·5</td><td>25·6</td></tr>
<tr><td>1000 × g</td><td>61·1</td><td>6·2</td><td>4·34</td><td>76·6</td><td>15·3</td><td>17·3</td></tr>
<tr><td>10,000 × g</td><td>87·3</td><td>3·5</td><td>3·48</td><td>65·9</td><td>15·7</td><td>19·9</td></tr>
<tr><td>50,000 × g</td><td rowspan="2">20·0</td><td rowspan="2">0</td><td rowspan="2">23·7</td><td>9·7</td><td>0</td><td>23·5</td></tr>
<tr><td>144,000 × g</td><td>0</td><td>0</td><td>18·6</td></tr>
<tr><td>Supernatant</td><td>0</td><td>0</td><td>0</td><td>0</td><td>0</td><td>0</td></tr>
</table>

A = Dichlorophenolindophenol reduction.
B = $NADP^+$ reduction.
C = $NADP^+$ reduction in the presence of the ascorbate-indophenol couple and DCMU.
Rates are expressed as $\mu$moles reduced/mg chlorophyll/hr.

Table IV. Effect of white light on the oxidation-reduction level of plastoquinone in sugar beet chloroplasts and digitonin fractions of chloroplasts

| | % of total plastoquinone in oxidized form | |
|---|---|---|
| | Dark (3 min) | Light (3 min) |
| Chloroplasts | 91·5 | 76·2 |
| | 52·9 | 31·1 |
| 1000 × *g* | 81·1 | 76·4 |
| | 76·2 | 65·6 |
| 10,000 × *g* | 73·6 | 54·7 |
| 50,000 × *g* | 100 | 91·4 |
| 144,000 × *g* | 40·9 | 49·5 |
| | 78·1 | 80·0 |
| Combined 144,000 × *g* + supernatant | 39·4 | 50·6 |
| | 50·0 | 63·5 |
| Supernatant | 44·0 | 44·0 |

When two values are given for a particular fraction, these have been obtained with fragments prepared from different batches of chloroplasts.

system 2 and the observed photoreduction of endogenous plastoquinone. When only system 1 is present we can only observe photo-oxidation of plastoquinone. These experiments therefore lend further support to the hypothesis that plastoquinone is situated between the two light reactions.

## Discussion

The reason why an overall photoreduction of plastoquinone is observed in chloroplasts and fragments which have both photochemical systems working whereas only a photo-oxidation is seen when system 2 is inhibited or removed, may probably be explained on the hypothesis that plastoquinone reduction is close to system 2. However, plastoquinone oxidation will require other carriers such as cytochrome *f* and probably plastocyanin which are themselves photo-oxidized by system 1 (Kok *et al.*, 1964). Since cytochrome *f* is only present in about one-tenth the amount of plastoquinone, on a molar basis, there could be a build-up of reducing equivalents in plastoquinone which could only slowly be transferred along the chain of electron carriers; thus in complete systems there would always be a tendency to observe a photoreduction of plastoquinone.

## Acknowledgement

This work was supported in part by a research grant from the Science Research Council.

## References

Amesz, J. (1964). *Biochim. biophys. Acta* **79**, 257.
Bailey, J. L. and Whyborn, A. G. (1963). *Biochim. biophys. Acta* **78**, 163.
Boardman, N. K. and Anderson, J. A. (1964). *Nature, Lond.* **203**, 166.
Crane, F. L., Ehrlich, B. and Kegel, L. P. (1960). *Biochem. biophys. Res. Commun.* **3**, 37.
Friend, J. and Redfearn, E. R. (1962). *Biochem. J.* **82**, 13P.
Friend, J. and Redfearn, E. R. (1963). *Phytochemistry* **2**, 397.
Good, N. E. (1960). *Biochim. biophys. Acta* **40**, 502.
Klingenberg, M., Müller, A., Schmidt-Mende, P. and Witt, H. T. (1962). *Nature, Lond.* **194**, 379.
Kok, B., Rurainski, H. J. and Harmon, E. A. (1964). *Pl. Physiol.* **39**, 513.
Krogmann, D. W., Jagendorf, A. T. and Avron, M. (1959). *Pl. Physiol.* **34**, 272.
Redfearn, E. R. and Friend, J. (1961). *Nature, Lond.* **191**, 806.
Redfearn, E. R. and Friend, J. (1962). *Phytochemistry* **1**, 147.

## Discussion

*H. Baltscheffsky:* You mentioned, I believe, that addition of $NH_4^+$ to chloroplasts caused an oxidation of the steady-state level for plastoquinone. Did you add ADP, orthophosphate or 2,4-dinitrophenol to the system under similar conditions as those in the ammonia experiment? I am thinking in terms of testing the possibility that a coupling site might exist at the plastoquinone level with the Chance–Williams crossover-method for identification of coupling sites in electron transport phosphorylation systems.

*J. Friend:* In early experiments we found no difference in the steady-state levels of plastoquinone measured in phosphate or *tris*, and additions of ADP to phosphate media had no effect either.

However, on reflection, it is possible that these chloroplasts could have been completely uncoupled, and it will probably be useful to repeat these experiments using chloroplasts in which phosphorylation is coupled to photoreduction.

# New Histochemical Study at Cellular Chloroplast Level: Characterization and Localization of Plastoquinones

L. Khau van Kien

*Département de Biologie, Service de Radioagronomie, CEN Cadarache, France*

In a previous paper (Khau van Kien, 1963), we tried to detect and localize plastoquinones in chloroplasts. The histochemical reactions used were based on the detection of phenolic and quinonoid functions and on the cytological localization in illuminated or dark chloroplasts. The phenol reactions (Millon's reaction, Molisch's argentaffin reaction, Champy-Coujard's reaction), the quinone reactions with iodate, dinitrophenylhydrazine, etc., allowed cytological localizations to be made in chloroplasts in connection with illumination and with the different forms of plastoquinone.

A specific enzymatic reaction, based on the fact that plastoquinones are coenzymes of succinic dehydrogenase and reduce nitro blue tetrazolium (N.B.T.), shows that the reaction decreases positively under illumination. Previous extraction of plastoquinone and thus the neutralization of succinic dehydrogenase make the reaction negative. Addition of plastoquinone to the medium restores the enzymatic activity.

The autoradiographic studies with $^{32}P$ demonstrated that $^{32}P$ is well incorporated in illuminated chloroplasts and only very little in dark chloroplasts. Extraction of plastoquinone results in the disappearance of $^{32}P$ from "isolated" and *in situ* chloroplasts but does not remove inorganic phosphorus.

All these reactions enabled us to localize plastoquinones within chloroplasts, i.e. in characteristic sites such as grana, chloroplast periphery and stroma.

In spite of these localization reactions, there may still be doubts regarding the characterization of plastoquinones. Our present study aims at clarifying this point with new histochemical and autoradiographic techniques of characterization, by means of biophysical and biochemical controls ($CO_2$ absorption, oxygen release), chromatographic separation on alumina columns of a [$^{14}C$]labelled product and subsequent u.v. check of the absorption spectrum, and finally by readdition of the [$^{14}C$]plastoquinone, which has been extracted and determined through its absorption spectrum.

## First Localization Technique

A first histochemical localization technique follows the pattern of the plastoquinone demonstration by means of neotetrazolium, after thin layer chromatography. It was used by Leister and Ramasarma (1959) and is well known to biochemists concerned with photosynthesis. Pure plastoquinone is chromatographed on slides and reduced by borohydride. After excess borohydride has been destroyed by hydrochloric acid, a buffered neotetrazolium solution is sprayed on to the slide; after slight heating in an oven, a violet colouration is observed, which localizes the plastoquinone.

We then proceed in the same way with the difference that, instead of using thin layer slides, we make stamped smears of dark *Tradescantia* and maize chloroplasts on histological slides. The preparations, fixed with formol vapours or by heating, are processed in the same way as chromatographic slides (same time and same reagents). The histological slides are heated moderately (56°) in order to avoid morphological and chemical alterations. They are left to cool and then mounted with Apathy syrup. This microscopical reaction, which follows the pattern of the biochemical demonstration, is expected to detect plastoquinones wherever they are located, with the same colouration. This is confirmed by microscopical examination. Chloroplasts show a distinct positive reaction, sharp at grana level and not so sharp on the periphery.

An improved variant of this technique for the characterization of plastoquinones is based on the fact that the chloroplast under illumination rapidly colours, compared with the dark control chloroplast. The reaction is made *in vivo*; instead of killing the chloroplast and reducing with borohydride, we use the reducing property of light with secondary reduction of neotetrazolium. Thus working on a living cell allows a dynamic localization of the plastoquinone according to illumination time.

Sections or fragments of fresh leaves are incubated between slide and cover glass in buffered neotetrazolium solution (pH = 7). The preparation is examined with an inversed Zeiss microscope for plankton under a strong illumination from above. After 15 min, a violet colouration develops, first in a few chloroplasts, then over the whole preparation. The colouration reaction is first noted on the periphery, then in the grana and finally in the stroma, with a fainter colouration. The peripheric localization is frequent. Some grana may colour first. This behaviour *in vivo* suggests different chemical states and different reactions of plastoquinones.

The control preparation, kept in dark during the same time, does not show any colouration. The reduction of neotetrazolium that can be observed in the illuminated preparation, is rapid; it is not due to the cytoplasm reducing power which is much slower. If the illuminated preparation is kept in the dark together with the control one, the macroscopic colouration of the latter is roughly found, one day later, to be darker than the illuminated preparation, which suggests the reaction is not specific; however, the microscopical examination shows that chloroplasts still were green whereas the cell cytoplasm contains formed elements: isolated grains, small chains or small rods which, all with a violet colouration, are nothing else but mitochondria. As we know, the latter are the seat of respiration, they contain reducing ubiquinones and their reducing power would appear later on. Improved results on plastoquinone localization have been obtained with the same leaf fragments which have been cut with a "cryotome" after inclusion into liver. The fine sections give more accurate localization than observation in the mass. The colouration observed with pure plastoquinone on chromatographic slides, killed chloroplast smears or living chloroplasts of fresh leaves with endogenous plastoquinone, respectively, favours the possible characterization of cellular plastoquinones according to illumination and to the rate of the appearance of the colours. The leaves used in the experiments were previously tested with respect to their $CO_2$ consumption and oxygen release, in the light and in dark. This new reaction *in vivo* with neotetrazolium is different from the enzymatic reaction with N.B.T. Contrary to the results obtained with the enzymatic reaction, this reaction increases with illumination time.

## Second Technique

The second plastoquinone characterization technique is based on the superposition, on a single histological slide, of the reducing power of barley chloroplasts containing reduced plastoquinone (positive Molisch's reaction), and the autoradiographical presence of [$^{14}$C]labelled substances in chloroplasts—illuminated for 15 min with $^{14}CO_2$—returned to normal air for 2 hr and extracted with the specific solvents of plastoquinone. For a better comparison of the superposition of both techniques, silver salts, resulting from Molisch's reaction and situated in the chloroplasts, are dipped into a gold chloride solution, which gives them a brown-orange colouration with a violet shimmer, different from the black silver grains situated in an upper level in the nuclear emulsion, which has been impressed by the radioactivity of [$^{14}$C]plastoquinone. Along with the histochemical and autoradiographical sampling, we extracted the [$^{14}$C]plastoquinones and related products, which will be considered later.

The Molisch's reaction was slightly modified in our experiments, in order to fit it to our material: we used either an ammoniacal solution of slightly alkaline silver nitrate (10%) or a buffered ammoniacal silver nitrate solution, pH=7·5. The reaction is carried out in an oven at 56°; a dark control sample is treated similarly. The illuminated sample rapidly turns black. The reaction is stopped when the dark control sample is just turning to grey. The time required is 30 to 40 min. The samples are washed, then processed for 1 hr with a saturated solution of sodium nitrite or with 10% formol. After washing, sections are made either directly with a "cryotome" or after inclusion in soluble wax, which results in more regular sections (5 $\mu$). A first series of these is directly autoradiographed. Another set is extracted with solvents in the same ways as for chromatography on alumina columns (5 min with pure light petroleum ether (LP), 3 min with LP containing 2% ethyl ether, 3 min with LP containing 5% ethyl ether, 3 min with LP containing 10% ethyl ether, 3 min with LP containing 20% ethyl ether). Then they are left to dry and the emulsion is spread on the slide. After proper exposure, the sample is mounted with balsam.

A microscopical examination of illuminated leaf preparations shows good coincidence between the histochemical Molisch's reaction and the autoradiographic response with $^{14}$C. The black silver grains are grouped above the chloroplasts, which are brown-orange coloured after gold chloride action. On the solvent-washed slides, chloroplasts have lost the greater part of their activity; only some residual extra-chloroplastidic cytoplasmic radioactivity is still to be found. It can be assumed that the loss in radioactivity occurring only in chloroplasts results from the disappearance of plastoquinone which is ten times more radioactive than carotenoids, vitamin K, tocopherylquinone, plastochromenol or similar products, which we isolated by chromatography and determined by their u.v. spectrum.

## Third Technique

Although counting gave us evidence that plastoquinone extracted from barley was highly radioactive and in spite of its disappearance from washed sections leading us to assume its presence in chloroplasts, we tried to get further conclusive evidence by adding back [$^{14}$C]plastoquinone. Using a 60 g sample of barley leaves which was left 15 min in contact with $^{14}CO_2$ under illumination and then returned to normal air, we could separate, besides carotenoids, vitamin K, tocopherylquinone, etc., with a low radioactivity (250 to 300 cts/min), the three plastoquinones A, B, and C (about 2800 to 3000 cts/min). An examination of these plastoquinones gave us evidence that they have different activities: 20 cts/min for one drop of plastoquinone A on a slide, 80 cts/min for plastoquinone B and 40 to 45 cts/min for plastoquinone C. After being purified, the three plastoquinones are mixed and lyophilized. This mixture was used for adding back. Normal barley leaves are kept in diffuse light during 2 to 3 hr. A fraction of these is fixed with formol vapour and cut with a cryotome; another fraction is included into carbowax and cut into 3, 5, and 7 $\mu$ sections. Some sections are kept as control samples, the others are washed with light petroleum (10 min), then with light petroleum containing 2% ethyl ether (10 min), and finally with light petroleum containing 20% ethyl ether (10 min). The [$^{14}$C]plastoquinones which are kept under a vacuum, are dissolved in 5 ml light petroleum. They are then incubated on slides, care being taken that the light petroleum does not evaporate (a cover-glass with four curved corners is used which allows light petroleum to be added as it evaporates). Readdition requires 30 min. The excess plastoquinone is washed away with light petroleum for 15 to 20 min, then left to dry. Autoradiographs are made with a nuclear Ilford G5 emulsion. After exposure, development and colouration, the slides are mounted with Canada balsam.

When the preparation in which [$^{14}$C]plastoquinone is added back to chloroplasts, from which endogenous plastoquinone had been extracted, is compared with a [$^{14}$C]labelled barley leaves preparation, identical localizations are found. However the radioactivity of the preparation in which

→

FIG. 1. Histochemical localization of plastoquinones (PQ) in chloroplasts. The three techniques are: (1) Characterization reaction of plastoquinone; (a) beginning of the reaction after illumination; (b) 15 min or 20 min after illumination; (c) control in the dark after 20 min; (d) control in the dark after a night. (2) Superposition of the histochemical Molisch reaction and the autoradiographic response of the $^{32}$P labelled compound; (e) without plastoquinone extraction; (f) after action of plastoquinone solvent. (3) Regrafting of PQ (A, B, C) on chloroplasts deprived of endogenous PQ (g), (h).

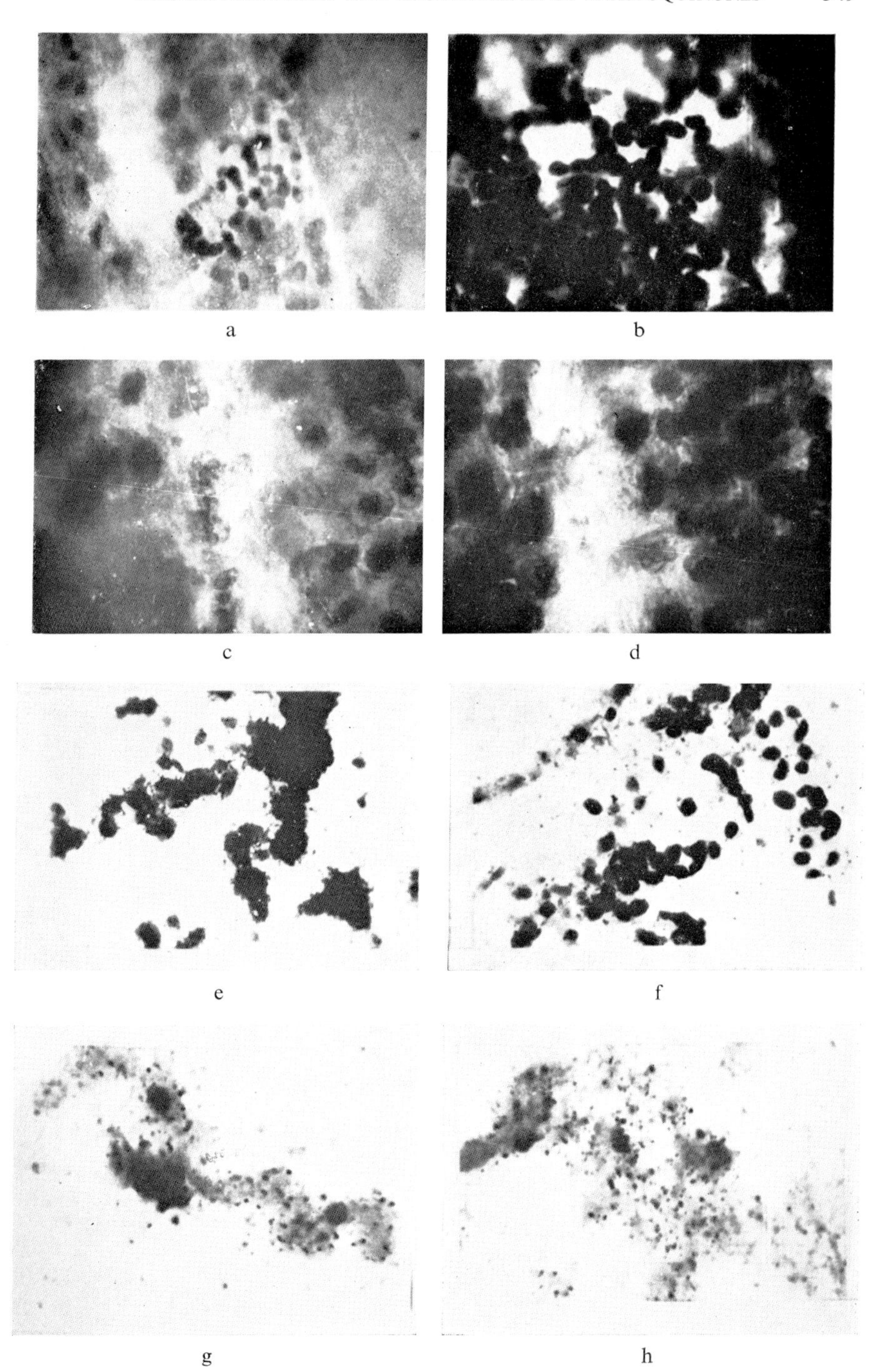

a b

c d

e f

g h

plastoquinone is added back is lower, which allows a more accurate localization to be made. The washed without any readdition controls show no activity.

Other variants using whole lyophilized leaves, the plastoquinone of which has been extracted from the fragment with subsequent readdition, yield similar results but the technique is longer, difficult, and results are not so accurate. Attempts at readdition to leaf preparations processed according to the Molisch reaction and then washed with light petroleum, yielded some analogous results but readdition is much more difficult.

In short, with both our former and new techniques, it seems we can affirm the existence of plastoquinones, visualized by different ways in chloroplasts. Detection, characterization and localization have been performed in fixed and living cells. Extraction with selective solvents has been completed by addition of chromatographically pure plastoquinones, by the study of their u.v. spectrum and by comparison of their radioactivity with that of related substances.

These different techniques (physical, biochemical, histochemical and autoradiographical) used jointly on a same preparation of the same material, allow a better characterization of plastoquinones. The results are summarized in Fig. 1.

## Summary

Attempts have been made to produce a more accurate characterization and localization of plastoquinones in living or fixed chloroplasts. Three cytological techniques are reported based on:

(1) The biochemical characterization of plastoquinone (technique of Leister and Ramasarma on thin layer chromatography slides, using neotetrazolium and borohydride or light);

(2) the superposition (on the same leaf chloroplast smear having absorbed $CO_2$) of an argentaffin reducing substance (Molisch's reaction) and of an autoradiographic [$^{14}C$] labelled substance, disappearing with the plastoquinone solvents;

(3) the extraction, purification and identification of [$^{14}C$]plastoquinones (u.v. spectrum or radioactivity) and readdition of these labelled plastoquinones to lyophilized chloroplasts deprived of endogenous plastoquinone.

These new techniques, together with the former ones, allow cellular plastoquinones to be characterized and localized.

## References

Khau van Kien, L. (1963). *In* "La Photosynthèse"—Colloques internationaux, 23/27 Juillet 1962, C.N.R.S. p. 423 Paris.

Leister, P. L. and Ramasarma, T. (1959). *J. biol. Chem.* **234**, 672.

# Oxidation-Reduction Potential of Plastoquinones

Jean-Marie Carrier
*Département de Biologie, Service de Radioagronomie, CEN, Cadarache, France*

It is now accepted that plastoquinones are involved in the electron transfer of photosynthesis. Three different types are known so far: A, B, and C, which are characterized by their chromatographic behaviour. Their u.v. absorption spectra are nearly identical and their chemical formulae are closely related: they are 2,3 dimethylated derivatives of parabenzoquinone with an isoprenoid chain, varying in length and, for B and C forms, probably more or less substituted. It has not been established yet whether the three of them are involved in photosynthesis and at what level, or whether one or several correspond to storage forms.

Their presence is essential to oxygen release and photo-phosphorylation by chloroplasts with a Hill oxidant (Bishop, 1961; Krogmann, 1961). More recently, Trebst and Eck (1963) studied photosynthetic reduction of $NADP^+$ by chloroplasts, the plastoquinone of which had been extracted with light petroleum, and they showed that this reaction could be regenerated by adding dimethylated benzoquinones with a lateral chain having at least one isoprenoid unit. Eventually Witt and his collaborators (Rumberg *et al.*, 1963) thanks to their flash differential spectrometric technique established that a plastoquinone is reduced after the intervention of an $h\nu_{II}$ photon and requires a second $h\nu_{I}$ photon for its reoxidation through a series of reactions, one of which, at least, is enzymatic. In the oxidation-reduction potential diagram attached to their scheme, plastoquinone has a definite position: level 0 at pH 7. This value, suggested in fact by Bishop (1961) after comparison with other substituted parabenzoquinones, has not been experimentally determined. More recently, Redfearn (1965) proposed the value of +120 mV, which is very close to that of cumoquinone. Consequently, we aimed at making *in vitro* potentiometric titrations. Indeed, as such potentials cannot be directly determined *in vivo* the *in vitro* reference was always used.

## Materials and Methods

The material used was obtained through extractions made in our laboratory from maple (*Acer pseudoplatanus*), fir (*Abies pectinata*) or holm oak (*Quercus ilex*) leaves. The light petroleum extract is separated on an alumina column with 7·5 % water; plastoquinone A is then obtained by elution with light petroleum ether and is then purified on another column. Plastoquinones B and C require the use of a stronger eluent (namely a mixture of light petroleum and ethyl ether, 100–2 and 100–5 respectively). Results concern plastoquinone A (also provided by Hoffmann La Roche, Basel, Switzerland) and plastoquinone C.

Substitutions on the parabenzoquinone nucleus make these materials insoluble in water. On the other hand, their solubility in 97·5% ethanol (volume by volume) amounts to 0·27 g/l at 20°, i.e. approximately $7 \cdot 10^{-4}$ equivalents/l for plastoquinone A and 0·41 g/l for plastoquinone C. A binary mixture of light petroleum and ethanol (25:75) increases the solubility ten times.

Measurements are made in two stages according to a classical method:
—measurement of the reference potential of Ag/AgCl electrode, versus hydrogen electrode in the cell

| $Pt/H_2$ | HCl | LiCl | Alcohol | Ag/AgCl |
|---|---|---|---|---|
| | 0·2 N | 0·2 N | × % v/v | |

—titration with titanous trichloride in the cell

| Ag/AgCl | HCl | LiCl | Alcohol | Quinone | Pt. |
|---|---|---|---|---|---|
| | 0·2 N | 0·2 N | × % v/v | y $N$ | |

The value is obtained by summing the half titration potential and the reference potential.

It is for convenience that the reference used was the silver chloride electrode: there is no junction bridge and consequently no non-reproducible junction potential. Besides, this electrode works reversibly in any water–ethanol mixture. On the other hand, its potential varies with the $Cl^-$ content of the medium, according to the following relation

$$E_{Ag/AgCl} = E_0 - \frac{RT}{n\mathscr{F}} \text{Log } a_{(Cl^-)}$$

where it can be seen that if $Cl^-$ activity increases, the electrode potential decreases.

The reference potential is also very dependent on any variation in the water content of alcohol (Fig. 1). In the water content of alcohol where we operate, the reference potential decreases very rapidly with the content approaching zero and it has to be measured before each titration. In all probability, this phenomenon is connected with the variations of the mixture dielectric constant according to its composition (Le Bas and Day, 1960), but studies initiated in this field show that the classical Born relation

$$E_{0(\text{ethanol+water})} - E_{0(\text{water})} = A + B\left(\frac{1}{(\epsilon)\text{ethanol+water}} - \frac{1}{(\epsilon)\text{water}}\right)$$

where $\epsilon$ is the dielectric constant of the indicated medium, does not account very well for experimental determinations (Feakins and French, 1957).

Owing to the two possible causes of variation in the reference potential, the reducing agent, and aqueous and HCl solutions (1·8 N HCl) of titanous trichloride, are diluted in the ethanol solvent. However, with a twenty-five times diluted reagent, the variation remains important (0·8 mV/ml) and is taken into account.

Titrations are performed by hand or by means of a Titrigraph type automatic Radiometer apparatus (Copenhagen).

Various quinones have been used to test the method: parabenzoquinone (after sublimation), menadione, synthetic vitamin $K_1$ (Hoffmann La Roche).

Before indicating the results obtained, the following points have to be noted: When diluted in ethanol, titanous trichloride is rapidly oxidized. It has to be

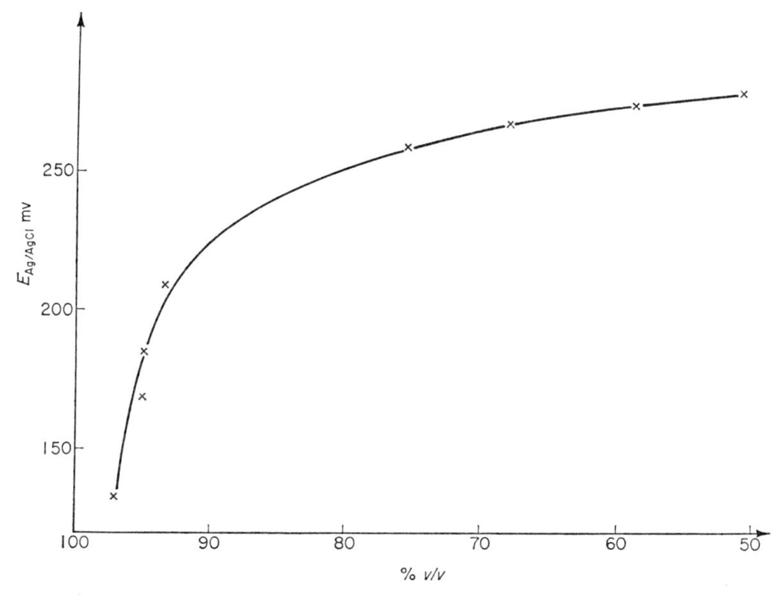


FIG. 1. Variation of the reference potential with the concentration of ethanol in water. Reference potential (mV) versus % of ethanol in water.

prepared immediately before each addition of reagent. In the automatic method, the reagent should be introduced into the apparatus all at once, at the beginning. It should be diluted in hydrochloric acid (1·8 N) where it remains stable. This introduction of water and hydrochloric acid involves an important variation in the reference potential, which has to be corrected.

## RESULTS AND DISCUSSION

### REFERENCE POTENTIAL OF THE Ag/AgCl ELECTRODE

The value found for the above concentrations in HCl and LiCl and for 97·5% ethanol is $+132$ mV $\pm 2$ mV at 20°.

Questions may be raised as to the validity of this potential, as no reference is made to the normal hydrogen electrode in aqueous medium. This problem remained unsolved so far but, with Conant and Fieser (1924), we prefer to define the oxidation-reduction potential of a quinone versus hydrogen electrode in the same solvent on the assumption that this potential directly measures the free energy of the chemical processes concerned.

Another question is that of pH, which has to be indicated with any redox potential value. There is no indicator electrode able to measure the pH in ethanol. A classical glass electrode can be used but the resulting pH value for the solution used is negative and is therefore not very significant. For want of an absolute reference, a comparison was made between the redox potential values measured on substances soluble in both water and ethanol, with the same concentrations in HCl and LiCl. $E_0'$ being the potential measured in 0·2 N HCl+0·2 N LiCl in the medium involved, and $E_0$ being the potential at pH 0, following results have been obtained:

| | $E_0'$ Ethanol 20° | $E_0'$ water 20° | $E_0$ water 25° |
|---|---|---|---|
| Quinhydrone | 696 | 693 | 699 |
| Phenazine methoxysulphate | 310 | 306 | |

The slight difference between values in water and ethanol indicates that the reference to hydrogen electrode in ethanol is consistent. On the other hand, the slight difference between $E_0'$ and $E_0$ in water allows us to place our results to pH next to zero.

## OXIDATION-REDUCTION POTENTIAL OF QUINONES

Before handling the plastoquinone samples, we tested the experimental technique on quinones of biological interest and known redox potentials.

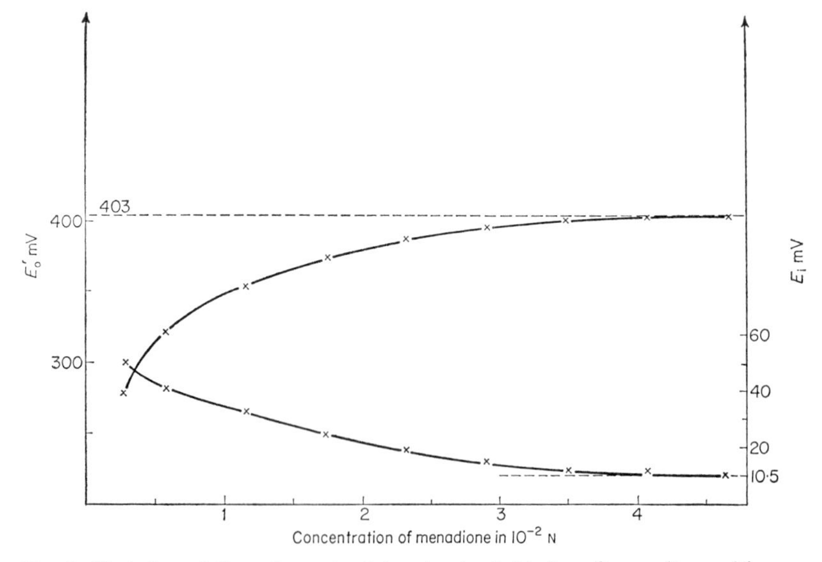


FIG. 2. Variation of the redox potential and potential index of menadione with concentration.

With a sample quinone (parabenzoquinone), the automatic titration method leads to a consistent result at all concentrations.

With a more complex quinone such as menadione (2-methylnaphthoquinone), the automatic method shows a potential variation increasing with the concentration (Fig. 2). At the same time, the potential index decreases

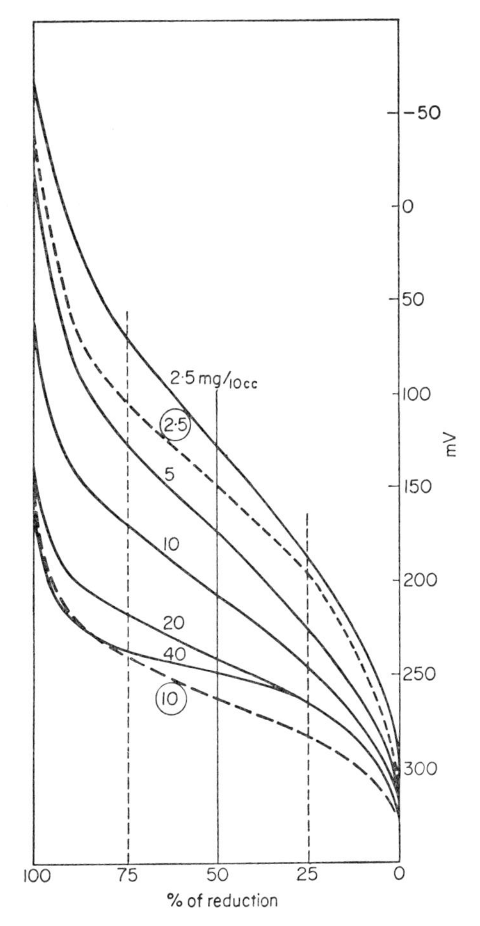


FIG. 3. Automatic titration curves of menadione; (——) ethanol + light petroleum (25%); (----) ethanol alone.

towards the theoretical value of bielectronic reaction. The highest concention leads therefore to the best result, i.e. 403 mV at 20° which is consistent to 5 mV with the literature value (Fieser and Fieser, 1935).

This variation indicates that reactions, at the indicator electrode level, are slower in ethanol than in water. The lower the concentration and the more the quinone is substituted, the more obvious this effect is. It is still more strongly marked in light petroleum (Fig. 3). The automatic method gives evidence of this phenomenon because the time which elapses between two additions of the reagent is too short. On the contrary, manual titration allows one to wait until the potential is equilibrated (i.e. 50 min) and therefore in the case of menadione, to measure the potential directly.

Conditions are different with quinones, such as vitamin $K_1$, with a long lateral chain, for which potential varies with concentration, even when operating by hand (Fig. 4). Here also the higher the concentration is, the better the result is. The value obtained (400 mV) is undoubtedly more consistent than that of 365 mV mentioned in the literature (Riegel *et al.*, 1940) where no indication of concentration is given.

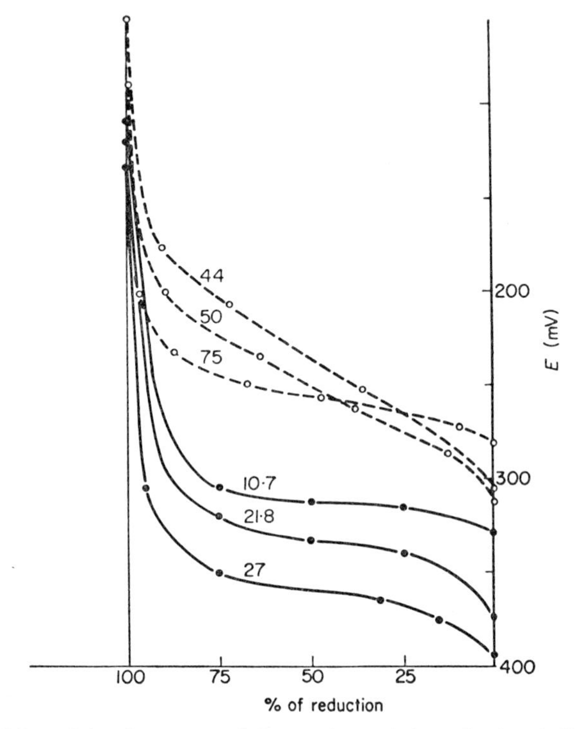


FIG. 4. Manual titration curves of plastoquinone A (●—●); vitamin $K_1$ (○—○).

The behaviour of plastoquinones is similar to that of vitamin $K_1$: potential varies to a great extent with concentration (Fig. 4). But in this case, the limited solubility does not allow measures in ethanol to be carried out in a wide range. That is the reason why a light petroleum–ethanol mixture is used (25–75). Such a medium results in lower values as can be seen with vitamin $K_1$ (Fig. 5) but the difference does not exceed 20 mV and decreases as concentration increases. Finally, the best result obtained for plastoquinone A is 499 mV, the actual value being expected to reach 20 mV more (i.e. 519 mV), because of the presence of light petroleum. Such an exact figure cannot be given for plastoquinone C, but it is interesting to note that the curve of its potential variations is parallel to that of form A, with a difference of 58 mV, so that its potential can be expected to be 461 mV.

The above results correspond to a pH value very close to zero. Arylated derivatives of parabenzoquinone are not ionized in the 0–7·5 pH range. Their potential dependence of pH therefore results in a curve parallel to that of the hydrogen electrode potential. Therefore the difference is obtained according to the relation:

$$E_7 = E_0 - 2{\cdot}3\frac{RT}{n\mathscr{F}}\text{pH}$$

i.e. 406 mV at 20°.

At pH = 7, the oxidation-reduction potential of plastoquinone A is located next to 113 mV and that of plastoquinone C next to 55 mV.

Figure 5 summarizes all the results.

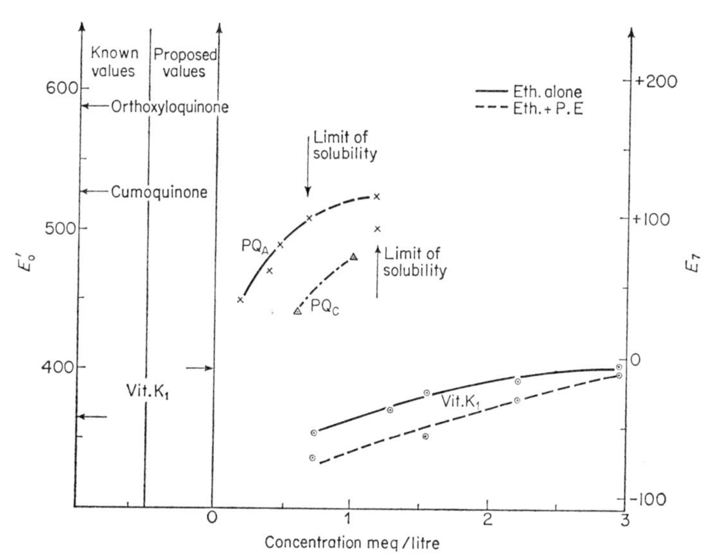


FIG. 5. Summary of measurements.

## References

Bishop, N. I. (1961). *In* "Quinones in Electron Transport" Ciba foundation symposium, p. 385.

Conant, J. B. and Fieser, L. F. (1924). *J. Am. chem. Soc.* **57**, 491.

Feakins, D. and French, C. M. (1957). *J. chem. Soc.* 2581.

Fieser, L. F. and Fieser, M. (1935). *J. Am. chem. Soc.* **57**, 491.

Krogmann, O. W. (1961). *Biochem. biophys. Res. Commun.* **4**, 4, 275.

Le Bas, C. L. and Day, M. C. (1960). *J. phys. Chem.* **64**, 465.

Redfearn, E. R. (1965). *In* "Biochemistry of Quinones" (Morton, R.A. ed.), p.175 Academic Press, London and New York.

Riegel, B., Smith, P. G. and Schweitzer, C. E. (1940). *J. Am. chem. Soc.* **62**, 992.

Rumberg, B., Schmidt-Mende, P., Weitkard, J. and Witt, H. T. (1963). *In* "Photosynthetic Mechanisms of Green Plants". Publication 1145 of Nat. Acad. Sciences, U.S.A.

Trebst, A. and Eck, H. (1963). *Z. Naturf.* **18b**, 694.

# Oxidation-Reduction Potentials in Relation to Components of the Chloroplast

R. Hill and D. S. Bendall
*Department of Biochemistry, University of Cambridge, England*

Leonor Michaelis and Mansfield Clark gave a solid foundation for the study of reversible oxidation-reduction systems by biochemists. Many of these studies have been made available in a remarkably simplified form by Dr. Malcolm Dixon (1949); his book on "Multi-Enzyme Systems" has become a classic. Some years ago we re-examined the determination of reduction-oxidation potential for some of the cytochrome *b* components in relation to the photosynthesis and respiration of certain plants. We did not publish this at the time and would like to take the opportunity of discussing our results now.

The method used in the majority of the cases has been the measurement of the ratio of oxidized to reduced form of a compound by spectrophotometry. The potential values are defined by the use of a known oxidation-reduction buffer (or rH buffer) which is in equilibrium with the substance under examination.

The limits of accuracy of our method based on rH buffers will be in terms of the oxidation-reduction potentials of the buffer mixtures, similar to the determinations of pH by using indicators. We might put the limit as 0·2 rH unit or about 6 mV. The determination of the characteristic potential of the buffer may be in error to the extent of $+0{\cdot}003$ V so that our present results will be taken as significant only to $\pm 0{\cdot}01$ V.

The determinations depend primarily on the characteristic oxidation-reduction potential ($E_0'$) of the rH buffer used. This has to be measured directly against a single half-cell whose potential is known in relation to the standard hydrogen electrode, for example, the calomel electrode or the quinhydrone gold electrode may be used. Previously the values of the two important rH buffers, ferri-ferrocyanide and ferri-ferro-oxalate had been taken from measurements in the literature. Clark, Cohen and Gibbs (1925) determined the value $E_0' = +0{\cdot}365$ V for the ferri-ferrocyanide system at pH 7 and 30°; Schaum and Linde (1903) had found the value 0·409 V at 25°. We had used the former value, whereas Nishimura (1959) used the latter for the *Euglena* cytochrome $c_2$.

The value for ferri-ferro-oxalate of $\pm 0$ V at pH 7 we originally took from Michaelis and Friedheim (1931). By direct measurement against a calomel half-cell we subsequently found that under conditions where phosphate buffer is used the $E_0'$ is more positive. Experimental data (from Bendall, 1957) are given in Table I. For cytochrome $b_6$ a correction is necessary for the value given previously (Hill, 1954). With the conditions as specified in our previous measurements the estimated correction is $+0{\cdot}06$ V, which is outside the limit of accuracy. Thus cytochrome $b_6$ would correspond to an $E_0'$ of $\pm 0$, as shown in Fig. 1, and our determination of cytochrome $b$ would be now corrected to $+0{\cdot}060$ V which is in reasonable agreement with the value of $+0{\cdot}077$ V obtained by Colpa-Boonstra and Holton (1959). The corrected value for $E_0'$ of the $b_6$ component of chloroplasts brings the potential of this nearer to that of the plastoquinones.

Table I. Oxidation-reduction potentials of the ferri-ferro-oxalate system at 20° measured against a calomel electrode containing saturated KCl ($E_h = +0{\cdot}247$ V)

| [Fe] (M) | [Pot. oxalate] (M) | [Phosphate] (M) | pH | $E_0^1$ (V) |
|---|---|---|---|---|
| 0·001 | 0·05 | 0·04 | 7·0 | +0·032 |
| 0·001 | 0·10 | 0·04 | 6·9 | +0·029 |
| 0·001 | 1·0 | 0·04 | 6·9 | +0·027 |
| 0·002 | 0·05 | 0·04 | 7·0 | +0·026 |
| 0·001 | 0·40 | 0·03 | 6·8 | +0·016 |
| 0·001 | 0·50 | 0·067 | 7·0 | +0·061[1] |

[1] Calculated value.

It is concluded that in the region $\pm 0$ V there are these two components. Again, near to $+0{\cdot}4$ V, there are the components cytochrome *f* (Davenport and Hill 1952), plastocyanin (Katoh, 1960) and P 700 (Kok, 1961). This leads to some comments in relation to the so-called "Z" scheme of electron or hydrogen transport in chloroplast systems.

Certain chlorophyll-deficient leaves, observed with a microspectroscope in strong light, gave evidence that cytochrome $b_6$ was reduced while cytochrome *f* was oxidized. This suggested that cytochrome components were involved in the photochemically induced H-transport against the thermochemical gradient. The net result of the light action is to produce oxygen and reduced coenzyme together with one equivalent of ATP, as was shown by Arnon (1959). This could all be simply explained in terms of the two light reactions: (1) oxidation of cytochrome *f* and reduction of coenzyme; (2) reduction of cytochrome $b_6$ and production of oxygen, as indicated in Fig. 1. Then, with a thermochemical reaction between reduced $b_6$ and oxidized cytochrome *f* the ATP could be formed, the reaction being analogous to oxidative phosphorylation in mitochondria. This interpretation was first proposed by Hill and Bendall (1960) and forms a useful basis for co-ordinating a variety of observations with the green plant system. In the diagram given by one of us (Hill, 1965) the connecting lines between the two light reactions were double. This was to suggest that between the two quasi-independent light absorbing systems hydrogen transport might in fact involve more than one path in parallel. There was no direct evidence for this; however, if this is assumed, then it may help to co-ordinate a variety of results. For example, if it is proposed that there is more than one photophosphorylation step these could be in parallel between the same limits of oxidation-reduction potential (for example, we have the pairs, cytochrome *f*:plastocyanin and cytochrome $b_6$:plastoquinones).

If the representation of oxidation-reduction potential levels is to include the behaviour when the system is not uncoupled but is able to phosphorylate, then there may be a modification of the potential values of certain components.

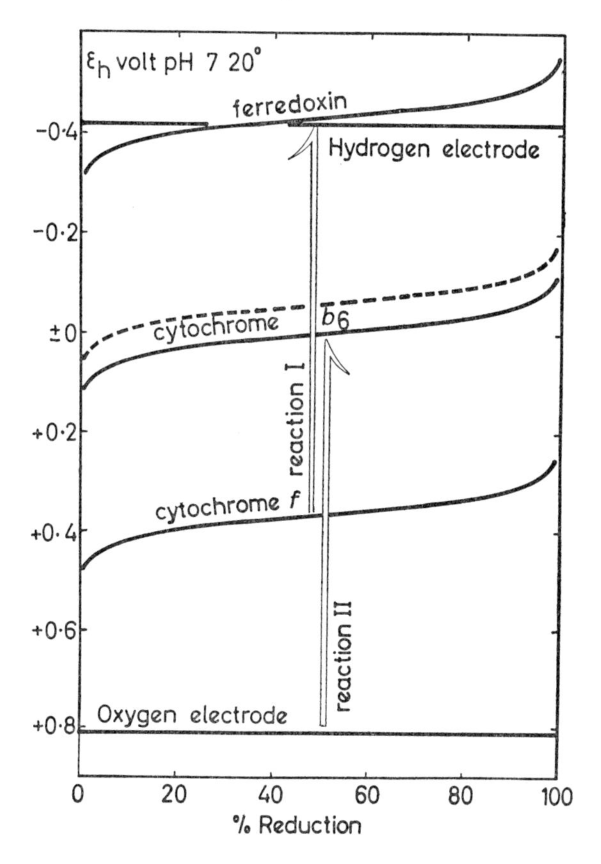


FIG. 1. Diagram to show the relation of the oxidation-reduction potential of cytochrome $b_6$ to those of cytochrome *f* and ferredoxin. The dotted line represents the former value given for cytochrome $b_6$. The two light reactions are indicated by the arrows representing photo-induced electron transport against the thermochemical gradient to the extent of 0·8 V.

We may refer back to the triose phosphate system of Warburg and Christian (1939) for relating the process of phosphorylation to oxidation-reduction potentials. If the system is normal the reduction of NAD in equilibrium with equal concentrations of phosphate ($P_i$), triose-phosphate (TP) and diphosphoglyceric acid (DPGA), represents an oxidation-reduction potential of −0·286 V at pH 7 (Krebs and Kornberg, 1957). If the system is "uncoupled" with arsenate then the NAD is completely reduced because the potential of the system

TP + phosphoglyceric acid (PGA) is below that of hydrogen (−410 mV, pH 7). The difference 410 − 286 = 124 mV for two equivalents of H represents a minimal estimate of the energy available for $P_i \rightarrow \sim P$. Hence it follows that in any H-transport system concerned with phosphorylation the oxidation-reduction potential of the crucial component will be different when the system is uncoupled. The triose-phosphate system becomes more positive when DPGA is a product of oxidation as compared with PGA. Thus we might look for a change in the apparent oxidation-reduction potential of some component of a hydrogen transport chain when the coupled and uncoupled systems are compared.

What we would like to do now is to attempt to describe a speculative "world" of the chloroplast. This, while having, let us say, no counterpart in the real "world", seems to be useful for co-ordinating a miscellaneous number of observations in a way which gives a sense of coherence. It seems that we are justified in assigning the conventional oxidation-reduction potentials, which refer actually to ideal solutions, to components of an electron transport chain where the components are part of the structure. This may be still somewhat of a puzzle from the theoretical standpoint. When it is attempted to relate the photochemical behaviour of the chemically active chlorophyll *a* component to the oxidation-reduction potentials we are still on dangerous ground. Yet in the case of reaction I there seems to be some justification for this. Bessel Kok (1961) has determined the oxidation-reduction potential of the pigment P 700 discovered by him. He has concluded that it is near to the potential of ferro-ferricyanide (+0·4 V) and that one equivalent is involved; the oxidized form is considered to be a free radical and is responsible for a light-induced EPR signal given by both whole cells and by chloroplast preparations. The work of Krasnovsky (1960) and his colleagues on the reduced chlorophyll system gives a potential of the order of −0·4 V. Chlorophyll *a* seems to have two oxidation-reduction potentials differing by 0·8 V; this is about the range of potential covered in the hypothetical "Z" scheme of hydrogen transport for each of the two photochemical reactions, but for reaction II the range is assumed to be at a more positive potential. Thus, taking for granted that chlorophyll has two different oxidation-reduction potentials separated by 0·8 V, we might form a rough picture of how a substance can be reduced simultaneously with a substance being oxidized when energy is supplied by absorption of light. It is possible to consider that absorption of the light could mediate an electron transfer between receptor and donor types of molecule whose respective characteristic oxidation and reduction potentials may differ by 0·8 V. The 0·8 V difference might be at any position on the rH scale—that is the characterististic potentials of the acceptor and donor molecule need have no fixed relation with the H-electrode. This would allow a close analogy between reaction I and reaction II. In these two reactions we represent the range as 0·8 V in each case. If reaction I was modified to have an electron donor of characteristic potential ±0 V, then it might be expected that an acceptor might be found which would have a characteristic potential of −0·8 V. It might be useful now to consider the structure of the porphyrin, chlorin or bacteriochlorophyll nucleus in a diagrammatic way.

A porphyrin might be regarded as a neutral hybrid structure between two oppositely charged structures, all three structures would have a complete system of conjugate bonds. The accepted structure for a neutral porphyrin has eleven double bonds. If two hydrogen atoms are removed then the oxidized conjugate structure will have twelve; if two hydrogens are added there is a completely conjugated system of ten double bonds. This very rough picture (Fig. 2) may help to co-ordinate a number of observations. The magnesium compounds, for example, can more easily occur in a reversibly oxidized form than in a reduced form. The free porphyrins can more easily occur in a reversibly reduced form rather than oxidized. The stable reduction products are neutral molecules in the present sense and have been characterized by Granick

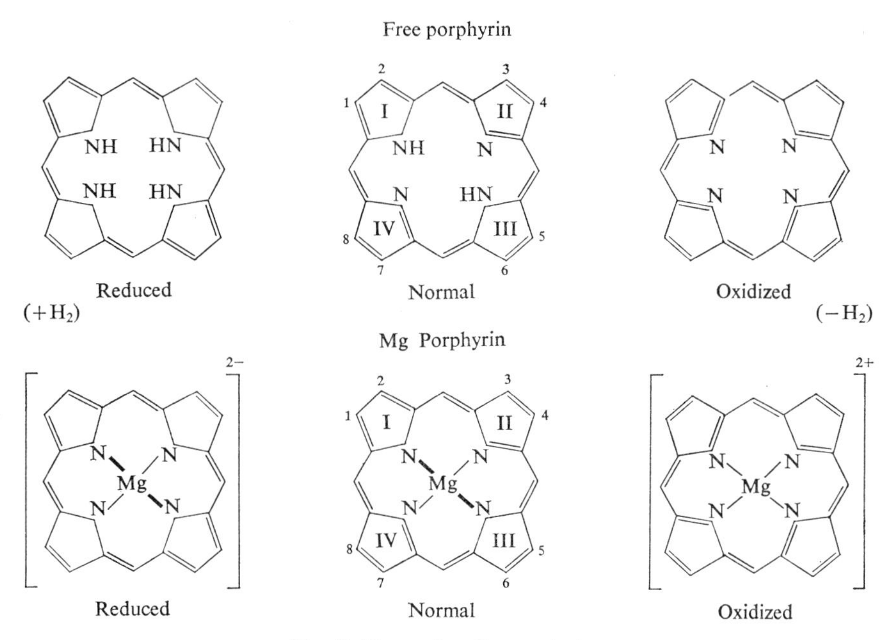


FIG. 2. For explanation see text.

and colleagues. The induced electron transport is more easily pictured with two molecules of chlorophyll in contact. There would seem to be little chance of actually isolating the hypothetical charged forms because they are defined only in terms of electron transport with pairs of other substances. This representation is virtually unaltered whether we consider the porphyrins, chlorins or bacteriochlorophyll, because positions 3,4 and 7,8 in rings II and IV are not affected by the proposed rearrangements of the conjugate system of double bonds. As is a fact of observation hydrogenation of positions 7,8, as in chlorophyll, and 3,4 and 7,8 as in bacteriochlorophyll, favours oxidation-reduction reactions.

The so-called "Z" scheme was put forward as a working hypothesis. A definite disproof of this kind of formulation would mark a great step forward in our understanding of the chemical change induced by absorption of light. It seems evident that if this scheme is taken literally and one photon is related to one hydrogen equivalent then ten or even twelve photons per $CO_2$ would be required for the formation of $O_2$ and carbohydrate if 3 ATP are necessary for this process. This seems to be against the best experimental data. The structure of a chlorophyll molecule is very different from that of an atom; can we propose one photon is related to two hydrogen equivalents? The energy of 1 mole quantum at 7000 Å is represented by about 1·78 electron volts. The difference in potentials between the couples chlorophyll : oxidized chlorophyll and chlorophyll : reduced chlorophyll is of the order of 0·8 electron volts. So that we could have a consistent scheme which conforms with the present measurements of oxidation-reduction potentials of known components.

In summary, the question here raised is whether the photo-induced electron transport depends on a single pair of substances, one being reduced, the other being oxidized, or whether a double pair of substances has to be present.

## References

Arnon, D. I. (1959). *Nature, Lond.* **184**, 10.
Bendall, D. S. (1957). Ph.D. Thesis, Cambridge University.
Clark, W. M. Cohen, B. and Gibbs, H. D. (1925). Studies on Oxidation-Reduction, VIII. *Public Health Repts.* (*U.S.*), **40**, 649.
Colpa-Boonstra, J. and Holton, F. A. (1959). *Biochem. J.* **72**, 4P.
Davenport, H. E. and Hill, R. (1952). *Proc. R. Soc.* B **139**, 327.
Dixon, M. (1949). "Multi-Enzyme Systems." Cambridge University Press.
Hill, R. (1954). *Nature, Lond.* **174**, 501.
Hill, R. (1965). *In* "Essays in Biochemistry" (P. N. Campbell and G. D. Greville, ed.), p. 121. Academic Press, London and New York.
Hill, R. and Bendall, F. (1960). *Nature, Lond.* **186**, 136.
Katoh, S. (1960). *Nature, Lond.* **186**, 533.
Kok, B. (1961). *Biochim. biophys. Acta* **48**, 527.
Krasnovsky, A. A. (1960). *Annu. Rev. Pl. Physiol.* **11**, 363.
Krebs, H. A. and Kornberg, H. L. (1957). "Energy Transformations in Living Matter." Springer-Verlag, Berlin.
Michaelis, L. and Friedheim, E. (1931). *J. biol. Chem.* **91**, 343.
Nishimura, M. (1959). *J. Biochem., Tokyo* **46**, 219.
Schaum, K. and von der Linde, R. (1903). *Z. Elektrochem.* **9**, 406.
Warburg, O. and Christian, W. (1939). *Biochem. Z.* **303**, 40.

# Photosynthetic Reactions with Pyridine Nucleotide Analogs: I. Isonicotinic Acid Hydrazide-NAD[1]

Peter Böger, Clanton C. Black and Anthony San Pietro
*Charles F. Kettering Research Laboratory*
*Yellow Springs, Ohio, U.S.A.*

## Introduction

The photoreduction of pyridine nucleotides by chloroplasts requires ferredoxin and a flavoprotein, ferredoxin-NADP reductase (San Pietro and Lang, 1956, 1958; Keister *et al.*, 1960, 1962; Tagawa and Arnon, 1962). It is well established that the reaction is specific for NADP; NAD is reduced at a greatly decreased rate (about 1 %) under comparable conditions. Furthermore, the reduction of NADP is accompanied by the formation of ATP in stoichiometric amount. In the experiments reported herein, the rate of ATP formation was approximately 150 μmoles/hr/mg chl[2] when NADP and ferredoxin were provided at optimal concentrations. The ferredoxin-NADP reductase was present in the chloroplasts, as prepared, in sufficient amount to catalyse this rate of reduction of NADP.

Although pyridine nucleotide analogs have been used to characterize the isomeric forms of certain enzymes (Anderson *et al.*, 1962; Cahn *et al.*, 1962; Chancellor-Madison and Noll, 1963); for example, lactic dehydrogenase, they have been seldom used in chloroplast reactions. Some years ago, San Pietro (1961) found that the 3-acetylpyridine analog of NADP was reduced by illuminated chloroplasts at a rate comparable to that of NADP reduction. In addition the 3-acetylpyridine analog of NAD behaved similarly to NAD.

Keister *et al.* (1960) used the 3-acetylpyridine and the deamino analogs of both NAD and NADP to study the pyridine nucleotide specificity of the ferredoxin-NADP reductase. They used as their assay the pyridine nucleotide transhydrogenase activity of the flavoprotein and measured the maximal velocity of the reaction with, and the Michaelis constant of, the various nucleotide analogs. Based on the kinetic measurements, they concluded that the flavoprotein was specific for NADP and this conclusion has been verified recently by Shin and Arnon (1965).

In a further investigation of the apparent involvement of an aldehyde in

[1] Contribution No. 209 of the Charles F. Kettering Research Laboratory. This work was supported in part by grants GM-10129 to A.S.P. and GM-12273 to C.C.B. from the National Institutes of Health, United States Public Health Service.

[2] The following abbreviations are used: ADPR, adenosine diphosphate ribose; INH-NAD and INH-NADP, isonicotinic acid hydrazide analogs of NAD and NADP, respectively; INH, isonicotinic acid hydrazide; P, phosphate; R, ribose; N, nicotinamide; PMS, *N*-methylphenazinium methyl sulfate (phenazine methosulfate); CMU, 3-(*p*-chlorophenyl)-1, 1-dimethylurea; CCCP, *m*-chlorocarbonyl cyanide phenylhydrazone; chl, chlorophyll; and EPR, electron paramagnetic-resonance.

photophosphorylation with vitamin $B_6$, Black and San Pietro (1963) observed that the 3-pyridine aldehyde analog of NADP-catalysed photophosphorylation in the absence of spinach ferredoxin. Thus, an investigation of the pyridine nucleotide specificity of photosynthesis was undertaken.

Primarily this paper reports experiments on several photosynthetic reactions with INH-NAD. Further reports will deal in more detail with photosynthetic reactions in the presence of other analogs. A preliminary report of these experiments has been presented (Black *et al.*, 1965).

## Methods

Preparation of chloroplasts and the ATP assay have been described (Black *et al.*, 1963). In control experiments, the glucose-hexokinase-Zwischenferment system gave results similar to the $AT^{32}P$ assay. Experiments were carried out routinely with chloroplast fragments prepared from spinach purchased from the local market. Some experiments, as noted, were conducted with chloroplast fragments from peas grown in the greenhouse at 20 to 25° for 14 to 20 days. Chromatophores were prepared by the method of Vernon (1963) and stored under argon in the refrigerator.

Table I. Photophosphorylation with various analogs[1]

| Analog | Final molar concentration[2] | $\frac{\mu\text{moles ATP}}{\text{mg chl.hr}}$ |
|---|---|---|
| Isonicotinic acid hydrazide-NAD | $5\times10^{-4}$ | 150 |
| Isonicotinic acid hydrazide-NADP | $3\times10^{-4}$ | 103 |
| Thionicotinamide-NAD | $10^{-2}$ | 34 |
| 3-Acetylpyridine-NAD | $5\times10^{-3}$ | 25 |
| 3-Acetylpyridine-deamino-NAD | $10^{-2}$ | 54 |
| 3-Pyridinealdehyde | $10^{-4}$–$10^{-2}$ | 5 |
| 3-Pyridinealdehyde-NAD | $10^{-3}$ | 100 |
| 3-Pyridinealdehyde-deamino-NAD | $5\times10^{-3}$ | 128 |

[1] The standard reaction mixture used throughout all experiments contained in $\mu$moles: *tris*-HCl buffer, pH 7·8, 40; $MgCl_2$, 2; $P+{}^{32}P$ (0·4 to 1·5 $\mu$c of $^{32}P$), 1; ADP, 1; chloroplast fragments containing 15 to 30 $\mu$g of chlorophyll; total volume, 1 ml. Reaction mixtures were illuminated in cuvettes of 1 cm light path with white light of approximately 24,000 ergs/cm²/sec at room temperature. Gas phase: air.

[2] Optimal concentration.

Pig-brain NADase was prepared as described by Kaplan (1955) and used to synthesize the pyridine analogs (Kaplan, 1957). The deamino analogs were also prepared by the method of Kaplan (1957). The concentration of the INH analogs was calculated from the optical density at 385 m$\mu$ in 0·1 N NaOH using a millimolar extinction coefficient of 4·4 (Zatman *et al.*, 1954). The pyridones of NAD were prepared according to a modified method of Leighton (1956). NAD, NADP, and the analogs listed in Table I were purchased from P-L Biochemicals, Milwaukee. The other NAD analogs tested were a generous gift of Dr. N. O. Kaplan, Brandeis University. INH is a product of General Biochemicals, Laboratory Park, Chagrin Falls, Ohio.

Phosphorylations under anaerobic conditions were carried out in Thunberg-type glass cuvettes. They were evacuated three times to a pressure of 5 mm of mercury and flushed subsequently with argon or air, respectively. Ferredoxin was not present unless stated otherwise.

Cytochrome *c* reduction was measured according to Keister and San Pietro (1963) except that KCN was omitted. The chlorophyll content was 4–5 $\mu$g/ml.

The preparation of spinach ferredoxin, ferredoxin-NADP reductase and the antibody against the latter protein will be described elsewhere.

EPR signals were measured with a Varian Model 4500 EPR spectrometer, equipped with a slotted cavity for illumination. Oxygen uptake was measured at 22° with a Clark type electrode, manufactured by the Yellow Springs Instrument Company, in connection with a Varian graphic recorder. Unless otherwise noted, the light intensity used in all experiments was saturating; that is, the rate of the reaction was independent of light intensity.

## Results

### PHOTOPHOSPHORYLATION WITH PYRIDINE NUCLEOTIDE ANALOGS

In recent years, Kaplan and his collaborators have synthesized a number of analogs as illustrated in reaction (1). In these syntheses NADase is used to cleave the nicotinamide component from NAD forming an enzyme-ADPR complex (reaction 1a). This complex can either undergo hydrolysis or react with certain pyridine derivatives, for example, INH (reaction 1b).

$$\underset{\text{(NAD)}}{\text{A–R–P–P–R–N}} + \text{INH} \rightarrow \underset{\text{(INH-NAD)}}{\text{A–R–P–P–R–INH}} + \text{N} \qquad (1)$$

$$\text{A–R–P–P–R–N} + \text{ENZ} \leftrightarrows \text{ENZ–R–P–P–R–A} + \text{N} \qquad (1a)$$

$$\text{ENZ–R–P–P–R–A} + H_2O \rightarrow \text{ENZ} + \text{ARPPR} \qquad (1b)$$

$$\text{ENZ–R–P–P–R–A} + \text{INH} \rightarrow \text{ENZ} + \underset{\text{(INH-NAD)}}{\text{A–R–P–P–R–INH}}$$

The pyridine nucleotide analogs, which gave measurable rates of photophosphorylation are listed in Table I. These rates, while representative, were not always obtained with each preparation of chloroplast fragments but varied by about 50%. This variation depended primarily on the batch of spinach. Furthermore, chloroplasts are subject to seasonal changes as reported by Heber (1960).

A number of analogs were assayed within the concentration range of $5 \times 10^{-5}$ to $2 \times 10^{-3}$ M. The following analogs catalysed a rate of photophosphorylation of less than 10 $\mu$moles of ATP/mg chl/hr: 3-amino-pyridine-NAD; 5-methyl-NAD; 5-amino-NAD; propylpyridyl-ketone-NAD; ethylnicotinate-NAD; ethylpyridyl-ketone-NAD; 4-methyl-NAD; *N*-methyl-nicotinamide-NAD; nicotinyl hydroxamic acid-NAD; nicotinic acid-NAD; butylpyridyl-ketone-NAD; benzylpyridine-NAD; 4-amino-5-imidazole-carboxamide-NAD; NAD; NADP; deamino-NADP; ADPR; and pyridones (presumably the 2- and 4-pyridones) of NADP.

Photophosphorylation with all analogs showed a marked dependence on concentration. The effect of analog concentration on the INH-NAD catalysed photophosphorylation, shown in Fig. 1, is representative of all analogs tested. The concentration at which optimal activity was observed varied with the analog (Table I). A similar type of curve has been reported for the PMS-catalysed photophosphorylation by Jagendorf and Avron (1958) with an optimal concentration between 1 and $6 \times 10^{-5}$ M. Our data indicate that the optimal concentrations for the INH analogs are some ten times higher. The optimal concentration of INH-NAD was the same under aerobic and anaerobic

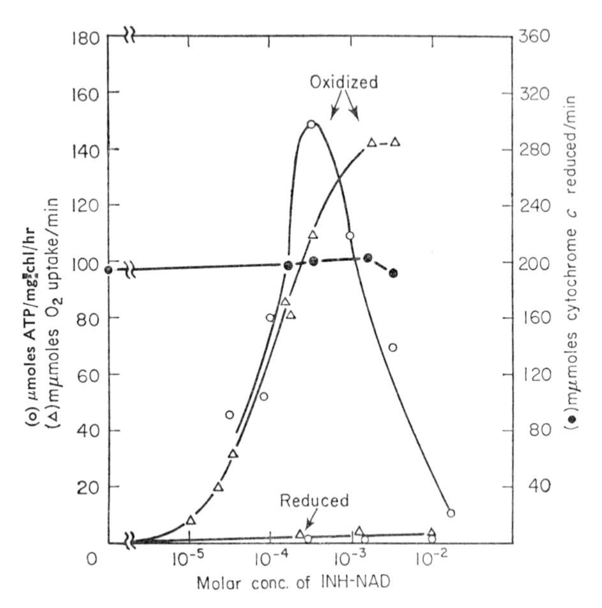


FIG. 1. Oxygen uptake and photophosphorylation with INH-NAD and the effect of the analog on cytochrome *c* reduction. Chlorophyll content was 115 µg chl/4 ml for oxygen uptake; the reaction mixture for photophosphorylation is given in Table I. Cytochrome *c* reduction was measured as described in METHODS. Gas phase: air. Curve labeled reduced: analog treated with borohydride.

conditions. Furthermore, the same optimal concentration and photophosphorylation rates were measured with pea chloroplast fragments. Similar data have been reported by Black (1965) for photophosphorylation with aromatic aldehydes.

No stimulation of photophosphorylation by INH-NAD with chromatophores of *Rhodospirillum rubrum* was detected in the absence or presence of succinate using 10 or 18-day-old chromatophores.

## PHOTOPHOSPHORYLATION ACCOMPANYING SYNTHESIS OF INH-NAD

To confirm that the photophosphorylation catalysed by INH-NAD was not due to an impurity, ATP formation was related to the synthesis of the analog.

Figure 2 shows the time course for synthesis of INH-NAD and the yield after 22 hr was approximately 85%. In accordance with the synthesis, the photophosphorylation rate increased with synthesis of the analog and decreased after the amount of INH-NAD surpassed the optimal concentration (Fig. 1).

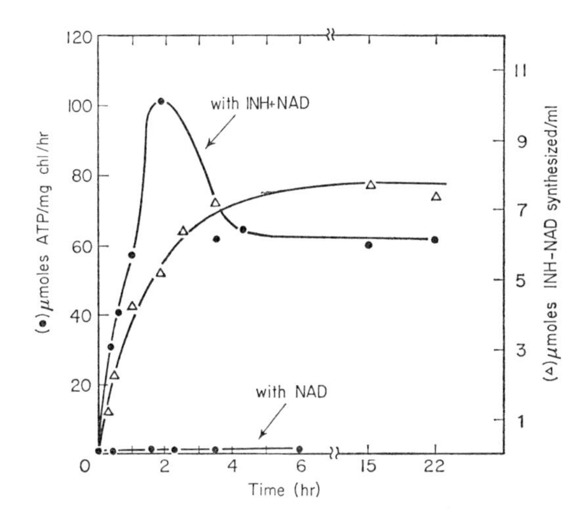


FIG. 2. Photophosphorylation and INH-NAD synthesis. Gas phase: air. At various times, a 0·1 ml aliquot was removed and used to measure photophosphorylation in the standard reaction mixture of 1 ml volume (see Table I).

There was no photophosphorylation observed when INH was omitted from the reaction mixture. These results indicate that the activity was due to INH-NAD and not due to impurities present in, or released from, the components of the reaction mixture used for synthesis and purification of the analog.

Table II. Influence of INH and NAD on INH-NAD catalysed photophosphorylation

| Catalyst in final molar concentration | μmoles ATP / mg chl.hr |
|---|---|
| INH-NAD, $10^{-4}$ | 41 |
| NAD, $5\times10^{-4}$ | 4 |
| INH, $5\times10^{-4}$ | 2 |
| INH-NAD, $10^{-4}$+NAD, $2\times10^{-4}$ | 42 |
| INH-NAD, $10^{-4}$+NAD, $5\times10^{-4}$ | 42 |
| INH-NAD, $10^{-4}$+NAD, $10\times10^{-4}$ | 35 |
| INH-NAD, $10^{-4}$+INH, $2\times10^{-4}$ | 41 |
| INH-NAD, $10^{-4}$+INH, $5\times10^{-4}$ | 40 |
| INH-NAD, $10^{-4}$+INH, $10\times10^{-4}$ | 41 |

Gas phase: air.

Furthermore, it was shown that neither NAD nor free INH catalyses photophosphorylation (Table II) and, in the concentrations listed, they do not substantially influence photophosphorylation catalysed by INH-NAD. Since the pyridones of NAD were inactive, it is unlikely that an oxidation product of INH-NAD is active. To rule out this possibility further, the analog was synthesized under argon or air. Both samples showed the same reaction kinetics and photophosphorylation activity based on INH-NAD concentration.

## REDUCTION OF INH-NAD

INH-NAD forms a peak at 385 mμ in alkaline solution but not after treatment with borohydride or dithionite in excess (a slight peak was observed at 340 mμ probably due to NADH). The disappearance of the peak at 385 mμ in alkaline solution is tacitly assumed here to indicate reduction of the analog,

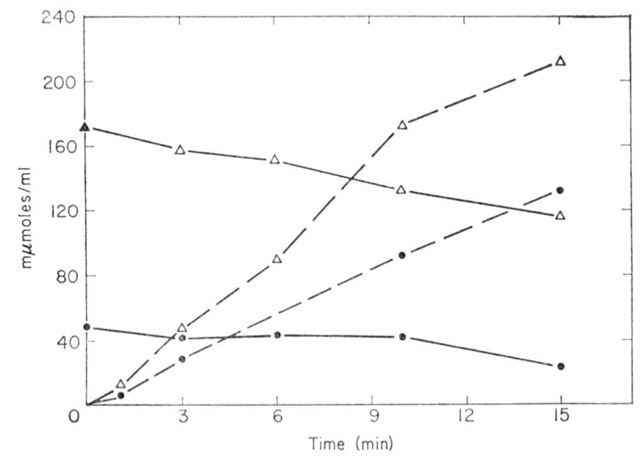


FIG. 3. INH-NAD catalysed photophosphorylation and accompanying decrease of INH-NAD. Gas phase: argon; 40 μg chlorophyll/ml. ●, 45 mμmoles INH-NAD at start, △, 170 mμmoles INH-NAD at start, ——INH-NAD, ----ATP.

although it is not known to which carbon atom of the pyridine ring the hydrogen is attached. With borohydride, the hydrogen atoms will presumably go to carbon-2 and-5 (Lyle et al., 1962; Sund *et al.*, 1964). The reduced analog was not reoxidized by air or $H_2O_2$ (pH 7 to 8). There was no reoxidation with ferricyanide, probably because the INH-NAD was destroyed by splitting off the $-NH-NH_2$ group (Wojahn, 1952).

Figure 3 demonstrates that prolonged illumination of INH-NAD with chloroplast fragments resulted in a slight reduction of the analog. This has not been possible heretofore with a number of dehydrogenases (Zatman *et al.*, 1954). Under aerobic conditions, the rate of reduction was decreased slightly (approx. 30%). In either case, subsequent treatment with air in the dark did not result in any detectable reoxidation of the reduced analog. Addition of ferredoxin did not increase the reduction.

When INH-NAD was reduced with borohydride and the excess borohydride oxidized with acetone, phosphorylation was not observed with the reduced INH-NAD either under aerobic or anaerobic conditions.

In one experiment, INH-NAD and chloroplast fragments were illuminated for 2 hr with saturating white light, while another aliquot was kept in the dark. After illumination, the first sample (L) contained 60% less INH-NAD than the dark control (D). The photophosphorylation rates obtained with aliquots of these two samples are shown in Fig. 4. The rates for both are linear and fall on the same line when plotted as a function of INH-NAD concentration. It is concluded that: (a) only the oxidized INH-NAD catalyses photophosphorylation; and (b) that the reduced form does not inhibit phosphorylation.

The reduced form of INH-NAD obtained by treatment with borohydride may not be the same as that obtained by illuminated chloroplast fragments.

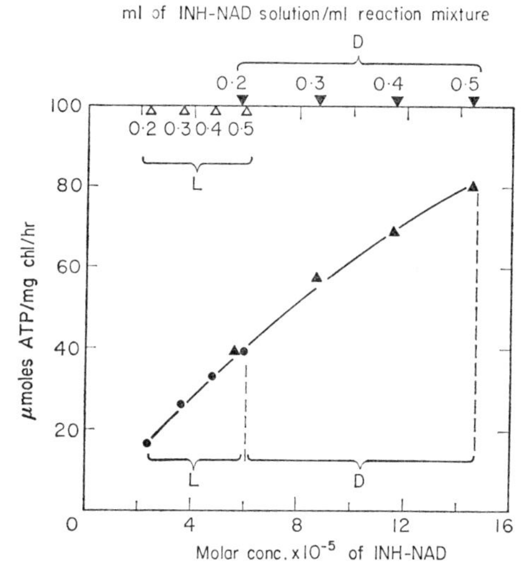


FIG. 4. INH-NAD catalysed photophosphorylation. INH-NAD ($2{\cdot}9 \times 10^{-4}$ M) and chloroplast fragments (32 μg chl/ml; 0·02 M *tris*-HCl buffer, pH 7·8) were illuminated for 2 hr. An aliquot was kept in the dark. After centrifugation, aliquots of the supernatant of the illuminated sample (L) and the dark one (D), upper abscissa, were made up to 1 ml with the standard reaction mixture as in Table I. See lower abscissa for final concentration. Gas phase: air.

However, both are similar in that: they have no peak at 385 mμ in alkali; they do not catalyse photophosphorylation; they are not reoxidized by air or $H_2O_2$; and they do not mediate oxygen uptake.

The polarographic half-wave potential of INH-NAD was found to be similar to that of NAD. The $E_0'$ for NAD is different when it is determined by thermodynamic equilibria in enzymatic redox reactions (Clark, 1960).

## INFLUENCE OF FERREDOXIN AND ANTIBODY TO FERREDOXIN-NADP REDUCTASE

The data in the first column of Table III demonstrate the catalysis of photophosphorylation by ferredoxin alone. In contrast to the previous findings of

Fewson *et al.* (1963) who used an antibody against a crude enzyme preparation, we could not confirm that the ferredoxin-catalysed photophosphorylation was inhibited by an antibody made against a highly purified ferredoxin-NADP reductase. The presence of the antibody did cause a 95% inhibition of photophosphorylation with NADP (Table II, column 2).

Table III. Photophosphorylation with INH-NAD and ferredoxin

| Additional components | Nucleotides in final molar conc. — | NADP $5\times10^{-4}$ | INH-NAD $3\times10^{-4}$ |
|---|---|---|---|
| −Ferredoxin | 2 | 3 | 116 |
| +Ferredoxin | 20 | 142 | 126 |
| −Ferredoxin+antibody | 3 | 5 | 144 |
| +Ferredoxin+antibody | 23 | 27 | 154 |

Final concentration of: ferredoxin, 0·23 mg protein/ml; antibody, 5 mg protein/ml. Phosphorylation rates are expressed as $\mu$moles ATP/mg chl/hr. Gas phase: air.

Photophosphorylation catalysed by INH-NAD, however, was not affected by the addition of ferredoxin or the antibody to ferredoxin-NADP reductase (Table III, column 3). The slight stimulation observed was due probably to a non-specific stimulatory effect of protein and was also observed by the addition of control sera.

### "CYCLIC" PHOTOPHOSPHORYLATION

INH-NAD supported a "cyclic" photophosphorylation (Fig. 3). After 15 min illumination, 0·045 mM analog yielded 0·130 mM ATP under anaerobic conditions. Under air, the photophosphorylation rate was 25 to 35% higher (see Fig. 6). The addition of an ethanol and catalase trap did not influence photophosphorylation under air or argon.

The reaction was linear during the routine experimental time employed (2–5 min.).

### OXYGEN UPTAKE

Oxygen uptake was demonstrable during ATP formation with INH-NAD as the cofactor (Fig. 1). It increased with increasing concentration of the analog and leveled off to a plateau. There was no decrease in oxygen uptake at higher concentrations of analog as was found for photophosphorylation. INH-NAD, reduced with borohydride, did not catalyse oxygen uptake with chloroplast fragments (Fig. 1).

### INHIBITORS

Both photophosphorylation and oxygen uptake were strongly inhibited by CMU as indicated in Fig. 5 and Table IV. Inhibition of the former was equal under air or argon and the concentration of the herbicide required to eliminate oxygen uptake was about the same as that needed to abolish photophosphorylation with INH-NAD or with ferredoxin plus NADP (*ca* $10^{-5}$ M). Antimycin A inhibited photophosphorylation to the same extent with both NADP and INH-NAD. Valinomycin had no substantial effect at concentrations up to $10^{-5}$ M.

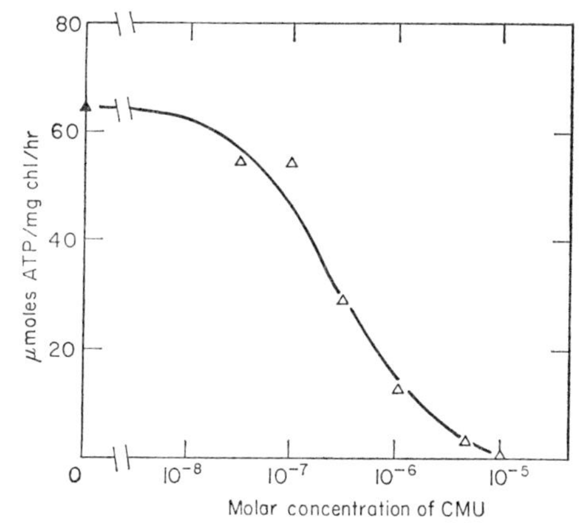


FIG. 5. CMU inhibition of INH-NAD-catalysed photophosphorylation. Gas phase: air.

Both the known uncouplers of photophosphorylation, CCCP and $NH_4Cl$, completely abolished ATP formation while oxygen uptake was stimulated slightly at concentrations of $5 \times 10^{-6}$ and $10^{-2}$ M, respectively (Table IV). This

Table IV. Influence of inhibitors of photophosphorylation on oxygen uptake

| Inhibitor | Final molar concentration | $O_2$ uptake in % of control |
|---|---|---|
| Control | | 100 |
| CMU | $1\cdot3 \times 10^{-6}$ | 33 |
| CMU | $2\cdot5 \times 10^{-6}$ | 0 |
| CMU | $5 \times 10^{-6}$ | 0 |
| CCCP | $5 \times 10^{-6}$ | 153 |
| $NH_4Cl$ | $2\cdot5 \times 10^{-3}$ | 112 |
| $NH_4Cl$ | $2 \times 10^{-2}$ | 116 |

No $O_2$ uptake occurred in the dark.

probably was due to an increased electron flow in the presence of uncoupling agents (Krogmann *et al.*, 1959). Additional information on inhibitors is given in the section on EPR signal.

### LIGHT INTENSITY

In Fig. 6 the phosphorylation rates are plotted against light intensity with the electron acceptors present in optimal concentrations. Both the INH-NAD and PMS catalysed photophosphorylations were inclined to reach a constant rate near the highest light intensity used. At lower light intensities, however, the efficiency was markedly different when PMS was compared with INH-NAD. Below 500 ergs/cm²/sec the rate with INH-NAD was twice as high as that with PMS. With increasing light intensities, the PMS rate was accelerated and

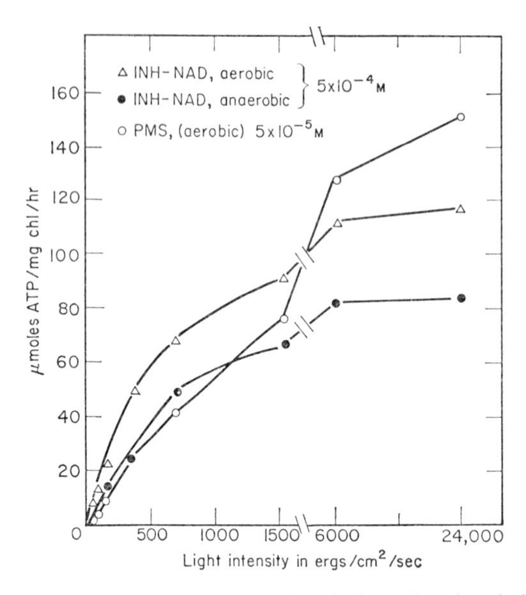


FIG. 6. INH-NAD and PMS-catalysed photophosphorylation.

exceeded that of INH-NAD above 6000 ergs/cm²/sec. With PMS, the rate increased almost linearly up to 1500 ergs/cm²/sec while with INH-NAD it was linear only to about 750 ergs/cm²/sec. As we would have expected, the phosphorylation rate under anaerobic conditions was lower than under air within the range of light intensity used.

## CHLOROPHYLL CONCENTRATION

The amount of ATP formed in INH-NAD catalysed phosphorylation was related in a linear fashion to chlorophyll concentration (Fig. 7). Higher chlorophyll contents than indicated were difficult to handle mainly because of sedimentation of the chloroplast material. With lower chlorophyll contents the rate of photophosphorylation with PMS was higher than with INH-NAD but vice versa with chlorophyll contents above 100 μg/ml. With higher chlorophyll contents, the light intensity in the sample will decrease. PMS catalysed photophosphorylation was linear only up to 40 μg chl/ml and then inclined (Fig. 7). According to Fig. 6, a decreased light intensity would affect the phosphorylation with PMS more than with INH-NAD. In the former case, light intensity became limiting whereas this situation was not observed for phosphorylation with the analog under the conditions of this particular experiment.

Linearity of photophosphorylation up to such high chlorophyll contents, as indicated here, was not found with other cofactors employing the same assay. With ferredoxin (Fewson *et al.*, 1963), phosphodoxin (Black *et al.*, 1963), and vitamin $B_6$ (Black and San Pietro, 1963), the rates were linear only up to about

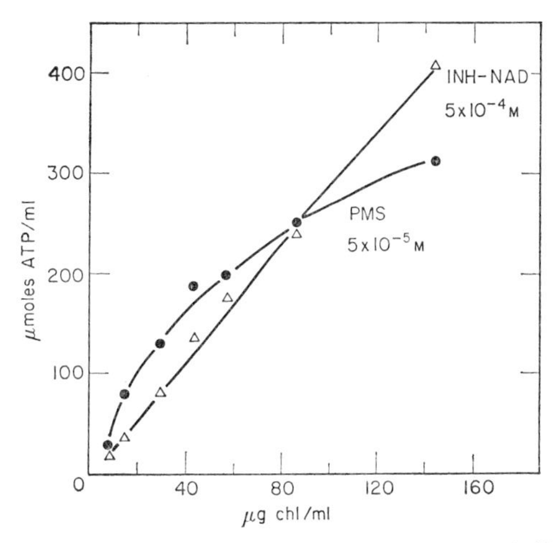


FIG. 7. INH-NAD-catalysed photophosphorylation and chlorophyll concentration. Gas phase: air.

50 μg chl/ml. The data of Allen *et al.* (1958) indicated linearity with flavin mononucleotide and vitamin $K_5$ below 100 μg chl/ml. However, one should consider the difficulties involved in comparing data when different methods are used.

## EPR SIGNAL

The EPR signals of illuminated chloroplast fragments in the absence of added cofactors were reported by Commoner *et al.* (1959, 1961) and Calvin and Androes (1963). We shall present some findings with chloroplast material and INH-NAD under phosphorylating conditions. Under the conditions described in Fig. 8, we could not detect a signal for chloroplast material itself even after prolonged illumination. A signal only became evident when the concentration of chloroplast material was doubled. We could neither observe a signal when ferredoxin and NADP were present in concentrations optimal for phosphorylation (0·24 mg ferredoxin/ml and $5 \times 10^{-4}$ M NADP) or when these concentrations were doubled, nor with $10^{-3}$ M NAD. With INH-NAD, however, a clear and light-dependent signal was observed (Fig. 8) which required variable periods of illumination to reach a detectable concentration. The signal (inset, Fig. 8) is different from the endogenous one of chloroplast fragments. It seems to be overlapped by at least five lines and has a broader signal width than that from chloroplasts alone. When the signal was established, the width remained constant. Therefore, the relative signal height represents the relative spin concentration.

The EPR signal increased during illumination and the final height was dependent on the concentration of the analog. It is interesting that the maximum amplitude was not reached at the concentration optimal for photophosphorylation but at a concentration of approximately $10^{-3}$ M. Longer illumination caused a decay of the signal. In the dark the signal disappeared in less than 15 sec. When the light was turned off for about a minute and turned on again, the signal reappeared immediately and with the same amplitude. Addition of $NH_4Cl$ ($2 \times 10^{-2}$ M) shortened the lag period (Fig. 8). When $5 \times 10^{-6}$ M CMU was included in the sample, the signal decreased by approximately the same amount as the concurrent photophosphorylation. Reduced

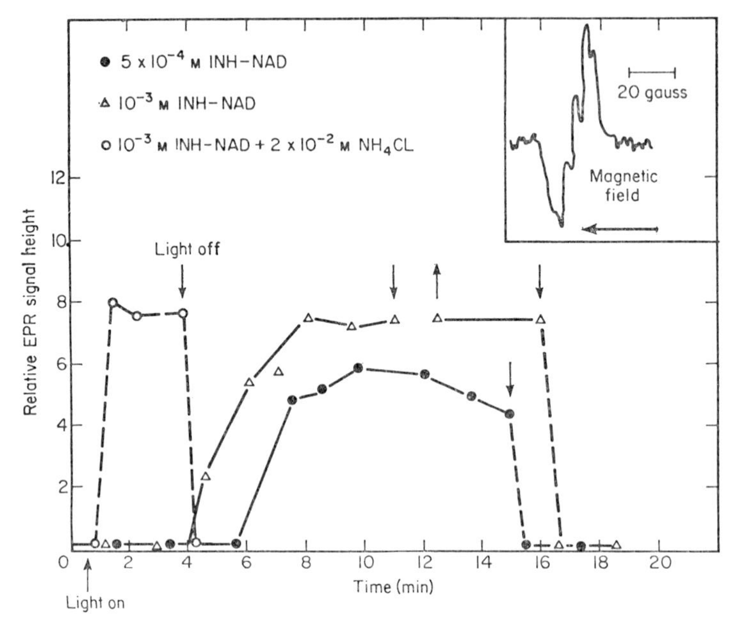


FIG. 8. EPR signal with INH-NAD under phosphorylation conditions. Modulation amplitude: 3 gauss; 170 $\mu$g chlorophyll/ml. Gas phase: air.

INH-NAD ($10^{-3}$ M) and chloroplast fragments gave no signal during 10 min. of illumination or in darkness. Also, neither light nor dark signals could be detected with $5 \times 10^{-4}$ or $10^{-3}$ M INH-NAD in the absence of chloroplast fragments.

## DISCUSSION

The reduction of INH-NAD by chloroplast fragments in the light cannot be related directly to ATP formation, which was three times in excess of the amount of analog present (Fig. 3). The reduced compound was stable in air for hours and did not catalyse photophosphorylation. Therefore, the low rate of reduction is neither due to re-oxidation by oxygen nor to re-oxidation by components of the electron transport chain of chloroplasts. On the other hand CMU inhibition of photophosphorylation with INH-NAD indicates that photophosphorylation requires electron flow. Reduction and photophosphorylation were not influenced by ferredoxin.

The possibility of oxygen evolution accompanying the reduction is masked by the substantially high oxygen uptake. The lack of oxygen uptake with the fully reduced analog indicates that the classical Mehler reaction may not occur (Good and Hill, 1955). The relatively small influence of anaerobic conditions on the INH-NAD catalysed photophosphorylation makes it improbable that the photophosphorylation is due exclusively to the electron flow caused by oxygen uptake. Although oxygen might participate in light-driven reactions which occur only when the analog is added to chloroplast fragments, this does not necessarily imply that oxygen will react directly with the INH-NAD. Oxygen may react with other components present in the chloroplasts or with the radical of INH-NAD.

Furthermore, oxygen may act as a "redox balance" for the electron transport chain, thus preventing an over-reduction of the components (Bose and Gest, 1963). As Whatley (1963) has shown, PMS-catalysed photophosphorylation is inhibited by CMU and dependent on oxygen, provided PMS is present in low concentration. It was not shown whether oxygen was consumed or acted catalytically. The explanation was based on a regulatory role of oxygen for electron flow. These findings indicate that inhibition of chloroplast reactions with CMU are of a diversified nature, and, together with the role of oxygen, cannot be adequately explained.

An EPR signal could be detected in the light with INH-NAD and chloroplast fragments. The decrease upon addition of CMU and the failure to observe the signal with the reduced analog indicate that the signal is related to electron flow in chloroplast material. A radical form of INH-NAD could be generated in the light and possibly react back with some component of the electron transport chain of chloroplasts in a cyclic manner. It would act catalytically and only be detected by EPR spectroscopy when it accumulates. Accumulation of the radical form of INH-NAD could occur when photophosphorylation diminishes (after approximately 10 min reaction time) or when an uncoupler is added (see Fig. 8).

The sharp decrease of photophosphorylation with higher concentrations of INH-NAD probably is due to an uncoupling effect of the analog itself. In support of this, the reduction of cytochrome *c* by chloroplast fragments was not inhibited by high concentrations ($10^{-2}$ M) of the analog. This indicates that electron flow is still operative. In addition, the rate of PMS (optimal concentration of $5 \times 10^{-5}$ M used) catalysed photophosphorylation was reduced by 85% in the presence of $10^{-2}$ M INH-NAD. Thus, it appears that the failure to observe catalysis of photophosphorylation at high concentrations of analog (Fig. 1) is due to an uncoupling effect of the analog at these concentrations. Additional points in favor of an uncoupling of ATP formation by high concentrations of INH-NAD are: the oxygen uptake and EPR versus concentration curves reach a plateau; and there is little effect of $NH_4Cl$ on oxygen uptake or the EPR signal.

## Summary

The isonicotinic acid hydrazide analog of nicotinamide adenine dinucleotide was synthesized and found to catalyse photophosphorylation at a rate of 150 μmoles ATP/mg chl/hr. Ferredoxin and the antibody to ferredoxin-NADP reductase had no influence on this photophosphorylation. The analog acts catalytically in that there was more ATP formed than analog present. The reduced form is stable under air and does not catalyse photophosphorylation, oxygen uptake or support an EPR signal. During ATP formation an appreciable amount of oxygen is consumed, up to half of the amount of ATP formed, but photophosphorylation is only slightly decreased by anaerobic conditions. EPR spectroscopy showed a light-dependent signal in the presence of INH-NAD and chloroplasts. It is suggested that a radical form of INH-NAD originates in the chloroplast system and it may be involved in electron flow.

## Acknowledgements

The excellent technical assistance of Mrs. Judy Christian and Mr. Michael Young is gratefully acknowledged. Gratitude is expressed to Mr. E. Thomas for assistance in the EPR studies.

## References

Allen, M. B., Whatley, F. R. and Arnon, D. I. (1958). *Biochim. biophys. Acta* **27**, 16.
Anderson, B. M., Ciotti, C. J. and Kaplan, N. O. (1962). *Biochim. biophys. Acta* **62**, 230.
Black, C. C. (1965). *Biochim. biophys. Acta* **94**, 27.
Black, C. C., Böger, P. and San Pietro, A. (1965). *Pl. Physiol.* Suppl. 40, xl.
Black, C. C. and San Pietro, A. (1963). *Archs Biochem. Biophys.* **103**, 453.
Black, C. C., San Pietro, A., Limbach, D. and Norris, G. (1963). *Proc. natn. Acad. Sci. U.S.A.* **50**, 37.
Bose, S. K. and Gest, H. (1963). *Proc. natn. Acad. Sci. U.S.A.* **49**, 337.
Cahn, R. D., Kaplan, N. O., Levine, L. and Zwilling, E. (1962). *Science, N.Y.* **136**, 962.
Calvin, M. and Androes, G. M. (1963). *In* "Microalgae and Photosynthetic Bacteria" (Japanese Society of Plant Physiologists, eds.), p. 319. University of Tokyo Press, Tokyo.
Chancellor-Madison, J. and Noll, C. R. Jr. (1963). *Science, N.Y.* **142**, 60.
Clark, W. M. (1960). "Oxidation-Reduction Potentials of Organic Systems", p. 493. Williams and Wilkins Co., Baltimore.
Commoner, B. (1961). *In* "Light and Life" (W. D. McElroy and B. Glass, eds.), p. 356. Johns Hopkins Press, Baltimore.
Commoner, B., Heise, J. J. and Townsend, J. (1959). *Proc. natn. Acad. Sci. U.S.A.* **42**, 710.
Fewson, C. A., Black, C. C. and Gibbs, M. (1963). *Pl. Physiol.* **38**, 680.
Good, N. and Hill, R. (1955). *Archs Biochem. Biophys.* **57**, 355.
Heber, U. (1960). *Z. Naturf.* **15b**, 653.
Jagendorf, A. T. and Avron, M. (1958). *J. biol. Chem.* **231**, 277.
Kaplan, N. O. (1955). *In* "Methods of Enzymology" (S. P. Colowick and N. O. Kaplan, eds.), Vol. II, p. 660. Academic Press, New York and London.
Kaplan, N. O. (1957). *In* "Methods of Enzymology" (S. P. Colowick and N. O. Kaplan, eds.), Vol. III, p. 873, 899. Academic Press, New York and London.
Keister, D. L., San Pietro, A. and Stolzenbach, F. E. (1960). *J. biol. Chem.* **235**, 2989.
Keister, D. L. and San Pietro, A. (1963). *Archs Biochem. Biophys.* **103**, 45.
Keister, D. L., San Pietro, A. and Stolzenbach, F. E. (1962). *Archs Biochem. Biophys.* **98**, 235.
Krogmann, D. W., Jagendorf, A. T. and Avron, M. (1959). *Pl. Physiol.* **34**, 680.
Leighton, S. B. (1956). Thesis, The Johns Hopkins University, Baltimore.
Lyle, R. E., Nelson, D. A. and Anderson, P. S., (1962). *Tetrahedron Letters*, **13**, 553.

San Pietro, A. (1961). *In* "Light and Life" (W. D. McElroy and B. Glass, eds.), p. 631. Johns Hopkins Press, Baltimore.
San Pietro, A. and Lang, H. M. (1956). *Science, N.Y.* **124**, 118.
San Pietro, A. and Lang, H. M. (1958). *J. biol. Chem.* **231**, 211.
Shin, M. and Arnon, D. I. (1965). *J. biol. Chem.* **240**, 1405.
Sund, H., Dieckmann, H. and Wallenfels, K. (1964). *Adv. Enzymol.* **26**, 115.
Tagawa, K. and Arnon, D. I. (1962). *Nature, Lond.* **195**, 537.
Vernon, L. P. (1963). *In* "Microalgae and Photosynthetic Bacteria" (Japanese Society of Plant Physiologists, eds.), p. 309. University of Tokyo Press, Tokyo.
Whatley, F. R. (1963). *In* "Photosynthetic Mechanisms of Green Plants" (Publ. 1145, National Academy of Sciences-National Research Council), p. 243. Washington, D.C.
Wojahn, H. (1952). *ArzneimittelForsch.* **2**, 324.
Zatman, L. J., Kaplan, N. O., Colowick, S. P. and Ciotti, M. M. (1954). *J. biol. Chem.* **209**, 467.

## Discussion

*O. Kandler:* What is the redox potential of the NAD analog.

*A. San Pietro:* The redox potential of the INH-NAD analog that pertains to thermodynamic equilibrium is unknown. It is possible to measure the polarographic half-wave potential of the analog but there is probably no relation between the polarographic half-wave potential and the potential that pertains to thermodynamic equilibrium (W. M. Clark [1960], "Oxidation-Reduction Potentials of Organic Systems", p. 493).

For example, we can consider the values reported for NAD. The $E'_0$, calculated either from equilibrium data [K. Burton and T. H. Wilson, *Biochem. J.* **54** (1953), 86] or determined by direct potentiometry [F. L. Rodkey, *J. Biol. Chem.* **175** (1948), 385], is reported to be about −0·320 volt at pH 7 and 30°. In contrast, the polarographic half-wave potential has been reported [B. Ke, *Biochim. Biophys. Acta* **20** (1956), 547; C. Carruthers and J. Tech, *Archs Biochem. Biophys.* **56** (1955), 441] to be about −0·936 to −0·980 volt versus the saturated calomel electrode (or −0·692 to −0·736 volt with respect to the normal hydrogen electrode).

As stated in our paper, we have measured the polarographic half-wave potential of the INH-NAD analog and it is similar to the values reported for the polarographic half-wave potential of NAD. However, no information is available at present concerning the potential of the analog that pertains to thermodynamic equilibrium.

# Energy Coupling at Different Coupling Sites in Photophosphorylation

Herrick Baltscheffsky

*Wenner-Gren Institute and Department of Biochemistry, University of Stockholm, Stockholm, Sweden*

## Introduction

In 1951 it was demonstrated that two cytochromes, cytochrome *f* and cytochrome $b_6$, occur in green tissues of plants (Hill and Scarisbrick, 1951). This discovery pointed towards the possibility that transport of electrons along a chain of electron carriers was of significance in photosynthetic energy transformations. Soon thereafter, Vernon (1953) found large amounts of a haem protein in extracts prepared from the photosynthetic bacterium *Rhodospirillum rubrum*. When photophosphorylation had been discovered, in chloroplasts from spinach (Arnon *et al.*, 1954) and in chromatophores from *R. rubrum* (Frenkel, 1954), much attention was given to the question of whether energy conversion in photosynthesis was mediated by electron transport and electron transport-coupled phosphorylation.

As has already become apparent from several previous reports in this section of the Conference, it is now well established that energy transfer in photophosphorylation is coupled to electron transport.

Similarities found between the respiratory chain in mitochondrial oxidative phosphorylation and the cyclic and non-cyclic electron transport chains operating in photophosphorylation soon after their discovery led to speculations and various considerations concerning the efficiency and the character of the photophosphorylation process. Did one or several coupling sites link electron transport with energy transfer reactions in photophosphorylation? Which, if any, of the multiple coupling sites in the respiratory chain of mitochondria had a counterpart in photophosphorylation systems?

During the last 5–6 years it has been possible to obtain some information about the actual number and the gross location of sites for coupling between electron transport and energy transfer reactions in photophosphorylation and about mechanisms involved in the energy transfer reactions, both in chromatophores from photosynthetic bacteria and chloroplasts from green plants. However, knowledge about these particular aspects of the transformation of light energy to biologically useful chemical energy in photosynthesis is still very limited.

In this presentation the number and the gross location of coupling sites in photophosphorylation of photosynthetic bacteria and green plants will be discussed. Attention will also be directed to certain results from our laboratory indicating that, in a photophosphorylation system, a basic difference may exist between the mechanisms of energy transfer from different coupling sites.

## Bacterial Chromatophores

The first evidence indicating that two separate coupling sites exist in bacterial photophosphorylation was given in 1960 (Baltscheffsky *et al.*, 1960; Baltscheffsky, 1961; Baltscheffsky and Arwidsson, 1962). Low concentrations of valinomycin, an uncoupling agent for oxidative phosphorylation, gave a partial inhibition of the "physiological" pathway for cyclic photophosphorylation in chromatophores from *R. rubrum*, but when this pathway for electrons was inhibited and phosphorylation was linked to a "by-pass"-pathway obtained by addition of the electron carrier phenazine methosulphate (PMS), no inhibition occurred. Detailed interpretations of these data, which indicate the existence of two coupling sites in cyclic photophosphorylation of the chromatophores, have been given earlier (Baltscheffsky and Arwidsson, 1962; Baltscheffsky and Baltscheffsky, 1963). It may thus suffice to give here only

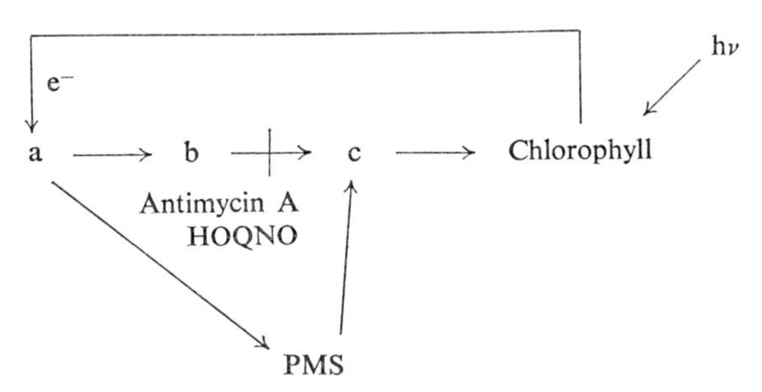


Fig. 1. Location of the two coupling sites in photosynthetic phosphorylation

the general scheme, which has been put forward to account for the gross location of the two coupling sites (Fig. 1).

Some independent support for the concept of two coupling sites in the "physiological system" and only one in the "PMS-system" was obtained in quantum requirement experiments, which were performed at the Johnson Foundation in Philadelphia, by Baltscheffsky *et al.* (1961) and later, with a more rapid and refined technique, by Nishimura (1963). A marked increase in quantum requirement was obtained upon substitution of the "physiological" system for the "PMS-system" in these studies.

The existence of two coupling sites in the "PMS-system" has recently been suggested (Horio and Yamashita, 1964). This was based on the demonstration of a biphasic inactivation-curve of photophosphorylation. The curve was obtained by treatment with increasing concentrations of Triton X-100. The present author finds it difficult to agree with this interpretation in the absence of a more pronounced effect than that described or any further support for the proposition.

Until very recently the detailed significance of the apparent difference in sensitivity to valinomycin between the energy transfer pathways emerging from the two coupling sites in photophosphorylation of chromatophores was only a matter of speculation. It was, for example, suggested as a possibility, that the insensitive pathway might be coupled to photochemical electron transport at the chlorophyll level, giving a unique mechanism of energy transfer, when compared with the mechanisms found in the sensitive pathway of chromatophores as well as in mitochondrial oxidative phosphorylation (Baltscheffsky and Baltscheffsky, 1963). It has recently been possible to obtain added experimental support for the concept that there is a basic difference between the energy transfer pathways linked to the two coupling sites in the cyclic electron transport chain of chromatophores and also, to obtain information about the location of this difference at the energy transfer level.

This support comes from the discovery (L. V. von Stedingk, in press) in our laboratory of a new and major light-induced pH-change in isolated chromatophores, which occurs in the absence of added orthophosphate, ADP

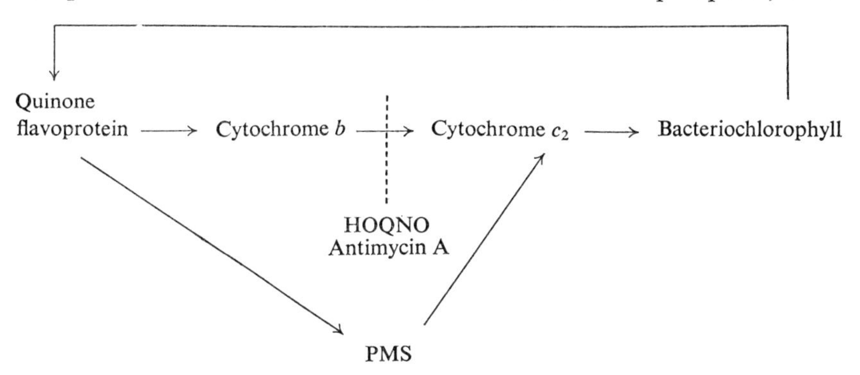


*Regions of the two coupling sites:*

Quinone flavoprotein ⟶ Cytochrome $b$ ⟶ Cytochrome $c_2$ = A ⟶ B

Cytochrome $c_2$ ⟶ Bacteriochlorophyll ⟶ Quinone flavoprotein = C ⟶ D

FIG. 2. Current views on electron transport in cyclic photophosphorylation in chromatophores.

and $Mg^{2+}$ and requires coupled electron transport. In this respect it resembles the earlier observed light-induced pH-change of isolated chloroplasts (Jagendorf and Hind, 1963).

As is described in more detail elsewhere (Baltscheffsky and von Stedingk, 1965), only the valinomycin-sensitive energy transfer pathway in the chromatophores seems to be connected to any appreciable degree with a light-induced increase of the pH in the medium. In the "PMS-system" practically no light-induced pH-change is obtained. In order to describe our interpretation of this result it may be useful first to give what may perhaps be regarded as a reasonable representation of current knowledge about electron transport in cyclic photophosphorylation of chromatophores, and about the location of the two coupling sites (Fig. 2). For the sake of simplicity the region of electron transport coupled to the valinomycin-sensitive energy transfer pathway may be designated A ⟶ B, and that coupled to the valinomycin-insensitive pathway C ⟶ D.

Energy transfer from the two coupling sites to ATP can now be visualized as is shown in the scheme in Fig. 3. In the present context it is sufficient to consider energy transfer from A → B in the region of X~I, which is described, with a question mark, as lacking, at least functionally, in the other energy transfer pathway. X~I is the well-known hypothetical energy-rich intermediate between electron transport and phosphate transfer in mitochondrial oxidative phosphorylation (Chance *et al.*, 1955) and appears to be the branching point for energy-requiring transport of ions in these organelles (Chance, 1963; Rasmussen *et al.*, 1965). The demonstration by Pressman (1965) that the uncoupling action of valinomycin in mitochondria is connected with its stimulation of $K^+$-transport (branching off at the X~I level) would appear to be highly relevant in connection with the observed distribution of the light-induced pH-changes in the energy transfer pathways of chromatophores. To put it briefly, the lack of both valinomycin-sensitivity and light-induced pH-change in the energy transfer pathway linked to the coupling site in the

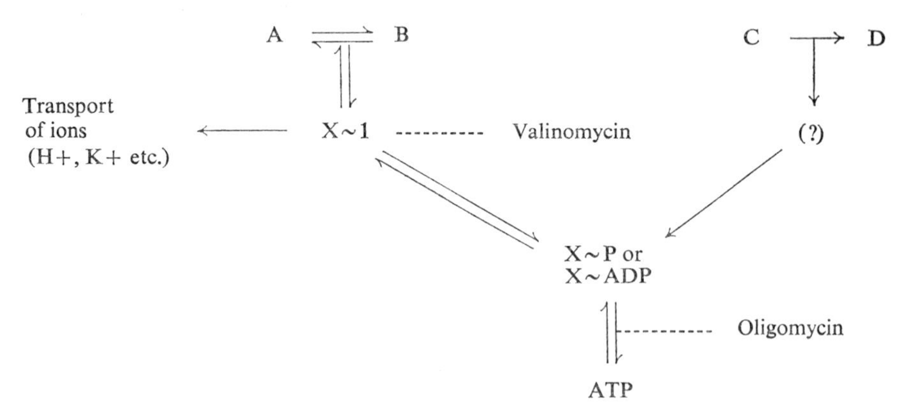


FIG. 3. Energy transfers from two coupling sites to ATP.

C → D region constitute two types of evidence for absence of a functional X~1-intermediate in this pathway.

All the evidence available so far thus seems to support the concept that two coupling sites exist in cyclic photophosphorylation of chromatophores from *R. rubrum*. Energy transfer from the coupling site in the region of dark electron transport only bears stronger resemblance to mitochondrial energy transfer than that from the coupling site in the region including the photochemical reactions. The latter energy transfer pathway appears by different criteria to be uniquely lacking the hypothetical intermediate X~I.

## Plant Chloroplasts

Investigators of photophosphorylation in plant chloroplasts have many possibilities to choose between various cyclic and non-cyclic systems for their study of mechanisms involved in the process. On the other hand, in spite of several attempts, it has not yet been possible to obtain preparations of isolated chloroplasts which give even a reasonably high rate of cyclic photophosphorylation in the absence of added electron carriers. Unfortunately any addition of such compounds may bridge not only a gap between two adjacent links in a cyclic, "physiological" electron transport chain, but also introduce a "by-pass" around a number of physiological carriers in the electron transport assemblies of the chloroplasts. In this way a physiological coupling site may well be by-passed and remain undetected. Thus the lack of an endogenous, functional, cyclic pathway for electron transport in isolated chloroplasts is a very serious source of difficulty in connection with the question about the number of coupling sites in chloroplasts from green plants.

Attempts to determine experimentally the number of coupling sites in chloroplasts have, however, not been lacking. Comparisons of the quantum requirement for formation of a molecule of ATP in systems where different electron carriers had been added were reported in 1961 by Arnon (1961a, b) and by Yin *et al.* (1961). According to Arnon, the quantum requirement for

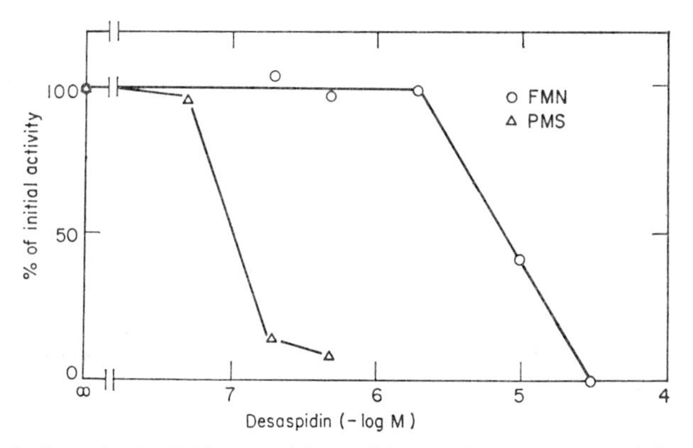


FIG. 4. Light-intensity 30,000 lux, aerobic conditions. Other reaction conditions and assays as previously described (Baltscheffsky 1965b). Courtesy *Acta chem. scand.*

formation of ATP in photophosphorylation with PMS as added carrier was much higher than with added menadione (vitamin $K_3$) or FMN, indicating more than one coupling or phosphorylation site in the two latter systems. However, Yin *et al.* did not obtain any significant difference between the quantum requirements of the three above-mentioned systems and concluded that each system contains only one coupling site. No explanation for this discrepancy between the results of the two groups has yet been given. Data from temperature variation studies have been interpreted by Hall and Arnon (1962) to suggest the actual operation of two coupling sites, but also in this case the results are only indicative of the possibility that more than one coupling site may exist. Whatley and Horton (1963) have further discussed the question of the possible existence and location of two phosphorylation sites in cyclic photophosphorylation of spinach chloroplasts. In recent review articles about chloroplasts, schemes with more than one coupling site have often been given, but a discussion of the experimental evidence for this has usually been lacking. Interestingly enough, Hall and Whatley (1967) in a very recent review article entitled "The Chloroplast", represent the electron transport chains in both cyclic and non-cyclic photophosphorylation as containing a single and common phosphorylation site, between plastoquinone and cytochrome *f*.

Our active interest in the number of coupling sites in spinach chloroplasts began when we tested the effect of desaspidin and found an unexpected and unusual relative response of two photophosphorylation systems to this agent (Baltscheffsky, 1964; Baltscheffsky and de Kiewiet, 1964). Desaspidin, which had been shown by Runeberg (1963) to uncouple oxidative phosphorylation, was found to inhibit photophosphorylation in the presence of PMS much more efficiently than in the presence of FMN (Baltscheffsky, 1965b). As is seen in Fig. 4, a factor of about 100 distinguishes the concentrations of the agent required for 50% inhibition. In order to obtain a better understanding of this great difference, it was considered to be important to test the action of

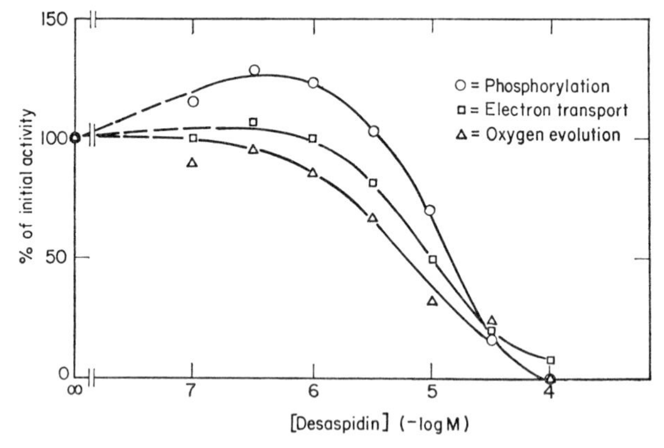


FIG. 5. Light-intensity 10,000 lux, anaerobic conditions. Other reaction conditions and assays as previously described (Baltscheffsky and de Kiewiet, in preparation).

desaspidin on other electron transport systems connected with photophosphorylation in the chloroplasts.

As has already been reported, it was found that, of three other systems tested, only the system where ascorbate served as electron donor, instead of water, was sensitive to low concentrations of desaspidin (Baltscheffsky and de Kiewiet, 1964). Figures 5 and 6 represent the two types of effect obtained. In Fig. 5 it is seen that only high concentrations of desaspidin inhibited non-cyclic photophosphorylation in the system $H_2O$/NADP and that both electron transport and phosphate esterification were inhibited. The same result was obtained when ferricyanide was added as terminal electron acceptor instead of NADP (+ferredoxin). In contrast to this situation, as is shown in Fig. 6, phosphorylation but not electron transport was inhibited by low concentrations of desaspidin in the system ascorbate/NADP, in the presence of DCPIP, to facilitate electron flow from ascorbate, and of CMU, to inhibit electron transport from the water-splitting light-reaction. It is seen that the concentrations required for this selective effect were similar to those which caused inhibition of photophosphorylation in the sensitive PMS-system.

Thus the picture emerged of a situation where two of five systems tested were sensitive to low concentrations of desaspidin and three were not. Similar results were obtained with another phlorobutyrophenone derivative, flavaspidic acid.

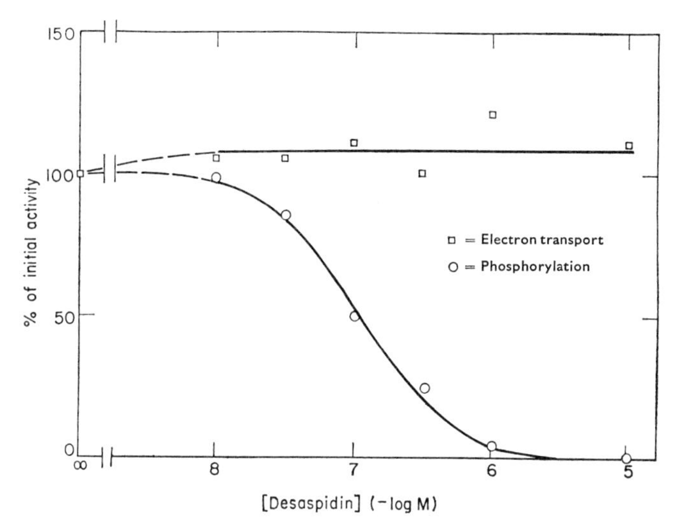


FIG. 6. Details as described under Fig. 5.

In order to corroborate the evidence for this difference between the five systems tested, experiments of the following kind were carried out. Photophosphorylation in a sensitive system was first blocked by addition of desaspidin.

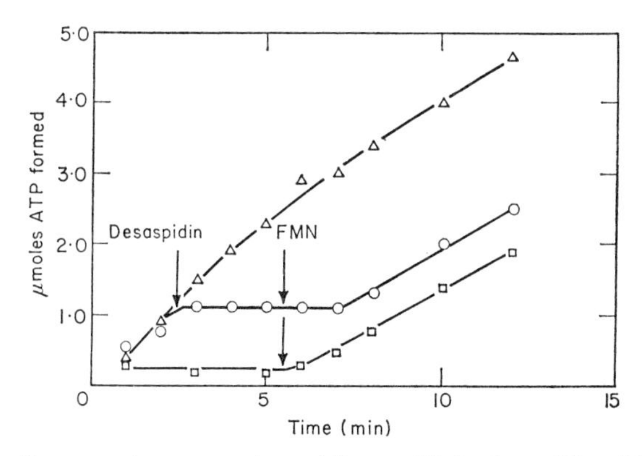


FIG. 7. In the three samples: △ = no desaspidin; ○ = $10^{-6}$ M desaspidin added after $2\frac{1}{2}$ min; □ = $10^{-6}$ M desaspidin added at zero time. Other details as in Fig. 4 and (D. Y. de Kiewiet and H. Baltscheffsky, in preparation).

Thereafter a suitable agent was added to the same sample of chloroplasts in order to restore photophosphorylation by introducing the functioning of a non-sensitive system. Figure 7 shows an example of this. It was found that the sensitive PMS-system and the non-sensitive FMN-system gave the expected pattern of desaspidin action (D. Y. de Kiewiet and H. Baltscheffsky, in preparation).

Our tentative interpretation of the results obtained with desaspidin have been given earlier in the minimum scheme shown in Fig. 8 (Baltscheffsky and de Kiewiet, 1964). It incorporates all the five systems tested into a simple and partially common reaction pathway. It accounts for the observed results by containing two distinct coupling sites with grossly specified location. It shows one coupling site with desaspidin-sensitive energy transfer in the cyclic electron

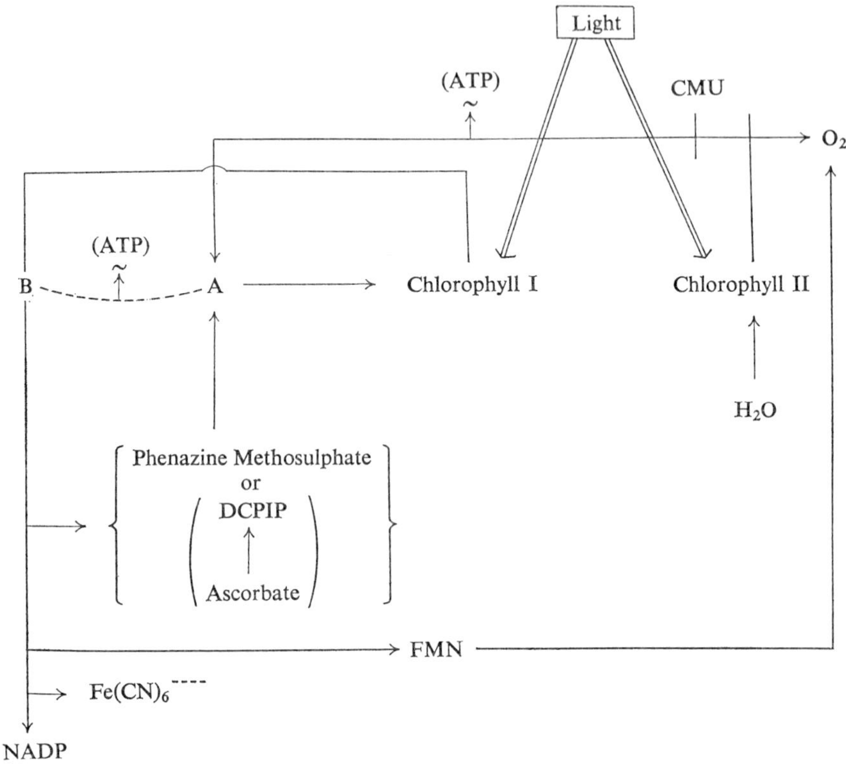


FIG. 8. Interpretation of results obtained with desaspidin. (Courtesy *Acta chem. scand.*)

transport pathway functioning after addition of electron carrier, and another coupling site from which emerge energy transfer reactions which are not sensitive to low concentrations of desaspidin and which is located in the non-cyclic region of electron transport.

The first evidence, from studies of differential effects of uncoupling agents, suggesting two coupling sites in photophosphorylation of chloroplasts, was given by Avron and Shavit (1963), who found that carbonyl cyanide *p*-trifluoromethoxyphenylhydrazone (*p*-$CF_3O$-CCP) inhibited ferricyanide-dependent photophosphorylation much more efficiently then PMS-dependent photophosphorylation. Our more extensive studies with desaspidin, which happens to give the opposite inhibition pattern for the two systems tested by Avron and Shavit, tend to support this concept and extend it to embrace five different systems, giving a coherent but still very preliminary picture of the location of the two suggested coupling sites. It is clear, however, that these studies must be supplemented with more detailed evidence, before a final judgement can be made in the question of whether chloroplasts contain one, two or maybe even more than two coupling sites in their physiologically functioning apparatus for energy transformation.

## Concluding Remarks

Our evidence from experiments with valinomycin, from determinations of the quantum requirement of ATP-formation and from measurements of light-induced pH-changes converges to indicate, very strongly, that the cyclic electron transport chain of phosphorylating chromatophores from *R. rubrum* contains two coupling sites for energy transfer reactions giving ATP.

The data obtained in several laboratories with chloroplasts from green plants, usually from spinach or swiss chard, concerning the number of coupling sites do not always unequivocally support the suggestions that multiple coupling sites occur also in this system. Studies of the quantum requirement for ATP-formation have, as was discussed above, resulted in quite conflicting views in this respect. Experiments where the temperature has been varied do not seem to allow conclusive interpretations in any direction. Data obtained with uncoupling agents such as *p*-$CF_3O$-CCP and desaspidin indicate that two coupling sites may exist also in chloroplasts. It may be appropriate to point out that our suggestion on the basis of results obtained with desaspidin that cyclic electron transport is linked to one of these two sites only and the non-cyclic electron transport emerging from the water-splitting reaction is associated with the other site does not necessarily imply any basic difference between chloroplasts and chromatophores with respect to number of coupling sites in the physiological cyclic electron transport chain. As long as the unfortunate situation prevails, that cyclic photophosphorylation in chloroplasts from green plants needs addition of electron carriers, be it PMS, FMN or ferredoxin, a "by-pass" around a coupling site may be introduced. Much more effort should thus be directed towards preparation of chloroplasts which would be capable of giving high rates of "endogenous" cyclic photophosphorylation, as do bacterial chromatophores. Only with such preparations will it be possible to obtain meaningful answers to questions about the number of coupling sites in cyclic photophosphorylation of isolated chloroplasts.

Some aspects of the information obtained from our studies relating to the number and the location of coupling sites in photophosphorylation may be summarized in four points:

| | *Chromatophores* (2 *sites*) | *Chloroplasts* (2 *sites*) |
|---|---|---|
| 1. | Valinomycin-sensitivity at one of two coupling sites. | Desaspidin-sensitivity at one of two two coupling sites. |

| *Chromatophores* (2 *sites*) | *Chloroplasts* (2 *sites*) |
|---|---|
| 2. Non-sensitivity in the PMS-system. | Sensitivity in the PMS-system. |
| 3. Non-sensitivity in electron transport-region including the light-reactions. | Non-sensitivity in electron transport-region including the light-reactions. |
| 4. Site by-passed by PMS appears to correspond to the second coupling site in mitochondrial oxidative phosphorylation. | Site by-passed by PMS appears to correspond to the second coupling site in mitochondrial oxidative phosphorylation. |

As has been pointed out earlier (Baltscheffsky, 1965a), the second of the three mitochondrial coupling sites appears to have its counterpart in photophosphorylation, both in chromatophores and in chloroplasts. In connection with the other listed properties of the two systems, it should perhaps be stressed that, in order to establish more firmly the degree of validity of our concept that the photophosphorylation systems of both bacteria and plants contain two coupling sites linked with different energy transfer mechanisms, more experiments are required. This is particularly the case in the isolated chloroplast system, where "endogenous" cyclic photophosphorylation is lacking and where only the data obtained with uncoupling agents provided the background for our suggestions.

In conclusion, basic similarities appear to exist between coupling and transfer of energy in the region including only dark electron transport reactions in photophosphorylation of chromatophores and chloroplasts and in oxidative phosphorylation of mitochondria. In the region of the electron transport chain which includes also the light reactions in photophosphorylation, a basic difference in the energy transfer has been demonstrated in chromatophores and some evidence for the existence of such a difference has been found also in chloroplasts. It may well be of significance that, when photophosphorylation and oxidative phosphorylation are compared, similarity with respect to transfer of energy is restricted to, or is particularly pronounced in, the region of similarity between the electron transport pathways.

## References

Arnon, D. I., Allen, M. B. and Whatley, F. R. (1954). *Nature, Lond.* **174**, 394.

Arnon, D. I. (1961a). *In* "Light and Life" (W. D. McElroy and B. Glass, eds.), p. 489. The Johns Hopkins Press, Baltimore.

Arnon, D. I. (1961b). *In* "Biological Structure and Function" (T. W. Goodwin and O. Lindberg, eds.), Vol. II, p. 339. Academic Press, London and New York.

Avron, M. and Shavit, N. (1963). *In* "Photosynthetic Mechanisms of Green Plants" (NAS.-NRC Publication 1145), p. 611. Washington.

Baltscheffsky, H., Baltscheffsky, M. and Arwidsson, B. (1960). *Acta chem. scand.* **14**, 1844.

Baltscheffsky, H. (1961). *In* "Biological Structure and Function" (T. W. Goodwin and O. Lindberg, eds.), Vol. II, p. 431. Academic Press, London and New York.

Baltscheffsky, H. and Arwidsson, B. (1962). *Biochim. biophys. Acta* **65**, 425.

Baltscheffsky, H. (1964). Federation of European Biochemical Societies 1st Meeting (London). Abstracts. p. A82.

Baltscheffsky, H. and de Kiewiet, D. Y. (1964). *Acta chem. scand.* **18**, 2406.

Baltscheffsky, H. (1965a). Federation of European Biochemical Societies 2nd Meeting (Vienna). Abstracts. p. 110.
Baltscheffsky, H. (1965b). *Acta chem. scand.* **19**, 1933.
Baltscheffsky, H. and von Stedingk, L.-V. (1965). *In* "Currents in Photosynthesis" (J. B. Thomas and J. C. Goedheer, eds.), Rotterdam.
Baltscheffsky, M. and Baltscheffsky, H. (1963). *In* "Bacterial Photosynthesis" (H. Gest, A. San Pietro and L. Vernon, eds.), p. 195. The Antioch Press, Yellow Springs, Ohio.
Chance, B., Williams, G. R., Holmes, W. F. and Higgins, J. (1955). *J. biol. Chem.* **217**, 439.
Chance, B. (1963). *In* "Energy-Linked Functions of Mitochondria" (B. Chance, ed.), p. 253. Academic Press, New York and London.
Frenkel, A. S. (1954). *J. Am. chem. Soc.* **76**, 5568.
Hall, D. O. and Arnon, D. I. (1962). *Proc. natn. Acad. Sci. U.S.A.* **48**, 833.
Hall, D. O. and Whatley, F. R. (1967). *In* "Enzyme Cytology" (D. B. Roodyn, ed.). Academic Press, London and New York.
Hill, R. and Scarisbrick, R. (1951). *New Phytol.* **50**, 98.
Horio, T. and Yamashita, J. (1964). *Biochim. biophys. Acta* **88**, 237.
Jagendorf, A. T. and Hind, G. (1963). *In* "Photosynthetic Mechanisms of Green Plants" (NAS.NRC Publication 1145), p. 599.
Nishimura, M. (1963). *In* "Bacterial Photosynthesis" (H. Gest, A San Pietro and L. P. Vernon, eds.), p. 201. The Antioch Press, Yellow Springs, Ohio.
Pressman, B. C. (1965). *Proc. natn. Acad. Sci. U.S.A.* **53**, 1076.
Rasmussen, H., Chance, B. and Ogata, E. (1965). *Proc. natn. Acad. Sci. U.S.A.* **53**, 1069.
Runeberg, L. (1963). *Societas Scientarium Fennica.* Commentationes Biologicae XXVI. 7. Diss., Helsinki.
von Stedingk, L. V., *Archs Biochem. Biophys.* (in press).
Vernon, L. P. (1953). *Archs Biochem. Biophys.* **43**, 492.
Whatley, F. R. and Horton, A. A. (1963). *Acta. Chem. scand.* **14**, S 140.
Yin, H. C., Shen, Y. K., Shen, G. M., Yang, S. Y. and Chiu, K. S. (1961). *Scientia Sinica* **10**, 976.

## Discussion

*J. F. G. M. Wintermans:* In view of results pointing to two sites of photophosphorylation differing in sensitivity to NAD, it may be recalled that light phosphorylation, and also photosynthesis, by whole *Chlorella* cells are sensitive to NAD only at low light intensities. Hence, it may be worth while to study these effects at various light intensities.

*H. Baltscheffsky:* The results concerning two sites of photophosphorylation differing in sensitivity to added uncoupling agents were obtained with valinomycin in chromatophores and with desaspidin in chloroplasts. The first results with desaspidin indicating that two coupling sites exist in spinach chloroplasts, and that energy transfer coupled to cyclic electron transport only was uncoupled by this agent, were obtained in our laboratory not very long ago and reported in 1964. I completely agree that it is important to vary light intensity—we have in fact already done a few such experiments—as well as the wavelength of light, among other pertinent variables.

# Reaction Kinetics of Photosynthetic Intermediates in Intact Algae

J. Amesz and W. J. Vredenberg
*Biophysical Laboratory of the University, Leiden, Netherlands*

On the basis of much evidence obtained in recent years, photosynthesis in algae and higher plants may be summarized by the following model:

$$
\begin{array}{c}
\qquad\qquad\qquad\qquad\qquad h\nu \qquad\qquad\qquad\quad ATP \nwarrow\!\!\smile\!\!\nearrow ADP+P_i \\
\qquad\qquad\qquad\qquad\qquad \downarrow \\
CO_2 \longleftarrow NADP \longleftarrow Fd. \longleftarrow X \longleftarrow \textit{system}\ 1 \longleftarrow P700 \longleftarrow Cyt. \longleftarrow PQ \longleftarrow \\
h\nu \\
\downarrow \\
\longleftarrow Q \longleftarrow \textit{system}\ 2 \longleftarrow Y \longleftarrow H_2O
\end{array}
$$

In this scheme X and Y are unknown primary reactants of the first and second light reaction, Fd. is ferredoxin, P700 a chlorophyllous pigment which is bleached upon oxidation, Cyt. is probably an *f*- or *c*-type cytochrome, PQ is a plastoquinone or a related substance, and Q is an intermediate which in the oxidized state quenches the fluorescence of the chlorophyll belonging to system 2. Phosphorylation probably occurs in the dark reaction between oxidized cytochrome and reduced plastoquinone. The horizontal arrows indicate the direction of electron or hydrogen transport. An extensive discussion of the evidence pertaining to the scheme can be found in recent reviews (Duysens, 1964; Trebst, 1964).

Although there is various evidence that P700, cytochrome, PQ and Q are redox catalysts in photosynthesis, and that they react in or near the chain between the first and second light reaction, quantitative evidence for this is still incomplete, and the kinetics of the reactions of these substances *in vivo* indicate that the scheme given above is too simple. In the following we will report a more detailed study of the reaction kinetics of these compounds in intact cells of the red alga *Porphyridium cruentum* and the blue-green algae *Anacystis nidulans* and *Schizothrix calcicola*. In these algae red ($> 650$ m$\mu$) and blue light are predominantly absorbed by system 1, and orange or green light predominantly by system 2.

## Methods

The algae were grown in the light of fluorescent lamps as described elsewhere (Hoogenhout and Amesz, 1965). *Porphyridium* was grown at 18° and

1000 lux in the medium of Jones *et al.* (1963). The blue-green algae were grown at 25° and 2500 lux, *Anacystis* in a slightly modified (Amesz and Duysens, 1962) medium C of Kratz and Myers (1955), *Schizothrix* (strain TX 27) in the medium described by Hoogenhout and Amesz (1965). During growth the algae were gassed with air enriched with 5% $CO_2$.

The apparatus for measuring absorbancy changes was the same as used in earlier experiments in this laboratory (see Amesz, 1964a). Fluorescence changes were measured by means of the same apparatus.

CCCP was a kind gift of Dr. P. G. Heytler, Wilmington, Del., to Prof. L. N. M. Duysens.

## Results and Interpretation

### Induction effects in P700 and cytochrome oxidation following a dark period

Figures 1, 2 and 3 show the kinetics of the light-induced reactions of P700 and cytochrome in a number of algae. It can be seen that illumination with red or blue light brought about an oxidation, but that, in confirmation of earlier

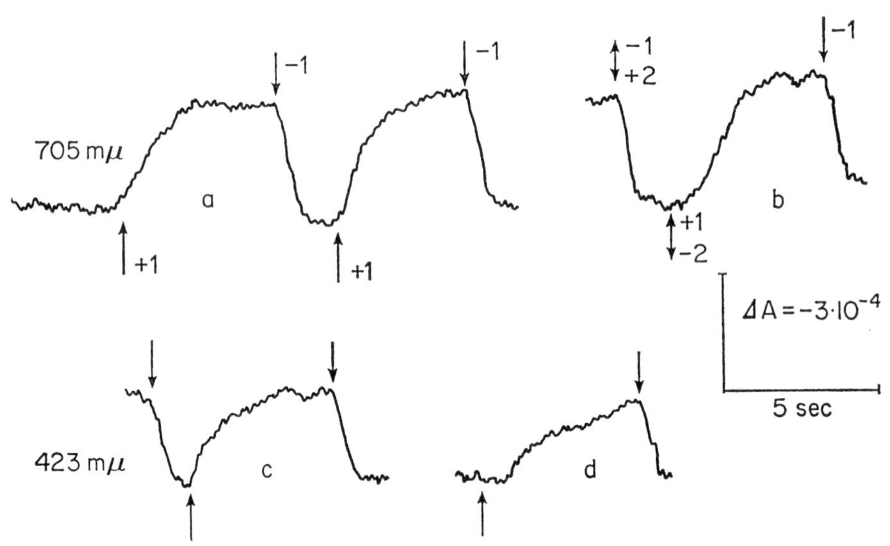


FIG. 1. The upper tracings show the kinetics of P700 in *Schizothrix* measured at 705 mμ. Upward and downward pointing arrows mark the beginning and the end of an illumination period. An upward tracing corresponds to a decrease of absorbancy and to an oxidation of P700. Tracing *a* shows the rate of oxidation in light of 410–440 mμ (1) of an intensity of $1{\cdot}7 \times 10^{-9}$ einstein/(sec cm$^2$) after a few min and after 2 sec darkness. In the second case the rate of oxidation is considerably higher. Tracing *b* shows the oxidation in light of 410–440 mμ (1) and reduction in light of 560 mμ (2) of $1{\cdot}2 \times 10^{-9}$ einstein/(sec cm$^2$). It can be seen that there is a retardation of the oxidation in blue light after previous green illumination. Tracing *c* and *d* give the kinetics of cytochrome oxidation, measured at 423 mμ, in light of 684 mμ, intensity $6{\cdot}0 \times 10^{-10}$ einstein/(sec cm$^2$). An upward tracing corresponds to a decrease of absorbancy and an oxidation of cytochrome. Tracing *d* was made after 3 sec darkness.

results (Vredenberg and Duysens, 1965), the rate at which oxidized P700 or cytochrome accumulated in the light depended upon the length of the preceding dark period. The rate of oxidation was much higher after a short than after a longer dark period. The extent of this induction effect after darkness was strongly dependent upon the intensity of illumination; it was most clearly visible at rather low light intensities. The amount of induction varied for different samples. After 1 min darkness it often took 20 sec or more to reach a steady state level of oxidized P700 or cytochrome upon illumination of a moderate intensity, whereas after a short dark period this level was reached

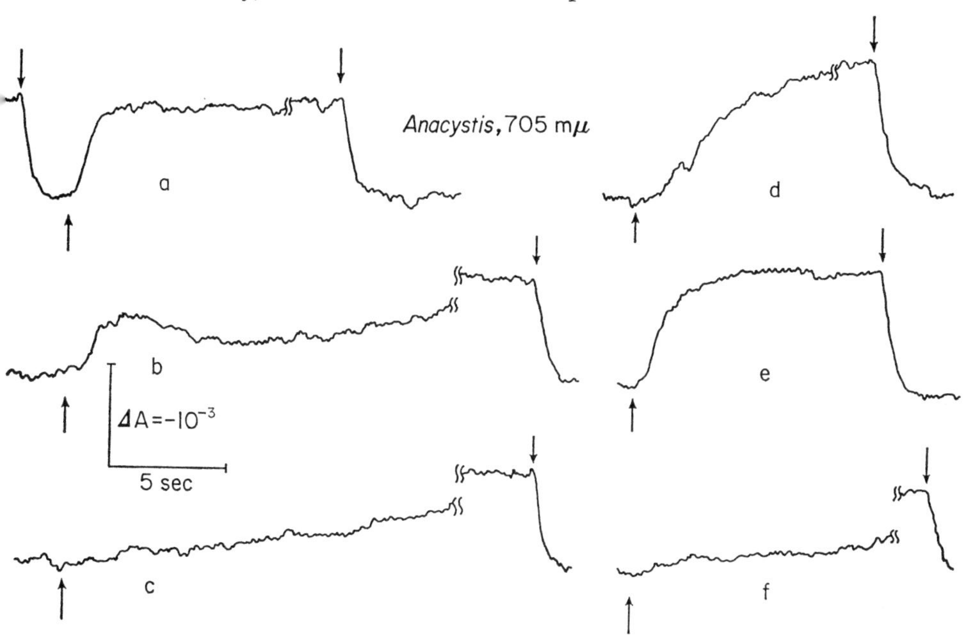


FIG. 2. Recorder tracings of P700 oxidation in blue light (410–430 mμ, $1{\cdot}1 \times 10^{-9}$ einstein/(sec cm²) in *Anacystis*. Tracings *a*, *b* and *c* were recorded after the algae had been in the dark during 2, 15 and 30 sec respectively, after a previous *blue* illumination of 1 min duration. Tracings *e*, *f* and *g* show the kinetics after 2, 5 and 30 sec darkness following a 1 min *orange* pre-illumination (620 mμ, $1{\cdot}2 \times 10^{-9}$ einstein/(sec cm²)). [See Fig. 1 for further explanation.]

within 1 or 2 sec in the light, but in some samples the effect was much less pronounced. In *Anacystis* (Fig. 2) the induction effect was maximal after about 30 sec of darkness (see also Vredenberg and Amesz, 1965).

Figure 3 shows kinetics of cytochrome reactions in *Porphyridium* after about 1 min darkness. A comparison of tracings *e*, *f* and *g* shows that the rate of reduction upon shutting off the light was considerably higher after a short light period than after several seconds of illumination, when the steady state level of cytochrome had been reached. This indicates that the low rate of accumulation of oxidized cytochrome is not caused by a low rate of the light-induced oxidation, but by a simultaneous high rate of reduction by a reduced compound. Upon continued illumination this rate of reduction then decreases.

Measurements of this type with *Anacystis* and *Schizothrix* were less conclusive because of the higher rate of the light-off reaction and the response time of the apparatus.

In some experiments it could be clearly observed (see tracing *b* of Fig. 2) that the rate of oxidation was relatively high immediately upon illumination, and that the level of oxidized P700 or cytochrome reached a maximum,

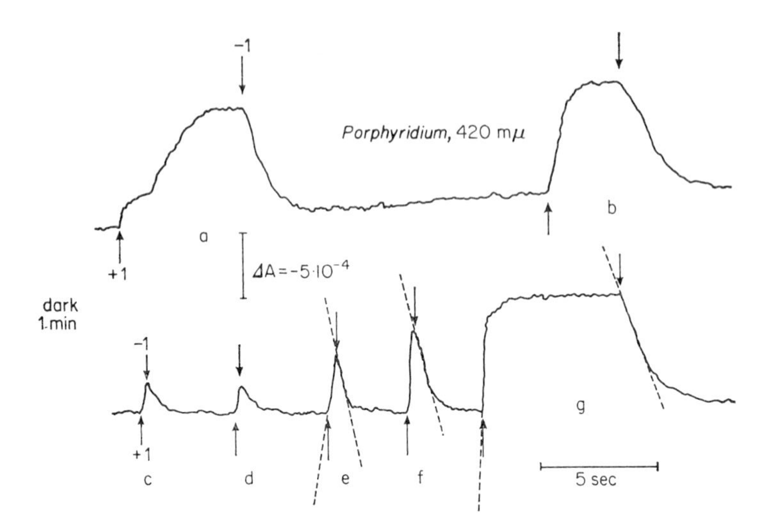


FIG. 3. Kinetics of cytochrome reactions in *Porphyridium*, after 1 min darkness. Actinic light: 688 mμ, $1 \cdot 3 \times 10^{-9}$ einstein/(sec cm$^2$).

decreased again, and only after a longer period of illumination increased again and reached its final steady state. This suggests that the dark reduction of P700 or cytochrome reached its highest rate only after one or more seconds of illumination and that the high rate of reduction which is induced by darkness is brought about by the light itself.

## INDUCTION EFFECTS CAUSED BY GREEN AND ORANGE ILLUMINATION

It has been shown (Duysens and Amesz, 1962; Amesz and Duysens, 1962; Vredenberg and Duysens, 1965) that in red and blue-green algae green or orange light causes very little oxidation of cytochrome or P700, and that it causes a reduction when applied during simultaneous red or blue illumination or in the presence of phenyl mercuric acetate. This can also be seen in tracings *a* and *b* of Fig. 4 which show that the rate of reduction of cytochrome in *Porphyridium* was considerably higher upon switching from red to green light, than upon switching off the red.

Figure 4 also shows that in *Porphyridium* the initial rate of cytochrome oxidation in red light was considerably lower after a preceding illumination with green or orange light than after a short period of darkness. For P700 a similar effect occurred, as illustrated in Figs. 1 and 2. The effect is probably caused by a high rate of simultaneous reduction, as indicated by tracings *c* and *d* of Fig. 4. These tracings show that the rate of reduction upon darkening is relatively high shortly after switching from green to red illumination in *Porphyridium*. This indicates that a reduced substance is formed in green light,

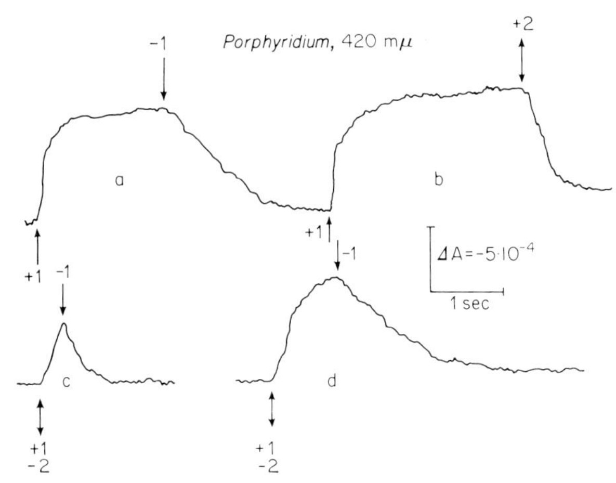


FIG. 4. Kinetics of cytochrome reactions in *Porphyridium* upon illumination with (1) red light (688 mμ, $2{\cdot}1 \times 10^{-9}$ einstein/sec cm²) and (2) green light (563 mμ, $4{\cdot}2 \times 10^{-10}$ einstein/sec cm²). See Fig. 1 for further explanation.

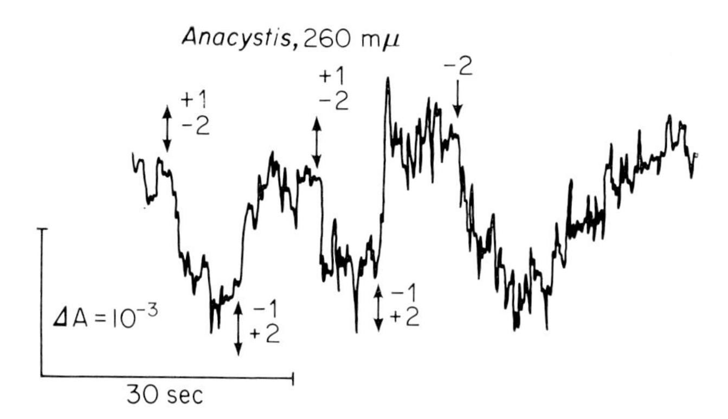


FIG. 5. Absorbancy changes at 260 mμ in *Anacystis* upon alternative illumination with (1) light of 684 mμ and $6{\cdot}2 \times 10^{-10}$ einstein/(sec cm²) and (2) light of 620 mμ and $7{\cdot}5 \times 10^{-10}$ einstein/(sec cm²). An upward tracing corresponds to a decrease of absorbancy and indicates a reduction of a plastoquinone.

which reacts with oxidized cytochrome. Figure 2 indicates that this induction effect is different from that induced by prolonged darkness. The steady state level was reached earlier and when a dark period of 5–8 sec was applied between the orange and red illumination, the induction practically disappeared and the effect of preceding orange light was nearly the same as that of red light after 5 sec of darkness or more.

Measurements of the ultraviolet absorbancy changes at 260 m$\mu$ in *Anacystis* and *Schizothrix* suggested that a plastoquinone, probably *Anacystis* plastoquinone (Henninger *et al.*, 1965), was the compound which was responsible for the induction effects in cytochrome and P700 oxidation after preceding green or orange light. Upon alternative illumination with light mainly absorbed by system 1 and system 2 the plastoquinone is alternatively oxidized and reduced (see Fig. 5 and Amesz, 1964b). The pool of plastoquinone reacting is about 1/150 of the amount of chlorophyll (Amesz, 1964b), i.e. large enough to cause appreciable induction effects in cytochrome or P700. In blue-green algae, this pool was in the reduced state in the dark and after turning off red light it returned slowly (i.e. in 20–30 sec) to the reduced state again (Vredenberg and Amesz, 1965). After turning off green or orange illumination it became oxidized after about 6–10 sec and afterwards slowly became reduced again (see Fig. 5). This could explain why the induction effect of P700 induced by orange light disappeared after 5–8 sec of darkness as mentioned above, and strengthens the evidence that plastoquinone is part of the photosynthetic chain between cytochrome and Q.

## EFFECTS OF INHIBITORS AND CHEMICAL AGENTS

DCMU (3,(3,4-dichlorophenyl)-1,1-dimethylurea) had little effect upon the induction of cytochrome and P700 in red light after darkness, as well as on the rate of reduction upon subsequent darkening. This indicates that system 2 is not involved in these processes. Carbonyl cyanide *m*-chlorophenyl hydrazone (CCCP), at a concentration of $10^{-5}$ M decreased the steady state in the light and enhanced the rate of reduction upon darkening of cytochrome and P700 in *Schizothrix* and of cytochrome in *Porphyridium*. 1 to $3 \times 10^{-3}$ M 1,1,1 trichloro-2 methyl-2 propanol (chloretone) gave about the opposite effect: it decreased the rate of the dark reduction of P700 and cytochrome in *Anacystis* and *Schizothrix* and inhibited the induction effect after darkness.

Phenazine methosulphate (PMS), in the concentration range of $10^{-7}$ to $1{\cdot}4 \times 10^{-6}$ M, and 0·7 to $2 \times 10^{-5}$ M 2,6 dichlorophenolindophenol (DCPIP) considerably enhanced the rate of dark reduction and decreased the steady state of oxidized P700 and cytochrome in *Anacystis* and *Schizothrix*. This agrees with the observations of Witt *et al.* (1963) on isolated spinach chloroplasts, and indicates that the reduced forms of these redox compounds react with cytochrome and perhaps P700, possibly in a "cyclic" reaction with the primary photoreductant of system 1.

PMS (even at $2{\cdot}7 \times 10^{-6}$ M) and DCPIP had no effect on the reactions of Q in *Schizothrix*. The kinetics of the increase of the yield of chlorophyll fluorescence in green light after preceding blue illumination, and of the subsequent decrease of fluorescence induced by additional blue light were not affected. The same applied in the presence of $1{\cdot}4 \times 10^{-3}$ M potassium ferricyanide and $1{\cdot}4 \times 10^{-6}$ M PMS or $2 \times 10^{-5}$ M DCPIP. These experiments indicate that Q and reduced Q do not react with PMS and DCPIP, and that reduced PMS and DCPIP cannot compete with the reduced products of system 2 in reducing the cytochrome. PMS had no effect on cytochrome kinetics in *Porphyridium*, possibly because it did not penetrate into the cell.

## Discussion

The results just described indicate that there are two different kinds of induction effects on light-induced P700 and cytochrome oxidation, which are both apparent as a retardation of the oxidation of these compounds in red light. One occurs after preceding illumination with light mainly absorbed by system 2, and is probably caused by a reaction of plastoquinol with oxidized cytochrome, the other is induced by prolonged darkness.

A reasonable explanation for the dark-induced induction could be given by a "cyclic" chain of reactions of cytochrome and P700 with the photoreductant of system 1, coupled with ATP production. This would also explain why the highest rate of dark reduction of cytochrome and P700 takes place after some light has been given, since light is needed for the production of reduced compounds for the "cyclic" reaction. The decrease of the reduction rate during continued illumination could then be due to a regulatory mechanism, e.g. controlled by the ATP/ADP ratio. There is evidence for the operation of a "cyclic" photophosphorylation *in vivo* (e.g. Forti and Parisi, 1963; Urbach and Simonis, 1964). A high ATP/ADP ratio brought about by red or green light then could decrease the reaction rate in the cyclic chain. In the dark this ratio then would decrease by utilization of ATP. In blue-green algae the induction after darkness presumably is partly, but, under appropriate conditions, to a smaller extent, also caused by a reaction of oxidized cytochrome with plastoquinol which has accumulated in the dark.

The experiments with DCMU and CCCP, an uncoupler of phosphorylation (Bamberger *et al.*, 1963), are consistent with the hypothesis outlined above.

Olson and Smillie (1963) also observed a slow initial rate of light-induced cytochrome oxidation in *Anacystis* and in *Euglena gracilis*, and explained this phenomenon by a cyclic electron flow. This gives only a satisfactory explanation for the low rate observed after prolonged darkness, at rather low light intensity, discussed above. However, the still relatively low quantum efficiency for e.g. cytochrome oxidation in *Anacystis* at room temperature at optimal conditions (rather high light intensity, short preceding dark time) cannot be easily explained by such a reaction, in view of the good efficiency for $NADP^+$ reduction in the same species at the same conditions (Amesz and Duysens, 1962).

## Summary

In the red alga *Porphyridium cruentum* and the blue-green algae *Anacystis nidulans* and *Schizothrix calcicola* the kinetics of light-induced cytochrome and P700 oxidation were studied. Two different induction effects upon illumination with red or blue light were observed. The first one occurred after darkness and is tentatively explained by a "cyclic" reaction with the primary photoreductant of light reaction 1, causing a retardation of the accumulation of oxidized cytochrome and P700. The second effect occurred after illumination with light mainly absorbed by the photochemical system 2, and is probably caused by a concomitant reduction of oxidized cytochrome by reduced plastoquinone.

## References

Amesz, J. (1964a). Thesis University of Leiden.
Amesz, J. (1964b). *Biochim. biophys. Acta* **79**, 257.
Amesz, J. and Duysens, L. N. M. (1962). *Biochim. biophys. Acta* **64**, 261.
Bamberger, E. S., Black, C. C., Fewson, C. A. and Gibbs, M. (1963). *Pl. Physiol.* **38**, 483.
Duysens, L. N. M. (1964). *Prog. biophys. mol. Biol.* **14**, 1.
Duysens, L. N. M. and Amesz, J. (1962). *Biochim. biophys. Acta* **64**, 243.
Forti, G. and Parisi, B. (1963). *Biochim. biophys. Acta* **71**, 1.
Henninger, M. D., Bhagavan, H. N. and Crane, F. L. (1965). *Archs. Biochem. Biophys.* **110**, 69.
Hoogenhout, H. and Amesz, J. (1965). *Arch. Mikrobiol.* **50**, 10.
Jones, R. F., Speer, H. L. and Kury, W. (1963). *Physiol. Plantarum* **16**, 636.
Kratz, W. A. and Myers, J. (1955). *Am. J. Bot.* **42**, 282.
Olson, J. M. and Smillie, R. M. (1963). *In* "Photosynthetic Mechanisms in Green Plants" p. 56. National Academy of Sciences—National Research Council, Washington, D.C.
Trebst, A. (1964). *Ber. dtsch. Bot. Ges.* **77**, 123.
Urbach, W. and Simonis, W. (1964). *Biochem. biophys. Res. Commun.* **17**, 39.
Vredenberg, W. J. and Amesz, J. (1965). *In* "Currents in Photosynthesis" Proc. of the 2nd Western-Europe Conf. on Photosynthesis, Woudschoten, Netherlands (J. B. Thomas and J. C. Goedheer, eds.). p. 349. Ad Donker, Rotterdam.
Vredenberg, W. J. and Duysens, L. N. M. (1965). *Biochim. biophys. Acta* **94**, 355.
Witt, H. T., Müller, A. and Rumberg, B. (1963). *In* "La Photosynthèse" Colloques Intern. du CNRS pp. 43–73. Paris.

## Discussion

*O. Kandler:* Do you know at what light intensity the oxidation of cytochrome or the reduction of P700 respectively is saturated?

*J. Amesz:* At an intensity higher than about $5 \cdot 10^{-9}$ einstein/(sec cm$^2$) an appreciable delay in the accumulation of oxidized P700 and cytochrome after prolonged darkness was not observed. This may suggest that above this intensity the cyclic reduction of these compounds becomes saturated. The light-induced oxidation of P700 and cytochrome is saturated at very high intensity, as indicated by experiments of Witt and of Chance and coworkers.

# Kinetic Studies on Photosynthetic Phosphorylation

A. H. Caswell

*Department of Biochemistry and Agricultural Biochemistry,*
*University College of Wales, Aberystwyth, Wales*

## Introduction

It has been recognized for some time that certain biological redox reagents interact with the photosynthetic phosphorylation process and cause a modulation of the rate of synthesis of ATP, which depends on the redox level of the exogeneous factor. Thus Vernon and Ash (1960) have shown that succinate inhibits phosphorylation in chromatophores, whilst fumarate has no such effect. Equally NADH inhibits, whilst $NAD^+$ does not influence the rate of synthesis. In a similar fashion ascorbate and dichlorophenol indophenol (Vernon and Ash, 1960; Bose and Gest, 1963) affect the redox balance. In the absence of a redox reagent chloroplasts exhibit low phosphorylating activity, but addition of certain reagents such as menadione or flavin mononucleotide cause substantial activation of photosynthesis.

The kinetic equations that follow have been derived to account in quantitative terms for the modifying effect that these reagents have on the cyclic electron transfer process. From the kinetic equations it is possible to predict the action of redox reagents and information is made available regarding certain parameters of photosynthetic phosphorylation.

## The Kinetic Equation

Consider the cyclic electron transfer process in its most general form as consisting of a cycle of transport between two endogenous reagents A and B. This is represented diagrammatically thus:

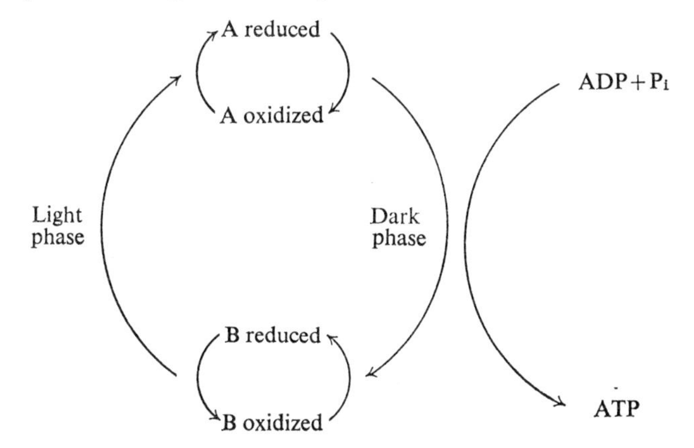

Illumination will cause reduction of A and oxidation of B but this is counteracted by oxidation of A and reduction of B, that accompanies ATP formation. The endogenous reagents will settle down to a steady state of redox level under the driving influence of light.

The following terminology is used in the mass action kinetic equations:

$[A]$ represents total concentration of endogenous reagent A
$[A_{red}]$ represents total concentration of A in reduced form
$[A_{ox}]$ represents total concentration of A in oxidized form
$[A_{red/ox}]$ represents ratio of concentration in reduced and oxidized form
$k_1$ is the rate constant for the light phase of the reaction
$k_2$ is the rate constant for the dark phase of the reaction

The rate of electron transfer is $\frac{de'}{dt} \propto \frac{d[ATP]}{dt}$

For light phase

$$\frac{de'}{dt} = k_1[A_{ox}][B_{red}] \tag{1}$$

For dark phase

$$\frac{de'}{dt} = k_2[A_{red}][B_{ox}] \tag{2}$$

Equations (1) and (2) can be coalesced to give rise to the rate equation

$$\frac{[A][B]}{de'/dt} = \left(\frac{1}{k_1} + \frac{B_{red/ox}}{k_2}\right)\left(1 + \frac{1}{B_{red/ox}}\right) \tag{3}$$

or

$$= \left(\frac{A_{red/ox}}{k_1} + \frac{1}{k_2}\right)\left(1 + \frac{1}{A_{red/ox}}\right) \tag{4}$$

Now the redox level of $A$ or $B$ is affected by interaction with an exogenous redox couple $R$, which is present in substantial excess of the quantity of endogenous reagent. It is the influence of $R$ on photosynthetic phosphorylation that is required.

$R$ interacts directly with, say, $A$ so that the redox potential of the two reagents is the same. That is:

$$E_A = E_A^0 - \frac{RT}{N_A F} \log_e [A_{red/ox}] \tag{5}$$

$$E_R = E_R^0 - \frac{RT}{N_R F} \log_e [R_{red/ox}] \tag{6}$$

but $E_A = E_R$, hence

$$\frac{\log_e [A_{red/ox}]}{N_A} = \frac{\log_e [R_{red/ox}]}{N_R} + (E_A^0 - E_R^0)\frac{F}{RT} \tag{7}$$

and thus

$$[A_{\text{red/ox}}]^{1/N_A} = \alpha\,[R_{\text{red/ox}}]^{1/N_R} \tag{8}$$

where $\alpha$ is a constant relating the two standard redox potentials. Substituting (8) in (4)

$$\frac{1}{\mathrm{d[ATP]/d}t} = \left(\frac{1}{k_1} + \frac{\alpha^{N_A}[R_{\text{red/ox}}]^{N_A/N_R}}{k_2}\right)\left(1 + \frac{1}{\alpha^{N_A}[R_{\text{red/ox}}]^{N_A/N_R}}\right) \tag{9}$$

and for $B$

$$= \left(\frac{1}{k_1} + \frac{\alpha^{N_B}[R_{\text{red/ox}}]^{N_B/N_R}}{k_2}\right)\left(1 + \frac{1}{\alpha^{N_B}[R_{\text{red/ox}}]^{N_B/N_R}}\right) \tag{10}$$

This is the equation that is used to interpret experimental results from the redox potential of exogenous reagent $R$. The equation is resolved by regression analysis for $\alpha$, $k_1$ and $k_2$ and the complicated quadratic equation that results is analysed by electronic computation.

## The Shape of the Redox Curve

The curve that results from varying the redox level of the exogenous reagent is illustrated in Fig. 1. The redox level or potential is altered by varying the

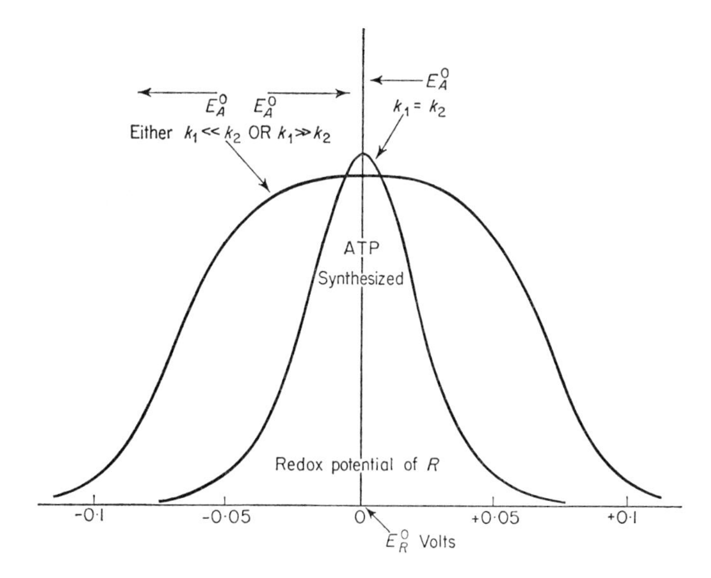


FIG. 1. Theoretical curve for varying redox level of the exogenous reagent added to chromatophores (for explanation of symbols, see text).

ratio of reduced to oxidized form, which is incubated with the chromatophores. The graphs are put in terms of redox potential which varies with the log $[R_{\text{red/ox}}]$. The following features should be noted:

(a) The solution for $\alpha$, $k_1$ and $k_2$ is a quadratic equation and so two values of each are elicited. These are such that equations (9) and (10) cannot be distinguished from each other mathematically and so it is not possible to estimate which endogenous reagent $R$ reacts with from a single experiment.

(b) Where $k_1 \gg k_2$ or $k_2 \gg k_1$, the curve has an extensive plateau region for optimum ATP synthesis whereas, if $k_1 = k_2$, the peak of the curve is sharp. Thus the length of the plateau is diagnostic of the ratio of $k_1$ to $k_2$.

(c) The values of $k_1$ and $k_2$ are functions of the conditions of incubation of the chromatophores and will vary from experiment to experiment; $\alpha$, however, relates the standard redox potential of $R$ to $A$ or $B$ and, within the limits of constant pH and ion concentration, $\alpha$ should be invariant.

(d) The maximum slope of the curve is diagnostic of $N_A$ or $N_B$. $N_R$ is known but $N_A$ and $N_B$ must be evaluated before the correct kinetic equation can be applied to the experimental data. This is achieved by examining the maximum slope, which varies within narrow limits depending on whether $k_1 \gg k_2$ or $k_1 = k_2$. However, the maximum slope varies considerably depending on whether $N_A = 1$ or 2.

(e) The assignation of $R$ to $A$ or $B$ can be determined directly if two graphs are obtained from an incubation experiment where $k_2$ has two different values. This is achieved by varying the concentration of ADP, $P_i$ or ATP, since $k_2$ is a function of these metabolites. If $k_2$ is reduced in one set of incubations by adding ATP and lowering ADP and $P_i$, then the two graphs obtained will fit one value of $\alpha$ better than the other. For example, if $R$ interacts with $B$, reduction of $k_1$ will cause a shift of the maximum towards more oxidized redox potential; if $R$ interacts with $A$, the shift is towards more reduced redox potential.

## RESULTS

Figure 2 illustrates the results obtained when bacterial chromatophores are incubated with varying concentrations of Na succinate/fumarate in a set of Thunberg tubes. The graph includes the peak of the curve but does not extend to the region of maximum slope. The "best fit" curve is based on the assumption that $N = 1$ and the parameters are determined, so that $k_1$, $k_2$ and the standard redox potential of the endogenous reagent are evaluated

$$k_1 = 450$$
$$k_2 = 1{\cdot}7$$
$$E^0 = -0{\cdot}058 \text{ or } +0{\cdot}075 \text{ volts}$$

Figure 3 shows the effect on the succinate/fumarate curve of reducing $k_2$ by reducing the concentrations of ADP and $P_i$ from 10 $\mu$moles to 1 $\mu$mole and adding 9 $\mu$moles of ATP. A shift of the peak towards more reduced redox potential is observed, which implies that succinate/fumarate interacts with $A$. Furthermore, the curve for low $k_2$ includes the area of maximum slope and this fits the provision that $N = 1$ but the slope is very much too slight to account for the shape that would be obtained were $N = 2$. This then is the justification for the assumption in Fig. 2. Figures 2 and 3 show that $R$ interacts with $A$ and

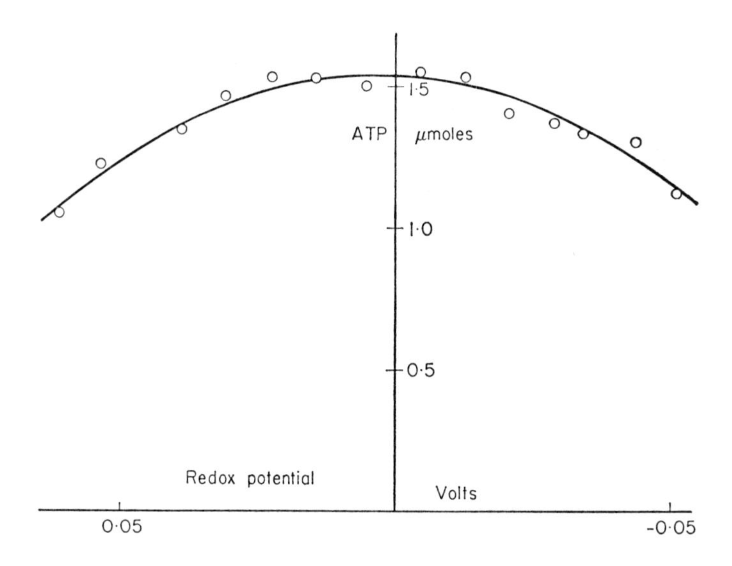


FIG. 2. Succinate/fumarate redox potential curve. Concentration of reagents: ADP 10 μmoles, $K_2HPO_4$ 10 μmoles, $MgCl_2$ 10 μmoles, Na succinate/fumarate 50 μmoles; total volume 3 ml pH 7·4.

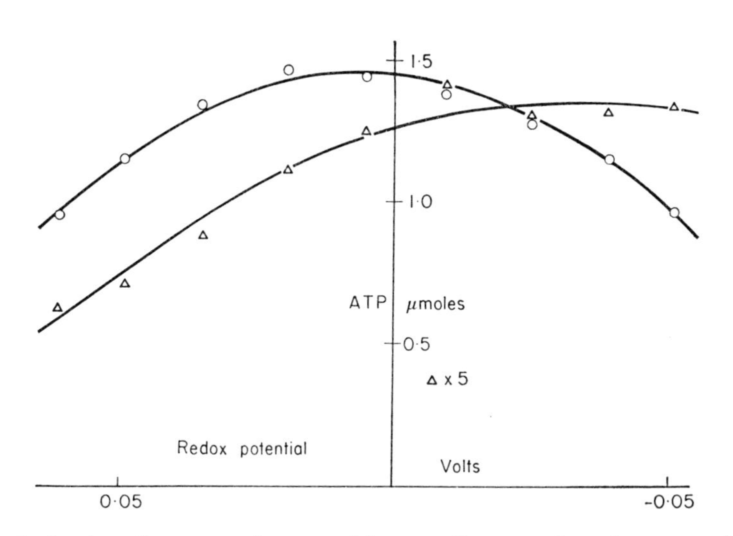


FIG. 3. Succinate/fumarate redox potential curve. Concentration of reagents: $MgCl_2$ 10 μmoles, Na succinate/fumarate 50 μmoles. ○, ADP 10 μmoles, $K_2HPO_4$ 10 μmoles. Δ, ADP 1 μmole, $K_2HPO_4$ 1 μmoles, ATP 9 μmoles.

that $E^0$ for $A$ is $+0{\cdot}075$ volts. The light phase is faster than the dark phase in accordance with the findings of other workers (Nishimura (1963) and they differ by a factor of about 270, where ADP and $P_i$ are present in 3$\mu$moles/ml.

Figure 4 is the redox curve for NADH/NAD$^+$. This curve includes the

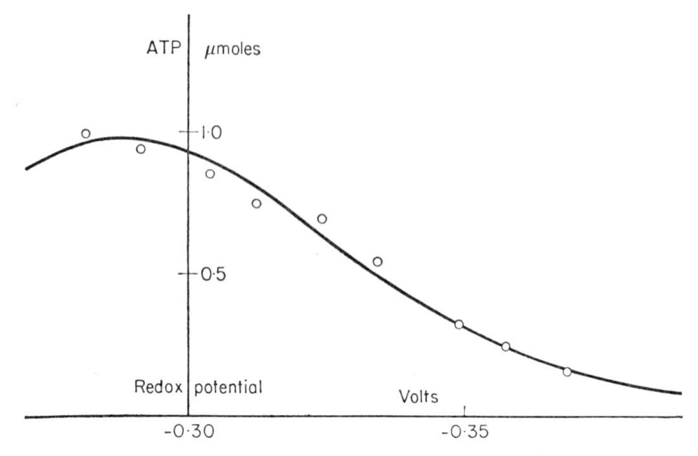


FIG. 4. NADH/NAD$^+$ redox potential curve. Concentration of reagents same as in Fig. 2 except redox reagent is NaNADH/NAD$^+$.

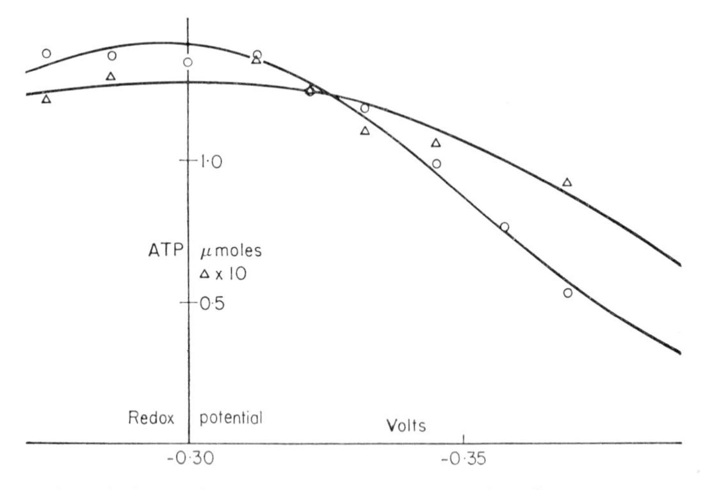


FIG. 5. NADH/NAD$^+$ redox potential curve. Concentration of reagents same as in Fig. 3 except redox reagent is NaNADH/NAD$^+$.

maximum slope in its field but not the full range of the peak. The "best fit" curve, obtained by computation, has been drawn in but the values of the parameters cannot be estimated properly, in view of the fact that the plateau is incompletely included.

Figure 5 illustrates the effect of reducing $k_2$. The curve for high $k_2$ is more complete than that in Fig. 5 and so the electron number N of the endogenous reagent can be estimated. A value of 1 is obtained. The curve for low $k_2$ admittedly does not provide a good fit, but a shift towards negative redox potential is discernible, commensurate with $R$ interacting with $A$. The values obtained for the parameters from the curve of high $k_2$ are

$$k_1 = 155$$
$$k_2 = 1{\cdot}7$$
$$E^0 = -0{\cdot}24 \text{ volts}$$

The value for $E^0$ is not estimated as accurately as in the succinate/fumarate curve, owing to the fact that the full plateau range is not included, and also NADH obtained commercially is contaminated with $NAD^+$ to an extent that is not readily discerned.

## Analysis

The most significant finding from these data is that succinate/fumarate acts on the reduced end of the transport cycle. This is unexpected in view of the standard redox potential of the endogenous reagent and also the fact that succinate can be used to photoreduce $NAD^+$ (Vernon and Ash,1 959; Frenkel, 1956). The appearance of two endogenous reagents acting at the reduced end of the scale and yet having entirely different $E^0$ is highly anomalous and might seem to imply that two light reactions occur in chromatophores. This is a feature for which no prior evidence is available and appears distinctly improbable. The finding of Horio *et al.* (1963) that incubation of chromatophores in the dark with NADH and fumarate does not give rise to equilibration of the redox potentials of these reagents, despite the fact that in the light photoreduction occurs, correlates with the finding above. Furthermore, the slight oxidation of NADH that occurs is found not to be accompanied by ATP synthesis so that the phosphorylating reaction cannot represent the barrier against equilibrium. This facet of succinate/fumarate does not appear to have a ready explanation.

The potential difference between the two endogenous redox reagents is 0·31 volts. This figure is unaffected by the value of $k_1$ or $k_2$ since the peak is shifted equally for each reagent towards the reduced end of the scale as $k_2$ is reduced with respect to $k_1$. This redox potential difference represents conditions at equilibrium. It compares with a value of 0·32 obtained by Horio *et al.* (1963). The value must be indicative of the energy released by the light reaction. A value of 0·35 volts is obtained as the standard free energy of hydrolysis of ATP. Although experimentally and physiologically conditions of ATP synthesis are not standard, the estimate of 0·31 volts is sufficient for synthesis of no more than one ATP molecule per quantum of incident light.

## Experimental

Chromatophores are released from *R. rubrum* cultured anaerobically by the method of Vernon and Ash (1959). Thunberg tubes containing the incubation medium are evacuated by a high vacuum pump and are incubated in a thermostatically controlled environment under conditions of saturating illumination provided by two 500-watt projector lamps. All reagents in the medium are adjusted to pH 7·4 and the following quantities are used: 2·0 ml chromatophores in 0·04 M *tris* buffer; 50 $\mu$moles Na succinate/fumarate or $NADH/NAD^+$; 10 $\mu$moles ADP; 10 $\mu$moles $K_2HPO_4$; 10 $\mu$moles $MgCl_2$; 0·5 ml [$^{32}P$]orthophosphate. The synthesis of ATP is estimated by the method of Avron (1960) modified for scintillation counting.

## References

Avron, M. (1960). *Biochem. biophys. Acta* **40**, 257.
Bose, S. K. and Gest, H. (1963). *Proc. natn. Acad. Sci. U.S.A.* **49**, 337.
Frenkel, A. (1958). *In* "Brookhaven Symposium of Biology", **11**, 276.
Horio, T. Yamashita, J. and Nishikawa, K. (1963). *Biochim. biophys. Acta*, **66**, 37.
Nishimura, M. (1963). *In* "Bacterial Photosynthesis", (H. Gest, A. San Pietro and L. P. Vernon, eds.) p. 201. Antioch Press, Ohio,
Vernon, L. P. and Ash, O. K. (1959). *J. biol. Chem.* **234**, 1878.
Vernon, L. P. and Ash, O. K. (1960). *J. biol. Chem.* **235**, 2721.

# Light-induced Changes of Chloroplasts Area in *Mnium undulatum*

J. Zurzycki

*Laboratory of Plant Physiology, University of Cracow, Cracow, Poland*

It is a well-known fact that isolated chloroplasts undergo structural changes under illumination. Recent investigations of Packer (1962, 1963), Itoh, Izawa and Shibata (1963), Izawa *et al.* (1963) and others have shown that illuminated chloroplasts show the phenomenon of shrinkage. This phenomenon consists in decrease of volume to 50–80% of the value before illumination. At the same time an increase of the flat area of chloroplasts is noticeable. The phenomenon of shrinkage is reversible, and the action spectrum of this process as well as the effects of inhibitors and cofactors suggest that the mechanism of shrinkage is connected with photophosphorylation.

In our previous study (Zurzycki, 1964) carried out on chloroplasts in intact leaf cells of the moss *Mnium undulatum* we have shown that there are some qualitative differences in the influence of the short and long wavelength range of the light on the area of chloroplasts. These differences are especially striking in strong light intensities, where the area of chloroplasts increases in red and strongly decreases in blue light.

In this report some further studies on the photostructural changes in chloroplasts of *Mnium undulatum* are presented. The experiments were carried out on intact leaves and on isolated chloroplasts. Isolation was performed in 0·4 M sucrose in M/15 phosphate buffer pH 7·3. The method of illumination and measurement of the area of chloroplasts was described in the earlier paper (Zurzycki, 1964). The results of planimetric measurements of the area of chloroplasts were expressed in % of the area of the same chloroplasts adapted to darkness.

The changes of area of chloroplasts illuminated with blue light are shown in Fig. 1 for intact and isolated chloroplasts. Isolated chloroplasts of *Mnium undulatum* manifest similar changes as the chloroplasts in intact cells, i.e. they increase their area after illumination with blue light of relatively low intensity and decrease this area under illumination with high intensity of the same quality of light. The changes of area of isolated chloroplasts run much quicker and are finished in $\frac{1}{2}$–1 hr of illumination, though much slower changes of chloroplasts in intact cells become stabilized after 5–8 hr.

However, light intensities necessary to produce the shrinkage of area of chloroplasts are rather high, the process of decrease of area seems to have no destructive character because it is fully reversible (Fig. 2). In about 3 hr after switching off the illumination the area of contracted chloroplasts returns to the former level.

The effect of light intensity on the changes in area of isolated chloroplasts is presented in Fig. 3 for blue and red light. As can be seen from the curves both quality of light cause the flattening effect, i.e. increase of the area in the region of intensities higher than 100 erg/cm$^2$ sec. The activity of blue light for this process is higher, while red light causes only a flattening effect until the highest intensities are used in these experiments, blue light of an intensity higher than

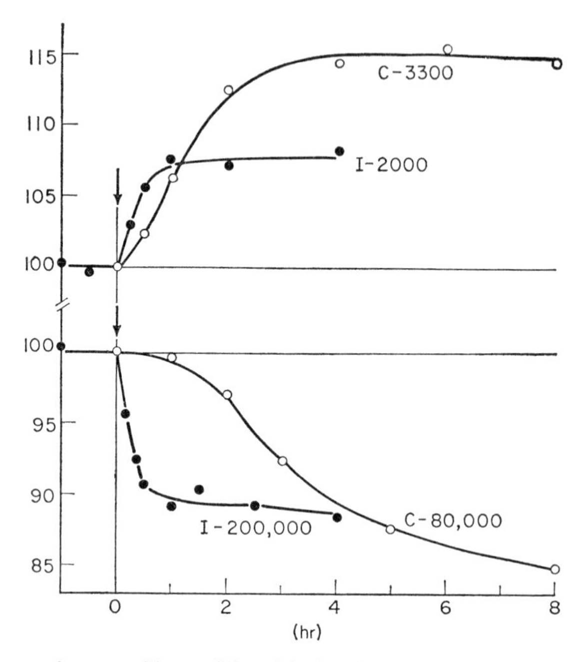


FIG. 1. Changes in area of intact (C) and isolated (I) chloroplasts under illumination by blue light (483 m$\mu$), X axis—time in hours, Y axis—area of chloroplasts in % of the value in darkness. At the time marked by arrows the exposure was applied. Figures by the curves show the light intensity in erg/cm$^2$ sec.

some 10,000 erg/cm$^2$ sec causes the shrinkage of area. The higher the light intensity the greater the degree of shrinkage.

It seems that there are two different processes connected with light which control the area of chloroplasts. The role of photophosphorylation in the structural changes of chloroplasts demonstrated by other authors (Packer, 1962, 1963; Itoh, 1963; Izawa *et al.* 1963) made us investigate the effect of some inhibitors of this process on the changes of chloroplasts area.

In Fig. 4 the dependence of the area of chloroplasts on the intensity of blue light in different concentrations of $NH_4Cl$ and *o*-phenanthroline is shown. The chloroplasts were pretreated with the inhibitors 1 hr before illumination. The influence of both inhibitors used is similar. In a suitable concentration the

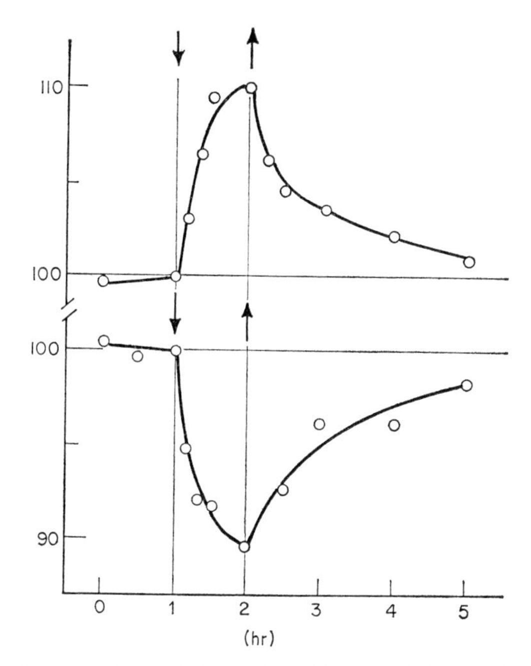


FIG. 2. Changes in area of isolated chloroplasts illuminated during 1 hr with blue light of the intensity 3000 erg/cm² sec (upper curve) and 200,000 erg/cm² sec (lower curve). Chloroplasts are kept in darkness before and after illumination. Time of illumination is marked by arrows.

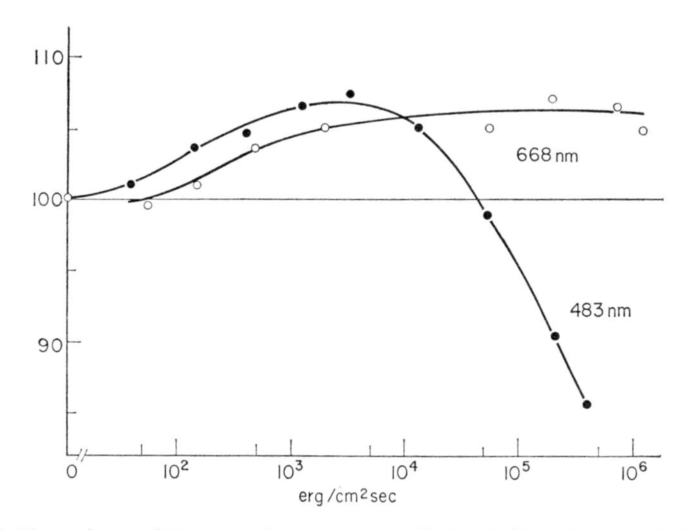


FIG. 3. Dependence of the mean change in area of isolated chloroplasts on the intensity of radiation as measured after 1 hr of illumination for two spectral region: 668 mμ and 483 mμ. X axis—light intensity in erg/cm² sec.

flattening effect of blue light is completely blocked while strong intensities of radiation cause still marked decrease of chloroplasts area.

The conclusion may be drawn that in the photostructural changes of

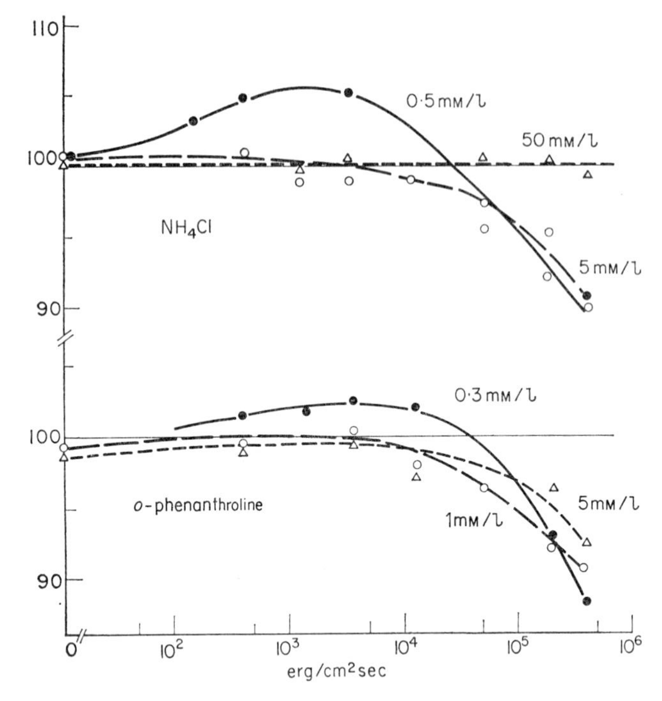


FIG. 4. Influence of two inhibitors: $NH_4Cl$ and *o*-phenanthroline, on the change of area of isolated chloroplasts caused by blue light of different intensities. X axis—light intensity in erg/cm² sec.

chloroplasts of *Mnium undulatum*, which may be demonstrated by the changes in area, two different processes may be distinguished:

1. The process of enlarging of an area caused by red light and by blue light of low and middle intensity. This process seems to be coupled with photophosphorylation.

2. The process of shrinkage of an area caused by strong blue light which overcomes the tendency to flattening. The last process, controlled only by the short wave part of the spectrum, is not sensitive to inhibitors of photophosphorylation.

## References

Itoh, M., Izawa, S. and Shibata, K. (1963). *Biochim. biophys. Acta* **66**, 319, 130.
Izawa, S., Itoh, M. and Shibata, K. (1963). *Biochim. biophys. Acta* **75**, 349.
Packer, L. (1962). *Biochem. biophys. Res. Commun.* **9**, 355.
Packer, L. (1963). *Biochim. biophys. Acta* **75**, 12.
Zurzycki, J. (1964). *Protoplasma* **58**, 458.

# Part VII. Biosynthetic Mechanism in Relation to Morphogenesis

# Biosynthesis and Morphogenesis in Plastids

Lawrence Bogorad[1]

*Department of Botany, University of Chicago, Chicago, Illinois, U.S.A.*

The production of lipids, proteins, nucleic acids, pigments, carbohydrates, and other components of plastids have been considered individually in this volume. The mature normal plastid is the properly integrated product of all this biochemistry and physiology. How is the production of these compounds related to plastid morphogenesis?

## Patterns of Plastid Development

Only mature plastids have been observed in many algae. In some of these organisms a fully formed plastid divides to give rise to two; each new plastid may then enlarge and divide again. The problem here is one of understanding growth and division, but it is difficult to analyse plastid development experimentally in these cases.

In a second type of plastid development small bodies grow into mature plastids. Spheres approximately 0·2 $\mu$ to 1 $\mu$ in diameter have been identified as plastid primordia in meristems of higher plants (e.g. Mühlethaler and Frey-Wyssling, 1959) and in some algae such as *Euglena* grown in darkness (Ben-Shaul *et al.*, 1964). In this alga there is no further plastid development without illumination but plastids continue to mature as long as light is provided. Particularly in higher plants, on the other hand, there is some further elaboration of structure in the small spheres even in darkness. The inner membrane of the primordial plastid appears to evaginate. Small vesicles are formed and extend into the center of the sphere. The entire structure enlarges, while at the same time more vesicles develop. If a higher plant is maintained in darkness, the characteristic proplastid containing protochlorophyllide and a crystal lattice-like body (the prolamellar body or Heitz-Leyon crystal) is elaborated. Ribosomes and sometimes even a few thylakoids are also visible in proplastids (Fig. 1). Prolamellar body-containing proplastids also have been seen in roots of illuminated plants. Only upon illumination do further modifications of this proplastid occur.

The subsequent discussion will be restricted largely to the development of plastids in higher plants.

[1] Research Career Awardee of the National Institute of General Medical Sciences, U.S.P.H.S.

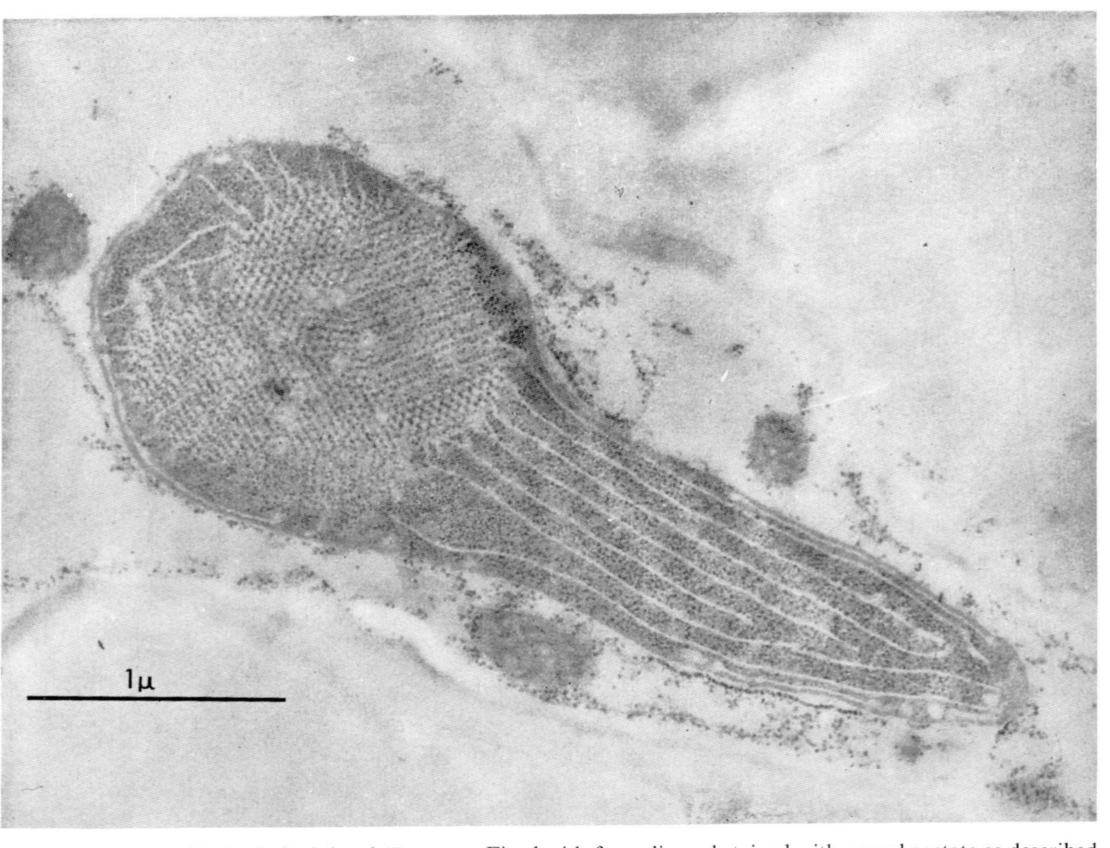


FIG. 1. Proplastid in leaf of etiolated *Zea mays*. Fixed with formalin and stained with uranyl acetate as described by Jacobson *et al*. (1963).

## Development of Plastids in Higher Plants

Among the higher plants, where development in darkness stops at the proplastid stage, there are two patterns of greening and further plastid development.

In both cases the protochlorophyllide which is present is converted into chlorophyllide by light. In some plants, and this depends upon age as well as species, chlorophyll formation and plastid maturation begin immediately upon illumination and continue at a rapid pace. In other cases, however, the initial transformation of the protochlorophyllide may be followed by a period of apparent relative inactivity, judged from chlorophyll accumulation (Virgin, 1955). For example, in 10 to 12-day-old etiolated maize leaves the chlorophyll content roughly doubles during the first 2 to 3 hr of illumination and then a phase of rapid chlorophyll formation begins. The final concentration of chlorophyll in fully greened leaves of maize is approximately three to four hundred times greater than that of the protochlorophyllide present before illumination. The separation of the greening process into three fairly distinct phases—protochlorophyllide transformation, the lag phase, and the linear phase—permits further analyses to be made of the controlling processes and the course of biosynthesis of plastid components.

Before examining chemical changes other than chlorophyll accumulation which occur during development of the proplastid into a mature chloroplast it is profitable to review the morphological alterations (von Wettstein, 1966, Vol. I). First, as the protochlorophyllide is converted into chlorophyllide, the crystal lattice-like structure of the prolamellar body dissociates, and in its place a group of vesicles, not so closely associated with one another, appears. This condition persists throughout the lag phase as long as plants are maintained in light, but von Wettstein and Kahn (1960) found that after illumination brief enough to cause dissociation of the prolamellar body the crystal lattice-like structure was reformed if the leaves were returned to darkness. In plastids of leaves illuminated after the end of the lag phase, the vesicles, still in about the same area previously occupied by the prolamellar body during the lag phase, disperse throughout the plastid as the linear phase of chlorophyll formation commences. These vesicles line up in rows across the plastid and appear to fuse with one another to form a number of primary thylakoids. Then, at various places along the primary thylakoids, additional layers of thylakoids arise appressed to the primary one. Grana appear to be built up in this manner.

The dissociation of the elements of the prolamellar body (tube transformation), measured as the proportion of proplastids with transformed prolamellar bodies, is closely correlated with the degree of conversion of protochlorophyllide into chlorophyllide in the proplastids as measured spectrophotometrically (Virgin *et al.*, 1963; Klein *et al.*, 1964). Further evidence of a close relationship between the photoreduction of protochlorophyllide and the morphological change is provided by the observation of Klein (1960) that tube transformation can occur at 3° under 400 ft-candle intensity. Gunning's elegant analysis of the structure of the prolamellar body (p. 655) should be helpful in leading to a better understanding of at least the physical basis of tube transformation.

The lag and linear phases seem to be controlled by two different photoreceptive systems.

It has been known for some time that the lag phase is mediated by the red: far-red reversible phytochrome system (Price and Klein, 1961; Mitrakos, 1961; Virgin, 1961). Exposure to red light initiates the lag phase but the promotive effect of this irradiation is eliminated by subsequent illumination with enough far-red light ($\sim$730 m$\mu$). After initiation by red light the lag phase can be completed in darkness; thus, if etiolated leaves which normally exhibit a lag phase in greening are irradiated for a short while and then are placed in darkness for a sufficiently long time—usually a few hours—they begin to form chlorophyll without appreciable delay upon reillumination.

The action spectrum for vesicle dispersal, the first morphological manifestation of the linear phase of greening, has been shown by Henningsen (p. 453) to be promoted strongly by light of about 450 m$\mu$; regions of lower effectiveness occur at 400, 550, and 660 m$\mu$. Chlorophyll accumulation during the linear phase slows and soon ceases if illumination is discontinued.

In view of Granick's demonstration that all the enzymes for protochlorophyllide formation except those for the synthesis of ALA ($\delta$-aminolevulinic acid) are active in proplastids, it is apparent that light is required for ALA formation during greening. The probability that continued protein synthesis is essential for ALA production during the linear phase is discussed in a later section.

If etiolated leaves are exposed to chloramphenicol (Margulies, 1962) or actinomycin D (Bogorad and Jacobson, 1964) before illumination, greening is inhibited. Thus, there is apparently a requirement for protein synthesis and for DNA-dependent RNA synthesis. Margulies (1964) had demonstrated that protein synthesis which normally occurs during greening fails to take place in bean leaves treated with chloramphenicol.

## THE COMPOSITION OF CHLOROPLASTS AND PROPLASTIDS

The obvious way to begin to try to understand the relationship between biosynthesis and morphogenesis in plastids would be to obtain a complete quantitative and qualitative inventory of the contents of plastids at various stages of development. By comparison of these inventories it should then be clear what compounds had to be made even though the relative importance of each compound in morphogenesis might not be revealed in this way. This is technically not a simple task and few data are available. Isolation of plastids in good condition is a relatively new and still primitive art. In general, the state of purity attained is not entirely satisfactory.

We can only guess at what may be in the 0·2 to 1 $\mu$ diameter plastid primordia. They probably contain DNA and protein; the membrane is probably composed of phospholipid as well as protein.

The proplastid of an etiolated leaf is much richer in morphology. Here DNA can be identified as long strands about 25–50 Å thick, ribosomes are present, and, as already described, the prolamellar body, which is probably composed largely of proteins, is the most conspicuous structural feature. Studies by Boardman and Wildman (1962) using fluorescence microscopy indicate that each proplastid has one or two localized centers of fluorescence; this suggests that the prolamellar body is the site of protochlorophyllide deposition.

The mature plastids of higher plants have varying numbers of grana and a conspicuous feature is the large amount of lamellar material which is not soluble when the plastid is disrupted osmotically. The lamellae are composed of about equal proportions of lipoidal and proteinaceous substances. With the electron

Table 1. Light stimulated changes in the composition of bean plastids[1]

| | Per Plastid | | | | |
|---|---|---|---|---|---|
| | Dry weight $\mu g \times 10^{-6}$ | Protein[2] $\mu g \times 10^{-6}$ | % of Protein in mature plastids | Lipid[3] $\mu g \times 10^{-6}$ | % of Lipid in mature plastids |
| Proplastids: | | | | | |
| 7-day-old plants | 3·2 | 1·8 | 34 | 0·6 | 30 |
| 12-day-old plants | 3·6 | 2·0 | 38 | 0·7 | 35 |
| Mature plastids: | | | | | |
| 12-day-old plants (7 days Dark + 5 days Light, 14 hr daily) | 7·9 | 5·3 | — | 2·0 | — |
| Plastids: | | | | | |
| 7-day-old plants + 90 min Light + 5 days Dark | 6·1 | 4·8 | 90 | 1·3 | 65 |

[1] Based on data from Mego and Jagendorf (1961).
[2] Based on N = 15% of protein.
[3] Not including chlorophyll.

microscope one can also detect a large number of ribosomes and some DNA in mature plastids.

Chemical analyses have been made of proplastids and mature ones. Mego and Jagendorf (1961) determined the dry weight as well as the gross lipid and nitrogen contents of proplastids and mature plastids isolated from bean leaves (*Phaseolus vulgaris* L. var. Black Valentine) in 0·5 M sucrose in 0·05 M phosphate buffer, pH 7, containing 2 to 4% formaldehyde to maintain plastid integrity. Some of their data are compiled in Table I.

It is estimated that on a dry-weight basis proplastids from etiolated Black Valentine bean leaves contain approximately 48 to 50% protein, 18 to 20% lipid; unidentified compounds make up the remaining 30 to 40%. Mature plastids from the same tissue contain about 67% protein and 25% non-chlorophyll lipids, both based on dry weight; these figures are not remarkably different from many presented by Menke and others during this conference. The data of Table I also show that there is little change in plastid composition between the seventh and twelfth day of growth of bean plants in darkness. However, upon illumination and maturation the dry weight roughly doubles, the total nitrogen increases by a factor of 2·7, and there is about a three-fold increase in the lipid content of each plastid. Mego and Jagendorf do not provide these data, but as pointed out above, in general, the chlorophyll content of mature plastids is roughly 100 to 300 times the amount of protochlorophyllide present in proplastids.

Obviously a thorough knowledge of the *kinds* of proteins, fat soluble materials, and other constituents of the proplastids and the mature plastid would be more useful than gross values in helping to understand structural elaboration and the mechanisms of control of plastid development.

A few of the enzymes present in proplastids are known to increase considerably during maturation. Among these are NADP-linked glyceraldehyde 3-phosphate dehydrogenase (Marcus, 1960; Margulies, 1964), ribulose 1,5-diphosphate carboxylase (Margulies, 1964), and photosynthetic phosphopyridine nucleotide reductase (Keister *et al.*, 1962). Of the enzymes involved in chlorophyll formation, at least some concerned with ALA synthesis must increase greatly, judging from the effect of supplying ALA to etiolated leaves (e.g. see Granick, p. 373). It may be significant that many of the enzymes which are subsequently seen in mature plastids can also be identified, although in lower concentrations, in proplastids. It is not understood why plastid development proceeds to the proplastid stage and stops in darkness.

Mego and Jagendorf also determined changes in plastid composition during an extended lag phase. Etiolated bean plants were exposed to light for 90 min and then returned to darkness for 5 days. The dry weight, nitrogen content, and lipid content of plastids from such leaves were compared with those of proplastids and of mature plastids from plants which had been illuminated for 14 hr each day over a 5-day period. These data are presented in Table I. The exposure to light for 90 min stimulated both protein and lipid formation—there is about twice as much of each of these classes of compounds in plastids from briefly illuminated as from unilluminated plants.

It is also striking that despite great morphological differences, the gross protein and lipid contents of mature plastids are not very different from those of plastids from plants which had been illuminated for one 90-min period 5 days before analyses were made. Mature plastids contained only about 10% more protein and 35% more non-chlorophyll lipid than proplastids from equally old plants which had been illuminated once for 90 min. In the mature plastid about 42% of the total protein and 90% of the lipid is in the insoluble lamellar material (Menke, Vol. I). (The difference in chlorophyll content would probably be about 100- to 150-fold.)

Are the grana and thylakoids formed by the rearrangement and/or association of lipids and proteins already present at the end of the lag phase? What kinds of proteins and lipids are made during the linear phase of greening? Do these include enzymes which catalyse the formation of new thylakoids from substances already present? Or, are many of the proteins and lipids accumulated during the lag phase destroyed and replaced by newly formed ones? These questions cannot be answered on the basis of evidence which is available currently, but it is clear from papers presented in this volume that many of the means for obtaining pertinent data are now available.

It is not known whether the subunits of the thylakoid "grow" from the protochlorophyllide holochrome by accretion—i.e. like a snowball—or whether the holochrome protein serves only as an enzyme for the conversion of protochlorophyllide. Does the chlorophyllide (or chlorophyll) molecule move from the holochrome protein to another protein which will become or already is part of the thylakoid?

The crucial questions of how light absorbed by one photoreceptive system or another promotes plastid development cannot be answered now, but the indications from inhibitor studies of the involvement of protein and nucleic acid synthesis in maturation of these subcellular organelles has stimulated research into certain events of the lag and linear phases in the hope that the site of light action can be established eventually.

## Ribonucleic Acid Metabolism During the Lag Phase

The observation that DNA-dependent RNA synthesis may be involved in chlorophyll formation and plastid maturation has led us to investigate RNA metabolism in developing plastids. Thus far, our investigations have been limited to events of the lag phase.

As has been pointed out repeatedly, proplastids in etiolated leaves contain DNA and large numbers of ribosomes (Fig. 1). The data of Mego and Jagendorf (1961) indicate that in proplastids of beans net protein synthesis is slight after a certain level of maturation has been achieved as long as the plants are not irradiated but is very great after brief illumination. The data on actinomycin D-inhibition of greening indicate that some sort of essential RNA is formed during the lag phase. The presence of ribosomes in the proplastid suggests that perhaps only informational RNA is required. Using currently available techniques it is possible not only to see whether additional RNA is formed, but to determine something of its character by sucrose density gradient or chromatographic analysis.

In these experiments, performed in collaboration with Dr. Ann Jacobson, we have supplied leaves of maize grown in darkness for 10–12 days with either $^{32}P$ as phosphate or with radioactive uridine. After letting detached leaves absorb radioactive material in darkness (the amounts we supplied were usually absorbed within about 30 min to an hour when the leaves are placed before a fan in a dark room) some were placed in the light (about 1800 ft-candles of fluorescent light) for periods of 30 min to 3 hr while others were maintained in darkness for equal lengths of time. The RNA was then extracted (by the phenol method) and the distribution of RNA as 260 m$\mu$ absorbing material and as radioactivity was determined after centrifugation in a sucrose density gradient (5 to 20% sucrose). We have generally obtained about the same kinds of patterns of incorporation of radioactivity into RNA during the lag phase whether we have illuminated leaves for 30 min and then returned them to darkness for another 90 min or exposed them to light continuously for an equivalent period, i.e. 120 min.

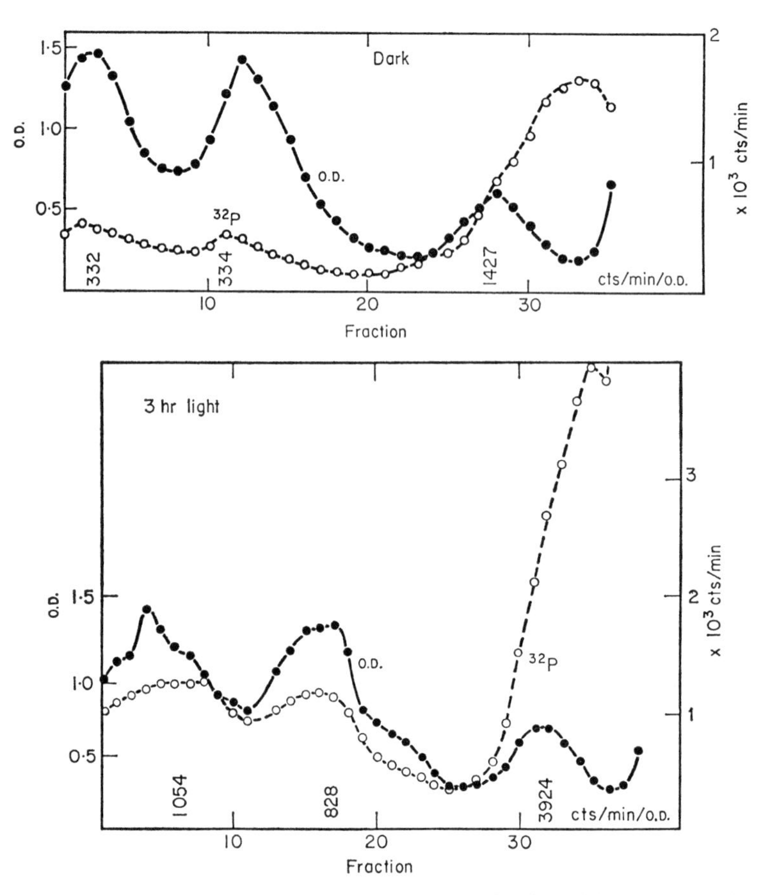


FIG. 2. Distribution of 260 m$\mu$ absorbing materials and radioactivity after centrifugation for 6 hr at 36,000 rev/min (Spinco Model L centrifuge, SW 39 rotor, at about 2°) in a gradient of sucrose (5 to 20% sucrose in 0·05 M *tris* pH 8·0, 0·1 M KCl) of RNA prepared from maize leaves supplied $^{32}$P-phosphate in darkness. "Dark": leaves maintained in darkness for 3 hr after administration of $^{32}$P. "3 hr light": leaves illuminated for 3 hr after administration of $^{32}$P. Fractions are numbered from bottom of gradient (densest portion) upward.

Figure 2 shows data obtained from an experiment of this sort comparing the radioactivity in RNA from maize leaves which had been illuminated for 3 hr and those which had been maintained continuously in darkness. It is apparent that in this experiment the specific activity of each kind of RNA we can separate according to size on a sucrose density gradient is at least three times higher when the source is illuminated leaves than when the RNA is obtained from comparable plants maintained continuously in darkness. The two heaviest RNAs, those nearest the bottom of the gradient, are of the ribosomal types; the RNA near the top of the density gradient (highest number fractions) is of the size range of transfer RNA.

These data on RNA from whole leaves do not demonstrate that RNA metabolism in the plastid is entirely or even partly responsible for the differences observed. In order to determine this, similar experiments were done, but

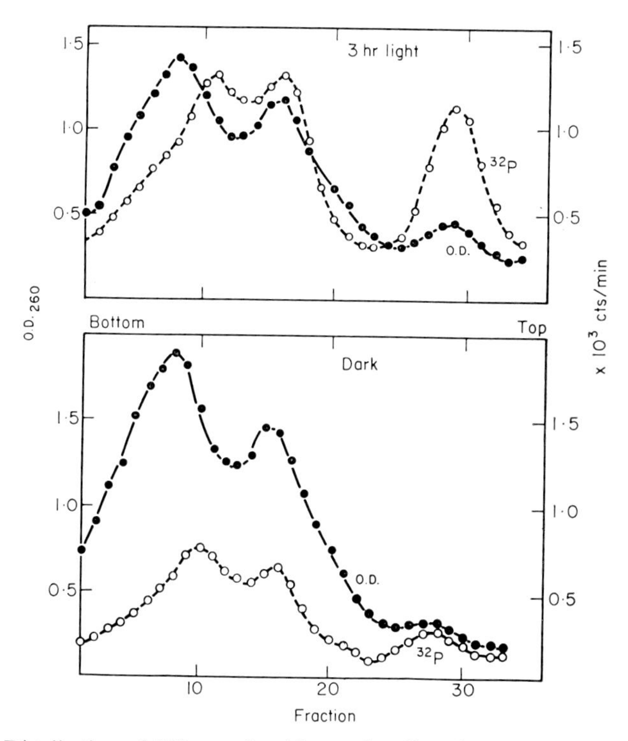


FIG. 3a. Distribution of 260 m$\mu$ absorbing and radioactive materials in sucrose after centrifugation (see Fig. 2). Upper figure: RNA from plastid fraction prepared from maize leaves which had been illuminated for 3 hr after administration of $^{32}$P-phosphate. Lower figure: RNA from plastid fraction prepared from maize leaves maintained in darkness for 3 hr after administration of $^{32}$P-phosphate.

the subcellular particles within the leaves were separated before their RNA was extracted. Again, leaves were permitted to absorb a tracer and were then illuminated or not. After this, leaves were ground (in 0·5 M sucrose, 0·5 M *tris* buffer pH 8, 0·001 M $MgCl_2$) and plastids were isolated first by differential centrifugation, and, in some experiments, further purification was achieved by centrifugation in a density gradient of sucrose. RNA was extracted from plastid fractions and analysed. In addition to this, after the plastids had been separated from the leaf homogenate by differential centrifugation, the supernatant fluid was used as a source of cytoplasmic ribosomes (isolated by the procedure of T'so *et al.* (1956)). Unfortunately, in our hands, techniques which were most suitable for isolating cytoplasmic ribosomes yielded poor-quality (for sucrose density resolution) plastid RNA and, conversely, techniques which were most effective for isolating chloroplasts and obtaining "good" RNA from them, led to formation of poor-quality cytoplasmic ribosomal RNA. (The preparative procedures differ in the molarity of the buffer used for grinding leaves.) Figure 3 shows the optical density and radioactivity profiles of RNA obtained from plastid and from cytoplasmic ribosome preparations. The curves for RNA from cytoplasmic ribosomes (undoubtedly contaminated with some chloroplast ribosomes) in this case are of poor quality for reasons

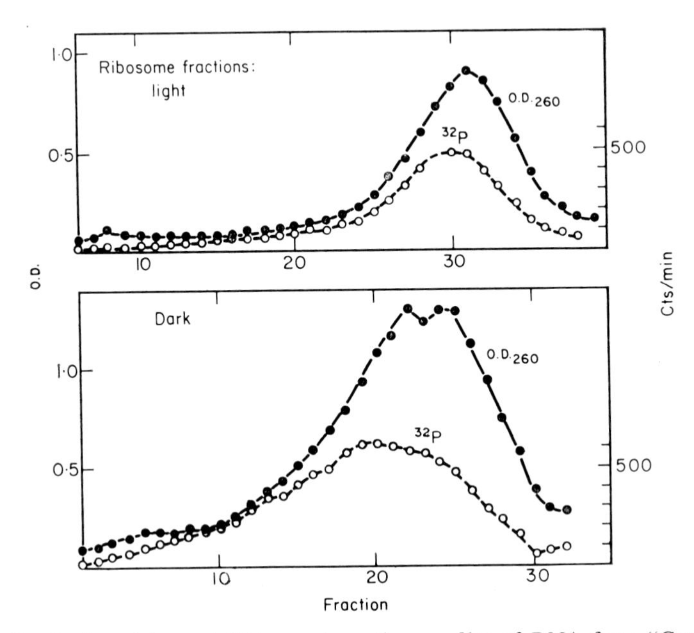


FIG. 3b. Radioactivity and 260 mμ absorption profiles of RNA from "Cytoplasmic ribosome fractions" of maize leaves. Each of these fractions was prepared from the corresponding homogenate from which plastids were prepared for determination of the RNA profiles shown in Fig. 3a.

which have been outlined, but calculations of the specific activity of total RNA reveals a small (~1·5-fold) difference between material obtained from illuminated and unilluminated leaves. On the other hand, the much higher specific activity of each kind of RNA from plastids of illuminated than unilluminated leaves is striking. The asymmetry of the radioactivity and the optical density patterns may indicate some differences in labelling rates of different RNAs.

If one or more specific types of messenger RNA are formed during the lag phase before massive RNA synthesis begins, we have not identified them in these experiments. It is apparent, however, that even if that does occur, all sorts of RNA are produced in larger quantities in plastids in illuminated than unilluminated maize leaves during the lag phase of plastid development.

Furthermore, if stimulation of nuclear RNA synthesis had occurred upon illumination, large differences in the specific activity of cytoplasmic RNAs from illuminated and unilluminated leaves should also have been seen. The small difference in specific activity of the cytoplasmic fractions may be partly or entirely attributable to contamination of these preparations with ribosomes from broken plastids.

This observation of an increase (i.e. difference in specific activity) in all of the kinds of RNA which we could separate led us to examine the possibility of a change in the level of RNA polymerase in plastids upon illumination of etiolated leaves. Using techniques essentially similar to those of Mans and Novelli (1961, 1964) for the assay of a soluble RNA polymerase in maize leaf extracts, we examined the activity of RNA polymerase in plastids isolated from illuminated leaves (i.e. a washed $1000 \times g$ fraction) and in the solution obtained by centrifuging the supernatant fluid from the chloroplast isolation at $27{,}000 \times g$

Table II. Maize leaf RNA polymerase

| Activity of $27{,}000 \times g$ supernatant fractions incorporation of [$^{14}$C]-ATP[1] | | |
|---|---|---|
| | +CT DNA | −CT DNA |
| Unilluminated | 652 cts/min[2] | 373 cts/min |
| 30 min Light + 90 min Dark | 647 cts/min | 318 cts/min |

[1] Reaction mixture = 0·8 ml. Components of reaction mixture: 1·8 μmole ea. CTP, GTP, UTP; 1·5 μc ATP (0·14 μmole); 0·02 mmole Mg Acetate; 0·04 mmole mercaptoethanol; 0·04 mmole *tris* pH 8·0; ±0·2 mg calf thymus DNA (CT DNA) and 0·1 mmole $NH_4Cl$, all in 0·4 ml *plus* 0·4 ml protein (enzyme preparation) in 0·5 M sucrose, 0·5 M *tris* pH 8·0, 0·001 M $MgCl_2$, 0·01 M mercaptoethanol. Incubation: 10 min, 30°. Protein per 0·80 ml reaction mixture: from unilluminated leaves 1·68 mg; from leaves illuminated for 30 min and returned to darkness for 90 min before harvest 1·78 mg.

[2] Cts/min/0·1 ml reaction mixture.

for 15 or 30 min. Mans and Novelli (1964) have studied soluble RNA polymerase in the supernatant of a maize shoot homogenate and demonstrated that activity in such preparations is stimulated by the addition of calf thymus DNA. We have assayed RNA polymerase in the presence of added DNA to obtain a measure of the enzyme independent of the amount of DNA carried along in the preparations. As shown in Table II there is essentially no difference in the activity of RNA polymerase in the supernatant fractions from illuminated and unilluminated leaves. On the other hand, striking differences are observed in the RNA polymerase activity of the $1000 \times g$ fractions from these two types of leaves (Table III).

The data for the $1000 \times g$ fraction are corrected to what is essentially RNA polymerase activity per plastid. The chlorophyll content per gram of leaf tissue was determined for unilluminated and illuminated leaves at the time of harvest. The chlorophyll content per gram of leaf tissue has generally been approximately 1·5 to 2·5 times greater in leaves illuminated for 2 hr (or illuminated for

30 min and then maintained in darkness for 90 min) than in unilluminated leaves. Observation with the light microscope reveals no conspicuous increase in the number of plastids; they are probably constant. (von Wettstein, Vol. I) reports no change in the number of plastids per bean leaf cell during 40 hr greening). The chlorophyll content of each suspension of plastids is also determined and equal numbers of plastids, based on the various chlorophyll determinations, are added to each incubation mixture (in some experiments variable amounts of plastid suspension were added and the activity was calculated back to permit comparison on an equal plastid basis). For a given plastid preparation, the amount of activity measured is proportional to the amount of chlorophyll (i.e. number of plastids) included in the reaction mixture.

Table III. Maize leaf RNA polymerase

| | Activity of "plastid" fractions incorporation of [$^{14}$C]-ATP[1] | |
|---|---|---|
| | −CT DNA | +CT DNA |
| Unilluminated | 166 cts/min[2] | 161 cts/min |
| 30 min Light + 90 min Dark | 419 cts/min | 410 cts/min |

[1] Reaction mixture = 0·8 ml. Components: 2 μmole ea. CTP, UTP, GTP; 1·4 μc ATP (0·14 μmole); 0·02 mmole Mg Acetate; 0·04 mmole mercaptoethanol; 0·04 mmole *tris* pH 8·0; ±0·2 mg calf thymus DNA (CT DNA) and 0·1 mmole $NH_4Cl$, all in 0·4 ml *plus* 0·4 ml of plastid preparation (washed and resuspended 1000 × *g* fraction) in 0·5 M sucrose, 0·05 M *tris* pH 8·0, 0·001 M $MgCl_2$, 0·01 M mercaptoethanol. Chlorophyll per 0·80 ml reaction mixture: from unilluminated leaves 2·0 mμg; from leaves illuminated for 30 min and returned to darkness for 90 min before harvest 4·1 mμg. These amounts of plastid suspension are calculated to be derived from almost equal amounts (fresh weight) of tissue, i.e. 0·63 mg (fresh weight) unilluminated leaf tissue; 0·60 mg illuminated leaf tissue. (Plastids were prepared from 58 g of each type of leaf tissue.) Incubation: 10 min, 30°.

[2] Cts/min/0·1 ml of reaction mixture.

In most experiments we have measured RNA polymerase activity of plastid and supernatant fractions by the rate of incorporation of [8-$^{14}$C]-ATP into an RNA-like material. (That is, the radioactivity is precipitable by cold trichloracetic acid, but is solubilized by RNase; incorporation of radioactivity occurs only when all four nucleotide triphosphates are included in the reaction mixture; incorporation is inhibited by actinomycin D; radioactivity from [$^{14}$C]-GTP is also incorporated and to about the same extent as that from [$^{14}$C]-ATP. In all these properties the system resembles those described for the soluble maize RNA polymerase by Mans (1964) and the chloroplast RNA polymerase preparations reported by Kirk (1964) and by Semal *et al.* (1964).)

The addition of calf thymus DNA to the test mixture in which maize leaf homogenate supernatant fluid was the source of enzyme-stimulated [$^{14}$C]-ATP incorporation, in confirmation of Mans' observation. On the other hand, the effect of addition of such DNA to reaction mixtures in which the RNA polymerase activity of $1000 \times g$ fractions was being measured was variable, and frequently had no effect. It should be remembered, however, that the $1000 \times g$ fraction can be rich in maize nuclear DNA.

If chloroplasts are isolated by sucrose gradient centrifugation of the $1000 \times g$ fraction, RNA polymerase activity is still detectable and differences between plastids from illuminated and unilluminated leaves are conspicuous (Table IV). RNA polymerase activity in these preparations has generally been found to be stimulated by the addition of calf thymus DNA. The greater effect of added DNA on the activity of plastids from unilluminated than from illuminated leaves may indicate that additional DNA synthesis may be stimulated by illumination, but this point must be checked by independent methods.

The light-induced increase in RNA polymerase activity is reduced in leaves

Table IV. RNA polymerase activity

"Purified" chloroplast fractions from maize leaves
$^{14}C$-ATP incorporated[1]

| | *Exp.* 1[2] | *Exp.* 2 | *Exp.* 3 |
|---|---|---|---|
| Dark | 10 cts/min | 11 cts/min | 29 cts/min |
| Light | 107 | 43 | 88 |
| Dark + CT-DNA | 25 | 20 | 58 |
| Light + CT-DNA | 149 | 71 | 103 |

[1] Cts/min/0·1 ml reaction mixture: 2 $\mu$mole ea., CTP, GTP, UTP; 1·1 $\mu$c ATP (0·14) $\mu$mole; 0·02 mmole Mg Acetate; 0·04 mmole mercaptoethanol; 0·04 mmole *tris* pH 8·0; $\pm 0{\cdot}2$ mg calf thymus DNA (CT-DNA) and 0·1 mmole $NH_4Cl$ all in 0·4 ml *plus* 0·4 ml plastids (from sucrose density gradient centrifugation) in 0·5 M sucrose, 0·05 M *tris* pH8·0, 0·001 M $MgCl_2$, 0·01 M mercaptoethanol. Incubation: 10 min, 30°. *Within* each experiment the quantities of plastids used in each reaction mixture were determined to be derived from equal amounts (fresh weight) of leaf tissue.

[2] Exp. 1: Light = 30 min light + 90 min darkness before harvest. Exp. 2 and 3: Light = 120 min light before harvest.

exposed to chloramphenicol during illumination. This suggests that light triggers the synthesis rather than the activation of this enzyme.

These data show that plastid RNA polymerase activity increases during the lag phase but do not reveal the sequence of events during the lag phase. That is, RNA polymerase may be the first enzyme whose production is stimulated by light or it is equally possible that increased production of RNA polymerase is concomitant with the accelerated synthesis of many proteins such as ribulose 1,5-diphosphate carboxylase, etc. Causal relationships in lag phase biochemistry and the primary effect of light on these processes remain to be determined.

In any event, during the lag phase RNA polymerase activity increases, all kinds of new RNA are formed, a relatively large amount of protein is synthesized, and much additional lipid is formed, but relatively insignificant structural changes occur in the proplastid. Tube transformation, the structural modification of the lag phase, is reversible in darkness. Again, the *sequence* of biochemical events is not known.

## THE LINEAR PHASE

Next, in plastids of higher plants major changes in the amount of chlorophyll formed and the number of thylakoids occur only under continuous illumination. Figure 4 presents some data on the consequence of returning dark-grown red kidney bean plants to darkness after they have been in the light for 6 hr (Gassman and Bogorad, 1965); no additional chlorophyll is formed but protochlorophyllide continues to accumulate for a few hours and then ceases.

Exposure of illuminated leaves to 0·005 M chloramphenicol (Fig. 5) or, with some delay (probably owing to absorption problems), to $5 \times 10^{-4}$ M puromycin arrests chlorophyll formation during the linear phase; thiouracil has a similar effect. Thus exposure of leaves to inhibitors of protein or RNA synthesis prevents greening.

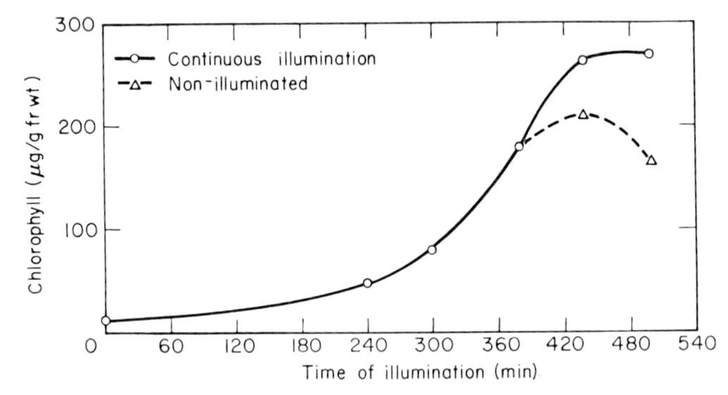


FIG. 4. The effect of light on chlorophyll synthesis during the linear phase of greening of red kidney bean leaves. Plants grown in darkness at about 25° for 8 days. Primary leaves excised and illuminated while on 0·2 M sucrose in Petri dishes. After 360 min illumination two sets of leaves were returned to darkness for analysis after 1 or 2 hr; before extraction of chlorophyll with 80% acetone these leaves were illuminated briefly to permit conversion to chlorophyllide or chlorophyll of the protochlorophyllide which accumulated. Photoconversion of protochlorophyllide in this manner simplified pigment analyses (Gassman and Bogorad, 1965).

Based on the same sorts of arguments presented above, it is apparent that continuous synthesis of at least some or some kind of protein is required for massive chlorophyll formation during the linear phase. The mimicking of removal of light by inhibitors of nucleic acid and protein synthesis suggests that light is acting here to stimulate these processes. Electron microscope observations of plastids in leaves treated with inhibitors during the linear phase have not yet been made.

How can one explain the relatively rapid cessation of protochlorophyllide formation when leaves are returned to darkness during the linear phase and the production of only small amounts of chlorophyll during the lag phase? Among possible mechanisms of control are: feedback inhibition, repression, and rapid turnover of the enzyme as well as of the informational RNA required for its production. If light accelerates protein synthesis in developing plastids, as it appears to do, and if the various inhibitors (e.g. chloramphenicol, puromycin, thiouracil) act in the same way in developing plastids as they do in bacteria, control of ALA synthesis by variations in the rate of formation and decay of an enzyme seems reasonable. The rate of decay of an enzyme in a

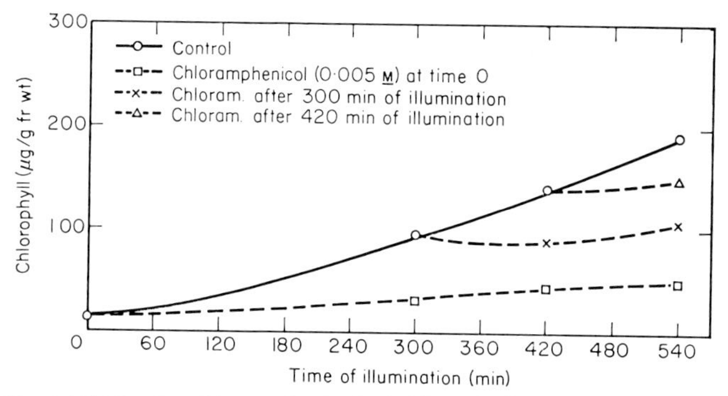


FIG. 5. The inhibition by chloramphenicol of chlorophyll synthesis by red kidney bean leaves during various stages of greening (Gassman and Bogorad, 1965). Leaves were illuminated throughout the periods of exposure to chloramphenicol.

constant environment is an inherent property of the molecule; the rate of formation can obviously be controlled by a number of factors. Unfortunately, there are few data available by which to assess this or almost any other suggestion. At this time it is not even clear what enzyme or enzymes catalyse the formation of ALA in the plastid.

## SUMMARY AND CONCLUSIONS

When higher plants are grown in darkness, plastids develop from spherical primordia lacking distinctive internal structures into proplastids, i.e. plastids which at maximum development characteristically contain a little protochlorophyllide and generally one or two crystal lattice-like prolamellar bodies. Proplastids in leaves which are illuminated form large amounts of chlorophyll and mature into chloroplasts.

The sequence of structural changes is well established, at least to the extent possible with current techniques of electron microscopy, but the order of biochemical events during plastid maturation is very uncertain.

Current values for total protein, total lipid, etc. per plastid of a single species must still be regarded as provisional because of difficulties in plastid isolation. But it is already clear that although these values when available for proplastids, proplastids late in lag phase development, and mature plastids give some gross ideas about metabolic activities in plastids progressing from one developmental stage to another they fail to reveal very much about the relationship between biochemistry and morphogenesis. Information about the nature of specific proteins and lipoidal materials in plastid thylakoids must be obtained and then the course of formation of each compound during development must be determined. In exploring the relationships between plastid metabolism and maturation, it is necessary to remain aware of possible contributions from nuclear activity and of the likelihood that the accommodation between plastid and nucleus may differ from species to species or even variety to variety.

Three different photoreceptors seem to be involved in plastid development: the phytochrome system appears to control the lag phase; the 450 and 550 m$\mu$ absorbing pigment(s) regulates vesicle dispersal; and, protochlorophyllide holochrome absorbs light for its own reduction to chlorophyllide.

Changes in RNA, RNA polymerase, and other enzymes can be detected in plastids during the lag phase, but we have no information about the causal relationships between light and the formation of these substances or the order of development of these materials—how is the production of one of them related to formation of the others?

The exploration of plastid metabolism and morphogenesis proceeds with an audacious and amusing naïveté. The chloroplast is in a sense a heterotrophic "organism" whose nutrition we do not understand except for a few aspects in a very small number of cases such as those described by Walles (1966 p. 633) and by students of the effects of mineral deficiency on plastid structure. Yet, even in these cases, the nutritional lack may affect the plastid indirectly by preventing the production in the cytoplasm of the true essentials for chloroplast growth. The problem of plastid nutrition is more difficult to approach experimentally than that of a free-living micro-organism. Despite this, the temptation to explore the problem is strong—structural changes can be seen, some poisons affect plastids without any apparent effect on the metabolism of the cell, and plastid development can be turned on and off by light! The answers seem tantalizingly close at hand. We are optimistic despite the experiences of our microbiological colleagues with ostensibly simpler problems.

## Acknowledgement

The research from the author's laboratory has been supported in part by grants from the National Science Foundation, the National Institutes of Health, U.S.P.H.S., and the Abbott Memorial Fund of the University of Chicago. This work was performed with skilled technical assistance provided by Miss Dagmara Davis and Mrs. Louisa Ni.

## References

Ben-Shaul, Y., Schiff, J. A. and Epstein, H. T. (1964). *Pl. Physiol.* **39**, 231.
Boardman, N. K. and Wildman, S. G. (1962). *Biochim. biophys. Acta* **59**, 222.
Bogorad, L. and Jacobson, A. (1964). *Biochem. biophys. Res. Commun.* **2**, 113.
Gassman, M. and Bogorad, L. (1965). *Pl. Physiol.* **40**, lii.
Jacobson, A. B., Swift, H. and Bogorad, L. (1963). *J. cell Biol.* **17**, 557.
Keister, D. L., Jagendorf, A. T. and San Pietro, A. (1962). *Biochim. biophys. Acta* **62**, 332.
Kirk, J. T. O. (1964). *Biochem. biophys. Res. Commun.* **16**, 233.
Klein, S. (1960). *J. Biophys. Biochem. Cytol.* **8**, 529.
Klein, S., Bryan, G. and Bogorad, L. (1964). *J. cell Biol.* **22**, 433.
Mans, R. J. and Novelli, G. D. (1961). *Archs Biochem. Biophys.* **94**, 48.
Mans, R. J. and Novelli, G. D. (1964). *Biochim. biophys. Acta* **91**, 186.
Marcus, A. (1960). *Pl. Physiol.* **35**, 126.
Margulies, M. M. (1962). *Pl. Physiol.* **37**, 473.
Margulies, M. M. (1964). *Pl. Physiol.* **39**, 579.

Mego, J. L. and Jagendorf, A. T. (1961). *Biochim. biophys. Acta* **53**, 237.
Mitrakos, K. (1961). *Physiol. Plantarum* **14**, 497.
Mühlethaler, K. and Frey-Wyssling, A. (1959). *J. biochem. biophys. Cytol.* **6**, 507.
Price, L. and Klein, W. H. (1961). *Pl. Physiol.* **36**, 733.
Semal, J., Spencer, D., Kim, Y. T. and Wildman, S. G. (1964). *Biochim. biophys. Acta* **91**, 205.
T'so, P. O. P., Bonner, J. and Vinograd, J. (1956). *J. biochem. biophys. Cytol.* **2**, 451.
Virgin, H. I. (1955). *Physiol. Plantarum* **8**, 630.
Virgin, H. I. (1961). *Physiol. Plantarum* **14**, 439.
Virgin, I., Kahn, A. and Wettstein, D. von (1963). *Photochem. Photobiol.* **2**, 83.
Wettstein, D. von and Kahn, A. (1960). *Proc. Eur. Reg. Conf. Electron Microsc., Delft* **2**, 1051.

# Use of Biochemical Mutants in Analyses of Chloroplast Morphogenesis

Björn Walles
*Department of Forest Genetics, the Royal College of Forestry, Stockholm, Sweden*

## Introduction

The chloroplasts of higher plants are known to develop from structurally undifferentiated precursors, proplastids, by growth in size and complexity. An extensive system of lipoprotein lamellae, in which the pigment-carrying grana constitute an essential part, is formed during this process. Gene mutations can block the morphogenesis of chloroplasts at different intermediate stages as demonstrated by electron microscopy. According to their degree of plastid development the chloroplast mutants can be arranged in a progressive sequence which begins with albina mutants having poorly differentiated proplastids and ends with viable pale green mutants with apparently normal chloroplast structures (cf. von Wettstein, 1959b, 1961; Eriksson, Kahn, Walles and von Wettstein, 1961). The majority of the chloroplast mutants can synthesize chlorophyll, which, however, is liable to photo-oxidation. The extent of photo-oxidation varies and is most pronounced in those mutants where the plastid differentiation comes to an end at an early proplastid stage. In order to resist bleaching the chlorophyll must presumably be bound to a thylakoid protein—i.e. constitute a holochrome complex—and be protected by carotenoids in the lamellar system.

Chloroplast mutations interfere with the normal development and function of chloroplasts by blocking the synthesis of individual plastid constituents. Some of the mutations might lead to loss of the ability to form certain proteins. Lack of individual structural proteins in plastids has so far not been verified experimentally. On the other hand deficiencies for more simple metabolites (amino acids, chlorophylls, carotenoids, thiamine) have been indicated in a number of chloroplast mutants after feeding tests or chemical leaf analyses (cf. Walles, 1963). In order to get information on the morphogenetic role of chloroplast constituents the plastid structures in a few of these biochemical mutants have been investigated. The experimental techniques employed have been described in detail previously (Walles, 1963, 1965). A brief review of the results is given here (cf. Fig. 1).

## Amino Acid Requiring Mutants

In green plant cells the majority of the protein—at least 60 % (Heber, 1960)—is found within chloroplasts. It is therefore to be expected that impaired synthesis of an amino acid would severely interfere with chloroplast development in mutant seedlings in spite of the fact that they initially are supplied with a certain amount of amino acids from the seeds. To investigate the occur-

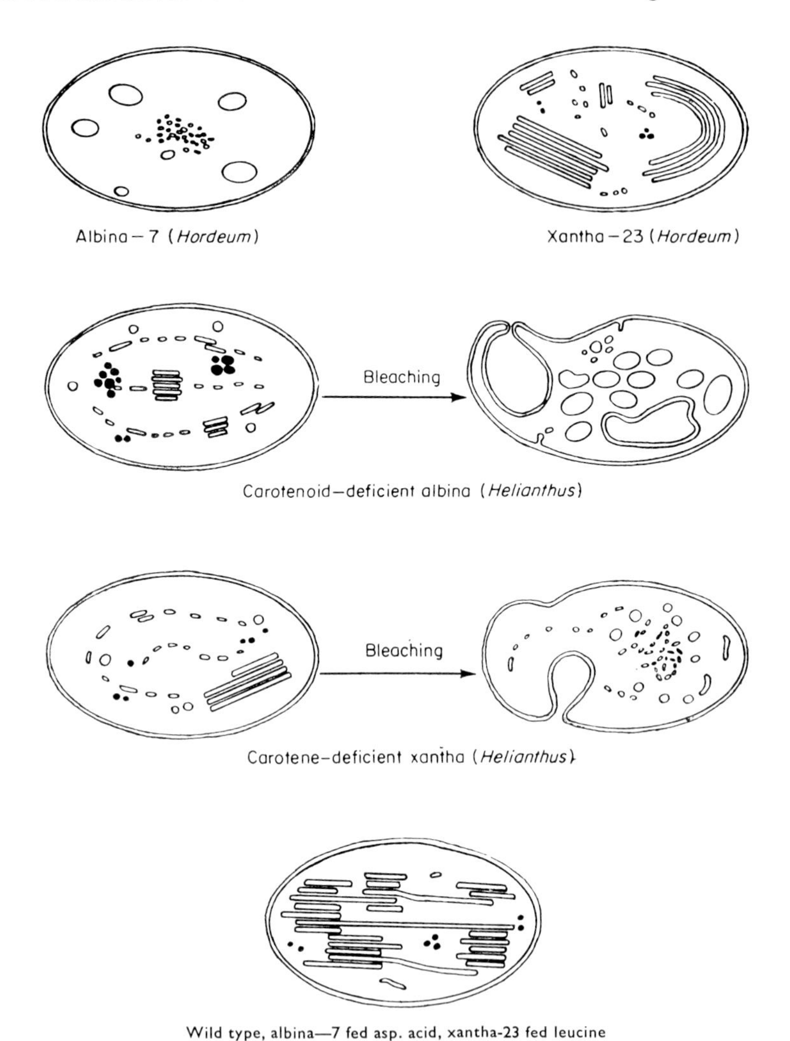


FIG. 1. Diagram of plastid organization of some chloroplast mutants in barley and sunflower.

rence of amino acid auxotrophs among lethal chloroplast mutants a test for amino acid requirements of mutant seedlings in barley was applied (Eriksson, Kahn, Walles and von Wettstein, 1961; Walles, 1963). The mutant seedlings were grown under aseptic conditions with their roots immersed in test tubes containing a Hoagland solution supplemented with an amino acid mixture

(Fig. 2). Pure Hoagland solution was used for the controls. Eventual greening and survival of treated plants was noted. The specific requirements of some mutants that were found to respond to the complete medium were determined by feeding individual amino acids.

Of thirty-seven tested chloroplast mutants of barley, six responded with greening on the supplemented medium. Four of these mutants (three of which

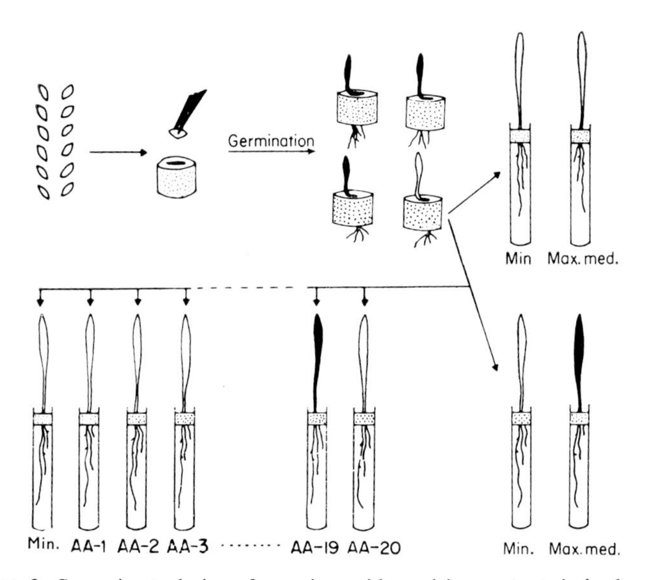


FIG. 2. Screening technique for amino acid requiring mutants in barley.

are allelic) are phenotypically xantha and require leucine, one is viridoalbina and requires aspartic acid and one is viridis with unknown specific requirements. Electron microscopical studies of plastid development and spectrophotometric chlorophyll determinations were made on seedlings of the aspartate-requiring mutant albina-7 and the leucine-requiring mutant xantha-23 (Eriksson, Kahn, Walles and von Wettstein, 1961; Walles, 1963).

## ALBINA-7

Seedlings of albina-7 are viridoalbina (white with pale green leaf tips) when grown at a temperature above 15° in greenhouse or laboratory environments. Under certain laboratory conditions (24 ± 1°, 2000 lux for 18 hr/day) a few of the mutant seedlings can become green spontaneously and survive. It was established, that the mutant offspring from heterozygous mother plants grown in the field during different years differ considerably in viability and chlorophyll content. The white leaf tissues of mutant seedlings have undifferentiated proplastids containing a few, often loosely aggregated vesicles (Fig. 3). The green tissues, including the green leaf tips of viridoalbina leaves, have structurally normal chloroplasts. Those albina-7 seedlings that can produce a

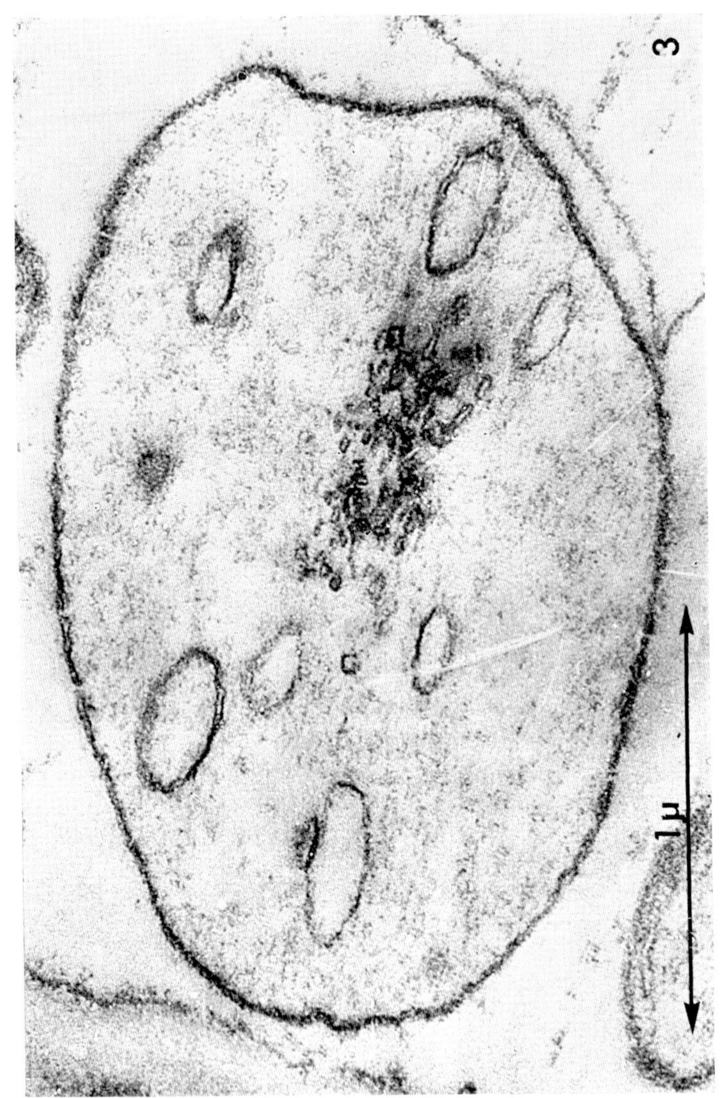


FIG. 3. Plastid from the white part of a primary leaf of albina-7. Fixed in 2% $KMnO_4$.

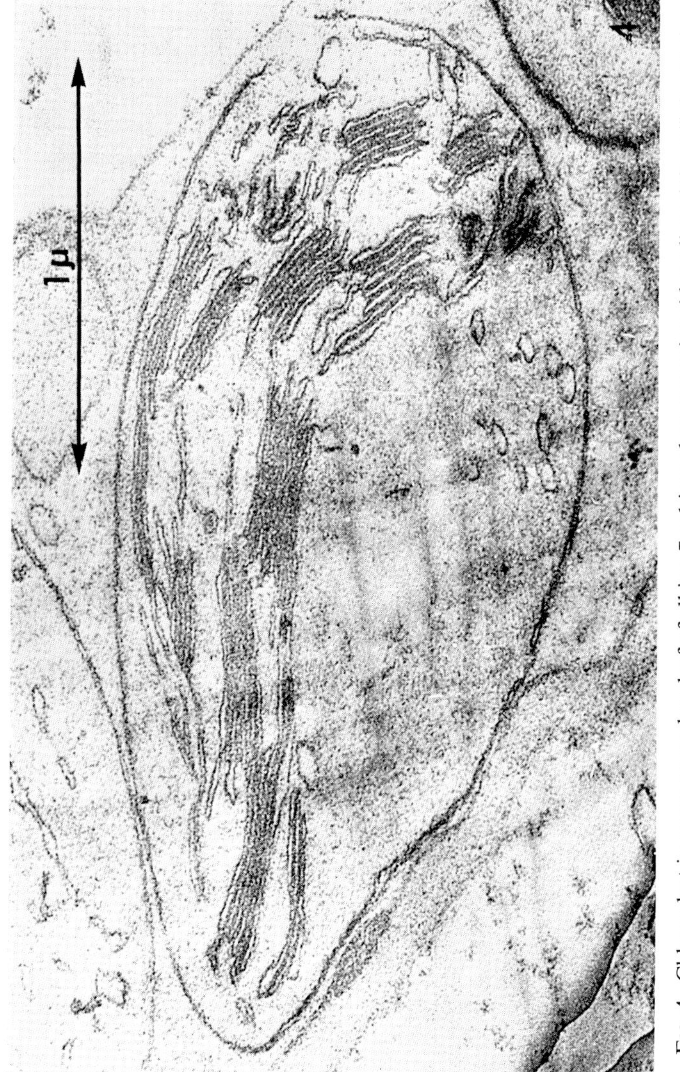


FIG. 4. Chloroplast in green secondary leaf of albina-7 cultivated on aspartic acid medium (50 mg/l) for 13 days. Fixed in 2% $KMnO_4$.

sufficient amount of green leaf tissue continue growth as autotrophs. The new leaves formed on such plants are entirely green. The chloroplasts of the green tissues can apparently carry out photosynthesis and—possibly as a consequence thereof—produce the substance they require for further growth and the formation of new functional chloroplasts. In this way sufficient initial greening will provide a compensation for the blocked biochemical pathway in mutant seedlings.

Aspartic acid given in concentrations of 25–200 mg/l raised the chlorophyll content of lethal albina-7 seedlings and made it possible for them to grow. Higher concentrations of the amino acid interfered strongly with the growth of the root system both in the mutant and in the wild type of barley. Norleucine, employed in the same range of concentrations as aspartic acid, was the only substance which was found to have the same capacity as aspartic acid to induce greening and survival of mutant seedlings. Other amino acids or asparagine had no effects in this respect. The green leaves of mutant seedlings treated with one of the efficient amino acids have mature chloroplasts with a normal structural organization (Fig. 4). Greening of the primary, initially white leaf is always restricted to the tip and the midribs but it is complete in the later formed leaves. Presumably supply of a proper substance is required from the very start of leaf growth in order to induce development of all proplastids into chloroplasts. After induction of sufficient greening of the first leaves, the supply of aspartic acid or norleucine is dispensable for further growth of the mutant plants. They grow in the same way as plants that have turned green spontaneously. The green albina-7 plants growing as autotrophs are markedly retarded in development as compared to wild type plants. The new leaves produced on the main shoot of these mutant plants are always green but the tillers formed are white with light green tips, i.e. their appearance is similar to that of untreated mutant seedlings. After having produced two or three such viridoalbina leaves the tillers eventually develop increasingly greener leaves. The photosynthetically active chloroplasts in the leaf tips might be able to produce the substance required for generation of more chloroplasts and greening of the tillers. Green mutant plants give rise to small spikes, which generally are sterile. The few seeds obtained give rise to typical albina-7 seedlings.

Albina-7 is considered a temperature-sensitive auxotroph requiring aspartic acid or a substance which can be substituted for by aspartic acid. The behaviour of the mutant plants indicate that the formation of aspartic acid during photosynthesis proceeds mainly by one pathway and during germination, early seedling and sideshoot growth by a different pathway which appears to be blocked by the mutation.

## XANTHA-23

Seedlings of xantha-23 are in laboratory tests (ab. 22°, 2500 lux for 18 hr/day) phenotypically viridis. Under these conditions the primary leaves of xantha-23 had a chlorophyll content of about 25 % of that found in the wild type.

The chloroplasts of xantha-23 seedlings contain many small and some very large grana (Fig. 5), which sometimes are cup-shaped or even of a concentric

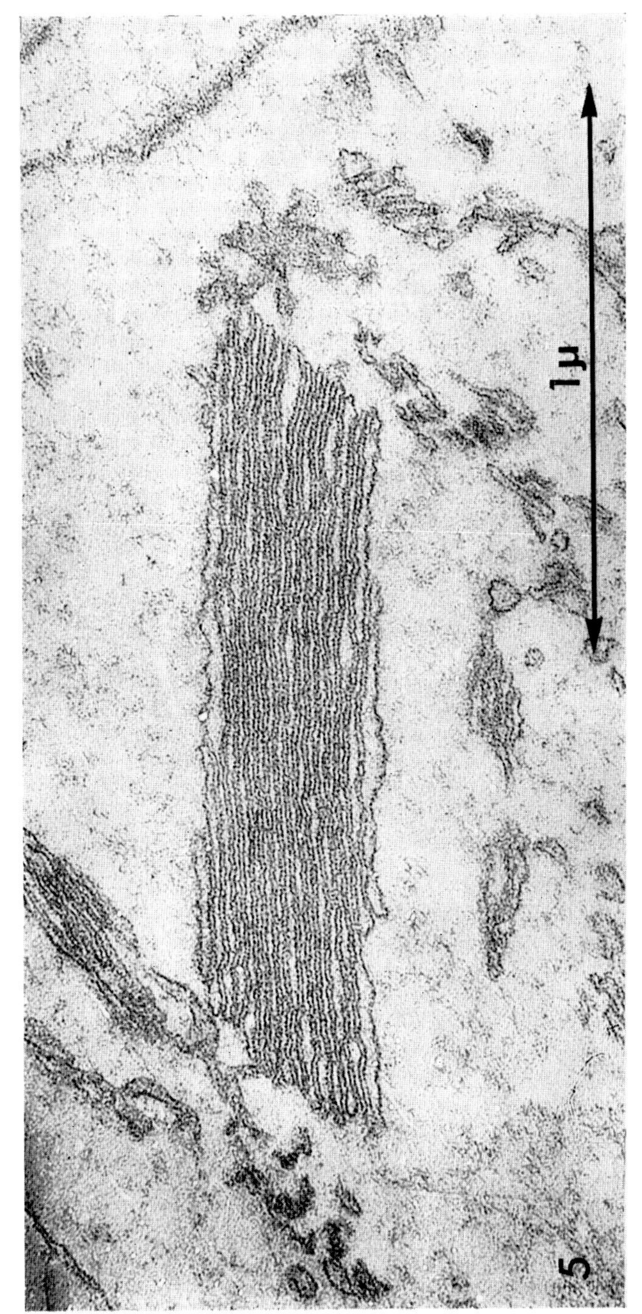


FIG. 5. Section through giant granum in 8 cm long secondary leaf of xantha-23. Fixed in 5% $KMnO_4$.

FIG. 6. Chloroplast in 3 cm long secondary leaf of xantha-23 grown for 6 days on leucine medium (1·6 g/l). Fixed in 5% $KMnO_4$.

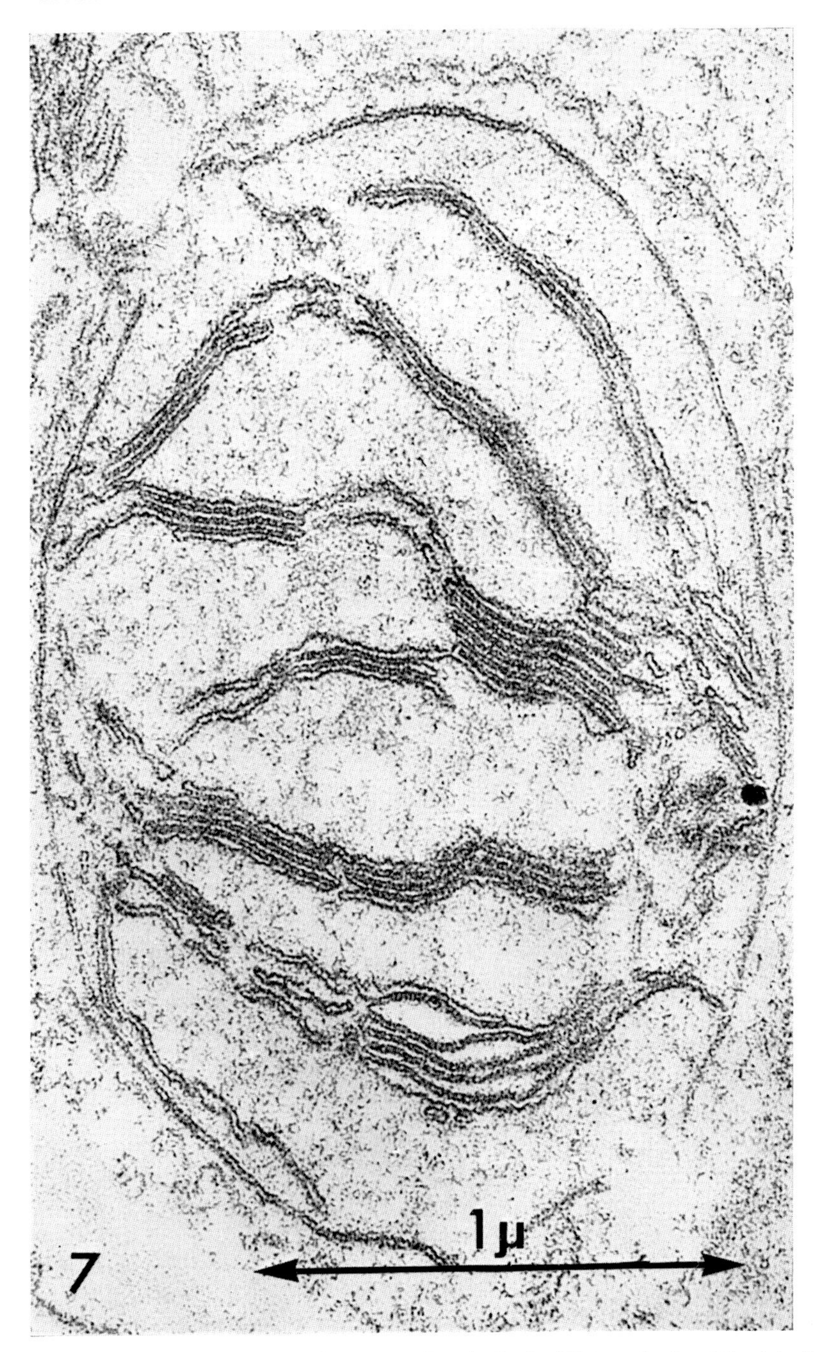


FIG. 7. Chloroplast in 1·5 cm long secondary leaf of wild type barley. Fixed in 2% $KMnO_4$.

shape. These giant grana have both a larger diameter and a higher number of aggregated discs than wild type grana (cf. Figs. 8–9). The individual grana piles are not arranged in the normal parallel manner but irregularly distributed so that adjacent grana frequently are seen lying at different angles to each other. There seem to be fairly few connecting membranes between the grana. The plastid organization in xantha-12 and its alleles (von Wettstein, 1960) corresponds to that of xantha-23. It was reported for mutants of the xantha-12 locus that the aggregation of the discs in their grana is less close than in normal grana and also in xantha-23 the spacing of the discs appears to be more remote

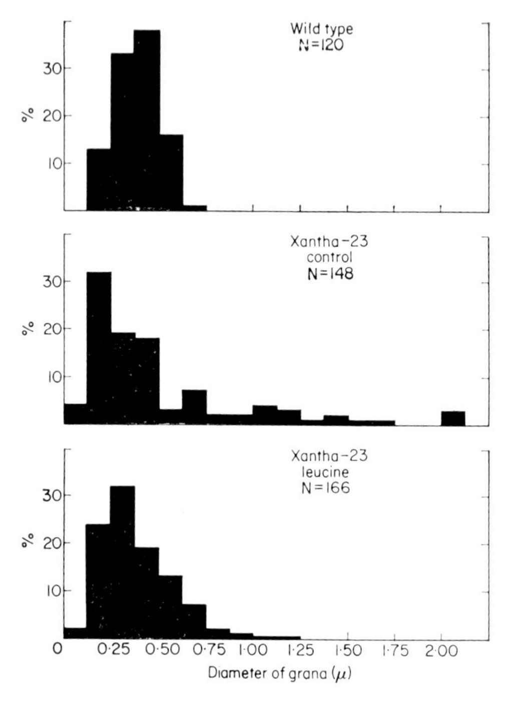


FIG. 8. Diameter of grana in secondary leaves of respectively wild type barley (8-day-old, 15 mm leaves), xantha-23 grown on minimal medium (15-day-old, 80 mm leaves) and xantha-23 grown on 1·6 g/l of leucine (14–15-day-old, 15 mm leaves).

than in the wild type. The thylakoid abnormalities of xantha-23 and -12 with its alleles may be interpreted as the consequence of the deficiency for a constituent required for formation of normal lamellar surfaces (cf. von Wettstein, 1960, 1961).

The chlorophyll content of the mutants increases considerably upon feeding with leucine, while other amino acids have failed to produce comparable responses. In primary leaves of xantha-23 growing on 2 g/l of leucine the chlorophyll content was raised to about 70% of that in the wild type. Also mutant seedlings of the xantha-12 locus increase their amount of chlorophyll when they are cultivated on leucine media (Walles, 1963). It was further found that leucine has a harmful effect on wild type barley and in higher concentrations inhibits the growth of both roots and shoots. Secondary leaves of 20-day-old wild type seedlings cultivated on 0·5 g/l of leucine had a fresh weight that was only half of that of untreated seedlings and a chlorophyll content that was 70% of that of untreated seedlings. In the same experiment the seedlings growing on media with still higher concentrations of leucine failed to develop secondary leaves.

The primary leaves of xantha-23 plants grown on leucine medium exhibited essentially the abnormal plastid structures typical for the mutant. This is most likely due to the fact that the leucine treatment was started 4 days after the

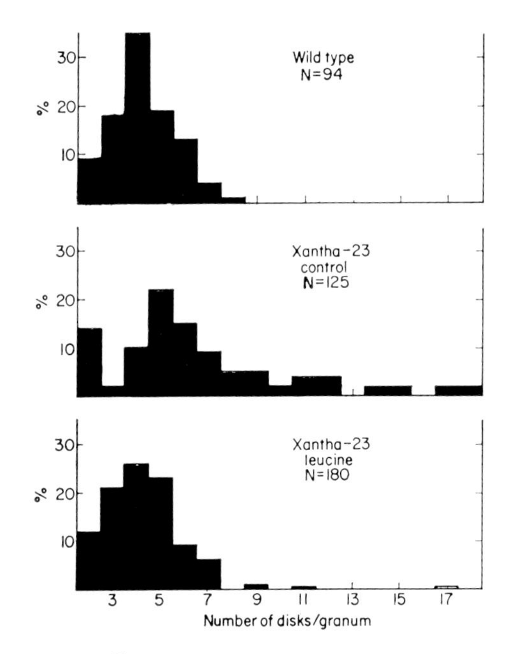


FIG. 9. Number of discs per granum in secondary leaves of the same plants as in Fig. 8.

germination of the seeds, and the primary leaves, about 2 cm long, had at that time accomplished development of plastids with aberrant grana. Supply of leucine can apparently not cure the abnormal plastid structures that have already been formed. It is possible, however, that in addition to the previous giant grana new normally constructed grana are developed in the chloroplasts under the influence of supplemented leucine. Such a process would be able to account for the increased chlorophyll content in the primary leaves. The chloroplasts in secondary leaves of treated xantha-23 seedlings developed after the seedlings had been transferred to the leucine medium. At a leucine concentration of 2 g/l (the highest concentration used) their content of chlorophyll was raised from the control value of 6% to about 75% of that of untreated wild type seedlings. In these leaves the chloroplasts differentiate in a normal way forming many small grana which are normally distributed in the organelle with a more or less parallel arrangement (Fig. 6; cf. Fig. 7 which shows a chloroplast of wild type barley). Measurements on electron micrographs revealed that the diameter of the grana and the number of discs per granum in secondary leaves of xantha-23 on leucine is comparable to what is found in wild type seedlings (Figs. 8–9).

## CHLOROPHYLL-LESS AND CAROTENOID-LESS MUTANTS

The structural role of chloroplast pigments can be evaluated by studies of seedlings that are deficient for certain of these pigments. In angiosperm seedlings chlorophyll synthesis can be blocked by darkness and gene mutation, and also the carotenoid synthesis can be arrested by gene mutation. The plastids of dark-grown normal seedlings produce crystalline prolamellar bodies composed of evenly spaced cross-connected tubes and some concentric discs. If the seedlings are exposed to light, their protochlorophyll is converted into chlorophyll *a* and drastic structural reorganizations take place (von Wettstein and Kahn, 1960; Eriksson, Kahn, Walles and von Wettstein, 1961; Virgin, Kahn and von Wettstein, 1963; Klein, Bryan and Bogorad, 1964). In an initial step following illumination the tubes of the prolamellar bodies are broken down into vesicles. The action spectrum of this tube transformation resembles that of protochlorophyll conversion (Virgin, Kahn and von Wettstein, 1963). In a second light-requiring step the vesicles are distributed into primary layers in the process of vesicle dispersal. This behaviour is most effectively induced by blue light (Henningsen, 1967). The third step requires prolonged illumination and comprises formation of thylakoids and their aggregation into grana. These processes are correlated with the rapid phase in chlorophyll synthesis.

### XANTHA-10

A few chlorophyll-less mutants have been reported in higher plants. The barley mutant xantha-10, which has a biosynthetic block between protoporphyrin IX and protochlorophyll (von Wettstein, 1959c), was analysed with regard to its plastid development (von Wettstein, 1958, 1959a, 1961; von Wettstein and Eriksson, 1964). Xantha-10 seedlings are unable to synthesize tubes and prolamellar bodies both in the dark and in the light. Consequently tube transformation and vesicle dispersal are absent. Large concentric double discs are formed but cannot be aggregated into grana because of the absence of chlorophyll.

### THE ALBINA MUTANT IN SUNFLOWER

Two lethal β-carotene-less mutants in *Helianthus annuus* (Wallace and Schwarting, 1954; Wallace and Habermann, 1959) have been studied to learn about the structural role of carotenoids (Walles, 1964, 1965, 1966). Under ordinary light conditions their chlorophyll is destroyed by photo-oxidation and the resulting leaf colour is dependent on what other pigments may be present.

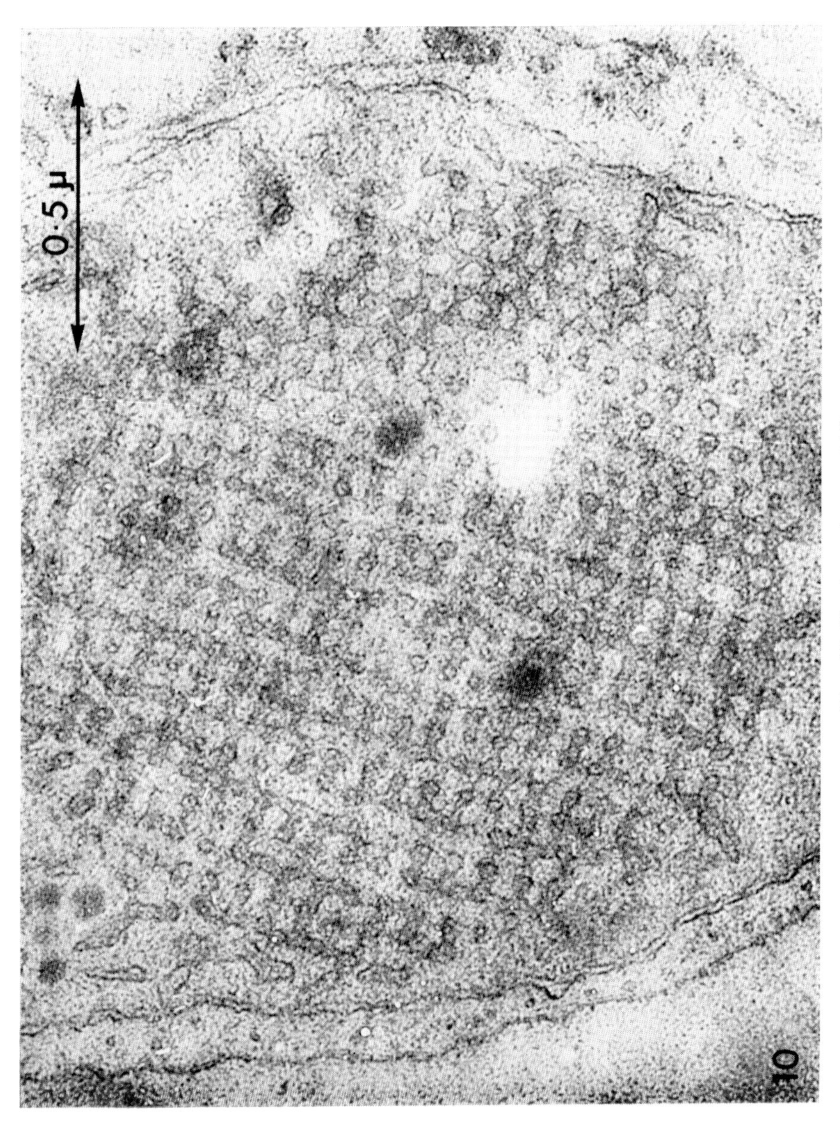


FIG. 10 (See legend on p. 646)

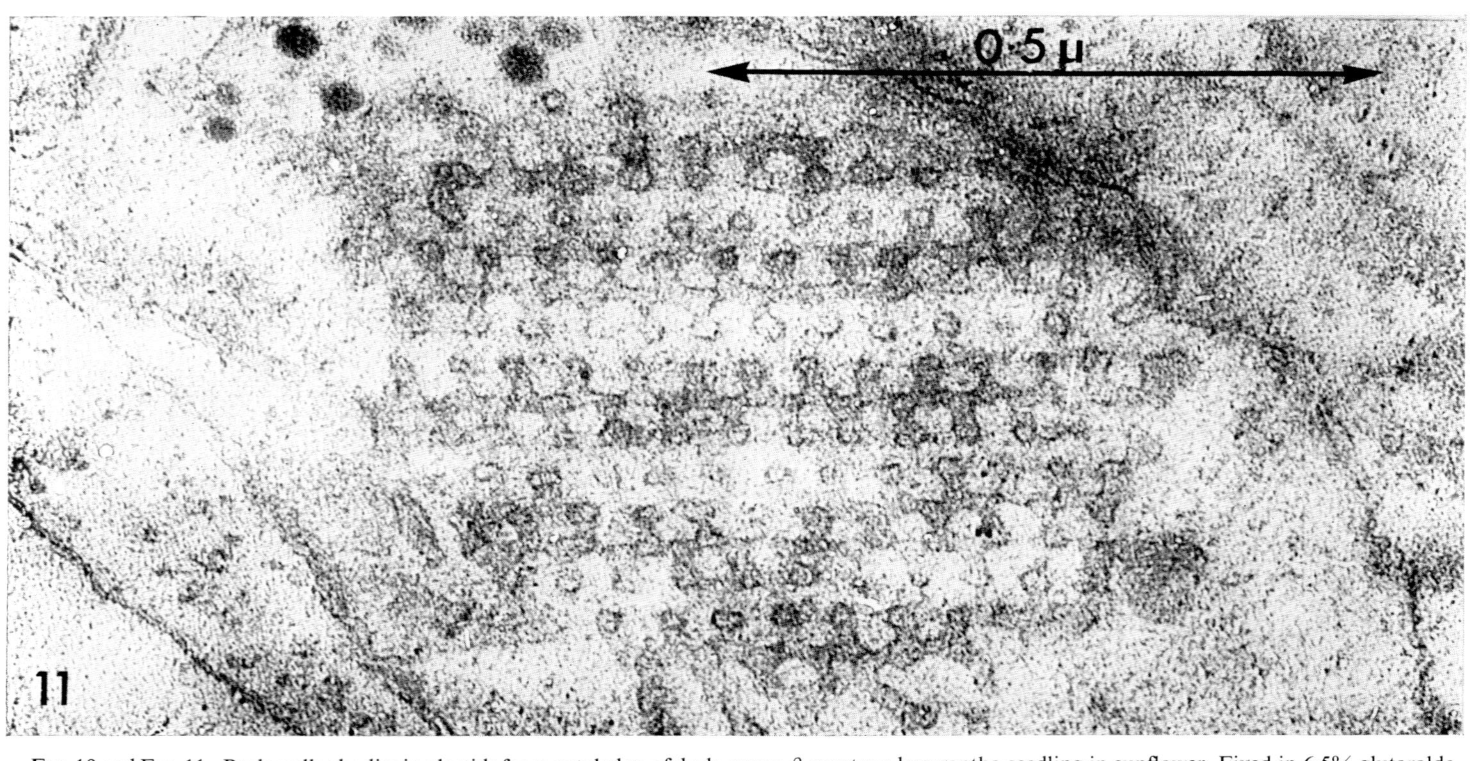


FIG. 10 and FIG. 11. Prolamellar bodies in plastids from cotyledon of dark-grown $\beta$-carotene-less xantha seedling in sunflower. Fixed in 6·5% glutaraldehyde and 2% osmium tetroxide. Post-stained with uranyl acetate and lead citrate.

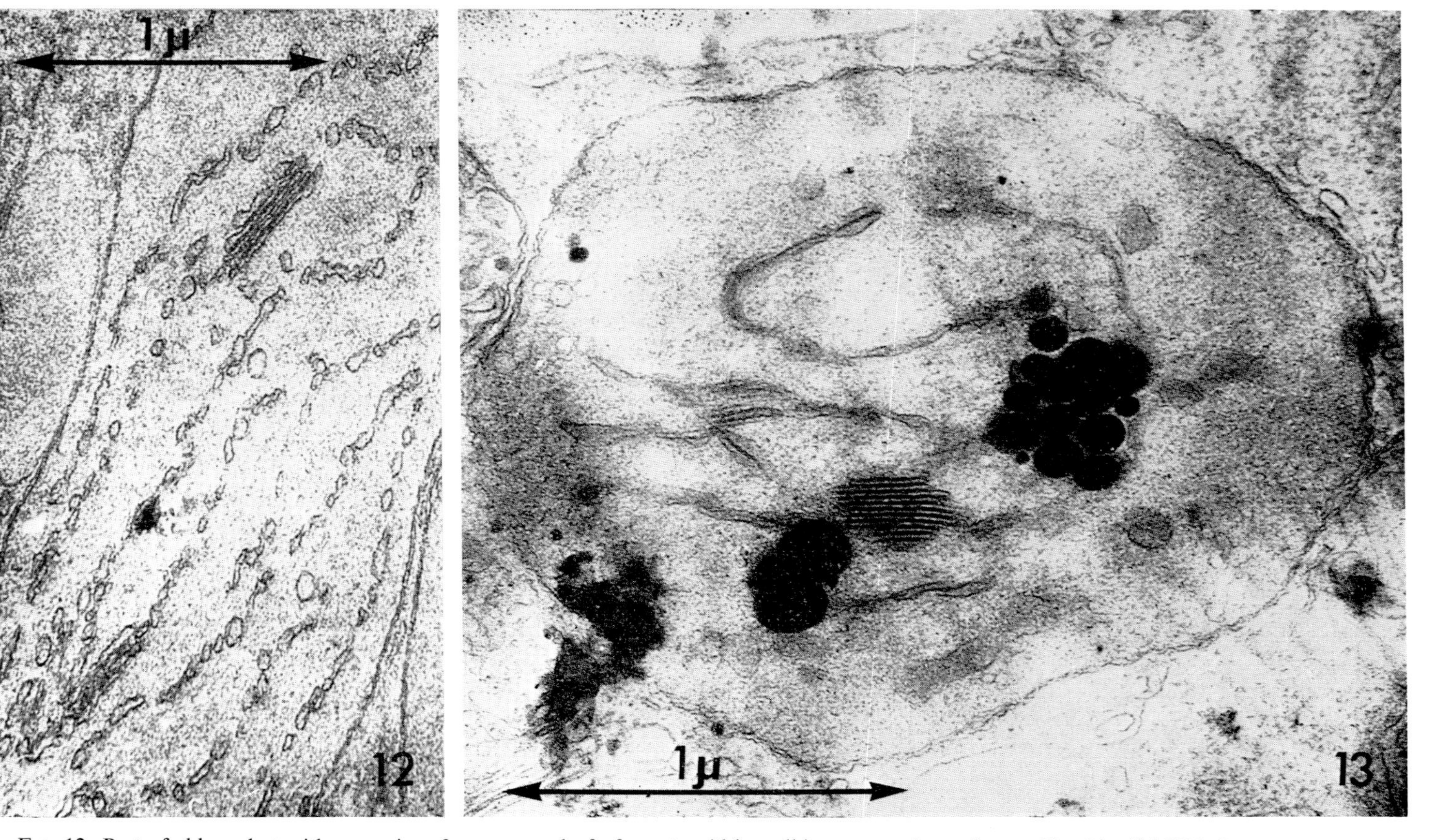


FIG. 12. Part of chloroplast with grana in a 3 mm green leaf of carotenoid-less albina mutant in sunflower. Fixed in 5% $KMnO_4$.
FIG. 13. Chloroplast with grana and groups of globuli from the same leaf as illustrated in Fig. 12. Fixed in glutaraldehyde and osmium tetroxide. Post-stained with lead citrate.

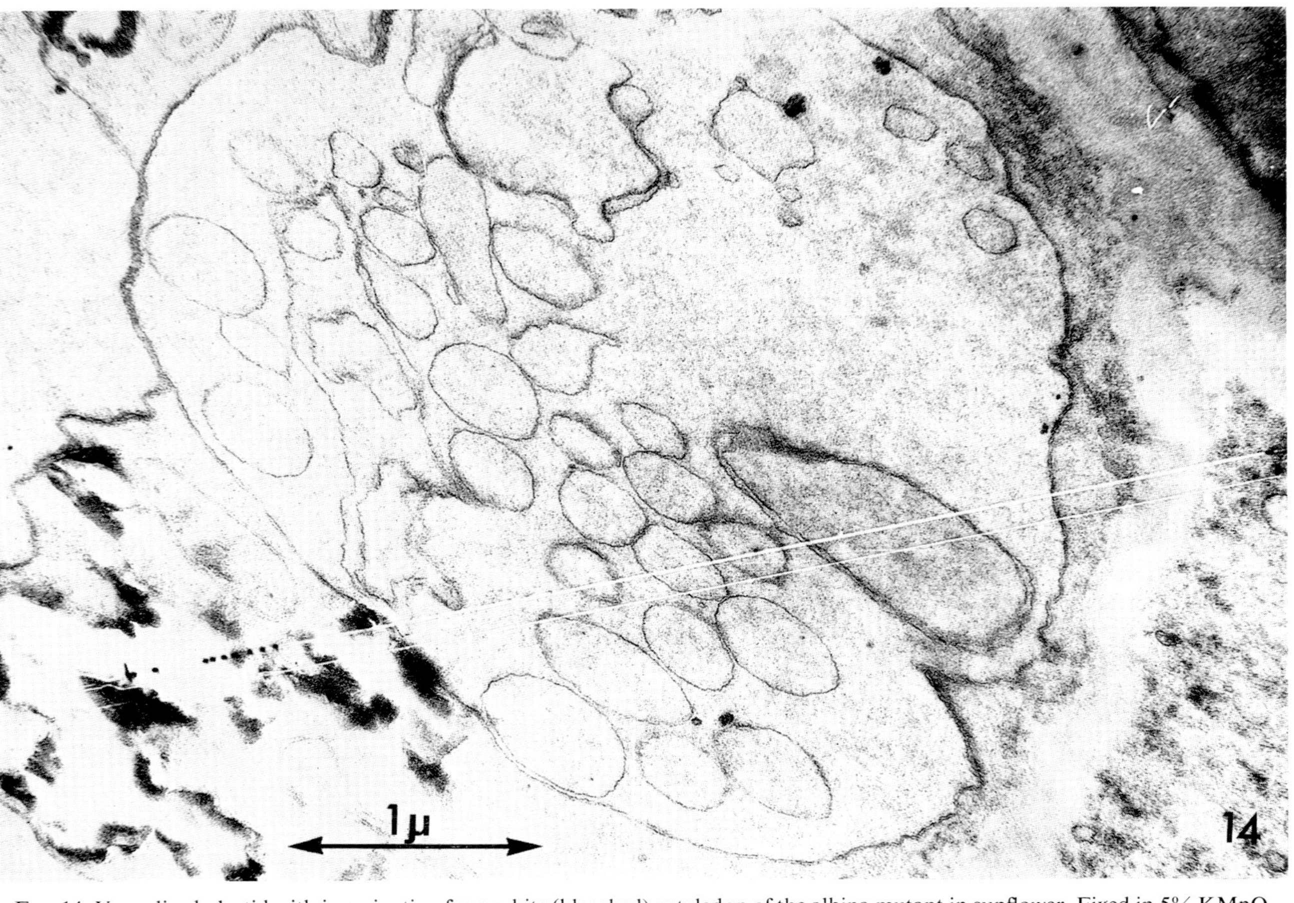


FIG. 14. Vacuolized plastid with invagination from white (bleached) cotyledon of the albina mutant in sunflower. Fixed in 5% $KMnO_4$.

The albina mutant is characterized by the absence of β-carotene and xanthophyll as indicated by absorption and derivative absorption spectra of intact leaves and acetone extracts (Habermann, 1960). The bleached seedlings are white, which also is the colour shown by dark-grown seedlings. The plastid structures were studied in seedlings grown in darkness (plants with protochlorophyll but no chlorophyll *a* or *b*), in weak light (plants with chlorophyll *a* and *b*) and in normal light (almost chlorophyll-free plants).

In darkness proplastids with typical crystalline prolamellar bodies (cf. Figs. 10–11) are formed. Their structure is indistinguishable from that of dark-grown wild type seedlings. The chloroplasts in unbleached leaves of mutant seedlings contain vesicles and discs arranged in primary layers. In a few places discs are aggregated into small grana (Figs. 12–13). The organization of the chloroplasts in the albina mutants is comparable to that of immature chloroplasts in greening wild type leaves. The chloroplasts of the mutant contain groups of large globuli (Fig. 13). Since the low number of grana hardly can incorporate all the chlorophyll present in unbleached leaves, the residual chlorophyll may be localized in the globuli. During bleaching of the mutant leaves their plastids usually lose their regular form and attain a more or less amoeboid shape. The discs swell and the grana disintegrate (Fig. 14). The globuli disappear.

From these results several conclusions can be drawn. Normal prolamellar bodies are synthesized in the plastids of dark-grown leaves in the absence of detectable amounts of β-carotene and xanthophyll. Of the light-induced and light-dependent structural alterations known to occur in plastids of dark-grown seedlings upon illumination, the process of tube transformation and the process of vesicle dispersal can proceed in the absence of carotenoids. In weak light formation of grana is possible to some extent without carotenoids.

## THE XANTHA MUTANT IN SUNFLOWER

In this carotene-less sunflower mutant the chlorophyll is also unstable, but since the mutant seedlings contain xanthophyll (Habermann, 1960) they have a yellow colour after bleaching. The mutant is therefore xantha. It was tested under the same set of environments as the albina mutant. In darkness the proplastids of the xantha mutant have crystalline prolamellar bodies (Figs. 10–11) and appear identical to those of dark-grown wild type seedlings. The chloroplasts of unbleached yellow-green leaves contain vesicles and discs and frequently a few grana of large size (Fig. 15). The chloroplasts of this mutant are more resistant to bleaching than those of the albina mutant. According to Habermann (1960) traces of chlorophyll are found in the bleached leaves of the xantha mutant. After bleaching of the leaves their plastids degenerate and become amoeboidic in shape (Fig. 17) but the grana may persist for some time (Fig. 16).

In confirmation of the conclusions derived from studies of the albina mutant the results from the xantha mutant show that synthesis of the typical structural material in plastids of dark-grown seedlings and the structural reorganizations of these plastids following illumination of the seedlings proceed in the absence

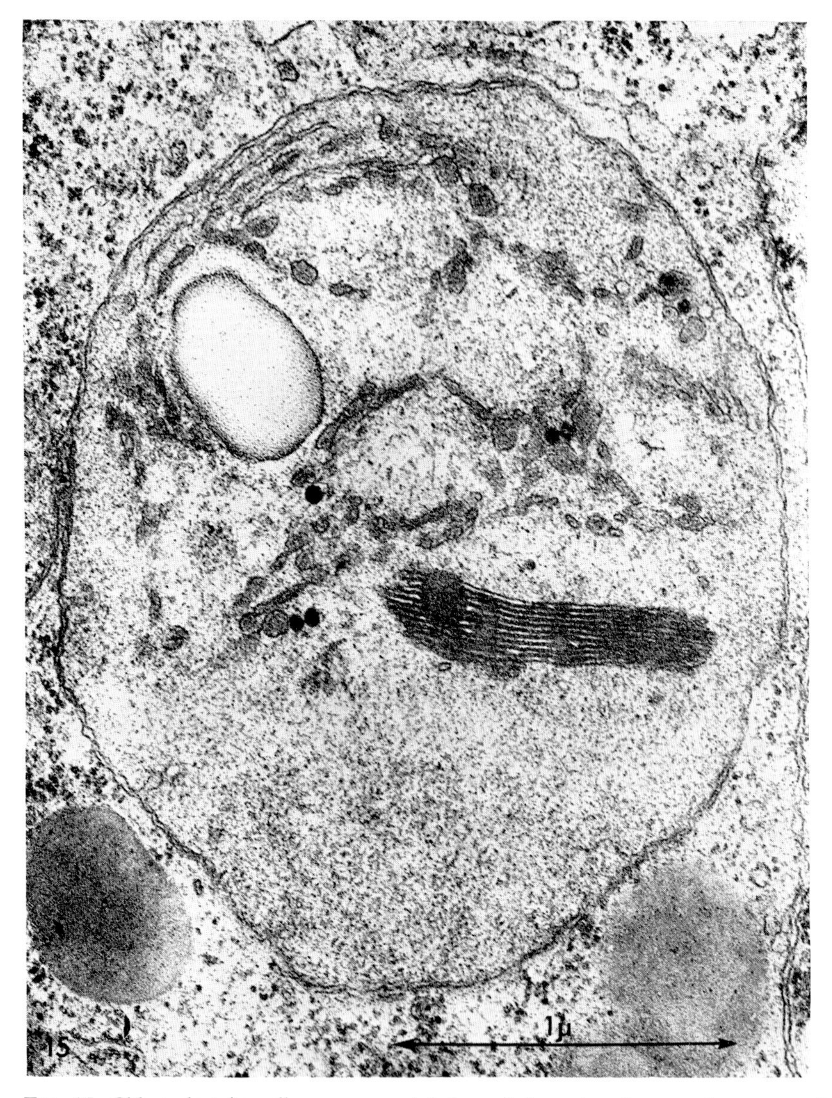


FIG. 15. Chloroplast in yellow green cotyledon of β-carotene-less xantha mutant in sunflower. Fixed in glutaraldehyde and osmium tetroxide. Post-stained with uranyl acetate and lead citrate.

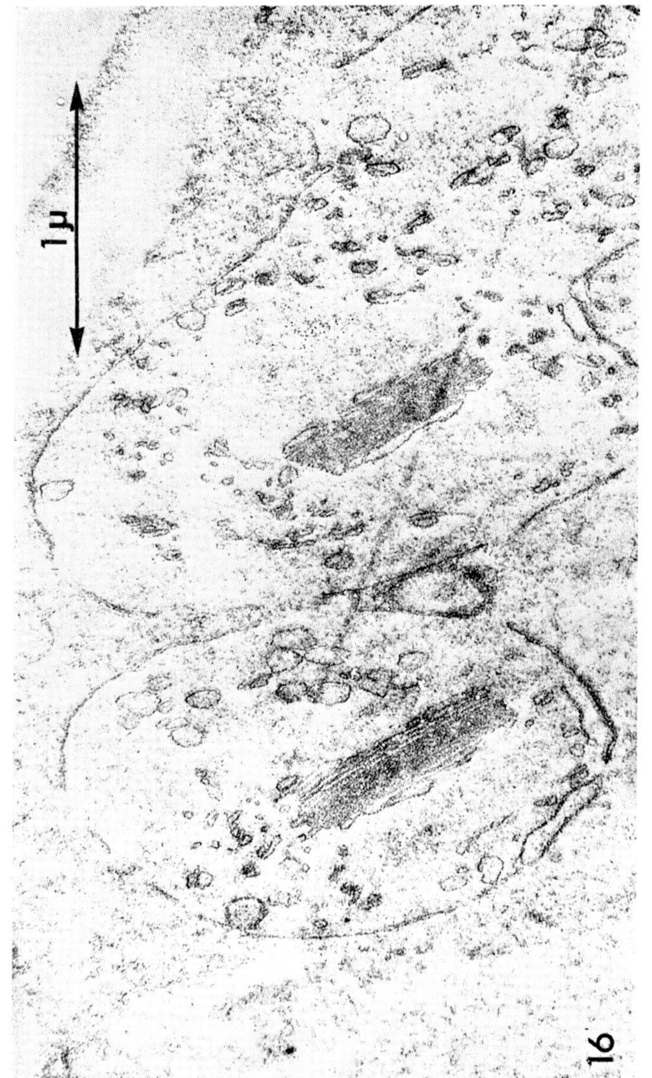


FIG. 16. Plastids in yellow cotyledon of the xantha mutant in sunflower. Fixed in 5% $KMnO_4$.

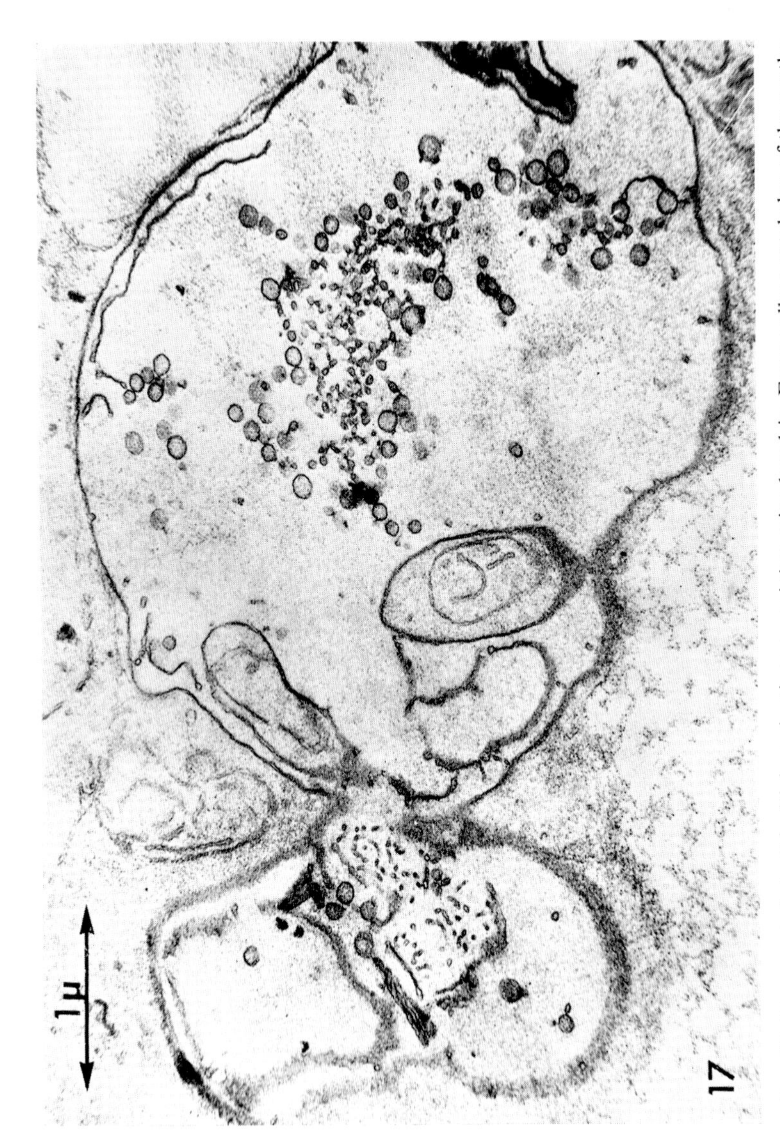


FIG. 17. Degenerated plastid with invaginations containing mitochondria. From yellow cotyledon of the xantha mutant in sunflower. Fixed in 5% $KMnO_4$.

of any detectable amount of $\beta$-carotene. This pigment is nevertheless necessary for the formation of a normal lamellar system, as well as for the protection of chlorophyll and the grana structure against photodestruction.

ACKNOWLEDGEMENT

Financial support from the Swedish Agricultural Research Council, the Swedish Natural Science Research Council and the United States Public Health Service (National Institutes of Health, G.M. 08877 and G.M. 10819) is gratefully acknowledged.

## References

Eriksson, G., Kahn, A., Walles, B. and Wettstein, D. von (1961). *Ber. dtsch. Bot. Ges.* **74**, 221.
Habermann, H. M. (1960). *Physiol. Plantarum* **13**, 718.
Heber, U. (1960). *Z. Naturf.* **15**b, 95.
Henningsen, K. W. (1967). *In* "Biochemistry of Chloroplasts" (T. W. Goodwin, ed.) Vol. II, Academic Press, London and New York.
Klein, S., Bryan, G. and Bogorad, L. (1964). *J. cell Biol.* **22**, 433.
Virgin, H. I., Kahn, A. and Wettstein, D. von (1963). *Photochem. Photobiol.* **2**, 83.
Wallace, R. H. and Habermann, H. M. (1959). *Am. J. Bot.* **46**, 157.
Wallace, R. H. and Schwarting, A. E. (1954). *Pl. Physiol.* **29**, 431.
Walles, B. (1963). *Hereditas* **50**, 317.
Walles, B. (1964). *Hereditas* **52**, 245.
Walles, B. (1965). *Hereditas* **53**, 247.
Walles, B. (1966). *Hereditas* **56**, (In press).
Wettstein, D. von (1958). *Brookhaven Symp. Biol.* **11**, 138.
Wettstein, D. von (1959a). *In* "Develop. Cytol." (D. Rudnick, ed.), p. 123. New York.
Wettstein, D. von (1959b). *J. Ultrastr. Res.* **3**, 235.
Wettstein, D. von (1959c). *Carnegie Inst. Wash. Yearbook* **58**, 338.
Wettstein, D. von (1960). *Hereditas* **46**, 700.
Wettstein, D. von (1961). *Can. J. Bot.* **39**, 1537.
Wettstein, D. von and Eriksson, G. (1964). "Genetics today." Proc. 11th Intern. Congr. Genet. The Hague 1963 III, 591.
Wettstein, D. von and Kahn, A. (1960). *Proc. Europ. Reg. Conf. Electron Microsc. Delft*, 1960 II, 1051.

# The Prolamellar Body

B. E. S. Gunning and M. P. Jagoe
*Department of Botany, Queen's University of Belfast, Northern Ireland*

## Introduction

Many of the membrane systems of plant and animal cells have been described in considerable detail, yet the processes and factors which determine their structural patterns remain obscure, even though these patterns may be highly characteristic and recognizable. One major difficulty is that most of the structures are not symmetrical in three or even two dimensions, and their irregular shape must presumably be governed by correspondingly asymmetrical forces. Pattern-controlling systems should be easier to recognize and investigate in cases in which the membrane configuration is precisely symmetrical. Such precision is rare, but is approached in a structure found in plastids in dark-grown leaves—the prolamellar body. The present communication aims at describing some of the properties of this highly ordered array of membranes and particles, and presents a hypothesis which suggests how its order might be achieved and maintained.

## Materials and Methods

Seedlings of *Avena sativa*, var. Victory, were used as plant material. Methods of fixation, embedding, and sectioning were as in Gunning (1965b), and sections stained with uranyl acetate and lead citrate (Reynolds, 1963) were examined using an Akashi TRS-80 electron microscope operating at 80 kV.

For pigment analyses the first leaves were cut off about 2 cm above sand level and immersed in boiling water for 1 min. The tissue was homogenized in a mixture of pure acetone and 80% acetone so as to give a final acetone concentration of 80%. The pigments were transferred to diethyl ether and estimated spectrophotometrically after the solution had been washed with water and dried over sodium sulphate. The equations of Koski (1950) were used.

To determine the percentage phytylation of the pigments an aliquot of the ether solution was shaken with 0·02 N potassium hydroxide (Holden, 1965), washed in water and dried over sodium sulphate. An equal aliquot was treated in a similar way, substituting water for potassium hydroxide. The ratio of the pigment in the first aliquot to that in the second was taken as a measure of the phytylation of the pigment in the original extract.

Etiolated plants, or plants that had been illuminated and then returned to darkness, were not exposed to any light until approximately 2 min before they were sampled and immersed in boiling water. They were at this point handled under a dim green safelight, but this exposure was shown to have no effect on the fine structure of the plastids, and did not convert protochlorophyll(ide) to chlorophyll(ide). The safelight consisted of a green fluorescent tube mounted behind Cinemoid filters (Strand Electric and Engineering Company, London) —three layers of dark green (no. 24), two layers of deep orange (no. 5A) and one layer of medium blue (no. 32). It is safe as regards the phytochrome system that sensitizes geotropic curvatures in *Avena* coleoptiles (Wilkins, 1965; M. W. Wilkins, personal communication), but it was found that (at least) prolonged exposures shortened the duration of the lag phase of greening in *Avena* leaves.

"Illumination" in the text refers to 600–1000 foot candles illumination provided by Philips VHO fluorescent tubes.

## Results

### The Structure of the Prolamellar Body

#### *The Membrane Lattice*

Figure 1 shows some of the features of plastids in etiolated leaves. There can be more than one prolamellar body in a plastid, and, in most cases, each prolamellar body is a compound structure, made up of adjoining masses of membranous lattice in "crystalline" form. These masses are usually oriented in different ways, so that when thin-sectioned, the prolamellar body as a whole presents a mosaic appearance. Depending on the orientation differences, the demarcation lines between the individual masses may be indistinct (e.g. Fig. 1, lower plastid) or well-marked dislocations (e.g. Fig. 4). All of these masses (except one type, see below) are thought to have the same basic structure, but they can look very different, depending on the angle between the plane of the section and the planes of symmetry of the lattice (see below). In a relatively thick section (Fig. 2) the crystalline arrangement can be much more obvious than in a relatively thin one (Fig. 3) where the membranes may even seem to be discontinuous in places.

There are usually a few thylakoids extending outwards from the prolamellar body (Fig. 3), even in material that has been grown and fixed in complete darkness. Occasional sections show that some of these thylakoids can connect with the inner layer of the plastid envelope (Figs. 1, 4, 5). Since the membrane surface is continuous both within the prolamellar body (the plastid stroma penetrates *between* its tubules) and outwards from it towards the envelope via these thylakoids, the space enclosed within the two layers of the envelope, within the thylakoids, and within the tubules of the prolamellar body itself is also continuous, and is not in direct contact with the plastid stroma. It has been shown (Gunning, 1965b) that after double fixation these membranes are 50–60 Å in thickness and possess the three layered construction of the "unit-membrane".

A detailed analysis of the structure of the prolamellar body membrane system has been presented elsewhere (Gunning, 1965b) and need only be summarized here. Figure 6 shows a model which can account for the more

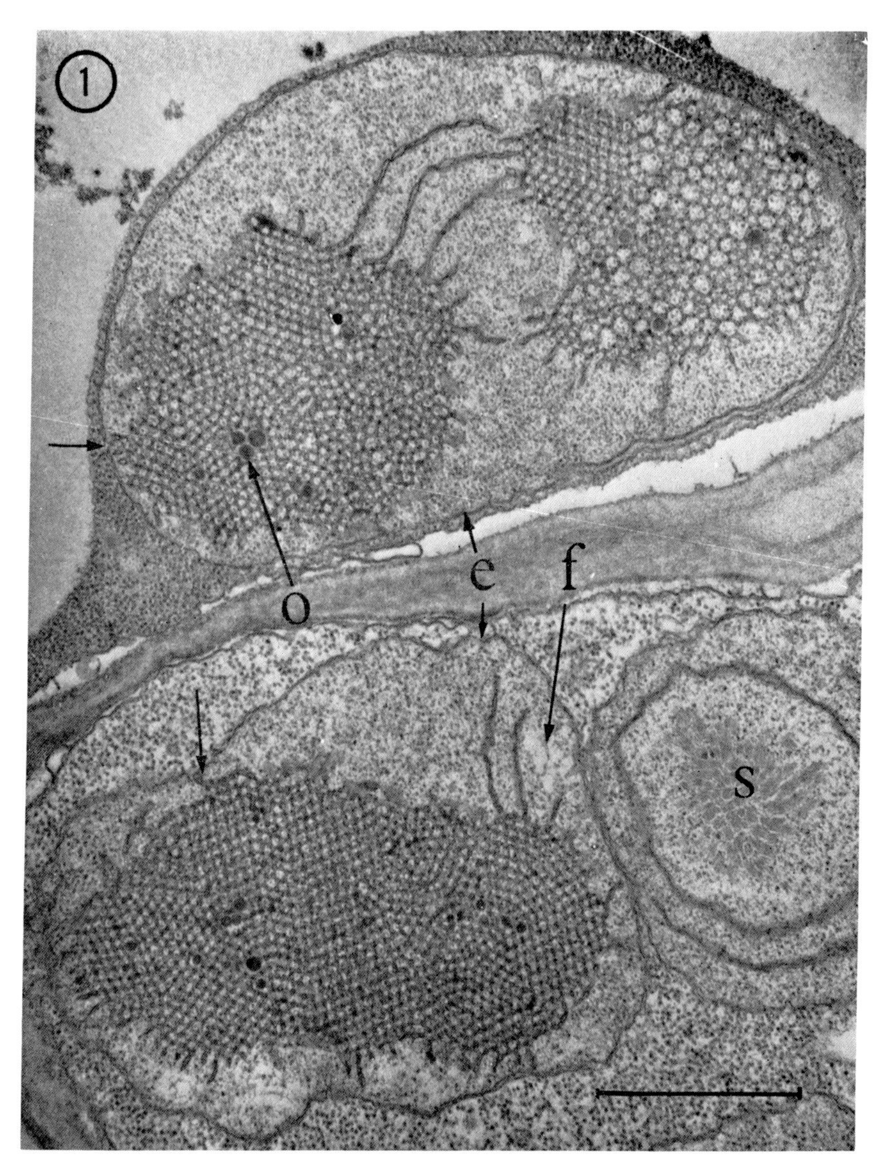


FIG. 1. Plastids in mature mesophyll cells in etiolated *Avena* leaf, showing prolamellar bodies in which membranes and particles are distinguishable. Arrows indicate connections between a prolamellar body and the inner layer of the plastid envelope (e). The upper plastid contains two crystal lattices; the one on the right consists in part of the "normal" system and in part of the "atypical", more open arrangement. Note that in this open lattice there are several particles in each space—compare the particle distribution in the other prolamellar bodies. s = stroma centre, f = fine fibrils in the stroma, o = osmiophilic globule.

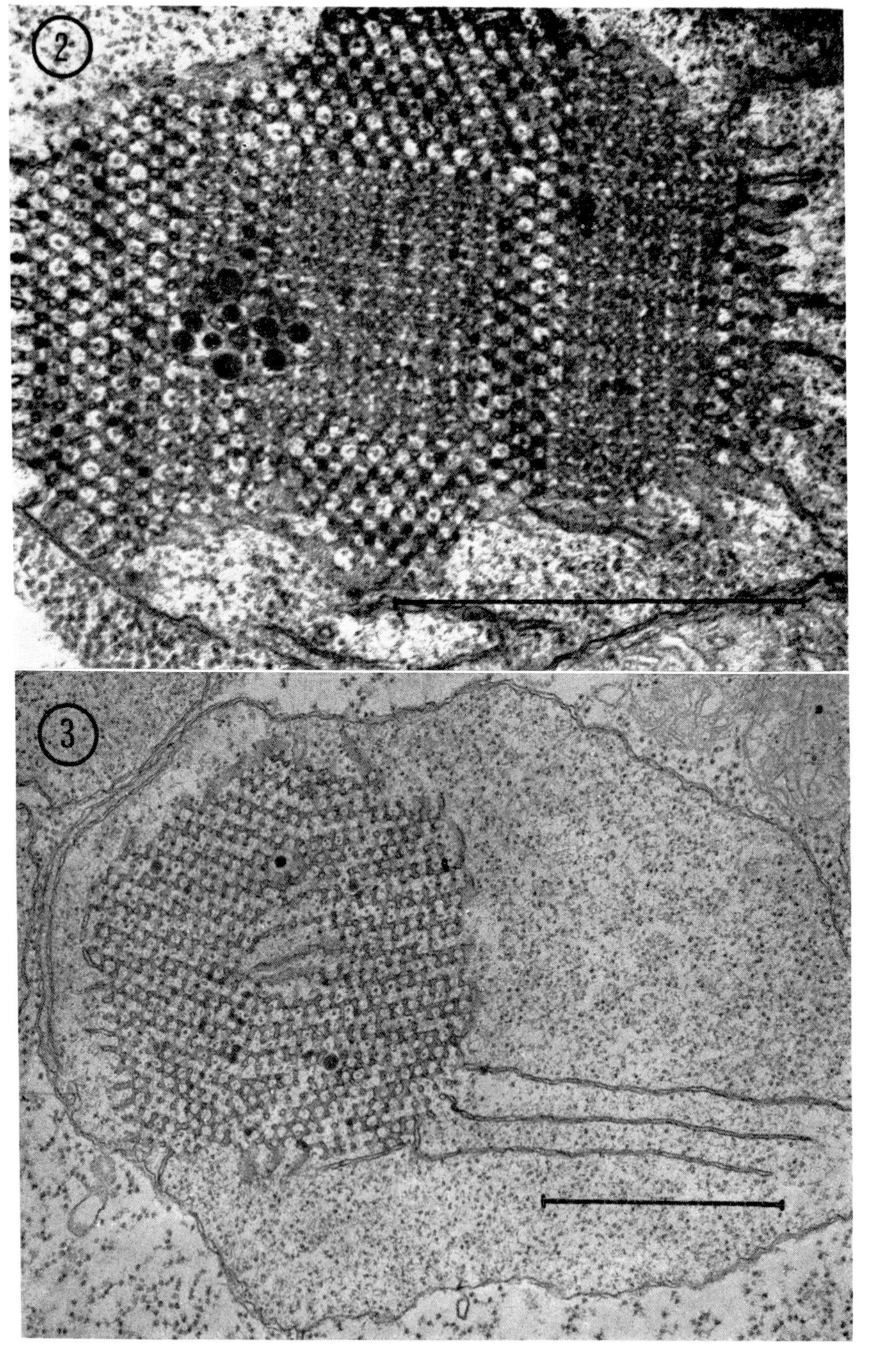


FIG. 2. A prolamellar body in a relatively thick section. Where hexagonal spaces are visible, there is usually a particle in the centre of each hexagon. The outlines of the membranous tubules and "nodal units" are indistinct.

FIG. 3. A relatively thin section, showing a prolamellar body with some thylakoids extending from it. Particles and membranes are more easily distinguished than in Fig. 2, however, the continuity of the membrane surface is less obvious.

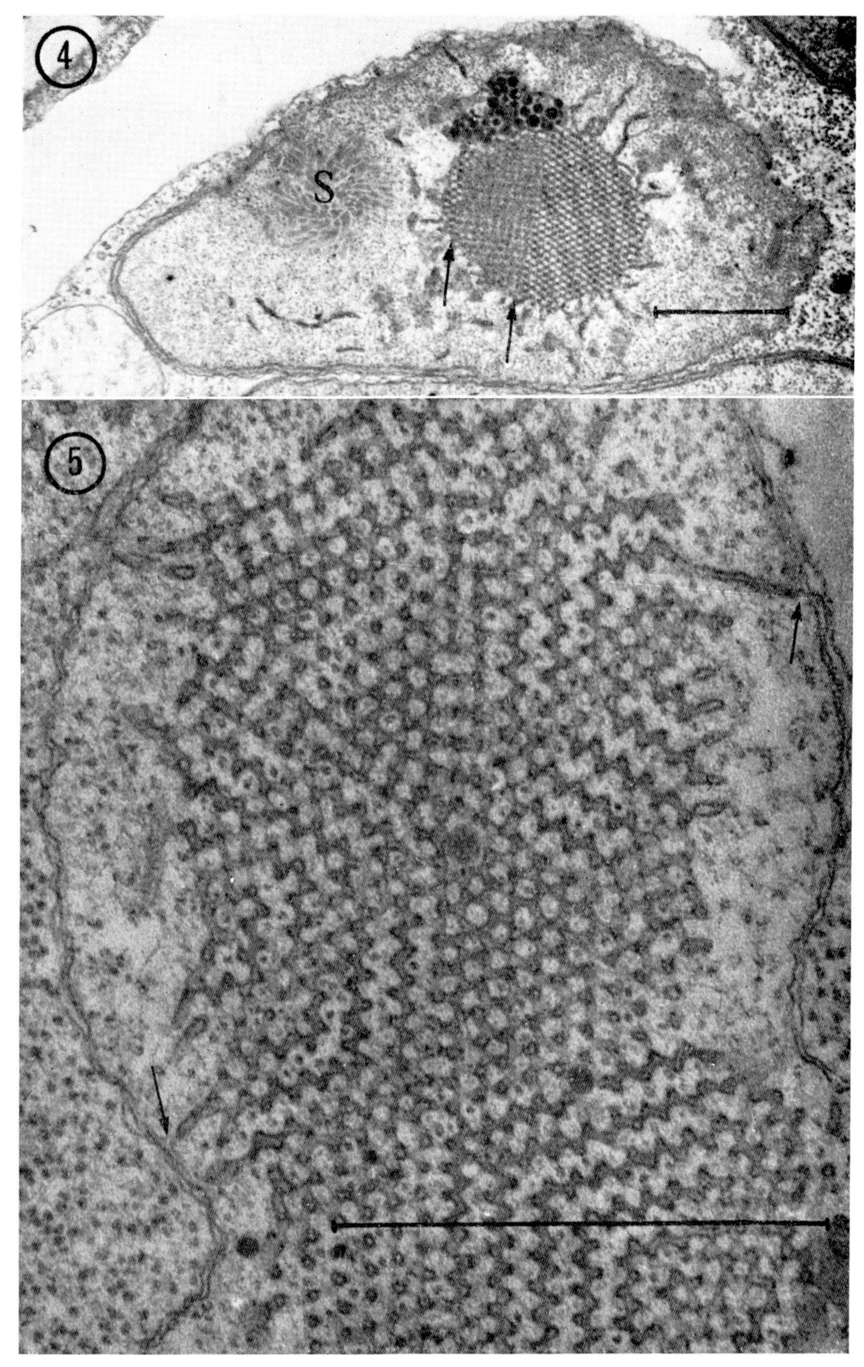


FIG. 4. A plastid containing a prolamellar body with (in this section) two well-defined dislocations (arrows). Other labelling as in Fig. 1.

FIG. 5. A compound prolamellar body sectioned to show how restricted areas can be uniform, with the structure as a whole having a mosaic appearance. "Waveforms", "hexagons", "tubules" and a small area of "rectangles" (see Gunning 1965b) are shown. Connections between the envelope and the prolamellar body are arrowed.

symmetrical profiles that are observed in thin section. Clearly, there are less regular portions of the lattice—e.g. in the neighbourhood of dislocations and osmiophilic globules (Figs. 1 and 2). The model is similar to the one presented by Granick (1961), which was based on von Wettstein's electron micrographs (von Wettstein, 1958, 1959). It consists of tubules which run in the three major

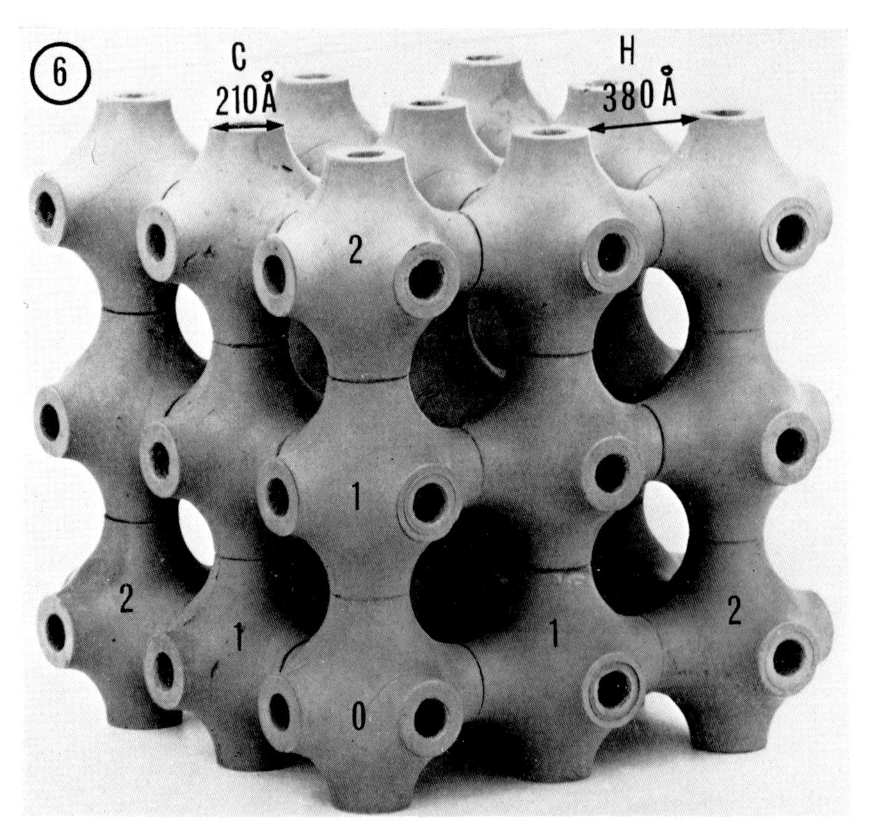


Fig. 6. Photograph of a model of the prolamellar body membrane system, consisting of twenty-seven six-armed "nodal units". The dimensions are indicated. The unit-membrane construction is not shown on the cut surfaces of the tubules. The units are numbered along the x, y, and z axes (see Fig. 7). Eight particles would be present in this portion of lattice, but these are not shown.

axes of a simple cubic lattice. Where three tubules meet and fuse at the corners of the cubical "unit cells" they are swollen so that their membrane surfaces are smoothly confluent. Perhaps the simplest description is that the membrane system is built up of six-armed "nodal units" which can be fitted together as in the model. In each unit, the membrane surface joining any two connecting arms which are adjacent in one plane delimits not a right angle, but a quadrant of a circle. The model photographed in Fig. 6 consists of twenty-seven such nodal units, and these enclose eight complete unit cells. Figure 6 also gives the dimensions of the units.

It is suggested that this structure is present in the crystalline portions of the prolamellar body. Face views of the lattice are very rarely encountered; much more frequently the section passes through the lattice in some plane other than a major plane of symmetry, and hence the membrane profiles observed are to some extent asymmetrical, and can be extremely varied. Figure 5 shows some of these variants in one section of a compound prolamellar body, and Fig. 7 indicates how the appearance of some of the profiles can be duplicated by sectioning the model in appropriate planes.

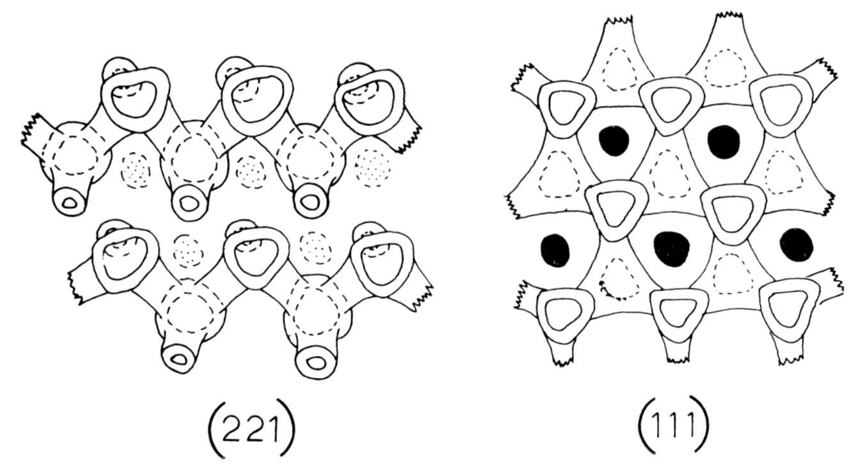


FIG. 7. The appearance of sections of the model in the (221) and (111) planes. These are equivalent to "waveforms", and "hexagons". The (221) sections do not pass through the centres of the unit cells, and so only portions of particles are included; the "hexagons" on the other hand include whole particles, since the centre of a unit cell lies at the centre of a "hexagon".

In *Avena*, there is at least one "atypical" lattice whose membrane configuration is not as depicted by the model. It is only rarely seen in mature leaves (but see p. 668), where it can occur in apparent isolation in a plastid, or connected to a "normal" type of lattice along a dislocation (as in Fig. 1, upper plastid, right hand side). It is a relatively open structure, and its lattice appears somewhat irregular.

## *The Contribution of the Plastid Stroma*

Between two-thirds and three-quarters of the total volume of a prolamellar body consists of material that is continuous with the plastid stroma. In thin sections, only one stroma component is clearly stained. It is a particle with a marked affinity for uranyl acetate, and in size and shape it seems identical to particles found in the stroma outside the prolamellar body. Evidence that these particles are in fact ribosomes has been presented elsewhere, for *Avena* (Gunning, 1965a, b; Brown and Gunning, Vol. I. p.365) and for other plants, especially *Zea* (Jacobson *et al.*, 1963).

These particles are precisely distributed within the prolamellar body, one per unit cell of the lattice. They can be seen in Figs. 1, 2, 3, 5, 17 and 18. Their location is best observed in sections which have cut diagonally through a unit cell to give a hexagonal arrangement of membrane profiles (Fig. 7). In such sections, the central point of the unit cell lies at the centre of the "hexagon". This (see Figs. 2, 5, and 15) is where the ribosome-like particles are found in the micrographs.

In many sections of prolamellar bodies, these particles do not appear in large numbers. This is because they are relatively small (about two-thirds the size of the cytoplasmic ribosomes outside the plastids) and are separated from one another by a distance that is usually greater than the estimated thickness of the section, so that the chances of including a detectably stained portion of a particle, or a complete particle, are much reduced from the theoretical maximum, which would be observed only when the section is oriented so as to provide face views of the unit cells. There would then be one particle in every 590 Å × 590 Å area.

## THE RESPONSE OF THE PROLAMELLAR BODY TO ILLUMINATION

There have been several descriptions of the process by which double lamellae develop in plastids when etiolated leaves are illuminated. Most of the studies have employed potassium permanganate fixation and most agree in that the prolamellar bodies are shown to dissociate into vesicles which disperse and ultimately fuse to form sheets of double lamellae. Our results, obtained through the use of glutaraldehyde–osmium tetroxide fixation, agree insofar as the prolamellar bodies are seen to become disorganized and give rise to sheets of double lamellae; however, this is accomplished by a process that does not involve vesicle formation and dispersal. In *Avena*, the timing would seem to be closer to that found in *Phaseolus vulgaris* by Klein *et al.* (1964), rather than to that reported by Eriksson *et al.* (1961) and Virgin *et al.* (1963).

### *The Surface Area of the Membrane in the Prolamellar Body*

A large surface area of membrane is stored in a compact form in the prolamellar body. It is of interest to know whether there is sufficient present to account for the observed production of lamellae during the initial response to illumination, that is, before the end of the lag period that lies between the initial conversion of protochlorophyll(ide) to chlorophyll(ide) and the onset of rapid production of chlorophylls (Bogorad, 1965). In *Avena* seedlings grown and harvested in our conditions, this lag period lasts for about 3 hr.

Since the dimensions are known, the surface area of a 6-armed unit of the lattice can be calculated. In the previous report (Gunning, 1965b) an estimate was obtained by assuming that the lattice consisted of alternating cubes (connecting arms) and cubo-octahedra (nodes) as in the Linde A-type molecular sieve (Duffett and Minkoff, 1964). A more accurate treatment is to consider each nodal unit as consisting of eight faces as in Fig. 8, each of which in turn is made up of three surfaces of revolution (as *abcd*), and a "triangular" portion (*cde*). *cde* is a curved surface, with its axis of symmetry equidistant from the three major axes. The discrepancy between its true area, and the area of its projection on a plane normal to this axis cannot be large.

$$\text{Projected area of } cde = \sqrt{3}\left[\left(\frac{C+H}{2}\right) - \frac{H}{2\sqrt{2}}\right]^2\left[1 - \frac{\pi}{4}\right]$$

$$\text{Area of } abcd = \frac{\pi H}{8}\left[\frac{\pi(C+H)}{4} - \frac{H}{\sqrt{2}}\right]$$

(Parameters "$C$" and "$H$" are shown in Fig. 6.) Using these expressions, the area per six-armed unit approximates to $7{\cdot}8 \times 10^{-3}\ \mu^2$.

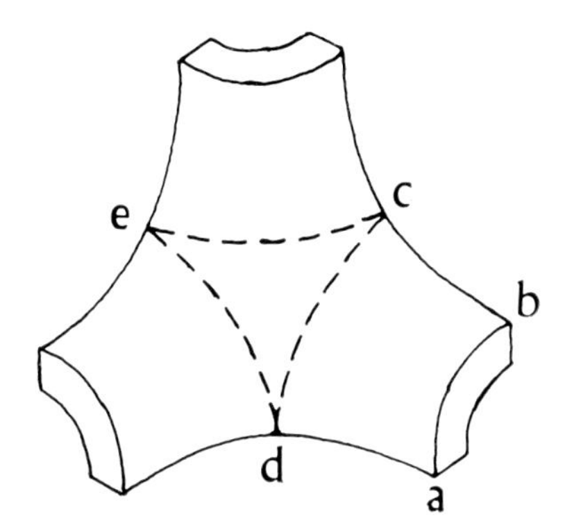


FIG. 8. One eighth of a six-armed "nodal-unit", showing how the surface can be subdivided in order to calculate its surface area.

It is not feasible to count the number of these units per prolamellar body. However, since a spherical prolamellar body of radius twelve unit cells (equivalent to 0·7 $\mu$) would contain about 7500, the number present is likely to range from 5000 to 10,000. This would provide (assuming that the membranes do not stretch or shrink) 40–80 $\mu^2$ of membrane surface, i.e., 20–40 $\mu^2$ of double lamella.

In other words, an "average" *Avena* prolamellar body contains 7500 unit cells, packs 60 $\mu^2$ of membrane into a sphere of radius 0·7 $\mu$, and could provide nine circular sheets of double lamella, each 2 $\mu$ in diameter. Alternatively, this is approximately equivalent to four concentric, spherical, double lamellae, separated radially from one another and from the plastid envelope by 0·3 $\mu$ intervals, all inside a spherical plastid of radius 1·5 $\mu$.

### *The Outgrowth of "Perforated Thylakoids"*

The outstanding difference between the glutaraldehyde–osmium tetroxide images and the micrographs published by other workers is that there is no stage during which the membranes dissociate into vesicles. Instead, the surface is continuous throughout the whole process of light-induced outgrowth of lamellae.

After 5 min light at 750–1000 foot candles, no further conversion of protochlorophyll(ide) into chlorophyll(ide) can be detected in *Avena* first leaves. The methods that we have used so far are not sufficiently precise to indicate whether *all* the original protochlorophyll(ide) is converted, or whether a fraction remains. Portions of leaf fixed at this stage contain plastids in which the order and regularity within the prolamellar bodies has been very largely lost without, however, destroying the continuity of the membrane surface

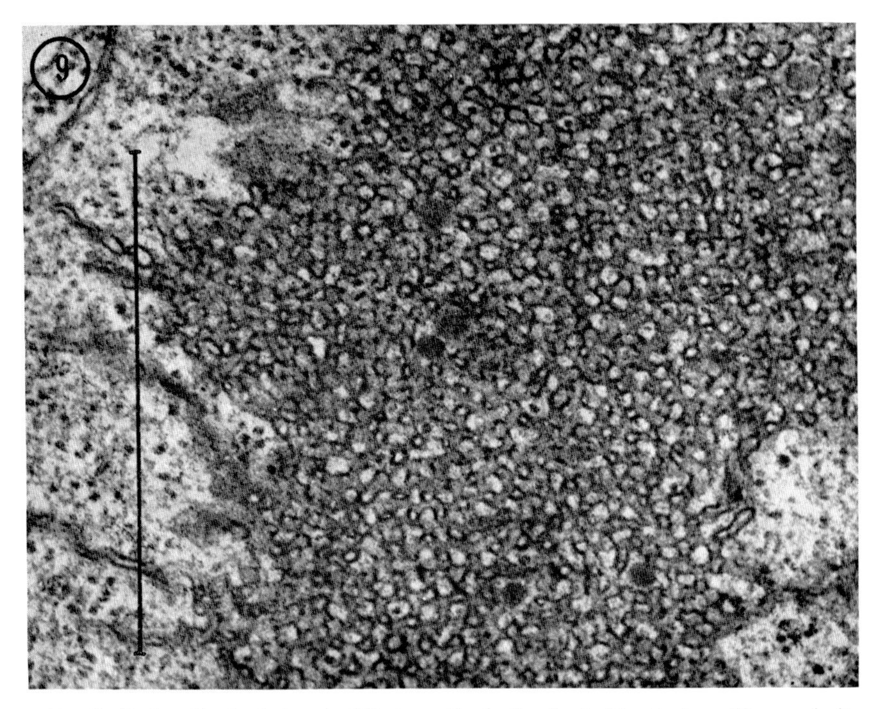


FIG. 9. Prolamellar body in a leaf that was fixed after 5 min illumination. The regularity of the lattice has been lost, but most of the connections remain.

(Fig. 9). It would appear that the balance of forces which locks the lattice into a stable and symmetrical framework is upset. Initially, comparatively few of the connecting arms between the "nodal units" are broken, but with continued illumination there is an increasing tendency for the connections to pinch off so as to generate two-dimensional sheets of double lamellae from the three-dimensional network of tubules (as shown in Fig. 13). Figures 11 and 12 are micrographs of adjacent serial sections giving surface and profile views of these sheets, and showing how they connect with the remnants of the prolamellar body. Figure 10 shows that they can lie in concentric layers.

These micrographs show how the "*H*" spaces (Fig. 6) of the prolamellar body survive as perforations in these sheets of membrane. It may be noted

(Fig. 13) that the presence of large numbers of perforations can give the sheets the appearance of rows of vesicles, especially when they are seen in profile (i.e. when they lie at right angles to the plane of the section). Certainly the "perforated thylakoids" must be equivalent to the sheets of vesicles described by other

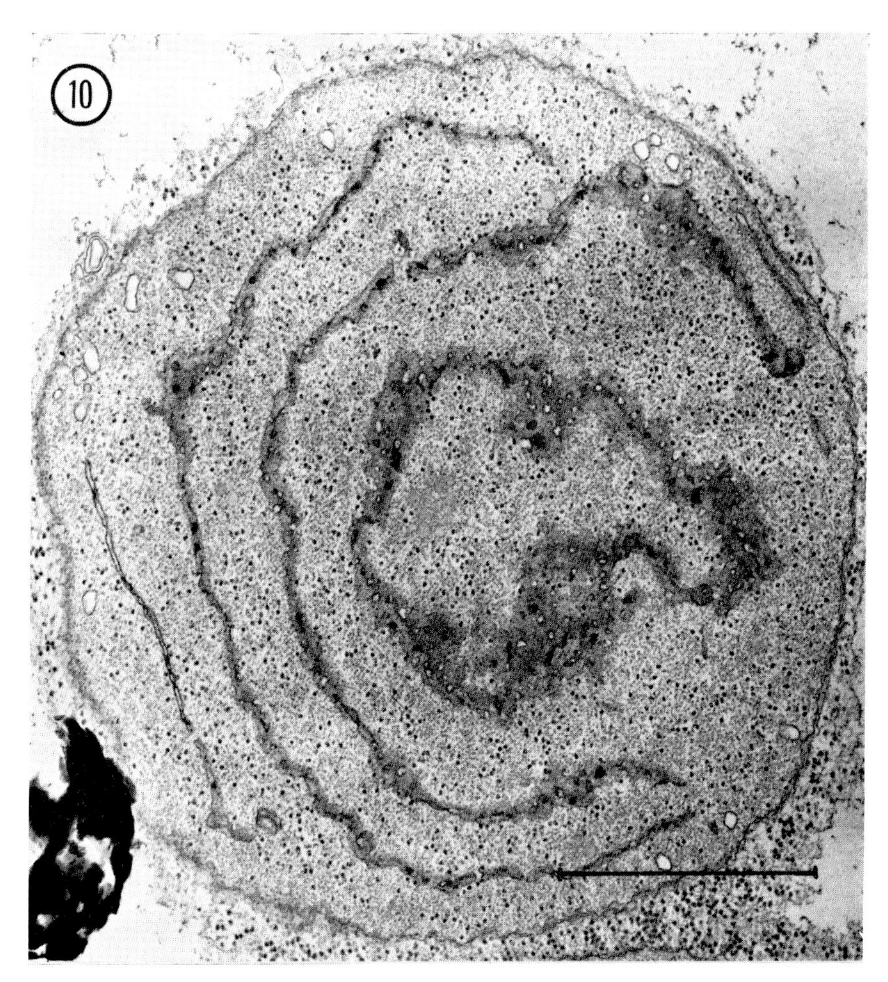


FIG. 10. Plastid fixed after 2 hr of light. The "perforated thylakoids" are in approximately concentric sheets. Many densely stained ribosome-like particles are seen in the stroma, amongst less well-stained particulate material. A few vesicles, probably formed by invagination of the inner layer of the envelope, are also present.

workers, although it is not possible to state definitely which image represents the true state of affairs. We have observed fragmentation into vesicles only in plastids which, because the double envelope was also fragmented, we have assumed were poorly preserved. It may be significant that breakdown and vesicularization of paired plasma membranes of ciliary epithelium cells has

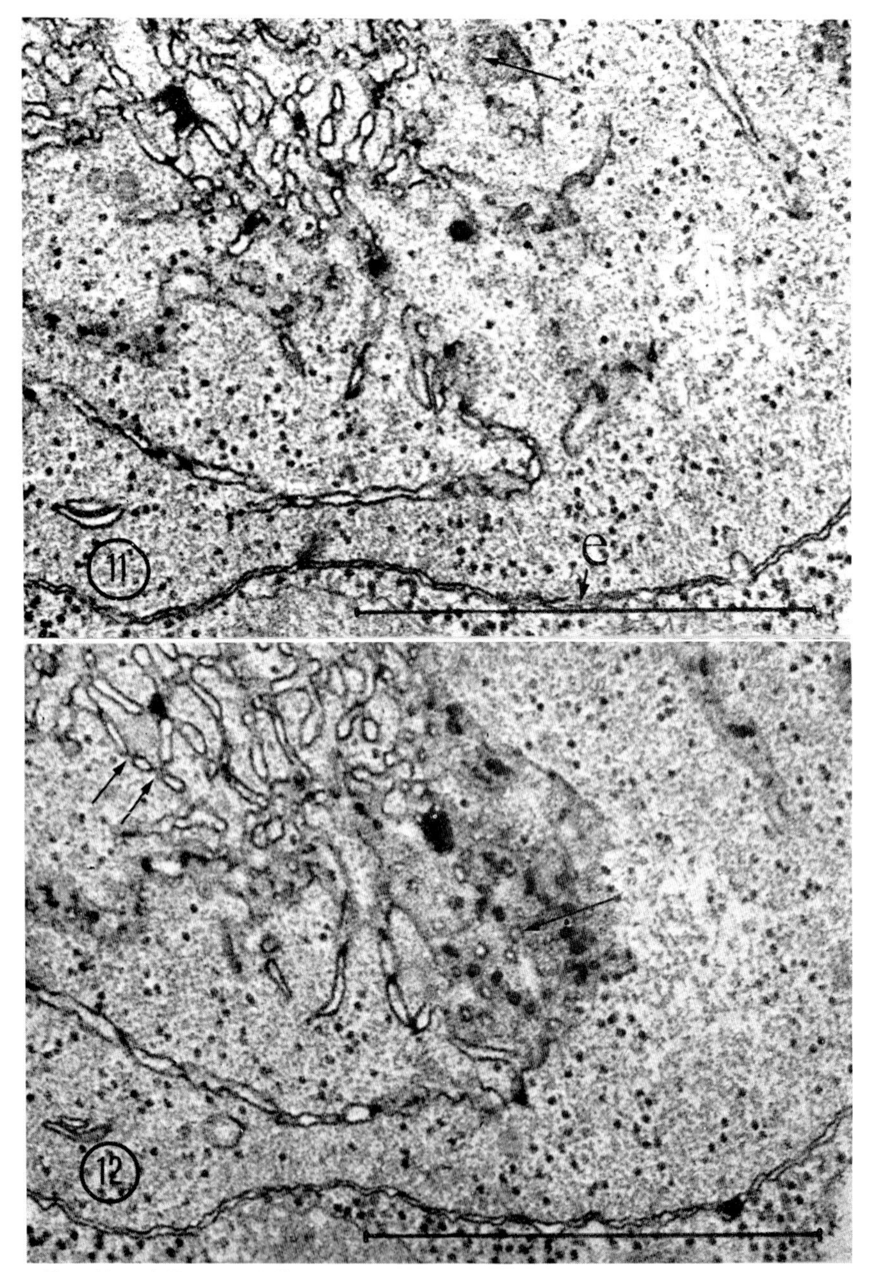


FIG. 11 and FIG. 12. Adjacent serial sections illustrating the connection between the remnants of the prolamellar body and the "perforated thylakoids". Some of the ribosome-like particles remain in the disorganized lattice; many more are present outside it in the open stroma. Perforations are seen in surface and side view (arrows). Other lettering as in Fig. 1. The leaf was fixed after it had been illuminated for 2 hr.

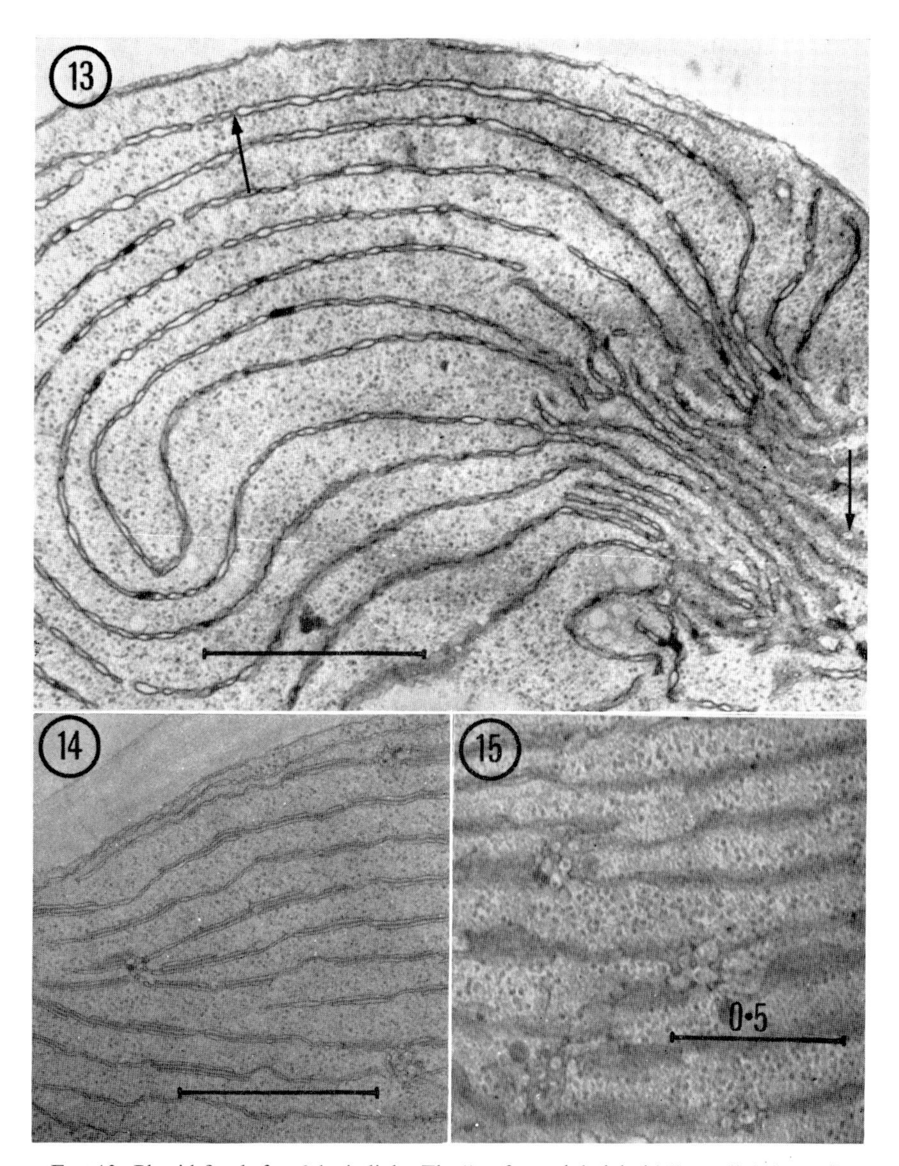


FIG. 13. Plastid fixed after 2 hr in light. The "perforated thylakoids" are slightly swollen, and although the pores can be clearly distinguished (arrows), the membrane surface resembles to some extent a sheet of separate, sometimes elongated vesicles. Derivation of flat sheets from the three-dimensional network is well illustrated where the section includes part of the disrupted prolamellar body (right hand side of plastid).

FIG. 14 and FIG. 15. Survival of portions of lattice in a plastid fixed after 10 hr illumination. Particles can be seen in the centres of the hexagonal spaces of the lattice. Grana are viewed obliquely and in profile.

recently been found to be a fixation artefact which can be prevented by use of glutaraldehyde prefixation (Tormey, 1964). A somewhat similar type of fixation damage has been reported in an examination of the stacked lamellae of retinal rod and cone cells (Eakin, 1965).

Although the total surface area of the "perforated thylakoids" can be estimated only very roughly, it can be tentatively concluded that by the end of the lag phase of greening, no large areas of membrane have been produced *de novo* in the plastids. This is indicated by counts of both flat sheets and concentric layers of "perforated thylakoids". One factor which was not taken into account, however, was the persistence of small portions of prolamellar body until well past the end of the lag period. This is illustrated in Figs. 14 and 15, which show grana alongside such portions in plastids that have been fixed after 10 hr of continuous illumination.

Perforations are transient, and characteristic of thylakoids that have recently been formed from a prolamellar body. The few thylakoids that have (presumably) been present for a considerable time in completely dark-grown plastids are not perforated (Fig. 3), and neither are the thylakoids found after 10 hr of illumination (Figs. 14 and 15). The disappearance of the pores may well be correlated in time with the end of the lag period, however, it is certainly not correlated with the time course of phytylation of the chlorophyllide produced during the initial illumination. In *Avena* (as in other species—Wolff and Price, 1957; Sironval *et al.*, 1965) this process takes only about 30 min. In oat leaves that have been infiltrated with actinomycin D or chloramphenicol development is arrested at the "perforated thylakoid" stage, although the pores may eventually disappear.

## The Formation of the Prolamellar Body

### Formation in Developing Mesophyll Cells

The base of the sheath of the first leaf of an etiolated oat seedling encloses the shoot meristem and very young leaves. Proplastids containing a few double lamellae are found in cells in and near the meristem, and the first sign of prolamellar body formation is the development of connecting arms between these lamellae. In large proplastids (or small plastids) in young mesophyll cells the prolamellar bodies are more obvious. Often they simply resemble small portions of a mature lattice, but, in contrast to the plastids in older mesophyll cells, there is a high proportion of the "atypical" open structures described on p. 661 (see Fig. 16, compare Fig. 1, upper plastid). Since it seems that the normal type of lattice can develop directly, the "atypical" system cannot be an obligatory developmental stage. Perhaps there is more than one form in which the prolamellar body components can crystallize, the less stable forms being eliminated through time in favour of the most stable.

Glutaraldehyde–osmium tetroxide fixation fails to reveal any vesicular stage in the outgrowth of lamellae from the prolamellar body (see p. 664). The same applies to the development of the lattice, and once again this contrasts with published descriptions, in which crystallization has been reported to be a relatively late event, prior to which the prolamellar body consists of a mass of vesicles and/or tubules, the latter sometimes regularly arranged in bunches

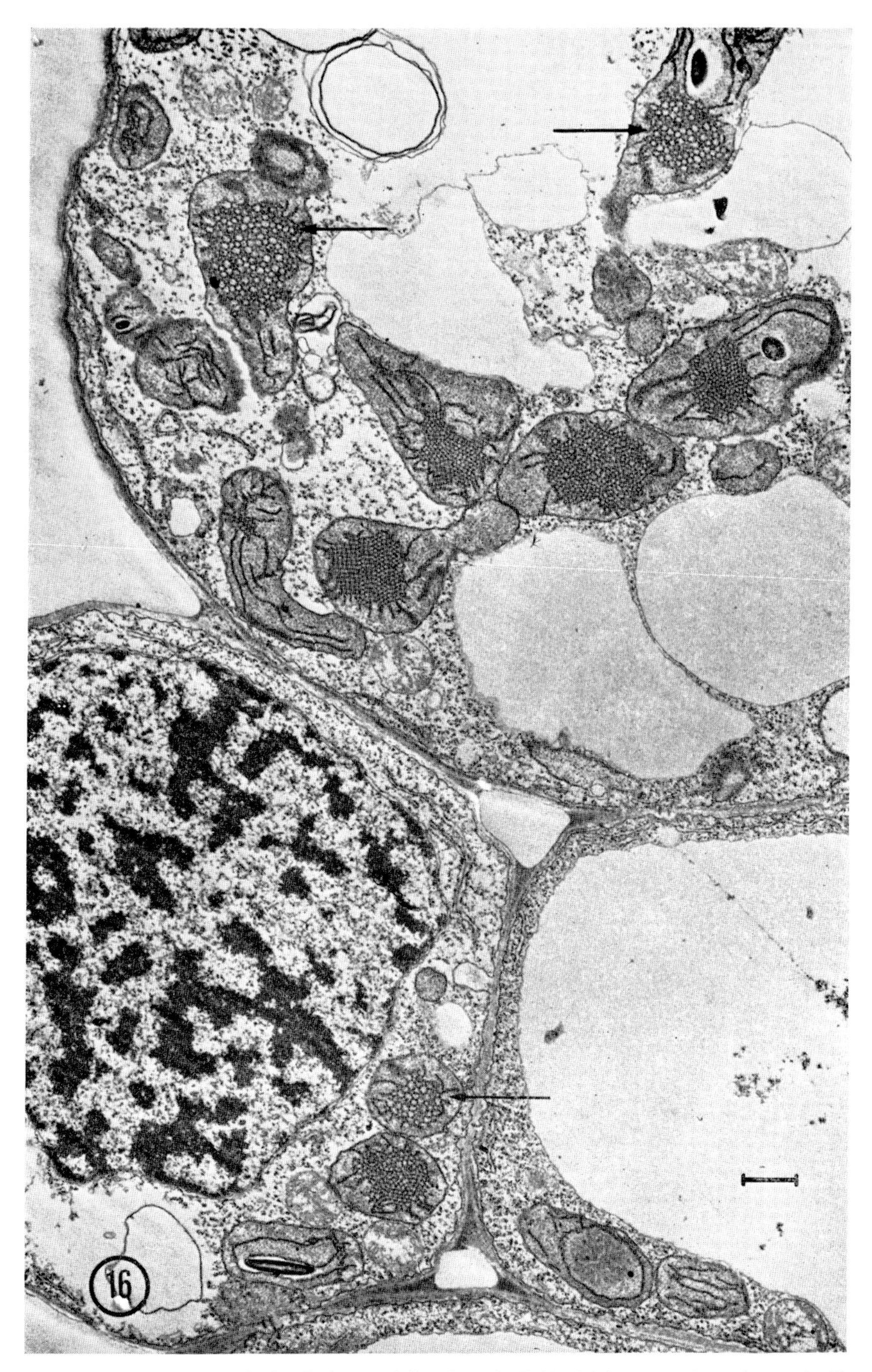


FIG. 16. Young mesophyll cells in an etiolated oat leaf. The high proportion of "atypical" prolamellar bodies (arrows) is illustrated; many of the plastids, however, contain normal lattices.

(Gerola *et al.*, 1960; von Wettstein and Kahn, 1960; Virgin *et al.*, 1963). It may be that such apparently non-crystalline stages are equivalent to the "atypical" open lattice found in *Avena*; alternatively *Avena* may behave quite differently from other species.

## RE-FORMATION IN DARKNESS FOLLOWING A PERIOD OF ILLUMINATION

Von Wettstein and Kahn (1960) have reported "reformation of tubules and prolamellar bodies" 2 to 6 hr after a briefly illuminated detached bean leaf is returned to darkness. According to their diagram no large, regular, crystalline lattice was formed. Klein *et al.* (1964) found neither progressive nor regressive changes in similar experiments, also performed on detached leaves. In intact *Avena* seedlings, on the other hand, prolamellar bodies will re-form readily after they have been disorganized.

Following a 5 min period of illumination, with subsequent sampling at 15 min intervals, re-formation of the crystalline lattice is first observed after 45 min in darkness. There is no synchronization, either within a leaf or within a cell, and plastids containing crystalline structures can be found alongside others which have not yet reached such a stage—e.g. Fig. 17. Meanwhile, protochlorophyll(ide) is accumulating in the leaf. In our conditions this protochlorophyll(ide) only became obvious after 60–90 min of darkness, following the initial 5 min in light. Although ultimately there was as much protochlorophyll(ide) present as in the original leaves prior to their first illumination, the oat seedlings evidently produce the pigment much more slowly than the wheat and barley plants studied by Madsen (1963).

The upper plastid in Fig. 17 contains a crystalline prolamellar body and comparatively few other membranes. If it is representative of the whole-leaf samples from which extracts were prepared, it contains some chlorophyll *a* (fully phytylated) and some protochlorophyll(ide). The appearance of the micrographs would suggest that the greater part of the re-formed prolamellar body is derived from membrane that was present in the original lattice. If this original lattice contained the original protochlorophyll(ide) (see Smith and French, 1963, and Klein *et al.*, 1964), the problem of the location of the chlorophyll *a* (which is present throughout the recrystallization process) derived from it arises. Is it built into the new lattice along with the old membrane?

With longer initial exposures to light, the situation can be more complex. After 5 hr in light, followed by half an hour in darkness, the plastids are no longer in the lag phase of greening and they contain both crystalline prolamellar bodies and small grana (Fig. 18). In such a case it is not possible to speculate on whether the new prolamellar body consists of pre-existing or of newly synthesized membrane (or both), or on the possibility that it contains pigments other than protochlorophyll(ide).

Re-formation of prolamellar bodies contrasts with the original formation process in at least one respect. The new structures are apparently organized directly, and at no stage is there a high proportion of the open "atypical" lattices such as is observed in young cells. Structurally the new prolamellar bodies seem to be identical to those originally present, as regards the distribution of both

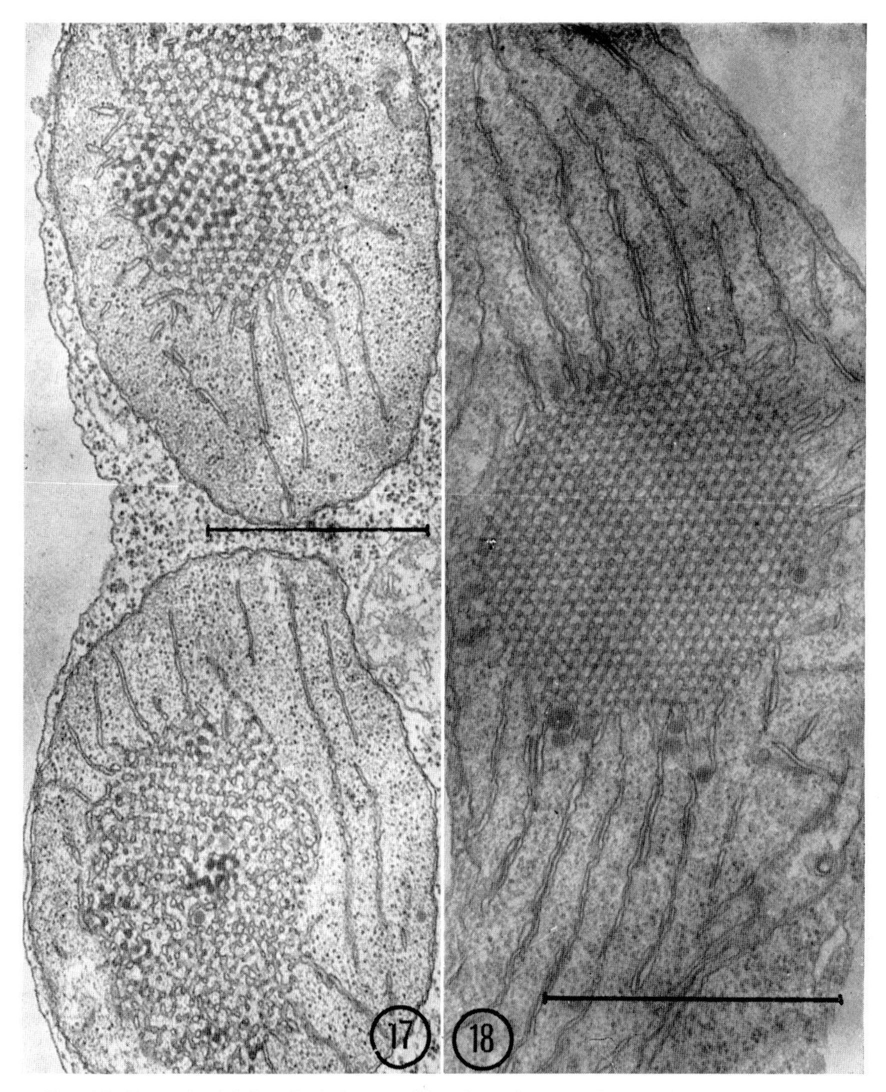


FIG. 17. Two plastids in a leaf that was illuminated for 5 min and then returned to total darkness for 45 min. Most of the membrane is present in the form of prolamellar bodies; one is crystalline, the other is only partly so.

FIG. 18. A plastid in a leaf that received 5 hr light followed by 30 min darkness. The prolamellar body (with particles visible) has re-formed, but thylakoids and small grana are also present.

particles and membranes (Figs. 17 and 18). This may be associated with the fact (Figs. 14 and 15) that portions of lattice can survive (albeit disorganized) for considerable periods in the light, and could act as centres of recrystallization.

## The Organization of the Prolamellar Body

The preceding information can be used as the basis for a hypothesis which accounts for the shape of the membrane surface in the prolamellar body, and indicates how its crystalline portions may be organized.

It is convenient to consider first of all an isolated simple cubic lattice of ribosome-like particles. In the *Avena* plastid these approximate to spheres of diameter 130 Å; the side of the unit cube is 590 Å in length. The hypothesis is that around each particle there is a field which is for some unknown reason unsuitable for the presence of membranes. If one particle is considered in isolation its field would be approximately spherical, with the particle at the centre. Contours can be envisaged within the sphere, such that the surface of a sphere of small diameter would represent a contour where the influence of the particle would be strong. A larger diameter sphere would represent a weaker, more distant influence. The particles, however, are in a lattice, and the fields around neighbouring particles would interact and merge with one another, at any rate in their outer, "weaker" regions. In a simple cubic lattice a contour representing a constant field strength would take the form shown in Fig. 19. Contours representing the stronger forces closer to the particles would not be distorted to the same extent, just as in the analogous case of electron orbits in a conductor with a simple cubic atomic lattice, where the inner orbits are approximately spherical, while in the outermost layer, lattice effects lead to the formation of a surface (the Fermi surface) on which electrons can flow throughout the whole lattice (Ziman, 1960), and which has exactly the same geometry as the surface shown in Fig. 19.

The geometry of this contour of constant property is the same as the geometry of the membrane lattice (compare Fig. 19 with Fig. 6). Both can be regarded as being composed of six-armed nodal units, and at any given point on their surfaces the principal curvatures are of opposite sign. By appropriate choice of dimensions, it is a simple matter to show that the two systems can be complementary in space. If the dimensions of the six-armed nodal unit of the field centred on a particle are as in Fig. 19, and if this surface is used as a template on which membranes become organized, then the membrane surface so formed consists of units with the complementary dimensions, as in Fig. 6. Fig. 20 shows how the two types of unit fit together.

In other words, it is suggested that there is, in a prolamellar body, a field centred on and generated by a lattice of ribosomes, which provides a template on which membranes either come to lie, or are actually formed. The perfection of the geometrical fit is the main evidence supporting the hypothesis, and the work of Dibble and Dintzis (1960) and Dibble (1964) may also be cited as indicating that the image of a ribosome as seen in the electron microscope after normal shadowing or staining techniques is a considerable underestimate of its

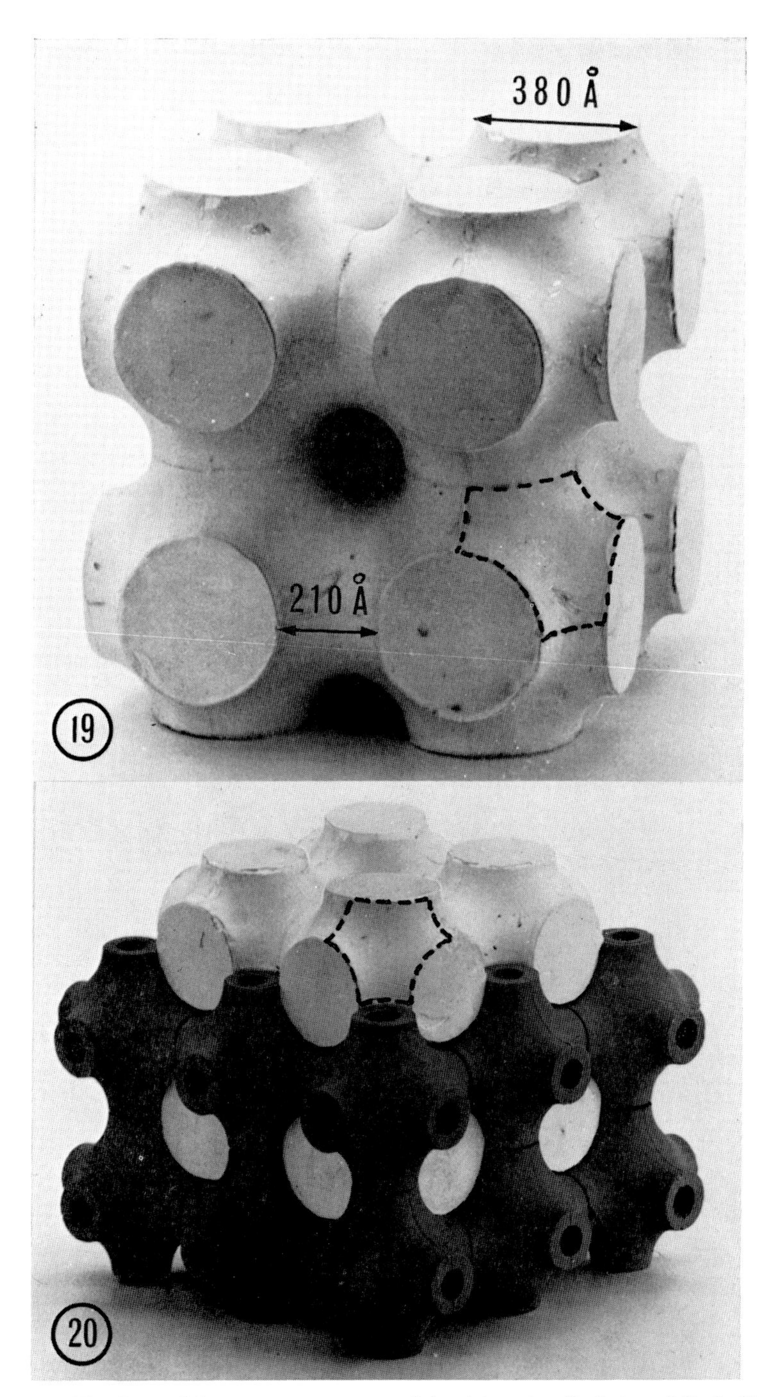


FIG. 19. The shape of the stroma component of the eight unit cells that are delimited by the membrane system illustrated in Fig. 6. It is suggested that this surface could be generated by ribosome-like particles which lie in the centre of each of the eight units.

FIG. 20. The lattice shown in Fig. 19 and part of that shown in Fig. 6 are shown combined here to demonstrate that they are complementary in space. The surface drawn in Fig. 8 would fit on to the portion of the "ribosomal field" surface that has been outlined in this figure and in Fig. 19.

*in vivo*, hydrated size. Forces which could lock the particles and membranes into a stable lattice are known, and it is not difficult to imagine a balance between forces of repulsion (due to double layer effects around, and possession of like charges by ribosomes) and forces of attraction (long range Van der Waals–London forces) between the particles. The continuity and space-filling shape of the membrane system (which is sealed—see p. 656) must provide some stability and rigidity, as could the symmetry of long range forces between the membranes and the particles. Kavanau (1965) has recently discussed the ways in which such forces can balance to bring about long range order in colloidal systems.

It is obviously not general for there to be forces which maintain a space between ribosomes and membranes—even within plastids (Brown and Gunning, Vol. I, p. 365). To that extent the prolamellar body membranes and particles would appear to be specialized. However, it is likely that such specializations can occur in other, quite different situations. The outstanding example here is the existence of a closely similar, if not identical, membrane lattice in crayfish oocyte mitochondria (Beams and Kessel, 1963). A further system which is analogous to a limited extent is found in down feather cells prior to the onset of keratinization (Bell *et al.*, 1965). It is an aggregate of four ribosomes arranged at the corners of a square and bears a strong resemblance to half of a unit cube of eight ribosomes, not only in its configuration but also in that it is inactive, long-lived, and is accumulated in the cells. The existence of these four-ribosome units suggests that in other systems of inactive ribosomes larger complexes could develop by aggregation.

Even if the hypothesis were to be proved correct, many problems would remain. How does darkness initiate re-formation of a prolamellar body? Is it the membranes, the particles, or some other plastid components which react to the absence of light? It is conceivable, for instance, that the lattice starts to form when there are large numbers of single ribosomes available, so that whatever controls the ratio of polyribosomes to single ribosomes indirectly determines whether or not it can develop. Indications that light could affect this ratio during the breakdown of the prolamellar body come from experiments on the effect of actinomycin D on greening (Bogorad and Jacobson, 1964) and observations on particle distribution (Brown and Gunning, Vol. I, p. 365). Similarly, the inactive, four-ribosome complexes of down feather cells give rise to active polyribosomes when keratin synthesis commences (Bell *et al.*, 1965).

Another problem is that of the location of the protochlorophyll(ide). The holochrome particle (Smith and French, 1963) is very large relative to the thickness of the unit membranes of the prolamellar body. Is the presence of protochlorophyll(ide) a prerequisite for lattice formation? This seems unlikely to be a general rule if the structure found by Beams and Kessel (1963) in crayfish mitochondria really is analogous. There is also (p. 670) the possibility that chlorophyll *a* can be present.

The nature of the forces that maintain the structure and stability of the lattice is unknown. However, the disruptive effect of rapid photoconversion of protochlorophyll(ide) to chlorophyll(ide) on a delicately poised equilibrium can be imagined. Weiss (1964) has indicated how disturbances could be propagated through a membrane-limited structure, and his ideas could well apply in the case of the effect of light on the prolamellar body. The ensuing controlled metamorphosis into, and dispersion of, sheets of "perforated thylakoids" is at present a much less understandable phenomenon.

## Acknowledgements

The support of the S.R.C. and the assistance of Miss M. Hayes and the staff of the Queen's University Electron Microscope Laboratory is gratefully acknowledged. We would like particularly to thank Professor D. J. Carr for his interest and for many helpful discussions throughout the course of the work.

## References

Beams, H. W. and Kessel, R. G. (1963). *J. cell Biol.* **18**, 621.
Bell, E., Humphreys, T., Slayter, H. S. and Hall, C. E. (1965). *Science, N.Y.* **148**, 1739.
Bogorad, L. (1965). *In* "Chemistry and Biochemistry of Plant Pigments" (T. W. Goodwin, ed.), Academic Press, London and New York.
Bogorad, L. and Jacobson, A. B. (1964). *Biochem. biophys. Res. Commun.* **14**, 113.
Dibble, W. E. (1964). *J. Ultrastr. Res.* **11**, 363.
Dibble, W. E. and Dintzis, H. M. (1960). *Biochim. biophys. Acta* **37**, 152.
Duffett, R. H. E. and Minkoff, G. J. (1964). *Discovery* **25**, 32.
Eakin, R. M. (1965). *J. cell Biol.* **25**, 162.
Eriksson, G., Kahn, A., Walles, B. and Wettstein, D. von (1961). *Ber. dtsch. Bot. Ges.* **74**, 221.
Gerola, F. M., Cristofori, F. and Dassu, G. (1960). *Caryologia* **13**, 179.
Granick, S. (1961). *In* "The Cell" (J. Brachet and A. E. Mirsky, eds.), Vol. 2, p. 489. Academic Press, New York and London.
Gunning, B. E. S. (1965a). *J. cell Biol.* **24**, 79.
Gunning, B. E. S. (1965b). *Protoplasma* **60**, 111.
Holden, M. (1965). *In* "Chemistry and Biochemistry of Plant Pigments" (T. W. Goodwin, ed.). Academic Press, London and New York.
Jacobson, A. B., Swift, H. and Bogorad, L. (1963). *J. cell Biol.* **17**, 557.
Kavanau, J. L. (1965). "Structure and Function in Biological Membranes", Vol. 1. Holden-Day Inc.
Klein, S., Bryan, G. and Bogorad, L. (1964). *J. cell Biol.* **22**, 433.
Koski, V. M. (1950). *Archs Biochem. Biophys.* **29**, 339.
Madsen, A. (1963). *Photochem. Photobiol.* **2**, 93.
Reynolds, E. S. (1963). *J. cell Biol.* **17**, 208.
Sironval, C., Michel-Wolwertz, M. R. and Madsen, A. (1965). *Biochim. biophys. Acta* **94**, 344.
Smith, J. H. C. and French, C. S. (1963). *Annu. Rev. Pl. Physiol.* **14**, 181.
Tormey, J. McD. (1964). *J. cell Biol.* **23**, 658.
Virgin, H. I., Kahn, A. and Wettstein, D. von (1963). *Photochem. Photobiol.* **2**, 83.
Weiss, P. (1964). *Proc. natn. Acad. Sci. U.S.A.* **52**, 1024.
Wettstein, D. von (1958). *In* "The Photochemical Apparatus, Its Structure and Function". Brookhaven Symp. in Biol. No. 2, p. 138.
Wettstein, D. von (1959). *In* "Developmental Cytology", 16th Growth Symp. (D. Rudnick, ed.), p. 123.
Wettstein, D. von and Kahn, A. (1960). *Proc. Europ. Reg. Conf. Electron Microsc. Delft.* **2**, 1051.
Wilkins, M. B. (1965). *Pl. Physiol.* **40**, 24.
Wolff, J. B. and Price, L. (1957). *Archs Biochem. Biophys.* **72**, 293.
Ziman, J. M. (1960). "Electrons and Phonons". Oxford Clarendon Press.

## Note Added in Proof

Detailed interpretations of *Phaseolus* prolamellar bodies have been published by Wehrmeyer (*Z. Naturf.* (1965). **20b**, 1270, 1278, 1288), since this manuscript was prepared. His work clearly indicates that following isolation of plastids and fixation with osmium tetroxide vapour, bean prolamellar bodies are constructed mainly of a system of tetrahedrally branched tubules. He did not observe a cubic lattice such as is unequivocally demonstrated by the serial sections presented by Gunning (1965b), nor did his methods visualize particles within the lattice. The relative proportions of tetrahedral, cubic, and "open" lattices remains to be elucidated. There could be variation between species and also during development within a species. Although the geometry of the tetrahedral system is different, the hypothesis that the membrane lattice is stabilized and possibly moulded by the presence of the particles could still apply to it as to the cubic system.

# The Lipid Compostion and Ultrastructure of Normal Developing and Degenerating Chloroplasts

B. W. Nichols, J. M. Stubbs and A. T. James
*Biosynthesis Unit, Colworth House, Sharnbrook, Bedford, England*

Reports from several laboratories regarding the nature of the fatty acid-containing lipids of numerous photosynthetic tissues have indicated the possibility of some basic difference between the lipid composition of cells capable of performing the Hill reaction (the various classes of algae and the photosynthetic tissues of higher plants) and those which cannot (the photosynthetic bacteria.

Photosynthetically cultured cells of freshwater and marine algae contain the same polar lipids which are present in the leaves and other chloroplast-containing tissues of higher plants, namely the galactosyl diglycerides, sulpholipid, cardiolipin, phosphatidyl-glycerol, -choline, -ethanolamine and -inositol (Benson, 1963; Lichtenthaler and Park, 1963). More particularly the major fatty acid-containing lipids of the chloroplast are the galactosyl diglycerides, sulpholipid and phosphatidyl glycerol (Benson, 1964; Nichols, 1963). Of the photosynthetic bacteria studied in several laboratories (Wood *et al.*, 1965; Benson *et al.*, 1959; Haverkate, 1965) none has shown a general lipid composition comparable to that of leaves or chloroplasts, and phosphatidyl glycerol is the only lipid common to all the photosynthetic cells studied (Table I).

It therefore seems permissible to argue that the four fatty acid-containing lipids present in the chloroplasts of algae and higher plants might be essential prerequisites for the Hill reaction, and three possible types of involvement can be postulated.

(1) Chemical involvement in some stage or stages of the electron transport mechanism for which it is reasonable to suggest the participation of the unsaturated fatty acids, since the other parts of the lipid molecule possess no groups capable of oxidation-reduction changes.

(2) As specific structural components maintaining the lipid-soluble pigments in the correct spatial alignment with one another and their associated enzymes.

(3) As relatively non-specific micellar elements which exclude water and in which the pigment–protein–enzyme complexes should be partially or completely embedded. Such water-free areas could allow the operation of electron transport mechanisms that would be inhibited by free water. The lipid would thus act as an organized medium of low dielectric constant.

From these three suggestions a few generalizations can be made.

(a) Chemical involvement of unsaturated fatty acids would require that

Table I. Polar lipids present in various photosynthetic tissues

| | No Hill reaction | | | | Hill reaction | | | | | |
|---|---|---|---|---|---|---|---|---|---|---|
| | Photosynthetic bacteria | | | | Blue/green algae | | | | | |
| Lipid | *Rps. palustris* | *Rps. capsulata* | *Rps. spheroides* | *R. rubrum* | *Anacystis nidulans* | *Anabaena variabilis* | *Euglena gracilis* | *Chlorella vulgaris* | Whole leaves | Leaf chloroplasts |
| Monogalactosyl diglyceride | 0 | 0 | 0 | 0 | / | / | / | / | / | / |
| Digalactosyl diglyceride | 0 | 0 | 0 | 0 | / | / | / | / | / | / |
| Sulphoquinovosyl diglyceride | 0 | 0 | / | 0 | / | / | / | / | / | / |
| Phosphatidyl glycerol | / | / | / | / | / | / | / | / | / | / |
| Phosphatidyl choline | / | / | / | 0 | 0 | 0 | / | / | / | 0 |
| Phosphatidyl ethanolamine | / | / | / | / | 0 | 0 | / | / | / | 0 |
| Phosphatidyl inositol | 0 | 0 | 0 | 0 | 0 | 0 | / | / | / | 0 |
| Unknown lipids | / | / | / | / | / | / | 0 | 0 | 0 | 0 |

photosynthetic systems of the same type would contain fatty acids of similar structures. Organelles produced in living tissues in the absence of light might be expected to contain lesser amounts of such components.

(b) The existence of specific interaction forces would require a similar lipid or group of lipids to occur in photosynthetic systems of a given type.

(c) The formation of complex micelles that provide water-free areas could be done by a variety of lipids. No particular association of either specific fatty acids or lipids would be expected within the different types of photosynthetic tissue.

## The Fatty Acids of Chloroplast Lipids

It is relevant to mention here a few established features of the fatty acid composition of the chloroplast lipids in various plants and algae. The photosynthetic tissues of higher plants and many algae contain a high proportion of linoleic and α-linolenic acids and these acids, particularly the latter, concentrate predominantly in the galactosyl diglycerides. Although there is sometimes only a comparatively slight difference in the fatty acid composition of these two lipids it is generally observed (Table II) that the monogalactosyl diglyceride fraction contains a greater proportion of the more highly unsaturated $C_{18}$ acids, usually α-linolenic acid. In some cases where the tissue contains a significant quantity of $C_{16}$ di- or tri-unsaturated acids, these acids usually predominate in the monogalactosyl diglyceride and are present in the diglycosyl lipid in only minor porportions. Usually the presence of such $C_{16}$ acids in the monogalactosyl diglyceride is "balanced" by a reduction in the proportion of the corresponding $C_{18}$ acid so that the total polyenoic acid content of the lipid is maintained at the "normal" level. Thus while in many leaves such as lucerne (alfalfa) (O'Brien and Benson, 1964) castor, lettuce and holly (Nichols, 1965a), the monogalactosyl and digalactosyl diglycerides contain about 90% and 80% α-linolenic acid respectively, in some tissues, namely spinach leaves (Allen *et al.*, 1964; Haverkate, 1965), moss (Nichols, 1965a) and *Chlamydomonas mundana* (B. W. Nichols, and B. J. B. Wood, unpublished data) the former glycoside contains less α-linolenic acid than does the latter but it nevertheless possesses a higher total trienoic acid content because of the presence of the $C_{16}$ trienoic acid. Similarly, when *Chlorella vulgaris* cells are cultured in the dark, the monogalactosyl diglyceride contains nearly all the $C_{16}$ dienoic acid synthesized by the cell, and the same cells when growing photosynthetically contain also a $C_{16}$ trienoic acid which appears exclusively in the monogalactoside.

These values indicate that if, as the results of Ferrari and Benson (1961) might suggest, digalactosyl diglyceride is synthesized in the chloroplast by galactosylation of the monogalactoside, then the enzymes concerned in this reaction must either be highly specific for monogalactosyl diglycerides of a distinctive fatty acid composition, or alternatively a certain amount of transacylation of these molecules is possible.

The phosphatidyl glycerol fraction from plant and algal photosynthetic tissues also shows a distinctive fatty acid composition quite different from that

Table II. Fatty acid composition of the monogalactosyl diglyceride (MGDG) and digalactosyl diglyceride (DGDG) fractions of some photosynthetic tissues

| | | 14:3 | 16:0 | 16:1 | 16:2 | 16:3 | 18:0 | 18:1 | 18:2 | 18:3 | 20:4 | 20:5 | 22:0 |
|---|---|---|---|---|---|---|---|---|---|---|---|---|---|
| Castor leaf | MGDG | 2·3 | 6·0 | | | | | t | t | 91·0 | | | |
| | DGDG | | 11·1 | | | | | | 3·7 | 85·3 | | | |
| Lucerne | MGDG | | 2·7 | | | | | | 1·7 | 95·0 | | | |
| leaf[1] | DGDG | | 14·0 | | | | 3·3 | | | 82·0 | | | |
| Runner | MGDG | | 2·3 | | | | | | 2·2 | 95·5 | | | |
| bean leaf[3] | DGDG | | 4·5 | | | | 1·0 | | 1·3 | 93·2 | | | |
| Holly leaf | MGDG | | 1·0 | | | | | | 2·3 | 96·7 | | | |
| | DGDG | | 13·4 | | | | | | 6·9 | 79·7 | | | |
| Spinach | MGDG | | t | | | 30·0 | | 1·0 | 1·0 | 67·0 | | | |
| leaf[2] | DGDG | | 6 | | | 3·0 | 1·0 | 4·0 | 3·0 | 84·0 | | | |
| Moss | MGDG | | 2·3 | t | 2·0 | 10·7 | 6 | 1·2 | 4·1 | 48·3 | 28·5 | 11·2 | 3·1 |
| | DGDG | | 6·2 | 1·8 | 1·2 | 1·7 | 1·6 | 2·5 | 4·6 | 62·2 | 13·8 | 4·4 | 2·6 |
| *Chlorella* | MGDG | | 3·7 | 14·3 | 21·4 | 1·8 | t | 18·4 | 29·9 | 8·0 | | | |
| *vulgaris* | DGDG | | 15·0 | 4·3 | 5·9 | 1·9 | 3·6 | 14·3 | 39·7 | 7·1 | | | |
| *Chlamydo-* | MGDG | 3·3 | 8·1 | 8·0 | 9·1 | 27·1 | t | 9·5 | 7·6 | 26·1 | | | |
| *monas* | DGDG | 1·1 | 11·5 | 8·5 | 14·0 | 17·8 | t | 16·4 | 10·0 | 20·0 | | | |
| *mundana* | | | | | | | | | | | | | |

[1] O'Brien and Benson (1964).
[2] Allen *et al.* (1964).
[3] Sastry and Kates (1964).

of the other chloroplast lipids, particularly the galactosyl diglycerides. Thus it usually contains a fairly high proportion of palmitic acid, and *trans*-$\Delta^3$ hexadecenoic acid, which in all such tissues so far studied is specifically combined in this lipid.

The distinctive fatty acid composition shown by each of the four chloroplast lipids when isolated from the same tissue suggests that they must all be synthesized from different diglyceride pools, or that they result from a very specific metabolism of the same diglyceride source.

Our more recent studies have been primarily directed towards an understanding of the qualitative and quantitative changes involved when tissues grown in darkness are then allowed to develop in the light over limited periods. Thus changes in lipid composition were investigated in castor leaves in which

Table III. Fatty acid composition of castor leaf lipids

| | 16:0 | $\Delta^9$ 16:1 | $\Delta^3$ 16:1 | 18:1 | 18:2 | 18:3 |
|---|---|---|---|---|---|---|
| Etiolated | | | | | | |
| Monogalactosyl diglyceride | 7 | 3 | — | 7 | 14 | 67 |
| Digalactosyl diglyceride | 14 | 2 | — | 8 | 21 | 51 |
| Phosphatidyl glycerol | 38 | 5 | — | 11 | 20 | 14 |
| Sulphoquinovosyl diglyceride | 29 | 5 | — | 15 | 12 | 27 |
| Phosphatidyl choline | 15 | 2 | — | 11 | 37 | 24 |
| After 20 hr illumination | | | | | | |
| Monogalactosyl diglyceride | 4 | — | — | 3 | 4 | 88 |
| Digalactosyl diglyceride | 15 | 2 | — | 7 | 7 | 65 |
| Phosphatidyl glycerol | 41 | — | 13 | 12 | 13 | 15 |
| Sulphoquinovosyl diglyceride | 39 | 3 | — | 6 | 10 | 33 |
| Phosphatidyl choline | 16 | 2 | — | 8 | 41 | 25 |

the plastids were undergoing the transition from the crystalline proplastid structure to the lamellar chloroplast configuration, as illustrated in Fig. 1. The changes in fatty acid composition of the leaf lipids which took place over this 16 hr period of illumination are indicated in Table III and are primarily an increased concentration of α-linolenic acid in the galactosyl diglycerides and the appearance of *trans*-$\Delta^3$ hexadecenoic acid in the phosphatidyl glycerol.

In *Chlorella vulgaris*, on the other hand, there are observed rather more dramatic changes in fatty acid composition, as illustrated in Table IV. When this alga is grown on a high nutrient medium either in the light or in the dark, the amount of photosynthetic growth is comparatively limited, and the cells contain only minor quantities of α-linolenic acid (thus contrasting with dark-grown castor leaf cells which contain a substantial quantity of this acid). If such cells are then transferred to a nutritionally poor medium (0·2 M phosphate buffer, pH 7·4) in the light then growth is largely photosynthetic and

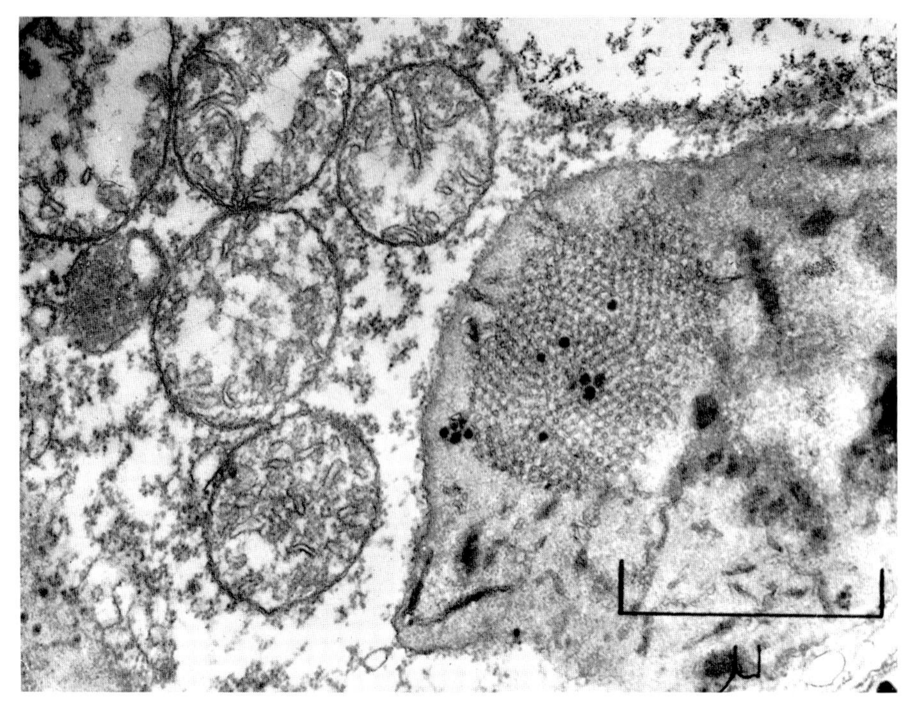


FIG. 1a.

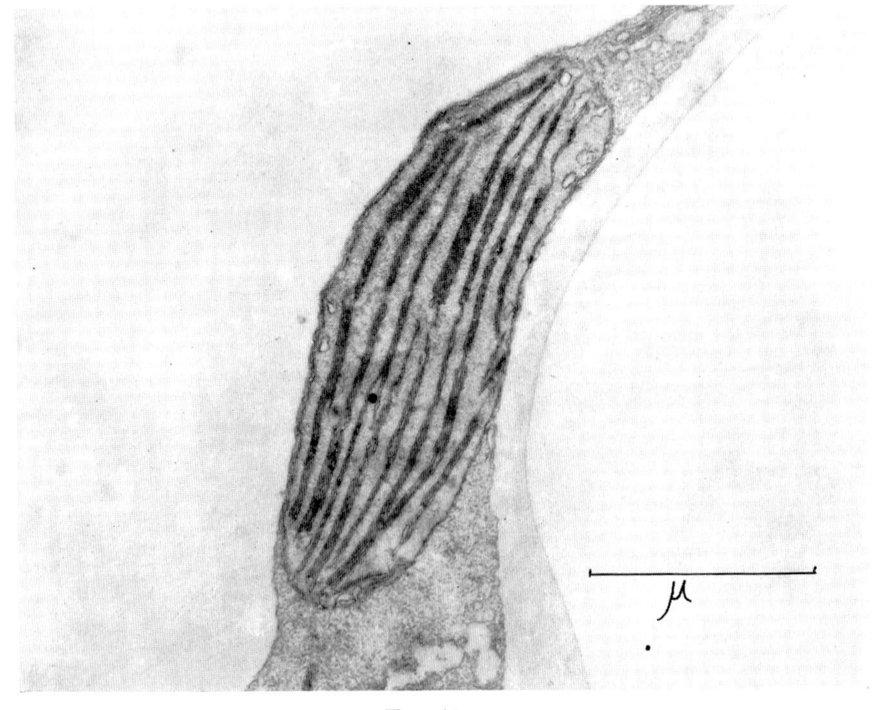


FIG. 1b.

there is a rapid synthesis of the polyenoic acids, which primarily concentrate in the galactosyl diglycerides, as well as the appearance of the *trans*-$\Delta^3$ hexadecenoic acid in the phosphatidyl glycerol fraction.

On the basis of the apparent ubiquity of α-linolenic in the photosynthetic tissues of plants and green algae and its absence from dark-grown *Euglena gracilis*, Erwin *et al.* (1964) suggested that this acid might be involved in electron transport related to the Hill reaction. Furthermore, we were unable to confirm the two references in the literature to plant photosynthetic tissues containing little or no α-linolenic acid, namely nettle leaves (Hilditch and Meara, 1944) and the needles of black pine (Tsujimoto, 1940). [See Table V.] In addition, we suggested (Nichols *et al.*, 1965b; Nichols, 1965b) the possibility of a relationship between the *trans*-$\Delta^3$ hexadecenoic acid and the Hill reaction on the basis

Table IV. Fatty acid composition of *Chlorella vulgaris* lipids

| | 16:0 | $\Delta^9$ 16:1 | $\Delta^3$ 16:1 | 16:2 | 18:1 | 18:2 | 18:3 |
|---|---|---|---|---|---|---|---|
| Partly etiolated | | | | | | | |
| Monogalactosyl diglyceride | 2 | 10 | — | 28 | 10 | 40 | 6 |
| Digalactosyl diglyceride | 11 | 7 | — | 9 | 17 | 42 | 9 |
| Phosphatidyl glycerol | 50 | 2 | — | 3 | 16 | 21 | 2 |
| Sulphoquinovosyl diglyceride | 36 | 4 | — | 3 | 13 | 32 | 5 |
| Phosphatidyl choline | 11 | 5 | — | 5 | 16 | 52 | 7 |
| After 24 hr illumination in buffer | | | | | | | |
| Monogalactosyl diglyceride | 7 | 6 | — | 22 | 5 | 35 | 26 |
| Digalactosyl diglyceride | 13 | 9 | — | 6 | 4 | 43 | 24 |
| Phosphatidyl glycerol | 46 | — | 16 | — | 8 | 25 | 6 |
| Sulphoquinovosyl diglyceride | 39 | 4 | — | — | 5 | 36 | 11 |
| Phosphatidyl choline | 16 | 7 | — | 3 | 9 | 52 | 10 |

of its absence from etiolated tissue (Nichols *et al.* 1965b; Haverkate, 1965) and its apparent ubiquity in photosynthetic tissues capable of performing the Hill reaction, namely in the leaves of spinach (Debuch, 1961; Allen *et al.*, 1964; Haverkate *et al.*, 1964), *Antirrhinum* (Debuch, 1961), clover (Weenink and Shorland, 1964), Castor, Holly, Moss (Nichols, 1965a) and in photosynthetically developed cells of *Euglena* (Haverkate, 1965), *Scenedesmus* (Klenk and Kripprath, 1962) and *Chlorella vulgaris* (Nichols, 1965b; Haverkate, 1965).

At this stage of our work, Holton and coworkers (1964) reported the absence of polyunsaturated fatty acids from the blue-green alga *Anacystis nidulans* (which performs the Hill reaction) although it was not clear whether or not they

←

FIG. 1. Electron micrograph of the proplastid bodies present in the etiolated primary leaves of castor (*Ricinus gibsoni*) (a), and that of a similar leaf after 20 hr illumination (b). Tissues fixed in osmic acid (Palade's fixative), embedded in epon 812, sectioned, and stained in lead citrate solution.

Table V. Fatty acid composition of the lipids from nettle leaves and the needles from black pine

| Leaf source | Fatty acid (%) | | | | | | | | | | | | |
|---|---|---|---|---|---|---|---|---|---|---|---|---|---|
| | 12:0 | 14:0 | 14:3 | 16:0 | $\Delta^9$ 16:1 | $\Delta^3$ 16:1 | 16:2 | 16:3 | 18:0 | 18:1 | 18:2 | 18:3 | 18:4 |
| Stinging nettle (*Urtica dioica*) | — | 3·0 | 1·8 | 11·7 | — | 2·0 | — | — | 1·2 | 2·2 | 15·3 | 66·1 | — |
| White dead-nettle (*Lamium album*) | — | 3·2 | — | 13·5 | — | 3·0 | — | — | 2·0 | 2·0 | 9·0 | 67·6 | — |
| Red dead-nettle (*Lamium purpureum*) | — | 1·2 | — | 16·8 | — | 2·6 | t | — | 2·0 | 2·0 | 11·1 | 62·9 | — |
| Black pine (*Pinus thunbergii*) | 6·0 | 3·7 | — | 16·6 | 2·1 | | 1·0 | 2·3 | 1·6 | 9·0 | 12·0 | 39·2 | 7·3 |

could have overlooked the presence of the *trans*-$\Delta^3$-hexadecenoic acid. Analysis of the fatty acid composition of the individual lipid fractions from this organism (Nichols *et al.*, 1965a) confirmed the absence of the hexadecenoic acid which was also found to be absent from another blue-green alga, *Anabaena variabilis*, previously studied by Levin and coworkers (1964) who demonstrated the presence of α-linolenic acid.

A study of the lipids present in these two algae showed a comparatively simple composition compared with that of other algae and the leaves of higher plants in that the major components were the four lipids we associate with chloroplasts, i.e. the two galactosyl diglycerides, sulphoquinovoso-diglyceride and phosphatidyl glycerol. *Anabaena variabilis* also contains an as yet unidentified lipid, which was also present in minor quantities in *Anacystis nidulans*.

A comparison of the lipid and fatty acid composition of numerous photosynthetic tissues now indicates that the basic lipid requirement for photosynthesizing cells capable of performing the Hill reaction is the presence of the

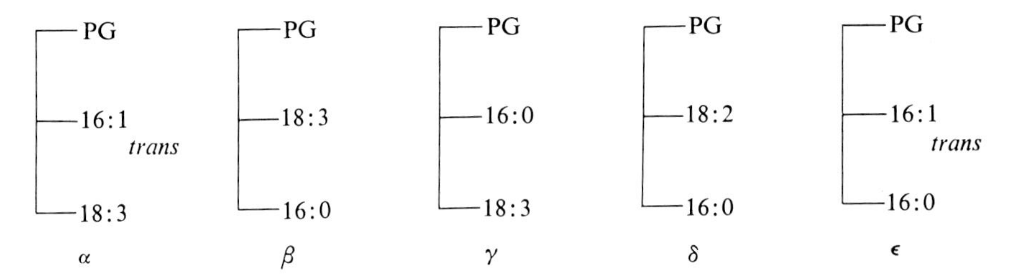


FIG. 2. Molecular species of phosphatidyl glycerol from spinach leaves (Haverkate, 1965).

galactosyl diglycerides, sulpholipid and phosphatidyl glycerol. These lipids are probably required for the maintenance of the highly organized structures present in the chloroplast, and their fatty acid compositions may be important only in providing molecules of a required geometry.

Haverkate (1965) has suggested that the phosphatidyl glycerol molecule is involved in the maintenance of membrane structure and that the *trans*-$\Delta^3$ hexadecenoic acid provides some special molecular geometry. We have considered the possibility that the membrane in question might be that surrounding the total chloroplast because of the absence of these membranes from blue-green algae which do not contain *trans*-$\Delta^3$ hexadecenoic acid. However, the proplastid bodies of etiolate leaves are also lacking in this acid and electron microscopy indicates that these particles are already surrounded by the membrane which eventually encloses the chloroplasts of green tissue and there is no evidence that this membrane undergoes any alteration during the light-induced transition of the crystalline proplastid structure into lamellae. There is therefore no direct evidence indicating which of the chloroplast membranes, if any, contain the phosphatidyl glycerol molecule although the concomitant synthesis of this lipid with that of chlorophyll and the formation of chloroplast lamellae suggests that it may be involved in the formation of lamellae membranes. Other evidence suggesting the importance of the phosphatidyl glycerol molecule in the formation of lamellae structures is the rapid synthesis of this molecule when dark-grown *Chlorella* is illuminated (J. A. Miller, quoted by Benson, 1963) and the rapid labelling and turnover of the fatty acids in this molecule during the first 8 hr of photosynthetic growth, which greatly exceeds that of the other lamellae lipids (B. W. Nichols, and A. T. James, unpublished work).

The excellent work of Haverkate (1965) has shown that the phosphatidyl glycerol of spinach leaves can be divided into at least three, possibly five, families of distinctive fatty acid composition (Fig. 2). Of these, a possible relationship between forms $\alpha$ and $\delta$ is suggested. The most notable effect of light on dark-grown photosynthetic tissue is the synthesis of the *trans*-hexadecenoic acid and an increased rate of synthesis of $\alpha$-linolenic acid. We know that the former fatty acid is produced by direct desaturation of palmitic acid (Nichols, Harris and James, 1965) whilst the latter is synthesized by desaturation of linoleic acid (Harris and James, 1965), and these precursor acids have been shown to be present together in the $\delta$-phosphatidyl glycerol by Haverkate. Possibly in dark-grown tissue phosphatidyl glycerol is in equilibrium with a pool rich in palmitic and linoleic (or oleic) acids which upon illumination of the cells are desaturated to form the hexadecenoic and octadecatrienoic acids. This desaturation could not take place directly on the phosphatidyl glycerol mole cule since the more unsaturated acids are esterified to the glycerol in the reverse positions from their precursors.

## Fine Structure of Flower Petals

We have been studying recently the fate of lipids when chloroplasts degenerate both during senescence and during the ripening of fruits and the maturation of flower petals. Steffen and Walter (1958) and Frey-Wyssling and Kreutzer (1958) have shown that the structures characteristic of the yellow plastids of Buttercup and Nightshade are homogeneous osmiophilic globuli of up to 1500 Å in diameter which first appear in young chloroplasts or leucoplasts. They are formed between lamellae and whilst increasing in size and number they destroy the lamellar structure until at maturity only these droplets remain lining the inner surface of the plastid membrane. In the dark yellow petal of Narcissus (Golden Harvest), however, electron microscopy has revealed large numbers of particles apparently consisting of concentric membranes and which seem to be derived from chloroplasts by a process involving the breakdown and rearrangement of chloroplasts by the sequence of events illustrated in Fig. 3. Thus while an immature green petal shows a preponderance of chloroplasts the mature yellow petal possesses the poly-membranous structure almost exclusively.

A study of the lipids of the mature petals of both buttercup and daffodil shows the absence of the *trans*-hexadecenoic acid and a reduced linolenic acid content, so that these acids appear to be lost along with chlorophyll during the maturation of the flower. During the maturation period, petal tissue readily incorporates labelled acetate into the lipid fatty acids, but the mature yellow

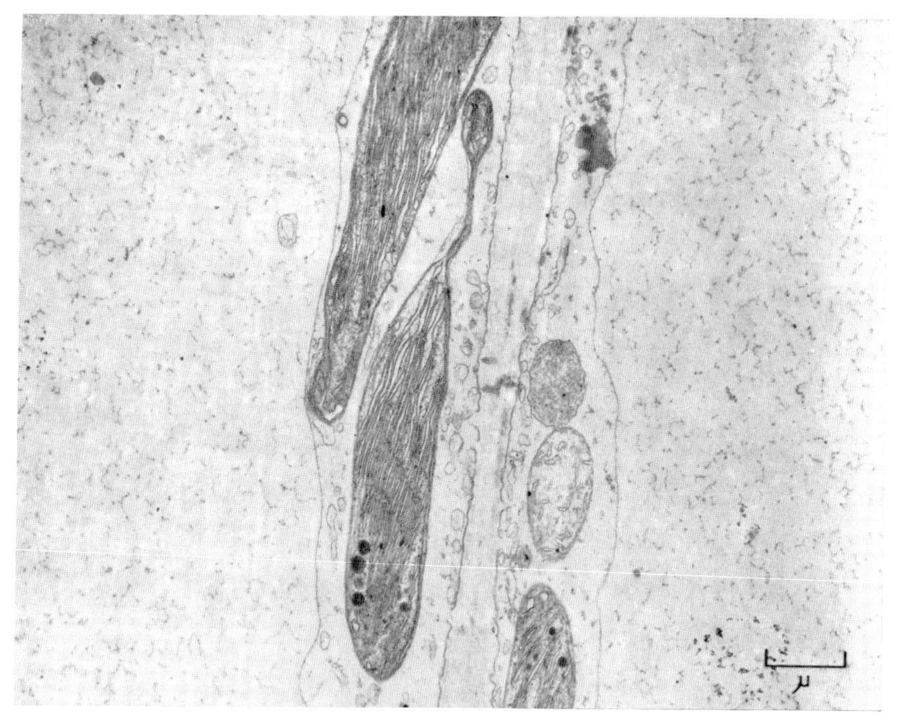


FIG. 3a

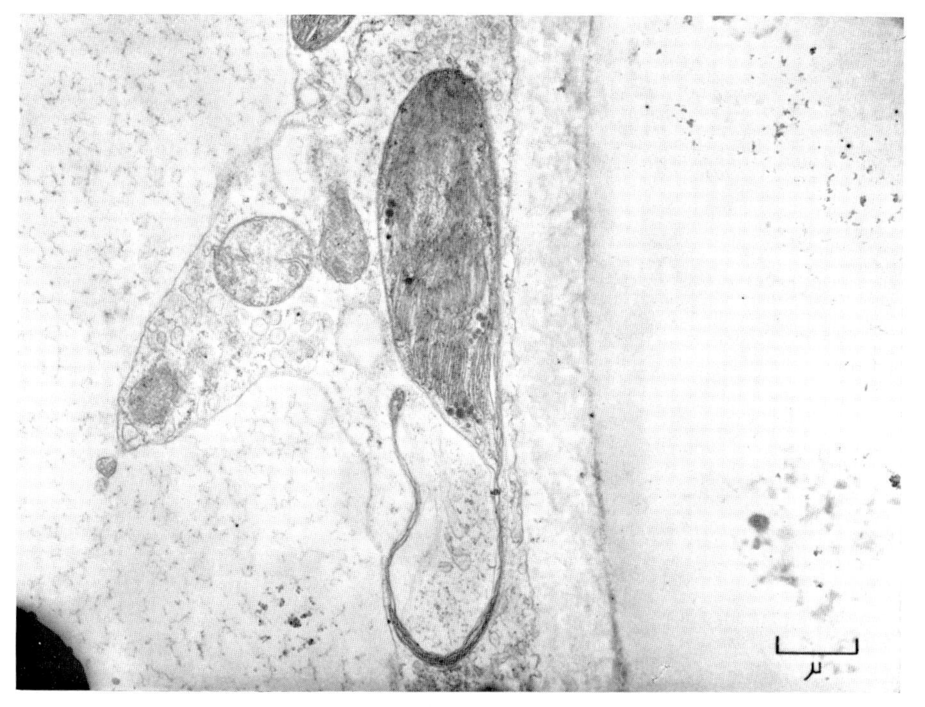


FIG. 3b (See legend on p. 689)

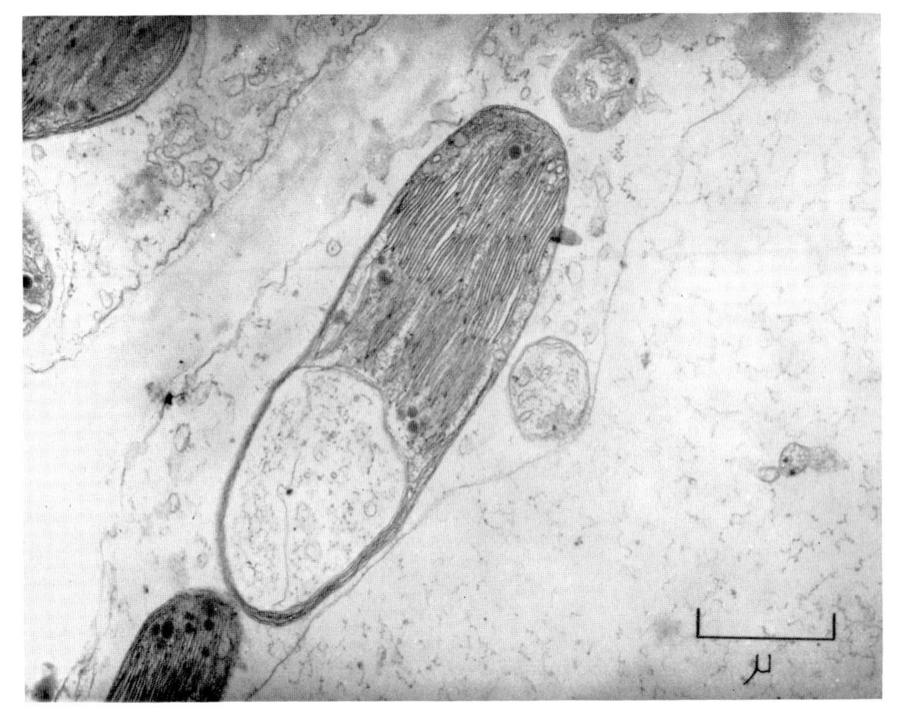


FIG. 3c

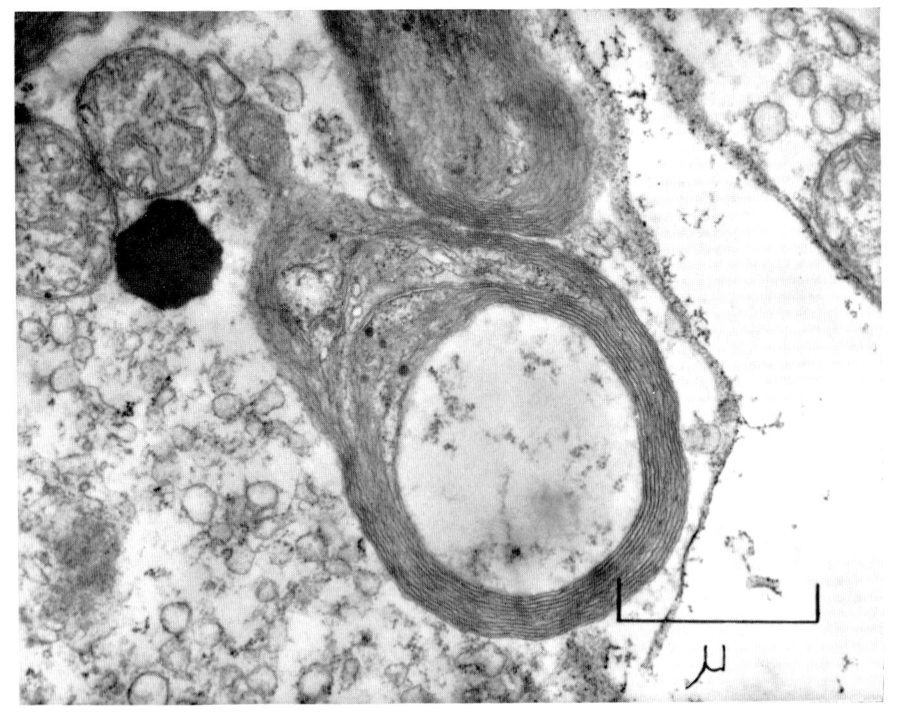


FIG. 3d (See legend on p. 689)

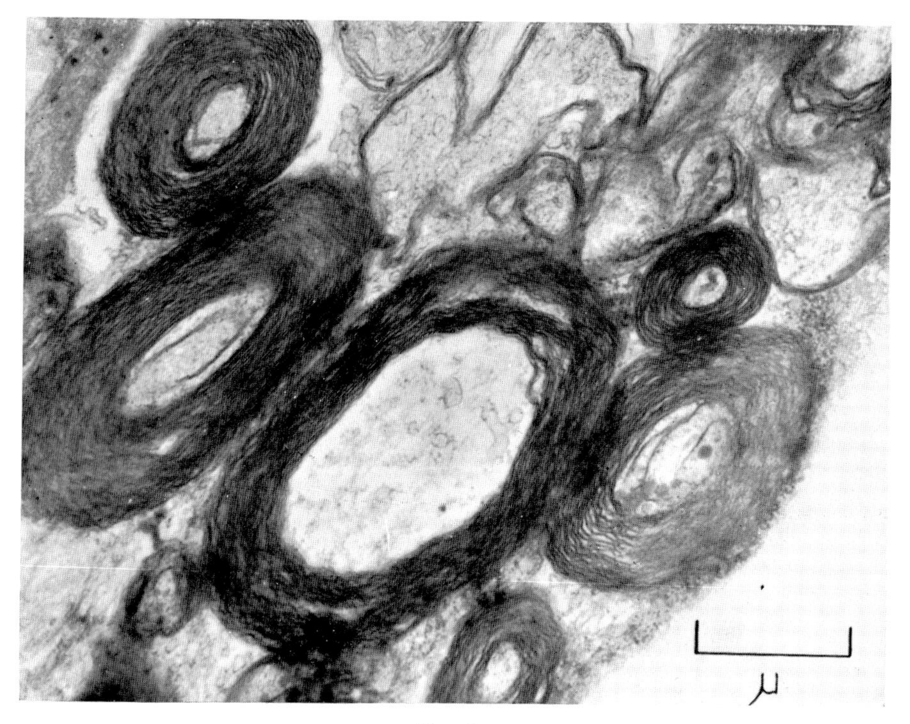


FIG. 3e

FIG. 3. The formation of plastid structures from chloroplasts in the trumpet of Daffodil (Golden Harvest). The chloroplasts develop protuberances (a), which lengthen and turn back towards the parent body (b), and finally fuse with it (c). General breakdown of the chloroplast lamellae then occurs with a concomitant thickening of the newly formed ring structure (d), until no trace of lamellae remains and the final particle consists solely of numerous (up to 25) concentric membranes (e). Tissues fixed in osmic acid (Palade's fixative), embedded in epon 812, sectioned, and stained in lead citrate solution.

petal containing the poly-membranous particles seems to be metabolically inactive and utilizes acetate only very slowly.

Consequently, the function of lipids in photosynthesizing and related systems would seem to be primarily one of maintenance of cellular structure, and the nature of the fatty acids present in these lipids is important mainly from the standpoint that they can help provide the required spatial configuration. This concept is supported by the fact that changes in lipid and fatty acid composition are greatest when changes in cellular structure occur.

## CONCLUSIONS

The work described here does not support the concept of involvement of any unsaturated fatty acid in the major pathways of photosynthesis, since the two such fatty acids whose biosynthesis is either light-induced or potentiated by light are absent from blue-green algae.

Phosphatidyl glycerol, the mono- and di-galactosyl diglycerides and the plant sulpholipid are found in all photosynthetic tissues capable of the Hill reaction but they also occur, though in different ratios, in protoplastids and "degenerated" chloroplasts whose fine structure is quite different from that of the mature chloroplast. Thus any specific structure-determining capacity of these four lipids must be due to their presence in a particular ratio.

There is, however, still the possibility that phosphatidyl glycerol containing *trans*-$\Delta^3$ hexadecenoic acid might be an important element in such a complex as the chloroplast lamellar membrane.

## References

Allen, C. F., Good, P., Davis, H. F. and Fowler, S. D. (1964). *Biochem. biophys. Res. Commun.* **15**, 424.
Benson, A. A. (1963). *In* "Mechanisms of Photosynthesis" (H. Tamiya, ed.). Pergamon Press, London.
Benson, A. A. (1964). *Annu. Rev. Pl. Physiol.* **15**, 14.
Benson, A. A., Wintermans, J. F. G. M. and Wise, R. (1959). *Pl. Physiol.* **34**, 315.
Debuch, H. (1961). *Z. Naturf.* **16**, 561.
Erwin, J., Hulanicka, D. and Bloch, K. (1964). *Comp. Biochem. Physiol.* **12**, 191.
Ferrari, A. A. and Benson, A. A. (1961). *Archs Biochem. Biophys.* **93**, 185.
Frey-Wyssling, A. and Kreutzer, E. (1958). *Planta* **51**, 104.
Harris, R. V. and James, A. T. (1965). *Biochim. biophys. Acta* **106**, 456.
Haverkate, F. (1965). Thesis "Phosphatidyl glycerol from Photosynthetic Tissues", Univ. of Utrecht.
Haverkate, F., de Gier, J. and van Deenen, L. L. M. (1964). *Experientia* **20**, 511.
Hilditch, T. P. and Meara, M. L. (1944). *J. Soc. Chem. Ind.* **63**, 112.
Holton, R. W., Blecker, H. H. and Onore, M. (1964). *Phytochemistry* **3**, 595.
Klenk, E. and Knipprath, W. (1962). *Z. Phys. Chem.* **327**, 283.
Levin, E., Lennarz, W. J. and Bloch, K. (1964). *Biochim. biophys. Acta* **84**, 471.
Lichtenthaler, H. K. and Park, R. B. (1963). *Nature, Lond.* **198**, 1070.
Nichols, B. W. (1963). *Biochim. biophys. Acta* **70**, 417.
Nichols, B. W. (1965a). *Phytochemistry* **4**, 769.
Nichols, B. W. (1965b). *Biochim. biophys. Acta* **106**, 274.
Nichols, B. W. Harris, P. and James, A. T. (1965). *Biochem. biophys. Res. Commun.* **21**, 473.
Nichols, B. W., Harris, R. V. and James, A. T. (1965a). *Biochem. biophys. Res. Commun.* **20**, 256.
Nichols, B. W., Wood, B. J. B. and James, A. T. (1965b). *Biochem. J.* **95**, 6P.
O'Brien, J. S. and Benson, A. A. (1964). *J. Lipid Res.* **5**, 434.
Sastry, P. S. and Kates, M. (1964). *Biochemistry* **3**, 1271.
Steffen, K. and Walter, F. (1958). *Planta* **50**, 640.
Tsujimoto, M. (1940). *J. Soc. Chem. Ind. Japan* **43**, 208.
Weenink, R. O. and Shorland, F. B. (1964). *Biochim. biophys. Acta* **84**, 613.
Wood, B. J. B., Nichols, B. W. and James, A. T. (1965). *Biochim. biophys. Acta* (In Press).

## Discussion

*D. I. Arnon:* We have been considering that the oxygen-evolving type of photosynthesis depends on the presence of chlorophyll *c* or some other "accessory" pigment in addition to chlorophyll *a*. In your talk, the oxygen-evolving type of photosynthesis has been linked to the presence of the four lipids. Is there any evidence that the presence of the four lipids in oxygen-evolving photosynthetic structures is more than a coincidence?

*B. W. Nichols:* The evidence for the essential participation of the four lipids in oxygen-evolving photosynthetic structures is at the moment circumstantial. It is possible that the ubiquitous presence of these lipids in such structures is only of evolutionary significance.

# The Ultrastructure of the Chromoplasts of Different Colour Varieties of *Capsicum*

J. T. O. Kirk
*Department of Biochemistry and Agricultural Biochemistry, University College of Wales, Aberystwyth, Wales*

B. E. Juniper
*Botany School, Oxford, England*

## Introduction

Chromoplasts, a form of differentiated plastid containing carotenoid pigments, are responsible for the yellow, orange or red colours of many kinds of fruit, flower petals and certain roots. Chromoplasts may develop from chloroplasts or amyloplasts and in the red pepper (*Capsicum annuum*) the chromoplasts of the fruit are formed from chloroplasts during ripening. Frey-Wyssling and Kreutzer (1958) have studied the ultrastructure and development of these chromoplasts. They reported that the mature plastids contained bundles of submicroscopic filaments: these filaments were thought to contain the carotenoids of the chromoplasts. Steffen and Walter (1958) found similar structures in the chromoplasts of the fruit of *Solanum capsicastrum*.

Different colour varieties of *Capsicum* fruit exist and in one of these Smith (1950) has found that the orange-yellow colour of the ripe fruit is controlled by a single gene, the yellow allele of which is recessive to the normal red allele. Different colour varieties must have different carotenoid compositions. We have set out to see what effect differences in carotenoid composition have on the fine structure of the chromoplasts. We have collected four colour varieties of *Capsicum*, red, orange, yellow and white, and in this paper we describe electron microscopical studies on the chromoplasts, and measurements of the carotenoid contents, of these different types of fruit.

## Methods

Three of the colour varieties, red, orange and yellow, were selected from seed obtained from S. Dobie and Son, Ltd., Chester, under the name "Ornamental Pepper (Capsicum)—Tall mixed". The white variety was kindly given to us by Mr. J. K. Burras, Superintendent of the Oxford Botanic Garden. The plants were grown in a greenhouse and fruits taken when ripe. Small pieces of the ripe fruit tissue, just below the skin were taken for electron microscopy, fixed at 0–2° for 2 hr in 2% glutaraldehyde buffered to neutrality with cacodylate, washed in pure buffer and post-fixed in 2% $OsO_4$ for 1 hr also at 0–2°. The tissue was then dehydrated through a progressive alcohol series, washed in propylene oxide and embedded in Epikote 812 resin. The blocks were cut on a

Cambridge "Huxley" microtome, stained with lead citrate and photographed in an A.E.I. E.M.6 electron microscope.

For the pigment analyses, an appropriate portion (0·182 g red, 0·312 g orange, 0·905 g yellow, and 1·018 g white) of pericarp was taken from the tip of the fruit, minced finely with scissors, ground for 1–2 min in a mortar with a pinch of $CaCO_3$, and then ground for a total of 10 min with two successive 3 ml quantities of isopropanol (Analar grade). The isopropanol supernatants were decanted into a centrifuge tube and the residue was then ground for 2 min with 2 ml of hexane (Hopkin and Williams "Spectrosol" for u.v. spectroscopy). In the case of the white tissue two further extractions with hexane were carried out. The hexane supernatants (plus further hexane to 5 ml) were combined with the isopropanol fraction. Enough water was added to cause separation of two layers and to drive all the pigments into the upper, hexane layer. After a brief centrifugation to ensure clean separation of the layers the hexane layer was quantitatively transferred to another centrifuge tube. The hexane fraction was washed with three successive 5ml quantities of water to remove isopropanol and was then dried with anhydrous $Na_2SO_4$. After appropriate adjustment of volume, the spectra were read against a hexane blank on the Perkin-Elmer 137 u.v. Spectrophotometer.

An accurate determination of the carotenoid content of any of these peppers would require complete separation, identification and estimation of each of the individual carotenoid components in each of the different colour varieties. However, very approximate values for the relative amounts of carotenoid in the different types of fruit may be obtained from the optical densities of the different extracts at their main absorption peaks. Data on the carotenoid composition of red *Capsicum* fruit are available (Cholnoky *et al.*, 1956; Curl, 1962), and on the basis of these data a very rough value of 1900 has been adopted for $E^{1\%}_{1cm}$ at 474 m$\mu$, for the red pepper carotenoid mixture in hexane.

Leaf carotenoids were extracted and saponified as described by Goodwin (1955). They were transferred to diethyl ether for determination of absorption spectra.

## Results

### Red Fruit

The chromoplasts in this particular variety of *Capsicum* are roughly ellipsoidal in shape; they do not have sharp, needle-like processes extending from them as do the chromoplasts in certain other *Capsicum* varieties. Like other plastids, the chromoplasts are bounded by a double membrane. This can be clearly seen in Fig. 1 (m), which shows two chromoplasts lying next to each other. The two most common types of structure within the chromoplast are fibres and osmiophilic globules (f and og, Figs. 1 and 2). Most of the fibres lie packed fairly close together, and parallel to one another in bundles. Figure 3 shows fibres cut transversely (tf) and obliquely (of). The fibres have diameters between 150 and 450 Å. Some of the fibres may be seen to have, either at the end, or somewhere in the middle, a swelling up to about 1000 Å in diameter (s, Fig. 2). The length of the fibres is somewhat difficult to ascertain since they are liable to pass out of the section. However fibres with lengths up to about

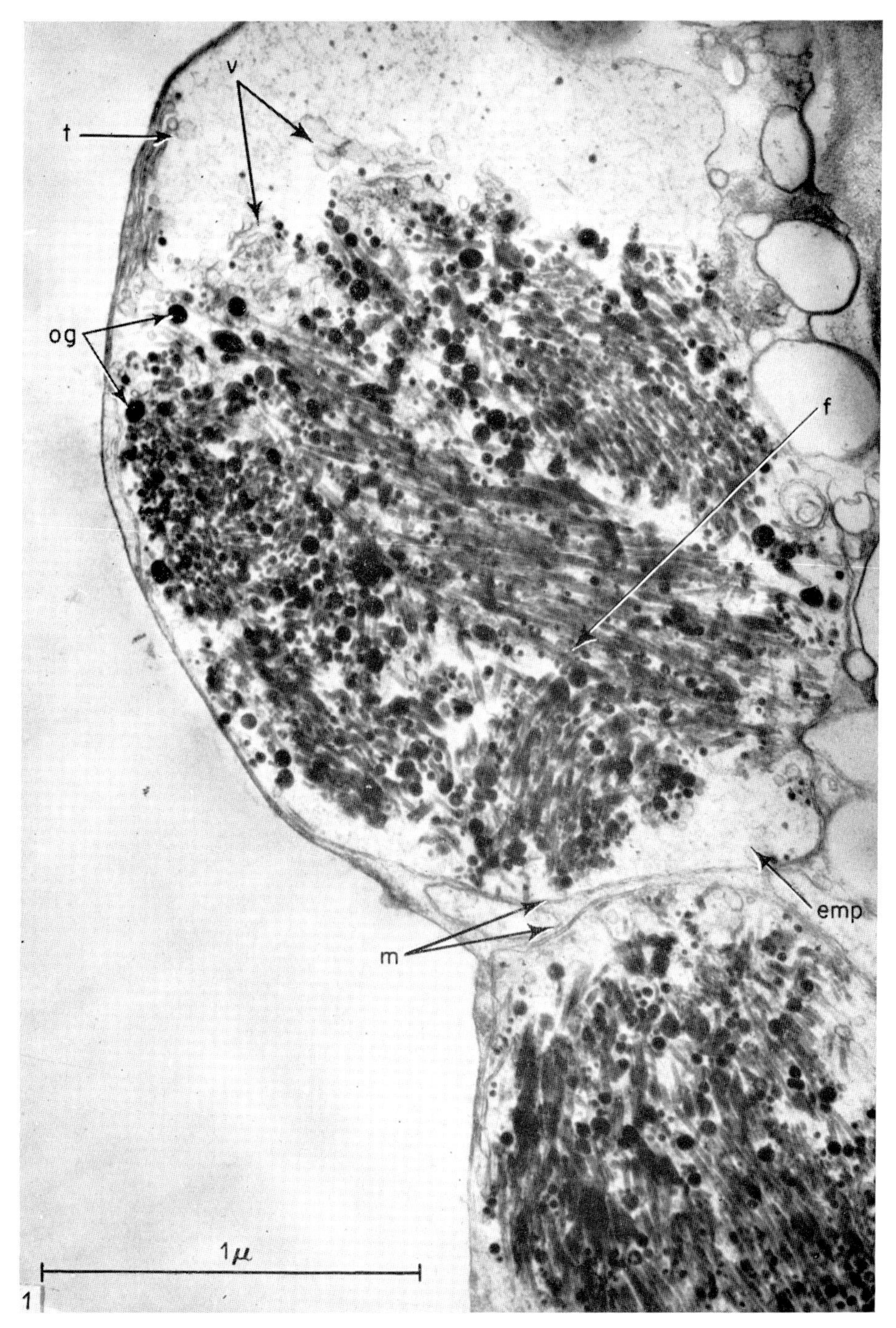


FIG. 1. Two chromoplasts in red fruit. m—double membrane bounding chromoplast. f—fibres. og—osmiophilic globules. v—vesicles. t—"tubules". emp—empty region.

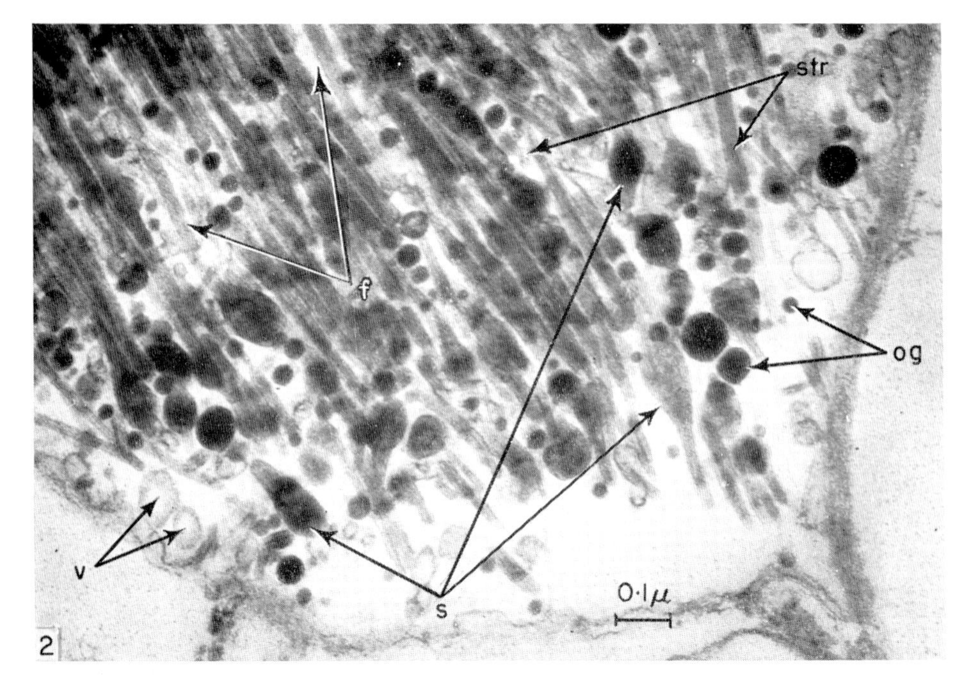


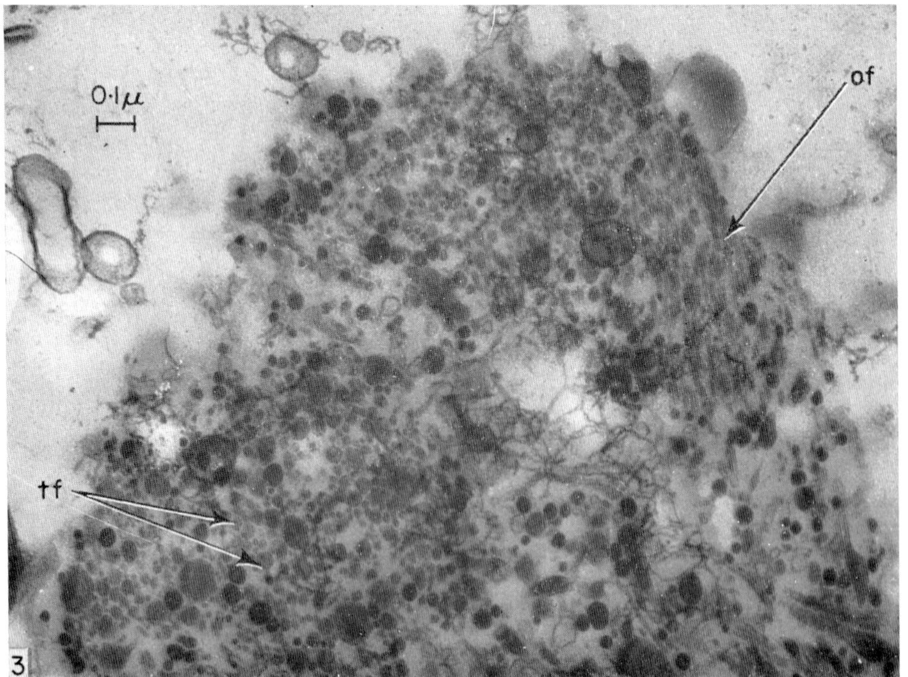


FIG. 2. Chromoplast of red fruit. s—swelling on fibre. str—striations along fibre. v—vesicles.

FIG. 3. Chromoplast of red fruit. tf—traversely-cut fibres. of—obliquely-cut fibres.

1·4 $\mu$ (14,000 Å) have been seen. On close inspection, the fibres may be seen to have faint longitudinal striations (str, Fig. 2); 3 or 4 dark lines, roughly 50 Å thick and 50–100 Å apart may be seen running along the fibre.

The osmiophilic globules have diameters of anything from 100–1000 Å. They are distributed more or less at random throughout the same region of the chromoplast which contains the fibres (Fig. 1), being commonly found within, as well as between, fibre bundles.

The chromoplasts also contain a number of membrane-bounded vesicles of irregular shape and variable size (v, Figs. 1 and 2). Short lengths of what appear in section to be tubules, or flattened sacs, are sometimes seen (t, Fig. 1). Most of the chromoplast volume is filled with fibres and globules but between the mass of fibres and globules and the chromoplast membrane there is a rather empty region (emp, Fig. 1) containing a certain amount of faint granular and fibrous material.

## ORANGE FRUIT

The chromoplasts of orange-coloured fruit differ from those of the red fruit in two main respects. First of all, no fibres are present (Fig. 4). However, there is still a large number of osmiophilic globules. Most of these occur within the same size range as the globules of the red chromoplasts, but occasionally very large ones, about 2000 Å in diameter, are seen. The globules are all clumped together; anything up to five separate clumps may be seen in one chromoplast. In general, the clump, or clumps, appear to occupy appreciably less than half the volume of the chromoplast.

The second main difference is that the orange chromoplasts contain membranous structures (t, Fig. 4) having the appearance in section of long tubules: in places these appear to break up into a series of vesicles. An alternative interpretation is that these structures are double-membraned sheets, and that the "vesicular" regions are places where there are holes through the sheets. The "tubules" follow no set path through the chromoplast: a given "tubule" may extend most of the way round a chromoplast; another may extend part of the way round and then double back on itself. These "tubules" are much longer than those seen in the red chromoplasts; however, it does seem possible that "tubules" present in the red chromoplasts might be hidden by the large amount of globules and fibres present.

Many orange chromoplasts also contain large granules similar in appearance to starch grains.

## YELLOW FRUIT

The chromoplasts of the yellow fruit are very similar in appearance to those of the orange fruit (Fig. 5). However, yellow chromoplasts with more than one clump of globules have not yet been seen.

## WHITE FRUIT

The "chromoplasts" of white fruit are similar to those of orange and yellow fruit to the extent that they contain osmiophilic globules but no fibres, and

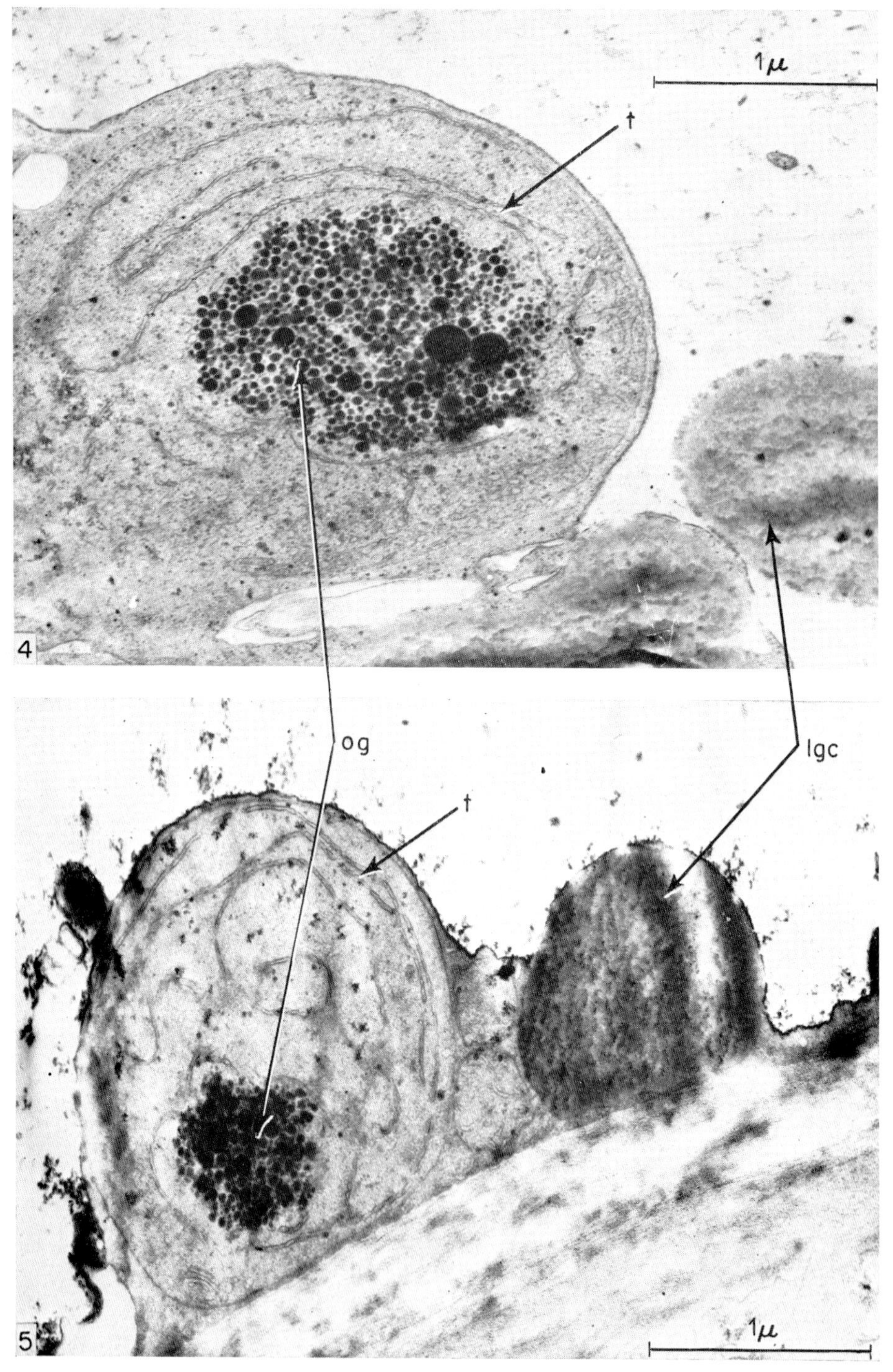


FIG. 4. Chromoplast of orange fruit. og—osmiophilic globules. t—"tubules". lgc—large granules in cytoplasm.

FIG. 5. Chromoplast of yellow fruit. og—osmiophilic globules. t—"tubules". lgc—large granule in cytoplasm.

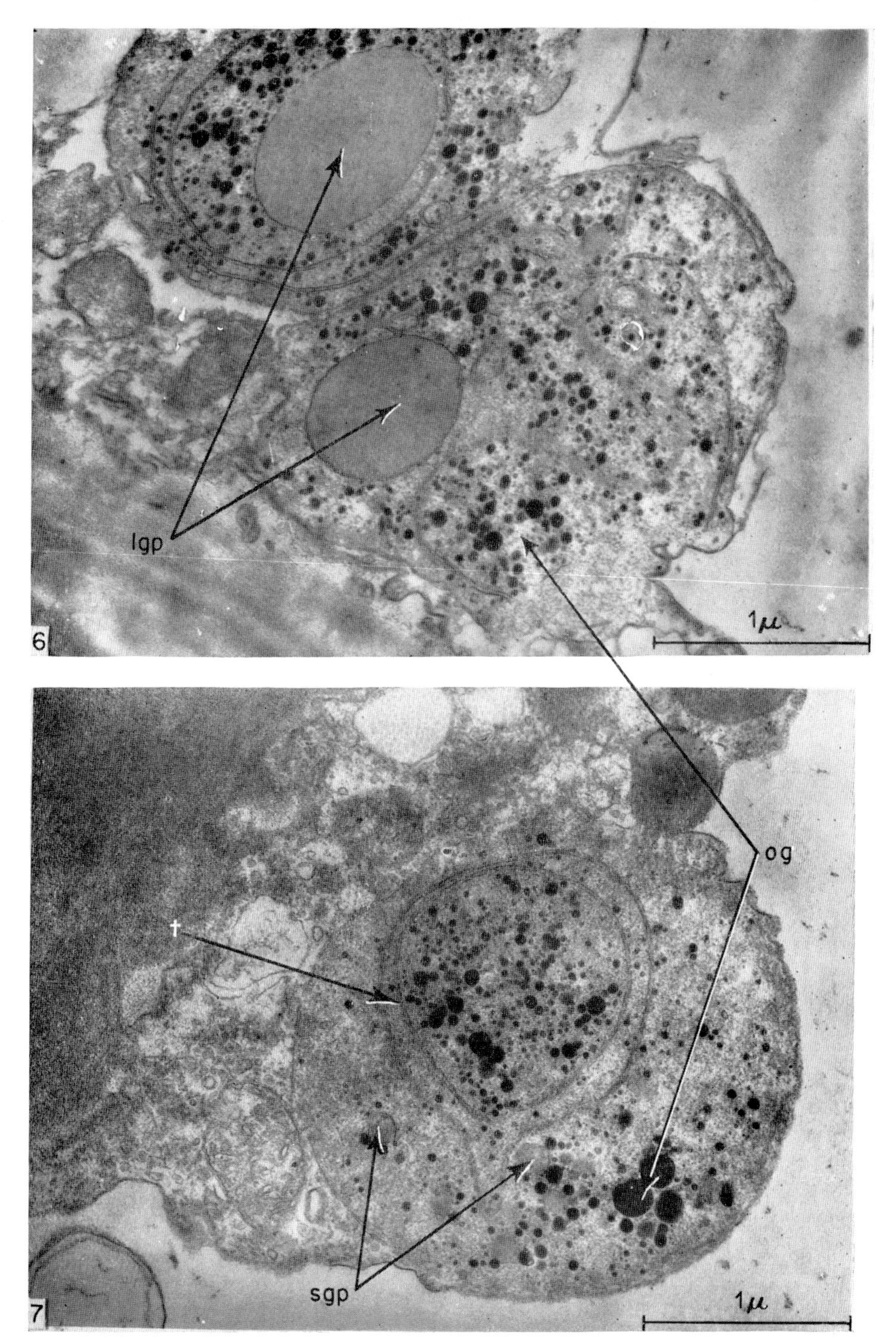


Fig. 6 and Fig. 7. Plastids of white fruit. og—osmiophilic globules. t—"tubules". lgp—large granules in plastids. sgp—small granules in plastids.

they also contain long tubules (t, Figs. 6 and 7). However, they differ from orange and yellow chromoplasts in that the globules in the white fruit plastids show no tendency to clump together: they seem to be distributed at random throughout the plastid.

Another difference is that many of the white fruit plastids contain a large egg-shaped granule about 1 $\mu$ (10,000 Å) long (lgp, Figs. 6 and 7); this stains lightly, but uniformly, with osmium and appears to be bounded by a membrane which in some places seems to be double. So far, not more than one of these large granules has been seen in any given plastid. These plastid granules are rather similar in size and general appearance to large granules which can be seen in the cytoplasm in all four fruit types (lgc, Figs. 4 and 5). Whether there is any relationship between these two kinds of large granules, or what their chemical nature is, is not known: they do not look very much like starch grains. The white fruit plastid also contains several granules of a type similar in appearance to the large granules, but much smaller, being about 1000–1500 Å in diameter (sgp, Fig. 7).

## CAROTENOID CONTENT OF FRUIT

The absorption spectra of the pigments of the four different kinds of fruit, in the visible and ultraviolet ranges are shown in Figs. 8 and 9. The red fruit have a carotenoid content somewhere in the region of 0·9 % of the fresh weight. The orange and the yellow types of fruit have, as might be expected, very much

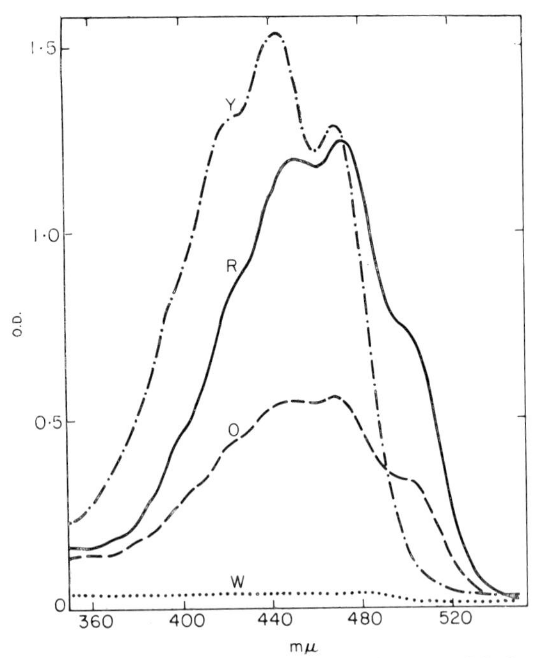


FIG. 8. Visible absorption spectra of *Capsicum* fruit pigments. Mg fresh weight of fruit corresponding to 1 ml of the final hexane solution: Red—7·3; Orange—113; Yellow—274; White—290.

less carotenoid than the red type; they both, in fact, appear to have something like one-thirtieth of the carotenoid level of the red fruit. The white fruit appear to have no coloured carotenoids.

The absorption spectrum of the pigments of the orange fruit in the visible range is quite similar to that of the pigments of the red fruit. The absorption spectrum of the pigments of the yellow fruit, on the other hand, is quite different from that of the pigments of the red fruit, suggesting a radically

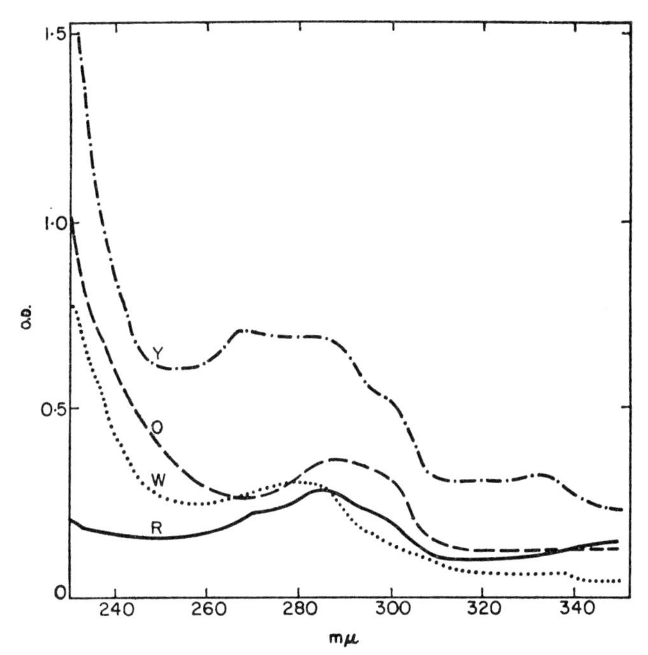


FIG. 9. Ultraviolet absorption spectra of *Capsicum* fruit pigments. Mg fresh weight of fruit corresponding to 1 ml of the final hexane solution: Red—7·3; Orange—113; Yellow—274; White—290.

different carotenoid composition. In the ultraviolet range, extracts of all four types of fruit have a rather diffuse peak, or combination of peaks, in the 260–310 mμ range.

## DISCUSSION

The formation of fibres in the red chromoplasts is associated with the synthesis of the characteristic chromoplast carotenoids: it therefore seems likely that the fibres contain at least some of these carotenoids. Since the orange and yellow chromoplasts contain osmiophilic globules but no fibres, then presumably the carotenoids are present in the globules. By analogy, we would therefore expect the osmiophilic globules of the red chromoplasts also to contain carotenoids. It might be possible to determine whether the globules and the fibres contain the same carotenoids by physically separating them and analysing them.

The swellings seen on some fibres are similar in size and general appearance to the osmiophilic globules with which they are mingled. It thus seems likely that, as suggested by Steffen and Walter (1958) for *Solanum* chromoplasts, the fibres actually develop from the osmiophilic globules, and that the fibres with swellings are, in fact transitional states in which the change from globule to fibre is not complete. In the absence of knowledge of the chemical difference, if any, between globules and fibres, it is impossible to say what the biochemical process underlying the structural change, is. The fibres might, for instance, consist of a carotenoid–protein complex, in which case conversion of globules to fibres might require the synthesis of a protein. In this connection, Frey-Wyssling and Kreutzer (1958) have already suggested that the fibres contain protein.

The determination of the nature of the biochemical lesion in the orange and yellow fruit must await a complete analysis of the carotenoids present. We are working on this at the moment. However, on the basis of the absorption spectra one might, perhaps, speculate that the orange fruit have similar carotenoids to the red, but in much smaller quantity; that is, there might be a low amount of an enzyme required for some step prior to the formation of the characteristic fruit carotenoids. The spectrum of the yellow fruit pigments, on the other hand, is quite different from that of the orange and red fruit, and in particular suggests that capsanthin and capsorubin are lacking. The yellow fruit might perhaps have a block in the capsanthin–capsorubin pathway. The white fruit apparently make no coloured carotenoids at all, and apart from a very low, diffuse, band in the 260–300 m$\mu$ region there is little indication of accumulation of colourless carotenoid precursors such as phytoene or phytofluene. The white fruit, therefore, presumably lack an enzyme or enzymes required for some very early stage in fruit carotenoid synthesis.

One of the interesting things about these *Capsicum* varieties with orange, yellow and white fruit is that despite their defects in fruit carotenoid synthesis, they can all make substantial quantities of leaf carotenoids, as the spectra in Fig. 10 show.[1] This is not so surprising in the case of the varieties with orange and yellow fruit, because the fruit can still, in fact, synthesize carotenoids. It is more surprising in the case of the white-fruited variety which appears to synthesize no fruit carotenoids. A possible explanation of this might be that fruit carotenoid synthesis involves a different series of genes, and hence of enzymes, from leaf carotenoid synthesis. An alternative, and more plausible explanation might be that the white fruit has a defect, not in a structural gene for a biosynthetic enzyme, but in a gene regulating fruit carotenoid synthesis. For instance, the regulating gene in the white fruit might form a faulty repressor, which could still repress the structural gene(s) for synthesis of the biosynthetic enzyme but which could no longer combine with whatever compound derepresses, or induces, fruit carotenoid synthesis.

The lack of fibres in the orange and yellow chromoplasts is, no doubt, related in some way to the difference between their carotenoid compositions and that of the red chromoplasts. Fibre development might, for instance, require the

[1] A similar observation was previously made by Mackinney, Rick and Jenkins (1956) on tomato varieties with different fruit carotenoid composition.

formation of a certain minimum level of carotenoids. A puzzling feature of the white fruit plastids is the presence of osmiophilic globules despite the lack of carotenoids: in this case the globules presumably consist of some other unsaturated lipid material. It is interesting that the globules in the white fruit plastids show no signs of aggregating together as they do in the orange

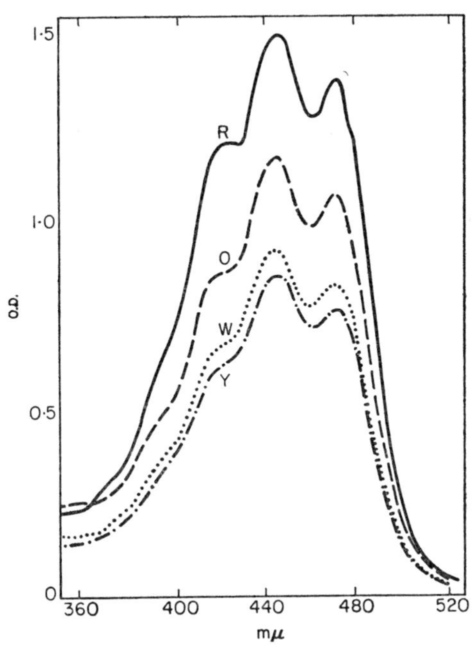


FIG. 10. Visible absorption spectra of *Capsicum* leaf carotenoids. Mg fresh weight of leaf corresponding to 1 ml of the final ether solution: Red-fruited type—24·2; Orange-fruited type—34·2; Yellow-fruited type—16·8; White-fruited type—17·9.

and yellow chromoplasts: this might perhaps be related to differences in the surface chemistry. The nature, origin and function (if any) of the long "tubules" in the orange, yellow and white fruit plastids are unknown.

## ACKNOWLEDGEMENT

The authors are indebted to the Science Research Council for financial support.

## References

Cholnoky, L., Györgyfy, C., Nagy, E. and Panczel, M. (1956). *Nature, Lond.* **178**, 410.
Curl, A. L. (1962). *J. Agric. Fd. Chem.* **10**, 504.
Frey-Wyssling, A. and Kreutzer, E. (1958). *J. Ultrastr. Res.* **1**, 397.
Goodwin, T. W. (1955). *In* "Modern Methods of Plant Analysis" (K. Paech and M. V. Tracey, eds.), Vol. III, p. 272. Springer, Heidelberg.
Mackinney, G., Rick, C. M. and Jenkins, J. A. (1956). *Proc. natn. Acad. Sci. U.S.A.* **42**, 404.
Smith, P. G. (1950). *J. Hered.* **41**, 138.
Steffen, K. and Walter, F. (1958). *Planta* **50**, 640.

# Resemblances Between Chloroplasts and Mitochondria Inferred from Flagellates Inhibited with the Carcinogens 4-Nitroquinoline *N*-Oxide and Ethionine[1]

S. H. Hutner, A. C. Zahalsky and S. Aaronson
*Haskins Laboratories, New York, New York, U.S.A.*

and Robert M. Smillie
*Plant Physiology Unit, School of Biological Sciences, University of Sydney, Sydney, Australia*

## Introduction

Molecular patterns and macromorphology hint at fundamental similarities between mitochondria and muscle fibers (Lehninger, 1964). This parallel is being extended to the chloroplast in such provocative detail that Duysens (1964) wonders whether "... the respiratory hydrogen transport chain [is] a modification of part of the photosynthetic hydrogen transport chain that generates ATP." Parallels between chloroplasts and mitochondria, notably in possession of a DNA–RNA–protein system, have been assembled (Gibor and Granick, 1964). Along these lines, Lieberman and Baker (1965) speculate that mitochondria were derived from microbial chloroplasts and have retained the "... reversed electron transfer and phosphorylation derived from primitive chloroplasts at the reduced end of the respiratory electron transport chain."

This paper describes explorations of the mitochondrion–chloroplast parallel, with emphasis on the chloroplast as an organelle strikingly sensitive to inhibition by certain carcinogens. Three kinds of experiments are joined: analysis of the disorder in electron flow in chloroplasts induced by 4-nitroquinoline *N*-oxide (NQO) (experiments done mainly by R. M. Smillie); cross-resistance studies based on the nutrition of intact organisms; and respirometric studies of resistant organisms.

## The Organisms: *Ochromonas danica* and *Euglena gracilis*

The main experimental organism, *Ochromonas danica*, is a voracious, vigorously photosynthetic chrysomonad flagellate. It avidly ingests particles, i.e., it is a phagotroph and hence by T. H. Huxley's criterion an animal (Hutner, 1961). Its permeation by dissolved materials, presumably as a concomitant of animality, exceeds that of photosynthetic bacteria, the common

[1] This work was aided by grants from the National Institutes of Health (GM 09103) and the Damon Runyon Memorial Fund for Cancer Research (DRG-827). Some of the early experiments on cell-free preparations were done at the Brookhaven National Laboratories. We are indebted to Dr. John M. Olson of Brookhaven for help in the experiments in which the Chance-type spectrophotometer was used.

green algae, and common fungi in respect to utilization of a variety of substrates and of quantitatively minor nutrients—a feature spelling advantages for discerning the mode of action of cytotoxic compounds (Aaronson *et al.*, 1964). Yet it can grow in inorganic media, carrying out a full photosynthetic $CO_2$ reduction aside from the minute dependence on fixed carbon represented by its absolute thiamine and biotin requirements. Gibbs (1962) observed that dark-grown cells had only 1–2% as much chlorophyll and 4–7% as much carotenoids as light-grown cells and no protochlorophyll was detected; it grew about as well in the dark as in the light on particle-free, rich media (generation time at 26°: 14 hr in the dark, 15 hr in light). The cells are easily broken up in a French press or by sonication (Ke, 1964). Heterotrophic (dark) growth, as implied by Gibbs' (1962) data, is as good as photosynthetic growth, given good aeration. It grows at neutrality or higher, also at pH's of 3·0 or lower (Table I). At these strongly acid media, auto-oxidation is lessened; free radicals, e.g. semiquinones, are more stable; and repression of ionization enhances the penetration of acidic inhibitors—an important consideration

Table I. *Ochromonas danica*: pH 3·6 medium (weight/100 ml final medium)

| | | | |
|---|---|---|---|
| DL-Malic acid | 0·04 g | Fe | 0·10 mg |
| $KH_2PO_4$ | 0·03 g | Mn | 0·30 mg |
| $MgCO_3$ | 0·04 g | Zn | 0·30 mg |
| $MgSO_4.3H_2O$ | 0·1 g | Mo | 0·037 mg |
| L-Glutamic acid | 0·3 g | Cu | 0·018 mg |
| $Na_2$ succinate . $6H_2O$ | 0·01 g | B | 0·0037 mg |
| $NH_4Cl$ | 0·05 g | V | 0·0037 mg |
| Glucose | 1·0 g | Thiamine HCl | 1·0 mg |
| $CaCO_3$ | 0·005 g | Biotin | 1·0 μg |

with NQO or the expensive alkyl hydroxyquinone *N*-oxides as lower concentrations are effective.

As with *Euglena*, the reversible light–dark, appearance–disappearance of the photosynthetic apparatus permits near certainty about whether a compound is associated with the photosynthetic apparatus. Thus Miyachi *et al.* (1966) showed that light-grown *O. danica* contained five to six times more chloroplast sulfolipid (sulfoquinovosyl diglyceride) than dark-grown cells.

Techniques for handling *O. danica* and *E. gracilis* have been detailed (Zahalsky *et al.*, 1963). The acid medium for *O. danica* (Table I) indicates the simplicity of its demands for dense heterotrophic growth. As a rough approximation, derived from exploratory surveys, its preferences for substrates are remarkably like those compiled for birds and mammals (Hutner and Provasoli, 1965). A biphasic inoculation medium (Zahalsky *et al.*, 1963) maintains vigorous cultures for at least 4 months, thus avoiding the sudden deterioration of chloroplast structure noted by Allen (1963) with rich liquid media.

The normal green Z strain of *E. gracilis* was used. Unless otherwise mentioned cultures were grown in the dark.

## Experimental Compounds

### 4-nitroquinoline *N*-oxide (NQO)

Earlier literature on the cytotoxicity of this potent carcinogen, whose structure obviously resembles the 2-alkyl-4-hydroxyquinoline *N*-oxides, has been summarized in our previous paper (Zahalsky *et al.*, 1963), our main results being as follows: NQO toxicity for *O. danica* and *E. gracilis*, and to some extent for *Rhodopseudomonas palustris* and *Corynebacterium bovis*, was annulled competitively and equally well by L-tryptophan, gramine or 5-methyl-DL-tryptophan. Secondary annullers included riboflavin, nicotinic acid, L-tyrosine, DL-phenylalanine and thymine. Naphthoquinones and reduced compounds also opposed NQO toxicity.

### ethionine (ETN)

This much-studied protein-synthesis paralyser and hepatic carcinogen becomes an ethylating agent by conversion to the *S*-ethyl homolog of adenosylmethionine. As reviewed by Farber (1963) and Farber *et al.* (1964), ETN is a methionine antagonist, also an adenine trap because of the pile-up of *S*-adenosylethionine. Furthermore, as adenine seems the rate-limiting precursor of ATP, ETN-poisoned cells tend to become ATP-depleted. ETN inhibits bacteriochlorophyll synthesis apparently by inhibiting the methyl transferase in biosynthesis (Gibson *et al.*, 1963). ETN inhibits chlorophyll synthesis in bean leaf discs (Lowther and Boll, 1960).

It was therefore of interest to see how ETN arrests photosynthesis. One approach, described later in the paper, was by testing for the development of cross-resistance between NQO- and ETN-resistant strains of *O. danica* and *E. gracilis*.

## Cross-Resistance between Nitroquinoline *N*-oxide and Ethionine

### aims

Compounds such as NQO may bridge the gap between two important kinds of carcinogens: the classical polynuclear-benzenoid tar compounds and azo dyes of the butter-yellow type, for the azo dyes are converted in the mammalian body into highly active *N*-hydroxy derivatives (Anderson *et al.*, 1964), and such typical tar carcinogens as benzpyrene are converted in the body into dihydroxy derivatives (Falk *et al.*, 1962). As there is substantial evidence that the 6-substituted, 8-aminoquinoline antimalarials are converted in the body into extremely active *o*-quinones (Alving *et al.*, 1962), it seems likely, by analogy, that the dihydroxy products of benzpyrene and the like are converted into *o*- and *p*-quinones, one or more of these products being the proximal carcinogen, interrupting the electron chain in chloroplasts and mitochondria much as does NQO. Superficially, ethionine (ETN) would appear to be different from both these classes of carcinogens. However, as noted, ethionine is converted into *S*-adenosylethionine in yeast and in higher animals. This raises the question of whether most organic carcinogens—at least in their activated forms—act primarily as alkylating or arylating agents, and if so, whether they affect the same site in the chloroplast. Should indeed the same site be attacked, a new question would arise: how equivalent or homologous is the sensitive site in the mitochondrion?

*E. gracilis* and *O. danica* grown in light in just sublethal concentrations of NQO are yellow-white. This arrest of the photosynthetic apparatus, unlike that seen with *E. gracilis* treated with streptomycin, barely sublethal high temperatures, u.v., and several other agents, is reversible. Conceivably, by finding antagonists to the lethality of NQO, the concentration range between chloroplast-arrested and killed cells might be widened, and under these conditions of diminished lethality permanent mass bleaching might be effected. We hoped also to induce "petites" in flagellates much as was done with some yeasts, thus extending the parallel betwen mitochondria and chloroplasts.

## LEVELS OF CROSS-RESISTANCE

Serial transfer of cells into increasing concentration of carcinogen served to obtain resistant cells. The following results are derived from cells grown in defined media. The same levels of resistance, in about the same number of passages was observed in the defined medium supplemented with a combination of yeast extract, liver extract, and an acid hydrolysate of casein, each at 0·01 %. At each step in the adaptation process, biphasic cultures (agar overlaid with distilled water, each phase containing the same concentration of drug) at the previous drug level were established to insure against cumulative toxicity leading to loss of strains. The levels of resistance attained (Table II) corresponded to the limits of solubility of L-ethionine, and was close to the limit for NQO. Tolerance tests were run in triplicate.

Table II. Toxicity of L-ethionine and 4-nitroquinoline *N*-oxide (NQO) (mg %) for sensitive and resistant flagellates

| | Ethionine | | NQO | |
|---|---|---|---|---|
| | $LD_{50}$ | $LD_{100}$ | $LD_{50}$ | $LD_{100}$ |
| *Euglena* (sens.) | 3 | 6 | 0·03 | 0·1 |
| *Euglena* (res.) | 750 | 1500 | 1·5 | 5·0 |
| *Ochromonas* (sens.) | 2 | 4 | 0·01 | 0·03 |
| *Ochromonas* (res.) | 750 | 1500 | 0·25 | 1·5 |

Viability was assessed by plating aliquots of treated organisms onto drug-free and drug-containing pH 3·6 agar plates. *E. gracilis* and *O. danica* remained indefinitely transferable up to the completely toxic concentrations; there was a very sharp cut-off at the fully toxic dose: such plates became completely sterile. No-drug control plates revealed plating efficiencies consistent with those for untreated, sensitive cells. For *E. gracilis*, the plating efficiency approached 100%; for *O. danica* 1–10%.

## NUTRITIONAL DISTORTIONS IN RESISTANT POPULATIONS

Resistant populations of *E. gracilis* and *O. danica* grew well in the standard defined media. The generation time of untreated *E. gracilis*, normally 12–16 hr was lengthened by 2–3 hr in resistant cells. This lag, also observed in resistant *O. danica*, was nearly abolished in media supplemented with commercial water-soluble liver preparations ("Liver L" and "15-unit" liver); the further addition of an acid-hydrolysed casein permitted full growth.

The competitive relationship in sensitive cells between L-ethionine *vs* the expected methionine–adenine combination persisted in resistant populations. But in resistant populations the solubility of ethionine limited extension of the dose-response curves.

## ANNULMENT OF NQO TOXICITY BY THERMOSTABLE COMPOUNDS

For NQO-resistant cells (studied in most detail in *O. danica*), L-tryptophan was kept at the level permitting half-maximal growth (i.e., $LD_{50}$). Restoration of near-maximal or maximal growth was obtained with the following (singly or in combination) in order of decreasing activity; tryptophan, kynurenine, phthiocol, tyrosine, riboflavin, uracil, and nicotinic acid. Crudes with high annulment activity included liver and yeast extracts. Sensitive cells responded to the same compounds as listed earlier in this paper; higher concentrations of these compounds had merely to be supplied for equivalent relief of inhibition at a given NQO concentration. Resistant populations derived from initially few-cell colonies did not acquire absolute requirements for any of the aforementioned metabolites. Preliminary results with tryptophan-requiring mutants of *Escherichia coli* support our previous hypothesis that NQO permeates the cell by a tryptophan-transport mechanism.

By use of minimal concentrations of methionine, similar annullers were identified for ethionine-resistant populations; these were adenine, uracil, tyrosine, and tryptophan.

## CROSS-RESISTANCE PATTERNS

Cross-resistance was also demonstrated by replica plating. With *O. danica* roughening of the plate with sterile velvet was necessary for initial colony growth on the agar surface. The velvet imprint provided depressions in which the organisms found the requisite moisture. The high motility and limited mucilage production of *O. danica* precluded use of pour-plate, streak or agar-overlay methods. As the colonies were not clearly clonal, and the results closely paralleled those obtained with liquid cultures, the nutritional results are not given in detail.

*E. gracilis* and *O. danica* were bleached on NQO plates. (Significant numbers of unbleached resistant colonies of both organisms also appeared on drug-containing plates). The bleached *Euglena* colonies remained bleached on subculture; the *Ochromonas* colonies tended to revert to brown. These populations with impaired photosynthesis were not investigated further because of desirability of using clonal cultures.

## INHIBITION OF $O_2$ UPTAKE IN SENSITIVE AND RESISTANT POPULATIONS

The growth experiments just outlined demanded appreciable quantities of inhibitor. As some highly interesting compounds are available in very limited quantities, more sensitive methods of testing for metabolic derangements were highly desirable. The use of polarography to measure $O_2$ uptake was therefore explored. The equipment used was not as sensitive as desired but as the results strikingly parallel those obtained by growth, a brief description is appended.

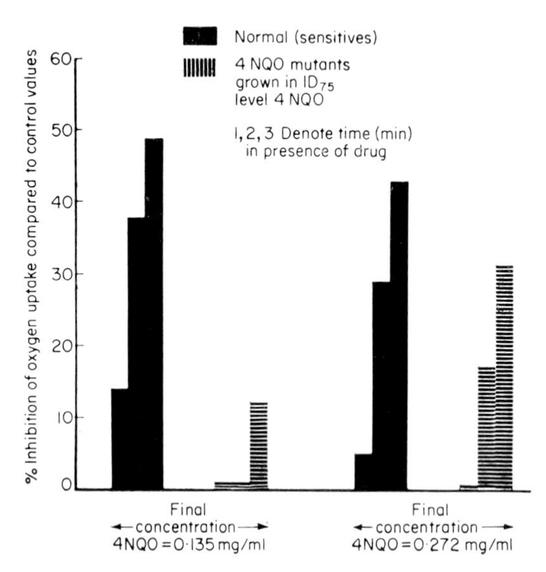


FIG. 1. Inhibition of $O_2$ uptake by sensitive and NQO-resistant populations of *O. danica*. $ID_{75}$ = concentration of NQO causing 75% growth inhibition. Each step interval = 1 min.

A Clark $O_2$ electrode with a steady potential of 0·6 V, seated in a water-jacketed glass vessel maintained at 25°, was connected to a Varian recorder calibrated to record $O_2$ uptake as $\mu$l $O_2$/min. Mass cultures of sensitive and resistant populations were grown heterotrophically in the dark on defined media. After 5 days' growth, the cells were harvested in the cold, packed-cell volume measured, and the cells resuspended in a substrate- and N-free basal medium. Cell densities were adjusted to 2·4 mg protein/ml (for *E. gracilis* a 1·5% cell suspension based on the packed-cell volume, and for *O. danica* 3·4–3·6 mg protein/ml, equivalent to a 2·0% suspension). Cell suspensions were bubbled with air for 20-30 min before starting a run. Eight ml of cell suspension were placed in the interior compartment of the water-jacketed vessel under constant mixing. Solvent effects were checked by appropriate controls. Time of onset of respiratory inhibition and time required for 100% inhibition were noted. Figure 1 summarizes the effect of NQO on the $O_2$ uptake by sensitive and resistant populations.

Na dodecyl sulfate at non-toxic doses (12 μg/ml) was used to enhance penetration of NQO. Sensitive cells pretreated with this surfactant for 1 min, followed by the standardized dose of 0·135 mg/ml NQO, completely stopped respiration in 3-4 min. This was approximately half the time for 100% inhibition by non-pretreated sensitive cells. Compounds acting as electron acceptors, i.e., phenazine methosulfate and K ferricyanide, added to sensitive cells previously exposed for at least 1 min to 0·062 mg/ml NQO, markedly enhanced respiratory inhibition. Complete inhibition of respiration was obtained with 3 min after addition of either compound (100 lambda of a 0·1 M solution).

Pre-incubation of sensitive cells in varying concentrations of L-tryptophan did not annul the inhibition of respiration. The only enhancement of annulment was a marginal opposition to inhibition when sensitive cells were pre-incubated with phthiocol. However, when phthiocol was added after sensitive cells had been exposed to 4NQO for 1–2 min, respiratory inhibition was enhanced.

## Action of 4-Nitroquinoline *N*-Oxide on the Photosynthetic Electron-Transfer System of Chloroplasts

### METHODS

All cells used in this phase of study of *Euglena* were grown autotrophically in an inorganic medium supplemented with thiamine and vitamin $B_{12}$ [basal medium of Hutner *et al.* (1956)]. The cells were grown in a continuous-culture apparatus at 25°; 5% $CO_2$ in air was circulated through the culture flask. Illumination was from a bank of white fluorescent tubes (approx. 700 foot candles). Cultures were started with a 10% inoculum and were harvested 3 days later.

Chloroplasts were prepared by disrupting cells in a French pressure cell at 10,000 psi. The cells were washed in 50 mM *tris* buffer pH 7·8 and suspended in the same medium before passing through the press. Few of the chloroplasts remained intact and a preparation consisting mainly of broken chloroplasts was collected by centrifugation between 1000 and 20,000 ***g***. The preparation was washed twice with 50 mM *tris* buffer pH 7·8.

Fd (ferredoxin) and cytochrome-552 were purified from extracts of autotrophic cells using DEAE-cellulose column chromatography (unpublished experiments). NADP reductase (pyridine nucleotide transhydrogenase) was purified from spinach by the procedure of Keister and San Pietro (1963).

Assays for photoreduction of $NADP^+$ and other compounds by isolated chloroplasts were performed as described (Smillie, 1962). Photo-oxidation of cytochrome-552 in cells and isolated chloroplasts was followed using a Chance-type double-beam spectrophotometer fitted with vertical optics. The cells were allowed to settle in the bottom of the cuvette for 5 min before measurements were begun. Illumination from a 500-watt projector was filtered through a Bausch & Lomb 700-mμ interference filter and a Corning cut-off filter (*C.S.* 2-64).

The *tris* buffers used in these experiments were freshly prepared and immediately frozen. They were thawed just before use. Fresh batches were prepared every 3 to 4 weeks.

## PHOTOREDUCTION OF $NADP^+$

It is believed that the photosynthetic electron-transfer system in intact non-bacterial cells terminates in the reduction of $NADP^+$. During this process $O_2$ is evolved. A rapid photoreduction of NADP by isolated chloroplasts was first demonstrated by San Pietro and Lang (1958). Chloroplasts and a soluble

Table III. NQO inhibition of the photoreduction of NADP by chloroplasts

| Additions | NADP reduction (mμmoles/min) |
|---|---|
| None | 0·00 |
| NQO | 0·00 |
| Fd | 0·68 |
| Fd, cytochrome-552, NADP reductase | 1·83 |
| Fd, cytochrome-552, NADP reductase, NQO | 0·01 |

The reaction mixture (0·8 ml) contained chloroplasts (12 μg chlorophyll), *tris* buffer pH 7·8 (16 mM), $MgCl_2$ (1 mM), $NADP^+$ (110 μM) and where indicated in the table, Fd (0·6 μM), cytochrome-552 (0·38 μM), NADP reductase 0·12 μM with respect to FAD, and NQO 1·2 μM. Fd was omitted from the control reaction mixture. Controls containing the complete reaction mixture were also run in the dark.

Table IV. NQO inhibition of the ferredoxin-dependent photoreduction of cytochrome *c* by chloroplasts

| NQO (μM) | Cytochrome *c* reduction (mμmoles/min) |
|---|---|
| 0 | 0·72 |
| 2·5 | 0·01 |

The reaction mixture (0·8 ml) contained chloroplasts (4·8 μg chlorophyll), *tris* buffer, pH 7·8 (16 mM), (1·0 mM), beef heart cytochrome *c* (40 μM), and Fd (0·6 μM). Fd was omitted from the control reaction mixture. Controls containing the complete reaction mixture were also run in the dark.

protein, photosynthetic pyridine nucleotide reductase (now ferredoxin), are required. The effect of NQO on the photoreduction of $NADP^+$ by chloroplasts supplemented with PPNR is shown in Table III. The chloroplasts and a purified preparation of PPNR were obtained from autotrophically grown *Euglena*; NQO at μ-molar concentrations completely inhibited the photoreduction of $NADP^+$.

Two of the components of the photosynthetic electron transfer pathway of *Euglena*, cytochrome *f* (-552) and a NADP reductase [the pyridine nucleotide transhydrogenase of Keister and San Pietro (1963)], are partly removed during isolation of *Euglena* chloroplasts. The rate of photoreduction of $NADP^+$ by isolated *Euglena* chloroplasts is considerably increased by the addition of purified preparations of these two components (Table IV). The stimulated rate of NADP reduction is inhibited by NQO.

## FD-DEPENDENT PHOTOREDUCTION OF CYTOCHROME *c*

Cytochrome *c* is reduced by illuminated chloroplasts (Holt, 1950), both in the presence and absence of Fd. However, Fd increases the rate of reduction (Davenport and Hill, 1960). It is not yet clear whether one or two reductive sites are involved (Keister and San Pietro, 1963). As shown in Table IV, the Fd-stimulated photoreduction of cytochrome *c* is inhibited by NQO.

## THE HILL REACTION

At least two distinct light-dependent reactions are involved in the photoreduction of $NADP^+$ (Emerson *et al.* 1957; Duysens *et al.*, 1961; Kok and Hock, 1961). In one reaction a photo-oxidant is formed and a photoreductant which reduces $NADP^+$ via Fd (photosystem I). In the other, a photoreductant

Table V. Effect of NQO on the Hill reaction

| Hill oxidant | NQO (μM) | Rate of reduction (mμmoles/min) |
|---|---|---|
| Cytochrome *c* | 0 | 2·2 |
| | 1·0 | 2·9 |
| | 2·0 | 3·8 |
| 2,6-Dichlorophenol indophenol | 0 | 1·5 |
| | 1·0 | 1·5 |
| | 2·0 | 1·5 |
| Ferricyanide | 0 | 25 |
| | 1·0 | 25 |
| | 2·0 | 24 |

The reaction mixture (1·0 ml) contained chloroplasts (2·4 μg chlorophyll), *tris* buffer, pH 7·8 (20 mM), and where indicated in the table, beef-heart cytochrome *c* (40 μM), 2,6-dichlorophenol indophenol (50 μM), ferricyanide (0·6 mM) and NQO. Control reaction mixtures were run in the light in the absence of the Hill oxidant and in the dark with the complete reaction mixture.

is formed and water is oxidized (photosystem II). These two systems are normally coupled but appear to be uncoupled in certain mutants of *Chlamydomonas* (Levine and Smillie, 1963) and can be studied separately. The formation of photoreductant can be followed by coupling with suitable oxidants such as cytochrome *c*, various dyes and ferricyanide (the Hill reaction). Table V shows the rates of reduction of various Hill oxidants by *Euglena* chloroplasts and the effect of NQO on these rates. NQO did not inhibit cytochrome *c* reduction but instead stimulated it. The reduction of 2,6-dichlorophenol indophenol and ferricyanide was not affected by NQO. The stimulation of cytochrome *c* reduction by NQO indicates that NQO may act as a Hill oxidant transferring electrons between photoreductant and cytochrome *c*.

## PHOTO-OXIDATION OF CYTOCHROME *f*

When photosynthetic cells are illuminated there is a rapid oxidation of cytochrome *f*. The photo-oxidation of cytochrome *f* is closely associated with the photochemical reaction of photosystem I and in fact still proceeds at liquid $N_2$ temperatures (Chance and Bonner, 1963). The action of NQO on the photo-oxidation of the cytochrome *f* (-552) of *Euglena* was examined in cells and isolated chloroplasts.

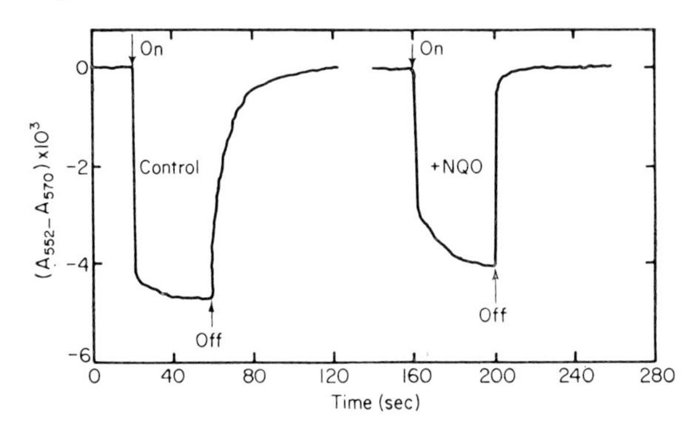


FIG. 2. Effect of NQO on the photo-oxidation of cytochrome-552 in autotrophic cells of *Euglena*. The reaction mixture (1 ml) consisted of *Euglena* cells (containing 0·3 mg chlorophyll), in growth media diluted 2:3 with water. NQO was 0·33 mM. Absorbancy changes at 552 m$\mu$ relative to those at 570 m$\mu$ upon illumination (700 m$\mu$, $8 \times 10^{-9}$ Einstein $cm^{-2}$ $sec^{-1}$) were measured using a Chance-type double-beam spectrophotometer.

Figure 2 shows the absorbancy changes at 552 m$\mu$ in cells in the presence and absence of NQO. In untreated cells there is a rapid decrease in absorbancy upon illumination and a much slower return to the original steady-state level when illumination ceases. These absorbancy changes are due to the oxidation of *Euglena* cytochrome-552 in the light and its reduction in the dark (Olson and Smillie, 1963). In the presence of NQO the photo-oxidation was less rapid than in untreated cells. However the most striking difference in cells treated with NQO was the rapid reduction of cytochrome-552 that occurred in the dark. A possible explanation of these results is that the NQO sets up a cyclic flow of electrons and short-circuits the normal electron flow between the photoreductant produced in photosystem II and cytochrome-552.

A photo-oxidation of cytochrome-552 has also been demonstrated in isolated chloroplasts (Olson and Smillie, 1963). In the cell-free system, reduction of cytochrome-552 following illumination is much more rapid than in intact cells and NQO has little effect on either the photo-oxidation or the light-independent reduction of cytochrome-552.

The light-dependent oxidation and reduction of cytochrome $b_6$ in *Euglena* chloroplasts at low and high light intensities respectively, previously reported by Olson and Smillie (1963) was not affected by NQO.

## REACTION OF NQO WITH NADP

The terminal enzyme of the pathway in chloroplasts leading to reduction of $NADP^+$ is a flavoprotein, NADP reductase (Keister and San Pietro, 1963; Shin *et al.*, 1963). This enzyme can be coupled by Fd to cytochrome *c* (Equation 1) (Lazzarini and San Pietro, 1962).

$$\text{NADPH} \rightleftarrows \text{NADP reductase} \rightleftarrows \text{NQO} \rightleftarrows \text{cytochrome } c \qquad (1)$$

NQO has no effect on the diaphorase activity (2,6-dichlorophenol indophenol used as the acceptor) or the transhydrogenase activity of NADP reductase (Fig. 3).

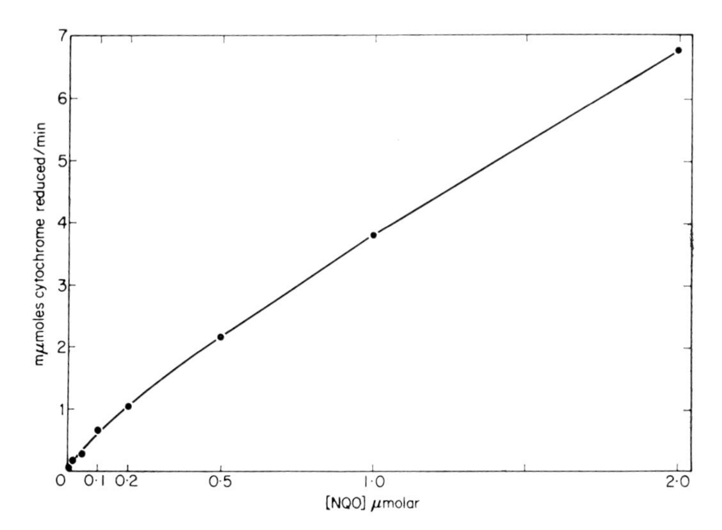


FIG. 3. Capacity of NQO to act as an electron carrier between flavoprotein and cytochrome *c*. Reaction mixtures contained *tris* buffer, pH 7·8 (25 mM) excess NADP-reductase purified from spinach chloroplasts (2), beef-heart cytochrome *c* (40 μM), NADPH (100 μM) and various concentrations of NQO.

## CONCLUSIONS

The data in Tables II and IV show that NQO inhibits the Fd-dependent photoreduction of NADP and cytochrome *c* by chloroplasts isolated from *Euglena*. The two light reactions of photosynthesis are not directly affected by NQO since neither the Hill reaction activity nor the photo-oxidation of cytochrome-552 in isolated chloroplasts is inhibited by NQO. The apparent photo-oxidation of cytochrome-552 in intact cells is decreased by NQO, but this can probably be attributed to a marked increase in the competing light-independent reduction of cytochrome-552. The fact that NQO causes a rapid reduction of cytochrome-552 in cells immediately following a period of illumination and further that it stimulates the transfer of electrons to cytochrome *c* in the Hill reaction suggests that NQO reacts with reducing equivalents formed in the light reactions. It is postulated that a cyclic flow of electrons is produced

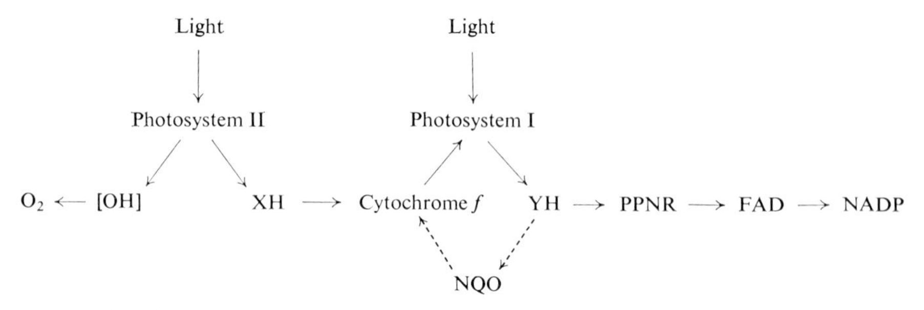


FIG. 4. Postulated site of action of NQO in photosynthetic electron transfer.

in photosystem I and that this disrupts the normal noncyclic flow to NADP (Fig. 4). NQO could react directly with cytochrome-552 to complete the cycle. The data of Fig. 2 accord with this suggestion. It is not clear whether NQO can react with reducing equivalents formed in photosystem II and if it does, whether a cyclic flow of electrons is produced. When cytochrome *c* is added to the isolated chloroplast system, the cytochrome *c* will continue to be reduced whether NQO is present or not since NQO can react directly with cytochrome *c* (Fig. 3).

## DISCUSSION

### INTERPRETATION OF CROSS-RESISTANCE: ROLE OF –SH GROUPS

The cross-resistance between NQO and ETN implies the alkylating and arylating carcinogens share common targets. First, NQO is not a chelator and so metals are not directly the targets. Colored compounds formed when 2-ethyl-4-hydroxyquinoline *N*-oxides reacted with $FeCl_3$, cited by Tappel (1960) as evidence for chelation, are probably a co-ordinated complex of the catechol type, the *o*-hydroxyl arising by oxidation. To lessen such auto-oxidations, media even more acidic are desirable. In preliminary experiments, media like those in Table I, supplemented with Mg, Ca, and Fe permitted normal growth at least down to pH 3·0; as expected, NQO, an acid, became more toxic presumably because penetration is favored as ionization is suppressed. A curious artefact emerged: Cu ($\sim$1·0 mg%) autoclaved with NQO in the glucose-containing medium conferred remarkable NQO-annulling power; when the glucose was autoclaved separately, Cu was inert.

Tissues probably reduce NQO to some extent to the 4-hydroxyaminoquinoline *N*-oxide; liver microsomes have a flavin-catalysed nitro reductase (Kamm and Gillette, 1963) but whether this is concerned with NQO reduction, and the role of the various quinone reductases, e.g., menadione reductase, in this reduction is unclear, as well as the extent to which these reductions are non-enzymatic. *Aspergillus niger* and other fungi reduce many organic nitro compounds. We find that *E. gracilis* carries out a virtually quantitative reduction of NQO under $N_2$; in one run the yield (estimated from amino determinations) was 132 mg from 138 mg NQO. The hydroxyamino derivative was less toxic than NQO to *A. niger* but equally mutagenic (Okabayashi *et al.*, 1964), and was definitely (though less) carcinogenic (Endo and Kume, 1963; Nakahara, 1964; Shirasu, 1965).

The easy interconversion of activated amino and *N*-hydroxy groups in the mammalian body is borne out by the ready interconvertibility of urethane and *N*-hydroxyurethane in the mouse (Mirvish, 1964). Instances are multiplying in the conversion of secondary and tertiary nitrogen to *N*-oxides, predominantly by microsomal enzymes. The 2-acetylaminofluorene carcinogens are likewise converted into more carcinogenic *N*-hydroxy metabolites (Miller *et al.*, 1964).

Quinones condense with –SH compounds (Wilgus and Chapman, 1964). This accounts for much of the toxicity of quinones for intact organisms such as bacteria (Geiger, 1946), fungi (Mace and Herbert, 1963), and for enzymes (Hoffmann-Ostenhof, 1963), judging from the efficacy of –SH compounds in annulling inhibitions. However, as reviewed by Sexton (1963), condensations with phenols, aldehydes, and indole compounds help bind quinones to proteins, sometimes more importantly than to –SH.

That the alkyl hydroxyquinoline *N*-oxides inhibit between cytochrome *b* and cytochrome *c* is amply documented by Kogut and Lightbown (1962); their paper reviews the older literature; more recently these inhibitors have been applied to Athiorhodaceae (Nishimura, 1963); to a micro-aerophilic vibro (Jacobs and Wolin, 1963); and to liver mitochondria (Howland, 1963). The consensus is that phosphorylation is blocked in the respiratory chain at Site 2 (Griffiths, 1965; Pinchot, 1965; Racker, 1965), which harmonizes with the present results with NQO.

Because of the conspicuous interaction of NQO with –SH groups, some workers have attempted to correlate carcinogenicity with avidity for SH; thus Okamoto and Itoh (1963) point out that NQO reacts with thioglycollate five to thirty times as rapidly as the noncarcinogenic 4-nitroquinoline. As a complication, Okabayashi *et al.* (1964) emphasize that the substitution reaction with –SH liberates nitrous acid—a mutagen. That similar reductions may take place in the chloroplast is suggested by findings by Wessels (1960) that illuminated chloroplasts reduce 2,4-dinitrophenol under anaerobic conditions and that the product can catalyse photophosphorylation. Mammalian tissues have an enzyme catalysing the replacement of labile aromatic nitro groups with glutathione; NQO reacts spontaneously with glutathione, however the enzyme speeds the reaction (Al-Kassab, *et al.* 1963). Neish *et al.* (1964) postulated that carcinogenicity in the azo-dye series depends on dye-binding to protein, and that carcinogenicity correlates with increase in liver glutathione.

To the evidence for an attack on critical proteins at their –SH groups (perhaps with a displacement of critical quinones bound to these –SH groups), and on alkylable sites on DNA, there is evidence that ribosomes as well as the transmethylating enzymes interposed between DNA and sRNA are targets for carcinogens. NQO produced nucleolar aggregates ("caps") in Hela and Chang liver cells (Reynolds *et al.*, 1963; Montgomery, 1963); similar caps were induced by actinomycin D (Reynolds *et al.*, 1964). The nucleolus is rich in transmethylases (Borek, 1964).

Since streptomycin appears to affect ribosomes (Brock, 1964; Gorini, 1964), bleaching of *Euglena* and metaphytes by streptomycin perhaps denotes an exceptional vulnerability of these chloroplast ribosomes to streptomycin as well as to other agents inducing permanent bleaching as distinct from temporary bleaching (the old term *etiolation* seems appropriate when applied to temporary bleaching caused by dark growth). The insensitivity of ordinary cytoplasmic ribosomes can be inferred from the remarkably wide gap between bleaching and killing doses of bleaching agents as reviewed by J. A. Schiff in this symposium (p. 341). The converse of impaired or destroyed chloroplasts accompanying presumed retention of ribosomal and mitochondrial function, might be something like the induction of petites in yeasts or, more drastically, the disappearance of mitochondria in yeast grown anaerobically. A correlation such as that between ability of yeasts to grow anaerobically and to form petites (Bulder, 1964) has not been tested in flagellates. Our efforts to induce petites or bleached strains in *Euglena* by treatment with acriflavin have failed: gradient plates invariably showed green colonies sharply up to the lethal zone.

A likelihood is that the 2-alkyl-hydroxyquinoline *N*-oxides are carcinogenic. These poisons had been isolated from pseudomonads as antagonists to the antibiotic action of streptomycin; antimycin was not an antagonist (Lightbown and Jackson, 1956). Curiously, ETN is normally synthesized by *E. coli* grown on methionine-free media (Loerch and Mallete, 1963). The microbial production of carcinogens and aralkylating agents is now a commonplace with the identification of anti-tumor, alkylating antibiotics such as mitomycin, the hepatocarcinogenic aflatoxin, and the carcinogenic, radiomimetic methylazoxymethanol derivatives isolated from cycads (Teas *et al.*, 1965). The existence of these compounds, as with colchicine, bespeaks the existence of intracellular compartmentation. The effectiveness of NQO and ETN in arresting photosynthesis would then, in this light, point to the chloroplast having tryptophan- and methionine-transport systems. Because of expense—cross-resistance and other experiments on growth consume much material—we have not tried to see whether alkyl-hydroxyquinoline *N*-oxides block bleaching of *Euglena* by streptomycin; we plan to try this with NQO. With refinement of the $O_2$-electrode technique, it should be less expensive to test for cross-resistance between NQO and alkyl-hydroxyquinoline *N*-oxides. Radiation resistance in *M. radiodurans* appears to depend on enhanced production of reducing material, presumably mainly –SH compounds (Bruce and Malchman, 1965) and on remarkable ability to repair DNA damage (Setlow and Duggan, 1964; review: Mosely, 1965). Does this hold for NQO- and ETN-resistant flagellates? Enhanced production of reducing material and enhanced DNA repair need not be mutually exclusive: the repair enzymes (Setlow and Carrier, 1964) might themselves be –SH enzymes.

Resistance to ethionine follows the –SH pattern. Strains of *Candia utilis* adapted to high concentrations of ethionine excrete forty times the normal amount of methionine in the medium, and intracellular methionine is increased (Musíliková and Fencl, 1964). Compounds with free –SH groups, e.g., *S*-methyl-$\beta$-mercaptopropionic acid, are excreted, among other methionine derivatives, by an ethionine-resistant *Neurospora* (Galsworthy and Metzenberg, 1965). That some factors in carcinogenesis by NQO and ETN apply to the tar carcinogens is made likely by NQO blocking benzpyrene carcinogenesis (Searle and Woodhouse, 1964) presumably by competing for the –SH groups necessary for attachment.

## ARALKYLATING AGENTS: METHYLATION OF RNA; RADIOMIMETIC AGENTS

Since there is cross-resistance between ethionine (an alkylating agent) and NQO (an arylating agent), the term *aralkylation* may be more apropos to designate common features in their action. The –SH groups of proteins in the electron chain of the mitochondrion and the chloroplast would appear especially vulnerable to certain aralkylating agents. Other likely sites of action are aralkylatable sites on DNA and RNA, notably those revealed by the occurrence of methylated purines and pyrimidines in RNA (yeast: Hall, 1965; Holley *et al.*, 1965; *E. coli*: Hurwitz *et al.*, 1965). The vulnerability of the 7-position of guanine to alkylating agents (Brookes and Lawley, 1964; Lee *et al.*, 1964) is said to account for some of the characteristic damage by radiomimetic agents. The cross-resistance between radiomimetic agents and (*a*) u.v. and nitrofurane derivatives (which also bleach *Euglena*) (McCalla, 1965); (*b*) u.v. and proflavin (Karrer and Greenberg, 1964); and (*c*) u.v. and $HNO_2$ (Zampieri and Greenberg, 1964), suggests experiments to see whether radiation-resistant organisms such as *Micrococcus radiodurans* have exceptional resistance to NQO and ETN.

The carcinogens formerly applied have been mainly of the tar series, and if higher plants lack the enzymes (predominantly microsomal in higher animals) for converting these compounds into the proximal carcinogens, the compounds would be inert. NQO has not yet been tried.

In surveying compounds annulling inhibitions by NQO and ETN, only water-soluble, autoclaved—i.e., thermostable—compounds were tried. We intend to test filter-sterilized or aseptically supplied lipid-rich crudes, and thus take advantage of the ability of *O. danica* to ingest these materials and digest them. Another limitation of these experiments is that populations, not clones, were used. Cross-feeding among different resistant types could have masked important targets. Annullers of the NQO-induced growth inhibition include riboflavin, and nicotinic acid, which suggests that NQO may interfere with riboflavin and pyridine nucleotide enzyme cofactors in the electron chain, resembling quinacrine (atebrine) in this respect.

Lovelock *et al.* (1962) suppose that the substances which disturb oxidative phosphorylation have an unusually high affinity for free electrons, and that the important step in carcinogensis is a disturbance of mitochondrial electron transport. A problem in testing so comprehensive a theory is the multiplicity of targets: *first*, in the functioning of mitochondria (and chloroplasts?)—assuming, mainly by analogy with the 2-alkyl-hydroxyquinoline *N*-oxides, that NQO paralyses mitochondrial function; *second*, in the functioning of the information system (DNA→RNA→protein) controlling the synthesis of nucleus and mitochondria and other quasi-autonomous cytoplasmic organelles. ETN causes a disaggregation of ribosomes in rat liver (Baglio and Farber, 1965). If in the chloroplast the ribosomes were the most ETN-sensitive targets, this could mean an appreciable advance in understanding carcinogenesis.

We end this discussion [heeding Racker's (1965) admonition that investigators of oxidative phosphorylation who used inhibitors should be more inhibited] by concluding that the cross-resistance between NQO and ETN offers an attractive way to pin down the mooted resemblances between chloroplasts and mitochondria.

The plant physiologist using carcinogens as useful tools for elucidating the functioning of chloroplasts, and the oncologist using chloroplasts as a convenient means of elucidating the mode of action of carcinogens, may be the same person.

## Summary

The water-soluble carcinogens 4-nitroquinoline *N*-oxide (NQO) and ethionine arrest photosynthesis in *Euglena gracilis* and the chrysomonad *Ochromonas danica*. Spectrophotometric studies of *Euglena* chloroplasts in the presence of NQO indicated that while the photo-oxidation of cytochrome-552 was not directly affected, the rate at which the cytochrome was reduced immediately following illumination was greatly increased; the Hill reaction and the light-dependent oxidation and reduction of cytochrome $b_6$ were not affected by NQO. An explanation of these results is that NQO sets up a cyclic flow of electrons and short-circuits the normal electron flow between the photoreductant produced in photosystem II and cytochrome-552.

By serial passage in increasing concentrations of drug-containing defined media, resistance of *O. danica* was raised between 25- and 50-fold for NQO and 350-fold for L-ethionine; resistance of *E. gracilis* was raised 50-fold for NQO and 350-fold for ethionine. Populations of resistant strains had almost complete cross-resistance, whether grown in NQO or ethionine, in growth and respirometric ($O_2$-uptake) tests. Likely metabolic targets considered for NQO were –SH groups of protein, alkylatable site on DNA, and ribosomes; these targets were postulated to be essentially similar in the mitochondrion and the chloroplast. Certain carcinogens were concluded to be useful reagents for exploring chloroplast function, and the chloroplast was viewed as a useful system for discerning the mode of action of carcinogens. The cross-resistance between NQO and ethionine was construed as meaning that these ostensibly different carcinogens have common features in their action, and that in turn the mitochondrion and the chloroplast shared equivalent sites sensitive to carcinogens.

## References

Aaronson, S., Baker, H., Bensky, B., Frank, O. and Zahalsky, A. C. (1964). *Devel Microbiol.* **6**, 48.
Al-Kassab, S., Boyland, E. and Williams, K. (1963). *Biochem. J.* **87**, 4.
Allen, M. B. (1963). Discussion *In* "First International Interdisciplinary Conference on Marine Biology" (G. A. Riley, ed.), p. 65. A.I.B.S. Washington, D.C.
Alving, A. S., Powell, R. D., Brewer, G. J. and Arnold, J. D. (1962). *In* "Drugs, Parasites and Hosts" (L. G. Goodwin and R. H. Nimmo-Smith, eds.), p. 83. J. and A. Churchill, London.
Anderson, R. A. Enomoto, M., Miller, E. C. and Miller, J. A. (1964). *Cancer Res.* **24**, 128.
Baglio, C. M. and Farber, E. (1965). *J. mol. Biol.* **12**, 466.
Borek, E. (1964). *Cold Spr. Harb. Symp. quant. Biol.* **28**, 139.
Brock, T. D. (1964). *Fed. Proc.* **23**, 965.
Brookes, P. and Lawley, P. D. (1964). *J. cell comp. Physiol.* **64** (Suppl. 1), 11.
Bruce, A. K. and Malchman, W. H. (1965). *Radiat. Res.* **24**, 473.
Bulder, C. J. E. A. (1964). *Ant. v. Leeuwen.* **30**, 442.
Chance, B. and Bonner, W. D., Jr. (1963). *In* "Photosynthetic Mechanisms of Green Plants", p. 66. Nat. Acad. Sci. U.S.—Nat. Res. Council: Washington, D.C.
Davenport, H. E. and Hill, R. (1960). *Biochem. J.* **74**, 493.
Duysens, L. N. M. (1964). *Prog. Biophys.* **14**, 1.
Duysens, L. N. M., Amesz, J. and Kamp, B. M. (1961). *Nature, Lond.* **194**, 510.
Emerson, R., Chalmers, R. and Cederstrand, C. (1937). *Proc. natn. Acad. Sci. U.S.A.* **43**, 133.
Endo, H. and Kume, F. (1963). *Gann* **54**, 443.
Falk, H. L., Kotin, P., Lee, S. S. and Nathan, A. (1962). *J. natn. Cancer Inst.* **28**, 699.
Farber, E. (1963). *Adv. Cancer Res.* **7**, 383.
Farber, E., Shull, K. H., Villa-Trevino, S., Lombardi, B. and Thomas, M. (1964). *Nature, Lond.* **203**, 34.
Galsworthy, S. B. and Metzenberg, R. L. (1965). *Biochemistry* **4**, 1183.
Geiger, W. B. (1946). *Arch. Biochem.* **11**, 23.
Gibbs, S. P. (1962). *J. cell Biol.* **15**, 343.
Gibor, A. and Granick, S. (1964). *Science* **145**, 890.
Gibson, K. D., Neuberger, A. and Tait, G. H. (1963). *Biochem. J.* **88**, 325.
Gorini, L. (1964). *New Scient.* **24**, 776.
Griffiths, D. E. (1965). *In* "Essays in Biochemistry" (P. N. Campbell and G. D. Greville, eds.), Vol. 1, p. 91. Academic Press, London and New York.
Hall, R. H. (1965). *Biochemistry* **4**, 661.
Hoffmann-Ostenhof, O. (1963). *In* "Metabolic Inhibitors" (R. M. Hochster and J. H. Quastel, eds.), Vol. II, p. 145. Academic Press, New York and London.
Holley, R. W., Apgar, J., Everett, G. A., Madison, J. T., Marquisee, M., Merill, S. M., Penswick, J. R. and Zamir, A. (1965). *Science, N.Y.* **147**, 1462.
Holt, A. S. (1950). *U.S. Atomic Energy Commission Document ORNL*-752.
Howland, J. L. (1963). *Biochim. biophys. Acta* **73**, 665.
Hurwitz, J., Gold, M. and Anders, M. (1965). *J. biol. Chem.* **239**, 3462.
Hutner, S. H. (1961). *In* "Microbial Reaction to Environment" (G. G. Meynell and H. Gooder, eds.), *Symp. Soc. gen. Microbiol.* **11**, 1.
Hutner, S. H., Bach, M. K., and Ross, G. I. M. (1956). *J. Protozool.* **3**, 101.
Hutner, S. H. and Provasoli, L. (1965). *Annu. Rev. Physiol.* **27**, 19.
Jacobs, N. J. and Wolin, M. J. (1963). *Biochim. biophys. Acta* **69**, 29.
Kamm, J. J. and Gillette, J. R. (1963). *Life Sci.* **4**, 254.
Karrer, P. W. and Greenberg, J. (1964). *J. Bacteriol.* **87**,536.
Ke, B. (1964). *Biochim. biophys. Acta* **88**, 1.
Keister, D. L. and San Pietro, A. (1963). *Archs Biochem. Biophys.* **103**, 45.
Kogut, M. and Lightbown, J. W. (1962). *Biochem. J.* **64**, 368.
Kok, B. and Hoch, G. (1961). *In* "Light and Life" (W. D. McElroy and B. Glass, eds.), p. 397. Johns Hopkins Univ Press: Baltimore, Md.
Lazzarini, R. A. and San Pietro, A. (1962). *Biochim. biophys. Acta* **62**, 417.

Lee, K. Y., Lijinsky, W. and Magee, P. N. (1964). *J. natn. Cancer Inst.* **32**, 65.
Lehninger, A. L. (1964). "The Mitochondrion". W. A. Benjamin: New York.
Levine, R. P. and Smillie, R. M. (1963). *J. biol. Chem.* **238**, 4052.
Lieberman, M. and Baker, J. E. (1965). *Annu. Rev. Plant Physiol.* **16**, 343.
Lightbown, J. W. and Jackson, F. L. (1956). *Biochem. J.* **63**, 130.
Loerch, J. D. and Mallette, M. F. (1963). *Archs Biochem. Biophys.* **103**, 272.
Lovelock, J. E., Zlatkis, A. and Becker, R. S. (1962). *Nature, Lond.* **193**, 540.
Lowther, R. L. and Boll, W. G. (1960). *Can. J. Bot.* **38**, 437.
McCalla, D. R. (1965). *Can. J. Microbiol.* **11**, 185.
Mace, M. E. and Herbert, T. T. (1963). *Phytopathology* **53**, 692.
Miller, E. C., Miller, J. A. and Enomoto, M. (1964). *Cancer Res.* **24**, 2018.
Mirvish, S. S. (1964). *Biochim. biophys. Acta* **93**, 673.
Miyachi, S., Miyachi, S. and Benson, A. A. (1966). *J. Protozool.* **13**, 76.
Montgomery, P. I. (1963). *Expl Cell Res.* (Suppl.) **9**, 170.
Mosely, B. E. B. (1965). *Sci. J.* **1**, 75.
Musílikovà, M. and Fencl, Z. (1964). *Fol. Microbiol.* **9**, 374.
Nakahara, W. (1964). *Arzneimittelforschung* **14**, 1842. (From *Carcinogenesis abstr.* (1965), 3, abstr. No. 62-235.)
Neish, W. J. P., Davies, H. M. and Reve, P. M. (1964). *Biochem. Pharm.* **13**, 1291.
Nishimura, M. (1963). *Biochim. biophys. Acta* **66**, 17.
Okabayashi, T., Yoshimoto, A. and Ide, M. (1964). *Chem. Pharm. Bull.* **12**, 257.
Okamoto, T. and Itoh, M. (1963). *Chem. Pharm. Bull.* **11**, 785.
Olson, J. M. and Smillie, R. M. (1963). *In* "Photosynthetic Mechanisms of Green Plants", p. 56. Nat. Acad. Sci. U.S.—Nat. Res. Council: Washington, D.C.
Pinchot, G. B. (1965). *Perspect. Biol. Med.* **8**, 180.
Racker, E. (1965). "Mechanisms in Bioenergetics." Academic Press, New York and London.
Reynolds, R. C., Montgomery, P. O. and Hughes, B. (1964). *Cancer Res.* **24**, 1269.
Reynolds, R. C., Montgomery, P. O. and Karney, D. H. (1963). *Cancer Res.* **23**, 535.
San Pietro, A. and Lang, H. M. (1958). *J. biol. Chem.* **231**, 211.
Shin, M., Tagawa, K. and Arnon, D. I. (1963). *Biochem. Z.* **338**, 84.
Searle, C. E. and Woodhouse, D. L. (1964). *Cancer Res.* **24**, 245.
Setlow, R. B. and Carrier, W. L. (1964). *Proc. natn. Acad. Sci. U.S.A.* **51**, 226.
Setlow, J. K. and Duggan, D. E. (1964). *Biochim. biophys. Acta* **87**, 664.
Sexton, W. A. (1963). "Chemical Constitution and Biological Activity", 3rd Ed., 517 pp. E. & F. Spon, London.
Shirasu, Y. (1965). *Proc. Soc. exp. Biol. N.Y.* **118**, 812.
Smillie, R. M. (1962). *Plant. Physiol.* **37**, 716.
Tappel, A. L. (1960). *Biochem. Pharmacol.* **3**, 289.
Teas, H. J., Sax, H. J. and Sax, K. (1965). *Science, N.Y.* **149**, 541.
Wessels, J. S. C. (1960). *Biochim. biophys. Acta* **38**, 195.
Wilgus, H. S. III and Chapman, D. D. (1964). *Org. Chem. Bull. Dist. Products Ind., Eastman Kodak* 36, No. 2, 4 pp.
Zahalsky, A. C., Keane, M. M., Hutner, S. H., Lubart, K. J., Kittrell, M. and Amsterdam, D. (1963). *J. Protozool.* **10**, 421.
Zampieri, A. and Greenberg, J. (1964). *J. Bacteriol.* **87**, 1094.

# Terpenoids and Chloroplast Development

T. W. Goodwin[1]

*Department of Biochemistry and Agricultural Biochemistry, University College of Wales, Aberystwyth, Wales*

## Introduction

The aim of this contribution is to present a hypothesis which goes some way to explain how the biosynthetic systems involved in terpenoid formation in developing chloroplasts are regulated so that just sufficient of the many terpenoids are synthesized to produce a functional chloroplast. A number of communications from our laboratory on certain aspects of this topic have already been presented to the Conference. I wish to draw these contributions and other data together in developing our general ideas on how regulation is achieved. Much elegant electron microscopy has also been discussed during the Conference and many aspects of the photomorphogenic changes which plastids undergo as they develop into functional chloroplasts are now reasonably clear. However, information on the regulation of the biochemical systems involved is almost non-existent. What follows represents one of the first efforts in this direction.

## Nature and Distribution of Terpenoids in Etiolated and Green Tissues

The terpenoids which are found in chloroplasts are the sterols (as reported by Mercer and Treharne, Vol. I) and carotenoids; whilst compounds in chloroplasts which contain a terpenoid side chain are the chlorophylls, vitamin $K_1$ and tocopherol and tocopheryl quinone (phytol side chain), and

Table I. Chloroplastidic and extraplastidic terpenoids

| CHLOROPLASTIDIC | EXTRACHLOROPLASTIDIC |
|---|---|
| Phytosterols | Phytosterols |
| Carotenoids | Pentacyclic triterpenoids |
| SIDE CHAINS OF: | SIDE CHAIN OF: |
| Chlorophylls | Ubiquinones |
| Vitamin $K_1$ | |
| Tocopherols | |
| Tocopherol quinones | |
| Plastoquinones | |

plastoquinone (solanesol-45 side chain); the main terpenoids found outside the chloroplast from our point pf view are sterols and pentacyclic triterpenes such as β-amyrin, whilst the main terpenoid derivative is ubiquinone (solanesol-50 side chain) (Table I). The structures of these compounds are given in Fig. 1.

[1] Present address: Department of Biochemistry, The University, Liverpool 3.

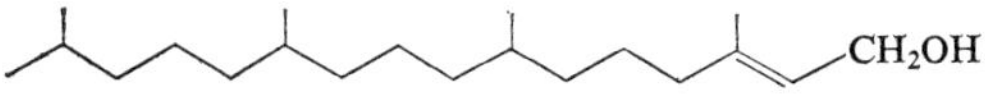

$CH_2OH$

Phytol

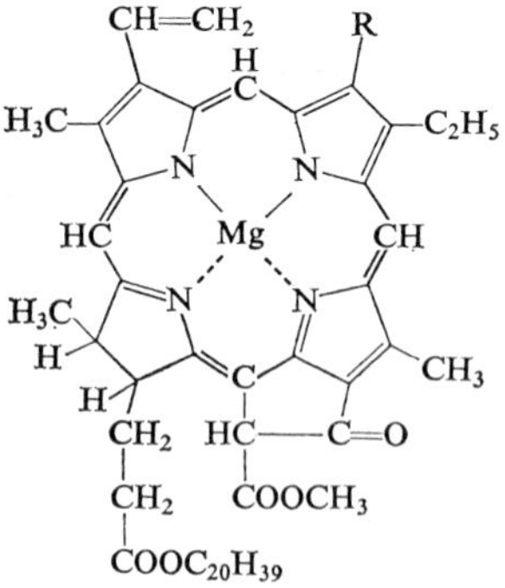

$CH{=}CH_2$ R H C $H_3C$ $C_2H_5$ N N HC Mg CH $H_3C$ N N H $CH_3$ H C $CH_2$ HC C=O $CH_2$ $COOCH_3$ $COOC_{20}H_{39}$

Chlorophyll *a*; $R{=}CH_3$
Chlorophyll *b*; $R{=}CHO$

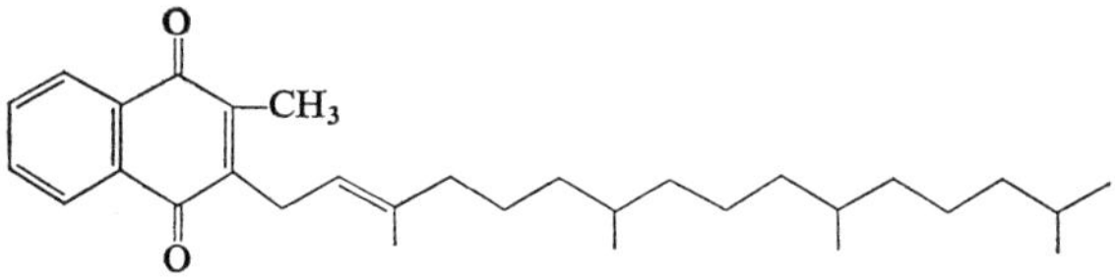

O $CH_3$ O

Vitamin $K_1$

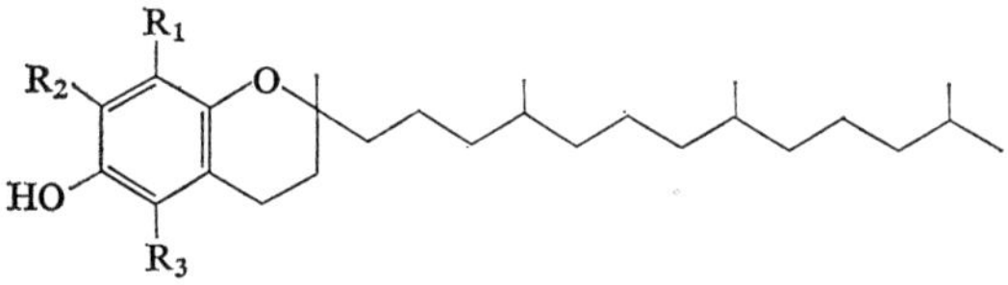

$R_1$ $R_2$ O HO $R_3$

α-Tocopherol; $R_1$, $R_2$, $R_3{=}CH_3$
β-Tocopherol; $R_1$, $R_3{=}CH_3$, $R_2{=}H$
γ-Tocopherol; $R_1$, $R_2{=}CH_3$, $R_3{=}H$

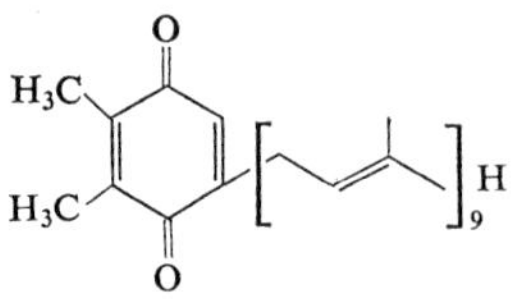

O $H_3C$ $H_3C$ O H 9

Plastoquinone

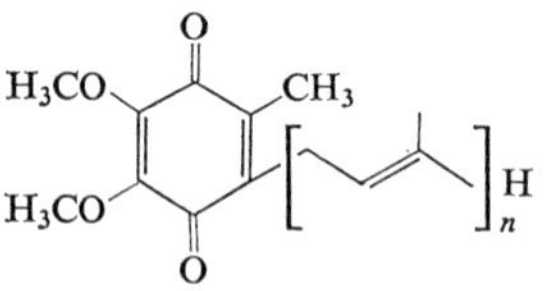

O $H_3CO$ $CH_3$ $H_3CO$ O H *n*

Ubiquinone, $n{=}6{-}10$

HO

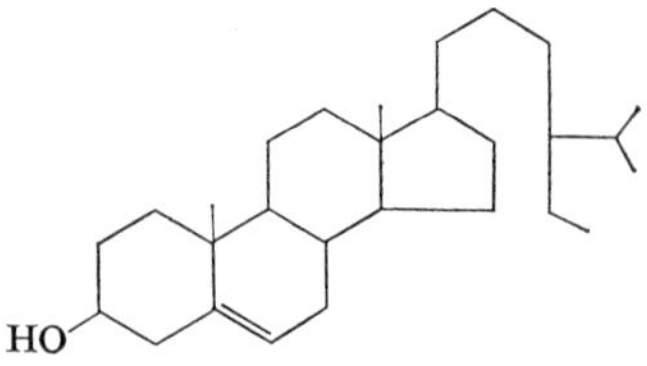

β-Sitosterol

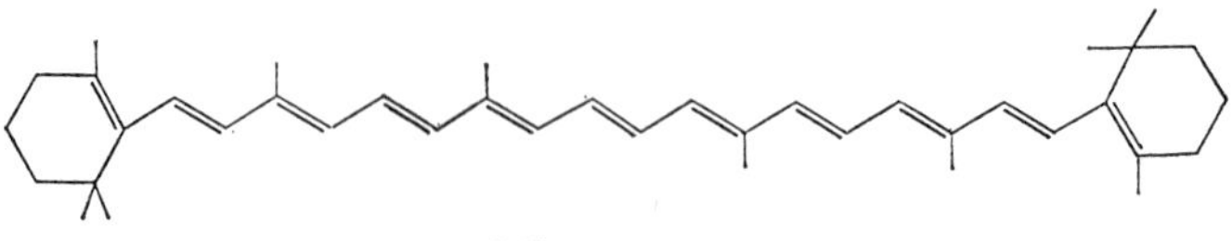

β-Carotene

FIG. 1. The structures of the major terpenoids and terpenoid-containing compounds in green plants and algae.

Etiolated seedlings of e.g. maize and etiolated *Euglena* all contain plastids, but they contain insignificant amounts of chlorophylls and little or no carotenoids, α-tocopheryl quinone, and only small amounts of plastoquinone and

Table II. Quinone levels in etiolated and green 6-day-old maize shoots[3]

| | Lipid [mg] | Chlorophyll [mg] | β-Carotene [μg] | Vit. $K_1$ [μg] | PQ [μg] | UQ [μg] | αTQ [μg] |
|---|---|---|---|---|---|---|---|
| [1] Etiolated | 41·4 | 0 | 19·2 | 9·45 | 28·7 | 38·0 | 0 |
| [1] Green | 72·8 | 10·05 | 77·4 | 12·24 | 227 | | 86·2 |
| [2] Etiolated | 323 | 0 | 149 | 74·4 | 223 | 304 | 0 |
| [2] Green | 503 | 69·6 | 523 | 82·8 | 1565 | 275 | 585 |

[1] Values calculated/g dry weight.
[2] Values calculated/100 shoots.
[3] Griffiths (1965).

vitamin K (Table II). They do, however, contain substantial amounts of sterol and ubiquinone. On illuminating etiolated seedlings of maize and *Euglena*, there is a rapid synthesis of all the terpenoid compounds, except the sterols and

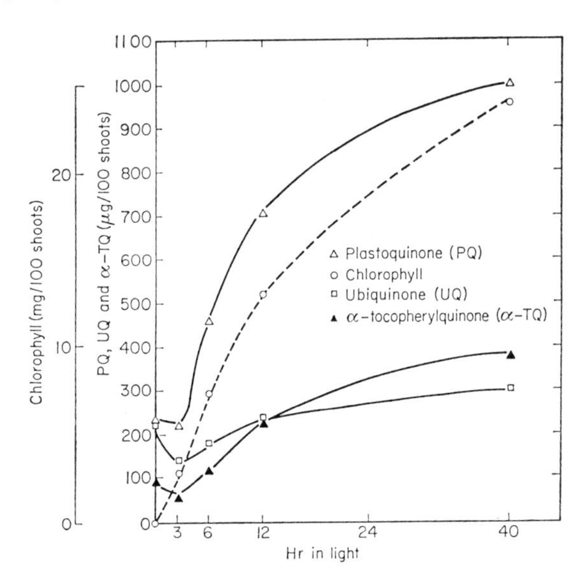


FIG. 2. Synthesis of terpenoid quinones in relation to chloroplast development (chlorophyll synthesis) in illuminated maize seedlings (Griffiths, 1965).

ubiquinone (Figs. 2 and 3). As the light-stimulated synthesis of the various terpenoids parallels the synthesis of chlorophyll, which can be taken as an indication of chloroplast development, they are obviously being synthesized

Table III. Intracellular distribution of chlorophyll, $\beta$-carotene, PQ, $\alpha$-TQ, and UQ in green maize.[1] Leaf material (120 g) homogenized for 30 sec in 160 ml NaCl buffer at 0°

| Fraction | | Percentage distribution | | | | | | | | | | | |
|---|---|---|---|---|---|---|---|---|---|---|---|---|---|
| | | Lipid | | Chlorophyll | | $\beta$-Carotene | | PQ | | $\alpha$-TQ | | UQ | |
| $\times g$ | Description | mg | % | mg | % | $\mu g$ | % | $\mu g$ | % | $\mu g$ | % | $\mu g$ | % |
| 3000 | "Chloroplast" | 36·7 | 34·8 | 2·20 | 71·1 | 80 | 64·5 | 246 | 68·3 | 97 | 68·2 | 15 | 24·9 |
| 20,000 | Mitochondria + Chloroplast fragments | 27·2 | 25·8 | 0·69 | 22·3 | 38 | 30·2 | 106 | 29·6 | 46 | 31·8 | 39 | 62·0 |
| 105,000 | | 19·3 | 18·3 | 0·21 | 6·6 | 6 | 5·3 | 0 | 0 | 0 | 0 | 8 | 13·1 |
| Supernatant | | 11·7 | 11·1 | 0 | 0 | 0 | 0 | 8 | 2·1 | 0 | 0 | 0 | 0 |

[1] Griffiths (1965).

alongside chlorophyll in the developing chloroplast. The synthesis of ubiquinone, which is a mitochondrial component and not a chloroplast component, is not stimulated under these conditions. The special position of the sterols is considered later (p. 731).

When the intracellular location of all the terpenoids in green maize was examined in the one experiment, the distribution was as expected, with one minor exception. The "plastid" terpenoids segregated with the chloroplasts but some plastoquinone always appeared in the supernatant fraction; ubiquinone segregated with the mitochondria (Table III); sterols were found in

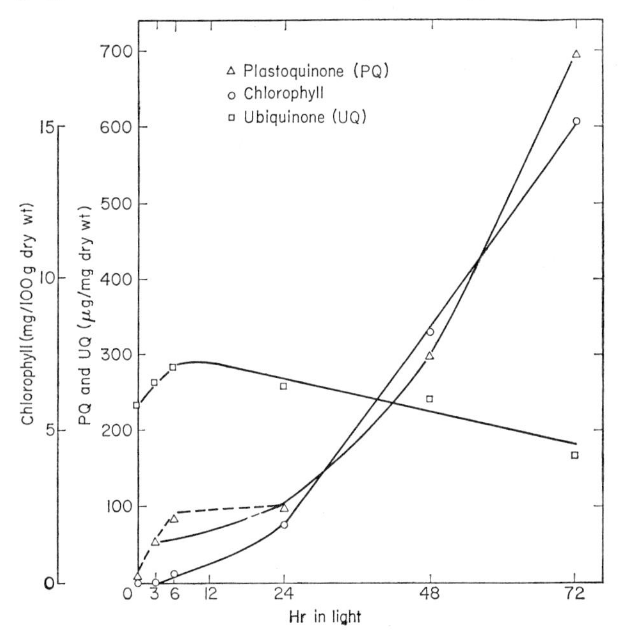


FIG. 3. Synthesis of terpenoid quinones in relation to chloroplast development (chlorophyll synthesis) in illuminated etiolated *Euglena*.

all fractions (see p. 731). In spite of their being located partly in the chloroplasts, synthesis of sterols is inhibited by illumination.

## THE PROBLEM OF REGULATION OF TERPENOID BIOSYNTHESIS

The biosynthetic pathways to all the terpenoids under discussion share a common route to C-15. Isopentenyl pyrophosphate, the common biological

Acetate ⟶ Acetyl-CoA ⟶ Acetoacetyl-CoA ⟶ Hydroxymethylglutaryl-CoA

↓

Isopentenyl pyrophosphate ⟵ Mevalonate

(I)

$$CH_2{=}C(CH_3)\cdot CH_2\cdot CH_2\cdot O\cdot P(O)(OH)\cdot O\cdot P(O)(OH)\cdot OH + H^+ \longrightarrow (CH_3)_2\overset{+}{C}\cdot CH(H)\cdot CH_2\cdot O\cdot P(O)(OH)\cdot O\cdot P(O)(OH)\cdot OH \xrightarrow{-H^+}$$

$$(CH_3)_2C{:}CH\cdot CH_2\cdot O\cdot P(O)(OH)\cdot O\cdot P(O)(OH)\cdot OH$$

$$(CH_3)_2C{:}CH\cdot CH_2\cdot O\cdot P(O)(OH)\cdot O\cdot P(O)(OH)\cdot OH \quad CH_2{:}C(CH_3)\cdot CH_2\cdot CH_2\cdot O\cdot P(O)(OH)\cdot O\cdot P(O)(OH)\cdot OH \xrightarrow{-PP_i}$$

$$(CH_3)_2C{:}CH\cdot CH_2\cdot CH_2\cdot \overset{+}{C}(CH_3)\cdot CH(H)\cdot CH_2\cdot O\cdot P(O)(OH)\cdot O\cdot P(O)(OH)\cdot OH \xrightarrow{-H^+}$$

$$(CH_3)_2C{:}CH\cdot CH_2\cdot CH_2\cdot C(CH_3){:}CH\cdot CH_2\cdot O\cdot P(O)(OH)\cdot O\cdot P(O)(OH)\cdot OH$$

(II)

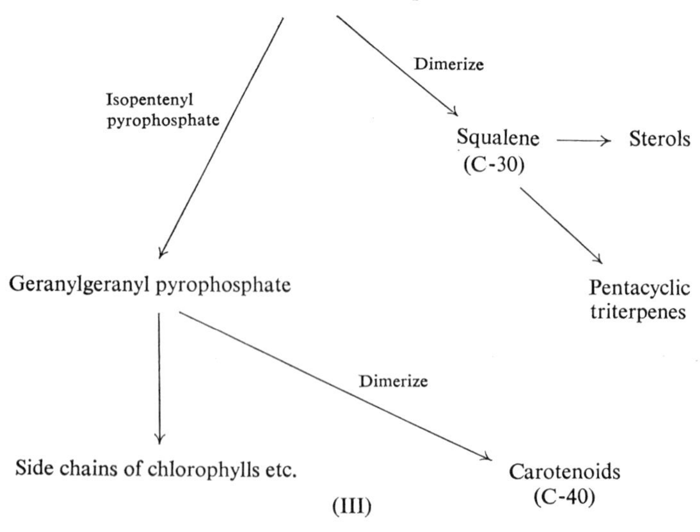


(III)

isoprene unit, discovered by Lynen is synthesized from acetate by a well-known pathway involving mevalonic acid (Scheme I). Isopentenylpyrophosphate is then isomerized to dimethylallyl pyrophosphate which acts as a starter for the enzymic chain elongation by the stepwise addition of molecules of isopentenyl pyrophosphate. The formation of geranyl pyrophosphate (C-10) is indicated in Scheme II. With the formation of C-15 the pathways divide according to the pathway outlined in Scheme III. The problem of regulation is basically how, when etiolated seedlings or *Euglena* are illuminated, the pathway is channelled at the farnesyl pyrophosphate stage away from synthesis of sterols and other triterpenoids and diverted into the synthesis of the terpenoids which are accumulating in the developing chloroplast.

## Proposed Hypothesis for the Regulation of Triterpenoid Synthesis in Greening Tissues

Evidence which we have obtained over the past 3–4 years, and which will be discussed in the next section, suggests that regulation is achieved by two mechanisms, probably functioning simultaneously: enzyme segregation and specific membrane permeability (Goodwin and Mercer, 1963; Goodwin, 1965). Firstly, it is assumed that enzymes concerned with the basic terpenoid reactions (up to C-15) are present both inside and outside the chloroplast whilst those concerned with synthesis of specific chloroplast terpenoids, e.g. phytol and carotenoids, are present only in the chloroplast, and those concerned with the synthesis of extraplastidic terpenoids are only synthesized outside the chloroplast. The problem of sterols which are present both inside and outside the chloroplast will be discussed later (p. 731). Secondly, it is assumed that the limiting membrane of the chloroplast is essentially impermeable, either way, to mevalonic acid, the first specific intermediate in terpenoid biosynthesis.

## Evidence in Support of the Hypothesis

If etiolated maize seedlings are excised from their roots, placed in water containing [2-$^{14}$C]mevalonic acid and illuminated for 24 hr then although the label is effectively incorporated into the terpenoids, it is located almost exclusively in squalene, sterols and ubiquinone; very little is incorporated into $\beta$-carotene and phytol although these were being actively synthesized during the exposure period. With $^{14}CO_2$ the situation is exactly the opposite, with the chloroplast terpenoids highly labelled and the sterols and ubiquinone relatively unlabelled (Tables IV and V). There thus appears to be a discrimination against mevalonic acid. This same pattern has been observed in many other tissues, e.g. oat, barley, french bean, and with mature etiolated tissues such as lettuce heart leaf. The same general pattern was observed with free solanesol (C-45) in tobacco leaves. This compound, which represents the side chain of plastoquinone, and which normally occurs in the free state only in minute amounts if at all, accumulates in considerable amounts in tobacco chloroplasts.

It is clear from the report of Więckowski and Goodwin (p. 445) that a small but significant amount of [2-$^{14}$C]mevalonate is incorporated into $\beta$-carotene in bean cotyledons and in pine seedlings; a superficial comparison of the specific activities of β-carotene and the sterols suggest that there is little difference between the incorporation of mevalonate into the two substrates; but it must be emphasized that most of the β-carotene was synthesized during the

Table IV. The incorporation of $^{14}CO_2$ and [2-$^{14}$C]MVA into the various chloroplastidic (β-carotene and PQ) and extrachloroplastidic (UQ, free and esterified sterols) terpenoids of illuminated etiolated maize shoots.[2] Shoots exposed to $^{14}$C-substrate for 24 hr with continuous illumination. Results expressed as per 100 shoots

| Terpenoid | [2-$^{14}$C]MVA (6 μc) | | $^{14}CO_2$ (200 μc) | |
|---|---|---|---|---|
| | Specific activity (d.$10^{-3}$/min/mg | Total activity[1] (d.$10^{-3}$/min) | Specific activity (d.$10^{-3}$/min/mg) | Total activity[1] (d.$10^{-3}$/min) |
| β-Carotene | 47·7 | 24 | 236 | 118 |
| Plastoquinone | 40·7 | 35 | 179·6 | 153 |
| Ubiquinone | 495 | 129 | 24·8 | 6 |
| Free Sterol | 96 | 2100 | 3·9 | 88 |
| Esterified Sterol | 294 | 500 | 18·3 | 31 |

[1] Values calculated assuming 500 μg β-carotene, 850 μg PQ, 260 μg UQ, 22 mg free sterol and 1·7 mg esterified sterol/100 shoots.
[2] Griffiths (1965).

Table V. Overall incorporation of [2-$^{14}$C]MVA and $^{14}CO_2$ into vitamin $K_1$, α-tocopherol and α-tocopheryl quinone in illuminated etiolated maize shoots[1]

| Compound | Concentration/ 100 shoots (μg) | Specific activity | |
|---|---|---|---|
| | | [2-$^{14}$C]MVA (d.$10^{-3}$/min/mg) | $^{14}CO_2$ (d.$10^{-3}$/min/mg) |
| β-Carotene | 300 | 37 | 210 |
| Vitamin $K_1$ | 150 | 26·4 | 115 |
| α-Tocopherol | 200 | 26·5 | 16 |
| α-TQ | 440 | 3·8 | 16·5 |
| 3β-OH-sterols | 17,000 | 102 | 4 |

[1] Griffiths (1965).

experiment whilst most of the sterol was already present when the experiment started. Thus the specific activity of the synthesized sterol should be much greater than that of the β-carotene. Evidence that this is so comes from the observation that in this type of experiment the sterol ester fraction has a very much greater specific activity than the free sterol fraction (Table VI).

In pine seedlings cotyledons which green up *in the dark* the mevalonate/$CO_2$ picture is the same as in bean seedlings; the full implications of these observations remain to be appreciated.

Direct evidence for a permeability barrier presented to mevalonate by the limiting membrane of chloroplasts came from the experiments of Rogers *et al.* (1966 and p. 283) who showed that mevalonate was not phosphorylated by intact chloroplasts prepared in non-aqueous medium. However, when the chloroplasts are ruptured by sonication the mevalonate kinase present inside the chloroplast is liberated and the mevalonic acid is converted into mevalonic 5-phosphate and 5-pyrophosphate.

When chloroplasts senesce in ripening fruit such as the tomato, and are converted into chromoplasts which synthesize large amounts of carotenoids,

Table VI. Incorporation of radioactivity into the total, free and esterified 3,β-hydroxy sterols.[1] Results expressed as per 100 shoots

| | Total sterol | | Free sterol | | Esterified sterol | |
|---|---|---|---|---|---|---|
| Sample No. | Wt. (mg) | Specific activity (d.$10^{-3}$/min/mg) | Wt. (mg) | Specific activity (d.$10^{-3}$/min/mg) | Wt. (mg) | Specific activity (d.$10^{-3}$/min/mg) |
| 0 | 19·7 | — | 15·2 | — | 1·21 | — |
| 3M[2] | 19·3 | 20·2 | 15·1 | 11·4 | 1·42 | 91 |
| 6M | 22·7 | 49·1 | 15·9 | 44·1 | 1·62 | 163 |
| 12W | 20·8 | 65·4 | 15·8 | 40·8 | 1·75 | 194 |
| 12M | 22·0 | 69·5 | 14·7 | 63·9 | 1·78 | 265 |
| 24W | 21·4 | 89·9 | 16·5 | 100 | — | 143 |
| 24M | 22·5 | 120·2 | 22·9 | 96 | 1·51 | 394 |
| $^{14}CO_2$ | 27·3 | 3·6 | 25·1 | 3·9 | 1·72 | 18·3 |

[1] Griffiths (1965).
[2] The numerals represent hr from zero time: M = MVA; W = $H_2O$; 12W means six hr after 6 hr treatment with MV4; 24W means 12 hr after 6 hr treatment with MVA.

permeability changes may also occur. Tomato slices incorporate [2-$^{14}$C]-mevalonate effectively into lycopene but no comparative measurements have yet been made on other terpenoids. Carrot root chromoplasts, in which the permeability seems to be different, incorporate [2-$^{14}$C]mevalonate into phytoene and α- and β-carotene as effectively as into squalene; but more work is also required in this area.

The impermeability of the chloroplast to mevalonic acid may, of course, be only an example of a much more generalized phenomenon. It has already been shown that neither the precursor of the nucleus of plastoquinone and ubiquinone, nor the methyl group of methionine, can be incorporated into plastoquinone although it is effectively incorporated into ubiquinone (Threlfall and Griffiths, p. 255.

It has also been possible to demonstrate that the incorporation of $^{14}CO_2$ into terpenoids of leaves depends on the state of development of the chloroplast, whilst the incorporation of [2-$^{14}$C]mevalonate is independent of chloroplast development (Fig. 4). If maize seedlings are illuminated for different times before being exposed for 3 hr to $^{14}CO_2$ or for 6 hr to [2-$^{14}$C]mevalonate, then during the early stages of illumination, chloroplast development, as indicated by chlorophyll synthesis, parallels the amount of $^{14}CO_2$ fixed. As the chloroplasts reach full maturity $CO_2$-fixation into the terpenoids falls, so that the turnover in mature chloroplasts become insignificant. The fixation of [2-$^{14}$C]-mevalonate is unchanged throughout this period. Clearly a new and important self-regulatory mechanism comes into play in mature chloroplasts which prevents $^{14}CO_2$ being unnecessarily fixed into chloroplast pigments and structural terpenoids. The mechanism of this regulation is unknown.

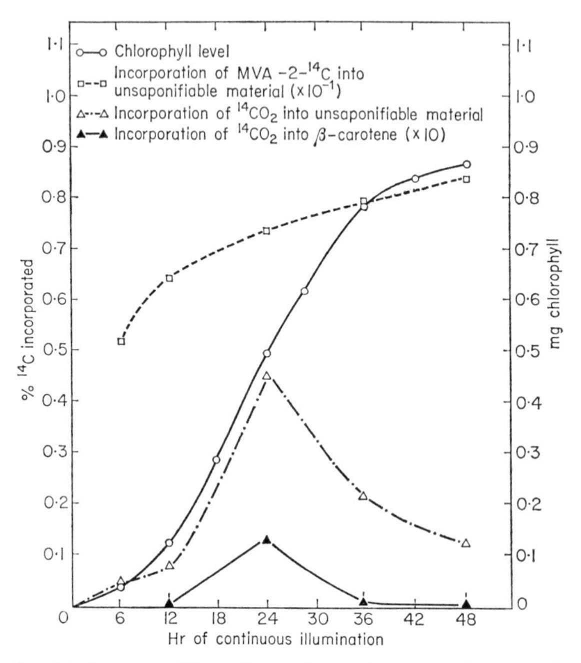


FIG. 4. Relationship between $CO_2$ and mevalonate incorporation into the unsaponifiable matter and the stage of chloroplast development in maize seedlings (Treharne, 1964).

The enzyme segregation which we envisage is illustrated in Fig. 5. Direct evidence in favour of a common basic pathway both inside and outside the chloroplast exists only for one key enzyme mevalonic kinase, which Rogers *et al.* (1965) showed to be present in both sites. Indirect evidence includes (a) the sterol precursors, squalene, lanosterol etc. are all extraplastidic (Goodwin and Mercer, 1963); (b) sterol synthesis goes on in the dark (Davies, 1963); (c) a cell-free system devoid of chloroplasts has been obtained which will incorporate [2-$^{14}$C]MVA into sterols (L. J. Goad, unpublished observations); (d) the carbons in the side chain and in the ring of plastoquinone synthesized in the presence of $^{14}CO_2$ have the same specific activity; this indicates that both are formed in the chloroplast, and that one component, e.g. the ring system, is not formed outside the chloroplast and transported there to be linked with the other synthesized inside the chloroplast.

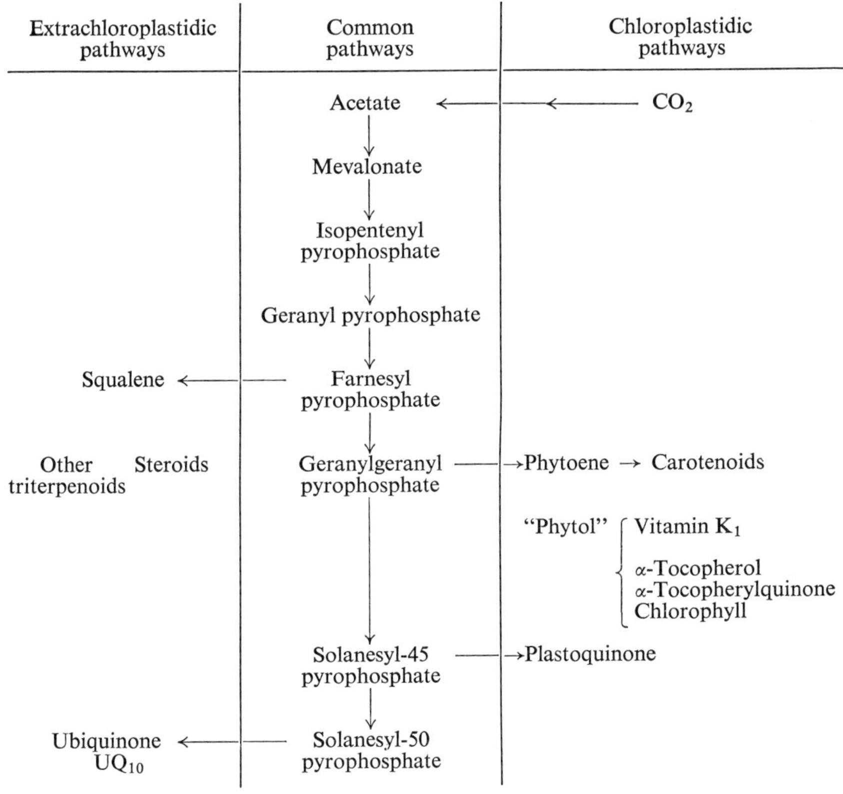


FIG. 5. Proposed scheme of segregation of enzymes concerned with terpenoid biosynthesis in developing seedlings. Centre column: enzymes present inside and outside the chloroplast; Right: enzymes present only or predominantly inside the chloroplast; Left: enzymes present only or predominantly outside the chloroplast.

## ORIGIN OF THE CHLOROPLAST STEROLS

The fact that sterols are present in the chloroplasts but synthesized outside raises interesting questions as to how they reach the chloroplast. If etiolated seedlings are illuminated then there appears to be a genuine redistribution of sterol; the observations cannot be accounted for by a synthesis of new sterol. This redistribution can also be observed after short exposures to red light; this is nullified by far-red light so the phenomenon appears to be phytochrome-mediated (Table VI).

A possible explanation is that the sterols are synthesized in the microsome fraction and then transferred to the chloroplast. The very high specific activity of the sterol ester fraction prompts the thought that it might be "transfer sterol". However, recent work by Treharne and Mercer (1966, Vol. I) suggests that there may be limited synthesis in the chloroplast. The very small amount of sterol which remains after vigorous washing has removed the phytosterols, is cholesterol; so perhaps the chloroplast has a limited capacity for sterol synthesis, but lacks the ability to alkylate the basic side chain to form the so-called phytosterols.

## CONCLUSION

The compartmentalization of terpenoid biosynthesis within the plant cell may be the way by which the plant regulates the production of terpenoid material to suit its particular requirements at various stages of development. The two main aspects of the regulatory system appear to be (i) specific enzyme segregation and (ii) specific membrane permeability of the plastid membrane to mevalonic acid.

During germination the extraplastidic synthesis of sterols required for membrane and lamellae formation occurs at the expense of endogenous food supplies from which mevalonate is synthesized; the mevalonate thus formed cannot penetrate the immature plastid to synthesize unnecessary pigments. As soon as a seedling emerges and is exposed to light, the plastids rapidly become functional by fixing $CO_2$ into mevalonate which is quickly used for the synthesis of chloroplast terpenoids which are incorporated into the developing lamellar structure of the chloroplast. Because of the impermeability, either way, of the limiting membrane of the chloroplast the mevalonic acid cannot pass into the cytoplasm and be utilized for synthesis of sterols and other triterpenoids which are not wanted at this time. Later, sucrose, formed in the chloroplast, can move out and provide the extraplastidic mevalonate required to form the sterols necessary for further growth of the plant.

## ACKNOWLEDGEMENT

Work from our laboratory reported here has been generously supported by the Science Research Council and by the Air Force Office of Scientific Research under Grant AF EOAR 63-24 through the European Office of Aerospace Research (OAR) United States Air Force.

## References

Davies, W. E. (1963). Ph.D. Thesis. University of Wales.

Goodwin, T. W. (1965). *In* "Biosynthetic Pathways in Higher Plants" (J. B. Pridham and T. Swain, eds.). Academic Press, London and New York.

Goodwin, T. W. and Mercer, E. I. (1963). *In* "Regulation of Lipid Metabolism" Biochem. Soc. Symposium, No. 24 (J. K. Grant, ed.). Academic Press, London and New York.

Griffiths, W. T. (1965). Ph.D. Thesis. University of Wales.

Rogers, L. J., Shah, S. P. and Goodwin, T. W. (1966). *Biochem. J.* **99**, 381.

Threlfall, D. R., Griffith, W. T. and Goodwin, T. W. (1964). *Biochem. J.* **90**, 56P.

Treharne, K. J. (1964). Ph.D. Thesis. University of Wales.

Treharne, K. J. and Mercer, E. I. (1966). *In* "Biochemistry of Chloroplasts" (T. W. Goodwin, ed.), Vol. I. Academic Press, London and New York.

## DISCUSSION

*M. Gibbs:* What I have to say is not directly pertinent to the subject matter but it may be of some interest. I noted that mevalonic acid is always supplied during a period of constant illumination. The lack of incorporation of this substrate into compounds found in the chloroplast may be due to its limited entry owing to constant illumination.

We found that the time of addition of arsenite or iodoacetamide was critical when investigating their effect on $CO_2$ fixation by whole chloroplasts. Thus arsenite or iodoacetamide added to whole chloroplasts in the light had no effect but added prior to illumination inhibited $CO_2$ uptake. On the other hand, arsenate inhibited $CO_2$ fixation even when added to the reaction mixture in the light.

Materials like egg albumin, avidin, protamine sulphate and bovine serum albumin show little effect on $CO_2$ fixation when added after the onset of photosynthesis. They either inhibit or enhance $CO_2$ fixation when added prior to illumination. If the period of photosynthesis is interrupted for a few minutes when these materials are present in the reaction mixtures, then they immediately exert an influence.

It would appear that the whole chloroplast during photosynthesis can block the entry of certain materials which can enter rapidly during the dark periods. Whether this occurs in the intact cell is not known. However, you could add evidence to this point by interposing a dark period in your experiments.

*T. W. Goodwin:* This is a possibility which we shall examine as soon as possible.

## Note Added in Proof

Addition of MVA to seedlings in the dark followed by illumination, did not significantly alter the general picture reported in this paper.

# Author Index

*Numbers in italic refer to the reference pages at the end of the chapters where the references appear.*

## A



## B

## D



## E

## H

## I



## J

# K

## Q



## R

T

## Y



## Z

# Subject Index

*Commonly accepted biochemical abbreviations are used whenever possible.*

## D

## F



## G

## H

I

## P

## Q

## R

## T

## U

## V



## W

## X



## Y



## Z

7970695
3 1378 00797 0695